I0038640

Metallurgy of Gold, by Sir Thomas K. Rose.

CYANIDE PLANT AND DUMP, KNIGHT'S DEEP AND NIGEL EAST, WITWATERSRAND.

THE
METALLURGY
OF
GOLD

with Application to
Mining and Mineral Processing

Sixth Edition

by

T.K. Rose
Associate of the Royal School of Mines

Revisions by

William Thomas Merloc, Ph.D.
Senior Chemist
Penrod Mining

Wexford Press
2008

PREFACE TO THE SIXTH EDITION.

FOLLOWING an old definition, which can hardly be bettered, a book on the metallurgy of gold should give an account of the extraction of gold from its ores and of adapting it for use. The book would then be useful to students and to metallurgists engaged in their profession. The most important function, however, which a book on metallurgy has to fulfil is to help those who are taking part in attempts to improve the existing practice. Progress in metallurgy depends on the capacity of metallurgists to apply their knowledge of physical science and engineering to the problems presented to them, and in this they are aided by a full understanding of the causes of the phenomena which they observe.

Hence, in addition to making an attempt to give, in as few words as possible, a complete picture of present practice, my aim in this edition has been to give full and accurate information as to the properties of gold, its alloys and compounds, and of the bearings of these on the work to be done. This has involved the rewriting of much of the work and a great expansion of certain sections corresponding with the rapid advance of science. References to the sources-of information are added in every case, to enable the reader to consult the original memoirs if he desires to do so.

A summary of the present general position in the working of placers, ore dressing, stamp milling and the cyanide process is given, with references to the lengthy treatises expressly devoted to each of these subjects. Some account is included of the treatment of gold ores in particular mills or districts, but this depends in great measure on local conditions and is best dealt with in separate text-books, such as the excellent one on *Rand Metallurgical Practice*. The chapters on the refining of gold and on assaying will, I trust, be found of value. Much care has, in particular, been devoted to the discussion of the electrolytic refining processes. Throughout the volume more attention is paid to the principles underlying practice than to the details of the machines employed, although descriptions are given of a number of machines which are typical of their class.

In the preparation of a portion of this edition I have been aided by my colleague, Mr. W. A. C. Newman, A.R.S.M., B.Sc., whose help has been specially useful in passing the work through the press. I am also indebted to many correspondents whose kind assistance has, I believe, been acknowledged in every case in the text, and last, but not least, to my publishers, who, with their usual thoroughness, have reset the work throughout on a larger page so as to compass in handy form the greatly extended text.

T. K. ROSE.

CONTENTS.

CHAPTER I.

THE PHYSICAL AND CHEMICAL PROPERTIES OF GOLD.

CHAPTER II.

ALLOYS OF GOLD.

CHAPTER III.

CHEMISTRY OF THE COMPOUNDS OF GOLD.

CHAPTER IV.

MODE OF OCCURRENCE AND DISTRIBUTION OF GOLD.

CHAPTER V.

TREATMENT OF SHALLOW PLACER DEPOSITS.

CHAPTER VI.

DEEP PLACER DEPOSITS.

CHAPTER VII.

ORE CRUSHING IN THE STAMP BATTERY.

CHAPTER VIII.

AMALGAMATION IN THE STAMP BATTERY.

CHAPTER IX.

OTHER FORMS OF CRUSHING AND AMALGAMATING MACHINERY.

CHAPTER X.

FINE GRINDING.

CHAPTER XI.

CONCENTRATION IN GOLD MILLS.

CHAPTER XII.

DRY CRUSHING.

CHAPTER XIII.

ROASTING.

CHAPTER XIV.

CHLORINATION.

CHAPTER XV.

THE CYANIDE PROCESS—CHEMICAL REACTIONS.

CHAPTER XVI.

THE CYANIDE PROCESS—GENERAL METHODS.

CHAPTER XVI.—*Continued.*

CHAPTER XVII.

THE CYANIDE PROCESS.

Special Methods and Examples of Practice.

SPECIAL METHODS.

EXAMPLES OF PRACTICE.

CHAPTER XVIII.

THE REFINING AND PARTING OF GOLD BULLION.

CHAPTER XIX.

THE ASSAY OF GOLD ORES.

CHAPTER XX.

THE ASSAY OF GOLD BULLION.

CHAPTER XX.—*Continued.*

CHAPTER XXI.

STATISTICS OF GOLD PRODUCTION AND CONSUMPTION.

BIBLIOGRAPHY.

LIST OF ILLUSTRATIONS.

b

THE METALLURGY OF GOLD.

CHAPTER I.

THE PHYSICAL AND CHEMICAL PROPERTIES OF GOLD.

Introduction.—From very early times the ancients were attracted by the beautiful colour, the brilliant lustre, and the indestructibility of gold.

Prof. Gowland points out [1] that on account of its wide distribution in the sands and gravels of rivers, and its distinctive appearance, it must have been the first metal to attract the attention of prehistoric man in most regions of the world. He also observes, however, that it could not have been used even for ornaments until the art of melting had been invented, and this could hardly have happened until man had passed the Stone Age culture and entered the Bronze Age.

No objects, he says, consisting of gold have been found with undoubted Stone Age remains. The earliest mining and metallurgical operations of which traces remain were those carried on in Egypt in dealing with the ores of gold. "The ancient mines are scattered over Upper Egypt, Nubia, and the Sudan," and consist of shallow pits in detritus, and trenches and shafts in hard rocks. The ore was broken by stone hammers, ground in stone mills or querns, and treated on inclined stone tables, on which the particles of rock were washed away from the gold. Shallow earthen dishes were used for the final washings, and the residual gold was melted with purifying fluxes in crucibles and cast into ingots. Remains of all the implements have been found, but their exact age is doubtful. [2]

Among the pictorial rock carvings of Upper Egypt there are several illustrations of the gold-extraction processes mentioned above. The earliest indications appear to be certain inscriptions on monuments of the Fourth Dynasty (4,000 B.C.), depicting gold washing. [3] Certain stele of the Twelfth Dynasty (2,400 B.C.) in the British Museum (144 Bay 1 and 145 Bay 6) refer to gold washing in the Sudan, and one of them appears to indicate the working of gold ore as distinguished from alluvial. [4]

In the code of Menes, who reigned in Egypt in 3,600 B.C. or about 2,000 years before Moses, the ratio of value between gold and silver is mentioned, one part of gold being declared equal in value to two and a half parts of silver, and it is, therefore, clear that the extraction of both metals from the deposits containing them must have been carried on before that time. A " gold bracelet found by Petrie on the arm of the queen of King Zer, successor of Menes, takes us back," almost as far, " whilst a small ingot of gold found

---

[1] Gowland, *J. Anthrop. Inst.*, 1912, **42**, 252-262. [2] Gowland, *loc. cit.*
[3] Wilkinson, *The Ancient Egyptians* (London, 1874), vol. ii., p. 137.
[4] Hoover, *Translation of Agricola* (London, 1912), p. 279.

by Quiball in a prehistoric grave at El-Kab demands an even more remote date " (Gowland).

In somewhat later times, in the collection of alluvial gold, the sands were washed down over smooth sloping rocks by means of running water, and the particles of gold, sinking to the bottom of the stream, were entangled and caught in the hair of raw hides spread on the rocks. Among the hides used were sheepskins, and hence originated the form of the legend of the Golden Fleece. Stripped of its heroic dress, this legend merely describes a successful piratical expedition about 1200 B.C. to win gold, which was being laboriously obtained from streams with the help of sheepskins or goatskins by the inhabitants of what is now Armenia. Similar expeditions have not been unknown in much later times, and the method of obtaining gold by washing river sand is still practised, with improvements in matters of detail, in many parts of the world. Metallurgists are almost proverbially conservative in their methods. Hides are even now occasionally employed to catch the gold, but sheep's wool, when used, is generally in the form of blankets.

At the present day, however, when auriferous sands are washed, the aid is also invoked of what Baron Born called in 1786 the " elective affinity " of mercury for gold when mixed with impurities. The ease with which gold-amalgam can be collected, in spite of its being less dense than gold itself, is due to the fact that it is wetted by mercury.

In the history of gold, it is also of interest to the metallurgist to remember that the earliest dawn of the science of chemistry was heralded by the study of the properties of gold, and by the efforts which were made to invest other matters with these properties. From the fourth to the fifteenth century, chemistry, which was first called " chemia " ($\chi\eta\mu\epsilon\iota\alpha$), and then " alchemy," was defined as the art of transmuting base metals into gold and silver, almost all the labours of philosophers being intended to aid directly or indirectly in solving this problem. At the end of this period, while Paracelsus was giving to chemistry a new aim—that of investigating the composition of drugs, and their effect on the human body—Agricola was reducing to order the numerous empirical facts which together made up the art of metallurgy, and although alchemy died hard, its era of usefulness may be said to have ended here. Gold has doubtless been the cause of many of the wars and marauding expeditions from which the world has suffered, but on the other hand it has been instrumental, in a far greater degree than most other commodities, in promoting the growth of civilisation, the efforts of the alchemists having laid the foundations of the science of chemistry, and those of the gold-seekers having resulted in the discovery of new countries, and in the spread of knowledge of all kinds.

For further notes on the history of the metallurgy of gold, see Chapters VII., Ore Crushing ; XIV., Chlorination ; XV., Cyanide Process ; XVIII., Refining ; XIX. and XX., Assaying ; and XXI., Statistics.

Colour.—The lustre and fine colour of gold have given rise to most of the words which are used to denote it in different languages. The word " *gold* " is probably connected with the Sanscrit word " *jvalita*," which is derived from the verb " *jval*," to shine. It is the only metal which has a yellow colour when in mass and in a state of purity. Impurities greatly modify this colour, small quantities of silver lowering the tint, while copper raises it. In a finely divided state, when prepared by volatilisation or precipitation, gold assumes various colours, such as deep violet, ruby and reddish-purple, the tint varying to brownish-purple and thence to dark brown and black. This purple colour has been supposed by some experimenters (viz., Guyton de Morveau, Büchner, Desmarest, Creuzbourg and Berzelius) to be due to the formation of a coloured oxide of gold of unknown composition, but Buisson, Proust, Figuier and, more recently, Krüss have shown that no oxygen can be obtained from this coloured material, and that it probably consists of metallic gold. Similar colours are seen in Purple of Cassius, and in Roberts-Austen's purple alloy of aluminium and gold, the colour in each case being probably due to a particular form of finely divided gold. "Faraday's gold," a ruby-coloured liquid prepared by the action of phosphorus dissolved in carbon bisulphide or in ether, or of formaldehyde on a cold dilute solution of chloride of gold, is a solution of colloidal metallic gold in water.[1] Faraday used a solution containing 0·6 of a grain of gold in a quart in preparing ruby gold. For further details as to colloidal gold, see Chap. III. Finely divided gold gives a faint blue tinge to light transmitted through the liquid in which it is suspended. The surface colour of small particles of native gold is often apparently reddened by being coated with translucent films of oxides of iron. Very thin films of gold are translucent, and appear green by transmitted light, while remaining yellow by reflected light. On heating, the green colour changes to some shade between ruby-red and violet, or disappears entirely owing to the breaking up of the continuous film into a network of metal through which white light passes. The green colour is restored by burnishing.[2] Molten gold is green, and its vapour is greenish-yellow.

Malleability and Ductility.—Malleability and ductility are possessed by gold at all temperatures to a far higher degree than by any other metal. A single grain of gold can be drawn out into a wire over 500 feet long, and leaves of not more than $\frac{1}{300000}$ of an inch in thickness can be obtained by beating. Faraday has shown that the thickness of these leaves may be still further reduced by floating them in a dilute solution of potassium cyanide, by which they are partly dissolved. Annealing is advantageous during the cold working of pure gold, but the temperature required is low, so that the gold beater's skin is uninjured.

Hardness.—Gold is softer than silver and harder than tin. Its hardness, according to Auerbach, is 2·5 to 3·0, and according to Rydberg 2·5, in the scale in which the diamond is 10 and talc 1.[3] The hardness of pure gold, however, like that of other metals, varies with its physical condition, as follows :—[4]

	Ludwik's Cone Machine.	Shore's Scleroscope, Magnifier Hammer.
Cast, ...	22	4·5
Hammered or rolled, . .	60-65	30-35
Annealed, ...	25	6

The scales are not the same. In the Ludwik scale, lead is 4·5, and quenched steel containing 0·9 per cent. carbon about 260. In the scleroscope scale, lead is 2 and steel about 175.

Tenacity.—According to Roberts-Austen,[5] pure gold when cast breaks

[1] Faraday, *Phil. Trans.*, 1857, p. 145 ; Zsigmondy, *Liebig's Annalen*, 1898, **301**, 29, 361.
[2] Faraday, *loc. cit.* ; G. T. Beilby, *Proc. Roy. Soc.*, 1903, **72**, 226 ; T. Turner, *Proc. Roy. Soc.*, 1908, **A, 81**, 301.
[3] *Landolt-Börnstein's Tabellen*, 1912, pp. 55-56.
[4] T. K. Rose, *J. Inst. of Metals*, 1912, **8**, 86 ; 1913, **10**, 150.
[5] Roberts-Austen, *Phil. Trans.*, 1888, **179**, 339.

with a load of 7 tons per square inch and an elongation of 30·8 per cent. on
a test piece 3 inches long. In Landolt's Tabellen,[1] the elastic limit of hard-
drawn gold wire is given as 14 kilos. per square mm. (8·9 tons per square
inch), and its tensile strength as 27 kilos. per square mm. (17·1 tons per square
inch). The tensile strength of annealed gold is 10 kilos. per square mm.
(6·3 tons per square inch).

Specific Gravity.—The specific gravity of gold is about 19·3. When cast
it is liable to contain cavities, by which the density is diminished, and after
compression the density is again higher, although it is not supposed that
the true specific gravity can be increased in that way. G. Rose (in 1848)
gave the density at 17·5°/4° (*i.e.*, gold at 17·5° compared with water at 4°),
as 19·28 to 19·31 when cast, and 19·31 to 19·32 after compression. Stas
gave cast gold 19·2860, after compression 19·3056. Roberts-Austen and
Rigg [2] gave cast gold 19·2945, after compression 19·3203, at 0°/4°. The
specific gravity of rolled sheet gold at 0°/4° is 19·2965,[3] that of soft annealed
wire at 20°/4° is 19·26, and of hard wire 19·25.[4] When crystallised from
solution the specific gravity is 19·431.[4] Henry Louis has shown [5] that the
specific gravity of unannealed "parted" gold (*i.e.*, the residue left after
boiling silver-gold alloys in nitric acid) is 20·3, its density being lowered by
the process of annealing. When precipitated by ferrous sulphate, its density
may be as high as 20·72 (G. Rose). The specific gravity of gold when precipitated from solution by oxalic acid is 19·49 (G. Rose).[6]

Taking the density of pure gold at 19·3, then

> 1 c.c. of pure gold weighs 19·3 grammes or 0·6205 oz. troy.
> 1 cubic inch weighs 316·25 grammes or 10·168 ozs. troy.
> 1 cubic foot weighs 546·485 kgrms. or 17569·9 ozs. troy.

The volume of 1 kilogramme of gold is 51·81 c.c., or 3·162 cubic inches.

> ,,   100 ozs. troy is 161·16 c.c., or 9·835 cubic inches.
> ,,   1 ton avoirdupois is 1·86 cubic feet.

Cohesion.—On heating, gold can be welded like iron below the point of
fusion, and finely divided gold agglomerates on heating without being subjected to pressure. Pressure alone is also sufficient to make gold dust cohere,
while a true flow of the particles of gold can be induced in the case of the pure
metal and some of its alloys.

Specific Heat.—The specific heat of gold is 0·0297 between − 188° and
+ 20°, 0·03103 at 18°, and 0·03114 at 100°.[7] It is 0·0345 at 900°, and 0·0352
at 1,020° (Violle).

Fusibility.—Gold fuses, after passing through a pasty stage, at a clear
cherry-red heat, just below the fusing point of copper and much above that
of silver. The metal expands considerably on fusing, and contracts again
on solidifying. The freezing point was given by Berthelot as 1,064°,[8] by
Day and Sosman as 1,062·4°,[9] and by Jaquerod and Perrot as 1,067·2°.[10]

---

[1] *Op. cit.*, p. 54.
[2] Roberts-Austen and Rigg, *Seventh Ann. Report of the Mint*, 1876, p. 44.
[3] T. K. Rose, *J. Inst. of Metals*, 1912, **8**, 111.
[4] *Landolt-Börnstein's Tabellen*, 1912, p. 164.
[5] Louis, *Trans. Am. Inst. of Mng. Eng.*, Chicago Meeting, 1893.
[6] G. Rose, *Pogg. Annalen*, 1848, **73**, 1 ; 1848, **75**, 403.
[7] *Landolt-Börnstein's Tabellen*, 1912, p. 751.
[8] Berthelot, *Compt. rend.*, 1898, **126**, 473.
[9] Day and Sosman, *Ann. Phys.*, 1901, **4**, [iv.], 99.
[10] Jaquerod and Perrot, *Landolt-Börnstein's Tab.*, p. 194.

[To face p. 5.]

SPARK SPECTRA OF GOLD.

5656
5251
5065
4812
4793
4608
4488

A
B

Fig. 1.— A, Gold, 990·99 line. B, Pure Gold, 1000·00 line.

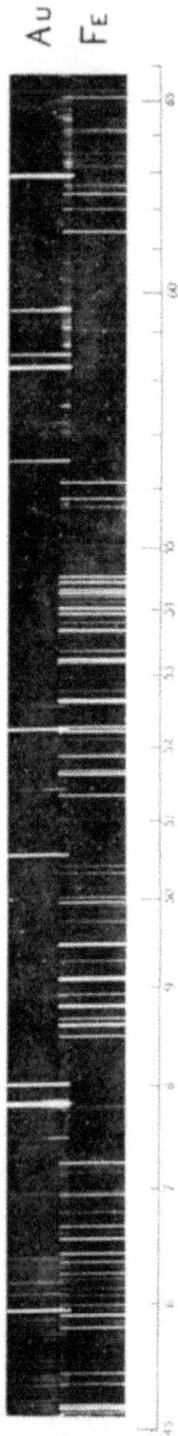

Au
Fe

Fig. 2.— Gold Spark in Air. Comparison with Iron.

Latent Heat.—The latent heat of fusion of gold is 16·3, and the normal lowering of the freezing point for 1 atom of impurity in 100 atoms of gold is 10·6°.[1]

Magnetism.—Gold is diamagnetic, its specific magnetism being 3·47 (Becquerel), if that of iron is taken as 100. Hanriot and Raoult [2] give the magnetic susceptibility of pure gold as not less than $- 0.234 \times 10^{-6}$.

Conductivity and Expansion.—Its electrical conductivity is 45.5×10^4 at 0°, that of silver being 68.12×10^4, and that of copper 64.06×10^4 (Dewar and Fleming). Its coefficient of thermal conductivity, K, is 0·7003 at 18°, that of silver being 1·006 (Jaegar and Diesselhorst). Its coefficient of linear expansion is 0·0000144 between 0° and 100° (Fizeau).

Atomic Weight and Volume.—Its atomic weight is 197·2 compared with oxygen = 16·00. The atomic volume of gold is 10·2.

Spectrum.—In the gold spectrum Huggins saw 23 lines. Mr. J. S. Clark, working under the direction of Prof. A. Fowler, has kindly taken new photographs of the spark spectra of gold, two of which are reproduced in Figs. 1 and 2. In Fig. 1 the spectrum of pure gold, which had been prepared at the Royal Mint, is compared with that of gold 999·96 fine, containing 0·04 per 1,000 of impurities, most of which probably consists of occluded gases. No difference was detected between the spark spectra of the two specimens. In this case the spark spectra were taken in air with self-induction, in series with spark gap, to suppress air lines. In Fig. 2 the arc spectrum of pure gold (as above) and that of iron are compared. In each case the spectra are slightly overlapping for comparison. In the part of the spectrum shown (wave length 4,500 to 6,500 $\mu\mu$), it will be seen that there are 13 principal lines in the gold spectrum, the wave lengths of some of these being 4,488, 4,608, 4,793, 4,812, 5,065, 5,231, 5,656, 5,835, and 6,276 $\mu\mu$ respectively. There are many other lines of shorter wave length down to about 2,300 $\mu\mu$, given by the gold spark both in air and in hydrogen.[3]

The Position of Gold in the Periodic Classification.—In Mendeléeff's Periodic Table gold occupies a position in the fourth long period, and among the " B " members of Group I. The " A " members of this group comprise the alkali metals, the companions of gold in the " B " series being silver and copper. The oxide type is R_2O, corresponding in the case of gold to aurous oxide, Au_2O.

Table I. shows the relative positions of gold, silver, and copper to the members of Group II., and also to the metals of the transition series in Group VIII.

It will be noticed that gold stands in the same relation to the heavy metals, osmium, iridium, and platinum, as silver and copper do to the two preceding series, forming a connecting link between the heavy metals and mercury, just as copper is a link between the iron metals and zinc, and silver between the palladium, rhodium, ruthenium series and cadmium. Like osmium, iridium and platinum, gold is dense, has a comparatively large atomic weight, and undergoes various degrees of oxidation, which are feebly acid or feebly basic. On the other hand, gold, like silver and copper, but unlike osmium,

---

[1] Roberts-Austen, *Proc. Roy. Soc.*, 1891, **49**, 352.

[2] Hanriot and Raoult, *Bull. Soc. chim.*, 1911, **9**, 1052 ; *Compt. rend.*, 1911, **153**, 182.

[3] For further information on the spectroscopic characteristics of gold, see Lockyer and Roberts, *Phil. Trans.*, 1874, **164**, [ii.], 495; and Frémy, *Ency. Chim.*, 1888, vol. iii., L'or, p. 40.

TABLE I.

		Group I. A. B.		Group II. A. B.		Group VIII.
1st long period,	even	K		Ca		Fe, Co, Ni
	odd		Cu		Zn	
2nd long period,	even	Rb		Sr		Ru, Rh, Pd
	odd		Ag		Cd	
3rd long period,	even	Cs		Ba		— — —
	odd	—		—		
4th long period,	even	—		—		Os, Ir, Pt
	odd		Au		Hg	

iridium, and platinum, is able to form compounds which conform to the type RX.

The compounds CuCl, AgCl, and AuCl are very much alike in physical and chemical properties ; they are insoluble in water, soluble in hydrochloric acid and ammonia, and also in potassium cyanide and sodium thiosulphate. Gold very easily forms higher halogen compounds of the type AuX_3, which may be readily converted into those of the lower type, AuX. This ease of conversion is a peculiar feature of the members of the odd series in the fourth long period.

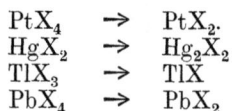

$$PtX_4 \rightarrow PtX_2.$$
$$HgX_2 \rightarrow Hg_2X_2$$
$$TlX_3 \rightarrow TlX$$
$$PbX_4 \rightarrow PbX_2$$

Table II. shows the relationships between Cu, Ag, and Au as regards density, atomic volume, and melting point.

TABLE II.

	Density.	Atomic Volume.	Melting Point, ° C.
Cu, . .	8·93	7·07	1,083
Ag, . .	10·49	10·21	961
Au, . .	19·26	10·11	1,063

The atomic volumes are relatively small, and the metals appear near the minima on the curve showing the variation of atomic volume with the atomic weight. These low atomic volumes are connected with the inertness of the elements as compared with the high activity of the alkali metals which occupy the maxima on the same curve. No gradation in the melting point is observable in the series, but the group forms an intermediate stage between the members of Group VIII., which have high melting points, and the easily fusible metals of Group II. B.

The affinity for oxygen in the series, copper, silver, gold, diminishes with rise of atomic weight from copper to gold. In the higher states of oxidation gold, unlike silver and copper, presents amphoteric properties—*i.e.*, acting both as acid and base. In the lower states of oxidation, however, copper, silver and gold are alike in exhibiting basic properties, although these in the case of gold are not very pronounced.

The facility with which the metal is precipitated from solution also increases with the atomic weight. Thus glucose precipitates cuprous oxide, and the pure metals silver and gold from solution, whereas ferrous sulphate, sulphur dioxide, oxalic acid and sulphuretted hydrogen precipitate the metal only in the case of gold. This diminution of chemical activity with rise in atomic weight is confined to the elements of Groups I. B, II. B, and VIII., and is associated with a diminution in electro-positiveness according to the electro-potential series.

Volatilisation of Gold.—Contrary to the belief of the older experimenters (Gaston Claves, and others), gold is sensibly volatile in air at temperatures not far above its melting point. Robert Boyle was unaware of this fact, but Homburg gilded a silver plate in 1709 by holding it over gold strongly heated in the focus of a burning mirror (*Encyclopœdia Britannica*, 1778, and *Gmelin's Handbuch*), and St. Claire Deville volatilised and again condensed gold when melting it with platinum. It has long been known that a discharge of high-tension electricity from gold points causes its volatilisation, and if the discharge is sent through a fine gold wire stretched on paper, it converts it into a purple streak of finely divided condensed particles of the metal. The rapid distillation of gold caused by heating it in a current of air of considerable velocity, such as that furnished by a blowpipe, by which the liquid is thrown into waves, may be shown at any time by heating a fragment of the precious metal of the size of a pin's head on a bone-ash cupel in the oxidising flame of a good mouth blowpipe. Almost immediately after the fusion is complete, a purple stain of condensed gold begins to form on the outer margin of the cupel. A piece of gold weighing 0·5 gramme loses half its weight in an hour, if heated on a cupel by a foot-blowpipe (the temperature attained being probably less than 1,300°), and only a few minute beads are observable, detached from the main button. Alloys of copper and gold disappear much more rapidly. No doubt some of the gold passes off as spray, but part of the loss is due to rapid volatilisation, and could not be correctly described as mechanical loss.

The volatility of gold, both when pure and when alloyed with silver and copper, was investigated by Napier,[1] who found that an alloy of 100 parts gold to 12 parts copper, if kept for six hours at a temperature just high enough to keep it melted, lost 0·234 per cent. of its gold contents, and at the highest temperature attainable in an assay muffle, it lost 0·8 per cent. in six hours. An increase in the amount of copper present caused an increase in the loss of gold. In the simple operation of pouring about 30 lbs. of a gold-copper alloy from a graphite crucible into moulds, fumes were given off, of which the part condensed in a wet glass beaker held above the crucible

[1] Napier, *J. Chem. Soc.*, 1857, 10, 229 ; 1858, 11, 168.

contained 4·5 grains of gold. Napier also found that gold does not appear to volatilise so readily when alloyed with silver only, as when copper is also present.

Makins found that gold volatilises sensibly along with silver and lead, when melted with these metals in a muffle in an ordinary bullion assay.[1] This has been confirmed at the Royal Mint, and the amounts were measured in 1910. The air led through the assay muffle in this case passed out into an iron pipe on its way to the flue, and the volatilised litharge condensed in large quantities in the first 2 or 3 feet of the pipe. The litharge condensed from 35 batches of gold bullion assays, of which there were 72 in a ·batch, was found to contain 70·7 grammes of lead, 0·210 gramme of silver, and 0·0012 gramme of gold. This corresponds to about 0·0005 milligramme of gold per assay, an amount which cannot be weighed on an assay balance. The temperature of the muffle furnace was about 1,050° C. In the cupellation of silver the temperature would, of course, be lower, so that the observed amount of loss by volatilisation of 0·08 per 1,000 of cupelled silver is higher than would occur in silver bullion assaying.

According to experiments made by the author,[2] the loss of gold on heating the pure metal rises with the temperature, being four times as great as 1,250° as at 1,100°, whilst it is insignificant at 1,075°, and probably inappreciable at lower temperatures. The nature of the atmosphere has also an effect on the rate of volatilisation, the loss in carbon monoxide being double that in air, and six times that in coal gas. A protective layer of charcoal would, therefore, increase the loss by volatilisation. The volatilisation of gold is also increased by the presence of any metallic impurity, even by the non-volatile metals, such as platinum. Lead and platinum have a very slight effect in increasing the volatility of gold; copper and zinc a more marked effect, while 5 per cent. of antimony or mercury causes losses amounting to about 2 parts per 1,000 of gold per hour at 1,245°. The metals which have most effect in reducing the surface tension of the liquid gold appear to increase its volatility in the greatest degree.

It has been found by Krafft and Bergfeld [3] that gold begins to volatilise at 1,070° in vacuo in a quartz vessel, and boils at 1,800° under the same conditions. It is estimated from these results that the boiling point of the metal at atmospheric pressure is 2,530°.[4] Calculated according to Wiebe's formula, the boiling point at atmospheric pressure would be 2,240°.[5] Richards has calculated the vapour tensions of gold for various temperatures on two assumptions—viz., (1) that the temperature interval between the first signs of vaporisation in a vacuum and boiling in a vacuum is equal to the interval between the latter temperature and the ordinary boiling point; and (2) that at equal fractions of the normal boiling point, expressed in degrees of absolute temperature, metals have the same vapour tensions. He has thus obtained the table on p. 9.

Thus, according to Richards, if the rate of volatilisation of gold is in direct proportion to the tension of its vapour, then, for example, it would be about 100 times as much at 1,387° as at 1,075°, and might be appreciable

---

[1] Makins, *Manual of Metallurgy*, 1873, p. 260; *J. Chem. Soc.*, 1860, **13**, 97.
[2] T. K. Rose, *J. Chem. Soc.*, 1893, **63**, 714.
[3] Krafft and Bergfeld, *Ber.*, 1903, **36**, 1670; 1905, **38**, 254.
[4] R. H. Richards, *Metallurgical Calculations* (1908), Part iii., p. 588.
[5] L. Meyer, *Modern Theories of Chemistry*, p. 134.

TABLE III.—VAPOUR PRESSURES OF GOLD.

Tension of Vapour in mm. of Mercury.	Temperature of Gold, °C.	Tension of Vapour in mm. of Mercury.	Temperature of Gold, °C.	Tension of Vapour in mm. of Mercury.	Temperature of Gold, °C.
0·0002	942	0·050	1,254	6·41	1,699
0·0005	987	0·093	1,298	14·84	1,788
0·0013	1,031	0·165	1,343	58·82	2,010
0·0029	1,075	0·285	1,387	195·	2,233
0·0063	1,120	0·478	1,432	451·	2,410
0·013	1,165	1·24	1,520	760·	2,530
0·026	1,209	2·93	1,611		

below its melting point. It is to be noted that this has not been observed directly.

The correctness of Richard's table has not been fully demonstrated by experiment, but it is not without some support. For example, in the cupellation observations given above, the tension of the vapour of gold (at 1,050°) would be, according to the table, about 0·002 mm. of mercury, and that of silver [1] 0·147 mm. of mercury, the ratio being 1 to 73·5. The volatilised metals, taking into account the densities of the vapours, should have been in the ratio of 1 to 40. The condensed metals were in the ratio of 0·0012 to 0·2100, but the ratio of gold to silver in the muffle was 1 to 2⅕. If equal quantities of gold and silver had been present in the muffle, the ratio of the metals condensed after volatilisation would have been presumably 1 to 81, or in fair agreement with the calculated ratio of loss by volatilisation.

Moissan found [2] that gold can be readily distilled when heated in the electric furnace. With a current of 300 ampères at 70 volts 59 grammes remained out of 107 grammes placed in a crucible, after six minutes. Copious fumes of a greenish-yellow colour were evolved. The gold was condensed as small regular yellow spheres, as brilliant yellow cubical crystals, as deep yellow leafy crystals, as filaments, or as a purple powder. In the distillation of gold-copper or gold-tin alloys, the residual ingot was richer in gold than the original alloy. The temperature attained was probably far above 2,500° C.

The loss of gold by volatilisation on melting its copper alloy is the common experience of mints. The melting loss is usually about 0·2 to 0·25 per 1,000, and after taking account of the amounts recovered from the ground-up crucibles, ashes, floor-sweepings and furnace bricks, it still amounts to 0·1, or 0·15 per 1,000. At the Royal Mint during the twenty years, 1870-1889, it did not exceed 0·140 per 1,000.[3] The loss depends on the temperature, the time of melting, the surface of metal exposed, but, above all, on the amount and nature of the draught passing over the metal, which sweeps away the volatilised gold, and enables the minute tension of vapour of the gold to come into play again and to renew the supply of gold vapour. In consequence of this, a cover of charcoal or other material on the gold should check the loss, and a cover to the crucible is also necessary. The effect of the cover on the composition of the atmosphere in contact with the gold must, however, be taken into account (see above, p. 8).

---

[1] Richards, *loc. cit.*   [2] Moissan, *Compt. rend.*, 1893, 116, 1429 ; 1905, 141, 977.
[3] T. K. Rose, *J. Chem. Soc.*, 1893, 63, 714.

The presence of volatile impurities in the bullion may also cause increased loss. Thus Hellot stated that if an alloy of 1 part of gold and 7 parts of zinc is heated in air, the whole of the gold comes off in the fumes,[1] but recent experiments show that if mechanical loss due to the violent boiling of zinc is avoided, the amount of gold carried off by zinc fumes is insignificant.[2]

It has also been shown by the author,[3] that tellurium does not cause volatilisation of gold at temperatures below 1,100°. Samples of an alloy of 78 per cent. of gold and 22 per cent. of tellurium were heated in a porcelain boat inclosed in a porcelain tube, through which a glass tube was passed. A current of water through the glass tube kept it cool. The alloys were heated for various lengths of time up to one hour at temperatures between 500° and 1,100°, in currents of different gases, air, carbon monoxide, hydrogen, and water gas (carbon monoxide and hydrogen in about equal volumes) being used in successive experiments.

In each case the whole or a part of the tellurium was sublimed and condensed on the cold tube, but the sublimates in only one case contained a trace of gold. In the other cases the whole of the gold was found still to remain in the boat. The exception was when a current of air was passed, the oxide of tellurium condensed on the cold tube in that case being found to contain 0·03 per cent. of the total gold originally present, while 99·96 per cent. of the gold was found in the boat.

A second series of experiments on a telluride ore from Western Australia, containing over 1,000 ozs. of gold per ton, gave similar results.

The losses incurred in roasting gold tellurides are probably in great part, if not entirely, due to fine dust being carried away mechanically, or to the absorption by the furnace bottoms of the very fusible mixtures which are formed.

The volatilisation of gold chloride and of gold in an atmosphere of chlorine is discussed below, see Chap. III.

Crystallisation of Gold.—Gold crystallises in the cubic system, occurring frequently in nature in the form of cubes, octahedra, and rhombic dodecahedra. Cubes and octahedra are often elongated, giving rise to rod-shaped crystals, and plates are also not uncommon. Twinning is frequent, giving rise to dendritic groups, tesselated surfaces and various fantastic and complicated forms. Some of these present the appearance of hexagonal pyramids or monoclinic prisms, and have even been described as such, but there is no reason to suppose that gold is anything but cubic and holosymmetric.[4] Cleavage is never exhibited. Single detached crystals are comparatively rare, and the crystals are usually attached end to end, forming strings, and branching, arborescent, or moss-like masses, which are composed of microscopic crystals, usually octahedra.[5] These forms occur frequently in quartz veins, but the single crystals, which are usually of larger size—viz., from $\frac{1}{4}$ to $1\frac{1}{2}$ inches in diameter—are mainly found in drift deposits. They are rarely perfect or of brilliant lustre, although such crystals were found at the Princeton Gold Mine, Mariposa County, California, but occur more frequently with

[1] Gmelin-Kraut, *Handbuch der anorganische Chemie*, vol. iii., p. 1039.
[2] "The Refining of Gold Bullion by Oxygen Gas." T. K. Rose, *Trans. Inst. Mng. and Met.*, 1905, **14**, 378.
[3] *British Association Report*, 1897, p 623.
[4] Miers, *Mineralogy* (1st edition, 1902), p. 302.
[5] See E. S. Dana, "On the Crystallisation of Gold." *Amer. J. Sci.*, 1886, **32**, 132, where other references are given.

rounded angles, raised edges, and cavernous faces, which are often marked with parallel striations, and possess little or no lustre (Fig. 3). The octahedra found in California are usually flattened parallel to two opposite faces, or elongated, or otherwise distorted. Still more frequently they are only partially developed, as in Figs. 4 and 5. In all these cases "the incomplete crystals have the appearance of a failure for lack of material " (W. P. Blake). Crystals of greater complexity, containing many modifying faces, occur chiefly in Siberia, Transylvania, and Brazil. The most common forms occurring naturally in Australia are the octahedron and the rhombic dodecahedron.[1]

Liversedge found [2] that, in the majority of cases, the gold embedded in massive quartz is remarkably free from any traces of crystalline form, and the larger the fragments of gold, the less crystalline form does it present.

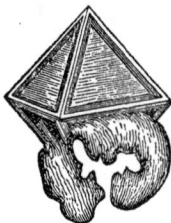

Fig. 3.

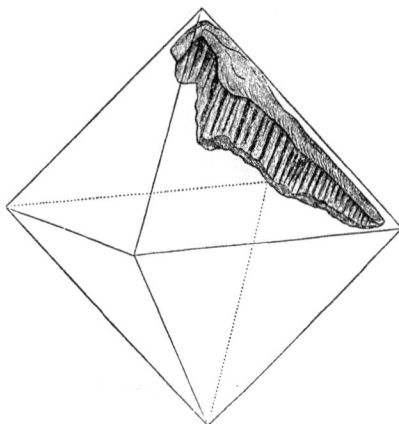

Fig 5.

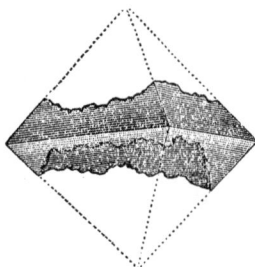

Fig. 4.

Figs. 3, 4, and 5.—Octahedral Crystals of Native Gold from California.

His observations enabled him to state that all well shaped crystals of gold appear to have been formed in what are now cavities, usually left by the removal of iron pyrite, or else in very soft matrices like iron oxides, clay, calcite, and serpentine. Crystallised gold is not usually met with in the quartz of the reef itself, but in the upper portions of the ferruginous and argillaceous casing of the reef and in the detritus near its outcrop.

On the other hand, polished and etched sections of nuggets [3] always show marked crystalline structure closely resembling that of a fused mass of gold. This is shown in Fig. 6, a section of a nugget from Coolgardie, Western

[1] For a full account of the crystalline forms of native gold, see a Paper by W. P. Blake, in *Precious Metals of the U.S.A.*, 1884, p. 573.

[2] Liversedge, *J. Roy. Soc. of New South Wales*, 1893, **27**, 299; *Chem. News*, 1894, **49**, 162.

[3] Liversedge, *J. Chem. Soc.*, 1897, **71**, 1125; Figs. 6 and 7 are reproduced from the Journal with permission.

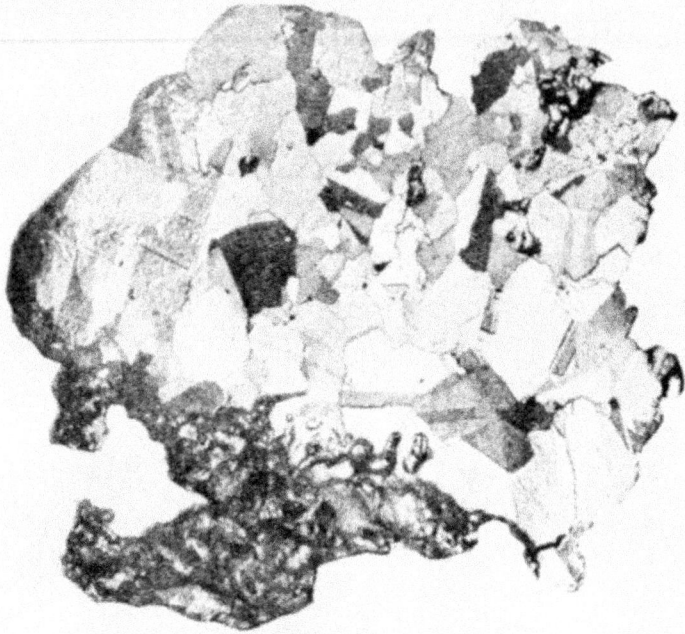

Fig. 6.—Gold Nugget, Internal Structure.   × 1½ diam.

Australia, which weighed 9·94 ozs., and consisted of gold 890, silver 105. Fig. 7 shows the outside of the same nugget. There are no traces of the

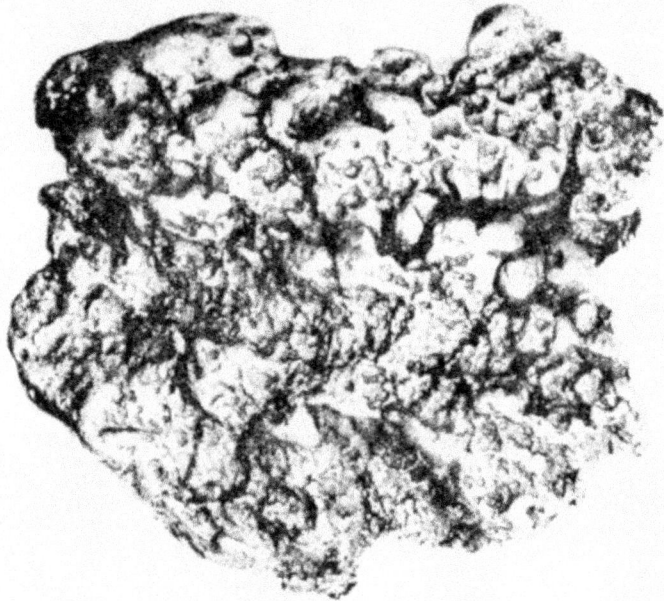

Fig. 7.—The same, Outside View.   × 1½ diam.

concentric structure which might have been expected if the nugget had been built up of successive coatings round a nucleus.

Artificial crystals can be obtained in several ways, but with some difficulty. Feathery crystalline plates are precipitated in the electrolysis of a solution of the chlorides of gold and ammonium. Formaldehyde in the presence of hydrochloric or nitric acid precipitates crystalline gold from solutions of gold chloride or bromide.[1] Crystals belonging to the cubic system are formed by the precipitation of gold from its solution as chloride by means of ether, phosphorus in ether, oxalic acid, ferrous sulphate, etc. When copper pyrite, mispickel, blende, etc., are used as precipitants, however, minute prisms, beautifully sharp and well-defined, are sometimes obtained.[2] The prisms are sometimes grouped in six-rayed stars, and six-sided plates may also be formed. Pseudo-hexagonal gold has also been observed by Blake, Chester, Dana, and others.[3]

By keeping an amalgam containing 5 per cent. of gold at a temperature of 80° for some days, and then digesting it at 30° with dilute nitric acid, bright crystals of gold can be obtained. These crystals are prismatic needles, and

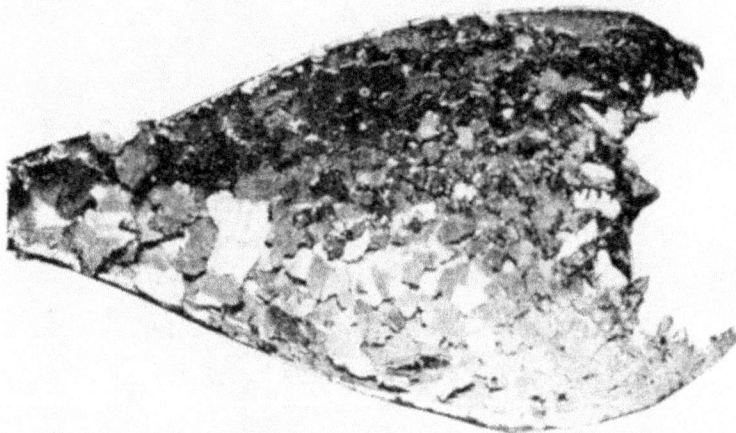

Fig. 8.—Gold Crystals.

are said by Chester to be regular hexagonal prisms with pyramidal terminations, and to contain 6 per cent. of mercury. In the Percy collection in the South Kensington Museum are some gold crystals found in the mercury troughs at the foot of the " blanket strakes," in an amalgamation mill. The troughs are placed so as to catch any stray particles of gold that may pass the blankets. As the amount of gold recovered in this way is very small, it is not worth while to clean out the troughs frequently, and in this case they had remained undisturbed for nine months, at the end of which time all amalgam was found to be crystallised. The mercury has been dissolved off by nitric acid, and the gold crystals remain. The smaller crystals are rather indefinite in shape, but amongst the larger ones (which are about half the size of a pea) are well-defined combinations of the octahedron, rhombic dodecahedron, and cube.

[1] Awerkieff, J. Chem. Soc., 1903, 84, [ii.], 218, 603.
[2] Liversedge, Chem. News, 1894, 49, 172.
[3] Loc. cit. ; Ditte, Compt. rend., 1900, 131, 143.

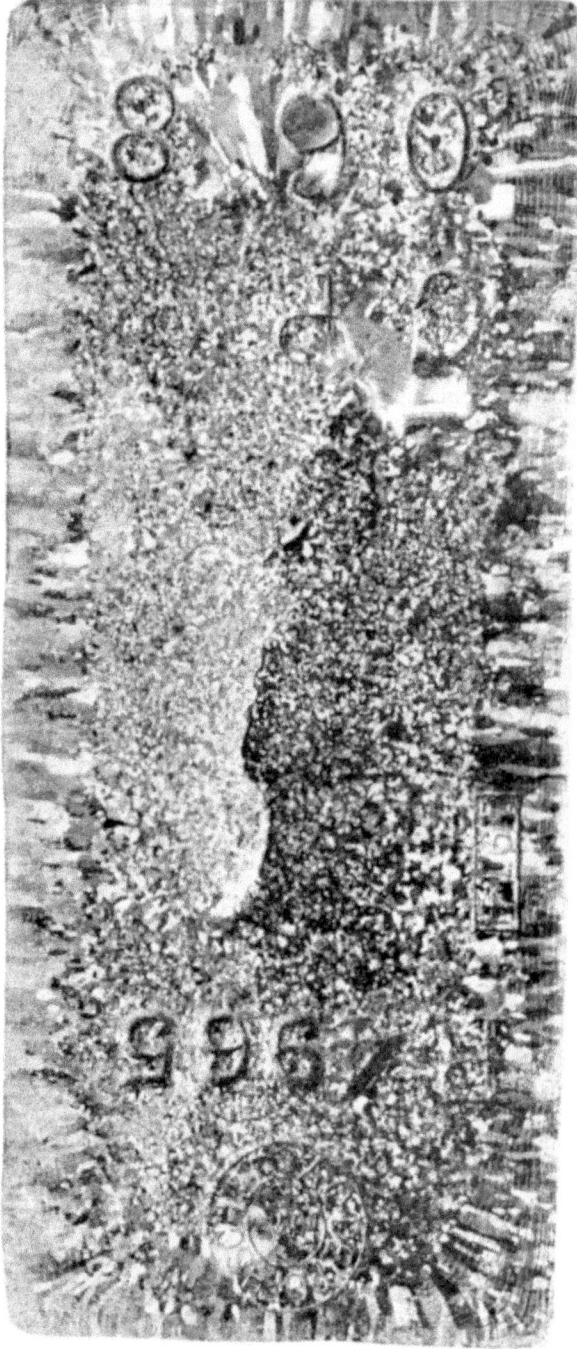

Fig. 9.—Cast Gold, Large Ingot (8 inches × 3½ inches), Etched.   (Upper surface, slightly reduced.)

When gold is cooled from fusion very slowly, fern-like structures in relief are formed on the surface of the metal, showing the rectangular arrangement of the axes. These raised portions are the parts of the metal first solidified, and, owing to contraction during solidification, they are left in relief above the general surface of the ingot. Sometimes faces and less frequently angles of octahedra are visible on the surface of the metal.[1] The purer the gold the more likely these crystals are to be observable. The presence of small quantities of copper reduces the size of the crystals. On pouring a partly solidified mass of pure gold from a crucible, Roberts-Austen obtained a shell consisting of an aggregate of well-formed crystals apparently octahedra.[2] These crystals are preserved at the Royal Mint, London (see Fig. 8, which is about half-size).

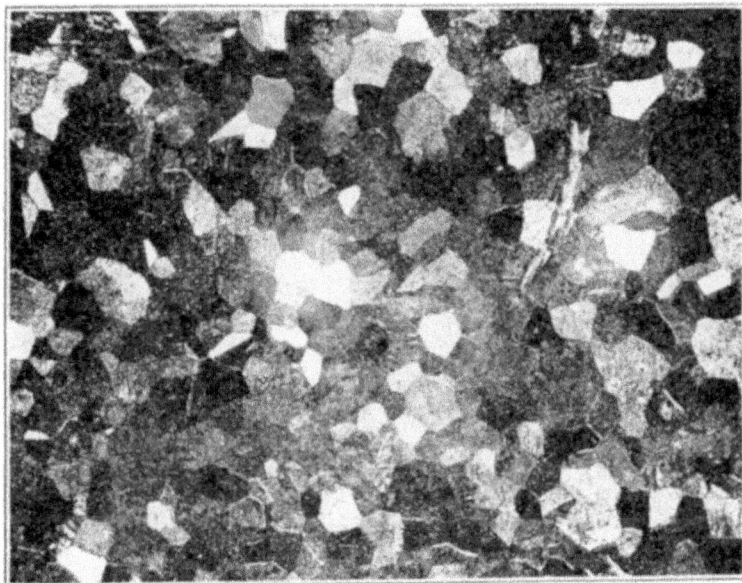

Fig. 10.—The same Ingot as Fig. 9. (Lower surface × 3·2.)

On solidification from fusion, whether quickly or slowly, gold, like other metals, sets in crystal grains or allotriomorphic crystals, consisting of irregular polygons of considerable size, which are larger as the rate of cooling is slower. These crystals may be seen without magnification by lightly etching the gold ingot with aqua regia. The best method of attack is to immerse the ingot in aqua regia diluted with an equal bulk of water at the ordinary temperature for 30 to 60 minutes. Photographs of cast pure gold etched in this way are shown in Figs. 9 and 10.

Fig. 9 is a photograph of the upper surface of a gold ingot, 999·9 fine by assay, and weighing 400 ozs. It was refined at the Royal Mint Refinery, and cast in an open mould. The difference in structure between the edge, where the

---

[1] See also Chester in *Amer. J. Sci.*, 1878, **16**, 29.
[2] *Encyclopædia Britannica* (9th edition), article "Gold."

metal was chilled by contact with the iron mould, and the centre is remarkable. It will be observed that the crystals in the inside of the free surface of the ingot are smaller than those outside. The surface of the ingot was not prepared by polishing or smoothing before being etched. The concentric lines near the edge are irregularities of surface, and pass through the crystals. Fig. 10 is a photograph of the lower face of the same ingot magnified 3·2 diameters.

The large ingot shown in Figs. 9 and 10 occupied a considerable time in cooling, and crystals bounded by straight lines were accordingly formed, the slow cooling producing the same effect as annealing. In small ingots which are cooled more quickly, smaller crystals are formed with irregular boundaries, as is seen in Fig. 11. On annealing, the irregular boundaries give place to straight lines, as in Fig. 12.

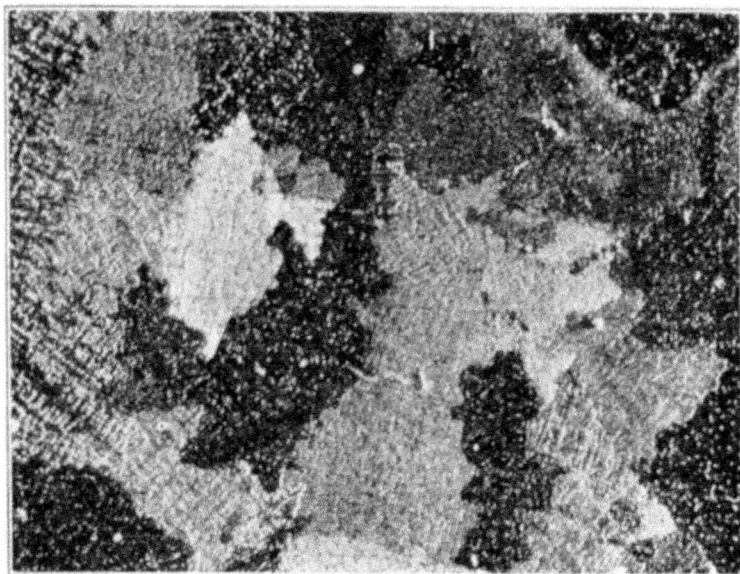

Fig. 11.—Gold, Small Ingot before Annealing (Etched).   × 100.

On rolling, such crystals are distorted by elongation, with the laminated effect shown in Fig. 13. On annealing, the metal is recrystallised, each large lamina breaking up into a number of small crystals, which appear and rapidly increase in number in particular laminæ, while others remain unaltered[1] (Fig. 14). The first appearance of recrystallisation, which is identical with softening, has been noted at 80° after 100 hours, and takes place in a few seconds at 200°. In course of time, or with rise of temperature, the new crystals increase in size, and obliterate the boundaries of the original crystal grains. Twinned crystals make their appearance, giving a characteristic banded structure, shown as certain narrow parallel-sided strips in Fig. 15.

---

[1] T. K. Rose, *J. Inst. of Metals*, 1913, **10**, 162.

Fig. 12.—The Same, after Annealing. × 100.

Fig. 13.—Fine Gold (Rolled and Etched). × 3.

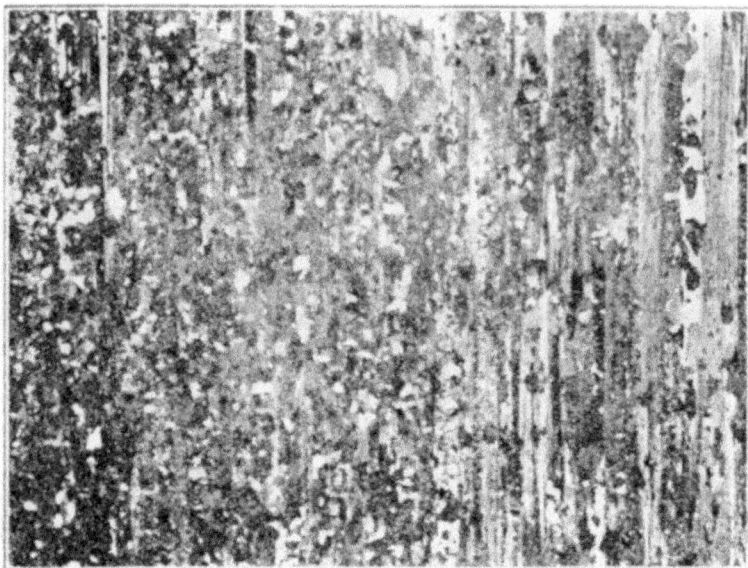

Fig. 14.—Gold (Rolled, Incompletely Annealed, and Etched).   × 4.

Fig. 15.—Fine Gold (Heated at 800° for 1 Hour).   × 7.

Dissolution of Gold.—Gold is readily soluble in aqua regia, or in any other mixture producing nascent chlorine, among such mixtures being solutions of (1) nitrates, chlorides, and acid sulphates—*e.g.*, bisulphate of soda, nitrate of soda, and common salt ; (2) chlorides and some sulphates—*e.g.*, ferric sulphate ; (3) hydrochloric acid and nitrates, peroxides such as permanganate, or chlorates ; (4) bleaching powder and acids, or salts such as bicarbonate of soda. Speaking generally, almost any chloride, bromide, or iodide will dissolve gold in presence of an oxidising agent. The action is much more rapid if heat is applied, or if the gold is alloyed with one of the base metals. The solution of gold in aqua regia takes place according to the equation—

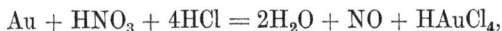

$$Au + HNO_3 + 4HCl = 2H_2O + NO + HAuCl_4,$$

and these proportions are most economical for employment.[1] They correspond to one part by weight of nitric acid of specific gravity 1·42 to four parts by weight of hydrochloric acid of specific gravity 1·2, or by volume, 1 part to 4·7 parts. The presence of silver in the gold retards the process, a scale of insoluble chloride of silver being formed over the metal, and the action may eventually be completely stopped if the percentage of silver present is large. Gold is also dissolved by liquids containing chlorine and bromine or a mixture producing bromine. The action is much slower than that of aqua regia, and subject to the same difficulties if silver is present ; heat assists the dissolution. Iodine dissolves gold only if it is nascent, or if dissolved in iodides or in ether or alcohol. Gold dissolves in hydrochloric acid in the presence of organic substances—e.g., methyl or ethyl alcohol, chloroform, glycerol, etc. The action is accelerated by heat.[2] Metallic gold does not dissolve in strong sulphuric acid unless a little nitric acid is added, when a yellow liquid is formed, which, when diluted with water, deposits the metal as a violet or brown powder. The mixture of nitric and sulphuric acids is a more rapid solvent for gold than nitric acid alone.

When gold is heated with concentrated selenic acid, H_2SeO_4, it dissolves with liberation of selenium dioxide and formation of a reddish-yellow solution of auric selenate, $Au_2(SeO_4)_3$; the action begins at 230°, but proceeds more readily towards 300°.[3]

According to Nicklès,[4] the easily decomposable metallic perchlorides, perbromides, and periodides are capable of dissolving gold, lower chlorides, etc., of the base metals being formed, and gold chloride, etc., produced. Some of these so-called persalts are, however, often regarded merely as solutions of chlorine, bromine, or iodine in the protosalts. Gold is soluble in ferric chloride [5] (see Table IV., p. 22), and in cupric chloride. In a re-examination of these observed facts in 1904,[6] Stokes found that the dissolution takes place readily at 200° C., thus—

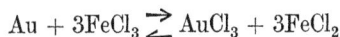

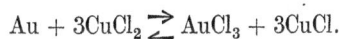

$$Au + 3FeCl_3 \rightleftharpoons AuCl_3 + 3FeCl_2$$

$$Au + 3CuCl_2 \rightleftharpoons AuCl_3 + 3CuCl.$$

Equilibrium is reached after a time, and no further action takes place, except on the addition of more ferric or cupric chloride or a rise in the temperature. A further addition of ferrous or cuprous chloride or a fall in the temperature causes some gold to be reprecipitated with the formation of ferric or cupric chloride.

M'Ilhiney, however, found [7] that hydrochloric acid in presence of air is without action on metallic gold, that ferric chloride is without action on gold unless oxygen is present, but that ferric chloride acts as an efficient carrier

[1] Priwoznik, J. Chem. Soc., 1911, 100, [ii.], 484.
[2] Awerkieff, J. Chem. Soc., 1908, 94, [ii.], 859.
[3] Lenher, J. Amer. Chem. Soc., 1902, 24, 354.
[4] Nicklès, Ann. Chim. Phys., 1867, 10, [iv.], 318.
[5] Napier, Phil. Mag., 1844, [ii.], 24, 370 ; Schild, Berg. und Hütt. Zeit., 47, 251.
[6] "A Treatise on Metamorphism," by Van Hise. U.S. Geol. Survey, 1904, 47, 1090.
[7] M'Ilhiney, Amer. J. Sci., 1896, 2, 293.

of chlorine in the presence of hydrochloric acid and oxygen. Stokes also observed [1] that gold is not appreciably dissolved in ferric sulphate unless chlorides are present at the same time, thus furnishing ferric chloride. These observations are of importance in considering the action of descending solutions in the belt of weathering in auriferous lodes.

Doelter has shown [2] that gold is somewhat readily soluble in a 10 per cent. solution of sodium carbonate, and also in an 8 per cent. solution of sodium carbonate containing excess of carbonic acid and containing sodium silicate. Becker has shown [3] that gold is easily soluble in sodium sulphide and in sodium sulphydrate.

Some other haloid compounds only attack gold in the presence of ether, in which case even hydriodic acid has a slight effect. Iodic acid has also been mentioned as a solvent for gold, but its action is very slight, much less, for example, than that of concentrated hydrochloric acid under similar conditions. A mixture of iodic and sulphuric acids dissolves gold when heated to 300° (Prat, also Victor Lenher). The effect of nitric and nitrous acids and of mixtures of them is described in Chapter XX. Alkaline sulphides attack gold slowly in the cold, and more rapidly if heated, producing sulphide of gold which is subsequently dissolved. Gold is also soluble in the thiosulphates of calcium, sodium, potassium, and magnesium, in the presence of an oxidising agent. According to H. A. White, [4] sodium thiosulphate slowly dissolves gold in the presence of air, and the action is greatly accelerated by ferric chloride and some other oxidising agents. He also found that ammonium thiocyanate alone would not dissolve gold, but that potassium thiocyanate in the presence of ferric chloride dissolved gold far more rapidly than was the case with thiosulphates. Fresh solutions of ferric thiocyanate dissolved gold slowly.

Spring has shown [5] that gold is soluble in hydrochloric acid if heated with it to 150° in a closed tube, and is subsequently reduced by the liberated hydrogen and deposited as microscopic crystals on the side of the tube. The author has found that boiling concentrated hydrochloric acid dissolves gold and maintains it in solution. Berthelot [6] found that the action takes place slowly in the cold in the presence of light and air, but not in the dark. Gold is also soluble in solutions of ferric and stannic salts in presence of hydrochloric acid, and is still more freely dissolved by a solution containing $CuCl_2$ and HCl, especially on heating. [7] Victor Lenher showed [8] that in presence of sulphuric acid many oxidising substances, such as telluric acid, manganese dioxide, lead dioxide, red lead, chromium trioxide, and nickel oxide, cause gold to pass into solution. In some cases phosphoric acid may be substituted for sulphuric acid. When gold is used as an anode, it is oxidised, and if the electrolyte is strong sulphuric acid, phosphoric acid or caustic soda or potash, part of the oxide is dissolved. [9] C. Lossen [10] has pointed out that if a solution of potassium bromide is electrolysed, the resulting alkaline

[1] Van Hise, loc. cit.

[2] Tschermak's Mineral. Mittheil., 1890, 11, 329 ; Van Hise, loc. cit.

[3] Becker, "Geology of the Quicksilver Deposits of the Pacific Slope," U.S. Geol. Survey, 1888, 13, 433 ; Van Hise, loc. cit.

[4] H. A. White, J. Chem. Met. and Mng. Soc. of S. Africa, 1905, 6, 109.

[5] Spring, Zeitsch. anorg. Chem., 1893, 1, 240.

[6] Berthelot, J. Chem. Soc., 1904, 86, [ii.], 569.

[7] M'Caughey, J. Amer. Chem. Soc., 1909, 31, 1261.

[8] Lenher, Eng. and Min. J., 1904, June 16, p. 963 ; J. Amer. Chem. Soc., 1904, 26, 550.

[9] Loc. cit. ; also Spiller, Chem. News, 1864, 10, 178.

[10] C. Lossen, Ber., 1894, 27, 2726.

solution containing hypobromite and bromate of potassium, is capable of dissolving gold. Gold is dissolved by aqueous solutions of simple cyanides in presence of an oxidising agent. This action is fully discussed in Chap. XV. Sulphocyanides, ferrocyanides, and some other double cyanides also dissolve gold, both at ordinary temperatures and on heating, but the action is very slow, even in presence of oxidising agents. Beutel has shown[1] that when potassium ferrocyanide is used, potassium aurocyanide is formed, and the resulting ferro-ions are oxidised by the air, giving ferric hydroxide. The solution formed is alkaline, and the reaction is probably represented as follows :—

$$3Au + K_4FeCy_6 + 2H_2O + O_2 = 3KAuCy_2 + Fe(OH)_3 + KOH.$$

Moir found[2] that finely divided gold dissolves slowly in acid (hydrochloric or sulphuric) solutions of thiocarbamide, and that the action is rapid in the presence of oxidising agents such as ferric chloride or hydrogen peroxide, especially if heated to 50°. The gold is very slowly reprecipitated by ferrous sulphate or stannous chloride.

The table on p. 22 gives the relative rate of dissolution of gold by a number of solvents. The gold used in each case consisted of a single " cornet " of " parted " gold, weighing about half a gramme, and consisting of gold 99·93 per cent., silver 0·07 per cent. Cornets offer a large surface to attack, as they consist of spongy gold, and those used did not differ in physical state. They were prepared by the method described under bullion assaying. The solutions were in excess, but there was no stirring or agitation.

Allotropic Forms of Gold.—Little is known of these. The marked influence of traces of other metals on the properties of gold has already been touched on ; from this and from the variations in colour and other properties the existence of several allotropic modifications of gold might be inferred.

Wilm[3] states that if gold is dissolved in dilute sodium amalgam under water, the aqueous liquid becomes dark violet, and when this is acidulated with hydrochloric acid, a black precipitate of pure gold is obtained. The black gold differs from the ordinary modifications in its extreme lightness ; moreover, it is soluble in alkaline solutions, and does not amalgamate with mercury or with sodium amalgam. When heated, it yields the ordinary modifications as a violet red powder. This form of gold appears, from Wilm's account, to resemble the black precipitate obtained on digesting certain aluminium-gold alloys with hydrochloric acid and that obtained by the action of water on potassium-gold alloys.

Julius Thomsen[4] stated that different allotropic modifications of gold are obtained by the reducing action of sulphurous acid on various solutions of gold compounds. The supposed allotropic forms obtained by the reduction of (a) neutral auric chloride, (b) auric bromide, (c) aurous chloride, bromide, or iodide have been designated respectively as gold, gold α, and gold β. Thus gold α gave out 3·2 calories in passing into gold. Van Heteren,[5] however, found that the potential differences between these samples of gold were no greater than between two samples of the first allotropic form, and consequently the forms were identical.

---

[1] Beutel, *J. Chem. Soc.*, 1910, **98**, [i.], 723.
[2] Moir, *J. Chem. Soc.*, 1906, **89**, 1345.
[3] Wilm, *Zeitsch. anorg. Chem.*, 1893, **4**, 325.
[4] Thomsen, *Thermochemische Untersuchungen*, vol. iii., p. 398.
[5] Van Heteren, *J. Chem. Soc.*, 1905, **88**, [ii.], 260.

TABLE IV.—RELATIVE RATES OF DISSOLUTION OF GOLD.

SOLVENT.	Time of Action.	Tempera-ture.	Amount of Gold dissolved per Cent. per Hour.
HCl and KNO$_3$, concentrated, . . .	20 hours.	15° C.	4·650
HNO$_3$ and NaCl, ,,	,,	,,	4·311
HCl and KClO$_3$, ,, . . .	..	,,	Completely dissolved in less than 20 hours.
HCl and K$_2$Mn$_2$O$_8$, ,, . . .	20 hours.	,,	0·862
HCl, ,, . . .	,,	,,	·003
,, ,, . . .	1 hour	70° C.	·150
HCl, 10 c.c. K$_2$Mn$_2$O$_8$, 0·2 gramme Water, 1 pint } (Etard's Mixture),	20 hours.	15° C.	·088
	1 hour.	70° C.	·850
NaCl, 1·5 gramme H$_2$SO$_4$, 1 c.c. K$_2$Mn$_2$O$_8$, 0·25 gramme Water, 1 pint, } (Black-Skeet Mixture), .	20 hours.	15° C.	·139
	1 hour.	70° C.	·620
H$_2$SO$_4$, concentrated, 95 parts, . . . {	20 hours.	15 °C.	·019
HNO$_3$, ,, 5 ,, . . .	1 hour.	70° C.	·010
NaCl, KNO$_3$, and a normal sulphate, strong solution, ... {	20 hours.	15° C.	None.
	1 hour.	70° C.	,,
NaCl, KNO$_3$, and an acid sulphate (KHSO$_4$), strong solution, ... {	20 hours.	15° C.	,,
	1 hour.	70° C.	1·830
NaCl, and ferric sulphate, strong solution, . {	20 hours.	15° C.	None.
	1 hour.	70° C.	0·340
HCl and H$_2$SO$_4$, concentrated, . . .	20 hours.	15° C.	None.
FeCl$_3$, concentrated, . . . {	,,	,,	,,
	1 hour.	70° C	4·060
Iodine and water in closed tube, . .	2 hours.	50° C.	None.
Iodine dissolved in KI, strong, . . .	20 hours.	15° C.	0·716
FeI$_3$ (prepared by dissolving excess of iodine in FeI$_2$), ... {	,,	,,	0·214
	1 hour.	70° C.	·450
HI, pure, ... {	20 hours.	15° C.	None.
	1 hour.	70° C.	,,
,, and ether, . . . {	20 hours.	15° C.	0·0007
	1 hour.	70° C.	·460
Iodine in alcohol, ... {	20 hours.	15° C.	·001
	1 hour.	70° C.	·120
,, in HI, ...	20 hours.	15° C.	4·950
Iodic acid, 5 per cent. solution, . . {	,,	,,	0·0005
	1 hour.	70° C.	·020
Chlorine, saturated solution in water (= 0·7 per cent.), ...	5 hours.	15° C.	1·152
	1¼ hours.	60° C.	3·592
Bromine, pure, ...	1 hour.	15° C	16·90
,, 5 per cent. solution in water, . {	,,	,,	8·30
	,,	50° C.	15·71
,, 1 per cent. solution, . . .	1¼ hours.	60° C.	5·17
,, 0·2 per cent. solution, . {	5 hours.	15° C.	1·162
	,,	50° C.	2·020
FeBr$_3$ (prepared by dissolving iron in bromine, expelling the excess of Br by heat, dissolving in water and filtering), . . . }	20 hours.	15° C.	0·003
	1 hour.	70° C.	·070
KCy, 1 per cent. solution, . .	1¼ hours.	15° C.	0·15
,, 5 ,, ,, . . .	,,	,,	·14
,, 25 ,, ,, . . .	,,	,,	·16
,, 1 ,, ,, . . .	,,	50° C.	·37
,, 1 ,, ,, . . .	,,	60° C.	·46
,, 1 ,, ,, . . .	5 hours.	100° C.	·26 Re-precipitated

Hanriot[1] studied the properties of brown gold, obtained by dissolving away the silver from silver-gold alloys with nitric acid, and concluded that it is a spongy mass formed of a mixture of ordinary or α gold, with a new modification, β gold. He leaves it as an open question whether the differences are due merely to the nature of the crystals, or whether they are sufficient to warrant the use of the name allotropic modification. The β variety has not been prepared in a pure state. It is always mixed with α gold, and also with traces of silver, copper, lead, iron, etc., as revealed by the spectroscope. It has a lower coefficient of diamagnetism than α gold, and is transformed into α gold at temperatures above 300°. Between 300° and 650° brown gold is metastable, and tends to change into ordinary gold with a contraction of 40 per cent. in the length of thin plates. This change is hastened by touching it with a plate of the α variety.

It has also been shown by Hanriot and Raoult[2] that brown gold is more soluble in nitric acid and in hydrochloric acid than yellow gold. They also found that brown gold is freely soluble in a hot hydrochloric acid solution of auric chloride, and that on cooling beautiful crystals of β gold are formed, consisting of a mixture of tetrahedra and rhombic dodecahedra. These crystals are very soluble in auric chloride.

Louis had previously found[3] that brown gold has a density of 20·3, and that it expands on being transformed by heat into the yellow variety.

The evidence in favour of the existence of amorphous or hard gold in cold-worked or polished specimens is similar to that in the case of other metals.[4]

---

[1] Hanriot, *Bull. Soc. chim.*, 4th series, 1911, **9**, 139, 339, and 1052.
[2] Hanriot and Raoult, *Compt. rend.*, 1912, **155**, 1085.
[3] Louis, *Trans. Amer. Inst. Min. Eng.*, Chicago Meeting, 1893.
[4] See G. T. Beilby, *Proc. Roy. Soc.*, 1907, **A, 79**, 463 ; *J. Inst. of Metals*, 1911, **6**, 5.

CHAPTER II.

ALLOYS OF GOLD.

Introduction.—Gold can be made to alloy with almost all other metals, but most of the bodies thus formed are of little or no importance to metallurgists. The binary alloys of gold with one other metal have been studied in a number of instances, but little systematic study has been devoted to gold alloys containing three or more metals, although the metallurgist has to deal chiefly with these. ⸰ In the following pages the binary alloys are described successively. As the effects of heat on alloys are most readily expressed in equilibrium diagrams, an explanation of these is given below.

A thermal equilibrium diagram (see Fig. 16, and also Figs. 17 to 24) expresses in diagrammatic form the temperatures at which physical changes take place in alloys. Each of the figures shows the equilibrium diagram of the series of binary alloys formed by gold with one other element. The principal physical change is that of melting or solidification. The *liquidus curve* joins the points, giving the temperatures at which alloys of different compositions begin to freeze or solidify. As the temperature falls (on a vertical line from a point on the liquidus curve) more and more of the alloy solidifies, and solid and liquid particles exist side by side. At length solidification is complete, and the curve joining the points, giving the temperatures at which this occurs, is called the *solidus curve*. Each alloy has a liquidus point and a solidus point. The interval between them is usually called the *pasty* stage. In a pure metal, or in a pure compound of two metals, or in a eutectic alloy (mother liquor which has the lowest melting point of the series), there is no pasty stage, and the liquidus and solidus points coincide. The solidus point is that at which a metal on being heated begins to melt.

The equilibrium diagram gives some further information. Fig. 16 represents a form of such a diagram, giving the solidus and liquidus curves for two metals, M and N, and all their alloys. Starting with a mixture represented by the point X, all the material is then molten. If it be allowed to cool gradually, the reduction in temperature will be represented by the line X P. At P crystals of composition represented by the point Q will separate, and temporary equilibrium will be established between these crystals and the melt at that temperature. By the separation of these crystals the melt becomes richer in N, and the freezing point of this remainder of the melt is lowered. On further cooling, the temperatures at which crystals begin to separate will tend to follow the line P E—the liquidus—while the composition of the crystals separating will tend to progress along the line Q T—the solidus—but such effects are attained only if the rate of cooling is extremely slow.

Similar conditions will prevail on the branch of the curve, which starts at the melting point of the pure metal N. At E, where the two branches meet, the melt will solidify as a whole. The crystals in equilibrium at the eutectic E—always provided that the condition of extremely slow cooling has been fulfilled—will have the compositions represented by T and T'.

By quenching an alloy from the molten state, as represented by X, the constitution at that point is preserved, and the separation into different kinds of crystals is wholly or partly prevented.

Besides the melting point, changes in the solid alloy (allotropic change, separation or union between the constituents, formation or decomposition of compounds, etc.) also occur. The curves joining the points, giving the temperatures at which these changes occur in solid alloys, are also given on equilibrium diagrams.[1]

Alloys of gold are generally prepared by melting their constituents together, but they may in certain cases be prepared by simultaneous precipitation from solution, as was shown by Mylius and Fromm in 1892.[2] Thus a gold-zinc alloy, containing equal weights of the two metals, and approximating in composition to $AuZn_3$, was obtained in the form of black, spongy flocks, by adding a solution of gold sulphate to water in which a zinc plate

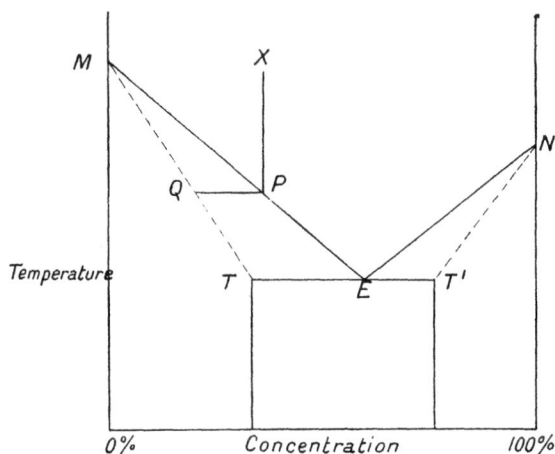

Fig. 16.—Thermal Equilibrium Diagram.

was placed. The gold-zinc slime obtained in the cyanide process (q.v.) may be compared with this. Gold-cadmium, similarly obtained, is a lead-grey crystalline precipitate, having the composition $AuCd_3$. If the gold-zinc alloy is shaken with a solution containing a cadmium salt, the gold-cadmium alloy and a zinc salt are obtained. Similarly, a copper plate, acting on solutions containing gold, yields a black, spongy compound of gold and copper; and gold-lead and gold-tin alloys in the form of black slimes are also readily prepared.

The diffusion of gold into other metals, both liquid and solid, was investigated by Roberts-Austen.[3] The rates of diffusion in liquid metals are of the

---

[1] For a more complete discussion of the equilibrium diagrams, see the chapter on "The Constitution of Alloys," pp. 91-115, in E. F. Law, *Alloys* (2nd ed., 1914); see also C. H. Desch, *Metallography*.

[2] Mylius and Fromm, *J. Chem. Soc.*, 1894, **66**, [ii.], 236; *Ber.*, 1894, **27**, 630.

[3] Roberts-Austen, *Proc. Roy. Soc.*, 1896, **59**, 281; Bakerian Lecture, *Phil. Trans.*, 1896, **A, 187**, 383.

same order as those of soluble salts in water, but the diffusion in solids is very slow. If gold is placed at the base of a cylinder of solid lead 70 mm. high, some is found to have reached the top in thirty days, the temperature being kept at 251°. The rate of diffusion is still measurable at 100°, but is almost inappreciable at ordinary temperatures.

The alloys which are most important to the metallurgist are those which gold forms with mercury, copper, silver, lead, and zinc.

Gold and Aluminium.—These alloys were investigated by Roberts-Austen,[1] and more recently by Heycock and Neville.[2] Several true compounds of the two metals were proved to exist, the most remarkable and stable being Au_2Al, a hard white alloy with a freezing point of only 622°, and $AuAl_2$, Roberts-Austen's beautiful purple alloy, which melts at about the fusing point of pure gold.

From the melting point of pure gold the freezing point curve falls steeply to a temperature of 545° C., and along this portion primary gold crystals first separate. At 545° C. a break occurs in the curve, corresponding to a compound having the composition Au_4Al (gold 96·7 per cent., aluminium 3·3 per cent.), which separates until the eutectic is reached at 527° C. (21·5 atomic per cent., or 3·6 per cent. by weight of aluminium). The curve rises from the eutectic to a second break at 575° C. (Au_5Al_2 or Au_8Al_3), containing about 5·1 per cent. by weight of aluminium, and finally reaches the first true maximum at an alloy with 6·3 per cent. of aluminium (Au_2Al). A second eutectic occurs at 570° (8·1 per cent. Al), and a third intermediate break at 685° C. ($AuAl$, containing 12 per cent. Al), whence the curve proceeds to the maximum denoting the compound $AuAl_2$, which contains 21·5 per cent. by weight of aluminium. This compound is always formed when mixtures of gold and aluminium are fused, and allowed to cool, provided that not more than about 90 per cent. of gold is present. The purple alloy is seen in patches on a white ground containing the excess of aluminium or of white compounds of gold and aluminium. A third eutectic occurs very near to the boundary of the curve for pure aluminium.

Gold and Antimony.—Gold is very easily dissolved by molten antimony. A piece of gold wire held in the heated material can be readily seen to be going into solution, forming radial stream lines.

A small amount of antimony lowers the melting point of gold considerably, and causes a long pasty stage. Hatchett[3] found that molten gold absorbs the vapour of antimony and becomes brittle. Roberts-Austen[4] found that an addition of 0·2 per cent. of antimony to gold reduced its tensile strength from 7·0 to 6·0 tons per square inch. The author[5] found that about 1·0 per cent. of antimony makes gold brittle, but that the brittleness is removed by annealing.

None of the alloys are malleable. They are all formed with contraction; they are hard, pale yellow or grey, and not easily attacked by acids, except aqua regia. Cosmo Newbery has shown that when the alloys are finely divided and ground with mercury they are slowly decomposed, yielding gold amalgam and black metallic antimony, mixed with antimonide of mercury.

---

[1] Roberts-Austen, *Proc. Roy. Soc.*, 1892, **50**, 367.
[2] Heycock and Neville, *Phil. Trans.*, 1900, **A, 194**, 201 ; *Proc. Roy. Soc.*, 1914, **90**, 560.
[3] Hatchett, *Phil. Trans.*, 1803, **93**, 43.
[4] Roberts-Austen, *Phil. Trans.*, 1888, **A, 198**, 339.
[5] T. K. Rose, 33rd *Annual Report of the Mint*, 1902, p. 73.

From Vogel's results [1] it appears that the liquidus curve consists of three branches, along which three different series of crystals separate. The two series containing most gold do not appear to consist of solid solutions. The third series of crystals corresponds to the formation of the compound $AuSb_2$, at 460° C., and 55 per cent. Sb, the point of maximum thermal effect in the reaction. At the same point there is a break in the liquidus curve. Vogel records that these alloys are strongly susceptible to undercooling, and that only by inoculation can solidification be promoted. In this respect they resemble the tellurium-gold alloys.

The compound $AuSb_2$ is of the same colour as antimony, is extremely brittle, and considerably harder than the individual components. There is a well-marked eutectic at 360° C. and 24 per cent. Sb.

Gold and Arsenic.—These resemble the gold-antimony alloys. They are readily fusible, hard, brittle, pale yellow or grey alloys, formed with contraction and decomposed by mercury like the antimony alloys. They are soluble with difficulty in aqua regia. Hatchett found [2] that the vapour of arsenic, acting on a suspended plate of gold at a red heat, forms a readily fusible alloy which drips off the gold, so that the residue of the gold plate is malleable and free from arsenic.

The alloys containing from 0 to 11 per cent. of arsenic were examined thermally by Schleicher. [3] The freezing point curve indicates a eutectic point at about 25 per cent. of arsenic, the melting point of which is 665°. The eutectic is still observable when only a small proportion of arsenic is present. Arsenic is evolved suddenly at the eutectic temperature on melting.

Gold and Bismuth.—Bismuth is an element which, when present in even the smallest amounts, is very injurious in the ordinary commercial applications of gold. It renders the latter very brittle and unworkable. Of all metals, it has the greatest effect in reducing the ductility of gold. Thus the author found [4] that fine gold containing 0·25 per 1,000 (1 part in 4,000) of bismuth is fragile, breaks under the hammer, and cannot be rolled. The same proportion of bismuth in standard gold (gold 916·6, copper 83·3) also causes brittleness, and differs from lead in the fact that the brittleness is not removed by annealing. Molten gold absorbs the vapours of bismuth and becomes brittle. All the alloys exhibit a strong tendency to segregation, so that when solid they are not uniform in composition.

It was observed by Arnold and Jefferson [5] that in gold containing 0·2 per cent. of bismuth, when somewhat quickly cooled from fusion, the geometry of the large primary crystals of gold is completely destroyed, and the structure may be described as consisting of irregular grains, forming a series of cells cemented together with walls which are probably formed of the gold-bismuth eutectic. The whole thus presents the appearance of a network. Slowly cooled specimens are similar in appearance, but the grains are larger. When broken by working, the fracture takes place along the cell walls, and any detached grain is perfectly malleable, probably consisting of pure gold, and can easily be beaten out into gold leaf. Similar effects were noted when lead or tellurium was substituted for bismuth.

Osmond and Roberts-Austen [6] found that in quickly cooled ingots

[1] Vogel, *Zeitsch. anorg. Chem.*, 1906, **50,** 145-157 ; *J. Chem. Soc.*, 1906, **90,** [ii.], 679.
[2] Hatchett, *loc. cit.*   [3] Schleicher, *Internat. Zeitsch. Metallographie*, 1914, **6,** 18.
[4] T. K. Rose, *33rd Annual Report of the Mint*, 1902, pp. 72-73.
[5] Arnold and Jefferson, *Engineering*, 1896, **61,** 176 ; also Andrews, *ibid.*, 1898, **66,** 541.
[6] Osmond and Roberts-Austen, *Phil. Trans.*, 1896, **A, 187,** 417-432.

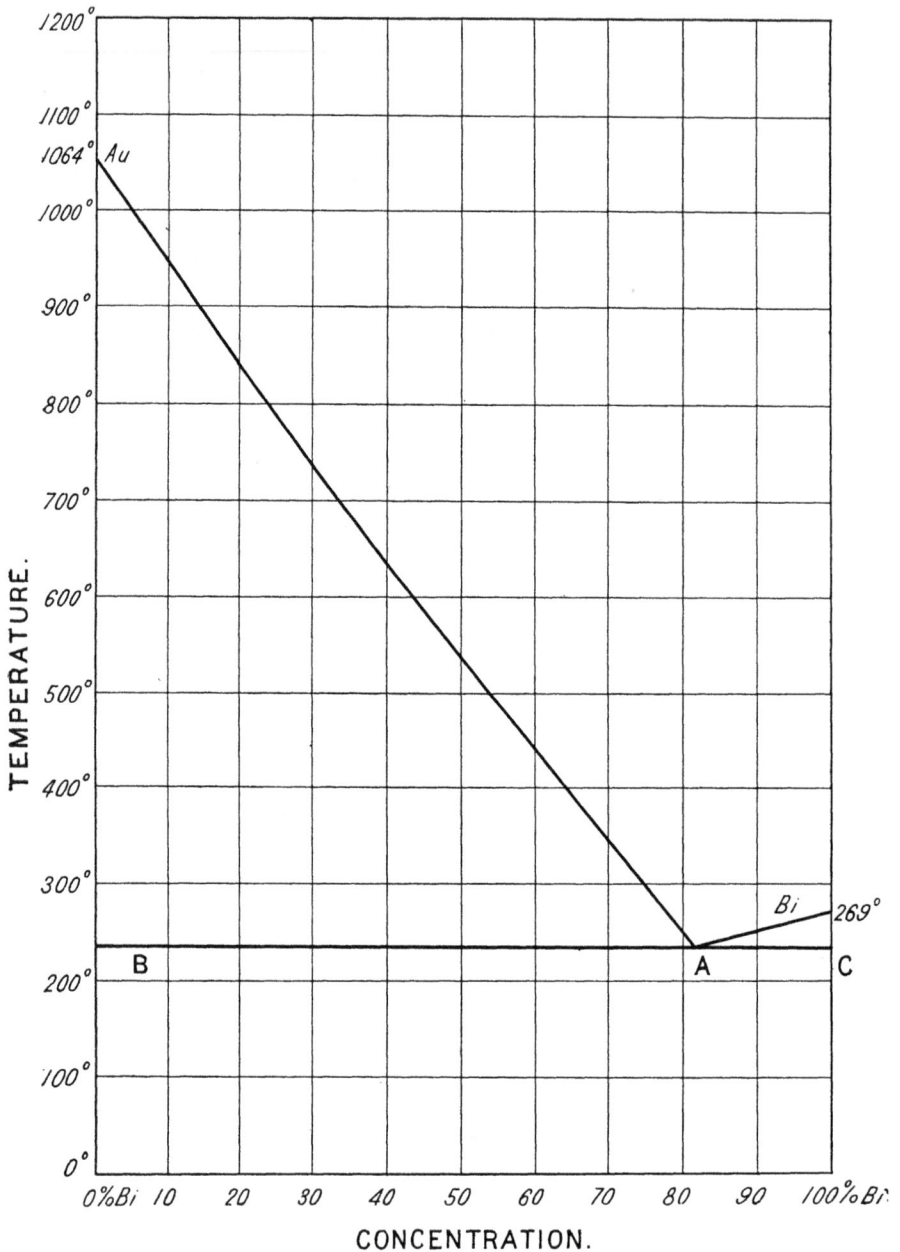

Fig. 17.—Thermal Equilibrium Curve of Gold and Bismuth.

containing 0·2 per cent. of bismuth the thickness of the joints between the crystals was 2·5 μ (0·0025 mm.). On annealing at 200° to 250°, the large grains were subdivided into small polyhedral grains.

The liquidus curve for this series of alloys was studied by Heycock and Neville [1] in 1890, but their researches only extended to alloys containing 1·8 per cent. gold, an amount which they found lowered the freezing point of bismuth by 4°. Roberts-Austen [2] showed that gold-bismuth alloys had a deep eutectic, and that the melting point of gold is regularly lowered by the addition of bismuth. The first additions of bismuth do not greatly lower the freezing point of gold, but give rise to a long pasty stage. In 1895 Roessler [3] melted a little gold with an excess of bismuth, and obtained on cooling a mixture of yellow and grey crystals. He isolated those crystals which were richer in gold, by means of nitric acid and by volatilising the bismuth under carefully controlled conditions, and found their composition to correspond very closely to the formula Au_3Bi.

It is interesting to note that MacIvor [4] has published particulars concerning a natural mineral, which is composed almost entirely of gold and bismuth, in the proportion expressed by Au_2Bi. This alloy is very malleable, and has been termed "Maldonite."

The thermal equilibrium diagram constructed by Vogel,[5] with some corrections in the solidus, is shown in the curve in Fig. 17. The liquidus curve has two branches, which meet in a eutectic point A at 82 per cent. bismuth, and at a temperature of 240° C. The eutectic horizontal B A C reaches to pure bismuth on the one side, and to gold containing less than 0·025 per cent. bismuth on the other side. This is proved by the brittleness of gold containing 0·025 per cent. of bismuth. No isomorphous mixture of gold and bismuth is known to occur. The action of gravity causes the bottom of a culot of any of the alloys to be enriched in gold. Vogel states that in these alloys the primary crystals of gold have a zoned structure, being yellow in the interior (pure gold) and white and poorer in gold on the exterior (probably nearly pure eutectic).

The structure of the eutectic alloy is quite altered by quenching from 400° C. The white element is replaced by a mixture of what appear to be yellow and black components arranged in lamellæ. The gold-bismuth alloys are all brittle, but are not harder than their constituents, and, with the exception of the gold-rich alloys, have the white colour of bismuth. The gold-rich alloys are pale yellow or greyish. They exhibit marked segregation, which is not surprising, considering the wide separation of the liquidus and solidus curves.

A peculiar effect has been observed on the surface of bismuth-rich alloys. After cooling, glittering spherical nodules of metal appear, which have been attributed to the well-known property of bismuth to expand on solidification. These globular masses have a composition approximating always to 18 per cent. gold and 82 per cent. bismuth, which is that of the eutectic. This phenomenon may be compared with the effect of heating gold-tellurium alloys or native tellurides.

Gold and Cadmium.—By immersing gold in a concentrated boiling solution of a cadmium salt, Raoult [6] obtained a compound formed by the combination of the precipitated cadmium with the gold. The constituents could be again

---

[1] Heycock and Neville, *J. Chem. Soc.*, 1892, **61**, 888.
[2] Roberts-Austen, *Phil. Trans.*, 1888, **A, 179,** 417.
[3] Roessler, *Zeitsch. anorg. Chem.*, 1895, **9**, 70.
[4] MacIvor, *Dana. A System of Mineralogy*, p. 15.
[5] Vogel, *Zeitsch. anorg. Chem.*, 1906, **50**, 145-157.
[6] Raoult, *Compt. rend.*, 1873, **76**, 156.

separated only on long continued boiling with acids. Mylius and Fromm [1] (1894) repeated these experiments and obtained similar results. They also demonstrated that a gold-cadmium alloy separates from gold chloride solutions by the action of cadmium. This separated product had a constant composition corresponding to the formula $AuCd_3$. Roberts-Austen [2] found that an alloy containing 0·202 per cent. of cadmium had a tensile strength of 6·88 tons per square inch, and an elongation on a 3-inch specimen of 44 per cent. The material drew out in the manner of a viscous solid such as pitch.

Heycock and Neville [3] in 1892 detected a new compound, AuCd, by distilling off cadmium from melts of gold and cadmium, and proving that the gold and the cadmium were present in the residue in the proportion represented by the formula AuCd. Vogel,[4] however, pointed out that this result was not beyond dispute, because the difference in the partial pressures of the combined and uncombined cadmium is small, and by continuing to heat the alloy AuCd more cadmium is volatilised.

The complete equilibrium diagram was constructed by Vogel in 1906 [5] from the results of thermal and microscopical analyses. The liquidus curve has four branches, with two breaks at 623° C. (30 per cent. Cd) and 493° C. (63 per cent. Cd) respectively. These breaks correspond to concentrations represented by the formulæ Au_4Cd_3 and $AuCd_3$. In all, there are four distinct series of crystals observable throughout the whole range—(1) solid solutions of cadmium in gold ; (2) practically pure cadmium ; (3) and (4) two intermediate series of mixed crystals, the second of which forms with (2) a eutectic alloy at 303° C. (87 per cent. Cd). The progression in the crystallisation may be summarised as follows :—From 0 to 18 per cent. cadmium a series of isomorphous mixtures of cadmium in gold separates, and at the latter stage becomes saturated. From 18 to 30 per cent. the primary separation of isomorphous mixtures is followed by the formation of the compound Au_4Cd_3. A second series of isomorphous mixtures then appears from 30 to 51 per cent. cadmium, and finally an alloy with the formula $AuCd_3$ appears up to 63 per cent. cadmium, after which practically pure cadmium separates out. Guertler [6] is of opinion that Vogel's compound Au_4Cd_3 is really of the composition AuCd.

This compound fuses at 623° C., is of a silver-grey colour, is very fragile, and exhibits a highly crystalline fracture. It is scarcely attacked by dilute nitric acid (1 : 1), but dissolves easily in hot aqua regia. The hardness reaches a maximum for the whole series in the periods 18 to 30 per cent. Cd., and 51 to 63 per cent. Cd., when it has a value of 4 (on Mohr's scale). The other alloys have a hardness almost equal to that of the constituent elements. The brittleness is greatest in the alloys containing 51 to 63 per cent. cadmium, which consists of saturated mixed crystals of AuCd and cadmium.

Alloys of gold and cadmium can be " parted " with nitric acid, and are dissolved by aqua regia. They may be also parted with dilute sulphuric acid.

In Chap. XX. will be found particulars as to the use of cadmium in the assaying of gold-silver alloys.

Gold and Cobalt.—Wahl [7] in 1910 studied the thermal, magnetic, and microscopic properties of these alloys. The experimental difficulties were great,

---

[1] Mylius and Fromm, *Ber. deut. chem. Ges.*, Berlin, 27 Jahrgang, 1894, **1,** 636.
[2] Roberts-Austen, *Phil. Trans.*, 1888, **A, 179,** 339.
[3] Heycock and Neville, *J. Chem. Soc.*, 1892, **61,** 888-914.
[4] Vogel, *Zeitsch. anorg. Chem.*, 1906, **48,** 333-346.   [5] Vogel, *loc. cit.*
[6] Guertler, *Metallographie*, p. 503.   [7] Wahl, *Zeitsch. anorg. Chem.*, 1910, **66,** 60-72.

owing to the ease with which cobalt oxide is formed on heating. This influences the results by depressing the true freezing points. The difficulty was overcome by heating the alloys in an atmosphere of nitrogen in porcelain tubes. The cobalt first attacks the porcelain, but the oxide present being thus removed, subsequent fusing does not cause any further attack.

Pure cobalt melts at 1,493° C., but can be easily supercooled, as much as 216° C. of supercooling having been observed. The cobalt-gold alloys exhibit a similar tendency to pass their melting point before becoming solid. Wahl obtained a rather unusual form of liquidus curve with many turning points by plotting atomic percentages against temperatures, but his results will probably be simplified on revision. Speaking generally, there are two branches to the liquidus curve, meeting at a eutectic at 997° C., and corresponding to a composition of about 90·1 per cent. of gold; but the accuracy of these figures is open to doubt, as few experiments were made. The saturated solid solutions (ends of the horizontal eutectic line) contain 3·5 and 94 per cent. gold respectively.

The whole of the alloys were found to be magnetic, the magnetisability decreasing, at first rapidly and afterwards more slowly, with increase in the gold content. The magnetisability also decreases with decrease in temperature, being three times as great at the eutectic temperature as when cold.

The transition point of cobalt has been determined by magnetic means to be about 1,140° C., and is found to remain practically unaltered by the addition of gold. α-cobalt comes into direct contact with the melt at the lower temperatures. Crystals which are rich in cobalt are cubic above 1,140° and isomorphous with β-cobalt.

Wahl, in his paper, describes the appearance of six-pointed stars in the eutectic, which he considers may be hexagonal cobalt crystals, but similar hexagonal crystals have been observed in the undoubtedly cubic metals, gold and copper.

The addition of cobalt to gold makes it brittle in a greater degree than the addition of nickel. Hatchett found [1] that standard gold becomes brittle if one-sixteenth part of cobalt is melted with it.

TABLE V.—GOLD AND COPPER.

	Percentage by Weight.		Atomic Percentage. Au.	Solidification Temperature, ° C.	
	Au.	Cu.		Beginning.	End.
1	100	..	100	1,063°	..
2	98·33	1·67	95	1,034°	1,011°
3	94·42	5·38	85	978°	949°
4	90·29	9·71	75	934°	916°
5	86·0	14·0	65·93	890°	887°
6	82·0	18·0	59·5	884°	883°
7	80·0	20·0	56·34	886°	884°
8	69·90	30·1	42·94	900°	894°
9	50·87	49·13	25·00	942°	920°
10	30	70	12·55	1,018°	980°
11	14·02	85·97	5·00	1,056°	1,030°
12	..	100	..	1,084°	..

[1] Hatchett, *Phil. Trans.*, 1803, **93**, 43.

Gold and Copper.[1]—These metals are miscible in all proportions when molten and solidify without segregation, forming solid alloys of sensibly

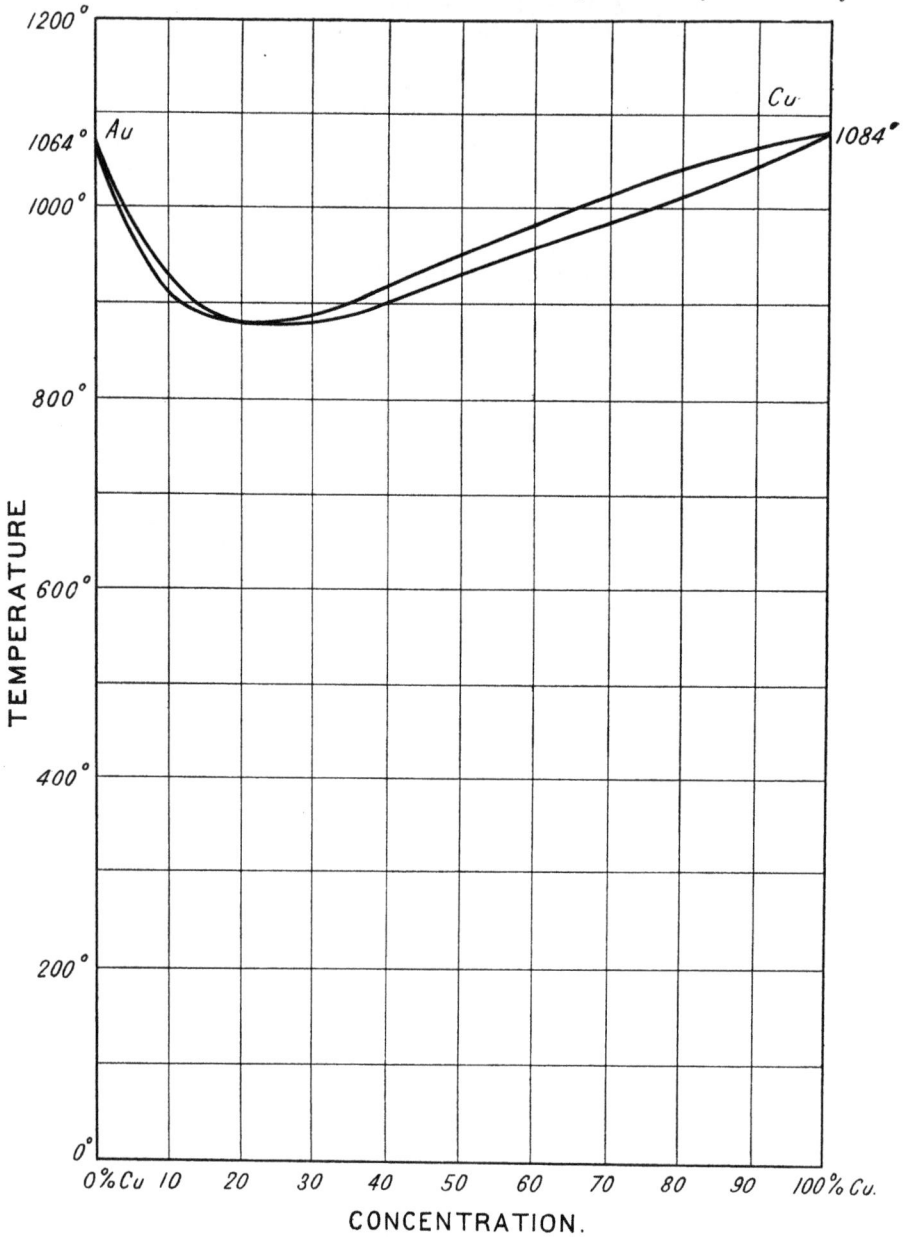

Fig. 18.—Thermal Equilibrium Curve of Gold and Copper.

---

[1] Roberts-Austen and Rose, *Proc. Roy. Soc.*, 1900, **67,** 105; Kurnakow and Zemczuzny, *Zeitsch. anorg. Chem.*, 1907, **54,** 149; *J. Russ. Phys. and Chem. Soc.*, 1907, **39,** 211.

uniform composition. The freezing and melting-point curves are shown in Fig. 18, the upper curve giving the temperature at which freezing begins, and the lower curve those at which it is completed. The difference between the curves corresponds to a pasty stage. The curve is constructed from Table V. (p. 31), which is after Kurnakow and Zemczuzny.

The freezing point of the British coinage alloy, containing gold 916·6, copper 83·3 parts per 1,000, is 949°, according to the curve in Fig. 18, or 951° according to Roberts-Austen and Rose. The freezing point of the alloy with gold 900, copper 100 parts per 1,000, is 931° by the curve. The alloy with the lowest freezing point contains gold 820, copper 180, solidifying at 884° without a pasty stage. This alloy is the only brittle member of the series, breaking with conchoidal fracture, and, being homogeneous, resembles a eutectic, but it is doubtful if it can be regarded as one. The other alloys are malleable and ductile, breaking with rough granular fracture after being nicked and bent backwards and forwards.

In the solidification of alloys rich in gold, cores are formed of solid particles containing a higher percentage of gold than the part remaining liquid, and successive layers solidify round the cores, each containing less gold than the previous one, until the last portions to be solidified form a network of copper-rich material. The grains vary in size with the velocity of cooling, being largest in slowly-cooled ingots. By long continued annealing the alloys would probably become completely uniform in composition, but with an increase in the size of the grains. After annealing the grains are nearly regular polygons. Figs. 11 and 12 (pp. 16 and 17) show standard gold before and after annealing.

In the solidification of alloys poor in gold, the cores are copper-rich, the succeeding layers containing more and more gold. For microscopic study, the alloys may be etched by aqua regia, or by ferric chloride in a hydrochloric acid solution.

According to D. M. Liddell [1] gold and silver dissolved in large amounts of copper tend to segregate, accumulating towards the bottom of a molten charge. His results on a slowly-cooled culot were as follows :—

TABLE VI.

	Silver, Ozs. per Ton.	Gold, Ozs. per Ton.
Top, . . .	53·14	5·86
Middle—Edge, .	53·08	5·82
Centre, .	59·92	6·18
Bottom, . .	69·58	6·72

With gold and copper only, in a quickly-cooled culot, the results were not so marked, and showed no definite law.

It has been pointed out by the author [2] that segregation takes place in standard gold made brittle by small quantities of lead or bismuth, the presence of 0·2 per cent. of either of these metals causing the centre of a sphere 3 inches in diameter to be enriched in gold to the extent of about one part per thousand. This is due to the fact, proved by Roberts-Austen, that a eutectic alloy of gold and lead remains molten after the remainder of the mass is solidified, and

---

[1] Liddell, *Eng. and Mng. J.*, 1910, **90**, 418.   [2] T. K. Rose, *J. Chem. Soc.*, 1895, **67**, 552.

is consequently driven towards the centre of the sphere, and that this fusible alloy contains much gold but very little copper. Jefferson and Arnold,[1] and also Andrews,[2] by micrographic studies, confirmed the view which follows from this, that brittleness in standard gold containing a little lead, bismuth, tellurium, or certain other impurities is due to the presence of films composed of such eutectic alloys separating the crystals of gold from each other, but Osmond and Roberts-Austen[3] were unable to observe these films in quickly cooled ingots.

Standard gold, however, is made too brittle for coinage by the presence of a far smaller amount of lead than 0·2 per cent. The addition of 0·05 per cent. of lead produces bars almost as brittle as those containing 0·2 per cent., and from researches by the author in the Royal Mint, it would appear that 0·01 per cent. of lead or even less renders gold unfit for coinage (see also the Gold-lead alloys).

The alloys of gold and copper are less malleable, harder, and more elastic than gold, and possess a reddish tint. Those with less than 12 per cent. of copper are easily worked ; when more than this is present they are more difficult to work owing to their hardness. Since no change of volume occurs when these alloys are formed, their densities may be calculated from those of gold and copper. The densities of gold-copper alloys as cast are given in the following table :—[4]

TABLE VII.

Gold.	Copper.	Specific Gravity at 15°.
Per 1,000.	Per 1,000.	
1,000	0	19·26
917	83	17·35
833	167	15·86
750	250	14·74
583	417	12·69
250	750	10·035
0	1,000	8·7

The densities of gold-copper alloys with large percentages of gold are given by Roberts-Austen as follows,[5] at 0° compared with water at 4°, the metal being in the form of discs compressed in a coinage press :—

TABLE VIII.

Gold.	Density.	Gold.	Density.
Per 1,000.		Per 1,000.	
1,000	19·30	923	17·57
980	18·84	916·6	17·48
969	18·58		(Broch)
959	18·36	900	17·17
948	18·12	880	16·80
938	17·93	861	16·48
932	17·79		

[1] Jefferson and Arnold, *Engineeering*, 1896, **61**, 176,
[2] Andrews, *ibid.*, 1898, **66**, 541.
[3] Osmond and Roberts-Austen, *Phil. Trans.*, 1896, **A, 187**, 417.
[4] Hoitsema, *Zeitsch. anorg. Chem.*, 1904, **41**, 65.
[5] Roberts-Austen, *7th Ann. Report of the Mint*, 1876, p. 44.

Recent determinations by the author at the Royal Mint of hard-rolled standard gold (gold 916·6, copper 83·3) give a density of 17·45 ± 0·005 at 0° compared with water at 4°. The density of a sovereign is sometimes as high as 17·48, owing to the presence of silver.

Many of the alloys have been used for coinage at various times. The Greeks and Romans, after electrum had fallen into disuse, employed the purest gold they could procure—viz., that from 990 to 997 fine. Under the Roman Emperors, however, copper was intentionally added, and in the two centuries preceding the fall of Rome very base alloys were used, some containing only 2 per cent. of gold or even less.[1] In the middle ages these base alloys were discarded, and the "byzant" of Constantinople and the "florin" of Florence were both nearly pure gold, while the first gold coins struck by the nations of Western Europe were also intended to be absolutely fine. The standard 916·6 or $\frac{11}{12}$ (i.e., 916·6 parts of gold in 1,000) was adopted by England in the year 1526, the standard of 994·8, which had been introduced in 1343, being finally abandoned in 1637. The 900 standard was introduced in France in 1794, and subsequently adopted in other countries. These two standards are now those most commonly used, the English standard being employed by Portugal, India, Turkey, Brazil, Chili, and Peru, and the French standard by all the other civilised countries except Egypt, which has a standard of 875. The Austrian ducat, however, has a fineness of 986$\frac{1}{9}$, and that of Holland a fineness of 983, but these coins are used only for trade purposes in Asia and Africa. Of all these alloys the 900 and 916·6 standards are those best adapted for coinage, keeping their colour fairly well, and resisting wear better than richer alloys. The 900 alloy is harder and wears better than the 916·6 alloy, but the difference is not great, the rate of wear depending less on such small differences of composition than on the mechanical and thermal treatment of the alloys during the operation of coining.[2] The alloys used in coinage formerly contained from two to twelve parts of silver per 1,000 in addition to the gold, but new coins now generally contain only one or two parts of silver per 1,000.

Gold-copper alloys tarnish on exposure to air owing to oxidation of the copper, and blacken on heating in air from the same cause. This oxidised coating may be removed and the colour of fine gold (not that of the original alloy) produced by plunging the metal into dilute acids or alkaline solutions, the operation being technically known as "blanching." The colour of some alloys may be improved without previous oxidation by dissolving out some copper by acids, a film of pure gold being thus left on the outside which can be burnished. French jewellers use a hot solution of two parts of nitrate of potash, one part of alum, and one of common salt for this purpose.

Nitric or sulphuric acid dissolves out the copper from gold-copper alloys under conditions similar to those under which it removes silver from silver-gold alloys. If the copper falls below 6·5 per cent. the alloy is not attacked by these acids (Pearce). Keller has shown [3] that when copper containing a small proportion of gold is dissolved in nitric acid, part of the gold is dissolved, and that more gold is dissolved from slowly cooled than from quickly cooled alloys. Aqua regia dissolves all the alloys completely.

---

[1] Roberts-Austen, Cantor Lecture, *J. Soc. of Arts*, Aug. 1884.
[2] Roberts-Austen, 14*th Ann. Report of the Mint*, 1883, p. 45, *et seq.*
[3] Keller, *Trans. Amer. Inst. Min. Eng.*, 1912, **43**, 582.

Gold and Iron.—Isaac and Tammann,[1] from a study of this series of alloys, have been unable to detect the presence of any well-defined compound. In the fused state the metals are miscible in all proportions, and form an interrupted series of isomorphous mixtures on solidification. The break in the formation of these solid solutions or mixed crystals occurs at 28 to 63 per cent. of gold. This break becomes greater at lower temperatures, owing to the transformation of the iron into another allotropic form. At 1,168° a second series of isomorphous mixtures is developed, owing to the reaction of the crystals containing 28 per cent. of gold with the fused mass. The minimum point on the liquidus curve (1,040°) is reached on the gold branch at a point corresponding to about 20 atomic per cent. of iron (or 5 per cent. of iron by weight). This composition agrees with the formula Au_4Fe. From this point, the liquidus curve rises steadily as the proportion of iron increases to 60 per cent., when the melting point is 1,405°. The curve then flattens, but continues to rise more gradually up to pure iron (1,520°).[2]

The transition temperature of iron is unaffected by the addition of gold. The hardness of the alloy containing 10 per cent. of gold is rather greater than that of pure iron, but beyond this point the hardness gradually diminishes, and an alloy containing 70 per cent. gold is considerably softer than iron. The alloys generally are hard, of high tenacity, and both malleable and ductile, so long as the proportion of iron does not exceed 80 per cent. All the alloys form with expansion and are hardened by quenching. The main difficulty in preparing the alloys is due to the high melting point of iron, but iron is gradually taken up by molten gold at a temperature of 1,100° to 1,200°, and more rapidly at higher temperatures. If cast iron, melting at 1,130°, is used, this difficulty is avoided. It is obvious that gold and its alloys when molten should not be stirred with an iron rod.

The alloys with 8 to 10 per cent. of iron are pale yellow, very ductile, and take a beautiful polish. Those with 15 to 20 per cent. of iron are used in France for jewellery under the name *or gris*. They are yellowish-grey, and, although hard, are not difficult to work. The alloy with 25 per cent. iron (*or bleu*) is also used in jewellery. Alloys with 75 to 80 per cent. of iron are silver-white in colour, and strongly magnetic.

Gold-iron alloys are not easily purified by cupellation, but the iron can be removed by sulphuric acid at 100°, if the proportion of iron is not too small. Nitric or hydrochloric acid may also be used. The action is slow. All the alloys dissolve completely in aqua regia.

Gold and Lead.—In 1905 Vogel[3] constructed the thermal equilibrium diagram of these alloys. Supercooling, amounting to 30° to 50°, was observed in the majority of the cases examined before solidification was induced either by vigorous stirring or by inoculation.

The liquidus curve has four distinct branches (see Fig. 19), and is easily followed. There are three eutectic points; at A, 215° C. (gold 14·8 per cent.), B, 254° C. (gold 27 per cent.), and C, 418° C. (gold 57 per cent.). Three series of isomorphous mixtures separate out in the whole range from gold to lead, and a consideration of the time required for the crystallisation of the eutectic shows (according to Tammann) the existence of two definite compounds

[1] Isaac and Tammann, *Zeitsch. anorg. Chem.*, 1907, **53**, 281-297; for earlier work on the influence of small quantities of iron on the properties of gold, see Hatchett, *loc. cit.*, and Wertheim, *Pogg. Annalen*, Ergzbd. 1848, **2**, 73.

[2] A diagram of the curve is given in T. K. Rose's *Precious Metals* (1909), p. 56.

[3] Vogel, *Zeitsch. anorg. Chem.*, 1905, **45**, 11-23.

of the two metals—namely, $AuPb_2$ (32 per cent. gold) and Au_2Pb (65·5 per cent. gold). The existence of these compounds has been confirmed by microscopic examination. $AuPb_2$ undergoes a polymorphic transformation

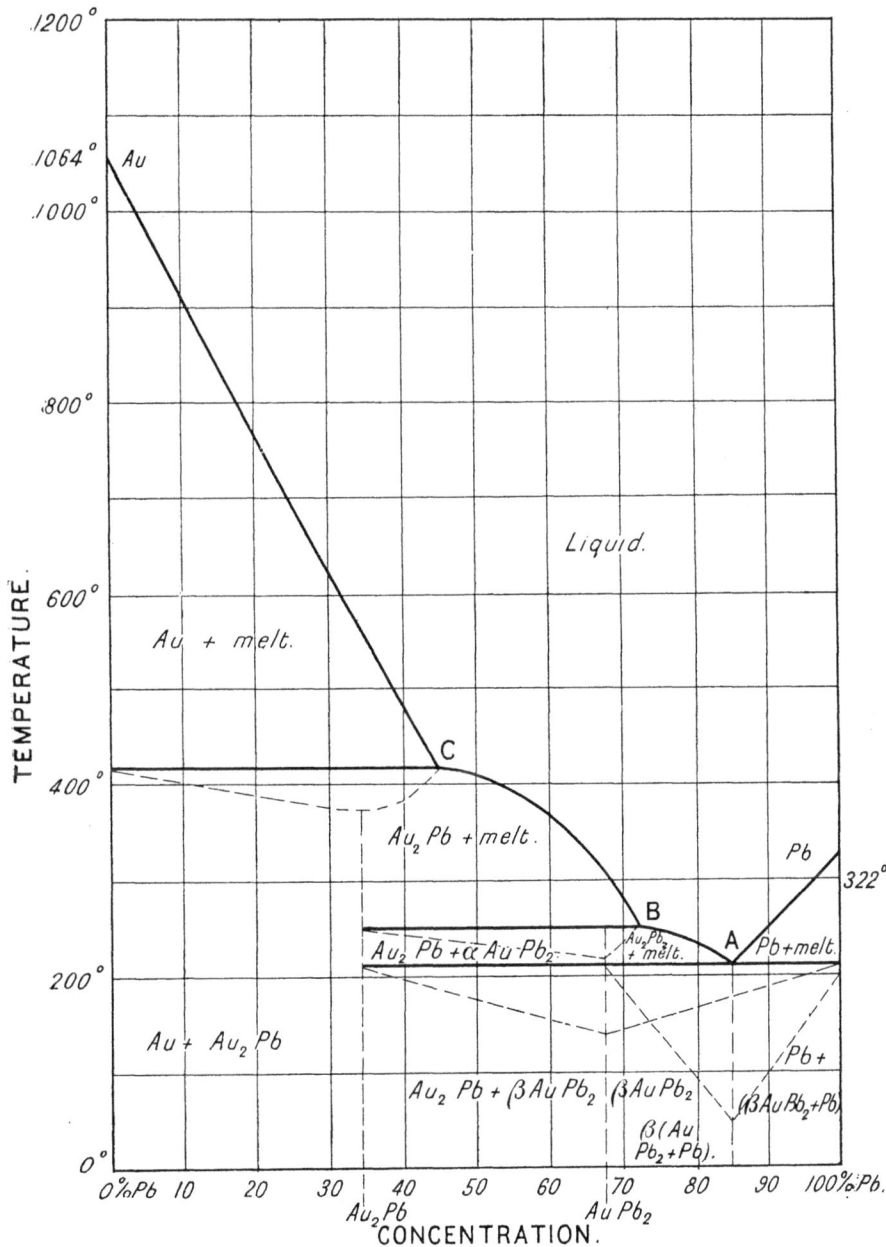

Fig. 19.—Thermal Equilibrium Curve of Gold and Lead.

at 211° C., and the conversion from the α to the β variety is reversible. This alloy forms long white needle-shaped crystals with rounded contours, while the crystals of Au_2Pb are larger, rhombic, and well-formed, thus easily distinguishable from the other constituents. Of the two compounds, Au_2Pb is the more brittle.

The appearance of three series of intermediate crystals in the case of the gold-lead alloys is particularly remarkable, and it distinguishes the latter from the neighbouring and analogous systems in the Periodic Table.

$$Au - Tl, \ Au - Bi, \ Ag - Pb \ and \ Cu - Pb.$$

All alloys containing less than 70 per cent. of gold solidify below a red heat.

In alloys containing 45 per cent. Pb three structural elements are observable—viz., (1) primary crystals of gold occurring in irregular forms ; (2) the compound Au_2Pb, which is almost completely surrounded by gold crystals ; and (3) a eutectic. Alloys with 45 to 72 per cent. lead do not show primary gold crystals, but large crystals of Au_2Pb are plainly visible in the section. These crystals are surrounded by a eutectic composed of the two compounds, Au_2Pb and $AuPb_2$. From 72 to 85 per cent. lead primary $AuPb_2$ crystals separate, and afterwards merely solid solutions of gold in pure lead.

The alloys, which are all brittle, range in colour from pale yellow through greenish-yellow to bluish-white. All are formed with expansion.

Considerable attention has been paid to the influence of small proportions of lead, bismuth, and tellurium on the chemical and physical properties of gold.[1] It is known that small quantities of lead increase the brittleness of gold considerably, and particularly is this the case with coinage alloys, 0·02 per cent. of lead being sufficient to cause cold-shortness.[2] In pure gold 0·15 per cent. of lead produces slight brittleness. None of the earlier observations resulted in any clear statement upon the significance of these mechanical effects and their relationship to the internal structure of the metal ; but it is proved that even with the very small percentages of lead, certain heterogeneous constituents (see Gold-bismuth alloys) appear in the structure, to which is attributed the rapid decrease in ductility and tensile strength of these alloys.

The vapour of lead, like that of bismuth, is absorbed by gold, making it brittle.[3]

In the smelting of lead ores,[4] the molten lead, which is formed in the incandescent mass, acts as a very efficient collector of any precious metals which may happen to be in the original ore. This mixture of lead, gold, and silver, called " base bullion," is collected in the Arent's syphon, or may be tapped from the furnace. Almost the whole of the gold may be collected in this manner, but only a proportion of the silver. Together with the base bullion, there are also formed in the lead-smelting furnaces matte (containing lead, copper, iron, and other metals in the form of sulphides), speiss (in which the iron and arsenic are gathered), and slag. In copper smelting, the copper

---

[1] Hatchett, *Phil. Trans.*, 1803, p. 43 ; Roberts-Austen, *Phil. Trans.*, 1888, **A, 179,** 339 ; *ibid.*, 1896, **A, 187,** 417 ; Heycock and Neville, *J. Chem. Soc.*, 1892, **61,** 909 ; Andrews, *Engineering*, 1898, **66,** 541 ; Jefferson and Arnold, *ibid.*, 1896, **61,** 176 ; T. K. Rose, 33*rd Ann. Report of the Mint*, 1902, p. 73.

[2] Rose, *loc. cit.*　　　　　　　　　　[3] Hatchett, *loc. cit.*

[4] For the extraction of gold from its ores by smelting with lead, see Collins, *Metallurgy of Lead*, 2nd edition.

matte there produced is the collector for the gold and silver, and it is found that here, too, the gold is absorbed more readily than the silver. In lead smelting, however, the lead is more efficient than the matte in extracting the gold, as its ability to take up this metal in solid solution is very marked. This will be seen to be in accordance with what has already been said concerning the gold-lead alloys.

The impurities in lead bars, with the exception of silver and gold, collect nearer the top than the bottom when the bars are slowly cooled, the differentiation being probably due to the differences in the specific gravities of the various constituents. Hofman [1] quotes figures to show that the precious metals concentrate more on the bottom when the bars are rapidly cooled. This liquation or, more correctly, segregation makes it necessary that due precautions should be taken in sampling such bars, in order to obtain a representative portion (see Sampling, Chap. XX.).

In the Pattinsonisation of lead containing gold, the eutectic alloy containing 14·8 per cent. of gold remains as the mother liquor, which cannot be enriched beyond this point. The gold is separated from the alloys by cupellation (q.v.), or by dissolving the lead in nitric acid. [2]

Gold and Magnesium.—According to Vogel and Urasow, [3] compounds are formed, the melting points of which are as follows :—

TABLE IX.

	Vogel.	Urasow.
$AuMg$, . . .	1,160° C.	1,150° C.
$AuMg_2$, . .	796° C.	788° C.
Au_2Mg_5, . .	. .	798° C.
$AuMg_3$, . .	830° C.	818° C.

The reaction between gold and magnesium is violent, and both investigators had recourse to special methods. Vogel melted the magnesium in an atmosphere of hydrogen and added the gold to the melt in small quantities. He found that alloys containing less than 27 per cent. of magnesium did not attack the porcelain tubes, even at 1,300° C., but that those containing more than this amount formed magnesium silicide very readily with the silica of the enclosing vessels. Urasow heated the components in the proportion required to form the compound $AuMg$ in a graphite crucible enclosed in an iron cylinder provided with a screw cap. Combination occurred at 700° C., and to the melt sufficient gold or magnesium was added under a layer of fused alkali chloride to form any particular alloy. The melting under these conditions was quiet.

Gold forms solid solutions with magnesium up to about 30 atomic per cent. of the latter metal (or about 5 per cent. by weight). The eutectic between pure gold and the point of formation of the compound $AuMg$ at 1,150° C. occurs at 827° C., corresponding to a concentration of 5·7 per cent. of mag-

---

[1] H. O. Hofman, *Metallurgy of Lead* (5th edition), p. 347.
[2] For full particulars of the Pattinson and zinc processes of separating gold and silver from lead, see Collins, *op. cit.*
[3] Vogel, *Zeitsch. anorg. Chem.*, 1909, **63**, 169-183 ; Urasow, *ibid.*, 1909, **64**, 375-396 ; Vogel and Urasow, *ibid.*, 1910, **67**, 442.

nesium. The eutectic range is from 5 to 5·7 per cent. of the base metal.
Between the maxima on the melting point curve corresponding to $AuMg_2$
and $AuMg_3$ the eutectic occurs at 770° C. and 70 atomic per cent. of magnesium.

The other two eutectics are as follows :—

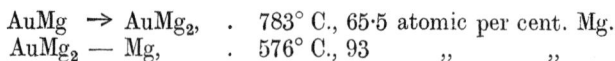

$AuMg \rightarrow AuMg_2$,   .   783° C., 65·5 atomic per cent. Mg.
$AuMg_2 — Mg$,   .   576° C., 93   ,,   ,,

Metallic magnesium absorbs but little gold in the crystallised state.

Magnesium-gold alloys containing up to 18 per cent. of magnesium retain
the yellow colour, but all others are silver-grey. Alloys with 33 to 66 per
cent. of the base metal are brittle, with a maximum hardness of 5, and have
a glassy fracture. They are stable in air at the ordinary temperature, and
only those tarnish which contain free magnesium. Swelling and disintegration
is noticed in alloys with 34 to 25 per cent. of gold at about 600° C. Dilute
acids attack all alloys containing more than 60 per cent. magnesium. Concentrated acid is required for 50 to 60 per cent. alloys, while below this concentration aqua regia must be used. Alloys rich in gold may be etched with
hydrochloric acid and bromine ; those poorer in gold may be treated by
simply rubbing with a wet chamois leather.

Gold and Mercury—Amalgams.—A piece of gold rubbed with mercury
is immediately penetrated by it and becomes exceedingly brittle. The ductility
is not always restored when the mercury is removed by distillation, a crystalline structure being often induced, perhaps because part of the mercury is
retained by the gold. A particle of gold " wetted " by mercury at once loses
its colour. A solid amalgam is formed, but is not readily dissolved in an excess
of mercury.

The amalgams of gold were studied by Böttger early in the last century,
and detailed descriptions by Sonnenschein [1] and Rammelsburg (1863, 1866)
are known, but these earlier workers devoted themselves mainly to quali-
tative work. Henry [2] had obtained an amalgam in which he showed
the solubility of gold in mercury to be 0·14 per cent., and also demonstrated the formation of a liquid phase in the series of alloys. Kasantseff [3]
isolated the crystals at various temperatures by passing the liquid amalgam
through a glass tube of 0·15 to 0·40 mm. diameter, and found the following
figures for the solubility :—

Temperature.						Percentage solubility.
0°,	.	.	.	.	.	0·110 per cent. gold.
20°,	.	.	.	.	.	·126 ,,
100°,	.	.	.	.	.	·650 ,,

These results are in agreement with those cited by Henry (0·14 per cent.)
and by Gouy (0·13 per cent.).

Tammann has studied a small portion of the liquidus curve corresponding
to the addition of minute quantities of gold to mercury, and finds that the
melting point of the latter is raised.

[1] Sonnenschein, *Zeitsch. der Deut. geologischen Ges.*, 1854, **6**, 243.
[2] Henry, *Phil. Mag.*, 1855, [iv.], **9**, 468.
[3] Kasantseff, *Bull. Soc. chim.*, 1878, **30**, [ii.], 20.

Percentage addition of gold.		Rise in M.P.
0·006 per cent. Au,	. . .	0·1°
·012 ,,	. . .	·1°
·025 ,,	. . .	·2°

The curve has a steep ascent, with which the sparing solubility of gold in mercury is in agreement.

The composition of the solid phase has been investigated by many workers, and the results obtained have been often at variance. Crookewitt[1] obtained a crystalline residue containing 32·75 per cent. gold by filtration of the amalgam through chamois leather. This composition corresponds to the formula $AuHg_2$. Henry repeated this experiment, and obtained a residue with 86 per cent. of gold by weight. Knaffl[2] noted that the quantity of crystals in the pulpy amalgam appreciably increases with a fall in temperature, and in more recent years Fay and North (1901), by centrifuging the amalgam, have succeeded in isolating a residue corresponding to the composition Au_4Hg.

Wilm[3] dissolved gold in sodium amalgam, and obtained results which were somewhat different from those of previous workers. The solution of gold in the sodium amalgam was boiled repeatedly with nitric acid, and crystals remained which contained 9·71, 11·45, 9·67, and 5·45 per cent. by weight of mercury. Henry[4] had also obtained a similar residue, containing 11·26 per cent. of mercury, which would correspond to the formula Au_8Hg. The contention that these crystals were not attacked by nitric acid is opposed by Kasantseff and Knaffl, but it has been more recently shown that with alloys having higher gold content than the above the attack by nitric acid is considerably reduced, which would tend to prove that in alloys containing up to about 10 per cent. of mercury the latter enters into a state of solid solution.

The maximum amount of mercury, however, which can be absorbed by solid gold is much greater. Thus a native amalgam of gold, Au_2Hg_3, is found in small yellowish crystals of specific gravity 15·47 in the native mercury of Mariposa, in California, containing gold 39·02 to 41·63 per cent., mercury 60·98 to 58·37 per cent. An amalgam of gold and silver occurring in white soft grains at Choco, New Granada, contains 38·39 per cent. gold, 5·00 per cent. silver, and 57·40 per cent. mercury.[5] By treating 1 part of gold with 50 parts of boiling mercury, keeping it heated for some days, allowing to cool, and carefully squeezing, Louis[6] obtained a hard alloy of gold and mercury, crystallising in silver-white, long delicate interlacing needles, and consisting of 41·43 per cent. of gold and 58·57 per cent. of mercury, thus corresponding exactly to the composition of native amalgam, but having a different crystalline form.

Schnabel and Louis[7] state that amalgam containing 90 per cent. of mercury is fluid, that with 87·5 per cent. is pasty, and that with 87 per cent. forms yellowish-white crystals. They give no evidence as to the truth of these doubtful statements. In practice in gold mills, the solid amalgam filtered

---

[1] Crookewitt, *J. prakt. Chem.*, 1848, **45**, 87 ; *Ann. Chim. Pharm.*, 1848, **68**, 289.
[2] Knaffl, *Dingl. Poly. J.*, 1863, **168**, 882.
[3] Wilm, *Zeitsch. anorg. Chem.*, 1893, [iv.], 325.
[4] Henry, *Phil. Mag.*, 1855, **9**, [iv.], 468.
[5] Rammelsberg, *Mineralchemie*, p. 10 ; Watt, *Dictionary of Chemistry*, 1864, **2**, 927.
[6] Louis, *Handbook of Gold Milling*, 1894, p. 82.
[7] Schnabel and Louis, *Handbook of Metallurgy*, 1905, **1**, 924.

off from the mercury by squeezing it through chamois leather usually contains about 1 part of gold to 2 parts of mercury, but it is doubtful whether this is a homogeneous mixture. The composition of a saturated solid solution of mercury in gold may, therefore, be regarded as unknown, with a presumption in favour of the view that mercury is present to the extent of not less than 60 per cent. This saturated solid solution is " wetted " by fresh mercury, but if more mercury is added the excess can be removed by filtering under pressure. Pasty amalgam is left in the filter bag, and becomes " drier " and less pasty the more the pressure is increased, until the whole of the liquid phase has been removed. The amount of pressure, however, does not alter the composition, either of the liquid or the solid phase.

On the application of heat, crystals of gold amalgam lose their contained mercury, but without change of their shape, and to some extent they retain their surface lustre. From the crystal structure it would appear that after heating, a porous mass remains, in which the interstitial spaces are the places from which the easily volatilised component—i.e., the mercury—has been removed. The original outer shape, however, is preserved. Similar phenomena of residual porous structure have been observed in the case of other amalgams, such as those of chromium and manganese.

From an observation by Souza[1] that at a concentration corresponding to the formula Au_9Hg a decided diminution in the rate of evaporation of the mercury occurred, it was concluded that there should exist a compound having this formula. But it was shown later by Merz and Weith[2] that these crystals still continue to give off their mercury very slowly after this particular concentration has been passed, and figures are given to prove their contention (see Table X.).

TABLE X.

Boiling Sulphur. 444° C.		Boiling Mercury. 353° C.		310° C.	
Time (Hours).	Per Cent. Mercury in Residue.	Time (Hours).	Per Cent. Mercury in Residue.	Time (Hours).	Per Cent. Mercury in Residue.
20	1·07	24	2·63	24	7·86
44	0·40	75	1·75	48	5·3
60	0·33	114	1·48	75	4·5
..	..	134	1·43	..	..

It has been noted as a general principle that the tendency to form intermediate series of crystals diminishes with rise in the atomic weight in the same periodic system. Such crystals do appear in the case of copper and silver amalgams, and in the former more readily than in the latter, so that it is not surprising that they are absent in the gold amalgams.

It has been shown by MacPhail Smith,[3] in studying the phenomena of diffusion that gold does not dissolve in mercury in the form of free atoms,

[1] Souza, Ber., 1876, 8, 1616 ; 9, 1050.
[2] Merz and Weith, Ber., 1881, 8, 1616 ; 9, 1050 ; 14, 1438.
[3] MacP. Smith, Ann. Phys., 1908, 25, [iv.], 252.

but as molecules of the type Au_4Hg, and that although intermediate crystals have not been obtained in the series, nevertheless compounds of Hg and Au do appear to be formed in the liquid phase.

The composition of the amalgam recovered in mills varies considerably. It may be regarded as a collection of little nuggets of gold, coated and partly saturated with mercury. The amalgam sinks to the bottom of the mercury. Coarse particles of gold are not saturated with mercury to their centres, although the outside layers of the particles consist of a saturated solution. The saturation of the interior, depending on the diffusion of mercury through solid amalgam, would probably not be complete for many days. The finer (*i.e.*, smaller) the particles of gold, the more nearly the interior approaches to saturation. It follows that coarse gold gives rich amalgam, and fine gold poor amalgam. The limits are pure gold on the one hand and a saturated solution of mercury in gold on the other, neither of which exist in practice. The practical limits are amalgams containing about 50 per cent. and 25 per cent. of gold respectively. These gold amalgams usually contain impurities in the shape of amalgams of silver and of base metals, as well as non-metallic substances. The actual value of squeezed amalgam consequently varies from £2 to a few shillings per ounce.

When amalgams are gradually heated, the mercury is distilled off by degrees, the action soon ceasing if the temperature is allowed to become stationary, and distillation recommencing if it is again raised. At 440° (somewhat below a red heat), an amalgam containing about three parts of gold to one of mercury is obtained, and at a bright red heat almost all the mercury is expelled, and if the heating has not been pushed too rapidly the vapours contain but little gold. According to Louis,[1] when properly heated the gold retains 1 to $1\frac{1}{2}$ per cent. of mercury, and the mercury vapours contain 0·005 per 1,000 of gold. The gold obstinately retains a little mercury, which is not driven off below the melting point of gold.

Gold and Nickel.—These alloys have only recently been systematically studied by Levin,[2] although Lampadius[3] showed that gold alloyed completely with nickel, this being contrary to the behaviour of the latter metal with silver. C. Hatchett[4] had also made observations upon alloys containing gold and nickel, with the view to testing their suitability for coinage purposes. Levin evolved the equilibrium curve from a study of the thermal properties of the series of alloys.

Alloys with 5 to 20 per cent. nickel and 50 to 70 per cent. nickel show a crystallisation interval, and the two branches of the liquidus curve intersect at a temperature of 950° C., corresponding to a composition of 53 atomic per cent. of the base metal. There is no evidence of the formation of a compound between the two metals, as no maximum was obtained in the curve nor any change in direction in either branch. On the one branch of the curve a saturated solution of nickel in gold separates, and on the other a saturated solution of gold in nickel. From the data he obtained, and the fact that the eutectic crystallisation is very marked only with the alloy containing 40 per cent. nickel, the curves for the other alloys showing only a break, Levin concluded that on what appeared to be the eutectic horizon no real eutectic alloy separated, but rather that there occurred a simultaneous solidification of two series of isomorphous mixtures. This conclusion was

[1] Louis, *loc. cit.*   [2] Levin, *Zeitsch. anorg. Chem.*, 1905, **45**, 238-342.
[3] Lampadius, *Schweigg. Journ.*, 1814, p. 174.   [4] Hatchett, *loc. cit.*

confirmed by microscopic examination. Alloys containing 5 to 10 per cent. of nickel are mostly homogeneous, while those with 20 to 90 per cent. of this metal show two contiguous structural elements, one easily and the other slowly attacked by nitric acid.

Alloys, too, whose nickel content is greater than that of the eutectic, are more readily attacked by nitric acid when they are slowly cooled than when they are quenched.

The transformation point of nickel-gold alloys lies very near the transformation temperature of pure nickel (323°)—i.e., the temperature of transformation of pure nickel from the α to the β variety is not appreciably altered by the addition of pure gold.

Similar results to the above have been obtained from a study of the magnetic properties of the alloys. At the transformation temperature of pure nickel, and passing from a higher to a lower temperature, there is formed a material of very slight permeability from one of much greater permeability.

Hatchett found [1] that the alloy, gold 916·6, nickel 83·3 parts per 1,000, is of the colour of fine brass, and breaks under the hammer with a coarse-grained earthy fracture, and that one-sixtieth part of nickel in standard gold makes it slightly brittle, but, according to Lampadius, the alloys are ductile, although hard and susceptible to polish. They range in colour from yellow to white, and are magnetic.

Gold and Palladium.[2]—Berzelius [3] records the finding of a natural gold-palladium alloy at Porpez, in Brazil, which contained 85·98 per cent. Au, 9·85 per cent. Pd, and 4·17 per cent. Ag. Mallet [4] also quotes a case of the occurrence in Brazil of a native alloy of almost pure palladium and gold of the following composition :—91·06 per cent. Au, 8·21 per cent. Pd, traces of silver and iron.

Gold containing 10 to 20 per cent. palladium is nearly white, and is hard and ductile. The 50 per cent. alloy is iron grey in colour, and is not ductile.

Gold and palladium form a continuous series of isomorphous mixtures, the equilibrium curve, therefore, being a simple curve concave to the axis denoting composition, and having only a small interval between the solidus and the liquidus. It is worthy of note that palladium also forms unbroken series of solid solutions with silver and copper.

The melting point of palladium is lowered 7° by the addition of 10 per cent. of gold, and that of gold is raised 207° by 10 per cent. Pd. The hardness increases with the palladium content up to 70 per cent. of that metal, but beyond this point it decreases. The electrical conductivity and the temperature coefficient sink to minimum values at 55 per cent. of gold, while the tensile strength has a maximum at 73 per cent. of gold. The thermo-electric power, as measured against platinum, has the greatest value for an alloy containing 60 per cent. of gold.

The capacity of palladium for occluding hydrogen is lowered by the addition of gold. Graham [5] showed that a 50 per cent. alloy exhibits an appreciable diminution in the power of palladium to absorb hydrogen as the following figures show :—

[1] Hatchett, loc. cit.
[2] Gubel, Zeitsch. anorg. Chem., 1910, 69, 38 ; Ruer, ibid., 1906, 51, 391-6.
[3] Berzelius, Pogg. Annalen, 35, 514.
[4] Mallet, Chem. News, 1882, 46, 216.
[5] Graham, Proc. Roy. Soc., 1869, 17, 500.

Volume of hydrogen absorbed
(in times own volume).

Pure palladium, ... 956·3
50 per cent. alloy, ... 459·9

More recently Berry,[1] by using a cell in which the cathode was composed of a palladium-gold alloy and the anode of pure platinum, has obtained results showing that 1 gramme of pure palladium absorbs 70 c.c. of hydrogen under normal conditions, and that this value decreases in a straight line to zero for alloys containing less than 25 per cent. of palladium.

Gold and Platinum.—Like palladium, platinum forms an unbroken series of solid solutions with gold (see Fig. 20). Observations have been undertaken by Erhard and Schertel (1879) and Dörinckel,[2] although no results were obtained for alloys containing over 60 per cent. of platinum. However, it is probable that no great variation from the curve given in the figure will be found on further investigation.

Barus[3] has investigated the electrical properties of the alloys, and his results support this view.

Additions of platinum to pure gold raise the melting point of the latter, and form malleable alloys. Up to 30 per cent. of platinum the yellow colour of gold persists, though in a diminishing degree, but alloys with a greater proportion of platinum assume the white colour of the latter element. An alloy of 75 per cent. Pt and 25 per cent. Au resembles grey cast iron in its hardness and brittleness.

It has been shown by Edward Matthey[4] that ingots containing 5 to 20 per cent. of platinum do not possess a homogeneous structure. Segregation occurs, and the platinum is concentrated in the interior.

The following table is a summary of the principal figures obtained by him from experiments on cast spheres of metal, 3 inches in diameter :—

TABLE XI.

	Alloy.	Maximum Difference in Assay between Centre and Outside.	Remarks.
1.	Au, 900 Pt, 100	Per 1,000. 58·9	
2.	,,	30·6	Heated in oil furnace, melted several times, and thoroughly stirred.
3.	,,	19·7	No. 2 alloy remelted.
4.	,,	27·2	No. 3 remelted in oxy-hydrogen flame in lime furnace.
5.	Au, 100 Pt, 900	6·9	Melted in oxy-hydrogen flame in lime furnace several times, and mixed thoroughly before pouring.
6.	Au, 250 Pt, 750	2·5	Max. diff. between intermediate and outside = 7·7 per thousand.

[1] Berry, *J. Chem. Soc.*, 1911, **99**, 463-6.　　[2] Dörinckel, *Zeitsch. anorg. Chem.*, 1907, **54**, 345.
[3] Barus, *Amer. J. Sci.*, 1888, **36**, 427.　　[4] Matthey, *Phil. Trans.*, 1892, **A, 183**, 629.

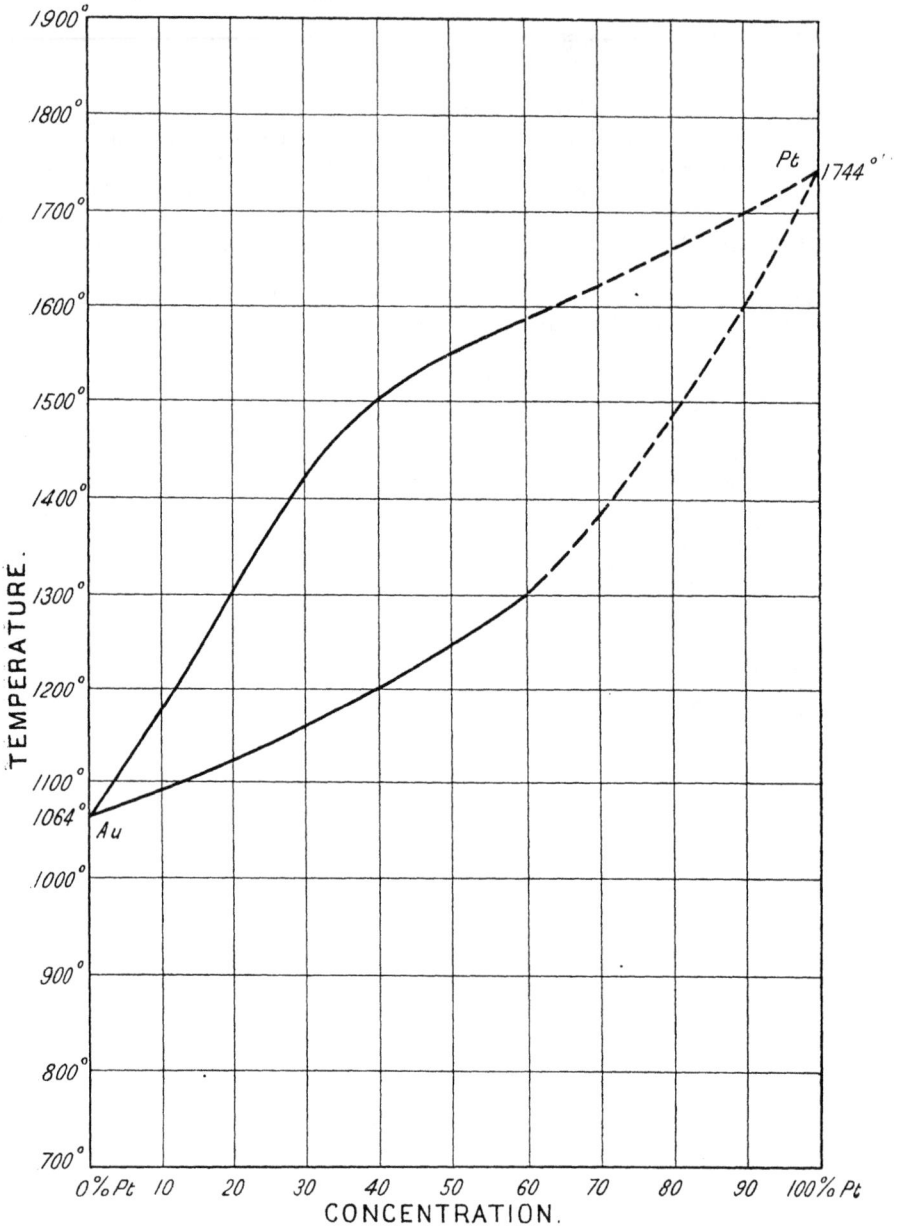

Fig. 20.—Thermal Equilibrium Curve of Gold and Platinum.

Table **XI.** shows that much depends on the temperature of casting, and that even where the platinum constitutes the major portion of the material, it is still concentrated in the interior of the sphere.

Platinum cannot be separated from gold by the ordinary methods of cupellation and parting. Both metals are dissolved by aqua regia, however.

Gold-platinum alloys are used in dentistry in the form of wire, and for some varieties of filling.

Gold and Silver.—Gold and silver are miscible in all proportions, forming an uninterrupted series of solid solutions (or isomorphous mixtures) with properties intermediate between those of the two metals. The first additions of silver do not perceptibly lower the melting point of gold, so that gold parted from silver by nitric acid, and containing about 999 parts of gold to 1 of silver, can be used instead of pure gold for calibration purposes in testing pyrometers, by observing the melting point, with very little error. The curve of equilibrium has been traced out by several observers [1] (see Fig. 21).

Jänecke's table of solidification points is as follows :—

TABLE XII.—SOLIDIFICATION POINTS OF GOLD-SILVER ALLOYS.

Composition.	Beginning of Solidification. ° C.	End of Solidification. ° C.
Gold, 100, .	1,064°	. .
,, 94, silver, 6, .	1,060°	1,053°
,, 88½, ,, 11½, .	1,054°	1,042°
,, 81½, ,, 18½, .	1,047°	1,035°
,, 73½, ,, 26½, .	1,036°	1,026°
,, 64½, ,, 35½, .	1,031°	1,018°
,, 54½, ,, 45½, .	1,015°	1,003°
,, 43½, ,, 56½, .	1,007°	997°
,, 32, ,, 68, .	993°	982°
,, 18, ,, 82, .	977°	970°
Silver, 100, .	961·5°	. .

The first additions of gold to silver raise its melting point.

The alloys are homogeneous, malleable, soft, and ductile. They are suitable for the material of trial-plates, as they are uniform in composition. The colour of gold is sensibly lowered by the addition of very small quantities of silver, and, on increasing the proportion of the latter, the colour changes by tints of a greenish-yellow (when from 20 to 40 per cent. of silver is present) to a faint yellowish-white (when 50 per cent. of silver is present), and white with a scarcely perceptible yellow tinge (when 60 per cent. of silver is present). The yellow colour finally disappears when some proportion between 60 and 70 per cent. of silver is present. The alloys containing small quantities (less than 20 per cent.) of gold are said to separate by gravity into two parts if kept for some time in a state of quiet fusion; an alloy containing one part of gold to five parts of silver sinks to the bottom, and slightly auriferous silver floats at the top.

The silver-gold alloys most used in jewellery are said to be *green gold* (silver 25, gold 75), *dead-leaf gold* (silver 30, gold 70), and the alloy containing

---

[1] Roberts-Austen and Rose, *Proc. Roy. Soc.*, 1902, **71**, 161 ; Raydt, *Zeitsch. anorg. Chem.*, 1912, **75**, 58 ; Jänecke, *Metallurgie*, 1911, **8**, 597 ; Heycock and Neville, *Phil. Trans.*, 1897, **A, 189**, 69.

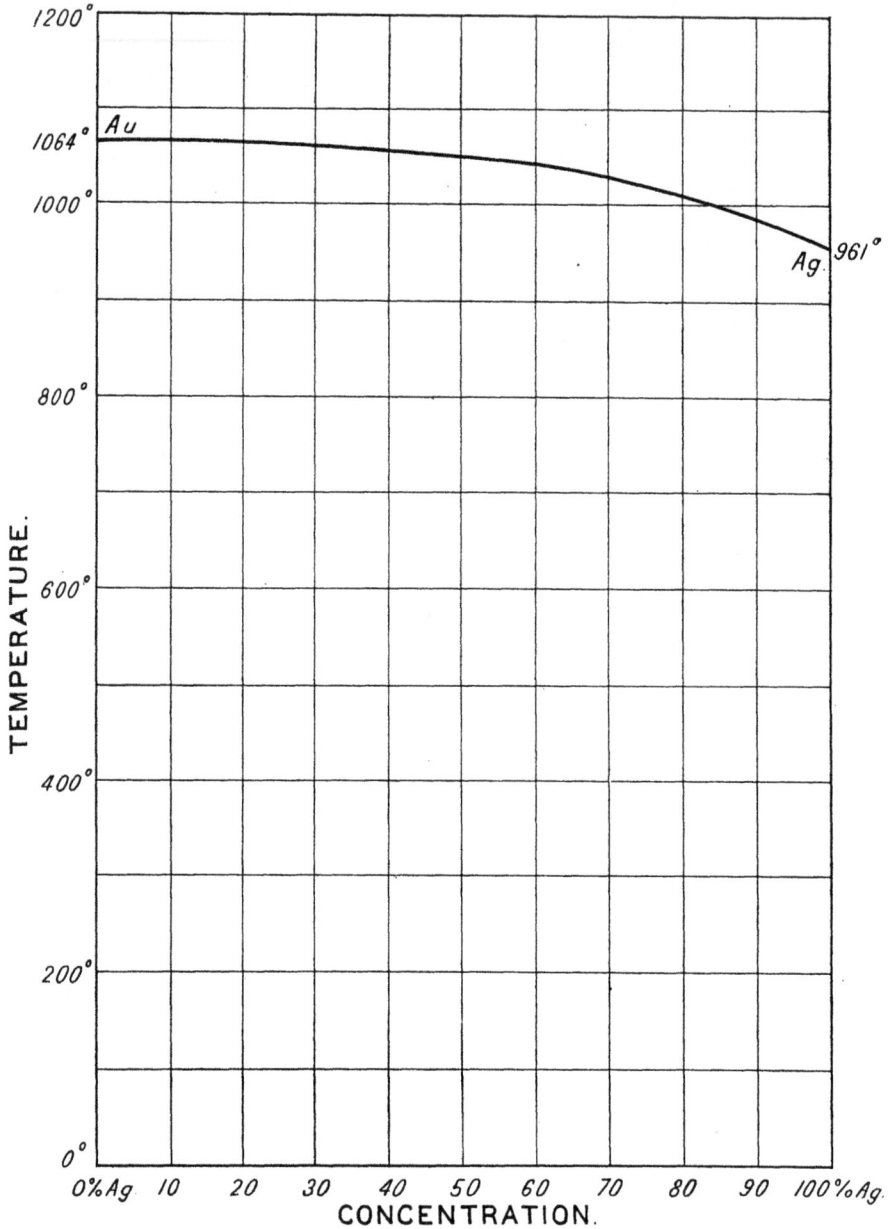

Fig. 21.—Thermal Equilibrium Curve of Gold and Silver.

40 per cent. of silver. Triple alloys of gold, silver, and copper are employed far more frequently by English jewellers than those last mentioned; of these, the alloys consisting of 22-, 18-, 15-, 12-, and 9-carat gold respectively,

can be Hall marked.[1] The relative amounts of silver and copper in these alloys is very variable, and the lower standards often contain zinc. The following are typical alloys :—

TABLE XIII.

		Gold.	Silver.	Copper, etc.
22 carat,	.	916·7	20	63·3
18 ,,	.	750	125	125
15 ,,	.	625	100	275
12 ,,	.	500	100	400
9 ,,	.	375	100	525

Other analyses are given by E. A. Smith.[2] The melting points of these triple alloys are given by Jänecke.[3]

Alloys of gold and silver were much used for coinage before the methods of parting became well known and inexpensive.

Electrum includes pale yellow alloys with from 15 to 35 per cent. of silver. It occurs native, and was much used for ornaments and coins by the Greeks and Romans, and by the nations which acquired their arts. It was the metal used for the earliest known coins, which were made by Gyges in Lydia about B.C. 727. Rods of electrum, containing gold 651 parts, silver 334 parts per 1,000, were used as money in Asia Minor at an earlier period.[4] The use of silver in the gold-copper coinage alloys was not discontinued until quite recently, all English guineas and the Australian sovereigns manufactured at Sydney up to the year 1871 containing some of it.

Both nitric and sulphuric acids attack silver-gold alloys, almost completely dissolving out the silver if it is present in amounts variously stated as at least 60 to 70 per cent., while, if the proportion falls below 60 per cent., some of the silver is left undissolved with the gold. Hydrochloric acid

TABLE XIV.—DENSITIES OF GOLD-SILVER ALLOYS.

Gold.	Silver.	Specific Gravity at 15°.
Per 1,000.	Per 1,000.	
1,000	0	19·26
917	83	18·08
879	121	17·54
843	157	16·96
750	250	16·03
667	333	15·07
500	500	13·60
333	667	13·00
250	750	11·78
167	833	11·28
0	1,000	10·45

[1] 22-carat gold contains $\frac{22}{24}$ by weight of fine gold, and so on. For details of these alloys see Gee, *Goldsmith's Handbook*, pp. 41-52.

[2] E. A. Smith, *The Sampling and Assay of the Precious Metals* (1913), p. 323.

[3] Jänecke, *loc. cit.*

[4] Schliemann's *Ilios*, p. 496. London, 1880. See Roberts-Austen, *J. Soc. Arts*, Aug. 1884.

4

scarcely attacks these alloys, and the action of aqua regia is soon arrested if the proportion of silver is considerable. They may be dissolved by a mixture of nitric acid and a concentrated solution of common salt.

The densities of the alloys of silver and gold are as in Table XIV.[1]

Tellurium and Gold.—For the gold tellurides occurring in nature, see p. 81.

By heating gold in the vapour of tellurium, Brauner[2] proved that the affinity between these substances is smaller than between tellurium and either silver or copper. He also showed the formation of what he considered to be a compound having the formula Au_2Te, which, when heated, lost some of its tellurium and became enriched in gold. The existence of such a compound, first asserted by Margottet, and here again published by Brauner, is contested by Coste,[3] who, from a microscopical study of the alloys, and also from experiments on the potential gradients obtained in solutions of nitric acid, obtained no evidence of its formation.

Lenher[4] has more recently studied the effect of heating this product at various temperatures and for various lengths of time. He found that after six hours' heating the exterior of the alloy remained golden, while the interior had assumed a much paler colour, approximating to white. He also found that by treating gold-tellurium alloys with sulphur chloride or nitric acid, pure gold was obtained. Moreover, the alloys precipitated gold from gold solutions, just as pure tellurium and the natural tellurides did. From these results Lenher concludes that both the natural and artificial tellurides are mixtures only, and not compounds. It may be taken as proved that compounds of gold and tellurium cannot be formed by precipitation, and that in any case they are easily decomposed.

Berzelius[5] believed he had obtained gold telluride by passing H_2Te into a solution of gold chloride, but more recent attempts to produce gold tellurides by the action of (1) excess of hydrogen telluride, (2) sodium telluride solution, on solutions of gold chloride have failed, pure gold being precipitated in every case. Dr. Tibbals[6] has also endeavoured to prepare a compound of the two elements by the reaction between sodium telluride and sodium sulphaurate solutions in an atmosphere of hydrogen, but obtained precipitates of varying composition with different concentrations of the reacting solutions, from which no definite conclusions could be drawn.

The bearing of the above observations on the natural occurrence of gold tellurides is interesting. Gold is transported from place to place in the form of a solution of one of its compounds, and these solutions, on meeting others containing tellurides, or on coming into contact with masses of telluride ores, throw down their gold. It may be also, as Van Hise points out,[7] that the solutions containing gold and tellurium respectively meet under reducing conditions, when both elements will be deposited, forming non-homogeneous mixtures. Actual observation does, in fact, show that the occurrence of tellurides is largely in the zone of reduction.

Concerning the common minerals containing both silver and gold in the

[1] Hoitsema, *Zeitsch. anorg. Chem.*, 1904, **41**, 66. These results agree closely with those obtained by Matthiessen, *Proc. Roy. Soc.*, 1859, **10**, 12.
[2] Brauner. *J. Chem. Soc.*, 1889, **55**, 391.
[3] Coste, *Compt. rend.*, 1911, **152**, 859-862.
[4] Lenher. *Economic Geology*, 1909, [iv.], No. 6.
[5] Berzelius, *Pogg. Annalen*, 1826, **8**, 178.
[6] Tibbals, *Bull. University of Wisconsin*, 1909, No. 274; *J. Amer. Chem. Soc.*, 1909, **31**, 902.
[7] Van Hise, *Treatise on Metamorphism*, p. 1122.

form of tellurides, it is probable that the two metals are carried in solution as chlorides along with an alkaline chloride, and that this mixture is met by another containing tellurides in association with sulphides, causing the deposition of more complex mixtures.

The thermal curve for the equilibrium between these elements has been studied by Rose,[1] von Pelabon,[2] and Pellini and Quercigh [3] (see Fig. 22), all of whom are in agreement that there is a maximum at the point A corresponding to a concentration of 44 per cent. of gold ($AuTe_2$), and a temperature of about 470° C. (Pellini and Quercigh). It will be noticed that this is the formula assigned to the natural mineral calaverite. It appears, therefore, to follow that the compound $AuTe_2$, whether natural or artificial, can only be formed on solidification from fusion. No other compound exists. The curve exhibits two eutectics, which occur at 39 per cent., B (Fig. 22), and 79 per cent. of tellurium, C, respectively. The corresponding eutectic temperatures are 450° C. and 416° C. (Pellini and Quercigh), or 432° C. and 397° C. (Rose).

The solubility of tellurium in gold crystals is very small, if any, and consequently its effects on the mechanical properties of gold are great. As with lead, minute quantities (0·025 per cent.) cause extreme brittleness in gold, rendering the commercial working of the latter impossible. (See Gold-bismuth and Gold-lead.)

Gold containing small quantities of tellurium is remarkable for the fact that on annealing at a low red or black heat it becomes more brittle than before annealing. This appears to point to the conclusion that tellurium is slightly soluble in solid gold (forms an isomorphous mixture) at high temperatures, and less soluble or insoluble at low temperatures. Bismuth-gold is unaltered on annealing, but lead-gold becomes less brittle, the lead appearing to pass into solution at the annealing temperature.[4]

On heating in air, the tellurium in alloys burns to oxide, leaving moderately pure gold, which remains brittle from the presence of tellurium.

Gold and Thallium.—Roberts-Austen,[5] alone and also in conjunction with Osmond, investigated the effect of small additions of 0·2 per cent. of thallium on the physical properties of solid gold. It is a little uncertain whether the measurements applied to the stable alloys, but it was shown that even very small quantities of thallium acted like lead, causing extreme brittleness in the gold, especially on heating. From a study of the equilibrium curve, as given by Levin,[6] this can be easily understood, for at very small concentrations of thallium the alloys contain it in the form of heterogeneous particles, which begin to melt at the eutectic temperature (131° C.), and which, moreover, are themselves very soft, and thus further diminish the strength of the alloy.

The full temperature concentration diagram derived by an application of Tammann's thermal method of analysis is of a very simple nature, being composed of two branches in the liquidus meeting at a single eutectic at 131° C. (27 per cent. Au). It affords no evidence of the formation of a compound

[1] T. K Rose, *Trans. Inst. Mng. and Met* , 1907-8, **17**, 285.
[2] Von Pelabon, *Compt rend.*. 1909, **148**, 1176-7.
[3] Pellini and Quercigh, *J. Chem. Soc.*, 1911, **100**, [ii.], 45 ; *Atti R. Accad. Lincei*, 1910, [v.], **19**, ii . 445.
[4] T. K. Rose. *33rd Ann. Report of the Mint* (1902), p. 73.
[5] Roberts-Austen, *Phil. Trans.*, 1888, **A**, **179**, 339 ; 1896, **A**, **187**, 417 ; *Contr. a l'étude, des alliayes*, 1901, 71; *Bull. Soc. d'encour*, 1896, [v.], **1**, 1136.
[6] Levin, *Zeitsch. anorg. Chem.*, 1905, **45**, 31-38.

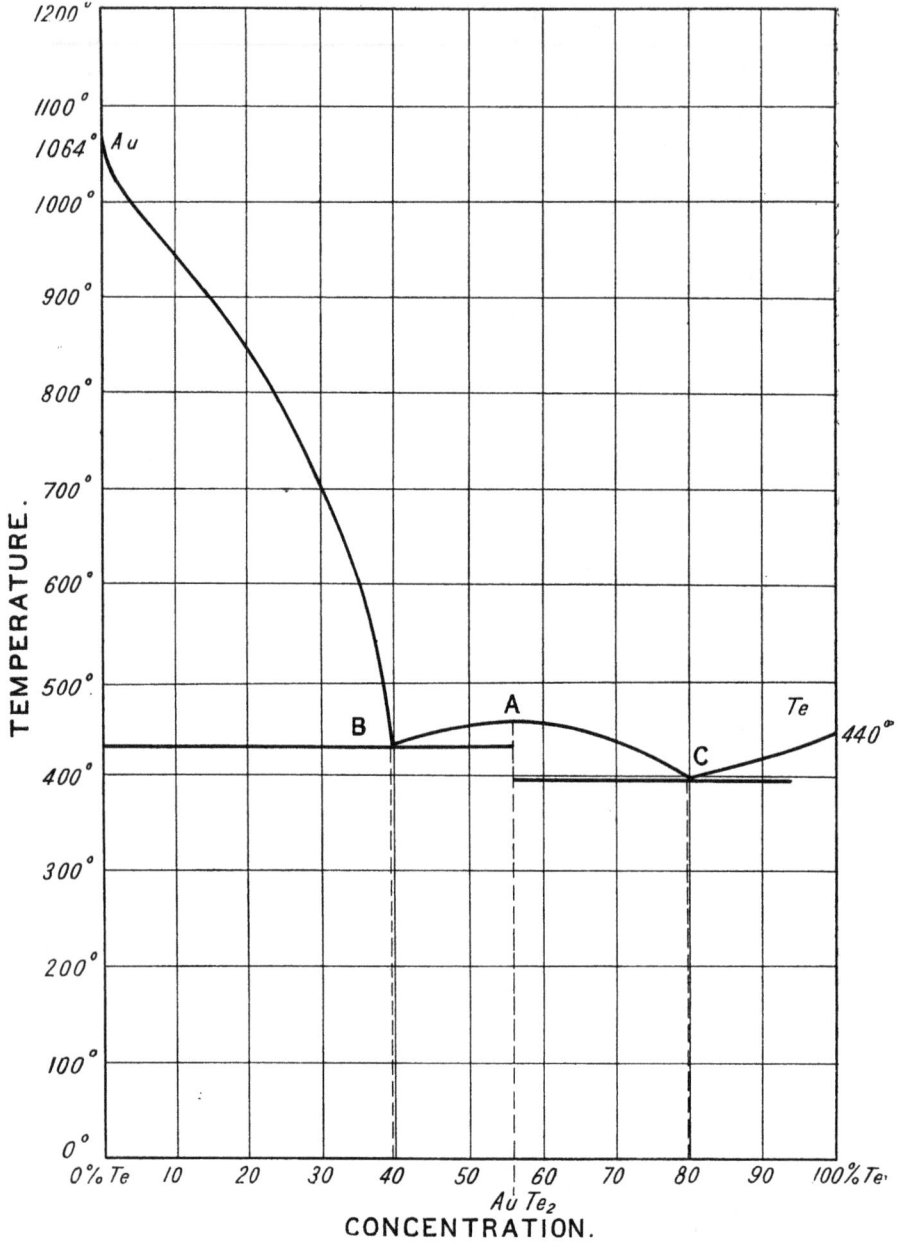

Fig. 22.—Thermal Equilibrium Curve of Gold and Tellurium.

and this is confirmed by microscopical examination. Levin observed that the yellow crystals of the gold show zoned colourings from the centre to the periphery of the individual crystals, whence it may be inferred that possibly quite appreciable amounts of thallium may be absorbed by gold on reaching the equilibrium state (*i.e.*, on annealing the brittleness should disappear). Gold dissolves in thallium crystals only to a small extent.

The thallium-gold eutectic melts at a lower temperature than any other alloy of gold except amalgam. A very small quantity of thallium causes a long pasty stage in the solidification of gold, terminating at 131°.

Gold and Tin.—Mathiessen,[1] Von Maey,[2] Le Chatelier,[3] and Guertler[4] have investigated the electrical and other physical properties of these alloys, while the electro-chemical properties have been observed by Raoult,[5] Mylius and Fromm,[6] Laurie,[7] and Le Chatelier. One interesting feature of the curve showing the relationship between the electro-chemical potential and the concentration is that on passing the point corresponding to a composition represented by AuSn there is a greater increase in the potential than one would be led to expect for this alloy. Heycock and Neville[8] studied the effect of small additions of gold—never exceeding 4 per cent.—on the melting point of pure tin. They found that 10 per cent. of Au lowers the melting point of tin by 18° C.

Vogel's[9] curve of thermal equilibrium is given in Fig. 23. He has also compared the curve with the curves of (1) electrical conductivity (Mathiessen), (2) electro-chemical potential (Laurie), (3) specific volume, to the first two of which reference has already been made. From the melting point of pure gold the liquidus falls very sharply to a eutectic at 280° C. (A, Fig. 23), and 20 per cent. tin. In this range there separate primary crystals of gold, surrounded by mixed crystals of gold and tin, together with small portions of eutectic. The curve rises again to a maximum, B, at 418° C. and 37·63 per cent. tin, denoting a compound with the formula AuSn, which separates in round dendritic crystals. Another eutectic appears at 217° C. and 90 per cent. tin, the point C, two breaks occurring intermediately at 308° and 60 per cent., and 258° C. and 80 per cent. tin. In addition to the compound AuSn, two others, $AuSn_2$ and $AuSn_4$, have been detected, and are formed on perieutectic horizontals at the two breaks at 308° C. and 258° C. respectively.

The compound AuSn is formed with slight dilatation of the volume of the reacting mixture. It has a metallic, silver-grey appearance, and is hard but brittle. Its electrical conductivity is greater than that of all other gold-tin alloys, except those with 95 per cent. or more of gold, and it is very resistant to acids.

$AuSn_2$ is distinguished by its characteristically large-sized crystals. At 308° it undergoes a transformation into AuSn and a fused alloy, as follows :—[10]

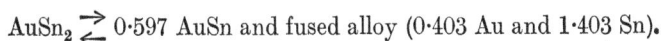

$$AuSn_2 \underset{\leftarrow}{\overset{\rightarrow}{}} 0.597 \text{ AuSn and fused alloy } (0.403 \text{ Au and } 1.403 \text{ Sn}).$$

$AuSn_4$ assumes a brown colour under the influence of nitric acid, whereas $AuSn_2$ remains unattacked.

[1] Mathiessen, *Phil. Trans.*, 1860, **150**, 161.
[2] Von Maey, *Zeitsch. physikal. Chem.*, 1901, **38**, 292.
[3] Le Chatelier, *Compt. rend.*, 1895, **120**, 835.
[4] Guertler, *Zeitsch. anorg. Chem.*, 1906, **51**, 397 ; *Phys. Zeitsch.*, 1908, **9**, 404.
[5] Raoult, *Compt. rend.*, 1873, **76**, 156.
[6] Mylius and Fromm, *Ber.*, 1894, **27**, 630.
[7] Laurie. *Phil. Mag.*, 1892, [v.], **33**, 94.
[8] Heycock and Neville, *J. Chem. Soc.*, 1891, **59**, 936.
[9] Vogel, *Zeitsch. anorg. Chem.*, 1905, **46**, 60-75.
[10] Vogel, *loc. cit.*

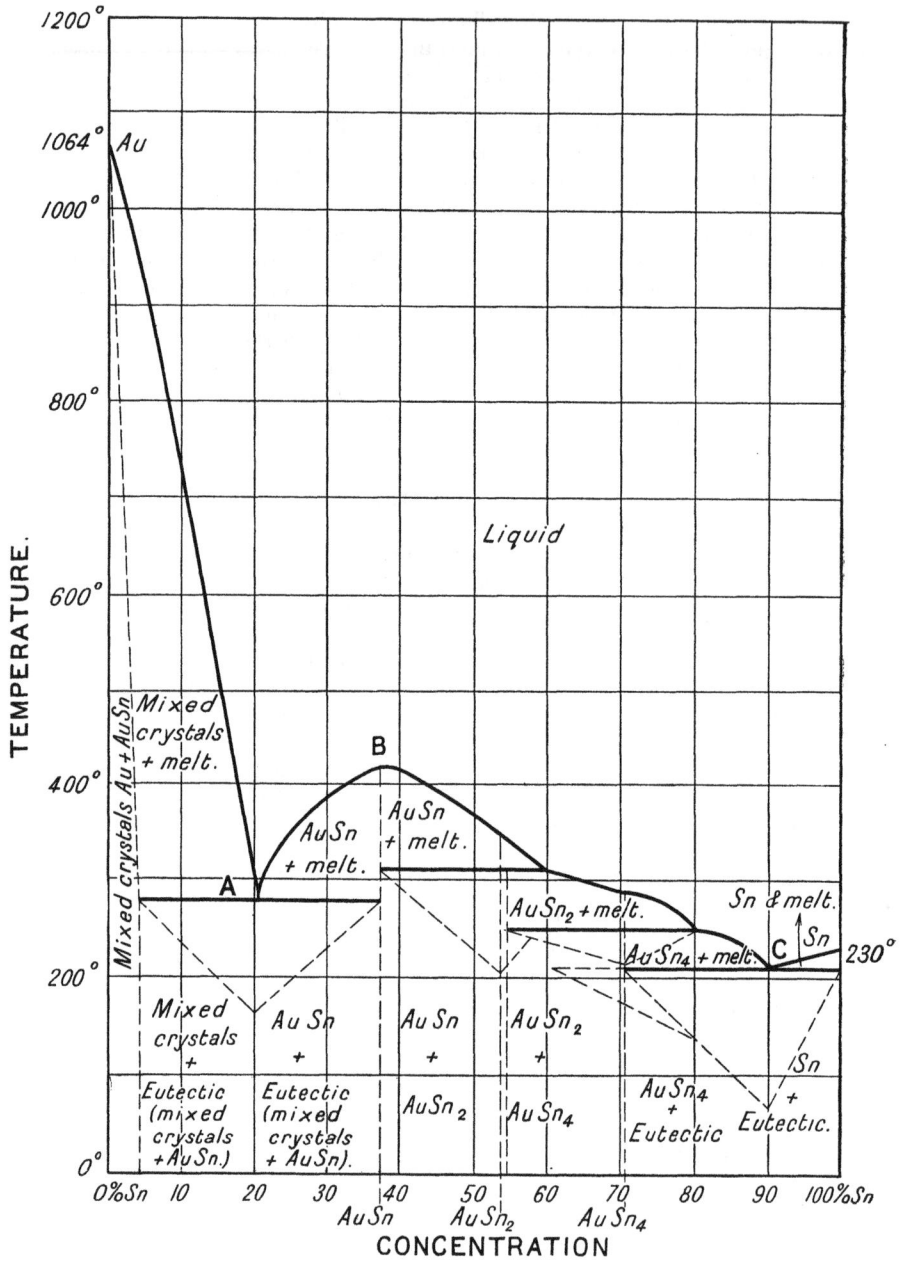

Fig. 23.—Thermal Equilibrium Curve of Gold and Tin.

In the intermediate series of crystals, as also in the case of pure tin, the formation of solid solutions is not observed to any great degree. Pure gold, on the other hand, takes up 5 per cent. by weight of tin in solid solution.

Under the usual cooling conditions, these crystals of tin dissolved in gold are yellow and rich in gold in the interior, while on the outside they are white and rich in tin, the whole crystals being enclosed within much eutectic. By long-continued heating at 800° C. the eutectic structure disappears, and the zoned portions in the crystals are replaced by a more homogeneous structure.

The solidus curve for the first series of solid solutions of tin in gold is not well defined. It has been found that the transformation temperature of tin is not appreciably altered by the addition of gold. Small quantities of tin are not harmful in their effects upon the mechanical properties of gold, as gold dissolves tin in solid solution.

With only a few per cent. of tin alloyed with it, gold loses its yellow colour, and assumes a grey appearance. Most of the alloys with comparatively large proportions of tin are very brittle, owing to the intermediate crystals which are formed according to the curve, possessing this property in a large degree. These two properties render tin-gold alloys very unsuitable for commercial purposes, in spite of the fact that such materials have a high resistance to chemical attack.

In a paper on the distillation of gold, Moissan [1] treats incidentally of the tin-gold alloys. 200 grammes of an alloy containing 40 per cent. gold were treated for three minutes in an electric furnace supplied with 500 ampères at 70 volts. At the end of this time the residue weighed 185 grammes, and had a composition of 41·08 per cent. Au, 59·72 per cent. Sn. After heating for four minutes the residue weighed 149 grammes, and contained 45·90 per cent. Au and 53·88 per cent. Sn. The boiling point of gold under atmospheric pressure, as calculated from its boiling point in the vacuum of cathode light, [2] is 2,530° C. This is higher than that of tin. When, therefore, alloys of tin and gold are distilled in an electric furnace, the residue becomes richer in gold than it was originally. If the vapours be allowed to escape into the atmosphere the tin burns to stannic oxide, and a substance remains having properties very similar to those of the Purple of Cassius. When freed from lime by hydrochloric acid the well-known purple colour appears. This fact is adduced in support of Debray's view of the constitution of Purple of Cassius, that the latter has no definite composition, but is merely a lacquer of tin oxide covered by fine gold. (See below, p. 73.)

Gold and Zinc.—The alloys of gold with zinc are of some metallurgical importance. In the Parke's process for the separation of the precious metals from base bullion formed by the smelting of lead ores, the principle involved is that gold and silver alloy more readily with zinc than with lead, forming a compound which, rising to the surface as a "crust," may be skimmed off and treated separately. [3]

When the impurities, such as copper and tin, have been skimmed as much as possible from the molten metal, zinc is added in predetermined quantities, according to the assay value of the bullion. The first crust to form on the surface contains practically all the gold, and a little silver and copper. Further additions of zinc are made in order to extract the remainder of the silver.

In 1906 Vogel studied these alloys. [4] The alloys were heated under charcoal in a stream of coal gas, to prevent oxidation of the zinc. According to

---

[1] Moissan, *Compt. rend.*, 1905, **141**, 977-983.
[2] Krafft and Bergfeldt, *J. Chem. Soc.*, 1905, **88**, [ii.], 144 ; *Ber.*, 1905, **38**, 254.
[3] See H. F. Collins, *Metallurgy of Lead.*
[4] Vogel, *Zeitsch. anorg. Chem.*, 1906, **48**, 319-332.

his results the curve has a eutectic point at 15 per cent. zinc (A, Fig. 24) (or according to the author [1] 14·4 per cent. zinc), and has a maximum at

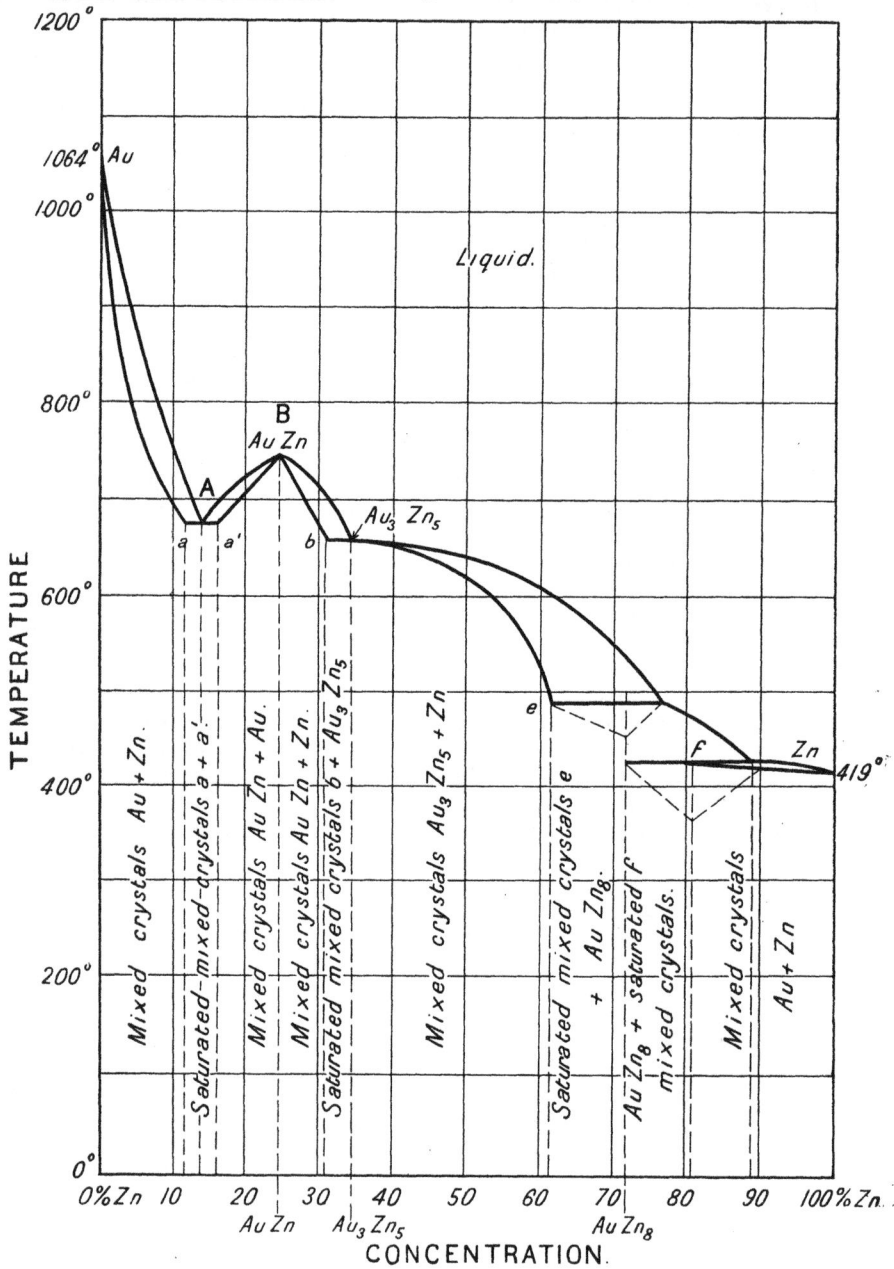

Fig. 24.—Thermal Equilibrium Curve of Gold and Zinc.

[1] T. K. Rose, *The Precious Metals* (1909), p. 46.

·25 per cent. (B) corresponding to a compound of the formula AuZn. Three breaks occur at 33, 77, and 88·5 per cent. zinc, and it is probable. that there exist also two other compounds with the formulæ Au_3Zn_5 (or according to the author $AuZn_2$[1]) and $AuZn_8$.

From 0 to 12·5 per cent. zinc a series of isomorphous mixtures separates, and then up to 16 per cent. zinc the primary separation of saturated crystals is followed by the appearance of a eutectic, the second component of which is a new series of isomorphous mixtures containing the compound AuZn. The new series containing AuZn separates alone between 16 and 25 per cent., and is succeeded at the latter point by a third series of isomorphous mixtures, which separate up to the point corresponding to a concentration of 31 per cent. of zinc. Here the compound Au_3Zn_5 (or $AuZn_2$) begins to form, and at 35·6 per cent. of zinc enters into a new series of mixed crystals, which reach up to 61 per cent. of the base metal. The range 61 to 72·6 per cent. zinc is occupied by the formation of the compound $AuZn_8$, after which isomorphous mixtures containing the pure metals separate.

These alloys have also been examined by the author, as already stated. He found the compounds to be AuZn solidifying at 750°, $AuZn_2$ solidifying at 650°, and $AuZn_8$. His other observations are as follows[2]:—The alloys with less than 14 per cent. of zinc are pale yellow, and about as hard as gold. They increase gradually in brittleness as the percentage of zinc increases. The first additions of zinc rapidly reduce the melting point of gold, giving rise to a long pasty stage during solidification. As the proportion of zinc increases from 14 to 25 per cent. the colour changes gradually from pale yellow to a reddish-lilac tint, the colour of the compound AuZn (cf. AgZn and AgCd), which forms lustrous crystals and is brittle. The alloys containing from 25 to 30 per cent. of zinc consist of lilac-coloured polygonal crystals of AuZn set in a white matrix. The compound $AuZn_2$, containing 39·8 per cent. of zinc, is a hard, white, homogeneous alloy, which will scratch steel, but is as brittle as glass. It breaks with a conchoidal fracture, giving lustrous surfaces. From 40 to 72 per cent. of zinc the alloys become less brittle and less lustrous. With above 72 per cent. of zinc the alloys resemble zinc in appearance, but are more brittle. The alloy $AuZn_8$, containing about 72 per cent. of zinc, is a hard white substance, solidifying with surfusion at 470°. It presumably occurs in the zinc crusts which are formed in Parkes' process.

Segregation in Gold Alloys.—The subject of segregation generally, including that observed in gold alloys, has been discussed in the volume introductory to this series.[3]

Segregation in the case of gold-copper alloys has been discussed above, p. 33. It has been shown by Matthey,[4] as mentioned above, on p. 45, that, when gold ingots containing members of the platinum group are cooled from a state of fusion, an alloy rich in the more fusible element (gold) falls out first, driving the less fusible constituent to the centre. Thus the assay of an outside cut of such an ingot gives a result too high in gold, sometimes by several per cent. It has long been known, moreover, that iridium and osmium become concentrated towards the bottom of the mass. The reason

---

[1] T. K. Rose, *op. cit.*   [2] T. K. Rose, *op. cit.*
[3] Roberts-Austen, *An Introduction to the Study of Metallurgy*, 1910, p. 76. The word "segregation" is used in this section as denoting the partial separation of the constituents of an alloy in such a way as to render the alloy non-uniform in composition.
[4] Matthey, *Proc. Roy. Soc.*, 1892, **51**, 447; *Phil. Trans.*, 1892, **A, 183**, 629-652.

for this is that, at the temperature of fusion of gold, these refractory elements, either free or alloyed with gold, sink in the molten metal and are left in the state of small crystalline particles.

Matthey has also further investigated[1] the question of segregation, in connection with the alloys of gold, silver, lead, and zinc produced in cyanide mills. He found that one such ingot weighing about 120 ozs. contained 662 parts per 1,000 of gold at the bottom corner, and only 439 parts at the top. In another case, when 16·4 per cent. of lead and 9·5 per cent. of zinc were present, the standard fineness of an ingot weighing 400 ozs., as shown by actually separating the whole of the precious metals, was, gold 614·0, silver 75·8, and its true value £1,028, while the value as deduced from the average of fourteen assays made on it (gold 576·0, silver 90) would have been only £965. Seven dip assays made on this ingot varied from 562 to 622 fine. Other cases of irregular distribution were even more remarkable. By experiments with synthetic alloys of gold and zinc, Matthey found that gold tends to liquate towards the centre of the mass, but only in a slight degree, the centre of a 3-inch sphere of an alloy containing gold 90, zinc 10, being only about 1 to $2\frac{1}{2}$ parts per 1,000 richer than the outside. Lead acts similarly, but with greater effect, the centre being about 29 parts per 1,000 richer in gold than the outside when 30 per cent. of lead is present, but the combination of 15 per cent. of lead and 10 per cent. of zinc is still more powerful in causing segregation, the sphere being found to contain 657 parts of gold at the top, 785 in the centre, and 790 at the bottom, gravity thus playing a part. The addition of silver, however, if it amounts to not less than about two-thirds of the quantity of zinc and lead taken together, appears absolutely to prevent any liquation from taking place, an alloy containing approximately, gold 55, zinc 7, lead 18, silver 20, being practically homogeneous.

---

[1] Matthey, *Proc. Roy. Soc.*, 1896, **60**, 21.

CHAPTER III.

CHEMISTRY OF THE COMPOUNDS OF GOLD.

Compounds of Gold.—Gold is characterised chemically by an extreme indifference to the action of all bodies usually met with in nature. Its simpler compounds are formed with difficulty, and decompose readily, especially when heated. The result is that gold is found in nature chiefly in the metallic form, and the mineralogist has, therefore, few compounds to consider. Gold also forms complex compounds, especially with cyanogen and sulphur. These do not exhibit the ordinary reactions of gold, and are in particular not so readily reduced. The gold in them is not present as an elementary ion, but forms part of complex ions.

Gold forms two series of simple compounds, having the general formulæ AuR and AuR_3, while doubtful compounds corresponding to AuR_2, AuR_4, and AuR_5, have been stated to exist by Thomsen, Prat, Figuier, and others. The two undoubted series are denominated *aurous* and *auric* respectively.

Compounds of Gold with the Halogens.—Gold forms two series of compounds with the halogens, the general formulæ being AuR and AuR_3 respectively. Supposed compounds having the formula AuR_2 have been described, but are probably mixtures of the series denoted by AuR and AuR_3. All these bodies are very unstable, existing throughout only low ranges of temperature, whether in the dry state or in aqueous solution. The chlorides are the most easily formed and the least unstable, the bromides coming next. The heats of formation of the compounds from their elements, according to Thomsen,[1] are given in the subjoined table in calories per gramme-molecule, where one calorie is equal to the quantity of heat required to raise the temperature of a kilogramme of water from $18°$ to $19°$:—

TABLE XV.

	Chloride.	Bromide.	Iodide.
AuR, solid, .	$+ 5\cdot81$	$- 0\cdot08$	$- 5\cdot52$
AuR_3, solid, .	$+ 22\cdot82$	$+ 8\cdot85$	?

CHLORIDES OF GOLD.

Gold Monochloride or Aurous Chloride, $AuCl$.—This salt is prepared by heating auric chloride to $190°$ in air for ten hours.[2] It can also be prepared

---

[1] Thomsen, *Landolt-Börnstein's Tabellen*, 1912, p. 869.
[2] T. K. Rose, *J. Chem. Soc.*, 1895, **67**, 902.

in solution by running a solution of $HAuCl_4$ into an excess of a solution of sulphurous acid. (See " Preparation of Pure Gold," Chap. XX.) The gold chloride is decolourised, showing that the auric chloride has been reduced to aurous chloride. After some time the solution becomes turbid and metallic gold is precipitated. Victor Lenher[1] found that colourless AuCl is formed in solution by reduction by means of sulphurous acid of the double chlorides of gold and any one of the following metals :—Sodium, potassium, copper, magnesium, zinc, cadmium. The reaction is quantitative, according to the equation :—

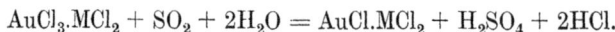

$$AuCl_3.MCl_2 + SO_2 + 2H_2O = AuCl.MCl_2 + H_2SO_4 + 2HCl.$$

In this way sulphurous acid can be used as an agent for the volumetric determination of gold. An arsenite can be used instead of sulphurous acid, thus :—

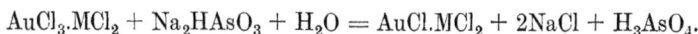

$$AuCl_3.MCl_2 + Na_2HAsO_3 + H_2O = AuCl.MCl_2 + 2NaCl + H_3AsO_4.$$

Diemer states that the double chlorides require a considerable excess of sulphurous acid for complete reduction with the precipitation of metallic gold. One molecule of NaCl, KCl, or NH_4Cl to one molecule of AuCl is sufficient to ensure the formation of the double salts, but AuCl.NaCl is rather unstable. The double salt is more stable if an excess of NaCl is present. About 40 molecules of $CaCl_2$ can take the place of 1 molecule of NaCl in preventing the separation of metallic gold by the action of sulphurous acid in slight excess.

Aurous chloride is non-volatile and unaltered at ordinary temperatures and pressure by dry air, even when exposed to light, but begins to decompose at temperatures above 160°, and the decomposition is complete if it is heated at 175° to 180° for six days, or at 250° for one hour. It combines with chlorine at the ordinary temperature, forming auric chloride.[2] Its density is 7·4. Water converts aurous chloride into a mixture of gold and auric chloride, thus $3AuCl = 2Au + AuCl_3$. The same reaction takes place slowly at the ordinary temperature in the absence of water. Aurous chloride is a citron-yellow amorphous powder. It is soluble in ammonia, and on the addition of HCl the white crystalline unstable substance $AuNH_3Cl$ is precipitated (Diemer). It is soluble completely in solutions of potassium cyanide and sodium chloride, and with separation of some gold when acted on by a solution of potassium bromide.[3]

Auro-Aurichloride, Au_2Cl_4.—A dark red compound having this composition is said, by Thomsen, to be obtained by heating finely divided gold in a current of chlorine to 140° to 170°. According to Krüss and Lindet it is merely a mixture of AuCl and $AuCl_3$. It yields gold and $AuCl_3$ if brought in contact with water, ether, or alcohol.

Auric Chloride or Gold Trichloride, $AuCl_3$—*Preparation.*—Trichloride of gold can be prepared, according to Debray,[4] by heating finely divided gold in a current of pure dry chlorine in a glass tube at a temperature of 300°. The chloride formed sublimes at this temperature and is deposited in the cooler part of the tube in fine red prisms and needles crystallising in the

---

[1] Lenher, *J. Amer. Chem. Soc.*, 1913, **35,** 546; see also Diemer, *op. cit.*, 552.
[2] T. K. Rose, *loc. cit.*
[3] Lengfeld, *J. Chem. Soc.*, 1902, **82,** [ii.], 27 ; *J. Amer. Chem. Soc.*, 1901, **26, 324.**
[4] Debray, *Compt. rend.*, 1869, **59,** 985.

triclinic system. Krüss tried to repeat Debray's experiments but did not confirm his results.[1] He found that $AuCl_3$ was formed at 140° to 150°, and was converted to AuCl at about 180° in spite of the presence of an excess of chlorine. At 220° the AuCl was completely decomposed, leaving metallic gold, while up to this temperature a little $AuCl_3$ continued to be sublimed, but towards 300° all action ceased, and the gold remained unattacked by the chlorine at all higher temperatures.

The exact point at which gold ceases to be attacked by chlorine and the rate of volatilisation of the chloride are of great importance in connection with the loss of gold on roasting auriferous materials with salt.[2] The matter was, therefore, investigated by the author,[3] with the following results :— Gold unites with chlorine if placed in the gas at atmospheric pressure at all temperatures up to a white heat, but the action is slight at a red heat. The absorption of chlorine by gold with the formation of chlorides at first increases in rapidity as the temperature rises above ·15°, and reaches its maximum at about 225°. The fact that gold is attacked by chlorine, and that the chloride is subsequently volatilised at all temperatures between 180° and 1,100°, was proved by means of Deville's hot and cold tubes, which enable part of the sublimed chloride to be collected. The rate of volatilisation at various temperatures is as follows :—

TABLE XVI.

Temperature, °C.	Percentage of Gold Volatilised in 30 Minutes.	Temperature, °C.	Percentage of Gold Volatilised in 30 Minutes.
180°	0·007	580°	0·60
230°	0·35	590°	0·58
300°	2·32	805°	0·50
390°	1·82	965°	1·63
480°	0·88	1,100°	1·93

For purposes of comparison, it may be added that when gold is heated in air or coal gas, no gold is volatilised below 1,050°, and only about 0·02 per cent. in thirty minutes at 1,100°,[4] or about one-hundredth part of that volatilised in chlorine at the same temperature.

The amounts volatilised vary according to two different factors—(1) The vapour pressure of gold trichloride, $AuCl_3$, which of course increases continuously as the temperature rises ; and (2) the pressure of dissociation of the trichloride, which also rises continuously with the temperature, but not at the same rate as the vapour pressure. The increase of vapour pressure tends to raise, and that of the pressure of dissociation to decrease, the amount of gold volatilised as chloride. The vapour pressure increases more rapidly than the pressure of dissociation at temperatures below 300°, and also above

[1] Krüss, *Ber.*, 1888, **20**, 212, 2364.
[2] See Prof. Christy's experiments, Section on "Roasting," Chap. xiii.
[3] T. K. Rose, *J. Chem. Soc.*, 1895, **67**, 881.
[4] T. K. Rose, *J. Chem. Soc.*, 1893, **63**, 717.

900°, but less rapidly at intermediate temperatures. Hence the curve (Fig. 25) showing the variation of volatilisation with temperature is irregular, passing through a maximum near 300°, and a minimum at a point somewhere below the melting point of gold. The first-named change in the direction of the curve probably occurs at the melting point of the chloride—namely, 288°. The second change is perhaps caused by the change of sign of the heat of formation of $AuCl_3$; when this becomes negative, the pressure of dissociation of the compound would decrease, in accordance with the law of van't Hoff and Le Chatelier. However this may be, it is certain that when gold is heated in chlorine at atmospheric pressure, auric chloride is formed and volatilised at all temperatures above 180°, up to, and probably far beyond, 1,100°.

If gold is treated with liquid chlorine in a sealed tube at ordinary temperatures, it is converted into a crystalline red mass.

The usual method adopted for the preparation of auric chloride is to dissolve gold in aqua regia, and then to drive off the excess of acid by heat, adding HCl if necessary to maintain its excess over the nitric acid, keeping

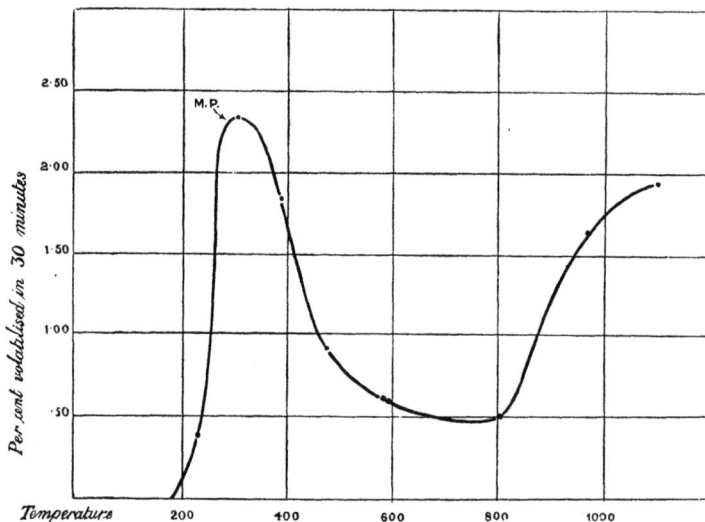

Fig. 25.—Curve showing Rate of Volatilisation of Gold Chloride at various Temperatures.

the temperature as low as possible. A brownish-red mass is thus formed, consisting of $AuCl_3$ mixed with more or less aurous chloride and hydrochloric acid. When heated at 95° to 100° until acid fumes no longer appear to be evolved, the resulting chloride solidifies at 70°. It consists almost entirely of $HAuCl_4$, but contains from 0·5 to 1·0 per cent. of AuCl. As noted below, pure anhydrous $AuCl_3$ solidifies at 288°. On taking up with water, aurous chloride is decomposed into gold and trichloride, but the hydrochloric acid can only be eliminated by shaking with ether, which withdraws auric chloride from its solution in water. If an attempt is made to drive off the hydrochloric acid by heat, a partial decomposition of the trichloride results.

Auric chloride exists both in the anhydrous state and in combination with two equivalents of water, $AuCl_3.2H_2O$, when it occurs in orange-red crystals. The anhydrous salt has a brilliant-red colour, crystallising in needles belonging to the triclinic system, and melting at 288° in chlorine under a pressure of two atmospheres. It can be prepared by drying the hydrated salt at 150°. The anhydrous and hydrated salts are both hygroscopic, and dissolve readily in water with elevation of temperature ; they are also soluble in alcohol and ether, and in some acid chlorides, such as $AsCl_3$, $SbCl_5$, $SnCl_2$, $SiCl_4$, etc.

Auric chloride is readily decomposed by heat. Löwe states [1] that 4 grammes of the trichloride, when heated in a porcelain basin on a boiling water bath, can be completely transformed into the monochloride, although not until after the lapse of several days. On the other hand, as has already been mentioned, Krüss states that the decomposition of auric chloride, in an atmosphere of chlorine, begins at 180°. According to the experiments of the author,[2] auric chloride is observed to suffer slow decomposition at as low a temperature as 165° in an atmosphere consisting of chlorine, about 1·6 per cent. being converted into monochloride in four hours at this temperature ; the decomposition is about five times more rapid at 190°. The decomposition in air can be readily observed at 100°, although it does not seem to be so rapid as was indicated by Löwe. In seven days only 6·6 per cent. of the trichloride was decomposed, the initial rate of decomposition being 0·041 per cent. per hour. At 165°, however, the initial rate of decomposition appeared to be 3·2 per cent. per hour, and the conversion into monochloride was complete in four or five days at 160° and in ten hours at 190°.

Solutions of gold chloride have also been shown to be decomposed by heat,[3] a solution of one part of the chloride in 15,000 parts of pure water yielding a precipitate of gold on heating for some hours. The solution was found to contain traces of hydrogen peroxide, the reaction being expressed by the equations—

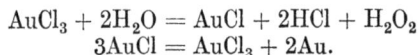

$$AuCl_3 + 2H_2O = AuCl + 2HCl + H_2O_2$$
$$3AuCl = AuCl_3 + 2Au.$$

A similar change, yielding hydrogen peroxide, was found by Sonstadt to take place when solutions containing 0·04 per cent. and also 0·007 per cent. of gold chloride were exposed to bright sunshine for several days.[4] The presence of free hydrochloric acid prevents this decomposition. Weak voltaic currents precipitate metallic gold from the solution of the trichloride upon the negative pole. Withrow has studied the rate of precipitation in the presence of potassium cyanide and of sodium sulphide, a rotating anode being employed.[5] The solution of trichloride of gold is also decomposed by many reducing agents, such as most organic substances, metals, and protosalts ; heating the solution in every case hastens the decomposition. The reduction by organic matter is assisted by the action of light, which is especially efficacious in the presence of starchy and saccharine compounds, or of charcoal or ether. In some cases the presence of hydrochloric acid prevents or retards the action, but in other cases this effect is not observable.

---

[1] Löwe, *Dingl. poly. J.*, 1891, **279**, 167.
[2] T. K. Ross, *J. Chem. Soc.*, 1895, **67**, 902.
[3] Sonstadt, *Chem News*, 1898, **77**, 74.
[4] Sonstadt, *Proc. Chem Soc.*, 1898, No. 198. p. 179.
[5] Withrow, *J. Amer. Chem. Soc.*, 1906, **28**, 1350 ; see also Neumann, *J. Chem. Soc.*, 1906, **290**, [ii.], 764.

Sugar added to $HAuCl_4$ in solution gives a blue coloration, and acids cause the gold to separate in the ordinary form.[1]   In direct sunlight ether deposits a bright mirror of metallic gold, but under ordinary conditions red colloidal gold (Faraday's gold) is set free.   Alkalies also quicken the action of organic matter, and it may be said that all organic compounds reduce gold chloride on boiling in the presence of potash or soda,[2] while Müller states that a mixture of glycerine and soda lye is one of the best precipitants for gold chloride, separating the metal completely in highly dilute solutions.   Priwoznik finds that a boiling solution of glycerine and sodium carbonate completely precipitates gold from its solution as chloride.[3] According to Krüss, if potash and soda are quite free from organic matter, they have no action on solutions of auric chloride, whether cold or hot.   If a small quantity of organic matter is present, sub-oxide of gold is precipitated ; if larger quantities are present, both metallic gold and sub-oxide are precipitated in the cold, but gold alone at boiling point.   Alkaline carbonates are without action on cold solutions, but if they are hot, then half the gold is precipitated as hydrate, while the other half remains in solution in the form of a double chloride of gold and the alkali.   Gold is also precipitated from solution by acetylene.[4]

The precipitation by means of charcoal is of especial importance in view of its adoption in practice.   Avery,[5] using a very pure charcoal prepared from cocoanut shell, made quantitative estimations of the HCl and CO_2 produced in this reaction.   He concluded that the reduction takes place in accordance with the equation proposed by König :—[6]

$$4AuCl_3 + 6H_2O + 3C = 4Au + 12HCl + 3CO_2.$$

Occluded gases, such as hydrogen and carbon monoxide, will also reduce gold, and König states that occluded gases are responsible for 95 per cent. of the precipitation.   The effect of the charcoal is diminished by use, but can be restored by heating the charcoal to redness in absence of air, or by the passage of a current of electricity, using the charcoal as the cathode. (See also " Precipitation of Gold," Chaps. XIV. and XV.)

It has been stated that a current of hydrogen gas will precipitate gold completely, especially on boiling the solution, but Krüss has proved that, if the hydrogen is quite pure, it has no effect either on cold or hot solutions. Sulphur, selenium, phosphorus, and arsenic all precipitate gold on boiling the solution of the trichloride.   Tellurium easily reduces gold chloride solution ; the precipitation is complete at the ordinary temperature, but the tellurium must be finely powdered, or it becomes coated with gold, and further action is prevented.   Silver telluride also reduces solutions of gold salts.   In these reactions tellurium tetrachloride is formed, thus :—

$$4AuCl_3 + 3Te = 4Au + 3TeCl_4.$$

[1] Vanino, *J. Chem. Soc.*, 1908, **94**, [ii.], 504.

[2] Frémy, *Ency. Chim.*, 16e Cahier, vol. iii., p. 74.

[3] Priwoznik, *J. Chem. Soc.*, 1912, [ii.], **102**, 562.

[4] Mathews and Watters, *J. Amer. Chem. Soc.*, 1900, **22**, 108.

[5] Avery, *J. Soc. Chem. Ind.*, 1908, **27**, 255.

[6] König, *Chem. News*, 1882, **45**, 215 ; *J. Franklin Inst.*, May, 1882.

The natural tellurides, calaverite, sylvanite, coloradoite, kalgoorlite, nag-yagite, hessite, and krennerite behave like tellurium in precipitating gold.[1] Many metals reduce chloride of gold, the action being, of course, most rapid in the case of the most highly electro-positive metals, such as zinc and iron. Lead sometimes gives fine dendritic plates of gold. Sulphuretted hydrogen precipitates sulphide of gold from both neutral and acid solutions, all traces of gold being readily removed from a solution by this reagent, whilst phos-phoretted, arsenuretted, and antimoniuretted hydrogen, as well as H_2Te and H_2Se, all precipitate metallic gold. The lower oxides of nitrogen, nitrous acid, and many other " *ous* " acids and oxides effect the same decomposition. Precipitation by potassium nitrite in presence of sulphuric acid gives gold nodules easily collected.[2] Sulphur dioxide is a convenient reagent, and is often used in the laboratory, being almost equally efficacious in cold and hot solutions. The reaction is

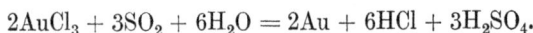

$$2AuCl_3 + 3SO_2 + 6H_2O = 2Au + 6HCl + 3H_2SO_4.$$

Various protosalts also reduce trichloride of gold. Ferrous sulphate is often used to detect the presence of gold in solution as chloride ; this reagent gives dilute solutions a pale blue colour by transmitted light, and brown by reflected light, owing to the formation of finely divided precipitated gold. The reaction is represented by the following equation :—

$$2AuCl_3 + 6FeSO_4 = 2Au + 2Fe_2(SO_4)_3 + 2FeCl_3.$$

To test a dilute solution for gold, a test tube filled with the liquid is held in the hand side by side with a test tube filled with distilled water and a few drops of a clear solution of ferrous sulphate are added to each. On looking down through the length of the test tubes from above, with a white surface as background, any slight changes of colour may be detected by comparison, and the liquid may also be compared with the original solution in a test tube. In this way, by a little practice, the presence of gold in the proportion of only $\frac{1}{720000}$ (1 dwt. per ton of water), or even less can be detected. The method was often used in the chlorination process, but it is better to use stannous chloride, $SnCl_2$. This substance gives a brown pre-cipitate of variable composition in concentrated solutions, but if mixed with the tetrachloride, $SnCl_4$, it gives a precipitate of Purple of Cassius. The reaction is very sensitive, and by its means a violet coloration by trans-mitted light can be obtained in a solution containing 1 part of gold in 500,000 parts of water, while by special means the presence of 1 part of gold in 100,000,000 parts of water can be detected, as described below.[3] The liquid supposed to contain gold is raised to boiling, and poured suddenly into a large beaker containing 5 to 10 c.c. of saturated solution of stannous chloride, and the liquids agitated so as to effect complete mixture. A yellowish-white precipitate of tin hydrate forms, which settles rapidly, and can be readily separated from the bulk of the liquid by decantation. If the solution originally contained at least 1 part of gold in 5,000,000 of water ($3\frac{1}{2}$ grs. per ton), the precipitate is coloured purplish-red or blackish-purple, according to the nature of the solution, and the condition of the precipitant. The colour can be seen without comparing it with other precipitates. If less gold than

[1] Lenher, *J. Amer. Chem. Soc.*, 1902, **24**, 355, 918.
[2] Jameson, *J. Amer. Chem. Soc.*, 1905, **27**, 1444.
[3] T. K. Rose, *Chem. News*, 1892, **66**, 271.

this is present it is better to compare the precipitate with one obtained by the use of boiling distilled water, and to increase the quantity of liquid used while adhering to the same amount of stannous chloride. In this way the presence of 1 part of gold in 100,000,000 parts of water (1 grain of gold in 6 tons of water) can be detected, the amount of liquid required in this case being about 3 litres. The gold is concentrated in the precipitate in which a distinct colour is caused by less than 0·05 per cent. of the metal.

Chlor-auric Acid or Hydrogen Aurichloride, $HAuCl_4$.—Gold trichloride in the presence of free hydrochloric acid forms this compound, which crystallises out on evaporation in vacuo in long yellow needles, having the composition $HAuCl_4 + 3H_2O$,[1] and since gold chloride unites with many other soluble chlorides to form double chlorides, the hydrochloric acid compound is regarded as an acid. It is more stable than gold trichloride. The chlor-aurates, having the general formula $M'AuCl_4$, or $AuCl_3.M'Cl$, are readily soluble bodies which can be crystallised, and which decompose with about the same readiness as chlor-auric acid. Double chlorides are also formed with organic bases. The sodium salt (sodium aurichloride), $NaAuCl_4.2H_2O$, is used for " toning " in photography. It contains 49·45 per cent. of gold, as made in England, but only 23 to 30 per cent. as made in the United States.[2] The acetates and succinates are also used in photography.[3]

On adding silver nitrate to a solution of hydrogen aurichloride a brown precipitate is obtained,[4] according to the equation :—

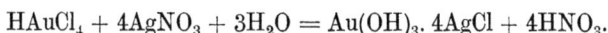

$$HAuCl_4 + 4AgNO_3 + 3H_2O = Au(OH)_3.\,4AgCl + 4HNO_3.$$

By the action of ammonia this is converted into fulminating gold (q.v.).

BROMIDES OF GOLD.

Aurous Bromide or Gold Monobromide, $AuBr$, is a yellowish-green powder obtained by heating the tribromide to about 115°. It is insoluble in water, but is decomposed by it, metallic gold and the tribromide being formed ; the change is especially rapid on boiling, and is hastened by the presence of hydrobromic acid. Aurous bromide is completely soluble in potassium cyanide ; and in ammonia, potassium bromide or HBr with partial separation of gold (Lengfeld).

Auro-auric Bromide, Au_2Br_4, is produced by the action of bromine on finely divided gold in the cold, some tribromide being simultaneously formed. Water breaks up this bromide into $AuBr$ and $AuBr_3$, and, according to some observers, it is only a mixture of these bodies.

Auric Bromide or Gold Tribromide, $AuBr_3$, is produced by the action of a mixture of bromine and water on gold, particularly on the application of heat. The action of bromine on gold, however, causes the formation of a film of $AuBr_3$, which prevents further action. The film is removed by shaking the mixture.[5] Auric tribromide resembles the trichloride in most of its properties. It is volatile at 300° in an atmosphere of bromine (Meyer). It crystallises in blackish needles or scarlet plates. It is deliquescent, but is

[1] Schmidt, *J. Chem. Soc.*, 1906, **90,** [ii.], 862.
[2] Kebler, *J. Soc. Chem. Ind.*, 1900, **19,** 1038 ; Johnson & Sons, *ibid.*, 1901, **20,** 210.
[3] Mercier, *Brit. Journ. Phot.*, **39,** 354.
[4] Jacobsen, *J. Chem. Soc.*, 1908, **94,** [ii.], 601 ; *Compt. rend.*, 1908, **146,** 1213.
[5] Meyer, *J. Chem. Soc.*, 1909, **96,** [ii.], 321 ; see also Lengfeld, *J. Chem. Soc.*, 1902, **82,** [ii.], 27.

far less soluble in water than gold trichloride. Concentrated solutions, which may contain about 1 per cent. of the tribromide, are nearly black in colour. Auric bromide suffers decompositions similar to those noted in describing $AuCl_3$, its solutions being still less stable than those of the chloride. A solution of gold tribromide is gradually decolourised by sulphur dioxide, being completely reduced to the state of monobromide before any precipitate of metallic gold is formed. It is prepared in a pure state by heating finely divided gold in sealed tubes with bromine and arsenic bromide, $AsBr_3$, to 126°. Gold tribromide forms intensely coloured brownish-red aqueous solutions, the presence of a mere trace of the salt in a solution being observable in this way. Double bromides exist analogous to the chlor-aurates.

The **Iodides of Gold** are of little interest to the metallurgist. Aurous iodide, AuI, can be prepared by the action of iodine on gold above 50°, the excess of iodine being removed by careful sublimation.[1] It is a white powder, turning green in air and decomposing at 190°. The tri-iodide is supposed to be formed if gold is acted on by a solution of iodine in potassium iodide. Rapid dissolution ensues, and the solution is fairly stable.

Gold Fluoride has been described by Lenher.[2]

CYANIDES OF GOLD.[3]

Cyanogen and gold unite in two proportions, forming aurous and auric cyanides, but the latter is only known with certainty in combination.

Aurous Cyanide, AuCy, is obtained by heating aurocyanide of potassium, $KAuCy_2$, with hydrochloric or nitric acid and washing with water. It is a lemon-yellow crystalline powder, insoluble in water, and unaltered by exposure to air. It is decomposed by heat, yielding metallic gold and cyanogen, and is soluble in ammonia, in yellow ammonium sulphide, in alkaline cyanides, and in hyposulphite (thiosulphate) of an alkali. It is unattacked by the mineral acids, except by aqua regia, but is decomposed when boiled with potash, metallic gold being thrown down and aurocyanide of potassium formed.

Potassium Aurocyanide, $KAuCy_2$, is obtained by crystallisation from its solution, which is prepared by dissolving metallic gold, auric oxide, or aurous cyanide in a solution of potassium cyanide. It is also formed by adding potassium cyanide to an acid solution of gold trichloride. The solution for electro-plating purposes may be prepared by precipitating a solution of gold chloride with ammonia and dissolving the fulminating gold in potassium cyanide, or by precipitating gold chloride with magnesia and dissolving the purified auric hydroxide in KCy, or by simply passing an electric current through a cyanide bath, with a gold anode. It forms a colourless solution in water, from which it can be crystallised as colourless, transparent rhombic octahedra. Cold water dissolves 15 per cent., and boiling water twice its weight of the salt. The aqueous solution, especially if hot, gilds copper or silver without the agency of a battery. Gilding is, however, generally effected by electro-deposition, using a gold anode. Gold is also precipitated from the solution by zinc and many other metals. Precipitates are also formed on the addition of salts of zinc, copper, tin, iron, or silver, no precipitates being

---

[1] Meyer, *Compt. rend.*, 1904, **139**, 733.
[2] Lenher, *J. Amer. Chem. Soc.*, 1903, **25**, 1136.   [3] See also Chapter XV.

formed if potassium cyanide is present in excess. According to Lindbom,
ferrous salts are without action on $KAuCy_2$, but oxalic acid, sulphurous
acid, or mercurous chloride, Hg_2Cl_2, precipitate aurous cyanide from hot
solutions.

Aurocyanides are decomposed by mineral acids, aurous cyanide being
precipitated, and hydrocyanic acid evolved. Iodine, bromine, and chlorine
are dissolved by $KAuCy_2$, and the iodine compound, $KAuCy_2I_2$, can be
crystallised out. The aurocyanides of sodium, ammonium, barium, calcium,
zinc, cadmium, and other metals have been prepared.

Potassium Auricyanide, $AuCy_3.KCy$, is formed by adding potassium
cyanide to a perfectly neutral solution of trichloride of gold, the precipitate
first formed being redissolved. The solution is completely decolourised,
and on cooling deposits colourless crystals of $AuCy_3.KCy.3H_2O$. These
effloresce in air, giving up two molecules of water; and, on heating, the
third molecule of water and some cyanogen are given off, aurocyanide of
potassium being formed, and this in its turn is decomposed at a slightly
higher temperature. Potassium auricyanide is somewhat soluble in cold
water, and readily soluble in hot water.

OXIDES OF GOLD.

Aurous Oxide, Au_2O.—This oxide is prepared by decomposing aurous
chloride, $AuCl$, or the corresponding bromide by potash in the cold (Berzelius)
when a violet precipitate forms, which is blackish when moist, but greyish
when dry. When freshly precipitated it is soluble in alkalies and in cold
water, forming an indigo blue solution, with brownish fluorescence, and
on warming the solution slightly the corresponding hydrate is precipitated.
It is also prepared by the action of nitrate of mercury on the trichloride,
and by boiling aurate of potash with organic compounds, such as citrates
or tartrates, or by boiling a solution of the trichloride with the potassium
salts of these acids. When prepared according to these methods, aurous
oxide always contains a certain proportion of metallic gold. The oxide may
be obtained pure by reducing brom-aurate of potassium at $0°$ by SO_2, passing
in the gas only until the solution becomes colourless, after which an
excess of gas would precipitate metallic gold. Aurous hydrate is then precipitated by potash, and, after being agglomerated by boiling, it is filtered,
washed with cold water, dried, and heated to $200°$ to expel the water of
hydration. At $250°$ it is resolved into gold and oxygen. Hydrochloric acid
decomposes aurous oxide into metallic gold and auric salts, slowly in the
cold, quickly at a boiling temperature; aqua regia dissolves the oxide,
but sulphuric and nitric acids are without action on it, while weak bases
at once decompose it.

An intermediate oxide, AuO, is prepared as a black powder by dissolving
metallic gold in aqua regia containing an excess of hydrochloric acid, then
adding an excess of carbonate of potash, and afterwards filtering and drying
the precipitate. It has been little studied, but the temperature at which
it decomposes has been fixed at $205°$ and its hydrate has been prepared.

Auric Oxide, Au_2O_3.—This, the best known oxide, is a black powder
when anhydrous, and is precipitated from solutions of auric chloride in
the form of a hydrate by the caustic alkalies, the carbonates of the alkalies,
and hydrates of the alkaline earths or zinc. The readiest method of

preparation of this compound is to add caustic potash, little by little, to a hot solution of gold chloride, until the yellow precipitate of auric hydroxide, $Au(OH)_3$, first formed is dissolved to a brown liquid which contains potassium aurate, $KAuO_2$. Then a slight excess of sulphuric acid or some sodium sulphate is added, the precipitate filtered off, washed and purified from potash by being redissolved in concentrated nitric acid, and reprecipitated by dilution with water. On drying this precipitate in vacuo, the hydroxide, an ochreous powder, results. It can also be prepared by heating a solution of gold chloride with magnesia and washing the residue with nitric acid. It is a yellow, olive-green, or brown powder (according to the method of preparation), and becomes brownish or black on drying. It dissolves in potash solution, and the resultant unstable potassium aurate can be used for electrogilding. If it is heated to 110°, oxygen begins to be given off; at 160°, AuO remains, and on heating for some time at 250°, metallic gold remains. Trioxide of gold dissolves in concentrated sulphuric and nitric acids, from which it is partly reprecipitated on boiling or on dilution, and these solutions are supposed to contain sulphates and nitrates of gold respectively. Double nitrates of gold and the alkalies have been obtained as crystals. Hydrochloric and hydrobromic acids dissolve the trioxide forming the haloid salts, but hydriodic acid decomposes it on boiling, giving iodine and metallic gold. Gold trioxide dissolves in boiling solutions of alkaline chlorides, giving aurates and chlor-aurates, while it also combines with metallic oxides to form aurates.

It is easily reduced by hydrogen, carbon and carbonic oxide, with the aid of very gentle heat. Boiling alcohol or hot alcoholic potash reduces it, yielding minute spangles of gold which were formerly used in miniature painting.

Aurates.—The aurates of potash and soda have the general formula $Au_2O_3.R'_2O$ or $R'_2Au_2O_4$ assigned to them. They are readily soluble, crystallisable compounds, and are formed when alkalies are added in excess to solutions of gold chloride. The aurates of calcium, magnesium, and zinc are insoluble in water, but soluble in hydrochloric acid. With organic matter they yield explosive powders (Meyer).

Fulminating Gold is a compound of auric oxide with ammonia, $Au_2O_3(NH_3)_4$, which is formed by precipitating gold chloride with ammonia or its carbonate, or by the action of ammonia on gold trioxide. When prepared by the former method its composition is variable, but the fulminate is always a high explosive decomposing with violence at 145°, or on being struck, and sometimes even spontaneously. It is decomposed without explosion by sulphuretted hydrogen, and by stannous chloride. It is a grey or buff-coloured powder, insoluble in water, but soluble in potassium cyanide, auricyanide of potassium being formed.

Sulphites of Gold.—Alkaline sulphites, or sulphur dioxide, which reduce gold trichloride easily, do not produce the same effect on a solution of an alkaline aurate. If sodium bisulphite is added to a boiling solution of sodium aurate ($NaAuO_2$) a yellowish precipitate is formed, soluble in excess of sodium bisulphite, and consisting of a double sulphite of gold and sodium, or *sodium aurosulphite*, having the composition $3Na_2SO_3.Au_2SO_3 + 3H_2O$. It is obtained pure by precipitating the corresponding barium salt with $BaCl_2$, and decomposing the precipitate with the minimum quantity of sodium carbonate. Double sulphites of potassium and ammonium with gold also exist. These salts are decomposed by acids, sulphite of gold being deposited, and also on boiling their aqueous solutions, but the addition of sulphuretted hydrogen or alkaline sulphides has no effect on them.

Thiosulphates of Gold.—The extraction of gold from auriferous silver ores, when these were treated by the ordinary "hyposulphite," or by the "Russell" process, depended on the formation of these compounds. The soluble double thiosulphates of gold with the alkalies and alkaline earths have the general formula $3R''S_2O_3.Au_2S_2O_3 + 4H_2O$. The double compounds of gold with sodium, potassium, calcium, magnesium, and barium are all known. The sodium salt is prepared by adding a dilute solution of gold trichloride little by little to a concentrated solution of sodium thiosulphate, when the following reaction occurs :—

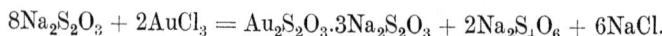

$$8Na_2S_2O_3 + 2AuCl_3 = Au_2S_2O_3.3Na_2S_2O_3 + 2Na_2S_4O_6 + 6NaCl.$$

The double thiosulphate may be separated by precipitation with strong alcohol, with which it is also washed, or it may be purified by repeated solution in water and precipitation with alcohol. Thus prepared, it consists of colourless crystalline needles, highly soluble in water, but almost insoluble in alcohol. The solution, which possesses a sweetish taste, decomposes under the influence of heat, the action being much more rapid when nitric acid is present ; metallic gold and sodium sulphate are formed. Gold, however, is not reduced from its solution as double thiosulphate by either stannous chloride, ferrous sulphate or oxalic acid, although sulphuretted hydrogen and alkaline sulphides give a black precipitate of Au_2S_3. The addition of hydrochloric acid or of dilute sulphuric acid does not immediately cause an evolution of sulphur dioxide and a deposit of sulphur, as in the case of ordinary thiosulphates. Since, therefore, the double sulphite of soda and gold does not present the characteristics of either aurous salts or of thiosulphuric acid, it has been suggested that it contains a compound radical and has a composition expressed by either $Na_3S_4O_6Au$ or $Na_3S_3O_6Au + 2H_2O$. The addition of any dilute acid soon effects the decomposition of this body in solution, gold sulphide being precipitated ; the reaction is accelerated by heat. This double thiosulphate exists in combined fixing and toning photographic baths.

The double thiosulphates of potassium, calcium, barium and magnesium present similar characteristics. If the barium salt is treated with the amount of sulphuric acid required by theory, a solution of the acid auro-thiosulphate, $3H_2S_2O_3.Au_2S_2O_3$, is obtained, but it cannot be crystallised. It has been supposed that the calcium salt is more easily formed than the sodium salt, and, therefore, that calcium thiosulphate was more suitable than sodium thiosulphate for use in the leaching process, whenever gold was present in perceptible quantities. According to a series of experiments conducted by Russell,[1] this was not the case.

Russell has demonstrated [2] that finely divided gold is soluble to a limited extent (*i.e.*, 0·002 gramme in 1,000 c.c. in 48 hours), in solutions of sodium thiosulphate of all degrees of concentration. The action depends on the oxidation of the gold by the air present in the solution, the soluble double thiosulphate, $Au_2S_2O_3.3Na_2S_2O_3 + 4H_2O$, and caustic soda being formed.

The formation of this thiosulphate by the action of the sodium salt on gold sulphide is far more complete and rapid. In twenty-four hours at 15°,

---

[1] Stetefeldt, *Lixiviation of Silver Ores*, New York, 1888, p. 90.   [2] *Ibid.*, p. 19.

0·066 gramme of gold, and in two hours at 65°, 0·117 gramme of gold were dissolved in dilute solutions.

Gold Carbide, Au_2C_2, is formed by passing acetylene into aurous thiosulphate. It is explosive when dry, and is decomposed by hydrochloric acid forming AuCl and acetylene. When treated with water, it yields gold and carbon.[1]

Gold Chromate can be obtained as crystals by treating silver chromate with auric chloride. Sodium aurochromate containing excess of chromate is obtained by mixing solutions of sodium aurate and chromate. An excellent photographic toning bath is formed, giving purple to bluish tones.[2]

Gold Selenate is formed by dissolving gold in selenic acid, see p. 19. It forms small yellow crystals, insoluble in water but soluble in sulphuric, nitric, or hot selenic acid. By the action of hydrochloric acid on it chlorine is evolved and auric chloride and selenious acid are produced. It is decomposed by heat with the production of metallic gold. On exposure to light it becomes dark green and afterwards bronze-coloured.[3]

Other Compounds of Gold.—Arsenates, alkyl gold chlorides, mercaptides, and other complex organic compounds have been prepared.[4] Certain thio-organic compounds of gold are soluble, and can be employed in the ceramic, enamel, and glass industries for the deposition of the finest layers of bright metal on various substances.[5] Textile fabrics, printed with a gold salt and then treated with a reducing agent, assume a beautiful grey colour. When the grey fabrics are subjected to heat between rollers, red, purple, or pink colours are obtained.[6]

SILICATES OF GOLD.

The existence of auro-silicates is now admitted without dispute, and gold has for centuries been used to impart colour to glasses, the method used being as follows :—A solution of chloride of gold is added to a mixture of sand with alkalies and alkaline earths or lead, and the whole is then fused, and colourless or yellow transparent silicates of gold thus formed. These are decomposed by being reheated gently to low redness, oxides of gold, or more probably metallic gold, being set free, and red or purple colorations thus obtained. The occurrence of silicates of gold in nature seems to be doubtful.

Experiments conducted by E. Cumenge [7] tend to show that the alkaline auro-silicates, obtained in the wet way, may have played an important part in the formation of auriferous quartz. The following conclusions have been established by these investigations :—

1. If an alkaline aurate, obtained by dissolving auric sesquioxide in caustic soda, is mixed with an alkaline solution of silicate of soda (soluble glass), the mixture may be concentrated by evaporation until it has attained

[1] Mathews and Watters, *J. Amer. Chem. Soc.*, 1900, **22**, 108.
[2] *J. Soc. Chem. Ind.*, 1900, p. 1038 ; Thorpe, *Dict. of Applied Chemistry*, 1912, vol. ii., p. 783.
[3] Lenher, *J. Amer. Chem. Soc.*, 1902, **24**, 355.
[4] For references to the original papers, see Thorpe, *Dictionary of Applied Chemistry*, 1912, vol. ii., p. 783.
[5] Pertsch, *Friedländer's Fortsch. d. Teerfarbenfabrikation*, 1894-97, 1324.
[6] Odenheimer, *J. Soc. Chem Ind.*, 1892, **11**, 600.
[7] Frémy, *Ency. Chim.*, vol. iii., L'or, p. 62.

a syrupy consistency without being decomposed. Auro-silicate of soda is, therefore, fairly stable, so long as there is an excess of alkali present.

2. The decomposition of this auro-silicate is effected by the addition to it of hydrochloric acid by which gelatinous silica is precipitated. This carries down a certain proportion of gold, which gives a rose colour to the white magma.

3. This decomposition may also be completely effected by the action of an aqueous solution of carbonic acid under pressure. Thus, if the syrupy, alkaline auro-silicate is introduced into a bottle of seltzer water, which is then hermetically closed, the decomposition can be seen to be gradually going on without the semi-fluid mass being dissolved, and the latter is replaced at the end of some days by coherent silica, which, on exposure to the air, assumes a white opaline appearance tinged with rose colour.

4. When gelatinous silica, obtained by the decomposition of an alkaline auro-silicate, is heated to redness in a current of steam, it assumes either a beautiful, unalterable rose colour, or a reddish tint with visible grains of gold, according to the proportion of precious metal present, and the conditions under which the precipitation has been effected.

For researches by Hatschek and Simon on the reduction of gold in gelatinous silicic acid, see p. 95.

SULPHIDES OF GOLD.

These compounds are prepared as brown or black precipitates by passing sulphuretted hydrogen through a solution of gold chloride. The exact composition of the precipitate varies with the temperature and degree of concentration of the solution, and the amount of free acid present. Levol and Krüss state that Au_2S is precipitated in the cold, but that only metallic gold and free sulphur are thrown down from boiling solutions. It seems probable, however, that free sulphur is usually formed in considerable quantities in both hot and cold solutions, and, as a general rule, definite compounds are not precipitated, variable mixtures of the three sulphides with free sulphur and metallic gold being formed. According to Ditte[1] and Antony and Lucchesi,[2] *aurous sulphide* is formed when H_2S is passed into an acidified solution of gold cyanide or chloride. It is a brown powder, soluble in water when moist. *Auro-auric sulphide*, Au_2S_2, is formed by passing H_2S into a cold neutral solution of gold chloride (Antony and Lucchesi; Christy). *Auric sulphide*, Au_2S_3 is obtained as a deep yellow precipitate by treating anhydrous lithium auri-chloride with H_2S at $-10°$ (Antony and Lucchesi). Similar precipitates are formed by alkaline sulphides and by sulphides of most of the heavy metals. The sulphides are soluble to some extent in a saturated solution of sulphuretted hydrogen, and are easily soluble in hot solutions of alkaline sulphides or alkalies, or alkaline sulphites, forming double salts so that precipitation from alkaline solutions is never complete. The sulphides are readily decomposed into gold and sulphur by the action of heat, the decomposition being complete at about 200°. Sulphide of gold is also dissolved at ordinary temperatures by potassium cyanide, and is slowly attacked by mercury with formation of mercury sulphide.

---

[1] Ditte, *Compt. rend.*, 1895, **120**, 320.
[2] Antony and Lucchesi, *Gazz. Chim. Ital.*, 1896, **26**, [ii.], 350.

When finely divided gold is heated with sulphur and potassium carbonate, a double sulphide of potassium and gold is obtained, which resists a red heat and is soluble in water. It is used for the production of Burgos lustre in gilding china.[1]

Purple of Cassius.—This body was discovered by Cassius of Leyden in 1683. It contains gold and oxide of tin, and is used to colour artificial gems, porcelain, enamel, glass, and glazes, various shades of violet, red, and purple being thus obtainable. Several methods of preparation are used, of which the following is that employed at the factory at Sèvres [2] :—Half a gramme of gold is dissolved in aqua regia composed of 16·8 grammes of hydrochloric and 10·2 grammes of nitric acid, and the solution is then diluted with 14 litres of water. To this solution is added, drop by drop, a solution of a mixture of dichloride and tetrachloride of tin, prepared as follows :—Three grammes of finely divided tin is dropped, little by little, into 18 grammes of aqua regia (constituted as above, with the addition of 5 c.c. water), the reaction is checked by cooling if it is too violent, and the solution of chloride of tin formed is allowed to cool. The precipitate of purple oxide thus obtained is finely coloured when it has been washed with boiling water. The purple precipitate obtained by Müller, by reducing chloride of gold with glucose in an alkaline solution containing tin oxide in suspension, and by various other methods not involving the use of tin,[3] differs from that prepared by the foregoing method in losing its colour at a red heat, while the true Purple of Cassius becomes brick red under such conditions. The true colour is seen when metallic tin acts on trichloride of gold, or when alloys of gold, silver, and tin are attacked with nitric acid. An alloy containing gold 2 parts, tin 3·5 parts, silver 15 parts is suitable.[4] Moissan obtained a finely divided mixture of stannic oxide, lime, and gold, having the colour and properties of Purple of Cassius, by distilling gold-tin alloys in an electric furnace made of lime.[5]

Purple of Cassius when dry is insoluble in alkalies, but when moist it dissolves in water and in the presence of very small quantities of acids and alkalies. When moist it is also soluble in ammonia. The solution precipitates gold on exposure to light. When dry, no gold is removed from the purple by mercury. It loses its colour at the melting point of gold, but without evolution of oxygen.

The constitution of Purple of Cassius, of which the composition, by analysis, is variable, has been the subject of much discussion, but has not yet been finally determined. Some chemists have considered it to be a compound containing hydroxides of gold and tin. Debray regarded it as a lake of stannic acid coloured by finely divided gold. Müller confirms Debray's views, showing that fine purple compounds can be made with gold and magnesia, lime, baryta, sulphate of barium, etc., the colour depending on the presence of finely divided gold and not on the other constituent. Schneider [6] considered that Purple of Cassius, at any rate in its soluble form, is a mixture of the hydrosols of gold and stannic acid, and Zsigmondy [7]

[1] Thorpe, *Dictionary of Applied Chemistry*, 1912, vol. ii., p. 782.
[2] Frémy's *Ency. Chim.*, L'or, vol. v., p. 63.
[3] Müller, *J. prakt. Chem.*, 1885, **30,** [ii.], 252.
[4] See also Schneider, *Zeitsch. anorg. Chem.*, 1893, **5,** 80.
[5] Moissan, *Compt. rend.*, 1905, **141,** 977.
[6] Schneider, *Zeitsch. anorg. Chem.*, 1893, **5,** 80.
[7] Zsigmondy, *Liebig's Annalen*, 1898, **301,** 29, 361 ; *J. Chem. Soc.*, 1898, **74,** [ii.], 522, 599.

regarded it as a mixture of colloidal gold and colloidal stannic acid. This investigator prepared a red solution of metallic gold in water by mixing formaldehyde rapidly with a feebly alkaline boiling solution of gold chloride. A solution of about 0·005 gramme of gold in 100 c.c. of water was thus obtained and concentrated by dialysis, until an intensely red solution containing 0·12 per cent. of colloidal gold was produced. Neutral salts, mineral acids, and alkalies precipitate the gold and an excess of alcohol changes the colour of the solution to dark violet, completely precipitating the metal, which retains the property of dissolving in water. If shaken up with mercury, these solutions are rapidly decolourised. The gold is also carried to the bottom by freshly precipitated lead sulphate and other precipitates. Gold purple of required composition and shade may be obtained by mixing solutions of colloidal gold and colloidal stannic acid and adding dilute acids or salt solutions. Precipitated Purple of Cassius prepared by Zsigmondy contained after washing and ignition 40·3 per cent. of gold and 59·7 per cent. of stannic acid.

Colloidal gold solutions (see Faraday's Gold, p. 3) can be readily prepared in many ways, and may be red, blue, violet, or green in colour.[1] An interesting method is by the passage of carbon monoxide gas through a solution of auric chloride containing from 0·002 to 0·05 per cent. of gold.[2] According to Steubing[3] and Gans,[4] the various colours of colloidal gold solutions are due to the difference in form of the particles. The size of the red particles, according to Svedberg[5] and Zsigmondy, varies in different samples from 1 $\mu\mu$ to about 20 to 30 $\mu\mu$. The solutions are decolourised when shaken with animal charcoal, barium sulphate, powdered porcelain, fibres of filter paper, etc., but are protected from this by gum arabic.[6]

---

[1] For details and references, see Thorpe's *Dictionary of Applied Chemistry*, vol. ii., p. 785.
[2] Donau, *J. Chem. Soc.*, 1905, **88**, [ii.], 462.
[3] Steubing, *Ann. Physik.*, 1908, **26**, [iv.], 329.
[4] Gans, *J. Chem. Soc.*, 1912, **102**, [ii.], 508.
[5] Svedberg, *ibid.*, 1909, **96**, [ii.], 645.
[6] Donau, *loc. cit.*

CHAPTER IV.

MODE OF OCCURRENCE AND DISTRIBUTION OF GOLD.

Dissemination of Gold.—The wide distribution of gold in minute quantities throughout the world was pointed out by J. R. Eckfeldt and W. E. Dubois, Assayers in the United States Mint, in 1861.[1] They found $17\frac{1}{2}$ grains of gold per ton of galena from Ulster, New York, and 12 grains of gold per ton of Spanish bar lead. A copper cent. of 1822 contained 1 part of gold in 14,500, and an English halfpenny a similar amount. Some metallic antimony contained 1 part in 440,000, bismuth 1 part in 400,000, zinc none. The clay deposit underneath the City of Philadelphia was found to contain 1 part of gold in 1,224,000, or 13 grains per ton. The wide dissemination of gold is further attested by a large number of specimens now in the Percy collection.[2] These consist of small specks of gold of different sizes which have been obtained from the most varied sources. Thus, samples of Pattinson's crystallised and uncrystallised lead, pig lead from all countries, lead fume, red lead, litharge, white lead, precipitated carbonate of lead and acetate of lead were all found to contain gold, which seems to be invariably present in galena. Moreover, it appears to be impossible to procure samples of copper in which gold cannot be detected, although the Lake Superior copper contains less than 1 part in 1,000,000 ; the bronze and copper coins of all nations are usually found to contain much greater quantities of gold than this. Similar evidence has been adduced which tends to show that all ores of silver, antimony, and bismuth contain gold.[3]

L. Wagoner[4] found minute quantities of gold and silver in a number of rocks taken from localities remote from veins or regions known to contain valuable minerals. The table on p. 76 gives his results, which are assays made by cyanide, and do not pretend to give the exact values of the rocks, but only the amounts extracted. By calcining and grinding, one of the samples showed 20 per cent. more.

The average amount of silver is seen to be about twenty times that of gold. The richest specimen contains over 17 grains of gold per ton.

J. E. Spurr[5] drew the deduction that igneous and metamorphic rocks contain more gold than sedimentary rocks, and that the order of richness of the sedimentary rocks in gold and silver may, on further investigation, prove to be (1) clays and shales, (2) sandstones, (3) limestones. The gold is no doubt derived in these cases from the sea, and is concentrated where much organic matter is present. It was found many years ago at the Royal Mint that coal from the North of England contained gold, and gold is also found

---

[1] Dubois, *J. Amer. Phil. Soc.*, June, 1861; *Amer. J. Numismatics*, Oct. 1885.
[2] See Percy and Smith. *Phil. Mag.*, April, 1853; Feb. 1854.
[3] *Loc. cit.*, and E. A. Smith on Bismuth, etc., *J. Soc. Chem. Ind.*, 1893, **12**, 316.
[4] Wagoner, *Trans. Amer. Inst. Mng. Eng.*, 1901, **31**, 798.
[5] Spurr, *ibid.*, February and May, 1902, p. 21.

TABLE XVII.

	Parts per Million.	
	Gold.	Silver.
1. Granite, California, ...	0·104	7·66
2. ,, ,, ...	·137	1·22
3. ,, ,, ...	·115	0·94
4. ,, Nevada, ...	1·130	5·59
5. Syenite, ,, ...	0·720	15·43
6. Sandstone, California, ...	·039	0·54
7. ,, ,, ...	·024	·45
8. ,, ,, ...	·021	·32
9. Marble, ,, ...	·005	·212
10. ,, Carrara, ...	·0086	·201
11. Basalt, California, ...	·026	·547
12. Diabase, ,, ...	·076	7·440

in coal of Cambrian age in Wyoming. A very rich deposit occurs in a bed of lignite in Japan.[1]

F. C. Lincoln [2] gives many records of the occurrence of " primary " gold (as distinct from gold subsequently deposited by infiltration, etc.), in rocks and waters. The amounts for igneous rocks range from 0 to 5 grammes per ton, the average gold content being 0·062 part per million, or 1 grain per ton. In sedimentary rocks, he estimates the average to be 0·015 part per million, and in sea water 0·028 part per million, or nearly half a grain per ton.

Gold in Sea Water.—The discovery of the occurrence of gold in solution in sea water was predicted by Percy,[3] and made by Sonstadt,[4] who states that it is far less than 1 grain per ton. Liversedge subsequently showed [5] that the amount of gold in sea water off the coast of New South Wales is from 0·5 to 1 grain per ton, or in round numbers from 130 to 260 tons of gold per cubic mile. Dr. Don [6] could not precipitate the gold from sea water either by charcoal, by insoluble sulphides, by metals or by a current of electricity. Wagoner [7] found 0·17 grain of gold per ton of water off San Francisco, and from 0·6 to 3·7 grains per ton of water from the depths of the Atlantic. In Chesapeake Bay, there was only 0·2 grain per ton. A statement has been made that about $\frac{1}{30}$ grain of gold per ton occurs in the water round the English coasts.

P. de Wilde [8] found from 0 to 64 milligrammes per ton of sea water, the water from the North Sea containing none. Friedrich [9] examined many samples of the salt deposits of Germany, the product of crystallisation from ancient seas, and found generally unweighable traces of gold in them. In six

[1] Gowland, *Non-Ferrous Metals*, p. 197.
[2] Lincoln, *Economic Geology*, May, 1911; *Mining Mag.*, July, 1911, **5,** 71.
[3] Percy and R. Smith, *Phil. Mag.*, Feb. 1854.
[4] Sonstadt, *Chem. News*, 1872, **26,** 159; *ibid.*, 1892, **65,** 131.
[5] Liversedge, Roy. Soc. of New South Wales, paper read on Oct. 2, 1895; see *Chem. News*, 1896, **74,** 146, 161, 166, 182, and 191.
[6] Don, *Proc. Amer. Inst. Mng. Eng.*, 1897, **27,** 564.
[7] Wagoner, *loc. cit.* and *Trans. Amer. Inst. Mng. Eng.*, 1907.
[8] P. de Wilde, *Archives des sciences physiques et naturelles*, Genf, 1905, pp. 559-580; *J. Chem. Soc.*, 1905, **88,** [ii.], 532.
[9] Friedrich, *Metallurgie*, 1906, **3,** 627.

samples gold was found in weighable quantities, varying from 3 to 13 milligrammes per ton. Liversedge,[1] however, had previously found 2·03 grains of gold per ton of Stassfurt salt, 1·7 grains per ton in rock salt from Cheshire, and larger quantities in kelp and bittern. The form in which gold exists in sea water is unknown, and the problem of its profitable extraction is still unsolved in spite of much patient research.

Gold Ores.—Gold is obtained—(1) From quartz veins (also called lodes, reefs, or leads) in rock formations. In this division may be included replacement-deposits, disseminations in rocks, and also, for example, the marine deposits accumulated in shallow water, such as the conglomerates of the Transvaal. (2) From placers or the alluvial deposits of ancient and modern streams. Modern beach deposits and loose sands or gravels generally, may be included in this section.

One of the most striking differences between the ores of gold and those of all other metals lies in the extremely small proportion which the desired material bears to the worthless gangue with which it is accompanied. Occasionally hand specimens in vein stuff are found containing several per cent. of the precious metal, but these are of quite exceptional occurrence, and have not the slightest economic importance. The greater part of the vein gold now being produced is derived from ores containing only about one part of gold in seventy or eighty thousand, whilst, under exceptional circumstances, a yield of one part in half a million parts of gangue may give handsome profits. Placer deposits are usually much less rich than this; the average amount of gold contained in those now worked does not exceed one part in one million, and in California deposits of gravel with only one part of gold in fifteen millions have proved susceptible of successful treatment by hydraulic mining on a large scale.

(1) *Vein Gold.*—In this case the metal, whenever it is present in visible grains or masses, has sharp angular edges, although it is usually not distinctly crystalline. It frequently penetrates the rock irregularly in various directions, and is completely interwoven with, and attached to the matrix, usually quartz, so that the metal cannot be separated from the rock without crushing the latter.

The gold in lodes is sometimes in the form of crystallisations, which are, however, exceedingly rare, and crystals of gold are still probably unknown to most miners, although they occur more frequently in placer deposits. Arborescent branching and dendritic masses of crystalline gold are more common than single crystals in both quartz lodes and placer deposits. The crystalline forms met with have already been described, p. 10. In the Transylvanian lodes, gold occurs chiefly in thin sheets or plates, often as much as from half an inch to two or more inches in breadth. Such plates are rarely thicker than a visiting card, and are generally covered with crystalline lines and markings, revealing a distinct geometrical structure. Gold also occurs in wire-like forms, sometimes penetrating crystals of other minerals, such as calcite and dolomite.

It frequently happens that the gold in lodes, etc., is in a state of fine division and is not visible without magnification.

The country rock with which auriferous quartz reefs are associated generally consists of slates or schists, especially hydromica and chloritic slates. Gold also occurs sparingly in similar veins in granite and gneiss.

---

[1] Liversedge, *J. Chem. Soc.*, 1897, **71**, 298.

Serpentine sometimes forms the walls of auriferous quartz lodes. Gold also occurs in the midst of rock formations without any obvious connection with quartz veins. Among such rocks are granite, aplite, gneiss, eurite, quartz-trachyte, syenite, andesite, basalt, diorite, gabbro, diabase, schists, porphyry and slate. These deposits seldom pay to work, and in that case can hardly be called gold ores.

Veins of auriferous quartz rarely occur except in association with eruptive rocks, such as dykes of diabase or diorite. So close is this association that we are led to believe that the eruptive rocks are the means by which the gold has been brought up towards the earth's surface, and from thence concentrated by slow aqueous action in the quartz-veins.[1]

J. E. Spurr writes [2]—" Native gold has been found in both basic and acid rocks. It has also been detected in the dark ferro-magnesian silicates of rocks of all degrees of acidity. The commercially-valuable concentrations of gold are generally connected, now with basalt or gabbro, now with diorite, now with phonolite, rhyolite or granite. They occur also in many different forms, as replacement-deposits in limestone, as disseminations in igneous and sedimentary rocks, as contact-deposits near intrusive masses, and in fissure-veins."

(2) *Placer Gold and Nuggets.*—Placer gold is usually in the form of small scales, but pellets or rounded grains also occur, the small pieces being called " gold dust," while large masses or nuggets are usually of a rounded mammilated form. The chief difference between the appearance of placer and vein gold lies in the fact that the former is always rounded, showing no sharp edges, even the crystals having their angles smoothed and rounded off. This has been pointed to by the advocates of the erosion theory of the origin of placer gold, as evidence in favour of their views, the roundness of the fragments being taken to prove that abrasion of the gold has been effected by attrition with water and grains of sand. " Quite commonly nuggets contain pieces of quartz, or they are sometimes embedded in quartz, where, in contact with the gold, the quartz is always angular and just such as occurs in reefs, and not rounded and water-worn, as would be the case if the gold had accumulated while in the alluvial deposits. Limonite frequently occurs associated with the gold of nuggets, sometimes with, at other times without, quartz." [3]

The largest masses of gold yet discovered have been found in auriferous gravel. The " Blanch Barkley " nugget, found in South Australia, weighed 146 lbs., and only 6 ozs. of it were gangue ; and one still larger, the " Welcome " nugget, from Ballarat, Victoria, weighed 2,195 ozs., or 183 lbs., and yielded gold to the value of £8,376 10s. 6d. It was found at a depth of 180 feet. The largest nugget ever found was " The Welcome Stranger," of 2,520 ozs. gross weight, containing 2,284 ozs. of gold. It was discovered in 1869 at Moliagul, Victoria, in a rut cut by a dray wheel, only a few inches below the surface of the ground, and was associated with 68 lbs. of quartz. In all, thirteen nuggets of over 1,000 ozs., and 1,327 of over 20 ozs. in weight are recorded as having been found in Victoria,[4] but many others were not recorded. In Russia a mass of gold was found in 1842 near Miask, weighing

[1] W. Topley, *Brit. Assoc. Report*, 1887, p. 512.
[2] Spurr, *Trans. Amer. Inst. Mng. Eng.*, February and May, 1902. "Igneous Rocks as related to Occurrence of Ores."
[3] E. J. Dunn, *Memoirs of Geol. Survey of Vict.*, No. 12, 1912.
[4] E. J. Dunn, *loc. cit.*

96 lbs. troy. The largest mass from California is given in the State Mineralogist's report as weighing 2,340 ozs., or 195 lbs., but no authentic cases seem to be on record of nuggets from this State weighing more than 20 lbs.

Minerals Associated with Gold.—The minerals most common in placer deposits are platinum, iridosmine, magnetite, iron pyrite, ilmenite, zircon, garnets, rutile and baryte ; wolfram, scheelite, brookite and diamonds are less common. Diamonds are associated with gold in Brazil, and also occasionally in the Urals and in the United States. In auriferous quartz lodes [1] the mineral most commonly associated with gold is pyrite. When pyrite is present, as is generally the case in auriferous quartz, the gold is almost always contained in the pyrite itself (see Gold in Pyrite, below). The pyrite is in rare cases replaced by pyrrhotine, which is found, for example, at Charter's Towers in Queensland and Passagem in Brazil. Pyrite is often accompanied by complex sulphides, which may themselves be auriferous, the most important being chalcopyrite.[2] Mispickel is found in a number of auriferous deposits in many parts of the world, notably at Pestarena, in California, the Urals, Brazil, Honduras, and Matabeleland. Auriferous stibnite is also far from uncommon, and is sometimes rich in gold, which is, however, irregularly distributed. Among the localities in which this is found may be mentioned the Murchison Range, Mashonaland, Armida in New South Wales, and Kremnitz in Hungary (De Launay). Galena and zinc blende are often found with gold ; and grey copper, tetradymite and many other sulphides are less commonly associated with it. Tellurides of gold are very widely distributed, occurring especially in Western Australia, at Cripple Creek in Colorado and in Transylvania. Gold is associated with selenium at Redjong Lebong in Sumatra. In gossans limonite occurs with gold as commonly as pyrite in unoxidised ores, and other metallic oxides, especially ilmenite and magnetite, although not containing gold, are often found with it. There must also be mentioned the frequent association of gold with calcite, and to a less extent with dolomite. Among other minerals occurring with gold are tourmaline, molybdenite, pyrolusite, tetradymite, uranium ochre, roscoelite, vanadinite, crocoite, wollastonite, gypsum and alunite.[3] The recognition of minerals as indicators of the presence of gold has been much discussed, and for particular localities valuable information may be obtained in this way. Speaking generally, however, there is no mineral, except gold itself, which infallibly indicates the presence of gold, and the absence of which denotes the absence of gold.

Gold in Pyrite.—As stated above, the sulphides present in auriferous quartz usually contain gold ; the gold in such an ore is usually in part quite free, disseminated through the quartz, in which visible grains of the metal often occur, and in part locked up in the pyrite, whence but little can in general be extracted by mercury. Dr. Don found that in many lodes in Australia, traces of gold at least were present wherever pyrite could be found and absent when no pyrite could be detected.

In seeking to explain the behaviour of gold in pyrite, when the ore in which it occurs is treated in various ways, various theories as to the form in which it exists have been propounded. According to one of these, the gold

---

[1] T. A. Rickard, *Trans. Inst. Mng. and Met.*, 1898, **6**, 194.
[2] De Launay, *The World's Gold* (London, 1908), p. 31.
[3] T. A. Rickard, *Trans. Inst. Mng. and Met.*, 1898, **6**, 194 ; Louis, *Handbook of Gold Milling*, p. 7, gives an almost complete list of minerals occurring with gold.

is supposed to exist in the pyrite in the form of sulphide combined with sulphides of iron, silver, copper, etc., and the refusal of the gold to amalgamate is explained in this way, auric sulphide not being acted on by mercury. Some observers have endeavoured to dissolve gold out of pyrite by the action of alkaline sulphides, and when, after many attempts, this was at length successfully accomplished, it was put forward as additional evidence that the gold must have been in the state of sulphide, although metallic gold is known to be soluble in these solutions.

The balance of evidence, however, seems to be in favour of the theory that gold exists in pyrite in the metallic state. Although the metal is generally invisible in undecomposed crystals of pyrite, it becomes visible when such crystals are oxidised either by air and water in nature, or by means of nitric acid, or by being roasted or subjected to deflagration with nitre. As a result of such decomposition, particles of bright, lustrous gold, angular and ragged in shape, but of considerable size, often become apparent. These particles may be separated from the oxides of iron by washing; and the use of nitric acid, followed by panning, is frequently resorted to in order to detect gold in pyrite. Moreover, although usually invisible, gold can sometimes be seen in unroasted pyrite. As long ago as the year 1874, Richard Daintree and J. Latta found specimens of cubical pyrite,[1] in which gold could be seen under a microscope gilding the cleavage planes of the crystals. Again, G. Melville-Attwood, on examining crystals of auriferous pyrite from California in 1881,[2] found that the faces of the crystals were gilded in some places, and that here and there little specks or drops of gold occurred, partially imbedded in the pyrite. These films were too thin to be detected by an ordinary lens, so that it did not seem surprising that such impalpable material could not be taken up by mercury. Louis Janin, Jr., more recently [3] found crystals of pyrite in a porphyritic gangue from the Republic of Colombia, which had gold in small globules on their surfaces. Lastly, it has long been known that crystals of pyrite are often found adhering to an amalgamated plate, the particles of gold on their surfaces having been amalgamated. It seems likely, in view of all these facts, that most if not all of the gold is in the metallic state, and its occasional refusal to amalgamate is not very surprising, when it is remembered how completely a thin coating of certain sulphurised compounds prevents amalgamation, and how readily sulphuretted hydrogen would be evolved from decomposing pyrite. It has been contended that the metallic gold is disseminated mechanically through the mass of pyrite, but the action of potassium cyanide, in dissolving the whole of the gold out of comparatively coarsely crushed pyrite, seems to point to the correctness of the view that the interior of the crystal is not auriferous, the deposition of the gold being superficial, so that the enrichment of the pyrite is confined to its crystalline faces, and possibly, but not probably, to its cleavage planes. Strong evidence of the richness of the outside of the crystals of pyrite in the Banket ore of the Transvaal is given by Caldecott.[4]

The following details [5] of a microscopical examination, by Prof. Morton,

---

[1] Daintree and Latta, *Proc. Royal Soc. of New South Wales*, 1874; "Iron Pyrites."
[2] Melville-Attwood, *Precious Metals in the United States*, 1881, p. 604.
[3] Janin, *Mineral Industry*, 1892, **1**, 249.
[4] W. A. Caldecott, *J. Chem. Met. and Mng. Soc. of S. Africa*, Dec. 1903, **4**, 196; *Trans. Inst. Mng. and Met.*, 1904, **14**, 51.
[5] Morton, *Mineral Industry*, 1892, **1**, 249.

of the condition in which pyrite is left after being leached with cyanide, confirms this view :—

" Upon the ordinary auriferous sulphide of iron, or arsenical pyrite, the solution of potassium cyanide acts readily, not by dissolving the sulphuret, but by attacking the gold upon its exposed edges, and eating its way into the cubes by a slow advance, dissolving out the gold as it goes. An examination with the microscope of the pyrite after the gold has been removed, suggests the method of the operation. A sample of very rich pyrite from a mine north of Redding, was treated with a weak solution, containing less than two-tenths of 1 per cent. of cyanide, for 168 hours ; the assay showed a complete extraction of the gold ; as the sulphurets showed no change in their appearance to the naked eye, some of them were placed under the microscope.

" There is no change visible in the form of the crystals as a whole ; along the fractured faces the mispickel looks clean and unaltered, showing the silvery-white colour and intense refraction of the arseno-pyrite. Upon the faces of the crystals appear dark lines, short, and parallel to each other. In places they are crowded close together ; in other parts they are at considerable distances, but always in parallel lines. The lines vary in length, being from four or five to over a hundred times their width ; the lines are very irregular and often broken. These lines are fissures in the pyrite, and extend so deep into it that the microscope does not reveal their depth. By using the higher powers the walls of one of the fissures were seen to be completely honeycombed, looking somewhat like two empty honeycombs set opposite each other ; evidently the mineral removed was crystallised along its contact walls at least. As the raw or untreated pyrite does not show any such fissuring, but, upon the contrary, shows a surface marked only by striation lines common to pyrite, I assume that the fissuring in the treated sample is caused by the solution acting upon some soluble mineral, probably gold, arranged in plates, occurring in groups, but which, by its colour and isomorphism and the extreme tenuity of its lines, is undistinguishable from the mass of pyrite enclosing it."

H. L. Smyth, W. Lindgren, and W. J. Sharwood [1] examined a number of specimens in which gold was associated with pyrite, and found that under the microscope, the gold could be seen at the surface of the pyrite or deposited in minute cracks in it, but never entirely enclosed within it.

Tellurides.—Certain tellurides, existing as distinct mineralogical species, contain gold as an essential constituent. These are described below.[2]

Calaverite is a bronze-yellow gold telluride, usually containing 2 or 3 per cent. of silver, occurring in Western Australia and in certain mines in California and Colorado. It is named after Calaveras County, California, where it was first found (Rickard). E. S. Dana gives the analysis by Genth as gold 40·7 to 40·9 per cent., tellurium 55·9 to 56 per cent., silver 3 to 3·5 per cent., corresponding to the formula $(AuAg)Te_2$.[3] The compound $AuTe_2$ can be formed by fusing the two elements together, but not readily in the wet way. The following analyses of calaverite are given by Charleton :— [4]

---

[1] Smyth, Lindgren, and Sharwood, *Mng. and Sci. Press*, July 14, 1906 ; Aug. 15, 1906 ; and Jan. 26, 1907.

[2] See T. A. Rickard, *Trans. Inst. Mng. and Met.*, 1899, **8**, 74-78 ; A. G. Charleton, *Gold Mining and Milling in W. Australia*, pp. 104-107 ; Carnot, *Compt. rend.*, 1901, **132**, 1298.

[3] T. A. Rickard, *Trans. Amer. Inst. Mng. Eng.*, 1900, **30**, 712.

[4] Charleton, *loc. cit.*

TABLE XVIII.

	Gold.	Tellurium.	Silver.	Total.
1. From Kalgoorlie (E. Simpson).	41·37	57·27	0·58	99·22
2. From Kalgoorlie (J. C. H. Mingaye).	41·76	56·64	0·80	99·20
3. From Kalgoorlie (P. Krusch).	58·63	37·54	2·06	99·91
4. From Cripple Creek (W. F. Hildebrand).	38·95	57·27	3·21	99·43
5. From Cripple Creek (F. C. Knight).	40·14	56·22	3·63	99·99
6. From Cripple Creek (Genth).	40·70	55·89	3·52	100·11

The densities were as follows :—

No. 1, ... 9·311
No. 2, ... 9·377
No. 6, ... 9·043

No. 3 contained also 0·29 per cent. of copper, 0·09 per cent. of iron, 0·07 per cent. of nickel, 1·13 per cent. of selenium and 0·10 per cent. of sulphur. The names given are those of the analysts.

Calaverite can be most readily distinguished from pyrite, according to Rickard, by its being easily cut by a knife and by rarely occurring in any other than a massive form.

Krennerite is similar in composition and appearance to calaverite but is orthorhombic. It occurs in Transylvania and Cripple Creek.[1]

Sylvanite, called also *graphic tellurium*, is a telluride of gold and silver, supposed to correspond to $(AuAg)Te_2$. It sometimes contains antimony and lead in addition. It is usually steel-grey to silver-white, but is sometimes nearly brass-yellow in colour. The arrangement of the crystals sometimes bears a resemblance to writing characters, whence the name graphic. The following analyses agreeing in composition with sylvanite by F. C. Smith are from the Black Hills, South Dakota, and from Cripple Creek respectively[2] :—

	Au.	Ag.	Te.	Total.
I.,	7·64	32·39	59·96	100·00
II.,	5·61	34·23	60·16	100·00

It occurs also at Kalgoorlie, in Transylvania, and in Calaveras County, California. Analyses of Kalgoorlie sylvanite given by Charleton prove it to contain gold 25 to 28 per cent., silver 13 to 15 per cent., and tellurium 56 to 60 per cent.[3] T. A. Rickard states that sylvanite is the characteristic telluride of Cripple Creek, and of Boulder County, Colorado, just as calaverite is of Kalgoorlie.[4]

---

[1] Chester, *Amer. J. of Sci.*, 1898, **5,** [iv.], 375.
[2] F. C. Smith, *Trans. Amer. Inst. Mng. Eng.*, 1897, **26,** 485.
[3] Charleton, *loc. cit.*
[4] Rickard, *Trans. Amer. Inst. Mng. Eng.*, 1900, **30,** 712.

Petzite is a telluride of silver, Ag_2Te, in which the silver is partly replaced by gold. A specimen from the Golden Rule Mine, according to Genth, contained tellurium 32·68 per cent., silver 41·86 per cent., and gold 25·60 per cent. It occurs in West Australia, Transylvania, Chili, California, Colorado, and Utah. It is black or steel-grey in colour, and is slightly harder and more brittle than sylvanite (Rickard). Charleton quotes [1] the following analysis of Kalgoorlie petzite, made by F. W. Grace :—The specimen had a specific gravity of 9 and a hardness of 2·5 to 3. The composition was gold 24·64, silver 40·47, tellurium 34·60, mercury 0·29.

Nagyagite or Foliated Tellurium, a sulpho-telluride of gold and lead, with a few per cent. of antimony, is remarkable for being foliated like graphite, which it also resembles in its colour, a blackish lead-grey, and in having a hardness of from 1 to 1·5 only. Its density, however, is above 7. It occurs in Transylvania, and contains, according to one analysis, tellurium 32·2, lead 54·0, gold 9·0 to 13·0 per cent. According to Louis it corresponds to the formula $Au_2Pb_{14}Sb_3Te_7S_{17}$.

Kalgoorlite is an iron-black mineral with sub-conchoidal fracture occurring at Kalgoorlie, Western Australia, and also in Colorado and Transylvania. Its analysis corresponds to the formula $HgAu_2Ag_6Te_6$.[2] The analysis by J. C. H. Mingaye gave gold 20·72, silver 30·98, mercury 10·86, tellurium 37·26, copper 0·05, sulphur 0·13. The proportion of mercury is very variable. It is massive and brittle, with a brilliant metallic lustre (Charleton).

Coolgardite is a sesquitelluride, $(AuAgHg)_2Te_3$, from Kalgoorlie, described by Carnot.[3] According to Spencer [4] and Liveing,[5] kalgoorlite and coolgardite are not true mineral species, but mixtures of coloradoite, $HgTe$, with the other tellurides. At Kalgoorlie, at the Cripple Creek district, Colorado, in Transylvania, in Boulder County, Colorado, and in many other localities, the value of the ore depends on the tellurides of gold contained in it.

At Cripple Creek, where the deposits are of late Tertiary age, it is the rule to find the tellurides in crystallised form, whereas at Kalgoorlie they are almost without exception in massive form. The age of the Kalgoorlie deposits is pre-Cambrian.[6] The oxidised portions of telluride ores contain finely divided yellow amorphous gold or *mustard gold*, and sometimes lemon-yellow tellurium-ochre or dioxide of tellurium (Transylvania and Boulder County), which is usually, however, converted into tellurite of iron, discovered by Knight.[7] The dull-looking gold from the oxidised ore of Mount Morgan, Queensland, has been recognised as having been derived from tellurides. Mustard gold occurs at Kalgoorlie in yellowish splashes like yellowish clay, and can be distinguished by burnishing (Rickard). The telluride ores of Kalgoorlie and Cripple Creek contain pyrite disseminated throughout them. Among the minerals specially associated with tellurides of gold, fluorite and roscoelite are especially notable, according to Rickard.[8] Fluorite characterises the telluride ores of Cripple Creek and Boulder County and roscoelite, a brownish-green hydro-mica containing vanadium, is found at Kalgoorlie, and

[1] Charleton, *op. cit.*, p. 107.
[2] Pittman, *Records Geol. Survey, N.S.W.*, 1898, **5**, 203.   [3] Carnot, *loc. cit.*
[4] Spencer, *J. Chem. Soc.*, 1903, **84**, [ii.], 378 ; *Mng. Mag. and Pacific Coast Miner*, 1903, **13**, 268.
[5] Liveing, *Eng. and Mng. J.*, 1903, **75**, 814.
[6] See also Lindgren, *Mng. and Sci. Press*, April 13, 1907, p. 472.
[7] T. A. Rickard, *Trans. Amer. Inst. Mng. Eng.*, 1900, **30**, 708.
[8] Rickard, *loc. cit.*

especially in the tellurides of Boulder County. Calcite is characteristic of the Kalgoorlie telluride ores. Sylvanite, petzite, calaverite, and nagyagite occur in Canada.[1]

Composition of Native Gold.—Native gold always contains silver, which occurs in varying proportions, the colour becoming paler with the increase of silver. In addition, small quantities of copper and iron are almost always present, and other metals are also sometimes found. The finest native gold yet found is that from the Pike's Peak Mine, Cripple Creek, which was 999 fine.[2] The mean fineness of Colorado gold is, however, only about 800. The gold from the Mount Morgan Mine, Queensland, was formerly stated to be 997 fine, but as the gold had been extracted by chlorination, more silver may have been present originally. The finest Russian gold was that formerly obtained at Katerineburg in the Urals, and yielded gold 989·6, silver 1·6, copper 3·5, and iron 0·5 (G. Rose). Gold dust from West Africa has been found to contain 978·1 of fine gold, the remainder being silver. The gold found in the British Isles varies from 800 to 900 fine, the remainder being silver ; the specimens from the district around Dolgelly are sometimes a little over 900 fine. The gold in the Andes is often no more than 600 or 700 fine, but from the Darien Mine it is about 990. Gold from Brazil varies from 737 to 978 fine.[3]

Native gold generally contains more than 99 per cent. of gold and silver taken together, and from 0 to 1 per cent. of copper, iron, etc. Native gold is usually coloured yellow owing to the predominance of gold, but nearly white electrum occurs native in a number of localities, and the proportion of silver, according to Phillips,[4] may exceed half the weight of the mixture, and certainly reaches 39 per cent. Native silver is usually free from gold, but a nugget of silver from Bolivia was found by D. Forbes to contain 2·5 per 1,000 of gold.[5]

According to Gowland,[6] placer gold in Japan ranges from 620 to 904 in fineness, vein gold in Japan from 566 to 926, and placer gold in Korea from 764 to 873.

Placer gold is usually finer than that derived from lodes, containing a smaller percentage of silver. The average composition of the placer gold formerly obtained in California was given as [7] gold 883·6, silver 112·4, base metals 4·0. Australian placer gold averages about 950 fine.

Gold is occasionally found alloyed with copper, and sometimes also with iron, bismuth, lead, mercury, tin, antimony, palladium, or rhodium. Rhodium gold from Mexico was found to be of the specific gravity 15·5 to 16·8 and contained 34 to 43 per cent. of rhodium. Palladium gold is called porpezite. The native alloy of palladium, gold and silver from Porpez contains 85·98 per cent. of gold, 9·85 palladium and 4·17 silver (Berzelius). A native amalgam of gold is found in small yellowish crystals of specific gravity 15·47 in the native mercury of Mariposa in California, containing gold 39·02 to 41·63 per cent., and the rest mercury. An amalgam of gold and silver is found in small soft white grains at Choco, New Granada. It contains

[1] Cairnes, *Can. Mng. J.*, April 1, 1911, p. 265.
[2] Furman, *Colliery Manager and Metal Miner*, Oct. 1896, p. 89.
[3] *Metallurgie*, 1906, p. 160.
[4] Phillips, *Gold and Silver* (1867), p. 2.
[5] Percy, *Silver and Gold* (1880), p. 190.
[6] Gowland, *Non-ferrous Metals* (1914), p. 198.
[7] *Report of the U.S. Census*, 1880, **13**, 352.

38·39 per cent. of gold, 5·00 of silver and 57·40 of mercury.[1] *Maldonite,* from Maldon, Victoria, contains gold 64·5, bismuth 35·5 (Louis).

Geographical Distribution of Gold.—It is pointed out by De Launay[2] that many goldfields have been worked out and abandoned, and that the existing goldfields will similarly lose their importance in course of time. All the districts from which the ancients derived their gold have long since ceased to yield any appreciable amount of the metal, and the days of activity have been almost forgotten. As De Launay says, " A day will, perhaps, even come in the near future when the Californians will be astonished on hearing that their ancestors gathered gold in their rivers, as might be the inhabitants of the Adour basin, the Malaga district, the plain of Granada, or the Po Valley on reading that placer working once existed in their country." [3]

Among the richest goldfields in ancient times were those of Armenia, Chaldæa, Asia Minor and Egypt. Of these, Asia Minor included the Pactolus, from which were drawn the riches of Croesus. Thrace, Arabia and some of the Ægean Islands were also celebrated before the Christian era. The Romans obtained their gold from Spain, the Pyrenees and the Alps, from Dalmatia and, above all, from Transylvania, which is still a goldfield.

Coming to the geographical distribution of gold at the present time, in the British Isles gold is found in some of the streams of Cornwall and in lodes and river gravels near Dolgelly and in other parts of Wales, in Sutherlandshire, and near Leadhills in Scotland, and in the County of Wicklow. On the Continent of Europe, gold is most abundant in Hungary and Transylvania, where the gold occurs in quartz lodes contained in eruptive rocks of tertiary age, chiefly propylite, porphyry, diorite and granite. The minerals occurring with the gold are galena, blende and pyrite. Gold mines have long been worked in Tyrol. In the German Empire, the gold obtained is chiefly derived from the smelting of argentiferous galena, in which small quantities of the more precious metal are contained. In Italy the only important mines are those of Pestarena and Val Toppa in North Piedmont, near Monte Rosa. Gold is also found in the sands of the Rhine, the Reuss, the Aar, and other rivers, and in small quantities in Sweden and Finland. A little gold comes from Spain (Corunna, etc.), and Servia.

The gold-bearing districts of Russia are (1) the Urals, (2) Eastern and Western Siberia, whilst an insignificant amount is also derived from Finland and from the Caucasus. The gold was formerly derived chiefly from lodes both in the Urals and in Western Siberia, but is now almost entirely produced by the placers of Siberia.[4]

In India, almost all the gold now being produced is derived from the quartz lodes of the Colar goldfield, Mysore, in Southern India, in which work was begun in the year 1880. A little gold also comes from the Presidencies of Madras and Bombay. In China, Korea, and Japan considerable quantities of gold are produced; little is known of the methods used and of the amount produced in China. Some gold is obtained from the auriferous sand of Bokhara. Among other gold-producing districts of Asia and the adjoining islands may be mentioned Annam and other countries in French

---

[1] Rammelsberg, *Mineralchemie,* p. 10.
[2] De Launay, *The World's Gold,* p 81, *et seq.*
[3] See also T. A. Rickard, " Persistence of Ore in Depth," *Trans. Inst. Mng. and Met.,* Nov. 1914.
[4] See *Mineral Industry,* 1892, p. 203, and Levat, *L'or en Siberie Orientale* (Paris, 1897).

China, the Malay Peninsula, the islands of Borneo, Celebes, Sumatra, Mindanao in the Philippines, and New Guinea.

From the United States a large percentage of the total gold production of the world is obtained. The chief producing States are California, Colorado, Dakota, Montana, Alaska, Arizona, and Utah, but smaller amounts come from many other States. The produce is now far more from lodes than from placer deposits, and in the treatment of auriferous quartz and pyritic ores almost all the known methods of treatment are applied in different localities. Canada (Yukon Territory, British Columbia, Nova Scotia and Ontario) produces large quantities of gold, and gold ores are also found in various parts of Mexico, Colombia, Bolivia, Chili, Venezuela, Brazil, Peru, and the small States of Central America. The production of several of these countries was formerly much larger than it is at the present day, the reduction being especially marked in the cases of Brazil and Venezuela.

Gold is somewhat widely distributed in Africa, the chief sources of production in former times being the placer deposits of the Gold Coast and Abyssinia. The discoveries of auriferous conglomerates in the Transvaal since 1884 have converted that region into the most important gold-producing country. Rhodesia and West Africa are large producers, and some gold comes from Egypt, Madagascar and the Soudan.

Gold is found in all the Colonies of Australia, and in Tasmania and New Zealand. In Western Australia, Kalgoorlie or East Coolgardie is by far the richest goldfield. Other important districts are Murchison, Mount Margaret, North Coolgardie, East Murchison and Coolgardie. The chief gold-producing districts of Queensland are Charters Towers, Rockhampton (where the Mount Morgan mine is situated), Croydon and Gympie. Almost all the gold is produced from the quartz mines, the placers having been practically exhausted. The chief producing districts in Victoria are Ballarat, Sandhurst (Bendigo), Beechworth, Maryborough, Castlemaine, Gippsland, and Ararat.

The table given in Chap. XXI. gives the quantities of gold now being produced from all the more important goldfields of the world.

Geological Age of Gold Deposits.—It has been demonstrated above that deposits of gold occur in many different kinds of rock and in most countries of the world. At each centre of gold mining the conditions are different from those existing elsewhere, and even within the narrow limits of one particular district the strata bearing the precious metal are often of very divergent geological and mineralogical constitution. Murchison,[1] in attempting to assign all gold deposits to one geological horizon, was led into error.[2] He compared the gold-bearing rocks in the Urals, which are distinctly of Silurian age, with some samples sent from Australia. These, by correlation of fossils and minerals, were also proved to be Silurian. As a result, Murchison put forward a theory that all rocks which contained gold would be found to have their genesis in the Palæozoic era. That gold in large amounts was eventually discovered in Victoria seemed to add to this belief, but Whitney's researches in California robbed the theory of much of its truth, for the rocks there were found to be Jurassic. Murchison's original theory has in no way been confirmed by subsequent gold dis-

---

[1] Murchison, "Siluria."
[2] T. A. Rickard, *Amer. Mng. Cong.*, Oct. 25, 1906 ; *Mining Reporter*, 417 ; *J. Chem. Met. and Mng. Soc. of S. Africa*, 1907, **7**, 311.

coveries, for they have been found to be distributed throughout rocks of all ages.

Thus Rickard [1] gives the following table, showing the varied distribution of gold, as illustrated by the principal mining districts of the world :—

TABLE XIX.

Period.	Rock.	District.	Region.
Quaternary.	Andesite.	Monte Christo.	Washington.
Tertiary.	Eruptive.	Cripple Creek.	Colorado.
Cretaceous.	Sandstone.	Verespatak.	Transylvania.
Jurassic.	Amphibolite Schist.	Mauposa.	California.
Triassic.	Limestone.	Raibl.	Carinthia.
Permian.	Conglomerate.	Stupna.	Bohemia.
Carboniferous.	Shale.	Gympie.	Queensland.
Devonian.	Conglomerate.	Witwatersrand.	Transvaal.
Silurian.	Slate and Sandstone.	Bendigo.	Victoria.
Cambrian.	Slate and Quartzite.	Waverly.	Nova Scotia.
Algonkian.	Schist.	Homestake.	South Dakota.
Archæan.	Granite and Schist.	Lake of the Woods.	Ontario.

The prominent goldfields are associated mainly with the intermediate igneous rocks and with granites and tonalites. The auriferous Archæan schists are principally derived from molten magmas, and the ancient veins have shared the metamorphism of these schists. Gold is often found in association with andesitic rocks of tertiary age, but what is the precise connection between them is a question to which no direct answer has been given. No important goldfield now being worked occurs in sedimentary rocks of age more recent than Cretaceous, although there are many igneous formations belonging to the Tertiary period. Generally speaking, the sequence of the gold-bearing strata throughout the ages is complete.

Nevertheless, some generalisations may be made by a comparison of the petrological characters of formations on various continents, and sometimes, though not in every case, by observations upon the contained fossils. This latter mode of correlation is not applicable in the case of Archæan rocks, which are unfossiliferous.

T. Lindsley [2] classifies gold deposits in two broad divisions—(1) Tertiary deposits ; (2) deep-seated deposits. He discusses both classes from the point of view of persistence in depth. The tertiary formations exhibit a depreciation of primary precious metal-contents with depth, while in the deep-seated type there is a much less marked decrease.

Malcolm Maclaren [3] has attempted to trace out the relationships existing among the rocks in the gold-bearing areas, and has broadly divided them into two large classes—Primary and Secondary. By a Primary auriferous province he signifies " one which shows no prior state of combination and has had no former locus in space." Such are the auriferous sulphides and many free gold deposits. A Secondary province is " one which has been derived from some earlier deposits, either from sulphide or telluride ores or from

[1] Rickard. loc. cit.
[2] Lindsley, Eng. and Mng. J., May 23, 1914, p. 1043.
[3] J. M. Maclaren, Gold, p. 42.

gold quartz veins." The following table shows Maclaren's[1] general classification :—

CLASSIFICATION OF AURIFEROUS DEPOSITS.

Primary.

Connected with extrusion of intermediate or basic igneous rocks (andesites or diabases)	Archæan	Schistose rocks : West Australian (Kalgoorlie, etc.), India (Colar), (Hutti), Rhodesia.
	Pre-Cambrian	Arising from intrusion of diabase and diorite dykes through Archæan schists : West Australia, India (Dharwar), South Africa (Pilgrim's Rest, Witwatersrand, and Barberton), Guianas, Appalachian Fields, and Eastern Canada.
	Tertiary	Andesitic goldfields : Northern Chili, Peru. Colombia, Mexico, California (Bodie), Nevada, Utah, Colorado, Unalaska, Japan, Sumatra, Celebes, New Zealand, Transylvania.

Connected with extrusion of acid rocks of granodioritic type	?	—Urals.
	Permo-Carboniferous	—Eastern Australia and Tasmania.
	Jurassic	—Western North America : Alaska, Oregon, and California.

Secondary.

Deposits produced or modified by chemical agencies	Free gold in original sulphide and telluride veins	*a.* Arising from decomposition of auriferous sulphides and tellurides by acid waters or by tellurides below zone of oxidation. *β.* Arising from decomposition of sulphides and tellurides in zone of oxidation.
	Placer gold	—In part.

Deposits produced by mechanical agencies	Placer gold in part.

A. Primary Deposits connected with Extrusion of Basic Igneous Rocks :—

(1) *Archæan.*—If the Archæan group in Maclaren's classification be studied, it is seen that all the members are schistose in nature, although it is often difficult to determine whether they have originated under sedimentary or igneous conditions. It is in the igneous amphibolites of this period that the oldest known forms of auriferous deposits occur. The rocks are unfossiliferous, and can only be correlated from observations on their petrological characteristics. The best defined group of Archæan rocks are found in countries bordering on the Indian Ocean. In the Lake Superior district of Canada similar formations are associated with large copper deposits.

In India there occurs a complex series of Archæan schists and highly metamorphosed rocks called Dharwars,[2] consisting of boulder beds or conglomerates, pebbly grits, quartzites, limestones, argillites and chloritic schists. They stretch from Bombay through Mysore, and are apparently

---

[1] J. M. Maclaren, *Gold*, 1908, p. 44. The table is reproduced by permission of the *Mining Journal.*

[2] J. M. Maclaren, *Rec. Geol. Survey*, India, 1906, **34**, 96.

sedimentary. In some places their structure suggests formation in running water. The most characteristic rock in the whole series, and one which is observed in all the regions associated with Archæan gold deposits, is a well-banded contorted hæmatite and magnetite quartz rock. Its origin, though obscure, may be due to (1) silicification along the shearing planes, or (2) metamorphism of the ferruginous silicate and carbonate bands in depth. In the Dharwar rocks themselves there appear to be two periods of vein formation and gold deposition, the older of which is said to be of dynamic metamorphism. The rocks have been subjected to great strain and assume a very dark colour, due to total reflection from internal stress surfaces. Gold is found in the interior of the lumps of ore, and this fact has been advanced as evidence of the simultaneous deposition of gold and quartz. In some cases the horn-blende schist has been penetrated by intrusive granites, and here the walls of the dyke have become changed into diorite. This would seem to have an important bearing on the occurrence of gold in seemingly metamorphosed igneous rocks.

In West Australia there are Archæan rocks existing under similar conditions to those found in India. Gneissoid granites overlaid by greenstone schists (amphibolites and hornblende) are the primary rocks of the country. The characteristic magnetite is present and runs parallel with the foliation and direction of the main schistose belts. Two distinct types of formation are noticeable in this district—(1) lode formation, as at the main foci of gold ores in the State—viz., Kalgoorlie, Kanowna and Peak Hill: the auriferous rock runs imperceptibly into the barren country rock, and the limits can only be determined by assay values; (2) Quartz veins of the blue and white varieties.

Gold-bearing Archæan rocks are not so common in South Africa as in other parts of the world. The economically important gold deposits occur in rocks of more recent date. The Barberton series in Swaziland may be correlated with the Archæan schists in Australia and India, which have been described already. They consist of chlorite and talc schists, argillites and, finally, the characteristic hard, banded hæmatite. The latter withstands denudation more than the other beds, and thus stands out in relief across the country. Further North, in the Murchison Range, and also in Rhodesia, similar rocks have been identified, in some cases associated with amphibolitic and quartzitic schists. The Buluwayo schists in Rhodesia have been referred to basic igneous intrusions.

In America, the Apalachian, South Dakota and Brazilian fields belong to the Archæan period. The sedimentary rocks in the first area consist of the Talladega series of slates, quartzites, conglomerates and dolomites, while the igneous rocks are complex green schists, diorites and gneisses. In some districts in Carolina the gold is accompanied by tetradymite. The Homestake Mine in Dakota lies in a field of highly metamorphosed Archæan schists impregnated with auriferous pyrite, and containing many lenticular masses of gold-bearing quartz. Some doubt has arisen as to the true age of the rocks in Brazil, but they are more usually referred to as Archæan. There is no mention of igneous intrusions in the Brazilian formations. The gold is in lenticular masses or in auriferous lines of iron ore in itaberite.

The Otago district in New Zealand possesses large alluvial deposits. It is probable, however, that they may be younger than Archæan, and have been said to be Carboniferous or Devonian. Maclaren points out that they may be either due to lateral secretion from contemporaneously deposited alluvial gold, or they may be connected with the granitic rocks intruded during the period of the Middle Mesozoic lift in the Southern Alps.

(2) *Pre-Cambrian.*—Auriferous rocks of Pre-Cambrian age occur in India, West Australia, the Rand, America, and the Guianas—*i.e.*, the same districts as those already mentioned in which gold is found in Archæan surroundings. Indeed, the Pre-Cambrian gold-bearing strata are usually found either in, or together with Archæan rocks. Two important exceptions to this generalisation have been noted at Nullagine in West Australia, and in the Witwatersrand in the Transvaal, where conglomerates occur which are porous and offer but slight resistance to the passage of auriferous solutions.

In India the younger Pre-Cambrian doleritic dykes are found to penetrate the Archæan Dharwar series, and may be correlated with the Cheyair group of the lower Cuddapah system. Where subsequent movements have burst the Archæan fissures, it is possible to find older blue and younger white quartz side by side. The latter often contains graphitic matter, which has probably been taken up from a carboniferous band during its intrusion.

In West Australia there is again the penetration of vertical doleritic dykes into older schists. Acidic dykes occurring in the same district are barren. In the North it is found that where doleritic dykes are cut by other dykes and faults, they become auriferous, and the gold is found for a few feet on either side. In the intersection of white quartz veins with doleritic intrusions there is said to be a close parallel with the Banket reefs of the Rand, where the auriferous conglomerate occurs as lenticular masses and contains gold in white quartz veins, and also interspersed in the matrix. The veins are much richer than the conglomerate.

The theory of the origin of gold on the Rand is discussed below (see p. 94). Doleritic dykes break through the Barberton series, forming banded shoots similar to those in West Australia. The geological age of the Witwatersrand series is not certainly known, though they are reputed to be older than the Devonian rocks of the Cape system, and younger than the Swaziland system. Maclaren states that there is no available evidence to prove whether the gold present in the infiltering solutions was derived from a diabasic magma or whether it was dissolved from presumably underlying Archæan schists. He considers the latter assumption the more probable.[1]

The American fields in which the gold is present in rocks of Pre-Cambrian age are almost identical with those previously mentioned. In South Carolina the country rock consists of an altered Archæan muscovite schist, which is often metamorphosed for some distance from the intruding Pre-Cambrian dyke. It seems probable that the controlling factor in the formation of the gold deposits in this dyke has been the heat of the igneous magma. In Colorado the hornblende schists have replaced mica schists in many instances. The Klondike region, according to Tyrrell,[2] is underlain by Pre-Cambrian schistose rocks which have been very much folded. Outliers of sedimentary rocks of Eocene age occur round the margin of the district, indicating a depression of the surface at that period. After Eocene times the land was raised and has suffered no subsidence since.

(3) *Tertiary.*—The Tertiary goldfields are andesitic in character, ranging from the Eocene to the Pliocene eras. They are distinctly modern, following the lines of modern volcanic activity, the " Pacific Circle of Fire." This circle passes from Valparaiso through Mexico and the Western States, including

---

[1] Maclaren, *op. cit.*, p. 57.　　　　　　[2] Tyrrell, *Econ. Geol.*, 1907, **2**, [iv.], 343.

such well-known fields as Nevada, Utah, Colorado. There is a break in British Columbia, but the line is continued at Unga Island (Alaska), and then follows round through Japan, Sumatra, Celebes, and Burma to New Zealand and Fiji. Two exceptions only of any importance are known, and these are in the Transylvanian and Hungarian goldfields, which were probably formed in the same period as the Mediterranean volcanoes of Mesozoic age.

It is characteristic of the Tertiary andesitic goldfields that the andesite tends to become propylite, in which chlorite and epidote have replaced the ferro-magnesian silicates, and quartz, chlorite and epidote have replaced the felspars. Tellurides are also characteristic, and are usually set in a matrix of quartz.

Maclaren considers that Tertiary andesitic goldfields have probably derived their gold from Pre-Cambrian deposits, either by leaching or by fusion. Lindgren,[1] however, is of the opinion that the gold in Tertiary lodes is derived from the leaching of Pre-Cambrian placers by magmatic or meteoric waters, which have been caused to circulate after the intrusion of the andesites.

In America the Tertiary fields pass from Mexico through Western North America, and belong to the Miocene period. At Pachuca the precious metal occurs only in andesitic tuffs, overlying Cretaceous sediments which probably were broken up in the middle Tertiary period. The Comstock and Cripple Creek regions belong to this period. At Cripple Creek, Pre-Cambrian schists are occasionally found in the granite in association with diabasic dykes. Oligocene and Miocene volcanic rocks prevail in the auriferous area, and the earliest of these consist mainly of andesitic breccias, and in a few cases of lavas and tuffs. It is probable that hydrothermal agencies have been active in the region. Phonolitic breccias and dykes succeed the andesitic rocks, and nepheline basalts have been formed by the intrusion of basic magmas into the former. The veins contain fluorite, which is unusual in andesitic rocks. Tellurides occur in the porous breccias.

According to Rickard,[2] the rocks in Colorado are Post-Cretaceous and Pre-Pliocene. Following the Cretaceous period in this district there occurred an interval of volcanic upheaval and a large outpouring of lava forming brecciated rocks in surrounding water, dykes traversing older strata, and volcanoes as at Cripple Creek and Silver Cliff. Volcanic eruption was very vigorous in Eocene times, and continued into the Miocene. Finally, it was succeeded by a period of intense underground thermal activity. So far as is known, no igneous rock associated with auriferous deposits in this district is older than Tertiary.

The belt of vulcanism is continued in Japan and Formosa, the gold areas in which are not now very productive, though they were greatly exploited by the Dutch and Portuguese merchants in the fifteenth and sixteenth centuries.

Although in New Zealand volcanic eruptions are almost unknown at the present day, geysers and hot springs are still found. There is abundant evidence of the junction of two great Pacific axes of folding and faulting in the Hauraki Peninsula. The gold-bearing veins occur in breccias of andesite and dacites, in which propylitisation has been extensive, and which probably form the Upper Eocene beds, the oldest eruptions of Tertiary age in this

---

[1] Lindgren, *Econ. Geol.*, 1911, **6**, 247.
[2] Rickard, *J. Chem. Met. and Mng. Soc. of S. Africa*, 1907, **7**, 311.

district. Obscure rocks of Mesozoic and Palæozoic age underlie these formations. The Waihi mine occurs in this field.

B. Primary Deposits connected with Extrusion of Acid Rocks.—These occur in three principal regions—(1) Urals, (2) Western North America, and (3) Eastern Australia, which contain the chief placer gold deposits of the world. They all present a characteristic mingling of igneous rocks. It is probable that in Tertiary times the auriferous areas were elevated, since the gold occurs free and in coarse grains, due to the exposure to washing and sorting agencies.

(1) *Palæozoic.*—The deposits in the Urals are supposed to be of late Palæozoic age.

(2) *Jurassic.*—The Californian belt in Western North America includes Rossland, Oregon and Alaska. Lindgren[1] has shown that there is a close relationship between the veins in this area and the metamorphic series of early Palæozoic-Jurassic ages. They consist of altered slates, sandstones, limestones and quartz porphyrites. The gold quartz veins are closely connected with the grano-dioritic rocks. The Great Mother Lode is one of the richest deposits in this belt.

(3) *Permo-Carboniferous.*—The Eastern Australian fields exhibit intrusions which are older than Triassic, and it is probable that the maximum enrichment of the auriferous rocks occurred in the Carboniferous period. In the Snowy River and Mitta Mitta valley, Lower Devonian porphyrites are found, giving evidence of early activity. The gold deposition in the Mount Morgan, Gympie, Lucknow, Wood's Point and Walhalla districts is referred to the Permo-Carboniferous period. The character of the rocks at Mount Morgan is sedimentary, which is, however, an exception to the general rule that usually sedimentary rocks of Ordovician and Silurian age are found in the South, while igneous rocks occur in the North. The latter contain much graphite which was probably assimilated by the magmas from and adjacent Carboniferous formation during intrusion. At Ballarat the origin of the gold is connected with granitic dykes intruded in late Palæozoic times into the vertical Ordovician slates and quartzites. Tertiary basic dykes penetrate through the Silurian slates. It is stated, however, that these are not associated with the deposition of the precious metal, but that the original lode formation began at the time when the neighbouring granite was extruded at the end of the Silurian period, and before the Devonian sediments were formed.

Auriferous quartz veins are visible at the summits of parallel anticlines—saddle reefs—at Bendigo. The country rock consists of Ordovician black clay slates.

C. Secondary Deposits.—These are correlated in point of time with the primary deposits from which they have been derived. Subsequent action, however, may not have occurred immediately on the formation of the primary rocks, and processes such as filtration and the chemical action of gases may proceed long after the original deposit has been laid down. Thus it becomes increasingly difficult to assign secondary rocks to their respective geological periods, and correlation must be based on a study of the primary country rock, which may be found in the vicinity.

Placer Deposits.—According to Maclaren, all the important placers were formed in Tertiary times, and are thus comparatively recent. No

---

[1] Lindgren, *Bull. Geol. Soc. Amer.*, 1895, **6**, 225.

placers the gold of which has been derived from the primary contents of igneous rocks have proved of much value. A few belong to the Cretaceous and Eocene periods, but the greater proportion has been referred to the Recent, Pliocene and Miocene. The conglomerates of Tallawang, New South Wales, are most probably Permo-Carboniferous. In some instances considerable doubt has arisen as to the true geological horizon of placer deposits. Thus Lindgren [1] describes the conglomerates at Mine Hill, California, as Jurassic, and Dunn [2] places those at Klamath, Oregon, in the Cretaceous period. Fairbanks [3] says that in both cases the gold has been deposited by infiltration of solutions, and that the previous deductions as to geological age are in error. As has been mentioned above, the two great goldfields in the Witwatersrand and at Homestake are believed to be of Pre-Cambrian and Archæan origin. Usually it is found that placers have been formed subsequent to any period of great gold deposition, but it has been pointed out by Lincoln [4] that Cambrian placers are rare, although the Pre-Cambrian period is known to have been one of active lode formation.

Origin of Gold Ores.—The origin of mineral veins, including those in which gold is contained, has long been discussed by geologists. The old theory that the quartz of veins was originally in a molten condition and was ejected from below into fissures is no longer maintained, although in 1860 H. Rosales brought forward evidence in its favour as far as the Victorian lodes are concerned. One of the theories now strongly advocated is that the materials forming the veins have been transported in aqueous solution and precipitated where they occur. In certain cases, superheated vapours may have played a part, and there is also the theory of magmatic segregation to be considered. One view is that the solutions found their way downwards from above, and the *ascensional* theory and the *lateral secretion* theory have both been advocated. The last-named theory found its principal supporter during many years in Prof. F. von Sandberger, who pointed out that the gangue of many lodes varies in composition if the nature of the rocks through which they pass is changed, and claimed to have proved by analysis that the materials forming vein-stone are derived from the the adjacent country rocks. He stated, moreover, that such minerals as augite, hornblende, mica, and olivine, which are essential constituents of crystalline rocks, contain small quantities of the heavy metals occurring in veins.[5] Although Sandberger did not try to detect gold in the silicates, this metal is not likely to be an exception. Prof. A. Stelzner objected to these conclusions, urging that small quantities of the sulphides of the heavy metals were probably mechanically mixed with the crystals of minerals which Sandberger analysed in the belief that they were pure. Stelzner advocated the retention of the ascensional theory, which alone affords a satisfactory explanation of the difference in composition observable in neighbouring lodes passing through the same rocks, and apparently formed at different periods. The two theories are, however, not contradictory, and perhaps neither need be entirely rejected, the solutions being supposed

---

[1] Lindgren, *Amer. J. Sci.*, 1894. **48**, 275.
[2] Dunn, 12*th Ann. Rep. Cal. State Mineralogist*, 1894, p. 459.
[3] Fairbanks, *Eng. and Mng. J.*, April 27, 1895, p. 389.
[4] Lincoln, *Econ. Geol.*, 1911, **6**, 247.
[5] Sandberger, *Untersuchungen über Erzgänge*. Wiesbaden, 1882 and 1885. Useful abstracts are given in Phillips' *Ore Deposits* and in Le Neve Foster's *Ore and Stone Mining*.

to pass more or less freely in the plane of the lode after they have been impregnated.[1] For the origin of placer gold, see p. 96.

In 1897, Dr. Don announced,[2] as the result of several years' work in Australia and New Zealand, that the gold in those regions if present in country rock is invariably contained in pyrite. He found that the amount of pyrite and its richness in gold diminished with increase of distance from an auriferous lode, and the gold soon disappeared. In the various minerals of igneous rocks, augite, hornblende, mica, etc., no gold could be detected, although by his method the limit of accuracy was 0·05 grain per ton. He concluded, therefore, that the gold in these lodes was not derived from the country rock adjacent, but that the latter was impregnated with gold from the lode by solutions rising from a depth greater than that of any of the rocks exposed at the surface. L. Wagoner[3] doubts the accuracy of these statements of Dr. Don, finding that they do not agree with his results when examining the rocks of Western America. He finds gold in rocks apparently free from sulphides.

J. E. Spurr[4] traces certain quartz veins directly to igneous magmas as the result of a process of magmatic segregation, the veins being the outward extensions of pegmatite or other igneous dykes. The magmatic concentration is succeeded by aqueous concentration acting underground by dissolution and precipitation, and on the surface chiefly by mechanical means, in both cases producing ore bodies.[5]

The conglomerate gold ores of the Rand have been considered by some authorities to be placer deposits.[6] Gregory contended[7] that the banket was a marine placer in which gold and black sand (magnetite with some titaniferous iron) were laid down in a series of shore deposits. The gold was in minute particles, and it was concentrated by the wash to and fro of the tide, which swept away the light sand and silt, while the gold collected in the sheltered places between the larger pebbles. The black sand deposited with the gold has been converted into pyrite by infiltrating waters, and at the same time the gold was dissolved and redeposited *in situ*.

The infiltration theory of the origin of the gold in the Rand conglomerates, however, now receives more general acceptance.[8] According to this view, after the deposition and cementation of the pebble beds, the gold and pyrite have been introduced, dissolved in solutions flowing freely through the banket, which offers less resistance to their passage than the adjacent country rock, the process being thus similar to the mineralisation of ordinary veins. The deposition of pyrite from solution appears to have taken place

[1] See also Posepny "On the Genesis of Ore Deposits," *Trans. Amer. Inst. Mng. Eng.*, 1893, **23**, 197.

[2] Don, "The Genesis of Certain Auriferous Lodes," *Trans. Amer. Inst. Mng. Eng.*, 1897, **27**, 564.

[3] Wagoner, *Trans. Amer. Inst. Mng. Eng.*, Nov. 1901.

[4] Spurr, *Trans. Amer. Inst. Mng. Eng.*, Feb. and May, 1902.

[5] See also among other papers, W. H. Weed, "Ore Deposits near Igneous Contacts," *Trans. Amer. Inst. Mng. Eng.*, Oct. 1902; also "Observations on Gold Deposits," by C. W. Purington, *Eng. Mng. J.*, 1903, **50**, 855 and 893, where are references to other papers. A full discussion of the subject is given in Thomas and Macalister's *Genesis of Ore Deposits*, and by F. H. Hatch, Presidential Address, *Trans. Inst. Mng. and Met.*, 1914, **23**, xli.

[6] See G. F. Becker, *The Witwatersrand Banket*, 18th Ann. Rep. U.S. Geol. Survey, part v., 1897. E. T. Mellor, *Mng. and Sci. Press*, May 6, 1914.

[7] Gregory, *Trans. Inst. Mng. and Met.*, 1907, **17**, 2-41; also Becker, *loc. cit.*

[8] See Hatch, *Types of Ore Deposits*, San Francisco; *Mng. and Sci. Press*, Dec. 27, 1913, p. 1019; C. B. Horwood, "The Rand Banket," *Mng. and Sci. Press*, 1913, **107**, 563, 604, 647, etc. Further references are given in the articles mentioned.

by replacement of other material, such as silica, which has been removed in solution. The mechanism of the deposition of the gold appears to be less certain.

An important suggestion as to the deposition of gold in quartz reefs generally has been made by Hatschek and Simon, based on their work on siliceous gels.[1] They placed in test tubes silicic acid solution mixed with gold chloride, and after the solutions (sols) had become gelatinous (forming jellies or gels), solutions of reducing agents were poured on the top, or the test tube placed in a large vessel containing a reducing gas. Among the reducing agents tried were oxalic acid, ferrous sulphate, sulphur dioxide, carbon monoxide, illuminating gas, graphite, charcoal and crude petroleum. In some cases the reducing agent was diffused in the gel and a solution of gold chloride allowed to come in contact with it. It was found that in a gel consisting of silicic acid in which a gold salt is uniformly distributed a reducing or precipitating agent has the following effects :—

(a) If the reducing agent is hypotonic in relation to the salts in the gel (i.e., if the concentration reckoned in molecules is less in the reducing agent than in the gel), the gold will leave the gel and will deposit in aggregates, usually of crystalline form, at the area of contact of the reducing agent with the gel.

(b) If the reducing agent is hypertonic in relation to the salts in the gel, the gold will be precipitated within the gel, in crystals or reddish-brown amorphous particles, not always uniformly throughout the gel, but often in distinct layers parallel to the surface of contact, the distance between one layer and the next one increasing with the distance from this surface. Carbon also appears to be precipitated by hydrocarbons, recalling the fact that graphite is often present in gold and silver deposits.

The frequent banded structure of auriferous quartz is brought to mind by these results, and the occurrence of gelatinous silica in reefs has been occasionally noted. Chloride solutions frequently occur near the surface, and might penetrate downwards in lodes, carrying gold in solution. As pointed out by Sulman, however,[2] solutions rising from below are more likely to contain alkaline sulphides in which gold might be dissolved, and the enrichment of siliceous gels in veins might be carried out by the action of such solutions. There is no reason to suppose that the effects observed by Hatschek and Simon are limited to chloride solutions, although these may have played an important part in the secondary enrichment in gold of ore deposits.

Von Veimarn [3] had previously found that gold chloride was slowly reduced in silicic acid without the presence of a special reducing agent. He mixed together dilute solutions of $NaAuCl_4$ and Na_2SiO_3, and observed a gradual change in colour from yellow to colourless, then rose, lilac and blue. He supposed that an unstable silicate of gold was formed and afterwards decomposed and the gold reduced spontaneously. A precipitate of silicic acid formed after about a year contained disseminated gold in particles too fine to be seen under the microscope, even as ultramicroscopic appearances.

The deposition of gold by naturally occurring sulphides has been the

---

[1] Hatschek and Simon, Trans. Inst. Mng. and Met., 1912, 21, 451.
[2] Sulman, op. cit.. p. 466.
[3] Von Veimarn, Zeitsch. Chem. Ind. Colloide, 1913, 11, 287 ; Mng. and Sci. Press, 1913, 107, 309.

subject of much research. Among the more recent studies may be mentioned those of A. D. Brokaw [1] and of Chase Palmer and E. S. Bastin, [2] who proved that nearly all the sulphides and arsenides which commonly occur in ore deposits are capable of reducing gold from a solution of its chlorides, although at different rates of velocity. The apparent preference of gold for chalco-pyrite and tetrahedrite rather than for pyrite in deposits carrying these three minerals may be due to differences in their reducing power.

Origin of Placer Gold.—The origin of the gold in placers was for long a vexed question. It was formerly accepted without question that the erosion of auriferous quartz lodes existing at higher altitudes furnished both gravel and gold. In support of this it was urged that the same districts which furnished auriferous gravels abounded in quartz veins at higher levels, while Whitney pointed out that in California numerous lodes were intersected by the valleys and were still to be seen in the bed-rock. On the other hand, the fact that some nuggets found in drift deposits in Australia and California are much larger than any masses of gold encountered in veins, and that the placer gold is of superior fineness, are difficulties in the way of accepting this theory. Moreover, Egleston states that nuggets as large as a man's fist have been found embedded in the midst of fine sand, whither they could not have been carried by the action of running water, but authentic instances of such finds seem to be lacking, nuggets usually occurring in coarse gravel, among boulders. It is further declared by the opponents of the erosion theory, that if a small quantity of soft material like gold mixed with lumps of hard quartz were washed down by water, then, long before the quartz could be reduced by grinding to the condition of grains of sand, the gold would be worn down to such a fine state of division that none of it could lodge in the river bed at all. In opposition to this contention, it may be urged that the extreme malleability of fine gold would make this comminution very slow, and that, for example, scales of the metal have their edges blunted and thickened by the pounding action of dry sand moved by the wind, instead of having them worn away. Moreover, nuggets often include more or less quartz.

In 1864, in order to account for these and other facts, A. C. Selwyn, of Victoria, suggested a theory of solution in which it is supposed that the gold disseminated through the rocks and drifts is dissolved by percolating waters which contain acids and salts in solution, and is reprecipitated around certain centres. Selwyn considered that the waters capable of dissolving gold must have acquired this property by passing through the beds of basalt, etc., overlying the drifts, inasmuch as large nuggets occur in districts where basaltic eruptions have taken place, while, where these are absent, the gold is very fine, and nuggets can scarcely be said to exist. The fact has long been known that gold is soluble in certain dilute solutions of salts, likely to be met with in nature, such as a mixture of nitrates with chlorides, bromides or iodides, or as the haloid ferric salts. This has been firmly established by the researches of Skey, [3] Daintree, [4] Egleston [5] and others. Gold is also soluble in alkaline sulphides which may arise from deep-seated sources, although

---

[1] Brokaw, *Mng. and Sci. Press*, 1913, **107**, 309.

[2] Palmer and Bastin, *loc. cit.* ; *Bull. Amer. Inst. Mng. Eng.*, 1913, p. 843.

[3] Skey, *Trans. N.Z. Inst.*, 1870, p. 225 ; 1872, p. 370.

[4] Daintree, *Trans. Amer. Inst. Mng. Eng.*, Aug. 1893 ; *Trans. Inst. Mng. and Met.*, 1912, **21**, 456.

[5] Egleston, *Trans. Amer. Inst. Mng. Eng.*, 1880, **8**, 451.

not known to occur in surface waters. The precipitation of gold from these solutions around nuclei consisting of particles of gold, pyrites, etc., by organic matter present in the liquid, has been studied, and efforts made to form nuggets similar to those found in nature, without much success. This, however, is not surprising, since the conditions in nature, including almost unlimited time and immense quantities of exceedingly dilute solutions, cannot be reproduced in the laboratory. Among other pieces of evidence against the erosion theory which have been cited, may be mentioned the fact that some gold placers occur at higher levels than any quartz veins yet discovered or likely to be discovered ; also that nuggets are said to have been found embedded in decomposed rocks in positions to which they could not possibly have been carried by running water, so that these nuggets at least must have been formed by accretion. The prevalent belief among diggers that the tailing from sluicing operations grows in richness so as to be worth working over again after a few years is explicable on either hypothesis. This belief does not seem to have much basis on fact. Speaking generally, no such enrichment occurs, except as the result of natural sluicing, by which most of the tailing is removed and the residue is left richer than the average of the whole mass originally present.

The exponents of the erosion theory have pointed out that the fineness of the placer gold may be accounted for by supposing that the impurities (silver, copper, etc.), formerly present in the native gold have been dissolved away by natural waters, in which they are much more soluble than gold is. One of the difficulties in this view is that solvents cannot extract the silver from a mass of metal containing, say, gold 900, silver 100 unless the gold is also dissolved. The gold protects the silver from attack unless the alloy can be " parted." [1] A more reasonable view is that both gold and silver are dissolved, and a greater proportion of the silver than of the gold carried away, and less of it reprecipitated. It has been suggested by T. A. Rickard [2] that the reprecipitation may occur in gossans, and data are certainly required as to the relative fineness of gold in the oxidised and unoxidised portions of the same lode in given mines.

The existence of large masses of gold in placer deposits was accounted for by Whitney by assuming that the upper portion of the lodes, now washed away, were richer, and contained larger masses of gold than the remains of the lodes now left, but Liversedge has shown [3] that this assumption is not necessary. Some nuggets, too, have been found showing undoubted signs of erosion by water, but these are rare. Liversedge has also adduced evidence (loc. cit.) that, even if the small particles of gold found in placers have grown by accretion, nuggets cannot have appreciably increased in size. The suggestions made to account for the great richness at bed-rocks—viz., that gold has " settled " through the quicksands, or that the gold solution has remained longest in contact with the sand nearest bed-rock—are not wholly satisfactory, but may be supplemented by some such explanation as that given below, p. 132.

The view is now generally accepted that placer gold has almost invariably resulted from the erosion of older auriferous deposits, but there is still some

---

[1] See also Lindgren, " Tertiary Gravels of the Sierra Nevada," U.S. Geol. Survey, 1911, p. 68 ; Mng. and Sci. Press, 1913, 107, 310.

[2] Rickard, Mining Mag., April, 1911, 4, 256.

[3] Liversedge, " Origin of Gold Nuggets," Proc. Roy. Soc. of New South Wales, Sept. 1893.

room for doubt. If it be accepted that the gold in quartz lodes has generally been carried thither in solution by underground waters and deposited therein, either as the result of a reduction in temperature and pressure of the solutions, or by being precipitated by the action of reducing agents, and if it is further to be accepted that redissolution and reprecipitation of gold occurs during the oxidation of lodes to form gossans, there seems to be no reason to stop there. It is undoubted that driftwood has been found in placer deposits containing gold, which has replaced or been deposited on the woody fibres. The same or similar influences which have been at work on gossans could be doing similar work in drift deposits. Although nuggets are crystalline and not concentric in structure, gold deposited from solution is also frequently crystalline. The net result of experiment and discussion appears to be inconclusive. Neither the erosion nor the precipitation theory for the derivation of the gold in placers has been satisfactorily disproved, except perhaps in particular instances.

The gold in the gravels of the Klondike region differs from ordinary placer gold. It is often in large nuggets, including much quartz, and is usually rough and but little water-worn.[1] It is of low standard, and has resulted from the erosion of auriferous rocks by glaciers which have carried the broken material into the valleys and left them as moraines. In this region the grains of gold are flat, roughly elliptical plates, more or less smooth on both surfaces. This shape would not result from travel of the gold along with the gravel down the creek beds, but rather from the pressing and polishing action of the gravel, as it passed over the gold, flattening out the grains and elongating them in the direction of the passage of the gravel over them.[2]

---

[1] J. B. Tyrrell, *Amer. Geol. Soc.*, Dec. 1898.
[2] " Report of Supt. of Mines, Ottawa," 1902, [vi.], p. 17.

CHAPTER V.

TREATMENT OF SHALLOW PLACER DEPOSITS.

Introduction.—The deposits grouped together under the name of " placers " comprise sands, gravels, or any loosely coherent or non-coherent detrital beds containing gold. They have accumulated owing to the action of running water, in the beds of rivers, or on the adjoining inundation plains, or on sea beaches. They fall naturally into two groups, between which no strict line of demarcation exists. These are—

(1) Shallow or modern placers, which are usually in or near existing rivers, and have not yet been covered by other deposits. In certain Arctic districts, the gravel beds have been formed by the action of glaciers.

(2) Deep level or ancient placers, which now lie buried beneath an accumulation of debris or coherent rock, the rivers by which they were formed having often been deflected into other channels by more or less extensive changes in the physical geography of the district in which they existed. Beach deposits occur in each subdivision.

In this chapter, the first of these groups will be considered.

In the past, the greater part of the gold derived from all sources has probably been obtained from shallow placer deposits, but this has ceased to be the case for some time past as far as current production is concerned. Shallow placer deposits now yield only an insignificant proportion of the total output of the world, although, for example, 90 per cent. of the Russian output is from alluvial deposits (Perret). They contain metallic gold in fragments of all sizes, ranging from the finest dust to nuggets weighing thousands of ounces. Auriferous sands are found in the beds of most rivers which flow during any part of their course through a region composed of crystalline rocks. If the rivers have rocky beds, gold may be found in the crevices, caught in natural riffles, and the whole may subsequently be covered by beds of sand.

Gold also occurs in river bars and banks, in river " flats," or inundation plains, in the dry beds of streams which only flow after heavy rains (" gulch diggings "), in terrace gravels on the sides of valleys high above the present level of the water (" bench diggings "), and on the sides and tops of hills (" hill diggings "). The last two subdivisions are evidently ancient rather than modern deposits. The gravels may contain boulders of any size, up to several feet in diameter, or may shade off into fine sand, while sandy clays, especially if on the bed-rock, are frequently very rich. In the Urals, the placer deposits often consist of heavy clays, while others are formed of waterworn fragments of auriferous quartz, talcose and chloritic schists, serpentine, greenstone, etc. Gold occurs under very various conditions in these deposits. It may occur in the grass roots on inundation plains, or near the surface of the gravels in river beds, or dispersed through the whole thickness of a stratum. More commonly, however, the lowest part of the superficial beds,

just above " bed-rock " (the country rock of the district), is richest. In
hollows, cracks, and crevices of the bed-rock, or, if it is soft and decomposed,
in the substance of the upper part of the rock itself, to the depth of 1 or 2 feet,
gold occurs in the greatest quantities. In pipeclays just above bed-rock,
in Victoria, it was not uncommon to find 12 ozs. of gold or more in a single
tubful of " dirt," and similar rich bed-rock deposits were found in early
days in California. The depth at which bed-rock is found varies greatly ;
it may crop out at the surface, or it may be buried beneath hundreds of feet
of gravel, and great variations occur even in a single district. In Siberia,
however, the thickness of the gravel is usually less than 3 feet thick, and is
rarely more than 12 feet.

Methods of obtaining Gravel from Shallow Placers.—The different
varieties of gravel obtained from placer deposits are generally treated alike,
but the mode of " winning the dirt " varies with the necessities of the case.
On flats and bars, the surface gravel, if rich enough, is loosened with pick
and shovel, and then washed. If only the part just above the bed-rock will
pay for treatment, it is reached by " stripping," or, if covered by too great
a thickness of barren material, shafts are sunk, and short levels run from the
bottoms of them in all directions. If the nature of the ground permits of
it, tunnels are run without shafts, and the rich gravel is then followed from
the surface, wherever it is found. This system was much practised in the
early days in California, although now seldom to be seen in operation ; it
was called " coyoting," from the coyote, which lives in holes in the ground.
When water was encountered in the shaft, it was drawn out by a bucket,
until it came in too fast, when the claim was abandoned. In California,
in somewhat later times, efforts were made to reach the gold in the river
beds, by deflecting the streams of water from their courses, and in other
ways. These methods will be briefly described under the head " River
Mining." In Siberia and in the Yukon district the gravels are perennially
frozen, and are thawed with steam jets or wood fires set at the bottom of
shafts and at the ends of the drifts, and are then taken out with pick and
shovel, or by other methods.[1]

Methods of Washing the Gravel.—*The Pan.*—When the existence of
gold in the placers of California and Australia first became known, the diggers
were not acquainted with any apparatus which was well adapted to extract
the metal. The household pan was used everywhere to wash the gravel,
and although, in its original form, it was a difficult implement to use efficiently,
it has retained its place in both countries for prospecting and also for washing
small quantities of rich material, however they may have been obtained.
The pan (see Fig. 26) is usually made of stiff sheet-iron, is flat-bottomed
and circular, with a base of 8 to 10 inches in diameter. The sides slope outwards at an angle of about 45° to the bottom, and the depth of the pan is
from $2\frac{3}{8}$ to $3\frac{1}{2}$ inches. A riffle is a useful addition, formed by the thickening
or bulging inwards of the side, situated about half-way up the latter and
running about half-way round the pan. Grooves to act as riffles are also
sometimes added in the angle of the pan and on the sides. The inside of the
pan is kept smooth and free from rust. The method of using this pan embraces
several operations. First, it is filled to about two-thirds of its capacity with

[1] For methods of work on frozen gravels in Siberia, see " Winter Gold-mining in
Siberia," by C. W. Purington, *Mining Mag.*, 1912, **6**, 50 ; also Perret, *Trans. Inst. Mng.
and Met.*, 1912, **21**, 647.

pay-dirt, of which it then contains about one-tenth of a cubic foot. It is then placed at the bottom of a water-hole or convenient stream, and the dirt is thoroughly broken up with both hands, care being taken not to leave any lumps of clay. As soon as the contents of the pan are reduced to the consistency of soft mud, the pan is grasped with both hands a little behind its greater diameter, inclined away from the operator, raised until the dirt is only just covered with water, and shaken sideways, while a slight oscillatory circular motion is also imparted to it. The mud and fine sand are soon obtained in suspension in the water, and gradually pass over the far edge, which is lowered more and more, until little but the stones, coarse particles of sand, black sand, and gold is left. The larger stones lie on the top and are removed by hand. The final stage consists in lifting the pan with a little water in it, and by a movement of the wrist, something like that used in vanning, causing the material to be spread out by the water in a comet shape, in the angle of the pan. The separation of the gold is also sometimes effected by merely running the water round the angle of the pan. The " colours "— *i.e.*, yellow specks of gold—are seen at the extreme head of the comet, and

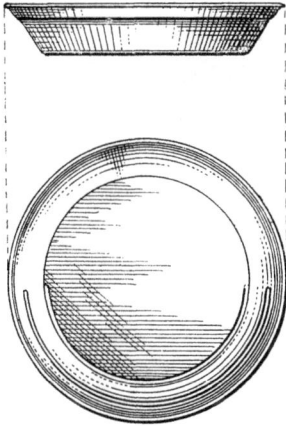

Fig. 26.—Miner's Pan.   Fig. 27.—Batea.

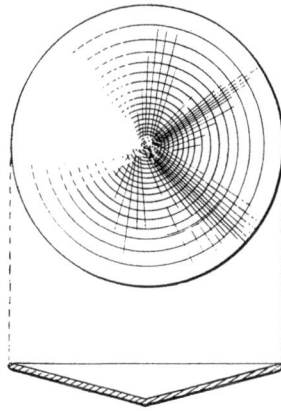

also occur in the succeeding inch or two, mixed with the black sand, while the quartz-sand forms the remainder of the tail and is scraped or washed off. The gold is separated from the black sand by (a) amalgamation with mercury, or (b) drying and blowing away the black sand, a wasteful process. Liquid amalgam is readily separated from sand, and the mercury is then driven off by heat (see below, p. 110).

The Batea (see Fig. 27) differs from the miner's pan in not having a flat bottom. It is of wood turned in a lathe, about 12 inches in diameter, conical, or more rarely basin-shaped, and about $1\frac{1}{2}$ inches deep in the centre, so that the angle at the apex is about 150°. The gold collects at the lowest point and clings to the wooden surface under conditions when it would slide over iron.

The batea consequently is more rapid and effective in obtaining a " prospect " than the pan, especially when the gold is fine, but is less frequently used in the United States and Australia. The best material for the batea is mahogany cut with the direction of the grain vertical to the surface of the implement. It had its origin in South America, and is especially favoured by the negro race.

The *Lotok*, also made of wood, is a Chinese form of batea (Purington).

An ancient form of bowl used by the Moravians for the same purpose as the pan is shown in Fig. 28, which is from Agricola.[1] The bowl was shallow and smooth, and painted black, so that the gold might be more readily seen. In the same illustration is seen a trough or sluice box for washing gold sands, which has a number of cross riffles, and resembles a long-tom.

Fig. 28.

A, Sluice ; B, Box with perforated bottom ; C, Bottom of inverted box ; D, Open part of it ; E, Iron Hoe ; F, Riffles ; G, Small Launder ; H, Bowl in which settled material is taken away ; I, Black bowl in which it is washed.

Prospecting Trough.—This instrument is used in the Far East, especially by the Chinese, Malays, Annamites, etc. It is made of wood, and is shaped in the form of a very flat reversed roof-top, the angle between the long sides being about 150°. In place of a circular movement of the water an alternating rocking motion is used, the water flowing up and down. The instrument is easily handled, but is very slow.

Horn Spoons cut out of black ox-horns have been used by prospectors, especially to finish the work begun by the pan. The surface holds the gold well and shows "colour" very readily.

---

[1] Agricola, *De re Metallica*, 1556, Lib. viii., p. 257.

The Cradle or *Rocker* was introduced in California soon after the first rush to the diggings took place in 1849. It consists of a rectangular wooden box, about 3 feet long and 18 inches wide, resting on two rockers (D, Fig. 29) similar to those used for infants' cradles. The shape of the walls is shown in Fig. 29, which is a section of the apparatus. The method of using it is as follows :—

The gravel is shovelled into the riddle-box, A, the bottom of which consists of $\frac{1}{2}$-inch mesh screen ; the workman sits by the side of the machine and rocks it with one hand, while he pours on water by means of a dipper filled from a water-hole with the other. The dirt is disintegrated and carried through the riddle, and falls on the apron, B, which consists of blanketing, canvas or wood. Here some fine gold is caught, and the dirt then passes out from back to front over the bottom, which is slightly inclined towards the front, and the coarse gold, black sand, etc., is caught in two or three riffles, C, consisting of transverse strips of wood each of about 1 inch in height, to which mercury is sometimes added to assist in retaining the gold. The rocking motion not only assists in the disintegration of the dirt, which is effected by the water, aided by the stones, but also prevents the sand from packing behind the riffles ; in the event of this happening gold would pass over the surface of the sand and be lost. Consequently the rocking should be quite continuous, since, after every pause, the sand in the riffles must be

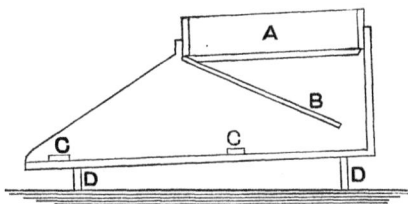

Fig. 29.—Cradle or Rocker.

stirred up before recommencing. It is, therefore, desirable for two men to work together at the cradle, one to carry the gravel and charge it into the hopper, and to remove the large stones from the latter by hand, while the other man rocks the cradle and pours on water. It requires three or four parts of water to wash one part of gravel, and it is, therefore, better to carry the ore to water than to carry water to the ore. When a clean-up of the cradle is desirable, the riddle is removed, the apron is taken out and washed in a bucket, and the accumulations behind the riffles are scraped out and panned. Most of the fine gold in the gravel is lost by the cradle, and two men working together can only wash from 3 to 5 cubic yards per day, according to the nature of the material. According to Richards,[1] the cradle is also used in cleaning-up sluices and quartz mills.

The Long-Tom, said to have been an importation from Georgia, was first used in Nevada County, California, in the latter part of 1849.[2] It consists of a sluice-box or trough (A, Fig. 30) about 12 feet long, 20 inches wide at the upper end, and 30 inches at the lower end, and 9 inches deep, with an inclination of about 1 inch to the foot. The lower end of the trough is cut

[1] Richards, *Ore Dressing*, 1903, p. 723.
[2] A. J. Bowie, *Hydraulic Mining in California*, p. 204.

off at an angle of about 45° and closed by a screen of sheet-iron, B, in which
a number of half-inch holes are punched, so that the fine dirt is allowed to
pass through while the stones are retained. Below the screen is the upper end
of the riffle-box, C, which is usually about 9 feet long, 3 feet wide, and at about
the same inclination as the upper trough. It is fitted with several riffles,
which are sometimes supplied with mercury. In working, a stream of water
enters at the upper end of the sluice-box, into which gravel is continually
shovelled, while a man breaks up the lumps with a fork, removes the large
stones, and puddles the lumps of clay. Two to four men can work at one
tom, and wash about five times as much in a day as can be done by one
or two men with the cradle. Only the coarse gold is caught, and the machine
is only suitable for washing small quantities of rich dirt where there is a
plentiful supply of water. The material caught by the riffles is scraped
out occasionally and panned, but the riffle-box is too short for close saving
of the gold.

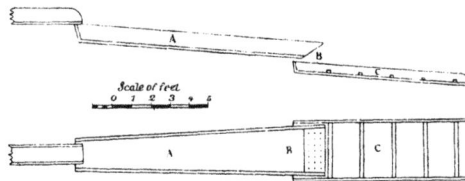

Fig. 30.—Long-Tom.

The Puddling-tub.—When water is scarce, as was the case in many places
in Australia where rich gravels were found, the long-tom is inadmissible,
and the puddling-tub is resorted to. This is particularly well adapted for
washing clays, and is still used to disintegrate lumps of clay encountered
in sluicing operations. It consists of one-half of a barrel which has been
sawn in two ; into this dirt is dumped and stirred up with water by means
of a rake, until all the clay is held in suspension in the water, when a plug
a few inches from the bottom is removed, and the slime run off. The operation is repeated until the tub is filled with gravel and sand to the level of
the plug-hole and this residue is then shovelled out and washed by the pan,
the cradle, or by sluicing. Large boxes were used in Australia in this way
in early days, the rakes being worked by horse or steam-power ; in 1860
no less than 3,958 boxes, worked by horses, were in use in Victoria alone.[1]

The cradle, long-tom and puddling-tub are now little used in districts
where work has been carried on for a long time except by the Chinese, who
gain a precarious livelihood with their help by washing over the heaps of
tailings accumulated from sluicing or hydraulic mining operations in Australia
and California.

In new districts, however, for a short time after shallow deposits have
been discovered, they are still extensively used. At Klondike and at Cape
Nome, for example, thousands of diggers used them, latter day improvements
on the old cradle being especially numerous.

The Siberian Trough.[2]—In Siberia, in the Urals and in the valleys of
the Obi, the Yenisei and the Lena, individual workers still exclusively use

[1] Philips, *Metallurgy of Gold and Silver*, 1867, p. 139.
[2] For further particulars see the account given by Cumenge and Fuchs in Frémy's *Ency.
Chim.*, L'or, Part iii., 1st Section, 12 ; and also Levat, *L'Or en Siberie Orientale*, Paris, 1897.

a trough, which differs from the long-tom mainly in requiring more constant attention on the part of the operator, and which resembles the old German buddle. The trough consists of a rectangular box open above and at one end.

When sandy gravels are being treated, the bottom of the disintegration box (A, Fig. 31), which is about 40 inches square, is made of a perforated screen of wood or sheet-iron, having holes of from ½ inch to 1 inch in diameter. The dirt is shovelled into this box, and, contrary to cradle-practice (see p. 103), if the gold is present in fine flakes, mercury is added here also, the amount depending on the richness of the auriferous material as determined by assay, the proportion used, however, being never more than 10 of mercury to 1 of gold. Water is directed upon the charge in the box, either by pipes from a reservoir or more often by pumping, and the fine material is carried through the screen and falls on to the table, B, while the pebbles are collected by hand and thrown away. If clay is being treated, no screen is used ; the lumps are puddled in the box, and the mud carried over by an overflow of water. The table is slightly inclined, about 20 feet long, and, for the greater part of its length, is about 20 inches wide. It is furnished with five riffles, of which two (C) near the top are about 2 inches high, while the others (D, E) are of less height. The disintegration of the sand is completed on the table with the aid of a small rake continually used by the workman. When disintegration is complete, the stream of water is diminished in amount, and

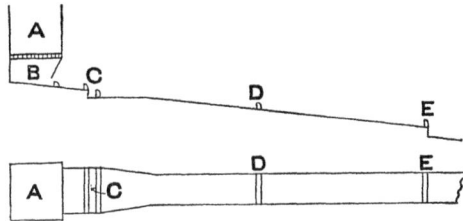

Fig. 31.—Siberian Trough.

the workman continues to rabble the sands which have accumulated above the riffles, pushing the contents of the lower riffles up the table again, until the water runs clear, and little except pyrite is left behind the riffles where the so-called "grey concentrate" accumulates. This is either concentrated further on the same table, or removed and worked on a smaller table. In either case the stream of water is still further reduced, being graduated so as to carry away the last particles of quartz, together with all materials of moderate weight, such as garnets, rutile, tourmaline, etc., and even all the fine pyrite. If mercury has not been added previously, it is sprinkled on before this last operation, unless the gold is very coarse, when no mercury is added at any stage of the proceedings.

The "black concentrate," thus obtained, consists entirely of amalgam, magnetite, and the large grains of pyrite. The final operation, by which the amalgam is separated, is the most difficult, and requires the greatest amount of skill on the part of the operator. The material is worked on the same table with very little water, with the aid of a small rake, or more often with the hand of the workman, who kneels down by the trough for the purpose. Finally, all the pyrite having been washed away, the magnetite is removed with a magnet, and the amalgam collected. The tailing from the black concentrate is treated over again, together with the grey concentrate.

The apparatus just described treats about 500 lbs. of sand at one time, and can be worked by one man, but usually gives employment to four people (frequently three of those are women), who can treat about 5 tons of sand per day. The degree of success attained depends largely on the skill of the workman ; in Siberia and Russia the art is handed down from father to son, certain families devoting their whole lives to the work during many generations. These workmen attain such a degree of dexterity in the use of the trough, that practically the whole of the valuable contents of the gravels treated are extracted by them, but the work is only suited to those who are content with small earnings.

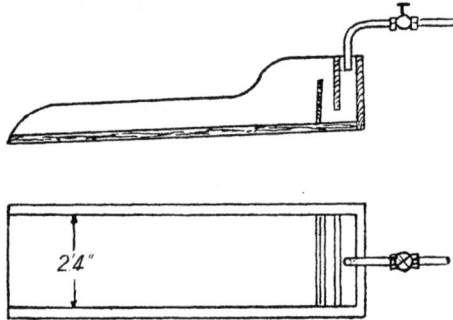

Fig. 32.—Washherd.

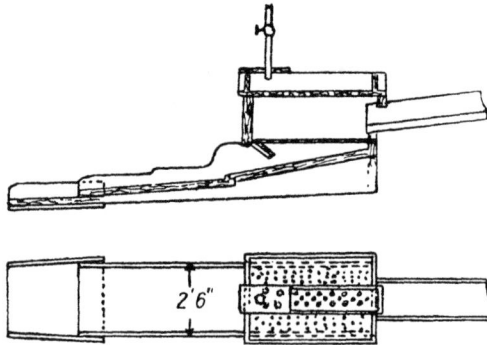

Fig. 33.—Stanok.

The washing of the samples obtained in prospecting work is effected in Siberia usually by means of the *Washherd* (Fig. 32), and in the Urals by means of the *Stanok* (Fig. 33).[1] The illustrations are self-explanatory.

A description of the washing trough or *butara*, which resembles a long-tom, and is used in the Urals, with illustrations and figures as to working costs, etc., is given by J. P. Hutchins.[2] The grade of this trough is 4 feet in 12 feet, an unusually steep grade, for which there is no good reason.

---

[1] Leon Perret, *Trans. Inst. Mng. and Met.*, 1912, **21**, 660. Figs. 32 and 33 are reproduced with the permission of the Institution of Mining and Metallurgy.
[2] Hutchins, *Mining Mag.*, 1914, **10**, 52.

The Sluice.—The sluice has replaced all these implements for washing the gravel from shallow placers, where water is abundant.[1] Sluices are constructed of " boxes," each of which resembles the upper part of the long-tom. The bottom of each box is made of rough boards, about 12 feet long, cut 4 inches wider at one end than at the other ; the total width is usually from 16 to 18 inches, while the sides are 8 or 10 inches high. It has often been pointed out that sluice boxes are usually made too narrow, so that the current is too deep, and fine gold is lost, while the flow is unduly checked by the sides, thus creating undesirable eddies. The box is held together with nails, and no attempt is made to render it water-tight, as the swelling, caused by the absorption of water, and the filling up of chinks with sand and clay soon stops all leakages. The narrow end of the box fits into the wide end of that next below it, and so a sluice, made up of hundreds of boxes, can be readily put up or taken down and moved to another locality.

In all but the simplest and cheapest sluice boxes, extra strips of wood are affixed to the sides so as to protect them from the wear caused by the grinding action of the stones and gravel carried through by the current. When worn thin, these strips are replaced. The bottom is similarly protected

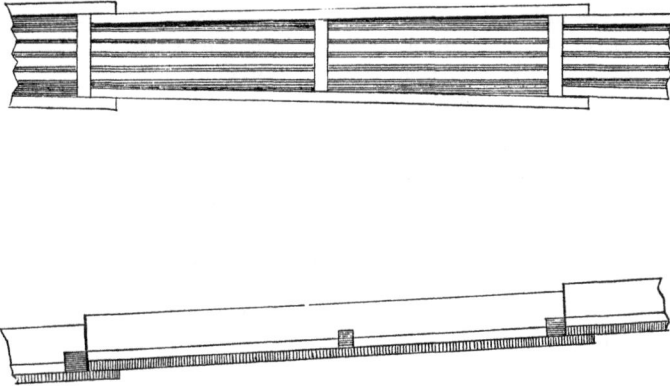

Fig. 34.—Sluice Boxes.

by riffle bars, whose main function is to catch the gold. Many different kinds of riffle bars are used.[2] The simplest are strips of cut wood or fir poles 3 inches in diameter and about 6 feet long. They are usually placed longitudinally, and are wedged in the boxes at a distance of 1 or 2 inches apart by means of transverse bars, so that two sets of riffles are placed in each box in the manner shown in Fig. 34, which represents the whole of one box and parts of two others. The depressions or riffles proper thus formed between the bars are well adapted to intercept all heavy particles that pass down the sluice, such as gold, mercury, amalgam, pyrite, etc., which gravitate to the bottom of the stream. Sometimes the riffles are placed transversely and sometimes for a short distance in zig-zag fashion. The latter arrangement does not retain anything, but affords a better chance of amalgamating the gold, which, together with the mercury (in this case fed in constantly),

---

[1] The large-sized sluices used in hydraulic mining are described in Chap. vi.

[2] See *Mining Mag. and Pacific Coast Miner*, Feb. 1905, **11**, 123. Article on "Washing Plants and Riffles," by C. W. Purington.

slides down the inclined riffles from side to side of the sluice, having sunk to the bottom by virtue of its high specific gravity.

When all is ready, a stream of water is turned into the head of the sluice, where the gravel is shovelled in also. The amount of gravel shovelled in per man depends on the height of the lift and the nature of the soil, as well as on the labourer. It varies from about 3 to 12 cubic yards per diem. The first gravel sluiced fills up most of the riffle depressions, leaving enough inequalities of surface, however, to intercept and retain the mercury, etc.

The length of the sluice varies with the consistency of the gravel, the fineness of the gold, the capital available, and the fall of the ground. It must be sufficient to complete the disintegration and then to catch the gold. The length may be adjusted by experiment ; if, in the clean-up, the lowest boxes yield much amalgam, an addition to the length is necessary, whereas if they yield none, the sluice may be shortened by the removal of one or more boxes. Even in the latter case, however, the tailings would almost certainly contain some fine gold. The grade of the sluice is measured by inches per box, so that a grade of " 12 inches " means one of 12 inches in 12 feet. The usual grade is about 6 inches per box, but it varies from 2 inches to 20 inches per box, depending—

(1) *On the fall of the ground*, since the sluice cannot be raised far above it, nor sunk deep into it, owing to the increased expense thereby occasioned.

(2) *On the nature of the gravel to be washed.*—Tough, tenacious, clayey, or cemented gravels require higher grades to effect their disintegration than loose material. Instead of being disintegrated, clay sometimes becomes aggregated into balls, which roll down the sluices, picking up particles of gold previously caught in the riffles, and these lumps of clay must be removed by hand and puddled. There must be sufficient grade to enable the water to carry away all but the largest stones, so as to avoid unnecessary hand-picking, but on the other hand, while coarse gold is readily caught, fine particles are lost if the current is too rapid.

(3) *On the quantity of water available.*—The reduction of the grade lessens the duty of the water, so that if the supply of the latter is short or costly, the grade is made as steep as possible, consistent with saving a fair proportion of the gold. A steep grade reduces the necessary length of the sluice, as disintegration takes place sooner. Since a steep grade, a rapid flow, and deep currents are best suited to effect speedy and thorough disintegration of the gravel, while a low grade, and slow and shallow currents are best adapted for saving the gold, the upper part of a sluice, for a sufficient distance to effect the complete disintegration of the gravel, is sometimes made of higher grades, or with narrower boxes than the lower part, which is occupied solely in catching the gold. When this is done additional supplies of water should be introduced at the point where the change is made, otherwise, the duty of the water being reduced, the sand packs in the angle where the grade is altered, and constant attention is required to prevent the stream from overflowing.

The requirements of disintegration and gold-saving are more often supplied by " drops," " mud-boxes," and " undercurrents." A vertical fall of the pulp constitutes a *drop*, which is arranged as follows :—The sluice terminates in a " grizzly," or inclined grating made of parallel iron bars placed longitudinally to the stream, and from 1 to 6 inches or more apart, according to the exigencies of the case. All the water and fine stuff pass through the grizzly and fall a distance of from 1 to 10 feet into a sluice below.

The larger stones or boulders roll down the inclined bars, and are shot over a precipice (if possible) or on to a steep slope outside the sluice, as, unless some arrangement for removing these rocks is made, they will accumulate until they can no longer roll off the grizzly. The higher the fall, the more effectively it acts in causing disintegration. Sometimes, near the head of a sluice, the grizzly is omitted from a fall, and the boulders are retained to help in breaking up the gravel. The chief disadvantage in permitting them to remain with the rest of the gravel lies in the fact that they wear out the sluice, and that much water is required to wash them down.

A *mud-box* is merely a wide part of the sluice, 2½ or 3 feet wide, and 12 or more feet long, at which a man is stationed to break up and puddle the lumps. A better, but more expensive, appliance is a trommel or a pan with a mechanical stirrer.[1]

The *undercurrent* is often used in conjunction with a drop. A grizzly with bars placed close together allows most of the water and fine material to pass through, while the coarse stuff is carried over and falls into the main sluice below. The fine material is carried off by a short sluice placed at right angles to the general direction of the main sluice, and is discharged into the upper end of a large broad box from three to ten times as wide as the sluice, and with its long diameter parallel to the main sluice. A number of check-boards help to distribute the stream evenly over the whole width of the box. This box, to which the name undercurrent is often given, although it properly belongs to the whole arrangement, is usually of higher grade than the sluice, having a fall of 10 or 15 inches per box, so that a broad shallow stream flows over its surface. It is plentifully supplied with riffles and mercury, and is intended to catch fine gold and amalgam. The tailing from the undercurrent is discharged into the main sluice below the drop.

Both grizzlies and undercurrents are used more frequently in hydraulic mining than in shallow placer sluicing, in which the large stones are usually removed by a man with a blunt pronged fork, who also either breaks up the lumps of clay or removes them and puddles them in tubs.

The Use of Mercury in Sluicing.—Mercury is added at the head of the sluice after washing has been in progress for a sufficiently long time for all leakages to have been stopped, and for the lowest depressions to have been filled in with sand. The mercury is sometimes sprinkled into the sluice, and sometimes poured into the riffles. The amount added varies with the richness of the gravel and the magnitude of the operations, enough being added to dissolve the amalgam formed. It is carried down the sluice and lodges in the riffles, the greater part being retained in the first few boxes. Fresh supplies are introduced every few hours at the head of the sluice, and sometimes at various points lower down the sluice also; in particular, mercury is added to the undercurrents, as it is especially valuable in catching the finer particles of gold which would otherwise be lost, whilst coarse gold can in great part be saved without mercury. Sometimes the latter is forced into the substance of the wooden riffles by driving an iron gas pipe into the wood, and filling it up with mercury, which is forced by the pressure of the column through the pores of the wood. The amalgam then forms on the surface and in the interstices of the blocks, and in cleaning-up this is scraped off. A better plan is to use amalgamated copper plates, which is

---

[1] Purington, *Mining Mag. and Pacific Coast Miner*, Jan. 1905, 11, 16.

now often done. These resemble the plates used in stamp batteries, described in Chapter VIII.

The Clean-up.—The length of each " run," at the end of which the boxes are cleaned-up, varies, according to the richness of the gravel, from a day to a whole season, but is usually a week. The upper part of the sluice, which retains most of the gold, is usually cleaned-up more frequently than the remainder. A clean-up is begun by discontinuing the supply of gravel, and letting the water continue to flow until it passes through the sluice quite clear. The first six or eight sets of riffle bars are then taken up, and the sand, mercury and amalgam washed down, all the latter being caught by the first riffle left in. It is scooped out thence by a wooden ladle or iron spoon into a bucket, and the rich sand is collected and panned. The next few riffle bars are now taken up, and so on, or alternatively the work may be begun on several sections at the same time. Lastly, the whole sluice is carefully searched over, and particles of amalgam or mercury picked out with spoons, penknives, etc., from every crevice where they have lodged.

The amalgam thus collected is stirred with fresh quicksilver in " amalgam kettles " or buckets, and the black sand and other foreign matter skimmed off. It is then strained through chamois leather or drilling, liquid mercury passing through and pasty amalgam being retained by the skin. The amalgam is well squeezed and then retorted. The retorts used in large well-conducted enterprises are similar to those in use in stamp mills, described in Chapter VIII. Amalgam obtained as the result of operations on a small scale, however, is often merely heated on a shovel over an ordinary fire, the mercury being driven off and lost.

Tail Race.—The tailing from sluicing operations on low ground which has not much fall is removed through a covered-in wooden sluice, or, better still, through a large iron pipe. The work proceeds in the up-stream direction, and the worthless material stripped from above the pay-gravel is thrown on the top of the tail race, which thus passes through a mound of earth and discharges into the open air lower down the valley. As the digging and sluicing progresses up-stream, the tail race is lengthened and the sluice boxes proper are conveyed further up the valley so as always to be near the auriferous material last uncovered. This method originated with the Chinese.

Ground Sluice.—In some cases boarded sluice boxes are not used, but a stream of water is conducted to a little trench cut in the pay dirt, which is soon enlarged by the action of the water, while the banks are at the same time shovelled or prised by the pick or crowbar into the sluice. The method is described by Agricola.[1] The gold is caught in the natural riffles afforded by the uneven wearing of the bed, or rocks may be added to arrest the gold, no mercury being used. Ground sluicing is only adopted where the supply water is precarious, or the season very short, so that violent rains cause floods that would sweep away sluice boxes, and then are succeeded by dry intervals during which the boxes would warp and crack. Only the coarse gold is saved, while the duty of the water is usually much less than in wooden sluices. After a time, usually when the water gives out, the auriferous material is collected from the sluice and washed in a long-tom or cradle.

Booming.—This method of sluicing is described by Pliny in his account of gold-washing in Spain in the first century. It is adopted when the water

---

[1] Agricola, *De re Metallica*, 1657 ed., Lib. viii., p. 270.

supply is insufficient for continuous operations. A dam with a light gate, capable of being easily lifted, is built just above the part of the valley where the auriferous gravel is situated. The water trickling down the valley accumulates behind the dam, and finally overflows at one point into a small rectangular box fastened to the end of a long lever. When full of water this box depresses its end of the lever and raises the dam-gate, so that all the accumulated water rushes out at once and scours the valley bottom. As the box falls it empties itself of water, and the dam-gate returns to its original position by its own weight. This device is usually employed in connection with ground-sluicing, but a line of sluice-boxes might be used through which the sudden flood could carry gravel piled just above the head of the series.

Tail Sluices are sometimes erected to intercept the tailing from one or several sluices with the object of collecting a further percentage of gold from the waste material. These tail sluices are made of much greater size than those described above, and in some cases pay for construction. The Kumara sludge channel, erected by the New Zealand Government to carry the tailing from the sluicing works of the district into the river, caught 957 ozs. of gold in the four years ending in 1891. This sluice is 3 feet 6 inches wide, and has a grade of 1 in 28, while the boxes which discharge into it are only 18 inches to 22 inches wide. Usually, in good work, the sluices are long enough to make the tailing too poor to be worked over again at a profit, except by the Chinese, until after it has been enriched by natural concentration in the rivers.

Fly Catchers were invented in Australia for the purpose of catching the fine particles of gold, which, successfully evading the riffles of all sluices, float down on the surface of the rivers. These devices consist of weirs constructed on piles driven into the river bed, and stretching across from bank to bank of the river. Boards covered with blanketing or coarse gunny-sacking are attached to the weirs and collect all particles floating on the surface of the water. At intervals the blankets are taken up and washed in a tank. These fly catchers soon pay for their cost of construction on many rivers, but are liable to be damaged by floods, and by being used as bridges by men and animals.

Dry-Blowing.—If no water can be obtained, it is sometimes profitable to concentrate pay-dirt by winnowing, tossing it in a pan until the lighter particles have been blown away, and finishing with mouth-blowing. It is, of course, a wasteful method of concentration. In West Australia the conditions are favourable to dry-blowing, winds being strong and constant and the air hot and dry. The method used there [1] is to slowly empty a pan full of dirt into an empty pan placed on the ground. This operation is repeated again and again, and is followed by tossing in a pan, by "panning" as though water were being used, and finally by mouth-blowing. The larger pieces of barren material are removed by hand at intervals.

Among machines used for the purpose, the simplest consist of flat screens supported on a frame and shaken by hand, the material falling through being winnowed by the wind. In other contrivances, a bellows is added worked by the same hand-mechanism by which the screens are shaken.

In some of the machines the blast of air is used to keep the sand partly in suspension, while it is moved by gravity down an inclined table which

---

[1] Rickard, *Trans. Amer. Inst. Mng. Eng.*, 1898, **28**, 502.

is furnished with riffles. The auriferous material must be quite dry or perfect disintegration cannot be accomplished. In a typical dry washer, the gravel is first made to pass through shaking screens, the mesh of which is adapted to the character of the material. The object of the screen is to eliminate the larger fragments, which are usually barren. The shaking screen delivers the material on to an inclined table formed of a wire-screen, covered with light canvas or some similar material through which are forced pulsating blasts of air. These sudden puffs throw up the sand and let it settle again alternately, and as a result the light material works down the table, while the gold is retained by the riffles, being too heavy to be tossed over them by the air.

In Edison's dry washer,[1] the auriferous sand falls vertically and is acted on by a horizontal current of air supplied by a fan, care being taken to avoid puffs and to keep the current constant. The air passes through a screen to eliminate eddies. The lighter material is deflected from its course more than the gold. As the gold is of different shapes and sizes, the gravel is carefully sized by screening. At a mill near Santa Fé, in New Mexico, where the gravel contained from 3 cents to 35 cents in gold per cubic yard, about 95 per cent. of the gold was saved. The iron was removed by a magnetic separator.[2]

In the State of Sonora, Mexico, dry-washing for placer gold is also practised.[3] Here Richards observes that cement gravel containing about $4 or $5 of gold per cubic yard is disintegrated in a Quenner disintegrator and passed through an air-blast machine consisting of an inclined table of burlap and calico on a wire screen. Puffs of air, 150 per minute, are forced upwards through the screen and cause the sand to jump the riffles and travel down the table while the gold is caught. The capacity of the table is 2 cubic yards per hour, and the value of the tailing is about 70 cents per yard. The gold is cleaned by tossing in a large wooden batea.

River Mining.—This method was formerly practised on the rivers of California, but is now superseded by dredging. An entire river was frequently deflected from its course so as to lay bare a section of its bed. This was usually done by building two dams from bank to bank, with their foundations on bed rock, the water being carried off in a wooden flume, starting above the head-dam and terminating below the foot-dam.

Sometimes, as on the American and Feather Rivers, tunnels were made to drain permanently large reaches and deliver the water at a lower point. Sometimes wing-dams were built out from the bank above and below the part of the river it was desired to work, and a third dam connecting their mid-stream ends was constructed parallel to the direction of the current. The space cut off was then pumped dry.

The river bed exposed by such methods was prospected and the pay-dirt when found taken out and washed, particular attention being paid to the surface of the bed-rock. The operations were usually terminated by the autumn floods.

River mining was probably subject to more uncertainty than any other branch of gold mining. The whole capital invested was often lost, and all works and machinery swept away by a flood before the pay-dirt was sighted,

---

[1] Chapman, *Eng. and Mng. J.*, May 9, 1903, p. 713.
[2] For a description of other dry or air concentrators, see Chapter xi.
[3] J. V. Richards, *Trans. Amer. Inst. Mng. Eng.*, 1910, **41**, 797.

while numerous instances are on record in which the alluvium on the river-bed, after having been laid bare at great expense, was not rich enough to pay for sluicing.[1]

Dredging.—This method of recovering gold from the gravel of river beds has of late years made remarkable progress, and its extension to the working of all flat placers, including those which are at some distance from the nearest stream, has completely changed the aspect of shallow alluvial mining. Large quantities of material can be worked at a low cost, and without filling the rivers with débris and causing damage to agricultural lands down stream. The consumption of water is small. Dredging is practised in many countries—e.g., New Zealand, Alaska, Klondike, Siberia, British Guiana, Colombia, S. America, West Africa, the Philippines, and, most of all, in California. A. Grothe[2] states that the first dredge was operated on the Clutha River in New Zealand in 1864, but even in 1891 the method had made but little progress, and was generally looked on with disfavour everywhere except in New Zealand. In dredging, gravel is raised from the bottom of the river and delivered into a barge (also called scow, pontoon or hull), and the material is there washed, the gold extracted, and the tailings sluiced back into the river or stacked on the bank. Gravel beds not in rivers are also dealt with. Dredges may be divided into three classes, according as (1) suction pumps, (2) continuous chain-bucket elevators, or (3) a crane and bucket or shovel are used to raise the gravel.

(1) *Suction Pumps.*—In this system, a centrifugal pump draws material through a large suction hose reaching to the bottom of the river. At Alexandra, New Zealand, as long ago as 1887, a Welman suction dredge was put into operation, and others were built for the ocean beaches. They were found suitable for the treatment of fine gravel, sand, etc., but not for coarse gravel and stones.[3] In 1891, a Welman suction dredge was in successful operation at Waipapa Creek, New Zealand.[4] The pump was 3 feet 6 inches in diameter, and the suction pipe, which was 13 inches in diameter, could be applied at any point in an area within a radius of 40 feet. The large stones were caught and separated from the fine stuff by a riddled hopper-plate. The gold was very finely divided, and was caught on plush mats which were washed every eight hours. The water for washing was supplied from a reservoir by means of an 18-inch pipe. Stones of 56 lbs. weight were lifted by this pump, but larger ones occasionally blocked the suction pipe.

It is difficult to regulate the relative amounts of gravel and water raised by suction pumps, the latter tending to be in great excess. The suction pipe is soon worn out, especially by coarse gravel, and the power required per ton of gravel is considerably greater than in the case of ladder-bucket dredges. Nevertheless, dredges of this type have been in successful operation on the Snake River, Idaho, where a suction pump was working in 1899 on the material of the bank. A pit, 20 feet deep, was first excavated, and the sides were then sluiced down by jets of water, and sucked up through a hose pipe until a hole 200 feet square had been made, when the dredging apparatus was moved and a second pit excavated, the tailing being discharged into the

---

[1] For a full account of river mining, with details of dam-construction, etc., see the article on the subject by R. L. Dunn in the *Ninth Annual Rep. of the Cal. State Mineralogist*, 1889, pp. 262-281. See also the *Eleventh Report*, 1892, pp. 150-153.
[2] Grothe, *Mineral Industry*, 1899, p 326.
[3] *New Zealand Mining Handbook*, 1906, p. 245.
[4] *Report of the Department of Mines, New Zealand*, 1891.

first pit.[1] The material handled ranged from fine sand to boulders 8 inches in diameter. Suction dredges, however, are not much used.

In 1909 proposals for the improvement of suction dredges were made by H. S. Granger.[2] They included the strengthening of the parts (pump-shell, runners and liners) which are most liable to wear and breakage, the provision of a cutter for loosening compact gravel at the intake, and, most important of all, an increase in the diameter of the suction pipe from the ordinary 10 or 12 inches to 24 inches, in order to allow larger stones to pass through. He instanced successful work done by suction dredges in harbour operations.

In Australia hydraulic sluicing with centrifugal pumps, which is not to be confused with suction dredging, is much used in reworking the old alluvial diggings, and has been found profitable.[3] In the Castlemaine and Beechworth districts many plants are reported at work. Work begins by excavating a hole to float the barge (dredge). A pressure of water of 60 to 70 lbs. is required to work the pump. The average yield of the old placer ground at Castlemaine was nearly 230 ozs. per acre. Boulders up to 60 lbs. in weight were raised and passed.

(2) *Chain-Bucket Dredges.*—This is, in the majority of cases, the most satisfactory and economical type, and the one in general use. Chain-bucket dredges are made in great numbers both in New Zealand and in Western America.

A full description of a number of types of bucket dredges with the necessary machinery is given by E. S. and G. N. Marks,[4] from which some of the following details are taken :—The simplest and cheapest form is the sluice-box dredge, on which the gold is caught in a long sluice run, into which the buckets tip their burden. The other type is the screen dredge, which is fitted with a revolving trommel, through which the material raised is washed on to tables. The diagram, Fig. 35, shows a Risdon screen dredge at work. In whatever way the gravel is washed the excavating is done by means of an endless chain of buckets passing round two tumblers, which are placed at the ends of a long ladder or frame. The ladder is of such a length that, when the buckets are working at the bottom of the water, it hangs at an angle of about 45°, as in this position the buckets give the best working results. The lower tumbler is, in modern dredges, an idler or sheave, not power-driven, and is now made round, not polygonal. It consists of ordinary cast steel or, better, of manganese steel. The ladder rollers, on which the chain runs, consist of chilled cast iron or steel. The upper tumbler is usually hexagonal, and acts as a positive driving sprocket for the bucket chain. Manganese, nickel, chrome, and other special steels are now much used in dredge construction[5] for the tumblers, buckets, rollers, spindles, etc., which are subject to heavy wear.

" The ladder is suspended at its lower end from a heavy gantry on the bow of the dredge hull, the raising and lowering of the ladder controlling the digging depth. The elevation of this ladder is controlled by a winch, called the *ladder hoist winch*, situated on the port side of the dredge. The

[1] Grothe, *Mineral Industry*, 1899, p. 328.
[2] Granger, *Trans. Amer. Inst. Mng. Eng.*, 1909, **40**, 496.
[3] F. D. Power, *Eng. and Mng. J.*, 1906, **81**, 759; Granger, *loc. cit.*
[4] E. S. and G. N Marks, *Trans. Inst. Mng. and Met.*, 1906, **15**, 453; see also Gardner and Shepard, *General Electric Review*, May, 1914, **17**, 436.
[5] Sibbett, *Eng. and Mng. J.*, 1914, **97**, 307.

lateral motion of the buckets is secured by swinging the entire dredge on one of the 'spuds' as a pivot. This is done by means of lines run from two

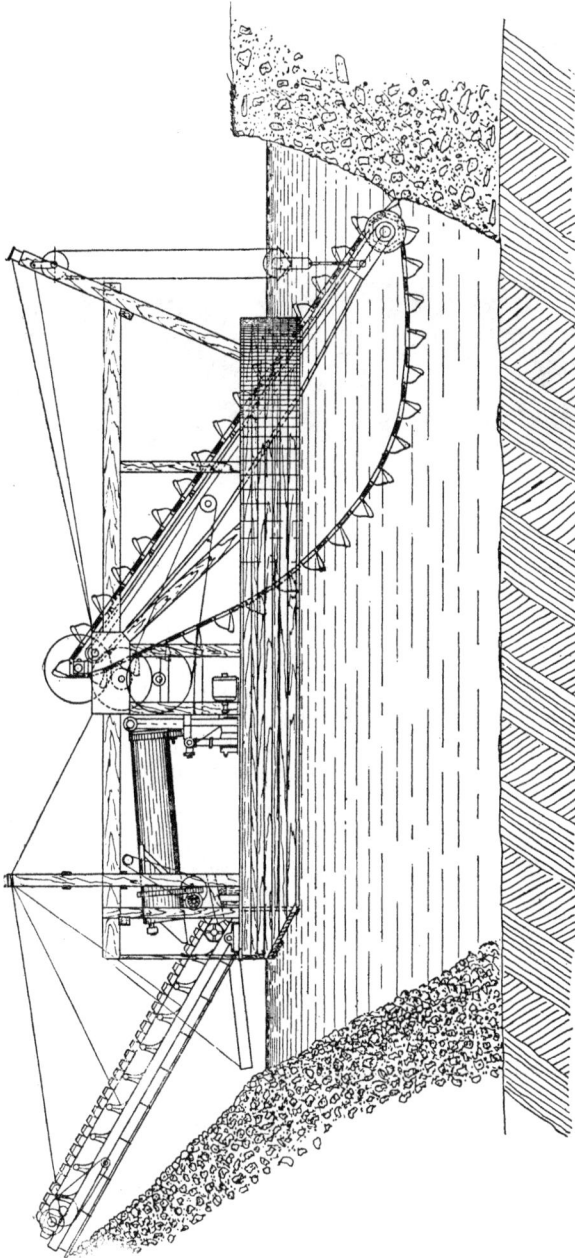

Fig. 35.—Risdon Gold Dredger.

drums, the ' swing winch,' on the starboard side, connected through sheaves on the port and starboard bow to ' dead men' on the banks. The continuously revolving bucket line is thus swung across the face of the bank for a width of from 190 to 300 feet, and at the extreme lateral limit of the swing the ladder is dropped and the side swing repeated. The revolving buckets thus terrace away the bank until bed-rock is reached, when the ladder is raised, the dredge stepped forward by swinging on alternate spuds, and the whole operation recommenced. These spuds, which are single sticks of lumber on the smaller dredges and of structural steel on the larger and more powerful dredges, are at the stern of the dredge and suspended from the stern gantry. On Yuba No. 14, for instance, these spuds weigh over 40 tons each, and are 60 feet long. By alternately raising and dropping these two

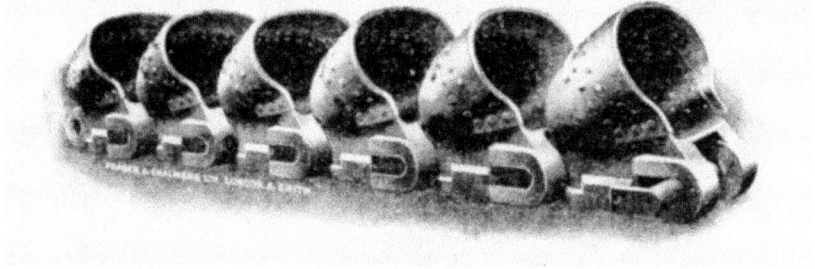

Fig. 36.—Close-connected Bucket Chain.

spuds as the dredge swings, it is stepped ahead much as a stiff-legged man might walk." [1]

The buckets are of steel plate, with a wide mouth and a shallow back to facilitate discharge. Their capacity varies from 3 to 16 cubic feet, but the tendency is towards larger buckets. In the Californian type of dredge the buckets are placed close together, following one another with no connecting link between (close-connected bucket chain, see Fig. 36), and in the

Fig. 37.—Open-Link Chain with Solid Links.

New Zealand type of dredge the buckets are spaced to form an intermittent chain (open-link bucket chain, see Fig. 37). The former type is generally better, especially in loose free-running ground, and the latter is said to be better in hard ground or when dealing with timber and boulders. A " save-all " sluice box is provided to catch the drippings from the buckets. A winch and drums are used for anchorage lines (when spuds are not employed) and for working the elevators. The power, formerly steam, is now electrical where possible. A heavy water pump is required.

---

[1] W. H. Gardner and W. M. Shepard, *General Electric Review* (New York), 1914, **17**, 438.

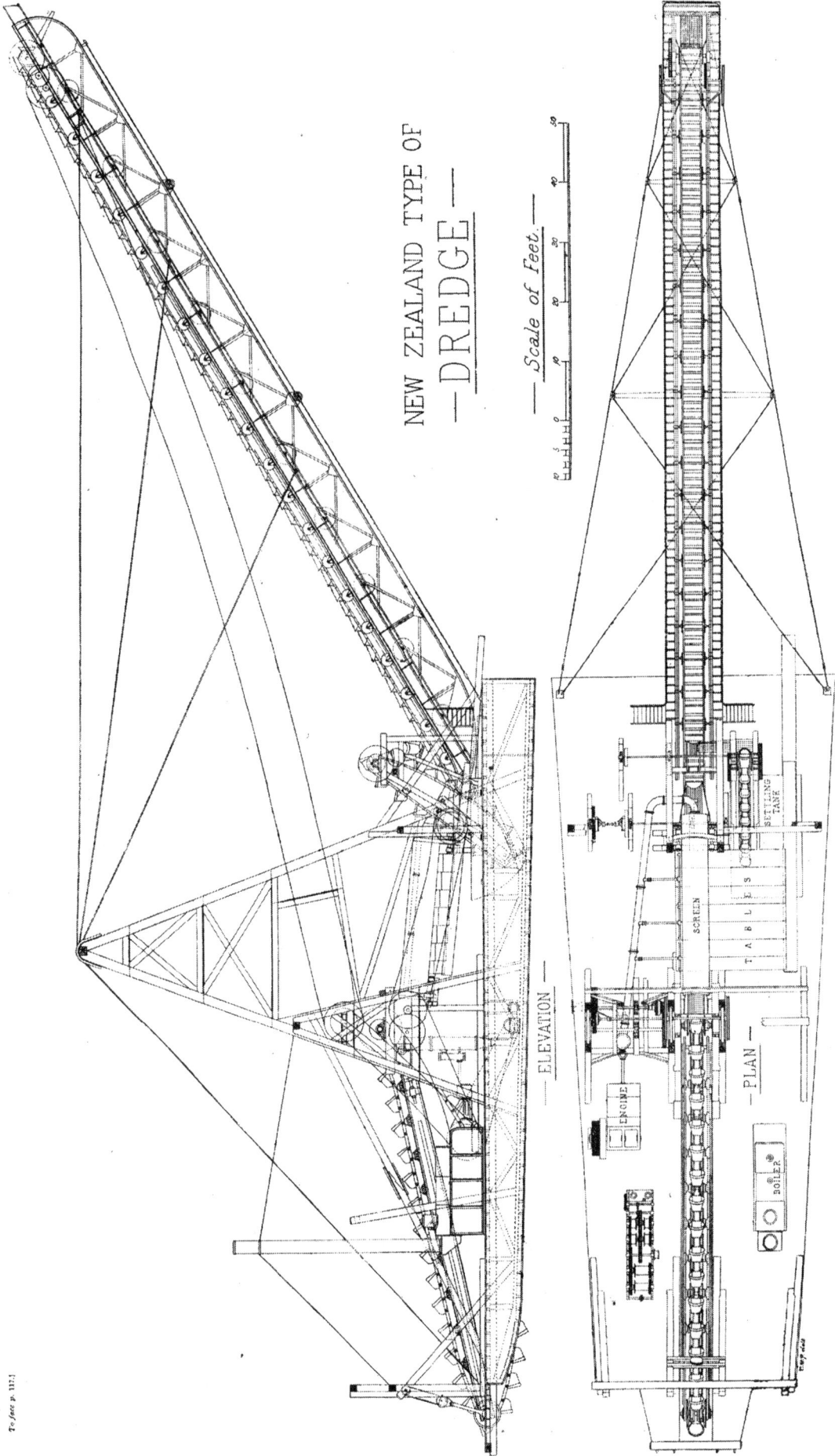

NEW ZEALAND TYPE OF
— DREDGE —

— Scale of Feet —

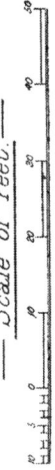

— ELEVATION —

— PLAN —

ENGINE

BOILER

SCREEN

TABLES

SETTLING TANK

Figs. 38 and 39.—Typical New Zealand Dredge.

To face p. 141.

" The ascending buckets, full of gold-bearing material, are dumped over the upper tumbler into a hopper, lined with heavy wearing bars, where it is subjected to high pressure sprays of water. The gravel is then delivered to the screen, a revolving cylinder sloping toward the stern and lined with perforated steel plates, in which it is continuously played upon by jets of water and is completely disintegrated. The sand, fine gravel and gold particles are washed through the perforations into the distributor under the screen which serves to properly distribute the mass on to the gold saving tables. These are in two banks, an upper and a lower, and are composed of fore and aft and thwart-ship sluices, lined with steel-shod sugar-pine riffles, where the gold is caught and amalgamated with mercury. The waste sand is delivered from the tail sluices at the stern of the dredge, some of it being deposited around the spud points to enable them to obtain a firmer hold, and the remainder dumped at some distance behind the dredge. The boulders and heavy gravel, which do not pass through the perforations, fall from the rear end of the screen into a chute and are thence delivered on to the stacker belt. This endless belt carries the material up a long stacker, which is hung from the stern gantry of the dredge, and is finally dumped far to the stern, forming the extensive ' rock piles ' characteristic of a dredging field." [1]

A typical New Zealand dredge is shown in Figs. 38 and 39. These figures are from drawings supplied to the Department of Mines of New Zealand by F. W. Payne, of Dunedin.[2] He also gave the following description of the plant.[3]

The hull is 119 feet long, 35 feet 6 inches wide at bow, and 50 feet wide at stern. Depth of hull forward 6 feet 6 inches, depth aft 9 feet 6 inches. The engine is 25 H.P. nominal. The ladder carrying the bucket-chain is of sufficient length to dredge 40 feet below water line. The buckets are of 7 cubic feet capacity, and run at a speed of ten buckets per minute. The ladder is raised and lowered by a line worked by a winch, and five other winches are designed for a head line and four side lines to moor the dredge to the bank and to move it from place to place. The buckets deliver the gravel into a revolving screen 31 feet long and 4 feet 6 inches in diameter, driven by friction rollers. Thence the coarse material is directed into the main tailing elevator-buckets, and the fine material passing through the screen is delivered on to the gold-saving tables. After running over these, the tailing is deposited in a settling tank, from which it is lifted by a supplementary elevator into the main tailing elevator. The main elevator is no less than 145 feet long between centres, and is capable of stacking tailing to a height of 80 feet above water line. The capacity would be about 120 cubic yards per hour with the buckets three-fourths full. The dredge was intended to work on Fraser Flat, near Alexandra, New Zealand.

The chain-bucket dredges in use in California bear a close resemblance to the type generally used in New Zealand, where most of the appliances originated. The following is a brief abstract of a description of a dredge placed at Oroville, on the Feather River, given by its designer, R. H. Postlethwaite [4] :—The hull is 80 feet long by 30 feet wide and 7 feet deep. The ladder, carrying the dredger buckets, consists of a heavy lattice girder, and the bucket belt travels on cast-steel rollers. There are 32 cast-steel

[1] Gardner and Shepard, loc. cit.
[2] Payne, Report of the Department of Mines, New Zealand, for 1899-1900, p. 43.
[3] Payne, loc. cit.
[4] Postlethwaite, Mng. and Sci. Press, Dec. 15, 1900, 81, 582.

buckets, each of 5 cubic feet capacity, the lips being made of nickel steel. The ladder runs at the rate of 12½ buckets per minute, giving a capacity of over 100 cubic yards per hour with the buckets three-fourths full. The gravel is delivered into a revolving screen 4½ feet in diameter and 25 feet long perforated with holes of ⅜ inch in diameter. The coarse material is conveyed away by a bucket elevator, but the fine stuff after passing over the gold-saving tables is sluiced into the river at a distance of several yards beyond the stern of the dredge. The dredge is moored by five lines and all power is supplied through electric motors. The crew consists of two men per shift, and about 70 H.P. is used in the various operations. The working costs of Risdon dredges in California are stated to be in certain cases about 4 cents per cubic yard of gravel, but are generally from 6 to 8 cents.

According to D. W. Brunton,[1] dredges built in 1909 were provided with close-connected buckets, of capacities up to 13·5 cubic feet, and capable of handling 10,000 cubic yards of gravel in twenty-four hours. There were bucket-ladders capable of digging 67 feet below the water line and 20 feet above it.

C. Janin, in describing a modern Californian dredge,[2] the Yuba No. 13, put in commission on August 10, 1911, says its dimensions are 150 feet by 58·5 feet by 12·5 feet, with an overhang of 5 feet on each side. The digging-chain is of plate-girder construction, designed to dig 65 feet below water level and is equipped with 90 buckets in a close-connected line, each of 15 cubic feet capacity. The weight of the chain and buckets is 700,000 lbs. The washing screen is of the revolving type, roller driven, and is 9 feet in diameter and 50·5 feet long. The gold-saving tables are of the double-bank type, and have a total area of 8,000 square feet. The conveyor stacker-belt is 42 inches wide and 275 feet long, on a stacker ladder (lattice-girder) 142 feet long. Nine motors are in use on the dredge, of a total capacity of 1,072 H.P. There are two steel spuds. The scow is constructed of wood, but steel scows are now generally preferred, especially in tropical climates. Even larger buckets up to at least a cubic yard in capacity are expected to be used.

The dredge described above handled 280,000 cubic yards per month on the Yuba River at an average operating and maintenance cost of 3·3 cents per cubic yard.[3]

The latest dredge, Yuba No. 14 (see Fig. 40), is described by Gardner and Shepard,[4] and was built on the Yuba River in 1912. Its hull, steel decks, housing and gold-saving tables are of steel, and it is electrically operated. The buckets are of 16 cubic feet capacity and the wearing parts are of manganese or nickel-chrome steel. The digging can be carried down to 70 feet below water level. The dimensions are similar to those of Yuba No. 13, and its weight is 1,994 tons.

The dredge is now often anchored by means of spuds, as in the cases described above, instead of by a number of mooring lines. In the former case the anchorage is more rigid, preventing the rocking of the barge, which interferes with the gold-saving tables.

(3) *Crane and Bucket Dredges.*—In these dredges a single large bucket or shovel, with a capacity of 1 or 2 cubic yards, is filled with gravel and hoisted

---

[1] Brunton, *Trans. Amer. Inst. Min. Eng.*, 1909, **40**, 557.
[2] Janin, *Trans. Amer. Inst. Min. Eng.*, 1911, **42**, 855.
[3] Gardner and Shepard, *Min. and Sci. Press*, June 27, 1914, **108**, 1053.
[4] Gardner and Shepard, *loc. cit.* ; also *General Electric Review*, *loc. cit.*

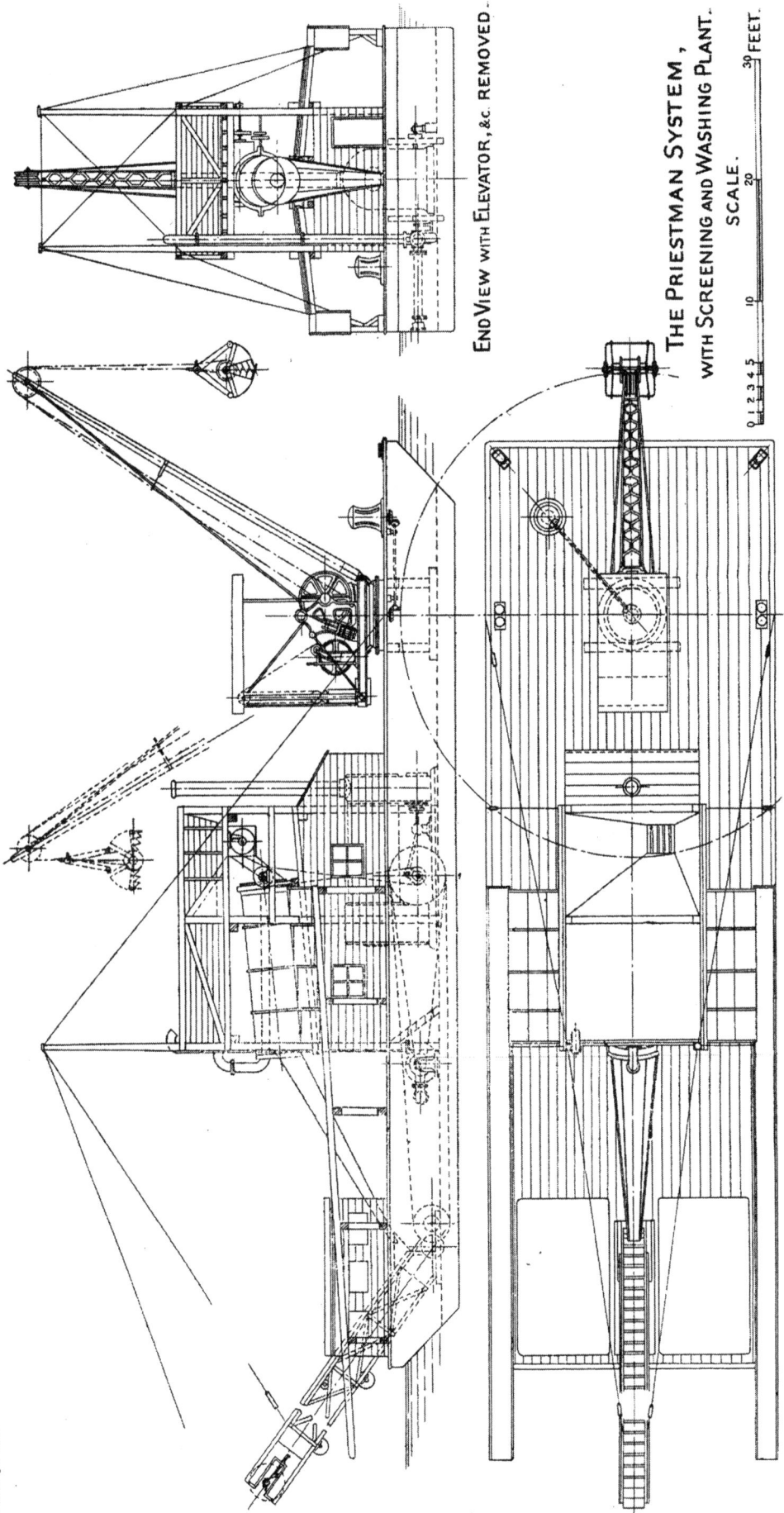

END VIEW WITH ELEVATOR, &c. REMOVED.

THE PRIESTMAN SYSTEM,
WITH SCREENING AND WASHING PLANT.

SCALE.

0 1 2 3 4 5   10   20   30 FEET.

Figs. 41, 42 and 43.—Crane and Bucket Dredge.

on board a barge by some form of crane. The bucket is sometimes of the Priestman grab type, opening at the bottom as it is dropped on to the gravel bed, and closing automatically when the chains begin to raise it. A large shovel is also sometimes used as in a case on the Fraser River, British Columbia, in 1896-1897.[1] Here there were two barges, one containing the dredging machinery, the other the gold-washing tables. The shovel made 90 trips per hour and raised 1½ cubic yards at a time. Boulders up to 5 or 6 tons in weight could be handled with ease. The Priestman grab bucket does not close completely, if a stone lodges between its jaws, and comes up partly empty, but in cases where large boulders, tree trunks, etc., are dispersed through the gravel, it might be used with advantage, instead of the ordinary ladder-bucket, which cannot deal with such obstructions. The wear and tear is said to be less with a Priestman dredge than with the bucket dredge.

A Priestman dredge is shown in Figs. 41, 42, and 43, which are from drawings supplied by Messrs. Priestman Brothers.

Fig. 40.—Yuba Dredge.

It is considered by some that " dipper-dredges " are useful under the following conditions :—(1) Where the ground is somewhat shallow ; (2) in small areas where a cheap dredge is required ; (3) where the material is rough, with boulders and stumps ; (4) where the ground is mixed with more or less clay, as the dipper will relieve itself notwithstanding the adhesiveness of the material (Janin).

In places unsuitable for dredges, *bucket-scrapers* are sometimes used.[2] This is a single bucket of about 1·5 cubic yards capacity on a crane, loaded by a drag line. The gravel is dumped by the bucket into a hopper, which feeds a trommel screen. The machine rests on rollers, by which it is moved on a plank track. It has been tried by Purington in Siberia.

Gold-Saving Apparatus.—The available space on a dredge is so small

---

[1] Grothe, *Mineral Industry*, 1899, p. 329.   [2] Janin, *loc. cit.*

that long sluices are out of the question, and it is, therefore, necessary to disintegrate the gravel thoroughly in revolving trommels or on shaking or travelling screens, and to make the gold-saving tables as wide as possible, so as to reduce the speed and depth of the stream of gravel running over them, especially where fine or scaly gold is abundant. The best size for the holes in the trommels is usually considered to be within a range from $\frac{5}{16}$ to $\frac{5}{8}$ inch. All material not passing through is regarded as worthless, and is not treated on the tables. In New Zealand the inclined tables are generally covered with coir matting or plush,[1] and in the United States with calico, covered with cocoa matting, on which is laid a sheet of expanded metal, the raised edges of which form a series of very effective riffles. Hungarian riffles, or bars somewhat resembling T-shaped rails, placed transversely, are also used.[2] Mercury is placed in these riffles, but mercury is not often used in dredge-washing.

A trommel described by E. S. and G. N. Marks[3] is 17 feet long and 54 inches diameter inside. It is constructed of steel plates with perforations ranging in four successive sections or rings from $\frac{5}{16}$ to $\frac{1}{2}$ inch in diameter. The pitch aft of the screen is 1 inch to the foot. Water is supplied by a 9-inch pipe perforated for 12 feet opposite the perforated rings of the screen with four rows of $\frac{1}{2}$-inch holes at 4-inch centres. The finer material undergoing treatment is washed through the screen perforations on to the gold-saving tables running athwartships. The tables in this case were 14 feet long and 12 feet wide, made in four widths of 3 feet each, stepped one above the other with division plates between them. The pitch of the tables is usually about 1 in 10, and they discharge into a shute which runs out over the stern of the dredge, either into the water or on to the bank. The oversize from the trommel is either passed direct to a tailings elevator set at an angle of about 30°, or is first treated in a sluice box fitted with calico, cocoanut matting and riffles to save gold, as in the case of the tables.

Marks[4] also describes a sluice box which may be used instead of a trommel and tables. The sluice may be 40 to 50 feet long and 3 to 6 feet wide, with a pitch varying from 1 in 8 to 1 in 10. The bottom of the sluice is laid with perforated plates and riffles overlying the cocoanut matting and calico throughout its entire length. In order to give variation in treatment as many different classes of riffles as possible, such as angle irons, venetians, perforated plates and crimped or diamond-shaped expanded metal, are used in the sluice. The riffles are prevented from becoming packed or choked. Undercurrents are sometimes used. (For description, see p. 109.)

On the Snake River in Idaho, the gold consists of thin flakes and is difficult to catch. The gravel is concentrated by passing through sluices lined with perforated steel plates. The fine material is then further concentrated on tables covered with burlap.[5] Similar methods have long been in use in New Zealand. J. P. Smith states[6] that fine gold is lost in most of the New Zealand dredges and gives a table, from which the following details are taken, of some of the most modern dredges in which fine gold is being saved :—

---

[1] J. P. Smith in *Rept. of Dept. of Mines, New Zealand*, 1899-1900, p. 51.
[2] For illustrations of these riffles, see Purington, *Mining Mag. and Pacific Coast Miner*, 1905, **11**, 21, 127.
[3] E. S. and G. N. Marks, *Trans. Inst. Mng. and Met.*, 1906, **15**, 471.
[4] Marks, *loc. cit.*, p. 479.
[5] Grothe, *Mineral Industry*, 1899, p. 331.   [6] J. P. Smith, *loc. cit.*

TABLE XX.

Dredge.	Width of Tables.	Material passed over Tables per Foot of Width per Minute in Cubic Inches.	Proportion of Sand to Water.
Fraser's Flat, ...	21 feet.	2,800 cubic inches	1 to 9·43
Olrig, ...	18 ,,	768 ,,	1 to 28·8
Leviathan, ...	81 ,,	674 ,,	1 to 17·2
Waimumu, ...	28 ,,	555 ,,	1 to 37
Mokoia and Buller Junction, .	15 ,,	1,152 ,,	1 to 33

The difference between the value of ground as determined in prospecting by treating samples in a cradle and that found in actual work with a dredge shows that some fine gold is lost in practice.

The sluice boxes or tables are cleaned-up weekly, the riffles or expanded metal being lifted out and the matting and calico washed in tubs. Most of the gold is found in the first few feet, which are sometimes cleaned-up every few hours. The washings or concentrate may be run through a small sluice box, 15 inches wide and 12 feet long, called a *streaming-down* box (Marks). It is laid with silk plush or green baize. The final product is amalgamated, sometimes in a clean-up pan. The original concentrate may also be treated by amalgamation, either by running it over amalgamated plates or by treatment with mercury in a clean-up pan or a rotating barrel.

Disposal of Tailing.—The tailing was formerly always discharged into the water again at some point beyond the stern of the dredge. It is now often piled on the bank to prevent any risk of mixing it with untreated gravel. Coarse and fine tailing is usually piled up separately, and when flat inland placers are being dealt with, as described below, the separated tailing occupies 50 or 60 per cent. more space than the undisturbed ground.[1] The bucket elevators for raising the tailing are shown in Figs. 38 to 42.

Treatment of Flat Inland Placers—" Paddock Dredging."—The advantages of dredges in handling gravel have appeared of late years so great that their use has been extended from the beds of rivers to all flat placer ground wherever situated. The dredge is placed in a reservoir which is filled with water. The machinery is then erected and dredging begun, the tailing being discharged at the stern, so that the reservoir and dredge move gradually forward together. By piling gravel round the reservoir and letting in more water, the dredge is raised and can be made to work its way up sloping ground, as at the Warm Springs, Idaho. According to Postlethwaite[2] clear water is kept flowing into the reservoir to prevent it from becoming too muddy for close gold saving. The amount of water required on a dredge is about 2,000 gallons per cubic yard dredged.[3]

Advantages of Dredging.—Investments in dredges have been popular in New Zealand of late years, for the reasons that results obtained have been good, the capital required is comparatively small, working expenses easily calculated and subject to few variations, and, if the claim turns out to be

---

[1] Grothe, *Mineral Industry*, 1899, p. 331.
[2] Postlethwaite, *Mining Mag. and Pacific Coast Miner*, Jan. 1905, **11**, 8.
[3] Richards, *Ore Dressing*, 1903, p. 1013.

barren, the dredge can be transferred elsewhere or sold. The Government Inspector, in fact, reported in 1900 [1] that in some parts of New Zealand no other form of mining would be entertained as an investment. Dredging on rivers is a far cheaper method of treating the gravel than any other, but any great extension in the future will doubtless be in inland work. The amount of water required is small, the risk of loss by floods or any kind of accident infinitesimal, and the labour costs trifling.

Difficulties in Dredging.—The difficulties of raising the gravel due to the occurrence of tree trunks or large boulders are obvious. Nevertheless, the chief difficulty is to save the fine gold in the gravel. The losses of gold are often large, though seldom considered. In one New Zealand dredge, it was found as the result of treating measured quantities of tailings that gold amounting to 2 grains per cubic yard of tailings treated was lost.[2] The yield on the dredge was about 3 grains per cubic yard, so that only 60 per cent. of the gold was saved. When there is an " overburden " or layer of clay above the auriferous material, it is found that lumps of clay and water carrying clayey matter in suspension are responsible for the losses of gold. The best method of avoiding these losses is to strip off the overburden of clay before attempting to raise and wash the auriferous gravel.

The bottom on which the gold-bearing material rests is also of importance. A hard bed rock prevents the gold lodged in its crevices from being removed by the dredge. Soft bed-rock partly decomposed *in situ* is, on the other hand, an excellent bottom for dredging. A soft tenacious clay underlying the gravel causes loss of gold in washing. A. F. J. Bordeaux [3] mentions the great difficulty in dredging in French Guiana due to a tenacious clay, which must be puddled by hand in the sluices to free the gold contained in, or picked up by, it. In a dredge the clay sticks tightly to the buckets and goes round and round in spite of all water jets. It is necessary to dislodge it with shovels.

One of the greatest sources of expense in gold-dredging is the repair of the bucket-line, including the tumblers. Stones carried between the tumblers and the buckets break or bend the latter. In 1905 such repairs amounted to one-third of the entire operating expenses, and was the cause of 39 per cent. of the time lost in stoppages.[4] Wooden hulls of dredges are destroyed by insects in tropical countries, and are now being replaced by steel.

Cost of Dredging.—The capital expenditure on a dredging plant in New Zealand varies from about £2,500 for a small dredge, with a theoretical lifting capacity of 60 cubic yards per hour, to £10,000 for a dredge capable of lifting 150 cubic yards per hour, and far more for the modern large dredges treating 250,000 cubic yards per month. The working cost of a small dredge is given as about 1 oz. of gold per day in New Zealand,[5] or about twopence per cubic yard of gravel treated. The cost of working in Western America is stated to be about four cents per cubic yard in a case where the power was supplied by steam from coal.[6] Working costs at Oroville, California, are given by Postlethwaite as ranging from 3·66 to 8·7 cents per cubic yard.[7]

---

[1] *Report on Dept. of Mines of New Zealand*, 1899-1900, p. 41.
[2] *Report on Dept. of Mines of New Zealand*, 1899-1900, p. 41.
[3] Bordeaux, *Trans. Amer. Inst. Mng. Eng.*, 1910, **41**, 587.
[4] Cal. State Mng. Bureau, Bulletin 36, 1905 ; *Trans. Amer. Inst. Mng. Eng.*, 1909, **40**, 504.
[5] *Loc. cit.*
[6] Grothe, *Mineral Industry*, 1899, p. 336.
[7] Postlethwaite, *loc. cit.*

The working cost on a dredge of moderate size is given by F. W. Taylor [1] as about £40 per week, and the return may be expected to be from 20 to 30 ozs. of gold per week, so that the profits are often large. A table of working costs in California in 1910 is given by C. Janin.[2] The costs varied from 2·30 cents per cubic yard working on fine gravel with easy digging by means of buckets of 13·5 cubic feet capacity to 9·60 cents per cubic yard with difficult digging with 5 cubic feet buckets. The cost in the case of the large Californian companies was 4 or 5 cents per yard in 1910, and about 5 cents in 1913. In French Guiana the cost is about 33 cents per yard.[3] In Siberia, where the industry was started in 1900, the working cost was given by Taylor as about 4·2d. per cubic yard in 1904. The working costs of the Phœnix (Victoria) Company are given by Marks as 2·6 pence per cubic yard.[4]

Method of Working Siberian Placers.—The methods and apparatus employed in Siberia differ so markedly from those which have been adopted elsewhere, that they are well worth a special description, although they cannot usually be applied to placers found in other parts of the world, owing to the difference in economic conditions. In California the valleys are narrow and the grade steep, so that watercourses are usually close to the auriferous deposits, and the sluices can be made of almost any length, while still conforming to the general slope of the soil. In Siberia the slope of the valleys is so gradual that the flow of the water is almost imperceptible, and often takes place through wide marshy tracts. The result of this is that the sluices must be short, being usually less than 100 feet long, and their upper ends are frequently raised on trestles. As a further consequence, also, the gravel must be excavated by hand and carried in waggons to the sluice, and the tailings removed in a similar way, the flow of water acting by gravity not being available for these purposes. Wire-rope haulage and narrow-gauge railway transport with the aid of locomotives are also used (Perret). The stripping of the overburden of barren material is also usually done by shovelling into two-wheeled carts drawn by horses. The introduction of mechanical excavators, such as steam shovels and drag-line (or scraper-bucket) excavators, has not been attended with success, according to Perret.[5] The excavation is made in benches or terraces, working up the valley. The height of each bench above the lower one is about 5 feet, and the gravel is picked down from the face of each bank and shovelled into carts by which it is conveyed to the washing establishments. The additional expense entailed by these causes is balanced by the low cost of labour in the country, and an incidental advantage lies in the fact that the washing apparatus can be placed outside the limits of the river during flood time, and so may remain for a number of years undisturbed, while all the gravel in the district is being washed. The workings are divided by Perret into open and underground workings, the latter being exploited preferably by drifting, when the fall of the ground admits of natural drainage, but shafts are also largely used. The workings are never of any great depth.

---

[1] Taylor, *Eng. and Mng. J.*, Jan. 14, 1904, p. 84.

[2] Janin and Winston, *Mng. and Sci. Press*, 1910, **101**, 151.

[3] *Trans. Amer. Inst. Mng. Eng.*, 1910, **41**, 586.

[4] Marks, *Trans. Inst. Mng. and Met.*, 1906, **15**, 492.

[5] Perret, *Trans. Inst. Mng and Met.*, 1912, **21**, 647-690. In this paper details are given of the methods of work, costs, etc.

There are three types of apparatus employed, each designed for washing a particular kind of deposit. They are :—

1. The Siberian sluice, which is used to wash light sand.

2. The Trommel, used for loamy sands.

3. The Pan, used for gravel which is cemented together by means of compact clay.

1. *The Siberian Sluice.*—The apparatus at Voltchanka, which may be taken as a type, consists of a head sluice and three secondary sluices, which are placed at right angles to the head sluice, and which leave it at different points and converge to a common centre, where the tailings are discharged. The head sluice begins at a height of 13 feet from the ground ; it is about 90 feet long by 2 feet wide, and has a fall of about 1 in 14. The sands are dumped from the waggons on to a wooden platform situated above the sluice-head, and shovelled into the latter, a stream of water being turned in at the same time. After passing through a grizzly, the gravel runs over a series of cast-iron cross-bar riffles, which form a number of rectangular depressions (or *pigeon-holes*) in the bed of the sluice, by which the disintegration is favoured. The stream then flows over an iron screen, through which a part of it falls into the first secondary sluice, while the remainder continues its course over more pigeon-hole riffles. This arrangement resembles the Californian undercurrent. A second and a third screen open on to the other secondary sluices, and the part of the gravel (consisting chiefly of small stones) which has resisted disintegration, and has not passed through the screens, is then let fall into a hopper, whence it is removed to the tailing waggons.

The secondary sluices are wider than the head sluice, and have a steeper grade, and the amount of water and auriferous material passed is, of course, much less in each one than in the principal sluice. The sands first pass over a number of transverse riffles, and then over about 30 feet of blanketing, the sluice being widened at the same time, and subdivided by longitudinal wooden partitions, and the grade being as much as 1 in 6, while small drops are introduced at intervals of a few feet. The third sluice has a more gentle inclination than the others. The tailings fall into a shallow sump, and are instantly lifted out of it by a bucket elevator or "tailings-wheel" operated by water power, and stored in a hopper, whence they fall into waggons, by which they are removed to the dumping ground.

No mercury is used in these sluices, and only the production of "grey concentrate" is attempted, this work being continued from 6 a.m. to 7.30 p.m. every day, after which the concentrate is collected from the riffles. It consists of gold in scales and plates, magnetic iron oxide, pyrite, rutile, together with some quartz, etc. It is treated with mercury in the Siberian trough or *washherd*, or on inclined tables, the method being that described on p. 105.

The apparatus described above treats 500 tons of gravel per day, the labour required being furnished by twenty men and ten horses. The gravel treated contains an average of from 12 to 15 grains of gold per ton, rarely falling below 6 grains per ton ; the exceptional richness of $3\frac{1}{2}$ dwts. per ton has been observed. The gold is chiefly found in the head sluice, where 70 per cent. is retained, 30 per cent. being caught on the secondary sluices. At a similar establishment at Tchernaia-Retchka, however, where the gold is less finely divided, 97 per cent. was caught on the head sluice, and only 3 per cent. on the secondary sluices. The amount of water used at Voltchanka is about six times the weight of the gravel. The cost of construction of the works was 70,000 roubles, or about £7,000.

Recently the wooden riffles used in Siberian sluices have been replaced by wire netting or expanded metal, laid on matting, similar to those used in America, Australia and other countries. The slope of the sluices is about 3 inches to the foot in Eastern Siberia, where the sluices are short (14 feet), giving a clean concentrate but a poor yield. In the Urals the slope is 1 to $1\frac{1}{2}$ inches per foot and the length 35 to 50 feet.[1]

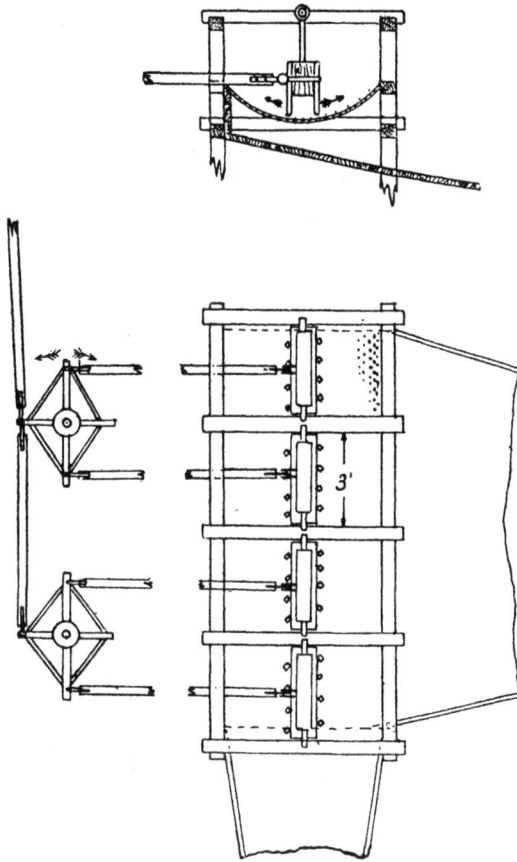

Fig. 44.—Boronka.

2. *The Trommel.*—Gravels which are too compact for satisfactory disintegration in the short sluices described above are subjected to a preliminary treatment by a trommel. At Berezovsk the trommel is of sheet iron of 9 mm. thick, having holes in it of about 1 mm. in diameter. The trommel is about 12 feet long, $3\frac{1}{2}$ feet in diameter at one end, and $4\frac{1}{2}$ feet at the other, and is set inside with denticulated plates of iron to assist in the disintegration effected by the water. The machine is driven by a water wheel, and is

---

[1] Perret, *Trans. Inst. Mng. and Met.*, 1912, **21**, 673.

sufficient for the disintegration of from 400 to 500 tons of gravel per day, requiring the expenditure of about 3 horse-power to drive it. The amount of water used in the trommel and on the tables is 67·5 litres per second, or about seven and a-half times the weight of the ore. The washing is effected on inclined tables only 30 feet long and 12 feet wide, and with a grade of about 1 in 4, placed with the incline at right angles to the length of the trommel. Near the head of the table, and stretching across it, is a deep trough-like depression, and below this there is a number of transverse riffles, in which grey concentrate is caught and treated as usual. The

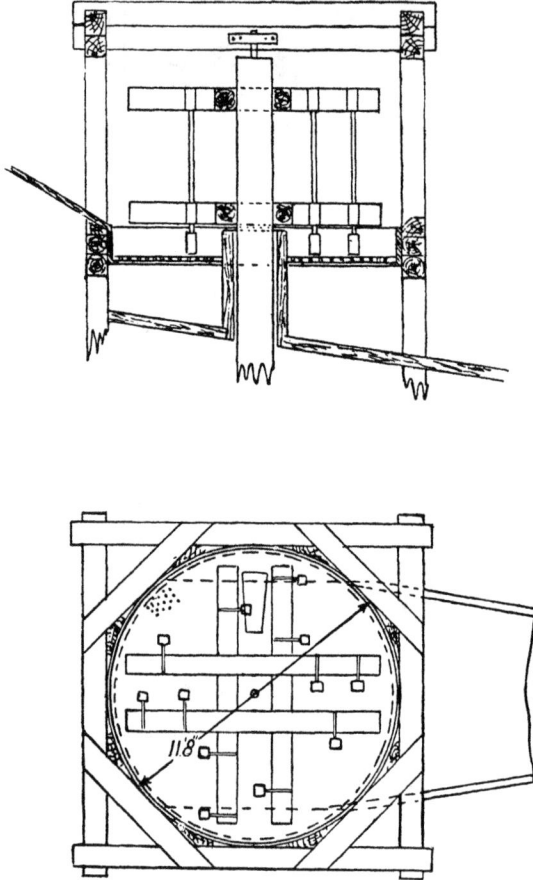

Fig. 45.—Chasha.

Berezovsk establishment employs twenty-five men and fourteen horses constantly; the trommel usually lasts for two seasons.

The *Boronka* (Fig. 44 [1]) consists of an arc-shaped concave cast-iron screen with ¾-inch holes, over which are suspended iron pendulums, which, when moving to and fro, mix with and disintegrate the material. It has a

---

[1] Perret, *loc. cit.* Reproduced with the permission of the Institution of Mining and Metallurgy.

capacity of 12 cubic yards per hour, and is used in small plants in the Urals as an auxiliary washer.

3. *Pan Washings.*—Sandy clays cannot be economically disintegrated in a trommel, and are, therefore, treated in a washing pan, or *Chasha,* which is similar to the puddling machine used in Australia (Perret). Fig. 45 represents a chasha in plan and elevation.[1] The pan usually consists of cast iron, and is from 8 to 16 feet in diameter, with vertical sides from 1 to 5 feet high. The bottom is of cast iron or sheet steel, and has numerous holes in it of about $\frac{3}{4}$ inch in diameter, widening downwards. The bottom is divided into 25 sectors, between which are deep groves for the collection of the pebbles. Through a circular opening in the centre of the pan there passes a revolving axis to which are suspended eight horizontal arms studded with vertical iron teeth, some of these being shaped like plough-shares. The revolution of these arms effects the disintegration of the sandy clays, which are fed into the pan together with water and puddled until fine enough to pass through the holes in the bottom, and the stones are removed at intervals by opening little gates placed opposite the radial grooves, or by a trap-door in the bottom. The disintegrated gravel falls from the pan on to concentration tables, similar to those used after disintegration in the trommel. At Berezovsk the pan is $11\frac{1}{2}$ feet in diameter and 5 feet deep, and the arms revolve at the rate of 25 turns per minute. As a rule, about 3 or 4 cubic yards of material are tipped into the chasha at a time, and after about five minutes' puddling the stones are discharged and another charge added. In this way from 25 to 30 cubic yards are washed per hour in a chasha of 12 feet diameter.[2] The water consumed, including that required for power, is about ten times the volume of the sand.

Beach Mining.—Beach mining is a comparatively unimportant form of shallow placer mining. The sea beaches on parts of the coasts of California, Australia and New Zealand contain small quantities of gold, which have been proved, in all cases in which the matter has been investigated, to be derived from the cliffs, in which the gold is generally less concentrated. Some streaks of black sand, however, in the " Gold Bluff," California, have yielded $135 or $6\frac{3}{4}$ ozs. per ton by actual working.[3] The waves of the sea wash down and partially concentrate the poor sands, and, under certain rather exceptional circumstances, as the tide goes out the surface of the beach is left covered with black sand, in which numerous specks of gold occur. This is carefully scraped up and transported inland to be washed, as sea water is not well adapted for the purpose, although it was used by one Californian company. The next tide usually washes away all the valuable material which has not been collected, or else covers it with barren sand. There is great difficulty in washing the black sand in California, as it consists largely of rounded grains of magnetite, the density of which is about 5·0, while the gold is in minute flakes and scales, which can be seen under the microscope to be oblong in shape, and thicker at the sides than in the middle (a shape due to continued pounding of a malleable material). This form is so easily moved and buoyed up by water that it is difficult to get a " colour " with the pan, and the amount caught by the mercury in sluices or long-toms is usually an insignificant proportion of the total assay value of the sand.

---

[1] Perret, *Trans. Inst. Mng. and Met.*, 1912, **21,** 671.   Reproduced with the permission of the Institution.
[2] Perret, *loc. cit.*   [3] *Prod. Prec. Met.*, *U.S.A.*, 1884, p. 557.

The industry is generally a languishing one, but the discoveries of rich deposits on the beach at Cape Nome, Alaska, resulted in great activity. In 1899, rockers of all sorts and descriptions were employed, and in 1900 when sluices and machines of various designs were used the production rose to £1,000,000.

Treatment of Shallow Placer Gravels by Steam Shovels and Amalgamated Plates.—A plant for the treatment of placer deposits, which are similar in nature to those worked in Siberia, has been devised in America, where labour-saving contrivances are indispensable in all such cases, owing to the high rate of wages. The apparatus consists essentially of a combination of a steam shovel or navvy, and an amalgamator consisting of a wide waggon-shaped wrought-iron trough loosely lined with silver-plated amalgamated copper plates, which form a series of steps or riffles at the sides of the trough. The material is elevated into a hopper by the excavator, and thence is charged into a revolving trommel placed inside the trough. Here the disintegration of the gravel is effected, and the fine material falls through into the trough, while the stones are discharged outside at the end of the trommel. At the bottom of the trough there is a water pipe, carrying water at high pressure, and in that pipe a series of jets pointing alternately forwards and backwards. The result is to give a series of eddies or whirlpools in the water in the trough, and the sand and fine gold is continually carried up to the surface near the middle line of the trough, and in descending again near the sides it comes in successive contact with several of the plates, which form a series of steps. The fine sand is eventually discharged at the end of the trough. It is stated by the Bucyrus Steam Shovel and Dredge Company, by which the machinery is manufactured, that such a machine erected in Montana had a capacity of from 600 to 800 cubic yards of gravel per day, all the water required being supplied by a pump raising 500 gallons per minute, the proportion being only three of water to one of gravel, which seems much too small. It is stated that gravel containing only 12 cents of gold (*i.e.*, about 3 grains) to the ton has been treated successfully by this machine, but exact records of continuous work done by it are wanting. It will be noted that special means must be adopted in each case for the handling of the tailings.

In 1903, a steam shovel plant was at work on Dominion Creek, Klondike.[1] The bucket shovel had a capacity of about 0·75 cubic yard, and excavated 800 cubic yards in a day of twenty hours. The ground could be cleared to a depth of 10 feet by this plant. The gravel was emptied into a hopper and sluiced thence to a trommel from which it passed to sluice boxes. The value of the gravel was $1·50 per cubic yard and the working costs were 13 cents per cubic yard.

---

[1] E. Haanel, *Eng. and Mng. J.*, April 11, 1903, p. 559.

CHAPTER VI.

DEEP PLACER DEPOSITS.

Nature and Mode of Origin of Deposits.—Both in Australia and California, besides the superficial placer deposits situated in or near the existing rivers, which in the deep cañons of the Klamath and other rivers in the extreme north of California attain a thickness of 250 feet, there exist auriferous gravels which bear no apparent relation to the present drainage of the country. These gravels often attain enormous thicknesses, and are in many places covered by volcanic rocks, consisting of basaltic lavas and tuffs, which are sometimes interbedded with gravel and loam. This latter circumstance shows that intermittent action of the volcanic vents, with long intervals of repose, has taken place. There has been some difficulty in accounting for the origin of these deep placers, and it has been ascribed in succession to the agency of the sea, of ice, and (for California) of a huge river flowing from north to south at right angles to the direction of flow of the existing rivers. None of these views is now entertained, and the " fluviatile " theory is generally accepted, the origin of the gravel being ascribed to the depositions of ancient rivers flowing in courses roughly parallel to those of existing rivers. The geological age of these ancient rivers has not yet been determined with certainty, but though they may be Pleistocene, the balance of palæontological evidence is perhaps in favour of Whitney's view that the deposits were formed in the Pliocene period.

The ancient Californian rivers probably had their sources at somewhat higher altitudes than those now existing, and had more uniform general grades, the slope of their beds corresponding more nearly to the general slope of the country. The existing rivers, on the other hand, have steep grades in the upper parts of their courses, followed by comparatively level stretches below. The old rivers, however, like their successors, had rapids, falls, and level stretches, the grade varying from 5 feet to 250 feet or more per mile. The Pliocene rivers ran in valleys which were broad and shallow in comparison with the present deep precipitous cañons, and the volume of water was in general much greater than that delivered by their representatives of to-day. The width of the valleys varied from 100 feet to fully 1½ miles (which is the width at Columbia Hill), and the depth must have been often over 1,000 feet. These valleys were already partly filled up by accumulations of gravel, when the outbreak of volcanic activity in many cases filled up the remainder, and the streams were deflected into other channels, which often lie close alongside the old cañons. These new channels have been excavated by the running water until they now lie much below the level of the beds of the Pliocene rivers, and consequently the gravels which were deposited in the old valleys now sometimes crown the highest ground in the district, the general level of the country having been greatly reduced in height. The new channels have been cut partly in the old country rock and partly in the

9

Pliocene auriferous gravels and their covering of volcanic rocks. Some-times the course of the present cañons cuts that of the old at several points owing to the sinuosity of both (see Fig. 46), in which A represents the modern river, and B the ancient one. The result is that sections of the old valley from bed-rock to surface are exposed in the sides of the cañons, usually at some height above the present level of the water, and it was at such points as these that the discovery of the existence of the deep placers was first made. The hard covering of basalt has served to protect the more friable gravels, which have been for the most part removed in those places where

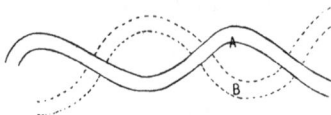

Fig. 46.—Intersection of Ancient and Modern Cañons.

the lava has been worn away or has never existed, so that the largest tracks of gravels still existent lie beneath the volcanic rocks.

Fig. 47 represents a section across two ancient channels (B, B) and a modern cañon, that of the American river. Here, A is the volcanic capping, which is 800 feet thick above the Red Point channel; B, B are the auriferous gravel channels; C, C are deposits of gravel on the "rims," containing gold in places; D is the bed-rock, consisting of dark-blue slates; E is a barren deposit of angular débris and boulders; F F are prospecting tunnels, which were put in at too high an altitude; F′ is the tunnel bored with the object

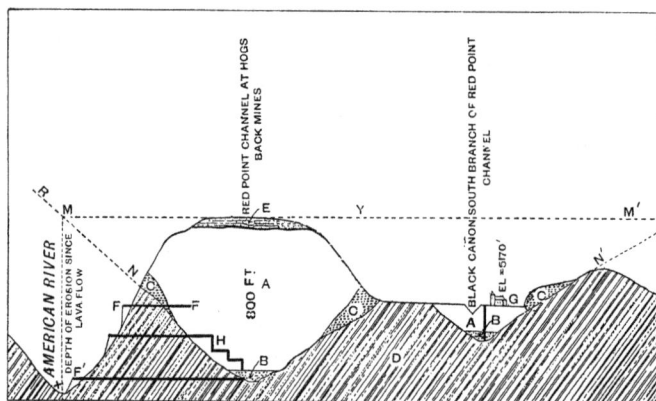

Fig. 47.—Section across Ancient and existing Cañons, California.

of reaching the bottom of the gravel deposit; H are prospecting winzes sunk in order to discover the position of the gravel. The space included within the dotted lines N M Y M′ N′ has been obviously denuded since the deposition of the volcanic cappings, the soft slate rims N M R and N′ M′ having been worn away, while the hard lava has resisted erosion. The vertical depth from M to the American river is about 1,800 or 2,000 feet.

This condition of things is that prevailing in California, but in Victoria the structure closely resembles that just described, with the exceptions that the old valleys were smaller and that the erosive action of the rivers since the deposition of the basalt has been comparatively slight, owing to the slight grades of the streams caused by the low elevation of the country and to the small amount of the rainfall. In consequence of this the basalt has usually not been worn through, and the " deep leads " or old river bottoms are often below the level of the present streams, so that although a larger proportion of the Pliocene gravel remains, it is more difficult and expensive to mine.

The shallow placers, at any rate in California, have resulted in the main from the erosion of these deep placers, the materials of which, having undergone a natural concentration in the ground sluices afforded by the river beds, furnished the wonderfully rich river-beds and bar deposits, which yielded so much gold between 1848 and 1860. The deep level gravels consist in slaty districts chiefly of quartzose sand, the fine materials furnished by the disintegration of the slate having been for the most part swept away, and the products of the quartz veins contained in the slate being left. These are the only gravels which pay for treatment. In granite districts, where the gravels are composed of more heterogeneous materials, and in cases where they consist of volcanic boulders and detritus, little or no gold is found. The lower parts of the gravels are often cemented into a conglomerate, called "cement," by infiltration of silica, oxides or sulphides of iron, or, rarely, carbonate of lime ; when the gravels are covered with lava, the whole thickness is in some cases converted into cement.

Distribution of Gold in the Gravels.—The gold is found chiefly either in contact with or just above bed-rock. If this consists of soft slate, and especially if the planes of cleavage are at a high angle to the horizon, particles of gold are often found in the natural riffles thus formed, and are disseminated through the rock to the depth of a foot or two. If depressions, potholes, or fissures exist in the old river bottom, they are usually very rich in gold. Where, as often happens, there is a channel, or " gutter," to adopt the Australian expression, cut by the stream in the lowest part of the valley, the gravel filling it is usually much richer than that found elsewhere. Such rich portions, often only a few feet wide, and of insignificant depth, but extending to considerable distances in the direction of the stream, are called " leads." Rich streaks also occur at various levels in the gravels, often resting on " false-bottoms," which consist of impermeable beds of clay or some similar material. Sometimes these streaks are richer than those encountered at bed-rock, as, for example, at the Paragon Mine, Placer County, California.

The amount and position of the gold vary, as in the case of the present rivers, with the grade, the shape of the valley, the volume of water, the amount of gravel being carried down, etc. " An underloaded current— i.e., a current charged with less detritus than it is well able to carry—is apt to cut its bed, and prevent the accumulation of gravel. A greatly over-loaded current will deposit too rapidly to admit of the concentration of the gold dust." [1] Under conditions intermediate between these extreme states, the current may be just strong enough to keep its bed clear from all accumulations except a small quantity of coarse gravel and the coarse gold, which is caught in the natural riffles, and thus all the conditions necessary to form a rich bed of pay-dirt may be present. If, however, the bed consists of granite or other rock which wears in smooth and rounded shapes, little

---

[1] Ross E. Browne, *Tenth Report Cal. State Mineralogist*, 1890, p. 448.

gold will be caught. Slates, consisting of layers of uneven hardness, wear irregularly, and afford a good gold catching surface. The conditions noted above as necessary to form rich gravels cannot be expected to have been prevalent over great distances. "An increase of grade or narrowing of the channel will cause an increase of velocity, and the same stream may be underloaded in a narrow steep section, and overloaded in a broad flat section."[1] The difference of velocity between the middle and sides of a stream, and between the inside and outside of a bend, may give the right conditions in one part of a river bed and not in another. Thus with high grades, rich gravels should occur in the less rapid, and with low grades, in the more rapid parts of a stream. Having regard to such considerations, the richer parts of existing rivers can be pointed out with little trouble. When, as in the Pliocene rivers, the beds are buried to a depth of hundreds of feet, the richer parts are more difficult to find.

The history of the Pliocene rivers began with a period when excavation exceeded deposition, when the rivers were underloaded for at least a portion of each year, and the channel was constantly being deepened. Some bench or terrace gravels were formed at this time, and being at the sides of an underloaded river tended to be rich. The river bed, although rocky and comparatively free from sand, would perhaps accumulate some coarse gold, which, as the channel deepened, was no doubt in part ground up into fine particles and carried off, but at the time when the excavation had reached its lowest point, some of this coarse gold would certainly be present. When the underloading of the stream ceased, whatever caused the cessation, a pause must in many cases have occurred before the gravel proved too much for the stream to carry. During this pause the conditions for gold catching were favourable, and hence rich gravels were formed on bed-rock in the gutters or channels. Then, as the streams became overloaded, sand and gravel accumulated rapidly, so that little concentration of the gold in them could take place. The rivers flowed over thick sand banks and, in consequence, frequently changed their courses. The sands, being deposited by overloaded rivers, of course, contained fewer and smaller boulders, and the thick masses of poor sand thus went on accumulating until the volcanic outbursts put an end to the process.

Materials occurring in the Placer Deposits.—In California, if quartz grains and silicified wood are excepted, the most abundant mineral is black iron-sand, which usually consists of magnetite, although menaccanite, a form of hæmatite in which part of the iron is replaced by titanium, also occurs. These minerals must have been derived from the lavas, as neither of them are known to occur in the quartz veins of the country. Platinum and its allies are usually present in more or less abundance ; thus iridosmine occurred to the extent of 1 in 100,000 of the gold in the early days, and increased afterwards to 15 or 16 times that proportion. It is still abundant in the beach sands of Northern California. Grains of native copper, nickel, and perhaps lead have also been detected, and a few diamonds occur, while garnets, small crystals of zircon and cinnabar are very abundant.

METHODS OF TREATING DEEP PLACER GRAVELS.

Both in California and Australia, when the first gold discoveries were made, the river beds and bars were at once explored, and soon afterwards

----

[1] *Ibid.*

the flats closely adjoining them. Subsequently, the bench gravels situated in the same valleys, and the side ravines and gulches, which remained dry during most of the year, were prospected and worked, owing to the rapid growth of the mining population and to the fact that the exhaustion of the shallow placers was already beginning to make itself felt. The result was that the exposed edges of the outcrop of some of the deep leads were found, and the pay-dirt followed into the hill-side by drifting. Then, as in many cases it was found that the gravels overlying the pay-dirt on hill-sides, although poor by comparison with the earth below them, nevertheless contained a small quantity of gold, the idea was evolved of breaking down the whole bank by jets of water, and passing all the material through the sluices. Hence arose the practices of drift and hydraulic mining,[1] of which the former is largely used in California, while the latter was much used in California prior to 1884 and in New Zealand and throughout the Western States of America, but cannot be applied in Australia or Siberia owing to the general flatness of the country. In Siberia, only shallow placer deposits are worked. In Australia, the deep leads are usually reached by shafts, since the surface of the country is not intersected by deep cañons, as in Western America. Drift [2] and shaft mining of deep leads will not be further referred to in this volume, attention being confined to the treatment of the gravel after it has been raised to the surface. In hydraulic mining, the breaking down of the gravel is so intimately connected with the extraction of the gold that a short description of the whole process is given.

Hydraulic Mining.—This method of working consists, as has been already stated, in breaking down banks of gravel by the impact of powerful jets of water, and passing the disintegrated material through a line of sluices, without the agency of hand labour. The chief requisites for the successful application of hydraulic mining are—

1. Large quantities of auriferous gravel, not less than 30 feet in thickness, and not overlaid by any appreciable thickness of barren material, which would necessarily be passed through the sluices with the pay-dirt. The gravel treated need not be rich, a mean yield of less than 1 grain of gold per cubic yard being often enough to furnish profits if the operations are on a sufficiently large scale.

2. A plentiful and uninterrupted supply of water throughout considerable portions of the year

3. Sufficient fall in the ground so that (a) the water may be delivered under the pressure of a head of from 100 to 300 feet, (b) the tailings can be easily carried away to a large dumping ground, which is most conveniently, either the sea or a large and rapid river.

Commencement of Operations.—In California, the naturally occurring banks or cliffs in the gravels in the sides of the gulches were first selected for attack. Later, some of the deposits occurring in those channels which are not intersected at favourable points by the present system of drainage were operated on. It is necessary in such cases to run a tunnel from the nearest cañon in the bed-rock to the lowest point in the gravel, this point

---

[1] The invention of hydraulic mining is ascribed to Edward Mattison, of Sterling, Connecticut, who used the method in 1852 on a small scale. For all details in connection with the industry, see *Hydraulic Mining in California*, by A. J. Bowie, New York, 1885. This has been largely used in drawing up the summary which follows.

[2] A detailed description of Drift Mining is given by Russell L. Dunn in *Eighth Repor Cal. State Mineralogist*, 1888, p. 736.

being found or guessed at by prospecting operations. The tunnels are often of great length, that at the North Bloomfield Mine, Nevada Co., Cal., for example, being 7,874 feet or 1½ miles long. One or more shafts are then sunk from the surface through the gravel to the tunnel, and washing operations are begun by ground-sluicing, letting the water and gravel fall down the shaft and run through the tunnel, in which the sluices are sometimes laid. The surface near the shaft is thus gradually lowered, or it may be terraced by hand labour, until an excavation is made of sufficient size to enable the ground to be attacked with the hose. Washing then proceeds regularly in this manner, the bank being broken down by jets of water, and the products being allowed to fall down the shaft and pass through the tunnel.

The Supply of Water.—The amount of water required by large undertakings is far more than can be obtained from the rainfall on the hills immediately round the mine. Thus the North Bloomfield Mine, in the season of 1877-8, used between sixty and seventy millions of gallons per day, or enough for the total supply of a city of three million inhabitants. The workings, moreover, must necessarily be considerably above the level of any large rivers in the neighbourhood, so that it is often necessary to construct huge reservoirs at convenient spots to store up the rainfall and melted snows of large districts, and to convey the water thence to the mine by ditches, flumes, or pipes. Wooden flumes are cheaper than ditches, but are more liable to be damaged. Sheet-iron pipes are made as much as 30 inches in diameter, and are cheap, durable, and easily repaired; they entirely prevent losses by leakage and evaporation, and deliver more water under similar circumstances than flumes, as they offer less fractional resistance to the flow.

Where it is necessary to cross side cañons or gulches the flume or iron pipe is carried over on a light trestle bridge, or the water is passed through an inverted siphon formed by a wrought-iron pipe which passes down one side of the valley and up the other, being filled from a head-box or reservoir, and delivering the water at a lower level on the other side. At Cherokee, Butte County, an inverted siphon pipe was used to carry the water across a ravine 873 feet deep. The diameter of the pipe was from 30 to 34 inches, and its greatest thickness (where there is a pressure of 384 lbs. per square inch) was 0·375 inch.[1] The Miocene Ditch Company, operating in the same County, carried their flume, 4 feet wide and 3 feet deep, round the face of a bluff 350 feet high, supported on L-shaped iron brackets made of bent T rails, soldered into holes previously drilled by men let down the face of the cliff by ropes.

The water is delivered at a convenient height above the workings into a "pressure-box," consisting of a small wooden reservoir from which the pipes take their origin. In many cases the reservoirs and ditches are owned by separate companies, or, as in New Zealand, by Government, and the water is sold to the miners by measure. The unit in New Zealand is a "government head," and in the United States a "miner's inch," following the system in vogue in Spain and Italy. The amount of water that will flow through an orifice 1 inch square, cut in a board 1 inch thick, under a head of water that varies with the custom of the locality, but is usually from 4 to 8 inches, is called a miner's inch. The amount of flow in twenty-four hours is called a "twenty-four-hour inch," and similarly there are

---

[1] J. H. Hammond, *Ninth Report Cal. State Mineralogist*, 1889, p. 125.

ten-hour and twelve-hour inches. The quantity of water in a miner's inch varies with the head of water used and the form and size of the orifice for delivery. Thus the amount delivered from an orifice 25 inches long and 2 inches wide is reckoned as 50 inches, although it will be more than fifty times as much as the delivery from an orifice 1 inch square. The twenty-four-hour inch under a head of 7 inches amounts to about 2,230 cubic feet.[1]

The water is conveyed from the pressure-box by pipes, which were formerly made of canvas hose, to which iron rings 3 inches apart were added for pressures of over 100-feet head. They are now made of sheet iron or steel. Sharp bends in them are avoided, as the flow of water is checked thereby. They are liable to collapse if the level of the water in them is reduced, and a partial vacuum formed inside ; hence, as in the case of all other sheet-iron pipes used in hydraulic mining, they are fitted with valves, which are constructed so as to freely admit air from without. The water is discharged through a nozzle called a " giant " or " monitor " (Fig. 48). The nozzle was at first a sheet-iron tube, having an aperture 1 inch in diameter, and was held in the hand. The size of the nozzle was gradually increased, until it reached a diameter of 11 inches. Such a stream, under a head of 200 feet, requires special appliances to control it, deflect it at will, and prevent the nozzle from " bucking."

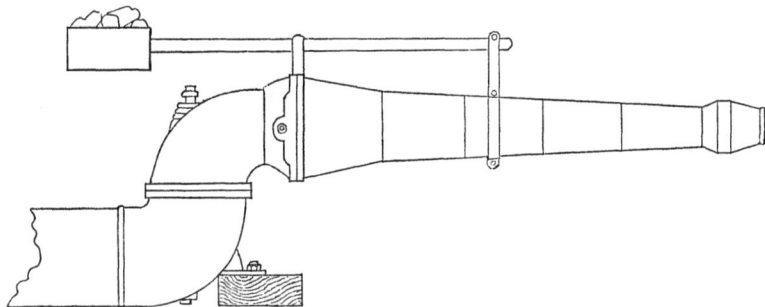

Fig. 48.—Hydraulic Giant or Monitor.

Breaking Down the Bank.—When the jet is first directed against the bank, the water spatters in all directions, then buries itself a little, and after a time, in loose ground, a " cave " takes place, the undermined bank falling down. By the method of undermining, the power of the giant is much increased, especially where hard and soft layers alternate. When large caves are about to take place the water is turned off, as otherwise the ground may run so far as to overwhelm the monitor and the workman directing it. The nozzle is placed as near to the bank as possible, consistent with the safety of the workers, so as not to waste too much of the initial velocity of the stream of water. Consequently, lofty banks are not advantageous, and if they exceed 200 feet, they are usually worked in terraces of 100 feet or so in height. In some parts of the Spring Valley Mine, however (see Fig. 49), a bank 450 feet high was worked in a single bench, and it was then not unusual for the runs of ground to bury pipes which were throwing 7-inch streams from a distance of 400 feet from the fall of the bank.

---

[1] For further details concerning miner's inch *vide* Art. by P. M. Randall in *Precious Metals in the United States*, 1884, pp. 558-572.

The jet is, if possible, delivered unbroken against the face of the bank, as its disintegrating power is thus kept at its maximum. However, in some

Fig. 49.—Hydraulic Mining.

cases, where it is cemented, the gravel is too hard to be economically broken down by the water alone, and blasting is then resorted to, a drift being run into the bank, and cross-cuts made at the end in which the powder is placed ; the drift is then filled up, and the charge exploded by electricity. It is more economical to blow out the base of the bank, as the upper part then falls by its own weight and can be broken up by the water. Sometimes arrangements are made to explode very large blasts ; thus, at the Blue Point Mine, Nevada County, a charge of 50,000 lbs. of powder was exploded in cross drifts at the end of a main drift 325 feet long in the year 1870, and 80,000 cubic yards of gravel were brought down, while at another mine, 3,500 lbs. of dynamite were exploded in 1872, and 200,000 cubic yards of gravel thrown down.

Washing the Gravel in the Sluices.—The sluices in which the gold is caught are constructed on exactly the same principles as those already described, but are larger and, though usually made of wood, are of more massive construction, in accordance with the great quantities of gravel to be handled and the continuous nature of the work. The sluices are commonly called " flumes," but it is better to restrict the use of this word to a conduit for carrying water only. The sluice boxes used in hydraulic mining, though, as usual, only 12 feet long, are as much as from 3 to 6 feet wide and from 2 to 3 feet deep ; they are lined with heavy planks on the sides, and the pavements are made of more durable materials than is usual in shallow placer sluicing, wooden blocks, rocks, or T railroad iron being most usually employed. The wooden blocks are from 12 to 30 inches square, and from 8 to 13 inches deep. They are usually made of one of the softer varieties of pine which " broom up " under friction, and thus present a better catching surface. The blocks are cut across the grain of the wood, and are set side by side across the sluice, each row separated from the next by strips of wood to which they are nailed, while they are also kept in position by the side lining which is placed upon them. The interstices in the block pavement act as gold catchers, and are filled with small stones, or, with less advantage, allowed to fill up with gravel when washing begins. On account of the rapid wearing away of the wood, much of the gold and amalgam caught is scooped out and carried off again. Wooden block riffles last only a few weeks when in heavy work, but are easy to take up and put down again in cleaning up ; they are discarded when worn so as to be only 4 to 6 inches thick.

Rock pavements are made of those boulders which are most easily obtained in the particular district. Basalt is generally used, oval stones of 15 or 18 inches long and from 9 to 12 inches thick being selected and placed on end, with a slight slant in the direction in which the current flows. They are held in place by wooden planks, which divide the sluice into compartments, so that if one stone works loose the pavement as a whole is not affected. The interstices as before are filled with gravel. Rock pavements are very durable, lasting from three to six months, but require more grade to the sluice, and occasion loss of time in cleaning-up and repaving the sluices. Consequently, they are never used near the head of a sluice, where cleaning-up is a frequent operation, but are often used for the lower parts of sluices, where they sometimes alternate with block riffles, and are especially suited for tail-sluices which are only cleaned-up once a year. Rock pavements cost less than other forms of riffles.

Iron riffles, which usually consist of T-iron rails, are placed longitudinally in the sluice, closely packed side by side. They present a large amount of space available for catching the gold and amalgam, last well, present little resistance to the current (so that the grade may be low while the duty of the water remains high), and are easily taken up and put down. They are, therefore, generally used at the head of the sluices. Though their first cost is higher than that of wooden blocks, they are more economical in the end, owing to the saving of time in cleaning-up and to their longer life. Egleston [1] instances the results of experiments made at the Morning Star Claim, California, where three sections of sluice, each 95 feet long, were laid at a distance of 300 feet from the face of the bank which was being worked. The first section was as usual laid with wooden blocks, the second with old rails, and the third with rocks. When the clean-up was made the middle section gave 9 ozs. more of gold bullion than both the others combined. If old rails cannot be had, strips of wood bound with iron are used, but are less durable and satisfactory.

At the Blue Spur Consolidated Gold Company's plant, Gabriel's Gully, New Zealand,[2] where the sluice is necessarily very short, most of the stones are first separated from the gravel, and the finer material is then passed over a sluice paved with transverse angle-iron riffles, placed with the hollow side facing down stream (Fig. 50, in which the arrow shows the direction of the stream); these iron riffles are placed 2 or 3 inches apart. Below the section containing these riffles, there is a false bottom to the sluice, formed by an iron plate perforated with small round holes through which some of the water and the finest particles of the gravel fall on to cocoa-nut fibre

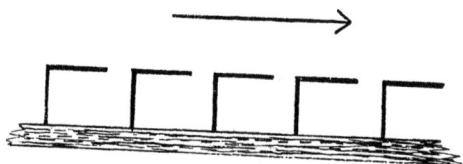

Fig. 50.—Angle-iron Riffles in Sluice.
Scale = ¼.

matting, laid on the true bottom of the sluice. Here the fine gold is caught, the principle being similar to that used in undercurrents.

The sluice is often divided into two by a median longitudinal partition, so that one side may be at work while the other is being cleaned-up or repaired, both sides being sometimes worked when water is very plentiful. There are usually unpaved rock-cuts above the sluice, leading to it from the places undergoing the process of piping. These rock-cuts are rarely supplied with mercury, and very little gold is usually caught there.

The length of the sluice, if capital is not lacking, depends on the cost of construction and of the maintenance, as compared with the value of the gold saved owing to the increased length of the system. The length may be diminished by a plentiful use of drops, grizzlies and undercurrents, all of which are described above under the head of shallow placer sluicing; they are made of proportionately large size in hydraulic mining. Coarse gold is, of course, soon caught, but fine gold may successfully evade all the riffles of a long sluice. The Spring Valley Mine had three parallel lines of sluices, each 2½ miles in length, and it was estimated that 95 per cent. of the gold

---

[1] Egleston, *Gold, Silver and Mercury in the United States*, 1890, **2**, 218.
[2] *Report of the Mining Commissioner of New Zealand*, 1891.

contents of the gravel was caught.[1] This length is unusual, the average not exceeding about 1,000 feet.

Grade of the Sluices.—The grade depends on the available fall of the ground and on the character of the material to be washed. The minimum is from 2 to 4 inches per box, such low grades being sometimes enforced by the nature of the ground, sometimes adopted from choice if the gravel is light, the gold fine, and water plentiful. With these low grades, however, disintegration is slow and incomplete; stones, unless they are small, cannot be sluiced; large ones block the sluices, and must be removed by hand, and the " duty " of the water, as regards sand, is greatly decreased. The 6-inch grade is that most generally used, but as much as 12 inches per box, or even more is required when water is scarce or the gravel coarse. Steep grades effect disintegration rapidly, thus shortening the length of the sluice, and enable all but the largest rocks to be sluiced, but less gold is then caught and a more plentiful use of undercurrents is necessary. It is considered necessary to have a sufficient depth of water to cover the largest boulders to be sluiced, but the deeper the water, the more difficult it is to save gold. Where poor or top gravel is being " piped," it is worked off as rapidly as possible, and with little regard to the percentage of gold saved.

The " duty " of the miner's inch—that is, the quantity of material washed down by an inch of water in twenty-four hours—varies with the grade and other conditions. It varies from about 1 to 5 or more cubic yards of gravel per inch.

The Use of Mercury.—Mercury is added daily during the run in gradually lessening quantities, the object being to keep the mercury uncovered and clean at the top of the riffles. The feeding is regulated by the appearance of the amalgam in the sluice, the additions being made in the riffles near the head-box and in the undercurrents. The loss of mercury is usually from 5 to 10 per cent. of the amount used per run. When cemented gravels are being treated, owing to the extra amount of trituration required, the loss may be as high as 30 per cent. These losses are the more serious, for the reason that amalgam is more easily lost than pure mercury, so that a heavy loss of mercury denotes a heavy loss of gold.

Cleaning-Up.—The process does not differ from that described under the heading of shallow placer mining. It is advisable not to defer the clean-up too long, as losses of amalgam are caused by the wearing of the riffles. Usually from 50 to 95 per cent. of the total yield of amalgam is caught in the first twenty or thirty boxes, which are cleaned-up frequently. The following table [2] shows the percentage yield of the various sections of the sluices, etc., at the North Bloomfield Mine, California, for the year 1877-8 :—

Total yield, ... \$311,276.20.

Near bank, from rock cuts in mine (all in gold dust, no
    quicksilver being added in the rock cuts), . . . 4·57 per cent.
Sluice in tunnel (1,800 feet), ... 86·26 ,,
Tunnel below sluice (6,000 feet), ... 4·50 ,,
Cut below tunnel (200 feet), ... 0·81 ,,
Tail sluices (300 feet), ... 1·21 ,,
From seven undercurrents, ... 2·65 ,,
                                             100·00 ,,

---

[1] Hammond, *Ninth Report Cal. State Min.*, 1889, p. 129.
[2] Hammond, *ibid.*, p. 131.

The first undercurrent caught five times as much as the sixth, and nearly three times as much as the seventh, which was of double size. The yield of the seventh ($947) induced the Company to add another undercurrent. This mine affords an example of the difficulty of catching fine gold. The gold loss was unknown, but was believed not to exceed 5 per cent. of the contents of the gravel.

The bullion obtained by retorting the amalgam from the sluices is finer than that from quartz mills, and is sometimes 990 fine in Australia, although California placer gold is often as low as 850 fine. The remainder is mainly silver, but copper, lead, iron, and some of the minerals existing in the gravel also occur. The amalgam from the head of the sluices yields finer gold than that caught lower down and in the undercurrents.

Tailing.—The tail-sluices usually terminate on the side of a cañon, in a river, or in the sea. The enormous amount of loose sand and gravel, delivered from the hydraulic mines of Placer County, California, and the neighbouring counties into the Yuba and Feather rivers prior to 1880, filled up their beds to such an extent that in rainy weather disastrous floods ensued, and much valuable agricultural land was buried beneath sterile drift deposits and rendered worthless. The farmers thereupon took action against the Mining Companies and obtained a perpetual injunction forbidding them to discharge their tailings into these rivers. The result has been to stop hydraulic mining in these districts, and the efforts to work the deep leads more extensively by drifting, or on the other hand, to impound the tailings by dams made of brushwood, or to return them to their original position, have not resulted in unqualified success. Consequently the gold winning industry has not been maintained on the extensive scale it had assumed prior to the action of the courts.

Treatment of Cement Gravel.—In many cases the gravel from deep leads won by drift or shaft mining is cemented by iron oxides or clay into a conglomerate which is too tough to be easily disintegrated in the sluices. It is then passed through "cement mills," which closely resemble the stamp battery to be described in the next chapter, the chief differences to be noted being in the facilities for delivery. Double discharge mortars are used, and the screens are very coarse, the mesh being usually about $\frac{3}{16}$ inch, but varying up to $\frac{1}{2}$ inch in diameter. One battery of ten stamps, each weighing 950 lbs., making 94 drops of 9 inches in height per minute, will crush about 40 or 50 tons of gravel in ten hours so that it will pass through a $\frac{3}{16}$-inch mesh screen. Mercury is put into the mortar, and most of the gold is usually caught there on amalgamated copper plates, but copper plates outside the mortar are also used as in quartz-milling, and rubbers are employed to brighten the gold. If well-arranged plates are laid down, the number of sluice boxes which can be added with advantage is very small, a length of from 50 to 300 feet being used, the former limit being most common. No attempt is made to save the auriferous magnetic sands and sulphides which these conglomerates usually contain.

In the case of the Morning Star Cement Gravel Mine, Placer County, California, quoted by R. H. Richards,[1] the weight of the stamps was 850 lbs., the height of drop 6 to 8 inches, the number of drops per minute 95 to 100,

---

[1] Richards, *Ore Dressing*, 1903, p. 211. *Can. Min. Rev.*, 1896, **15,** 255 ; article by J. B. Hobson, on "Cost of Milling in Cement Gravel Mines."

the size of screen holes $\frac{3}{16}$ inch (round), and the capacity per stamp per twenty-four hours 12 tons.

In cases where cement mills are not required, the gravel is washed in sluices which differ little from those already described. The boxes are not more than from 18 to 24 inches wide and deep, and the series is seldom more than 300 or 400 feet long. Iron riffles are most in favour. Where the amount of gravel to be washed is small, or the water is scarce, the gravel is allowed to accumulate for some time and the water stored in a tank or reservoir. It is in some cases a great advantage to keep compacted gravels exposed to the air during a few months before washing them, as they "slack" and disintegrate under the influence of the weather, and subsequently are more easily treated, while for a similar reason, tailing is sometimes impounded, and re-washed after some time has elapsed. The disintegration of cemented material, which has been "slacked" by exposure to the weather, is usually completed in a *cement-pan*. This is a cast-iron pan with perforated bottom, and with a gate in the side for the removal of boulders, which are mostly barren and are separated from the auriferous material by this system, instead of being crushed and mixed with it, as is the case when stamp-mills are used. In the pan, four revolving arms, furnished with plough-shares, break up the gravel, which is carried through the apertures in the bottom by a stream of water, and falls into the sluice. A pan of 5 feet in diameter and 2 feet in depth will treat from 40 to 120 tons per day, according to the nature of the gravel.

The Hydraulic Elevator.—In this machine a jet of water under high pressure forces water, gravel, and boulders up an inclined plane, and delivers them all at the head of the sluice, which may be as much as 100 feet above bedrock. The differences in construction between the machines made in Australia, New Zealand, and the United States are only matters of detail. They consist essentially of an upraise pipe, usually of wrought iron, having a diameter of from 12 to 24 inches, which

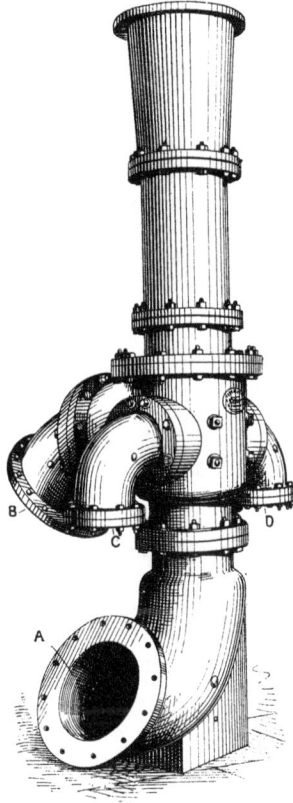

Fig. 51.—Hydraulic Elevator.

terminates below in an open conical funnel; a hydraulic nozzle, delivering water under the pressure given by a head of from 100 to 500 feet, projects into this funnel, and sand and gravel can also enter round the sides or through a special orifice. The inclination of the upraise pipe is usually from 45° to 65°. The top of the upraise pipe is turned over and terminates above a sluice, into which the gravel falls and is washed in the ordinary way. The Evans elevator made by the Risdon Iron Works, San Francisco, is shown in Fig. 51. A is the orifice through which the water for the jet is forced, B the main suction opening, and C and

D two auxiliary suction openings, which can be connected with pipes of any length and serve for raising water or fine material. The upraise pipe is now generally made with heavy steel plate, and the elevator itself of cast steel, instead of cast iron, the weight being much reduced in this way. The pipe is greatly contracted at the throat or lower end of the elevator, into which the nozzle discharges. The nozzle and the throat are sunk in a sump excavated in the bed-rock, and the gravel is washed down by any means (usually by a jet from an ordinary hydraulic nozzle) into this sump. The entrance to the upraise pipe is protected by a coarse grating, which prevents large stones, pieces of wood, etc., from entering it. The force of water is enough to complete the disintegration of the gravel during its passage through the upraise pipe, so that a short sluice is enough to effect the washing proper. If the excavation is carefully arranged, it may be kept funnel-shaped, so that the elevator, once placed in a sump, may be worked there permanently without being moved. When the pit is large enough, the washing may be done inside it, only the tailing being raised to the surface by the hydraulic elevator. The head of water required varies according to the vertical height through which the gravel must be raised ; a head of about 60 feet is required for every 10 feet of vertical upraise.

Wherever the necessary head of water is available, the hydraulic elevator is now recognised as a good method of working flat placers, river-bars, etc., or any deposits which are either below the water level of the district, or which have not sufficient fall for the disposal of the tailings by gravity. It is in wide use in California and New Zealand. The following instances of work in both countries may be given :—At the Blue Spar Consolidated Gold Mining Company, Gabriel's Gully, New Zealand,[1] tailing which has accumulated close to the sea on the foreshore, is sluiced in this manner. The vertical upraise is 60 feet, the angle of inclination of the upraise pipe being 63·5° ; about 480 tons of gravel are raised per shift, the head of water used being 400 feet, while the amount of water used in each elevator is seventeen government heads. The sluice is short, and has an inclination of only 3½ inches in 12 feet ; the upper parts are fitted with transverse, patent ⌐-shaped, angle iron riffles, in which the angle faces up stream (see Fig. 50). The lower parts of the sluice have a false bottom of wrought-iron plates, perforated with round holes ; beneath these plates is the true bottom of the sluice, covered with cocoanut matting in which fine gold is caught. The tailing is discharged into the sea.

At Quartz Valley, Siskiyou County, California,[2] on hard ground, where the elevator was first used, it took forty-three days to work out a piece of ground 300 feet by 250 feet, which was of an average depth of 18 feet. The bank was washed down with 600 miner's inches of water, and went to the elevator through a 30-inch bed-rock flume, which had a grade of 5 inches in 12 feet. The water and gravel were raised through a 20-inch elevator pipe without any contraction at the throat. It was set at an angle of 40°, and the pipe was 42 feet long, the vertical upraise being thus 28 feet. The force used was 1,000 inches of water with a head of 230 feet, delivered through a 6-inch nozzle, and the gravel was emptied into a sluice 6 feet by 3 feet, with a grade of 1½ inches in 12 feet. When 3,500 inches of water were running in this sluice, they could not carry off all the gravel raised by the

[1] New Zealand Mining Commissioners' Report, 1891, p. 65.
[2] Egleston, Silver, Gold and Mercury in the United States, 1890, vol. ii., p. 307.

elevator. The work was done without any delay from stoppage of the machine and there were no repairs, the wear of the elevator being very little.

In the Ruble hydraulic elevator, the fine gravel is separated from the coarse rock and boulders by means of a grizzly, while in transit up the elevator. The sluices operate on fine material only, and so gold-saving is facilitated.[1]

The use of the hydraulic elevator in Alaska is described by T. A. Rickard.[2] On the Discovery Creek, an elevator was raising gravel through 28 feet vertically, the angle being 80°. The head of water used was 940 feet and the nozzle $3\frac{1}{2}$ inches in diameter, 427 cubic feet of water being used per minute. The elevator was fed by a monitor, which washed gravel into a ground sluice where some gold was caught.

Cost of Placer Working.—In work on shallow placer deposits by individuals the results differ greatly, both according to the strength and skill of the worker and to the contents of the gravel. Under the best conditions of climate a strong, well-nourished, American digger may be able to raise by the shovel from 10 to 12 cubic yards of gravel per day, and throw it into a receptacle 3 feet above the ground. Native labour cannot be expected to effect so much, and in French Guiana it is reckoned that only about half a cubic yard of earth per man per day can be shovelled into the sluice. If the workman must wash the gravel, as well as raise it, much less can be accomplished. For an active man, it is a fair day's work to dig and wash from fifteen to twenty pans of dirt, the amount treated thus not exceeding about 10 cubic feet. On the other hand, with the cradle, the output may be from $1\frac{1}{2}$ to 2 cubic yards per man in a day, while with the long tom it may rise to 3 or 4 cubic yards per man. In the Siberian trough, only from 1 to $1\frac{1}{4}$ cubic yards can be treated by one worker per day, but a larger percentage of the gold is believed to be saved in this apparatus. The minimum contents in gold, which will make the gravel worth treating, depends on the cost of labour. In early times, in California and Australia, when the virgin shallow deposits were being worked, large sums were often realised by individual diggers, cases being on record in which 5 ozs. of gold were obtained from one pan of bed-rock scrapings lying under heavy gravel, and earnings of several hundred dollars per day were not uncommon. The results obtained on the Klondike in Canada are still more remarkable. Authenticated instances of 15 ozs. of gold per pan are recorded, and an average of $5 to $7 per pan was obtained on certain claims in 1897.

Concerted work, with the aid of the sluice, is much more effective ; in California, gravel containing about 1 pennyweight of gold per cubic yard is worked at a profit, the dirt being lifted into the sluice by hand-labour, and the tailings removed by sluicing with water ; at Ballarat, in Australia, where the gravel is raised to the surface from underground workings through a vertical shaft several hundred feet deep, and subsequently washed, 12 grains of gold per cubic yard of material pay for the treatment, while in Siberia, as stated below, the cost is even less. In French Guiana, the unhealthiness of the climate and the cost of supplies render it impossible to work gravel containing less than about 3 pennyweights of gold per cubic yard.[3]

In Siberia,[4] the distance of the workings from the nearest town, and the

---

[1] J. M'D. Porter, *Trans. Amer. Inst. Mng. Eng.*, 1909, **40**, 561.
[2] T. A. Rickard, *Min. Mag.*, 1909, **1**, 139.
[3] Frémy, *Ency. Chim.*, vol. v., L'or, part iii., p. 48.
[4] Frémy, *ibid.*, vol. v., L'or, p. 47.

traditions of the industry, require workmen to be hired by the year, or, in cases where no work is attempted in winter, for the season. The total cost of treatment of the gravels varies greatly with their geographical position. On the banks of the Lena, where the season only lasts for five months, it is estimated that the gravel must contain 2 *zollatniks* of gold per 100 *poods,* or 4⅓ dwts. per cubic yard ; but in the neighbourhood of Ekaterineberg, the deposits in the bed of the Pechma, a tributary of the Obi, are worked when they only contain from 8 to 9 grains per cubic yard.

The cost of working the perennially frozen placers is much more, but Levat gives an instance in which a bed of gravel, 9 feet thick, yielding about 1¼ dwts. per yard, and covered by 100 feet of barren material, was worked at Malamalski in the Trans-Baikal.[1]

The cost of hydraulic mining depends largely on the magnitude of the operations. With large quantities of water available at a cheap rate, and big banks of soft gravel, the cost has been reduced to from 2 to 3 cents per cubic yard in California, while the average cost there in 1884 was about 10 cents, and only in exceptional cases amounted to as much as 50 cents per cubic yard.

The cost of dredging in California may be put at 2½ to 10 cents per cubic yard of, say, 1½ tons.

---

[1] Levat, *L'or en Siberie Orientale*, vol. i., p. 146.

CHAPTER VII.

ORE CRUSHING IN THE STAMP BATTERY.

Primitive Methods of Crushing and Amalgamation.—In all countries, when the richest alluvial deposits have been worked out or have all been taken possession of, efforts have been made to extract the gold from the various hard, auriferous materials met with in veins. The earliest machines used for the purpose in Egypt were stone slabs, on which the gold quartz was broken by stone-hammers, hard stone rubbing mills with mullers for coarse crushing, grinding mills or querns from 18 to 20 inches in diameter for fine grinding, and stone inclined tables, on which the particles of rock were washed away from the gold. The washing was completed in broad, flat dishes.[1] Hollowed-out stone mortars for grinding gold ore have also been found in Wales, Central America, the Pyrenees, and Transylvania. Diodorus Siculus, the Greek historian, has given a detailed description of the method of gold-quartz reduction employed in the mines of Upper Egypt 2,000 years ago.[2] The ore was reduced to coarse powder in mortars, then finely crushed in hand-mills resembling the querns or hand flour-mills formerly in general use, and finally washed. In order to separate the pulp from the uncrushed lumps, the Egyptians, in common with other races in ancient times, employed sieves, but in extracting the gold from auriferous sands they used raw hides, on which the flakes of gold were entangled.[3] These devices closely resemble those still in use in many parts of the world.[4]

The date of the first use of mercury for amalgamation is unknown, but it has no doubt been used for this purpose for the last 2,000 years. The earliest mention of quicksilver itself appears to occur in the works of Aristotle,[5] who speaks of it as fluid silver, and in those of Theophrastus, about B.C. 300 ; but Diodorus, in the account just mentioned, does not refer to its use. Only a few years later, however, Vitruvius,[6] about B.C. 13, described the manner in which, by the help of mercury, gold was recovered from cloth in which it had been interwoven, and in Pliny's time the separation of gold from its impurities generally by the same means was well known.[7]   It is probable

---

[1] Gowland, "Huxley Mem. Lecture," *J. Roy. Anth. Inst.*, 1912, **42**, 256.

[2] *Diodorus Siculus*, **3,** 13. This description is quoted from Agatharcides, a Greek geographer of the second century B.C. Booth's translation (London, 1700, p. 89) is given in Hoover's *Agricola* (London, 1912), p. 279, A full translation is also given by B. H. Brough in his Cantor Lectures on Mine Surveying, *J. Soc. Arts*, March and April, 1892. See also T. K. Rose, *The Precious Metals*, 1909, p. 3.

[3] Beckmann, *History of Inventions* (London, 1846), vol. ii., p. 334.

[4] An interesting account of some primitive methods of treating gold quartz employed by the Chinese is given by Henry Louis in *Eng. and Mng. J.*, Dec. 31, 1892, p. 629, from which it appears that these methods bear points of resemblance to those of the Egyptians.

[5] Aristotle, *Meteorologica*, iv., 8, 11. Hoover, *Translation of Agricola*, p. 432.

[6] *Vitruvius*, lib. vii., cap. viii. For translation, see Hoover, *Agricola*, p. 297.

[7] Pliny, *Naturalis Historiœ*, lib. xxx., cap. vi., sect. 32. Quoted in full in Percy, *Metallurgy. Silver and Gold*, p. 559.

that this knowledge was never afterwards entirely lost, although the references to it in the Middle Ages are very scanty. For example, Geber,[1] in the thirteenth century, was aware that mercury would dissolve considerable quantities of gold and silver, but not earthy materials, and Theophilus, the monk,[2] in the eleventh century, carefully described the method of washing the sands of the Rhine on wooden tables, the final operation consisting in treating the concentrates with quicksilver for the removal of the gold. The extraction of silver by mercury on lines precisely similar to those used in the patio process is described by Biringuccio in his treatise (which was published in Italian in 1540),[3] and had apparently been a secret art for some time previously. For Agricola's account of the amalgamation of gold ores, see below, p. 150.

Mercury was introduced into Mexico as a means of extracting the precious metals by Bartolomé Medina in the year 1557,[4] and its use doubtless soon spread to Peru and the neighbouring countries. When Barba wrote in 1639 there were three amalgamation machines in use in Peru[5]—viz., the *mortar* (used in the Tintin process), the *trapiche* or Chilian mill, and the *maray*. The *mortar* was hollowed out in a hard stone, the cavity being about 9 inches both in diameter and in depth. The ore was triturated with water and mercury in this mortar with the aid of an iron pestle, while a stream of water, flowing through, carried away the crushed particles. The slimes were roughly concentrated, but most of the gold remained in the mortar with the mercury.

The *maray*, although equally primitive, was probably of greater capacity : it makes use of the principle employed in the bucking hammer. About the year 1825 Miers[6] saw it in Chili, where it is probably still in use. It consists of a flat or slightly concave stone, 3 feet in diameter, on which lies a spherical boulder of granite about 2 feet in diameter. This is rolled to and fro by two men seated on the ends of a long, wooden pole, which is firmly fixed to the boulder. Ore, water, and mercury are ground together in this machine, and then washed down.

The *Chilian mill* closely resembles the edge-runner mill of the present day, which has again come into some prominence for crushing gold ores and is also used for grinding and mixing mortar. The Peruvian *trapiche* had a similar circular bed of hard stone, but only one stone runner, which was driven by mules. The Chilian mill was used to prepare ores for treatment in the *arrastra*, which was not mentioned by Barba, and may perhaps be regarded as an outcome of the *trapiche*.

Among other ancient methods of amalgamating ores, the *tina* and *cazo* systems may be mentioned. The *tina* system, largely used in Chili, at least as late as the year 1853,[7] is a modification of that still at work in the Tyrol (see Chap. IX.). The old and long-abandoned Norwegian method of treatment, described by Schlüter in 1738,[8] and formerly practised at Kongsberg, was also, no doubt, derived from the same source. In Norway and Chili, however, the working was not continuous. Charges of crushed ore were

---

[1] Salmon's *Geber*, cap. xlvii. [2] *Theophili*, lib. iii., chap. xlix.
[3] Biringuccio, *De la Pirotechnia* (Venice, 1540) ; lib. ix., cap. xi., fol. 142. For translation, see Percy, *Metallurgy. Silver and Gold*, p. 560.
[4] Georg Agricola, *Bermannus* . . . übersetzt von *Friedrich August Schmid*, 1806, p. 49.
[5] Barba, *Arte de los metales*, lib. iii., cap. xxi.
[6] Miers, *Travels in Chile and La Plata*, vol. ii., p. 398.
[7] Percy, *Metallurgy. Silver and Gold*, p. 567.
[8] Schlüter, *Gründlicher Unterricht von Hütte-Werken*, p. 211.

introduced into wooden tubs (*tinas*) furnished with cast-iron bottoms and stirred up with water and mercury by means of revolving iron agitators, comparable to the mullers of the modern amalgamating pan. At Kongsberg only 80 to 100 lbs. of ore were treated at one time with about 40 lbs. of mercury, but in Chili a *tina* contained from 4 to 6 cwts. of ore and 150 lbs. of mercury, or more. After the agitators had been at work for about four hours making sixteen revolutions per minute, the mercury was drawn off, squeezed through calf-skin or canvas bags, and the amalgam removed from the bags and retorted. The *cazo* process, introduced in Peru in 1609,[1] was similar, but the bottom of the vessel was of copper and the pulp was boiled. These processes were used in treating ores rich in silver.

The *Kröncke* process [2] introduced in 1860 is a modification of the cazo process. In 1908 a plant was completed to operate on the gold ores of Hog Mountain, Alabama, by amalgamation with mercury in the presence of copper sulphate, salt, and iron, in a large 5- by 22-foot tube mill. The method is similar to those used in the Kröncke and Patio processes. The grinding, chemical treatment and amalgamation go on simultaneously. The pulp was subsequently passed over plates. This method was devised after treatment of the ore (red oxidised quartz) by cyanide had been tried and abandoned.[3]

The **Arrastra** was also one of the earliest crushing machines in use in America, being introduced at the same time as the Patio process—*i.e.*, about 1557—and until recently was in wide use in Mexico, although chiefly in the treatment of silver ores by the Patio process. It is now almost obsolete. It is a circular, shallow, flat-bottomed pit, 10 to 20 feet in diameter, surrounded by a circular wall of stones, and paved with hard, uncut stones. A typical arrastra is shown in Figs. 52 and 53 in plan and section.[4] Granite, basalt, and compact quartz are all used for the rock pavement, which is made 12 inches thick, and is either placed on a bed of well-puddled clay from 3 to 6 inches thick, or set in hydraulic cement, so that no chink or cranny remains into which the mercury or amalgam can find its way. In the centre a vertical shaft revolves, carrying two or four horizontal arms, to each of which is attached a heavy stone by thongs of bullock hide, or by chains. These drag stones weigh from 200 to 1,000 lbs. each, their forward ends being about 2 inches above the floor, whilst their other ends drag on it. They are moved by mules walking round outside the arrastra, or by water- or steam-power, the speed varying from four to eighteen turns per minute. Fig. 54 represents an arrastra of the simplest description; at the front the stones forming the edge have been removed, so as to expose a section of the rock pavement.

Ore of about the size of pigeon's eggs is introduced, enough water being added to make the pulp of the consistency of cream, and mercury is sprinkled over it after most of the grinding has been done. When the ore cannot be ground any finer, more water is added to dilute the pulp, the mule is driven slowly for half an hour to collect the mercury, and the pulp is then run out and another charge shovelled in. The pulp is run into large dolly-tubs, where more water is added, and the mercury and amalgam separated by settling. Most of the amalgam, however, remains in the arrastra, and is

[1] Barba, *Arte de los Metales. Loc. cit. ant.*
[2] See Collins, *Metallurgy of Silver* (1900), p. 69.
[3] T. H. Aldrich, jun., *Trans. Amer. Inst. Min. Eng.*, 1908, **39**, 578.
[4] Collins, *Metallurgy of Silver*, p. 40.

removed and dealt with only after many charges of ore have been treated. An arrastra of 10 feet in diameter takes a charge of 500 to 600 lbs. of ore, and treats about 1 ton per day of twenty-four hours. The amount of wear sustained by the grinding stones is equal to from 6 to 10 per cent. of the ore

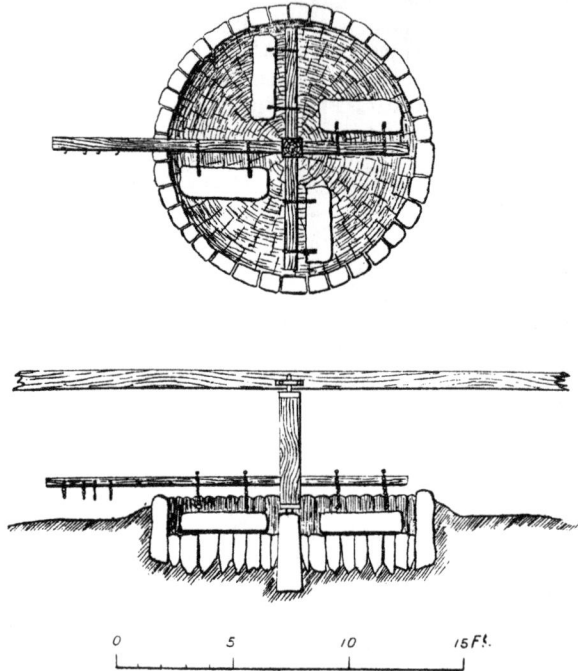

Figs. 52 and 53.—Arrastra (Plan and Section).

crushed. The output is so limited that the use of the arrastra has never been general outside Mexico, although it has been used in almost every district in the United States for a short time after the commencement of quartz mining in the particular locality.

Fig. 54.—Arrastra, General View with Front removed.

It is often stated that the results obtained, so far as the percentage of gold extracted from refractory ores is concerned, cannot be equalled by any other amalgamating appliance, and that the Mexicans, using the arrastra, formerly treated at a profit ores which hardly yield any gold to the stamp mill, or even to the amalgamating pan. In consequence of its power of saving gold and the cheapness with which it can be erected and worked, the arrastra is still valuable for prospecting. In preparing the bed for this purpose, every care must be taken that the surface is even, and all joints properly cemented, or else the mercury and amalgam will in great part find their way into the foundations.

The reasons for the high extractive power of the arrastra, when treating certain ores, are no doubt to be found in the extreme fineness to which the ore is reduced, and in the prolonged contact between the ore and the mercury, which is maintained while they are being ground together. Moreover, the grinding action of the dragged stones keeps the particles of gold bright and in a suitable condition for amalgamation, without exercising force enough to flatten and harden them. The relative advantages of grinding surfaces consisting of iron and stone are less certain, and probably vary with the nature of the ore in course of treatment. An instance is given by Colonel Harris,[1] in which stone was better than iron. In this case, at Cerro de Pasco, Peru, the old method of grinding ores in circular pans, by edge-runners of stone or granite, was found to entail a rapid wearing of the edge-runners, and, in order to remedy this, the runners were shod with iron. The returns at once fell off, and on careful trial it was found that the yield in the old machines was 15 per cent. more than in the new ones. Other similar instances are on record, but it must nevertheless be conceded that, with some ores, the presence of iron is necessary for good work, chlorides of silver and mercury being reduced by it, which would otherwise be lost in the tailings. The somewhat extravagant views often expressed, upholding the great superiority of the arrastra over its more modern rivals as an amalgamating machine for gold ores, are perhaps hardly justified, since a direct comparison with stamps or pans has been possible in only a few instances.

One of the most remarkable of these instances is that afforded by the experience at the Pestarena Mines, Val Anzasca, Italy. The ore of this mine contains about 20 per cent. of pyrite, and is of somewhat low grade, rarely containing as much as 15 dwts. of gold per ton. Efforts were at first made to treat it by means of stamps and amalgamated plates, with the result that only 65 per cent. of the gold was extracted. Better results attended the introduction of the Frankfort mill, a modified arrastra, driven in this instance by steam ; this mill is substantially a wooden pan, with dies and shoes of stone. The mercury was added to the pulp after it had been finely ground, and the amalgamation and grinding of the pulp subsequently kept up for seven hours. From ore containing 12·3 dwts. per ton, the mill extracted 10·2 dwts., or 83 per cent., for a considerable period of time, while in the year 1890 the average extraction was 79·4 per cent. There were twenty-eight mills, each of which treated 1,200 lbs. of ore per day. The stone bed was said to last ten months, and the shoes from six to eight weeks, the total cost of treatment being very low.

At Smartsville, and in other places in California, the arrastra was applied to crushing and amalgamating hard cement gravel obtained from a drift mine.[2] At Smartsville the cement is coarsely crushed by being passed through a Gates rock-breaker, and is then charged into four arrastras, each of which

[1] Harris, *Mng. J.*, December 24, 1892, p. 1466.
[2] *Eleventh Report Cal. State Min.*, 1892, p. 315.

is 12 feet in diameter and 3 feet deep and capable of containing from 5 to 9 tons of gravel. The grinding is effected by four blocks of diabase, each weighing from 600 to 1,000 lbs.; the rate of revolution is 14 times per minute, the time of grinding being one hour. A tablespoonful of mercury is fed in with each charge, and the total loss is only 10 per cent. of this. The pavement costs $40 and lasts for six months. At the end of the hour, a gate is opened in the side of the arrastra and the charge run into a sluice, 200 feet long, where the mercury and amalgam are caught by means of riffles. The capacity of one arrastra is 50 tons of hard cement per day, or 75 to 90 tons of soft surface-gravel per day. The cost is from 6 to 8 cents per ton, and more gold is extracted than when a stamp battery was used at a cost of from 20 to 40 cents per ton. One man attends to the four arrastras, and another to the Gates crusher.

Iron prospecting arrastras with stone drags are sometimes used.

The Stamp Battery.—The stamp battery evolved, no doubt, from the pestle and mortar was not introduced until a comparatively recent date. Beckmann[1] states that mortars, mills, and sieves were used exclusively in Germany throughout the whole of the fifteenth century, and in France stamps were unknown as late as the year 1579. Brough has suggested that the origin of stamp mills was probably due to the manufacture of gunpowder. It seems certain that in 1340 a stamp mill, used in connection with this industry, existed in Augsburg, and that Conrad Harscher, of Nuremburg, owned one in 1435. They were first applied to the gold industry at the beginning of the sixteenth century, a doubtful record stating that they were introduced into Saxony by Count von Maltitz in 1505. In 1512 Sigismund Maltitz, " rejecting the dry stamps, the large sieves and the stone mills of Dippoldswald and Altenberg, invented a machine which crushed the ore wet under iron-shod stamps."[2] In 1519 the processes of wet-stamping and sifting were established in Joachimsthal by Paul Grommestetter, who had some time previously introduced them at Schneeberg. The improvements gradually spread through Germany, and detailed descriptions and drawings of the apparatus were given by Agricola[3] in 1556, from which it appears that the dry-stamp batteries in use at that time consisted of three or four square wooden stamps with square iron heads, working on a flat surface without sides (*i.e.*, no mortar-box) and raised by levers or cams set in the axle of a water wheel. The wet stamps had a wooden mortar-box with an iron bottom, and an iron screen set in an opening at the end of the mortar through which the water flowed carrying the crushed ore. In some Hungarian mines, Bennett H. Brough[4] saw some primitive stamps in use, resembling those drawn by Agricola, weighing only 100 lbs. each, and having their heads made in some cases of hard blocks of quartzite. At that time, in cases where the conditions of water supply were favourable, these stamps were able to treat with profit an ore containing as little as $2\frac{1}{2}$ ozs. of gold to 50 tons of ore, and at Zell, in the Tyrol, they were able to treat a slaty material containing 1 oz. of gold to 50 tons of ore. Such economical work is seldom possible with the modern Californian stamp under the most favourable circumstances.

Agricola's exact description of the treatment of auriferous quartz in Germany in 1556 shows that the methods in use at that time were strikingly

---

[1] Beckmann, *History of Inventions*, vol. ii., p. 334.
[2] Agricola, *De re Metallica*, p. 246; Hoover's translation, p. 312.
[3] Agricola, *De re Metallica*, lib. viii., p. 220; Hoover's Translation, pp. 284, 287.
[4] Brough, *Proc. Inst. Civil Eng.*, 9th Feb. 1892.

Fig. 55.

A, Water Wheel; B, Main Axle; C, Stamp; D, Hopper in Upper Millstone; E, Opening
passing through its centre; F, Lower Millstone; G, Its Round Depression; H,
Its Outlet; I, Iron Axle of Millstone; K, Its Cross-piece; L, Beam; M, Drum,
with trundles on the iron axle; N, Toothed Drum on Main Axle; O, Tubs; P,
The Small Planks; Q, Small Upright Axles; R, Projecting Part of ore; S, Their
Paddles; T, Their Drums supplied with trundles; V, Horizontal Axle set into end
of main axle; X, Its Toothed Drums; Y, Three Sluices; Z, Their Little Axles:
A A, Spokes; B B, Paddles.

similar to those still employed in Transylvania and the Tyrol, which were
among the districts of which he wrote. Doubtless in these districts, the
methods have been handed down from generation to generation with little
change, while in other countries, where they were introduced hundreds of
years later, the changes have been rapid and striking. In those points in
which the older differed from the modern Tyrolean practice, it resembled
the practice of the ancient Egyptians, so that the origin of the methods
may, perhaps, be traced back to them. The wooden stamps, shod with
hard stone or iron, were arranged in sets of three or four, and raised by cams
to fall by gravity when released. The rock was shovelled dry into the mortar,
and coarsely crushed by the blows of the stamps.

Next, it was ground as fine as flour in a stone mill supplied with water,
and carried by the stream of water into the uppermost of three wooden
tubs, whence it overflowed in succession into the other two. Agricola's
illustration of the process is shown in Fig. 55. Revolving mechanical stirrers,
furnished with six paddles, kept in agitation the contents of the tubs and,
to quote Agricola, "separate even very minute flakes of gold from the
crushed ore. These flakes, settling to the bottom (*delapsa*), are drawn to
itself and cleansed by the quicksilver (lying in the tubs), but the water carries
off the dross. The quicksilver is poured into a bag made of leather or cloth
woven from cotton, and when this bag is squeezed, as I have described elsewhere, the quicksilver drips through it into a jar placed underneath. The
pure gold remains in the bag." [1] The same error of assuming squeezed
amalgam to be pure gold occurs in Pliny, but not in Biringuccio. Agricola
here expounds the theory of amalgamation still adhered to in Austria where
mercury is regarded merely as a useful means of collecting particles of gold,
which have already been separated from the crushed ore by their great
density. The Tyrolean bowls, still in use at Vöröspatak in Hungary and
in a few retired valleys in the Eastern Alps, do not differ essentially from the
tubs drawn and described by Agricola ; and, although wet crushing by the
stamps has been introduced, the mortar is even now seldom furnished with
screens in these mills.

Elsewhere the changes in stamp battery practice, introduced since
Agricola wrote his treatise, have been many and great, but they were not
introduced in Europe. In 1767 M. Jars saw stamp batteries in use in the
Hartz [2] which resembled those described by Agricola. Even then only a
single screen of brass wire 12 inches square delivered the product of three
stamps, and in several other districts in Germany screens had not been
adopted. The screen was completely protected from the splash of the stamps,
so that the sieving of the ore was very slowly effected by a current of water
flowing through.

The most important improvement has undoubtedly consisted in combining the operations of crushing and amalgamation by charging mercury
with the ore into the battery, and placing amalgamated copper plates in
the mortar, so as to catch gold. No mention of these practices appears to
have been placed on record before stamp batteries began to work in California in 1850, although they had possibly been adopted in Georgia before
that time. It remains to add that in more recent practice, especially on

---

[1] Agricola, 1657 ed., p. 233 ; Hoover, *Agricola*, p. 298.
[2] Jars, *Voyages Métallurgiques* (Paris, 1780), vol. ii., p. 309.

the Rand, the addition of mercury to the mortar has been discontinued and the principles of the more ancient practice re-introduced.

The use of the copper plate was probably suggested by the experience in Mexico and South America of the working of the Cazo process (invented by Barba in 1609, according to his own statement[1]), in which, as was well known, amalgam tended to adhere to the copper sides of the vessel unless the proportion of mercury to gold and silver present was less than four to one. Thus Baron Born wrote in 1786,[2]—"In new kettles . . . the inside becomes wholly and so perfectly silvered that it never can be cleaned . . . . The silvery coat is daily increased by slow and gradual apposition, and the crusts of amalgama, accumulating on the bottom and sides of the vessels, become gradually so thick that on emptying them they often fall off by their own weight as silver plates, which, when dry, show the laminated texture of their daily augmentation."

The knowledge of this behaviour of amalgam in the Cazo process must have been common to many who were engaged in exploiting the quartz veins of the West soon after their discovery, and the speedy application of this knowledge is exactly what might be expected from those quick-witted pioneers. Nevertheless, the exact date and locality of the introduction of the copper plate remains a matter for conjecture.

The copper plates fixed in the battery in the early fifties were about 4 inches wide and as long as the mortar, and were placed one on the " feed " side and one on the discharge side just underneath the screens. It was soon found that the plates worked better from the start if they were coated with mercury before they were placed in position, and this has now been invariably done for the last half century. Crushed ore, stones, water, and amalgam are flung violently against the plates, and the amalgam is retained in great part.

The scouring action of the pulp on the plates is, however, always great, and becomes more violent in proportion as the stamps are larger. The plate on the feed side, long ago condemned by many, is accordingly being discarded more and more, and that below the screens is curved away in such a manner that it cannot be struck directly by the splash from the stamp (see Fig. 66, p. 161). A more recent step in this direction was to line the mortar with cast-steel plates, furnished with slots for the purpose of catching amalgam (see Fig. 66).

As already mentioned, the tendency is now towards the discontinuance of inside amalgamation, and the abolition of gold-catching plates in the mortar. The introduction of coarser crushing in the battery, followed by regrinding, has, of course, assisted this movement.

Seventy years ago it was the practice, in districts other than Transylvania, to treat the pulp after it had left the mortar mainly by passing it over inclined tables covered with blankets, much in the same way as Jason may have seen the golden sands worked in Asia Minor. The sands accumulating on the blankets were washed off at intervals and ground in mills with mercury. In addition, the ore was frequently passed over or through baths of mercury, in imitation of the old Tyrolean practice.

Amalgamated copper plates over which the pulp flowed were tried, after

---

[1] Barba, *Arte de los Metales*, 1640, lib. iii.
[2] Baron Inigo Born, *New Process of Amalgamation*, translated by Raspe, p. 122 (London, 1791).

those placed in the mortars had been proved to be beneficial, but in Western America were at first almost everywhere rejected,[1] probably owing to the great depth of the stream of ore and water made to flow over them. When long afterwards, about the year 1870, this mistake began to be rectified, the value of the plates was soon recognised in California.

The pulp is now led over the surface of these inclined plates in a very thin stream, not more than a quarter of an inch deep. The pulp does not run down in a regular stream, but in a series of little wavelets which tumble over and over, and are supposed to bring every part of the pulp successively in contact with the amalgamated surface. The catching powers of the plates are thus supposed to be practically independent of the tendency of the particles of gold or amalgam to sink to the bottom of the stream. This theory is not accepted by the Austrian school (see above, p. 152), and it is certain that native gold is caught more easily in proportion as it contains less silver (and is of higher density), so that when the particles of metal consist of an alloy containing a large proportion of silver, and are, therefore, of comparatively low density, the yield on the plates is generally poor. In any case, however, whatever be the working hypothesis adopted, the amalgamated plate should theoretically be better adapted for its work than the Tyrolean mill and other machines using mercury baths, owing to the slight depth of the pulp on the plates, and the short distance through which the gold particles are compelled to settle before reaching a catching surface. The plates are wiped down with rubber or brushes as often as required (from once to twelve times a day), and the gold separated in the usual way from the excess amalgam thus collected.

The German stamp has a rectangular stem made of wood or, latterly, of iron, with an iron head, the total weight never exceeding 300 to 400 lbs. It was introduced almost unchanged into France, Cornwall, and, after the discovery of gold in 1849, into the United States, but has given place in all new districts to the Californian stamp, and need not be fully described here. In 1850, the first stamp mill seen in California was erected at Boston Ravine, Grass Valley. The stamps consisted of tree-trunks shod with iron, and the framework was constructed of logs.

The ordinary method of reduction and amalgamation of gold quartz in a stamp battery now consists of the following operations :—

1. The ore is broken down to a moderate size, usually to about 2-inch cubes (or, according to Caldecott, $1\frac{3}{4}$ inches on the Rand), by passing through the jaws of a rock breaker, or by hand hammers, the latter an almost obsolete method.

2. The ore is then fed into the mortar-box of a stamp mill, where it is pulverised to the required degree of fineness. Automatic feeders are now in general use. In wet crushing, a stream of water is introduced also, and the blows of the stamps splash the water and pulp against screens set in the side of the mortar, the finely divided ore being ejected in this way. In some cases the mortar-box is partly lined with removable amalgamated copper plates, by which some of the gold is caught and retained, small quantities of mercury being in this case usually fed into the mortar-box with the ore and water. Sometimes mercury is added when there are no inside plates.

3. On issuing from the battery, the pulp is allowed to run over a series

---

[1] See G. Küstel, *Nevada and California Processes of Silver and Gold Extraction*, p. 61. (San Francisco, 1863.)

of inclined, amalgamated copper plates, by which a further percentage of the gold is amalgamated and retained. Amalgam is also caught here when inside amalgamation is practised.

4. The residue or tailing is sometimes further treated by running over rough hides or blankets or riffled sluices, by which some particles of gold and pyrite are retained, or the pyrite is separated from the valueless sand by concentration on some form of vanner or jig. The concentrate is subjected to further treatment, usually either by cyaniding, smelting, or formerly by chlorination. In other cases the tailing is separated, by settling in water and decantation or by classifiers, into " sand " and " slime," which are cyanided without previous concentration. Sometimes the tailing goes to tube-mills for regrinding.

5. At intervals the gold amalgam is wiped or scraped off the copper plates, the excess of mercury separated by squeezing through filter bags of chamois leather or canvas, and the pasty amalgam thus obtained is retorted in order to distil off the mercury.

6. The retorted bullion is melted in crucibles with the necessary fluxes, and cast into bars, which are then sent to a refinery.

The following is a general description of the machinery employed in stamp-battery practice :—

Ore Bins.—For mixed ore as received from the mine, it is an advantage to instal large bins of sufficient capacity to hold several days' supply for the stamps, in case of breakdown or other delays in the mine. Bin floors were formerly made of planking, laid with the lengths in the direction of the slope, for, if placed transversely, the boards wear fast, and the ore packs at the edge of each one, with the result that its movement is impeded and must be assisted by shovelling. The slope should be at least 45° in order to enable the ore to move downwards by gravity, when the lowest portion is drawn from the shoot. The wear on the inclined bin floor is, however, very great, and the floor is now generally protected by steel plates or bars. The ore is discharged through a sliding door, which is also well protected against wear. Ore bins with flat bottoms have greater capacity, but may necessitate an additional handling of part of the ore. In this case the bottom of the bin can be made of wood throughout. The sills of the bins should be placed horizontally on terraced ground, not on the slope of the hill.

A shoot from the mixed ore bin door leads to a grizzly, through which the fine ore drops into the main battery ore bin. The larger pieces of rock are discharged either into a coarse ore bin or else upon the platform by the side of the rock-breaker and on a level with its mouth, into which it is shovelled by hand. The former course is preferable, as in that case the rock-breaker can be fed continuously by a gate in the coarse ore bin, which is opened and shut by a rack and pinion. By this arrangement there is a saving of labour, but the chief advantage is that the rock-breaker is thereby kept constantly at work. At the North Star Mill, California, it was found that when, by arranging for a continuous feed from the coarse ore bin down a shoot leading direct to the rock-breaker, the latter was in constant work, it absorbed 12 horse-power, as against 8 horse-power when in intermittent work, but its output was over 50 per cent. more.

Rock-breakers.—There are two classes of these machines in general use, viz. :—(a) Jaw crushers with reciprocating motion, and (b) Gyratory crushers. The Blkae and the Dodge crushers are representative of the former class, and the Gates and Comet crushers of the latter.

The *Blake Crusher* is shown in section in Fig. 56. The rock is crushed between the stationary jaw, B C[1], and the swinging jaw, D, which is pivoted at E, and moved by the eccentric, F, through the toggles, J K. The jaw plates, C[1], C[3], formerly consisted of chilled cast iron, and are now usually of manganese steel or some other special cast steel, suitable for resisting hard wear; they have longitudinal corrugations. The machine works at about 250 or 300 revolutions per minute. At each revolution the moving jaw is advanced about $\frac{1}{2}$ inch towards the other, and the lumps of rock which have dropped down between the jaws are broken; as the moving jaw recedes, the fragments slip lower down and are further crushed at the next advance, and this process is repeated until the ore is small enough to pass out at the opening at the bottom. The distance between the jaws at the bottom limits the size of the fragments, and this distance may be regulated at will by moving the wedge, L, or by changing the length of the toggles, J K. The jaws wear most at the centre and lower end, which is partly compensated for by reversing. The "Osborne" composite jaw, in use on the Rand, is designed to meet this difficulty.[1] The capacity of the machine is great, being about 300 tons of ordinary rock per day of twenty-four hours in the case of the

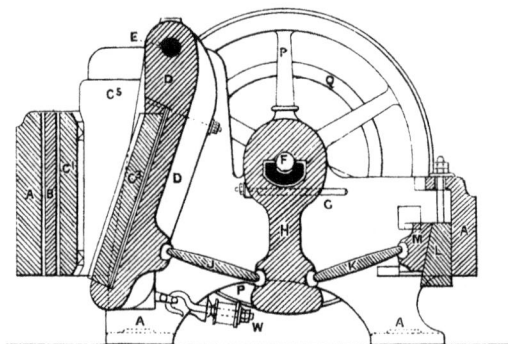

Fig. 56.—Blake Crusher.

machine whose dimensions at the mouth are 20 inches by 10 inches, when the lower edge of the jaws are set to approach within $1\frac{1}{2}$ inches of each other. The power required for this is stated to be 14 H.P. In large "coarse breakers" on the Rand 20 to 30 H.P. is required for a breaker dealing with from 40 to 60 tons of ore per hour.[2] Jaw breakers of this type are used on the Rand for coarse crushing to 4 or 5 inches maximum, as they make little "fines."

In the *Dodge Crusher*, shown in Fig. 57, the moving jaw is pivoted below instead of above. The effect of this arrangement is to make the product more uniform in size, and as there is little or no motion of the movable jaw at the delivery aperture, this may be made as narrow as desired, so that a finer product can be obtained, although it is at the expense of capacity. The Dodge crusher is more particularly recommended for fine crushing in concentration works, or where the product is to be subsequently passed through rolls.

Authorities differ as to the relative advantages of the two positions of

---

[1] C. O. Schmitt, *Rand Metallurgical Practice*, vol. ii., p. 32 (Griffin & Co., 1912).
[2] *Loc. cit.*

the pivot. R. H. Richards decides [1] in favour of placing the pivot above, assigning as a reason that the work of crushing is in that case more evenly distributed, whereas if the pivot is below, most of the power is used near

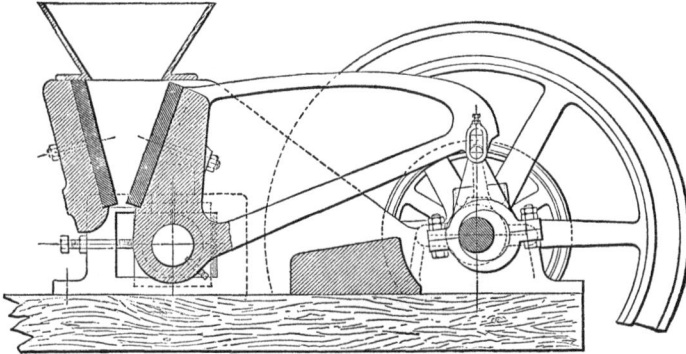

Fig. 57.—Dodge Crusher.

the throat or discharge aperture in recrushing particles which have already been reduced sufficiently (*i.e.*, " choke " crushing).

The *Gates Crusher* is shown in Fig. 58, and consists of a vertical spindle

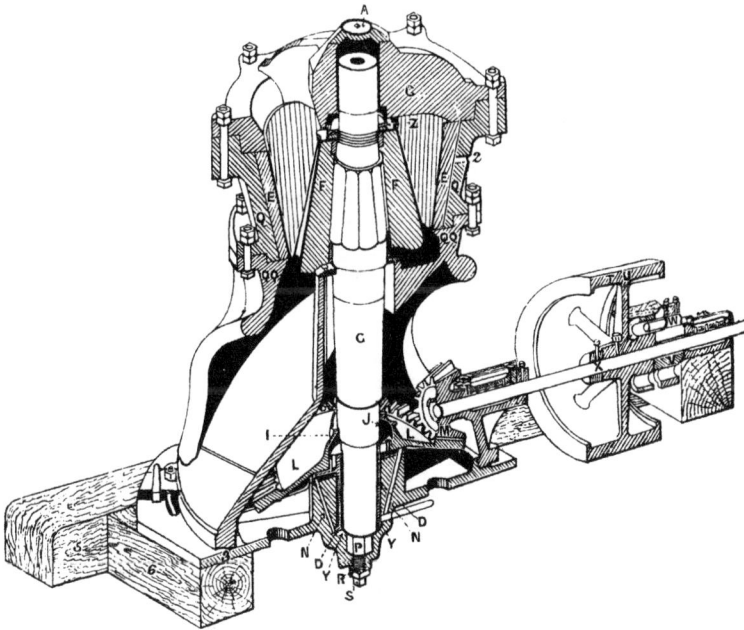

Fig. 58.—Gates' Crusher.

of forged steel, G, revolved by the bevel wheel, L, and a bevel pinion. The spindle is set eccentrically in the bevel wheel, and has a gyratory motion

---

[1] R. H. Richards, *Ore Dressing*, vol. i., p. 35.

when rock is being crushed. At the top of the shaft is a breaking head, F, and the shell surrounding this is lined with twelve concave pieces, E. These form the crushing faces, and were formerly of chilled cast iron, but now consist of manganese steel or other special hard cast steel. The faces wear rapidly and are frequently renewed.

The *Comet Crusher* is a similar machine which has come into use more recently. In the later form of this machine, the "Adjustable Comet," the size of the product can be varied by turning a hand-wheel. *The Hadfield ore breaker* is also widely used.

Gyratory crushers are of far greater capacity than machines with reciprocating motion, and have accordingly come into use where a large output is required. The largest size Comet crusher is said to have a capacity of from 100 to 200 tons of rock per hour.

On the Rand, gyratory breakers are used as "fine breakers," taking the product of the coarse breaker (see above, p. 156), after it has been passed through a trommel and sorted, and breaking it to a maximum of about $1\frac{3}{4}$ inches. The cost of fine breaking is given as follows :—

Maintenance, ...	0·2d. per ton.	
Labour, ...	·25d.	,,
Power, ...	·15d.	,,
Capital charges, ...	·12d.	,,
Total, ...	0·72d.	,,

The power required is $\frac{3}{4}$ to 1 H.P. per ton per hour.[1]

Position and Use of Rock-Breakers.—The aperture of a rock-breaker is placed on a level with the floor, so that the ore can be dumped down by the side and shovelled into the jaws, or the ore is fed direct into it through a shoot from the ore bins. It is now becoming customary to place the rock-breaker in a separate building distinct from the battery house.

A grizzly or screen of steel or iron bars, set 1 or 2 inches apart, is often employed to separate the fine material, which is passed straight to the stamps. Washing trommels are in use on the Rand. When the stamp battery is used only to crush the ore, which is subsequently treated in other machines, it is of great advantage to separate the fine product of the rock-breaker by sieving, instead of passing the whole through the stamps. This arrangement increases the output and prevents unnecessary sliming of the ore, thus greatly reducing the loss of sulphides when an attempt is made to save these by concentration.

The product of the rock-breaker is mixed as thoroughly as possible with the original fines, and led by means of a shoot direct to the automatic stamp feeders.

The efficiency and economy in crushing, attained by rock breakers, are so fully recognised that efforts have been made to use them to reduce gold quartz to a very small size before feeding it into the stamp batteries, with a view to increase the output. For fine crushing, multiple jaw crushers, on the same principle as the Blake, have been constructed, but have not passed into general use ; the use of a pair of rolls between the rock-breaker and the stamp battery has also been advocated. The usual size, however,

---

[1] Schmitt, *Rand Metallurgical Practice*, vol. ii., p. 51.

to which ore is reduced before it is fed into the stamp mill is $1\frac{1}{2}$ or 2 inches maximum. Any finer preliminary crushing is unusual, and, according to the researches of W. A. Caldecott and other members of the Mines' Trial Committee, S. Africa, the most economical size is about $1\frac{3}{4}$ inches.

The Stamp Battery.—California "gravitation" stamps are in general use at the present day for crushing gold ores. A stamp is a heavy iron, or

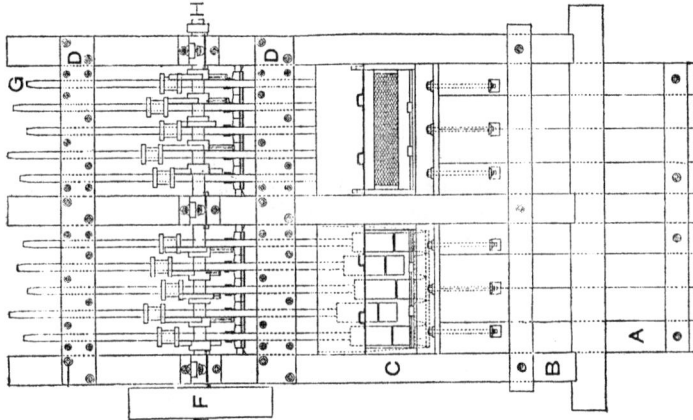

Fig. 60.—Stamp Battery—Front View.

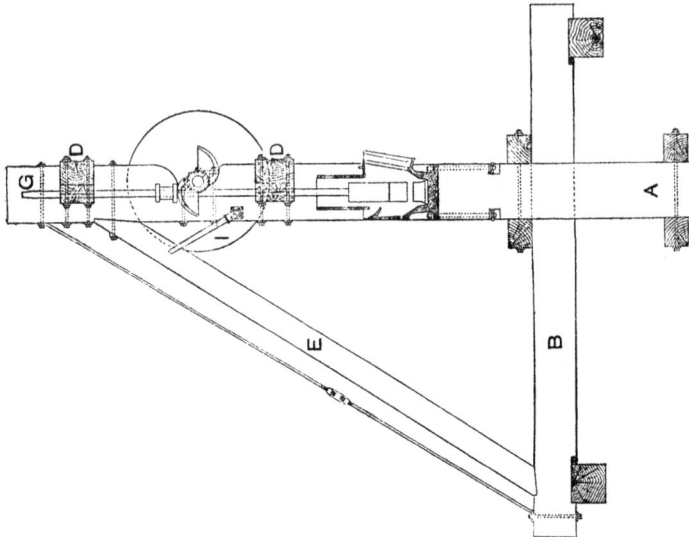

Fig. 59.—Stamp Battery—Side View.

iron and steel, pestle, raised by a cam keyed on to a horizontal revolving shaft, and let fall by its own weight. Stamps are ranged in line in groups of five stamps each, which have a mortar box in common. Fig. 59 represents the side view, and Fig. 60 the front view, of a ten-stamp battery, with the amalgamating tables removed to show the foundation timbers or mortar-blocks, A.[1] Figs. 61, 62, and 63[2] (see folder) give the details of a modern Transvaal ten-stamp battery (Simmer Deep).

The foundations are of the highest importance. If they are badly made through carelessness or false economy, the efficiency of the battery is greatly decreased, and it soon shakes itself to pieces. The blow of a stamp is partly employed in crushing the ore, and is partly expended in producing a concussion or jar acting on the framework and foundations. The amount of energy used up in the latter way depends largely on construction. In preparing the ground for the foundations, the earth is removed until bed-rock is reached if possible, and the latter is then carefully smoothed and covered with a layer of concrete. The wooden mortar-blocks of from 6 to 18 feet long are placed upright in this trench, and the space round filled up with sand, or, as in the Transvaal, solid masonry is built round the blocks. Concrete mortar blocks are now often used instead of wooden ones (see Figs. 61, 62, and 63). Heavy iron anvils or mortar blocks are also used on concrete foundations. These were introduced at the Village Deep Mill in 1903, but have been omitted in later mills. The *framework* is made of wood, iron, or steel. It consists of the massive cross sills, B (Fig. 59), on which rest the battery- or king-posts, C, and the braces, E. The cross sills rest on horizontal mud-sills, placed parallel to the line of stamps. The posts are held together by the guide- or tie-timbers, D. The mud-sills are shown below the cross sills. Frames of a number of different designs are in use. In recent mills on the Rand the king-posts and also the foundations generally have been made of reinforced concrete.

The Mortar.—The mortars are made of cast iron, but differ in shape according to the nature of the ore and the corresponding modifications made in the course of treatment. They weigh from $1\frac{1}{2}$ to $5\frac{1}{2}$ tons, being especially thick at the bottom where there is the greatest stress. An ordinary mortar is about 4 feet 7 inches long, 50 inches high, and 12 inches wide on the inside at the level at which the dies are set. The bottom is from 3 to 11 inches in thickness, and has a heavy flange cast on it by which it is bolted to the mortar blocks. For modern heavy stamps, mortars are larger and heavier,[3] the bottom being 15 inches thick. The mortar-blocks are tarred over, all cracks in them having been filled with sulphur, and are then covered with three thicknesses of blanket, carefully coated with tar on both sides. The mortar is placed on these blankets and securely bolted down. This arrangement lessens the chance of the mortar working loose, the jar being diminished. A sheet of rubber, $\frac{1}{4}$ or $\frac{3}{8}$ inch thick, is used instead of the blankets in many modern mills. Figs. 64 and 65 represent sectional elevations of the two chief types of mortars, Fig. 64 showing a mortar intended to be supplied with an inside amalgamated copper lining plate, *e*, on the screen side only, and Fig. 65 a mortar designed to have copper plates, *e, e*, placed both at the front and back. In both figures, *b* is the feed-opening through which the ore is introduced into the mortar; *c* is the bed on which the die is placed; *d* is the screen-opening. The chief difference between them is in the feeding arrangement; in the latter case the back plate is put in a recess, and is protected from the falling rock fed into the battery. Back plates are now

---

[1] For full details of construction of stamp-mills, see Richards, *Ore Dressing*, vol. i., pp. 144-230; also, *Rand Metallurgical Practice*, vols. i. and ii., from which many of the details in the preceding and following sections are taken.

[2] Schmitt, *Rand Metallurgical Practice*, vol. ii., p. 136.

[3] See *Rand Metallurgical Practice*—Smart, vol. i., p. 50; Schmitt, vol. ii., p. 80.

CAM SHAFT PLATFORM

FEEDER FLOOR

MILL FLOOR

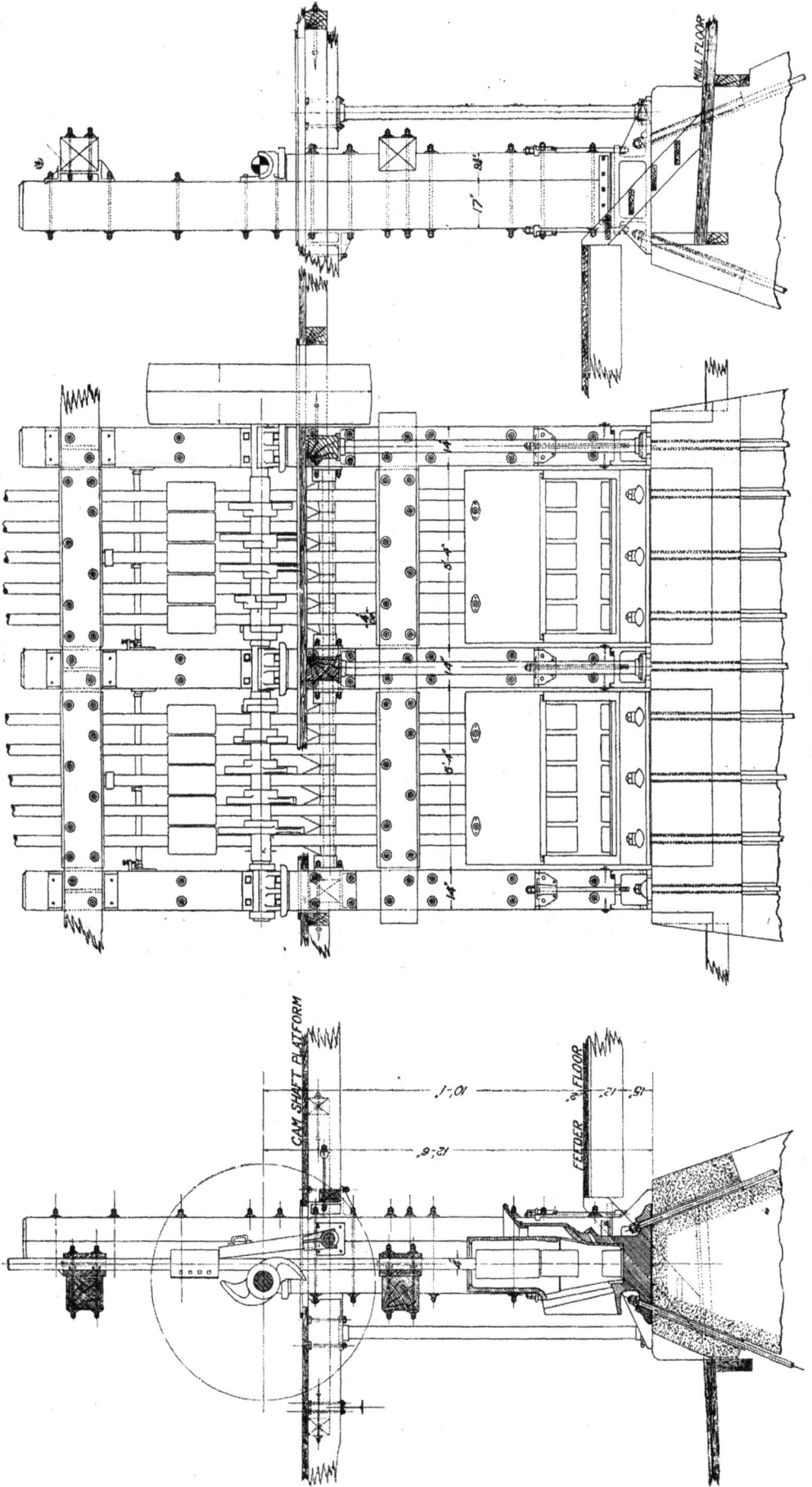

Figs. 61, 62 and 63.—Details of a Modern Transvaal 10-Stamp Battery (Simmer Deep).

not often used. The plates catch the coarse gold inside the mortar when the pulp is flung against them. Cast-iron or steel lining plates and false bottoms take up the wear inside the mortar and are renewed when worn out. They last a few months.

A section of the mortar used at the Croesus mill, in the Transvaal, is

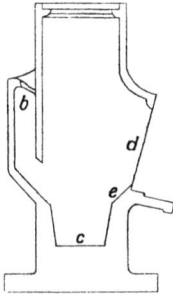

Fig. 64.—Mortar for Inside Amalgamation Plate on Screen Side.

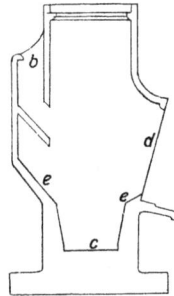

Fig. 65.—Mortar for Inside Plates on Both Sides.

shown in Fig. 66. a is a cast-steel lining plate with slots or recesses in it for collecting the amalgam; b is the feed opening; c the wooden blocks for carrying e, the copper plate; d is the screen opening; and f a steel plate. In more modern mortars in the Transvaal the back is vertical inside for about 16 inches (Smart).

The width of the mortar varies from 10 to 14 inches at the level of the

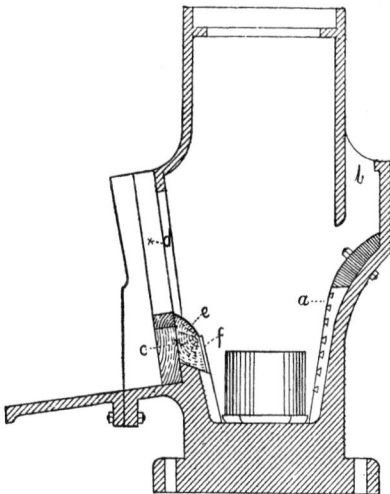

Fig. 66.—Mortar at Croesus Mill, Transvaal.

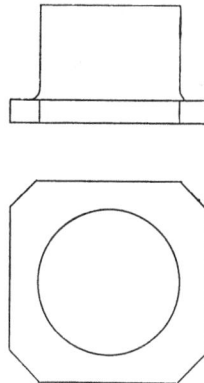

Fig. 67.—Die, Plan and Elevation.

bottom of the screens. As has been already mentioned, narrow mortars are best fitted for rapid discharge, but, if hard flinty ores are to be crushed, a narrow mortar causes frequent breakage of the screens, unless the discharge is deep—i.e., unless the bottom of the screens is a considerable distance above the surface of the dies. By this latter arrangement the output is

11

reduced, since, the nearer the screens are to the dies, the more rapid is the discharge.

The *depth or height of the discharge* is the distance, measured vertically, from the top of the die to the lower edge of the screen opening. It varies from zero to about 10 inches in modern mills, according to the nature of the ore and the general scheme of working. Adjustable battery-screens keep this height constant, in spite of the wearing of the dies. The screen-frame is supported on a wooden chuck-block, which is easily removable, and to which the copper plate is bolted. When the dies wear down, the chuck-block is replaced by one of less height, to which a suitable plate has already been fixed. The height of the die may also be regulated by periodically introducing liners below the die as it wears down. Smart points out[1] that, even though the surface of the die be level with the lower edge of the screen opening, there is actually about $1\frac{1}{2}$ inches height of discharge, due to the partially crushed ore banking against the lower edge of the screen, and in consequence there is nothing to be gained by employing a less height of discharge than, say, $1\frac{1}{2}$ inches. The greater the height of discharge the longer the ore is retained in the mortar and the finer the product, and, therefore, where coarse crushing is used as a preliminary to fine grinding in tube mills or other machines, the height of discharge is kept at about 2 inches. Where inside amalgamation is practised a greater height of discharge is maintained. The extreme height of discharge of 13 to 15 inches was formerly used in the Gilpin County Mills, Colorado.[2]

The *splash-box*, not shown in the figures, and now often omitted, is bolted to the outside of the mortar just below the screens. It is rectangular, consists of wood or iron, and is of the same length as the mortar. It receives the pulp as it passes through the screens, and distributes it evenly over the amalgamating tables by a number of spouts, usually three. Instead of the splash-box, a splash-board or a canvas shield is now almost universally employed. The lip of the mortar projects some inches, as shown in Fig. 66, and to it is bolted a cast-iron *apron* about 14 inches long, which often carries an amalgamated copper plate, the *apron-plate*. The pulp is dashed through the screen against the splash-board and falls thence on the apron-plate, which is often provided with a curved lip to prevent the splashes from reaching the amalgamated tables. The old form of mortar had its upper part, or housing, of wood, but, as mercury is lost through the smallest aperture, and it was difficult to make these wooden housings quite tight, mortars are now cast in one piece, including the housings. The roof of the mortar is made of 2-inch planking, through which holes are cut to admit the stems of the stamps and the water pipes.

When the mortar is in place, the *dies* are put into it, a layer of sand being often introduced first. The dies consist of two parts, the *footplate* or *base* and the *die* proper or *body*. Fig. 67 shows, in plan and elevation, one of the many forms of dies in use ; here the base is almost square, so as to fit the mortar ; it is 1 or 2 inches thick, and $9\frac{1}{2}$ to 10 inches square. The body is cylindrical, 5 or 6 inches high, and of the same diameter as the shoe. On the Rand it is generally 9 inches, or with stamps above 1,500 lbs. $9\frac{1}{4}$ inches in diameter. Shoes and dies are now generally made of forged steel, the dies being kept a little softer than the shoes. They last longer than the iron

---

[1] Smart, *Rand Metallurgical Practice*, vol. i., p. 71.
[2] T. A. Rickard, *Stamp Milling of Gold Ores*, 1897, 2nd edition, p. 16.

shoes and dies formerly in use. Chrome steel and manganese steel have also been used for shoes and dies. Sometimes steel shoes and iron dies are used. The *wear* of iron shoes and dies was formerly stated to be about 2 or 3 lbs. per ton of ore crushed in California. At the Robinson Mine, South Africa, the wear of steel shoes and dies is, according to Harland, 0·45 lb. per ton crushed for shoes and 0·30 lb. per ton for dies. According to I. Roskelley,[1] a 12-inch steel shoe weighing 225 lbs. is worn out on the Rand in about ninety days, during which from 450 to 500 tons of ore are crushed. The shoe is used until it is worn down close to the shank, and weighs only 35 or 40 lbs. when discarded. Forged steel dies,[2] with cylindrical bodies 6 inches high and weighing 120 lbs., last as long as the shoes, and weigh only about 40 lbs. when thrown out, if they have worn down evenly. These rates of wear correspond for shoes to about 0·40 lb., and for dies to about 0·17 lb. per ton of ore crushed. When the body is worn down to within from ½ inch to 1 inch of the base, the die is replaced. Dies wear more slowly than shoes, since they are protected by a layer of pulp, which is over an inch thick. The dies are all renewed together, as it is important that those in the same battery should be of equal height, otherwise one or more will become almost bare of ore, and a disastrous pounding result. If a die breaks, it is not replaced by a new one, but by one worn to the same extent as the others in the battery. Iron false-bottoms or chuck-blocks or sand packings are placed beneath partially-worn dies, so as to keep the height of the discharge constant.[3]

Dies require to be hardened and tempered very evenly, according to Roskelley,[4] to prevent uneven wear. A die worn down at one side causes a diminution of output, and may result in a broken stem. According to E. E. Aulsebrook,[5] the three centre dies in a mortar wear down at the back faster than at the front, on account of the size of the material which is fed at the back being larger than that which is crushed at the front. For this reason he recommends dies to be turned round once a month. He also states that dies often wear unevenly because the stamps are out of centre.

The *cam-shaft*, H, Fig. 60, is of wrought iron or, better, forged steel of the best quality, as it has to stand severe strains. Nickel steel has been tried. The shaft is usually about 6 inches in diameter. It is usual to have a separate cam-shaft for each five or ten stamps, which have thus a separate driving pulley. The advantage of this arrangement is that repairs can be done to one or more stamps without necessitating the stoppage of the whole mill, as used to be the case when there was only one cam-shaft. The cam-shaft is placed at a distance of from 5 to 10 inches from the stem-centre, and is 9 to 10 feet above the mortar bed. The bearings rest on supports often attached to the battery posts, generally on the discharge side. The cast-iron cam-shaft bearings and supports at the City Deep Mill are shown in Fig. 68.[6]

---

[1] Roskelley, *J. Chem. Met. and Mng. Soc. of S. Africa*, Feb. 1904, **4**, 410.
[2] *Loc. cit.*
[3] For the method of changing shoes and dies, see G. O. Smart, *Rand Metallurgical Practice*, vol. i., p. 65.
[4] Roskelley, *loc. cit.*
[5] Aulsebrook, *ibid.*, July, 1904, **5**, 12.
[6] Schmitt, *Rand Metallurgical Practice*, vol. ii., p. 119.

Fig. 68.—Cam-Shaft Bearing (City Deep).

The *cams* are made of cast steel. The double cam, illustrated in Figs. 69 and 70, is now in almost universal use, though single and treble forms have been employed. Sometimes cams are cast in two pieces which are bolted together, so that when one is worn out, it can be replaced without first removing the other cams on the shaft, but these sectional cams work loose, and are not much used. Cams are either right- or left-handed. Both are used in the same battery to equalise the cam thrust, the stamps being rotated in opposite directions. The shape of the cam face is the involute of a circle slightly modified at the end so as to stop the upward motion gradually. The radius of this circle is equal to the distance between the centres of the cam-shaft and the stem, which depends on the height to which the stamp is to be lifted, so that the curve of the cam varies with the drop.[1] A cam should last several years unless broken through being a faulty casting, or through carelessness in letting the stamp fall when hung up.

The old-fashioned keyed cam has now been replaced in many mills by the *Blanton cam*, which will take any position on the cam-shaft. To secure it two holes are bored in the cam-shaft, and two pins dropping into these hold fast a semi-circular tapering wedge or " bushing." The cam slips over the bushing and tightens itself in working. When it is necessary to take off a cam, a slight blow on the back edge with a hammer loosens it instantly. The " new " Blanton cam (see Fig. 71),[2] fastened by " rifling," is said to be found still more convenient than

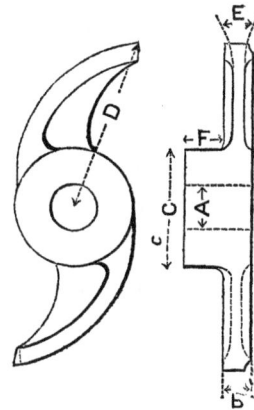
Figs. 69 and 70.—Cam, Front and Side Elevation.

the older form. The bushings are abolished, and the cam-shaft made with ten taper faces, so that its cross-section is like a ratchet-wheel with ten teeth instead of being circular. The bore of the cam has ten corresponding faces, and the cam slips on in any one of ten positions. It is then tightened sufficiently to hold it in position until put into operation, when it tightens itself further on the shaft in proportion to the work it has to do.

The cam-face works against the collar or *tappet*, shown in plan and section

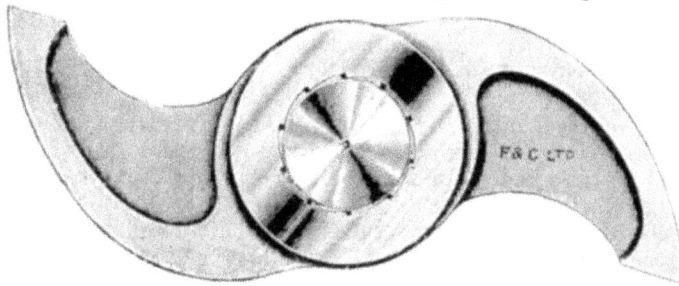
Fig. 71.—Cam with " Blanton " Rifling.

in Figs. 72 and 73, which is bored out to fit the stem of the stamp. The tappet is usually made of hard fine-grained cast steel, and is fitted with a wrought-iron gib, which is pressed against the stem by two or three keys behind it, thus binding the tappet firmly on the stem while, at the same time,

---

[1] For a full discussion of the cam curve, see H. Louis, *Handbook of Gold Milling*, 1894, pp. 479-491.
[2] Schmitt, *Rand Metallurgical Practice*, vol. ii., p. 113.

admitting of rapid adjustment to another position. With heavy stamps two or more gibs are used. The cam always strikes the tappet a heavy blow, and this could be diminished by a change in its design, such as that proposed in Behr's cam (see p. 171). The weight of the tappet should not exceed 12 to 15 per cent. of the total weight of the stamp.[1] Adjustment of the tappet on the stamp-stem is required every few days to allow for the wear of the shoes and dies.

The entire end surface of the tappet comes in contact with the cam-face, by which the stamp is raised and, at the same time, rotated. The effect of this is, that the shoe does not strike the ore in the mortar in exactly the same place twice in succession, and the wear of its face is made more uniform. The greater part of the revolution takes place during the raising of the stamp, but the latter does not quite cease to rotate as it falls, and a slight grinding action on the ore has been noticed by many observers. The amount of rotation varies with the fall, the extent to which the cam and tappet are greased, and the state of wear of their surfaces. A little grease is always added to reduce wear, but, if too much is present, the stamp does not revolve at all, while, according to J. H. Hammond, when the tappet is in the right condition, one revolution is effected in from four to eight blows, with a 6- or 8-inch drop. Other observers find the usual rate of rotation more rapid, and in Gilpin Co., Colorado, where the average drop is from 16 to 18 inches, the stamp makes from $1\frac{1}{4}$ to $1\frac{1}{2}$ revolutions at each blow, according to Rickard. Tappets last for four or five years; and, having both ends alike, they can be reversed when one end is worn out, and their worn and grooved faces can be planed down when necessary. Some millmen assert that tappets may be broken by the cam if keyed too tightly to the stem.

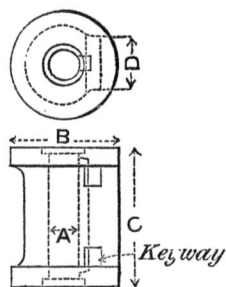

Figs. 72 and 73.—Tappet, Plan and Elevation.

As the cam-thrust is not applied at the centre of the stamp, there is always a considerable side pressure, which greatly increases the friction in the guides and wears out the latter, besides causing a loss of power. Moreover, another result of this is that the stamp tends to be inclined (not vertical) when it is released, and so the blow on the die is given slightly to one side— i.e., the side of the die on which the cam works. Consequently, there is a tendency for this side to wear down more quickly than the other. To obviate these disadvantages, cams have been introduced at Johannesburg with a wide hub, and the two blades set one at each end of the hub, so that they work on opposite sides of the stamp and cause it to revolve in different directions at each successive uplift.

The *pulley* on the cam-shaft (F, Fig. 60) is made of wood on cast-iron or cast-steel centres, which are keyed to the cam-shaft; if iron alone were used, the rapid succession of jars, caused by the dropping of the stamps, would soon cause the material to crystallise and break. A tightener pulley on the belt driving the cam-shaft is often used, by which the stamps can be put in motion or stopped without interfering with the driving power.

The stamp itself consists of three parts, the stem, the head or boss, and the shoe. The *stem* (G, Fig. 60) is from 12 to 18 feet long, and from 3 to

---

[1] Schmitt, *Rand Metallurgical Practice*, vol. ii., pp. 103 and 104.

$4\frac{1}{4}$ inches in diameter ; it is made of wrought iron or steel, or steel in the middle portion with wrought-iron ends, and has both ends tapered for a length of 6 or 8 inches to fit the heads, so that, if one end is broken off, the stem can be inverted and the other end used. The *head* and shoe are made of equal diameter—viz., about 8 or 9 inches. The head is of cast iron, or in more recent practice of hard fine-grained cast steel, about 15 to 24 inches long (or in later practice on the Rand 48 inches long, when the stem is reduced to 12 feet), and has a tapered socket at each end, the upper one for the stem and the lower for the tapered shank of the shoe. When these are driven into their respective sockets, into which a few strips of wood are inserted to keep the two metal surfaces from touching each other, a few blows by the stamp bind them securely together, no other fastening being necessary. Slots are provided at the base of the two sockets, through which wedges may be driven to force out the shoe or stem when necessary. Roskelley [1]

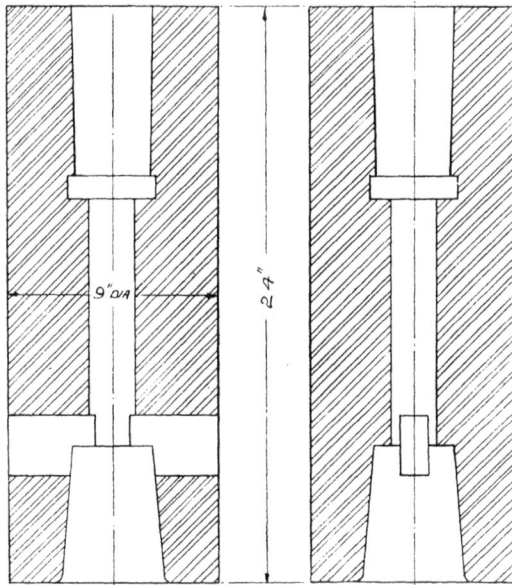

Fig. 74.—Improved Stamp Head with Axial Opening for removal of broken stems.

advocates a hole about $2\frac{1}{2}$ inches in diameter drilled through the axis of the head to facilitate the removal of broken stems. G. O. Smart has designed a head (Fig. 74) [2] in which the difficulty is overcome by having a hole drilled or cast in the head through the entire length, and with the aid of a stout rod and a hydraulic press the broken-off shank (or stem) is then removed with the greatest ease, instead of it being necessary to resort to dynamite, as was formerly sometimes the case. The head lasts several years, being rarely ruptured.

The *shoe* (Fig. 75) [3] consists of two parts, the shank, which fits into the

[1] Roskelley, *loc. cit.*
[2] Smart, *Rand Metallurgical Practice*, vol. i., p. 56.
[3] Schmitt, *Rand Metallurgical Practice*, vol. ii., p. 96.

head, and the shoe proper or butt. It is made of hard forged steel, with the shank made softer than the butt in the tempering. The diameter of the shank is about half that of the butt. The shoe is replaced when the butt, which is from 9 to 18 inches in length when new, has been worn down to about 1 inch in length. To keep the total weight of the stamp constant, several sizes of heads are sometimes used in one mill, the heavier heads taking partly-worn shoes. "Chuck-shoes" are inserted between heads and shoes with the same object. Shoes last about three months on the Rand, according to Roskelley, the wear being about 0·4 lb. per ton of ore milled. Compensating weights are affixed to the stem as the shoe wears.[1]

The relative weights of tappet, stem, head and shoe, which together make up the stamp, vary considerably. There is an advantage in increasing the diameter of the stem, as one of small diameter tends to spring and bend from the blow of the cam, or when the stamp falls, and to wear the guides rapidly. On the other hand, there is an advantage in diminishing the length of the stem, as, owing to its elasticity, the effect of the blow when the stamp falls is partly expended in compressing the stem momentarily. For the same reason it is advantageous to reduce the weight of the tappet and to increase that of the head. The stem weighs from 250 to 600 lbs., the tappet from 80 to 150 lbs., the head from 175 to 370 lbs., and the shoe from 100 to 250 lbs. The total weight of the stamp is usually from 650 to 1,350 lbs., but is sometimes as low as 450 lbs., and, for prospecting purposes, the weight is only from 100 to 300 lbs. The heaviest stamps yet made for the Rand are 2,000 lbs. in weight. The introduction of the long head on the Rand raised its weight to 900 lbs., or 44 per cent. of the total weight, while the stem fell from 46 to 27 per cent.[2]

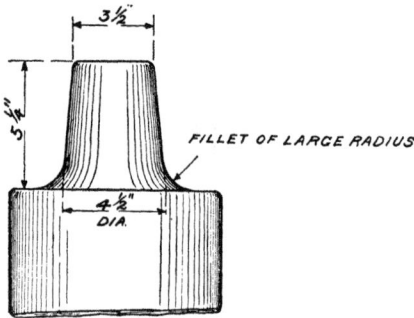

Fig. 75.—Shoe for Heavy Stamp.

The stamp stems are guided in boxes bolted to the wooden guide-timbers, which also serve to hold the battery posts together. There are two of these *guide-timbers* (D, Fig. 60), one within 2 or 3 feet of the top of the battery posts, and the other about 6 or 7 feet lower. The depth of each guide is about 15 inches, and the stems are fitted closely to the guides, metal guides being used occasionally (*e.g.*, the Ralok guide[3]), although wood is much more general. Wooden stem-guides wear the stems more rapidly than metal ones, in spite of a higher expenditure on lubricants. The guide-beams are sometimes pierced with large square holes in which bushes of wood, with the grain parallel to the length of the stamp, are placed fitting the stem exactly. In this way, the guide-beams themselves are preserved from wearing out. Sectional guides, consisting of a series of iron keys enclosing wooden bushings, are also used. In this case each stem has a guide to itself and the bushings can be renewed by hanging-up the one stamp without stopping the other stamps in the battery. The Ralok guide has this advantage also.

[1] Smart, *Rand Metallurgical Practice*, vol. i., p. 65.   [2] Schmitt, *ibid.*, vol. ii., p. 90.
[3] Schmitt, *ibid.*, vol. ii., p. 101.

Each stamp is provided with a *finger-bar* or *jack* (I, Fig. 59) made of wrought iron, or wood protected with iron, and carried on a separate *jack-shaft*, which is supported on cast-iron brackets attached to the king-post. The jack is for the purpose of raising the stamp and hanging it up out of reach of the cam. When this is to be done, a strip of wood or wrought iron (the *cam-stick*), an inch or more thick, is laid with one hand on the cam as it rises, and the stamp is thus raised an inch higher than usual, so that the jack can be slipped in under the tappet with the other hand. The stamp is thus suspended above the cam and can be repaired without stopping the others, while it can only be released in a manner similar to that in which it was hung up. Above the stamps there is a double rail, on which is a tackle block carriage (*crawl, crab*) ; by this the stamp, etc., can be lifted up for repairs. When the stamps are to be set up, the head is put on the die and the stem dropped into it, strips of wood or other packing being usually put into the head socket and the stem dropped into that. The stamp is then raised and dropped into the shoe, the shank of which is surrounded by strips of wood for packing. As already stated, the parts are soon wedged firmly together by raising and letting fall the stamp a few times.

The height of the " drop " of the stamps varies from 4 to 18 inches, and the number of drops per minute varies from 30 to over 100. These depend on one another to a great extent, an increase in the height of the drop being necessarily accompanied by a diminution in the number of drops per minute. With a drop of $8\frac{1}{2}$ inches, about 95 blows can be obtained, the tappet then just having time to fall after leaving one face of the cam, before the other begins to raise it. The actual height of drop, as pointed out by Smart, is the " set " height less the thickness of the ore on the dies after the stamp has fallen. As, within certain limits and under certain conditions, an increase of speed results in an increase of yield of pulverised ore, efforts have been made to raise the number of blows per minute. The subject will be returned to when the conditions for successful amalgamation are discussed.

D. B. Morison discusses the question of height of drop from the point of view of speed of crushing, in a valuable paper which should be consulted.[1] The time occupied in seconds in the fall of the stamp, if all friction, including the resistance of the air, is left out of account, is $\sqrt{\dfrac{2\,H}{G}}$, where H = height in feet, and G $= 32 \cdot 2$. For a fall of 8 inches this amounts to about 0·2 second, but owing to the friction between the stamp and the guides and the resistance due to the water, the actual time taken, according to Morison's experiments,[2] is about 0·225 second (see Fig. 76). The velocity at the time of impact and the force of the blow are correspondingly diminished, the latter being 17 per cent. less than it would be if the work were done without friction *in vacuo*, a mean result of twenty-four experiments. The time required in raising the stamp is somewhat greater, owing to the imperfect method necessarily employed. The shape of the cam causes the stamp to start upwards at a certain velocity immediately the cam meets the tappet, and to maintain the same velocity until near the end of the stroke. In a properly constructed cam, designed to give a drop of 8 inches, the vertical component of the velocity of the cam, is for 100 drops per minute, about 2 feet per second. If the

---

[1] "Gravitation Stamp Mills for Quartz Crushing," by D. B. Morison, *Trans. North-East Coast Inst. of Engineers and Shipbuilders*, 1896-7, Session xiii.
[2] Morison, *loc. cit.*

cam were suddenly removed from the tappet when moving at this velocity, the stamp would continue to rise against gravity for about $\frac{3}{4}$ inch, neglecting friction.

The inherent defect of the cam is that it strikes the tappet a tremendous

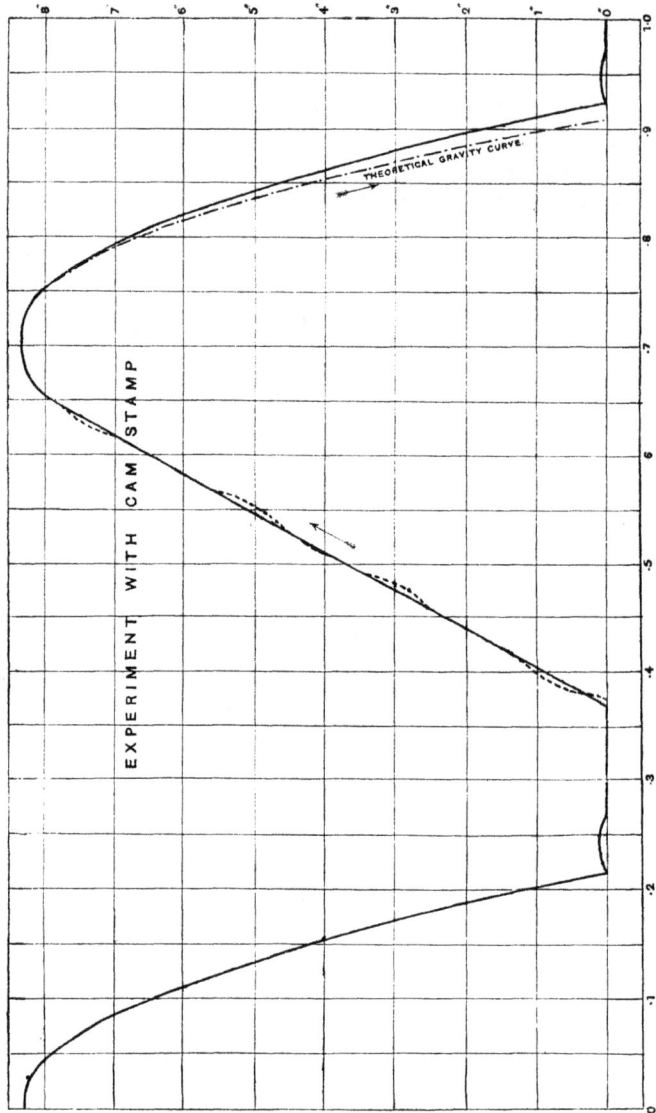

Fig. 76.—Curve illustrating Vertical Movements of Stamp.

blow in the effort to lift the stamp at the full velocity from the start, and it is this blow which is the cause of much of the intense noise and vibration felt in every stamp mill. In spite of the power of the blow, which tries the cam-shaft severely, the dotted curve in Fig. 76 shows that at first the move-ment of the cam is checked a little, probably in part by distortion of the cam, and that it rapidly recovers itself, and by elastic distortion in the opposite direction lifts the stamp faster than the normal rate, and these alterations succeed each other during the whole lift, after which the cam presumably recovers itself in the time that elapses before it strikes the next blow.

Fig. 76 shows a curve traced out by a pencil attached to a 900-lb. Sandycroft stamp, the ordinates representing vertical movement of the stamp and the abscissæ being time in tenths of a second.

Space-time diagrams have also been taken in actual practice on the Rand,[1] and show that at 100 drops per minute the total time of each cycle of 0·6 second is divided into rise 0·25 second, fall 0·22 second, rest 0·13 second. The Rand engineers continue, "the stamp, owing to the shock on the cam striking the tappet, bounds off at a greater velocity than that due to the speed of the cam, but the energy so imparted is not sufficient for the total rise ; hence the cam overhauls the stamp and imparts a fresh blow, this time less violent, but still resulting in a bound on the part of the stamp. This is repeated several times, until the full height is attained." The height of the rebound of the stamp on striking the ore varies with the thickness of the layer of ore on the die. Behr's cam, which is modified in shape so as to begin to lift the stamp gradually and afterwards to increase the speed of lift, reduces the shock and wear, but also reduces the number of blows per minute.[2]

Screens.—The screens are set in wooden or iron frames, which now usually slide in grooves cast in the mortar, and are keyed to it, but were formerly fitted into recesses and bolted. The joints are made tight by blanketing. Screens are made either of steel or brass wire-cloth, or of Russia sheet-iron or steel, or tin-plate, in which holes are punched, either round or consisting of long slots (from $\frac{1}{4}$ to $\frac{1}{2}$ inch long) ranged parallel or inclined to each other. The width or diameter of the holes ranges from about $\frac{1}{80}$ to $\frac{1}{12}$ inch or more, according to the nature of the ore and the method employed in its treatment ; the usual size is from about $\frac{1}{24}$ to $\frac{1}{40}$ inch. The relative advantages of wire-cloth and sheet-iron are not yet beyond dispute, and vary with the nature of the ore. Slots appear to be better suited for discharge than meshes, but, on the other hand, there is a great loss of discharge area in the use of punched iron. Thus a wire mesh screen, containing 18 holes to the linear inch, has 324 holes to the square inch, while a round-punched sheet-iron screen has only 140 holes of the same size per square inch. Wirewoven screening is universally employed on the Rand (Smart).

It must be borne in mind that a 20-mesh screen (20 holes to the linear inch) has apertures of only about 0·030 inch diameter, the exact size depending on the diameter or gauge of the wire. The "diameter" of the square apertures, the thickness of wire, the percentage of discharge area, and the number of holes per linear or square inch are all factors required to be known by the millman using any particular screen.[3]

Although iron is the material usually employed for screens, it is often preferable to use copper, as pyritic ores, if kept for any length of time after being mined, soon become oxidised and acidified, and the ferrous sulphate thus formed corrodes iron rapidly, whilst the water used is often more or less acid if it comes from mines. Copper is not attacked in the same way.

---

[1] *Rand Metallurgical Practice,* vol. ii., p. 107. [2] *Loc. cit.*
[3] For details of these, see G. T. Holloway, *Bull. Inst. Min. and Met.,* Feb. 1905 ; also Report of Committee, *J. Chem. Met. and Mng. Soc. of S. Africa,* June, 1906.

T. A. Rickard has shown [1] that the life of screens in the mills at Blackhawk, Colorado, diminishes as the creek is descended, the water becoming more and more impure. At the South Clunes United Mill, Victoria, iron-punched gratings lasted less than a week, but, on introducing copper plates containing 100 holes per square inch, the life of the screen was increased to a month, and 275 tons of ore were passed through it. At Blackhawk the iron screens last while from 80 to 430 tons are passed through them, according to the position of the mill. At Grass Valley the average is 200 tons, at Bendigo 134 tons, and at Otago, N.Z., only 40 tons. In California the brass-wire screens last from 10 to 14 days, corresponding to a passage of 120 to 140 tons, and the Russia-plate lasts from 15 to 40 days, the average being 30 days, corresponding to the passage of about 330 tons. The rate of wear of the screens depends greatly on their position, being more rapid with a shallow than a deep discharge, and more rapid in a narrow than a wide mortar. Pieces of iron or wood in the ore may cause the screen to break, and these should, consequently, be removed from the ore as far as possible. The battery is hung up now and then, so that a thorough inspection of all the screens may be made, and those that are broken replaced.

The screens were formerly set vertical, but they are now placed at an angle which varies somewhat but is never far from 10°, and this has been found to facilitate discharge. Smart, however, observes [2] that, with ore containing much wood or other débris, a vertical screen is advantageous in keeping the apertures from choking. In wet stamping, screens are usually placed on one side of the mortar only—viz., that opposite the feeding side. In cases where the discharge is required to be as rapid as possible, the screen area is sometimes increased, and double discharge (front and back) mortars have been made, but have not been used much, except for dry crushing, and screens at the ends of the mortars are used at Harriettville, Victoria (Rickard). Caldecott, however, has shown [3] that double discharge mortars do not give a higher stamp duty, and result only in an increase in the amount of water required. The area of the screen is usually from 3 to 4 square feet per battery, the height from the bottom to the top of the screen being from 8 to 10 inches. On the Rand, the screen openings are usually 52 inches long and 10 inches deep. According to H. T. Pitt [4] the output is increased by about 5 per cent. by making the screens 22 inches deep instead of 10 inches. Smart,[5] however, states that the portion of the screen area principally utilised is the lower 3 inches, the upper part of the screen receiving only a small amount of the splashed pulp. He considers that the ordinary discharge area is already ample in amount. Opinions are thus seen to differ as to the necessary amount of screen area to be used. It has been contended that the capacity of a battery is really limited, not by its crushing, but by its discharging power. Thus, by a number of experiments conducted some years ago at the Metacom Mill, California, it was shown that when crushed pulp instead of the unbroken ore was fed into the mortar, the rate of discharge was not increased. This was taken as a convincing proof that the discharge area is usually far too small. Nevertheless, double discharge is hardly ever used

[1] Rickard, *Eng. and Mng. J.*, Sept. 3, 1892, p. 224.
[2] Smart, *Rand Metallurgical Practice*, vol. i., p. 74.
[3] Caldecott, *Trans. Inst. Mng. and Met*, 1910, **19**, 57.
[4] Pitt, *J. Chem. Met. and Mng. Soc. of S. Africa*, May, 1904, **4**, 408.
[5] Smart, *Rand Metallurgical Practice*, vol. i., p. 74.

for wet crushing mills, the various objections that are made to it being summarised as follows :—

1. Inconvenience is caused in the arrangement of the copper plates, both inside and outside the battery.

2. So great a quantity of battery water must be used, that the pulp is too thin for efficient amalgamation on the plates.

3. The ore does not stay long enough in the battery to be effectively amalgamated, or, according to others, the time of stay and the amount of output is unaltered.

Order of Fall of Stamps.—The order in which the stamps drop is of considerable importance. If they were let fall in succession from one end to the other of the mortar, the pulp would be driven before them, so that the stamp which fell last would have its die covered by too deep a cushion of ore, while that at the other end would be almost bare. The result to be obtained is to keep the ore equally distributed through the mortar, so that each stamp shall do the same amount of crushing, although it is inevitable that the middle stamps should be more efficient than the end ones in discharging the ore. The order most favoured in California is 1, 4, 2, 5, 3, and that on the Rand is 1, 3, 5, 2, 4, whilst the orders 1, 5, 2, 4, 3 and 1, 5, 3, 2, 4 are also often used. Several other orders have their advocates, and are probably little inferior to the above for the particular ores on which they are employed. Since the end stamps are of less efficiency than the others, it has been argued that a larger number of stamps in one mortar would be advantageous, and at Clausthal, in the Hartz Mountains, there are usually from nine to eleven stamps in a battery,[1] placed close together, space being greatly economised in this way. Long and wide experience has, however, proved that the best number is five stamps in one mortar.

Feeding.—Ore is fed into the battery either by hand or by automatic machines. It is often asserted that really intelligent hand-feeding is better than the automatic method, since the stamps are not all equally efficient. The feeder on small mills is often expected to break down the big pieces of ore with a sledge hammer, a rock-breaker not being used, but this method of working may be safely set down as irrational and uneconomical, and the result usually is that large and small pieces go into the mortar together. In the United States and in the Transvaal self-feeders are universally employed in modern mills. The art of feeding consists in keeping the depth of pulp on the dies constant throughout the battery, as long as the work is carried on. This is much better done by automatic machinery than by hand, and it was found that by the introduction of the former in California the capacity of the stamps was increased by 15 to 20 per cent., while the wear of shoes and dies was decreased by 25 per cent., and that of the screens by 50 per cent. It is not difficult to discern the cause of the advantages, for, if the dies are insufficiently covered with ore, less crushing is done, while a greater concussion must be taken up by the stamp and by the die, mortar, etc. If the die is quite bare this concussion is so great that the stem may be bent or broken, and the shoe and die battered. On the other hand, if the ore is too deep in the mortar, there is so thick a cushion that much of the force is taken up in compression without crushing it ; whilst, besides the reduction of output, the head, under these circumstances, sometimes becomes detached from the stem, which is broken or battered by the next blow. The maximum

---

[1] Meinecke, *Proc. Inst. Civil Eng.*, 9th Feb. 1892,

capacity is obtained with " low feeding," the depth of pulp on the dies being about 2 inches or less.

One advantage of even feeding is that a larger proportion of gold is caught, owing to the more regular and even flow of the pulp over the plates, the danger of scouring being diminished.

There are many automatic feeders of different designs. *Hendy's Challenge Feeder*, shown in Fig. 77, is a typical machine, and is more widely used than the others. It is constructed so that the tray, A, below the sheet-iron hopper, B, is revolved in a horizontal plane by means of a gear-wheel placed below it. The ore is fed into the hopper and the amount passing to the tray, A, is regulated by a sliding door. The gear-wheel is set in motion by a friction grip, D, placed on the outside of the frame, and actuated through the lever, E, by the bumper-rod, G, against which the feed tappet strikes. This is a

Fig. 77.—Challenge Feeder.

collar clamped to the centre stamp-stem.[1] At each partial rotation a given quantity of ore is scraped off by the stationary wings or side plates, H, resting on the tray, A. This amount of ore is regulated by the condition of the mortar. The machine is especially adapted for very wet or sticky ores. In general one feeder is sufficient for a battery of five stamps (delivering up to 75 tons per day), more ore being fed to the middle stamps, where the most work is done, than to the end ones. The feeders are usually suspended from above instead of being supported from below.

---

[1] For G. O. Smart's improvement on the ordinary method of actuating the feeder, see *J. Chem. Met. and Mny. Soc. of S. Africa*, 1906-7, **7**, 133.

Water Supply.—The water is usually supplied to the stamps by horizontal pipes passing just above the top of the housing of the mortar-box, or through the front of the box above the head, with one or two feed pipes or, better, five to each battery. The Rand engineers have added [1] five 1-inch holes along the back of the mortar-box level with the feed chute, with the nozzles of the water pipes pointing so as to discharge the water against the faces of the dies from the back, in order to clean the dies of fine material and increase the output. In front of the battery there is sometimes another pipe of about half the size, to supply water to the tables to help carry off the pulp. The feed pipes are often pierced with pin-holes, so that the water is supplied as a number of fine jets, in order to keep the stems, etc., clean. The water pipes require to be cleaned out frequently (Smart). The water should be supplied under a constant head, but even then requires continual adjustment in amount to follow the changes in the screen used.[2]

The amount of water used varies from $1\frac{1}{4}$ to 14 gallons per stamp per minute, the average in California being about $2\frac{1}{2}$ gallons, on the Rand formerly about $5\frac{1}{2}$ gallons, but in modern practice, with an output of 20 tons per day per stamp, 12 or 14 gallons, and in Colorado formerly only about $1\frac{3}{4}$ gallons. In California, with fast-running rapid-discharge batteries, the amount of water per ton of rock crushed varies from 1,000 to 2,400 gallons, the mean being about 1,700 gallons, while in Colorado the average amount is as high as 2,500 gallons. On the Rand the amount of water per ton of rock varies from 4 to 10 tons, the average being 7 tons or 1,400 gallons (Caldecott) for the comparatively fine crushing formerly in use, and about $4\frac{1}{2}$ to 5 tons in modern practice with coarse crushing. Roskelley [3] gives 6 to $7\frac{1}{2}$ tons for the Robinson Deep in experimental runs. Besides varying with the method of crushing adopted, the amount of water varies with the nature of the gangue, clayey ores requiring more, while the large quantity required by sulphide ores is due to the deep discharge necessitated by the difficulty of catching the gold, as well as to the high density of the pulverised material, which renders it more difficult to convey in suspension over the plates. As a rule, the more rapid the output, the less water per ton of ore is required in the battery. Coarse crushing requires less water in the battery, but, on the other hand, more has to be added on the plates. The amount of battery water per ton is increased by over 20 per cent. by the employment of double-discharge mortars. The amount of water to be added on the plates varies with their grade, as well as with the density and size of the particles of crushed ore. It should be only just enough to prevent the pulp from accumulating on the plates, as any excess over this tends to check amalgamation and to scour the plates. Richards gives [4] the average amount of water used in 21 mills as 2·77 gallons per stamp per minute, or 6·68 tons of water per ton of ore crushed. The average duty of a miner's inch in a gold stamp-mill is given by P. M. Randall [5] as 12 tons of quartz if the head under which the water is supplied is 4 inches, and 15·88 tons if the head is 7 inches. This gives the proportion of the volume of water to that of ore as 11·1 to 1. This may be compared with the proportion of between 7 and 10 to 1 in Siberian

[1] F. T. Pitt, *J. Chem. Met. and Mng. Soc. of S. Africa*, 1907, **8,** 373 ; Caldecott, *Trans. Inst. Mng. and Met.*, 1910, **19,** 57 ; *Rand Metallurgical Practice*, vol. ii., p. 83.

[2] Schmitt, *Rand Metallurgical Practice*, vol. ii., p. 130.

[3] Roskelley, *J. Chem. Met. and Mng. Soc. of S. Africa*, Feb. 1904, **4,** 412.

[4] Richards, *Ore Dressing*, vol. i., p. 222.

[5] Randall, Article on Practical Hydraulics in *Sixth Report Cal. State Min.*, 1886.

placer working, and as much as 30, or even 50 to 1 in hydraulic mining. It may be mentioned that a ton of 2,000 lbs. of quartz occupies about 13 cubic feet when unbroken, and about 20 cubic feet after having been broken up, so that in a lode a cubic yard contains about 2 tons, and in the tailing heap only about 1⅓ tons.

Amalgamated Plates.—For particulars as to amalgamated plates, mercury traps, etc., see Chap. VIII.

Mill Sampling.—It is usual to take samples of the pulp issuing from the screens, and also after it has passed over the amalgamated plates. The method adopted at some mills [1] is to catch the whole stream falling from the lip of the mortar for a fraction of a second. This is done by a trough 6 inches wide applied by hand. Automatic sampling would be better. In other mills a portion of the stream of pulp is caught during the whole period of the run. Sampling troughs, basins, etc., are preferably enamelled inside to prevent contamination (Smart).

Mill Site.—This should be easily accessible by road, rail, and water, if possible ; moreover, it should be near both wood and water, and there should be a good fall of the ground. The least fall that is considered sufficient in California is 33 feet from the mouth of the rock-breaker to the floor on which the concentrators are placed, when rock-breakers are used, followed by stamps, copper tables, sluice plates, and two successive concentration tables. If a second concentrator is dispensed with, however, and space otherwise economised as far as possible, 29½ feet may be enough. There is also the question of foundations to be considered in connection with the nature of the subsoil, and also the relative positions of mine-shaft, mill and tailing dump, if any. Smart points out that a prevailing wind from the dump will carry dust into the mill and cause trouble with the bearings.

Arrangement of the Mill.—The general disposition of the machinery is shown in section in Fig. 78. This represents a mill in which the ore is delivered from the ore-cars through a grizzly on to the rock-breaker floor, and thence by a shoot to the automatic feeders of the stamp battery ; the pulp, after passing over the plates, is conveyed by sluices to the double row of "frue vanners" (described in Chap. XI.), which are shown standing back to back on the lowest floor.

The general design of a modern mill in the Transvaal, the Simmer Deep, is shown in Fig. 79,[2] which shows the back-to-back type of battery with mortar-boxes direct on concrete, and driven by one motor for every ten stamps.

Speaking generally, the mill building should be well ventilated and of ample size to allow space for the work of repairing the machinery and for exercising supervision, and for the storage of spare parts. Thus, for example, some space is required behind the feeders. The amalgamating tables should also be easily accessible, space being left to pass between them, and the same remark applies to the sluices and the tables or other appliances for concentration. H. T. Pitt [3] has suggested that the amalgamating tables should be separated from the battery by a space of several feet to allow space for sampling and for repairs ; the pulp would then be delivered at the tables by launders. On the Rand, the pulp in some mills passes direct

---

[1] *J. Chem. Met. and Mng. Soc. of S. Africa*, May, 1904, p. 412.
[2] Schmitt, *Rand Metallurgical Practice*, Fig. 56, vol. ii., p. 66.
[3] Pitt, *J. Chem. Met. and Mng. Soc. of S. Africa*, 1904, p. 409.

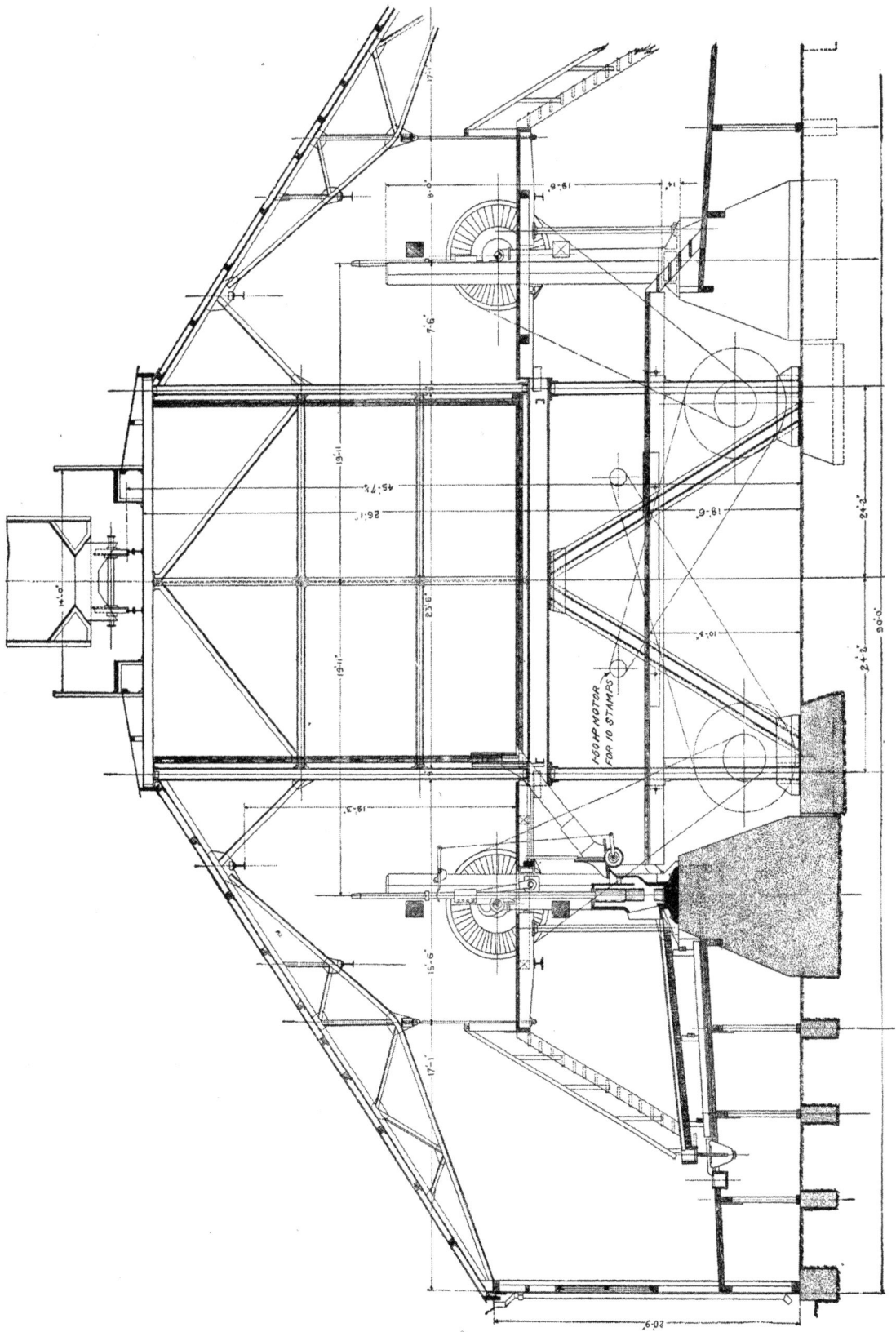

Fig. 79.—Modern Mill with Back-to-Back Type of Battery and Mortar-Boxes Direct on Concrete.

from the battery to the tube mills, the amalgamated plates being placed after the latter, and not in the same building as the battery. All shafts, bearings, etc., should also be easily accessible, so that oiling, relining, and repairs may be readily done.

The whole of the machinery is contained in a strong building of timber or steel framing covered by corrugated iron, to protect it and the work-

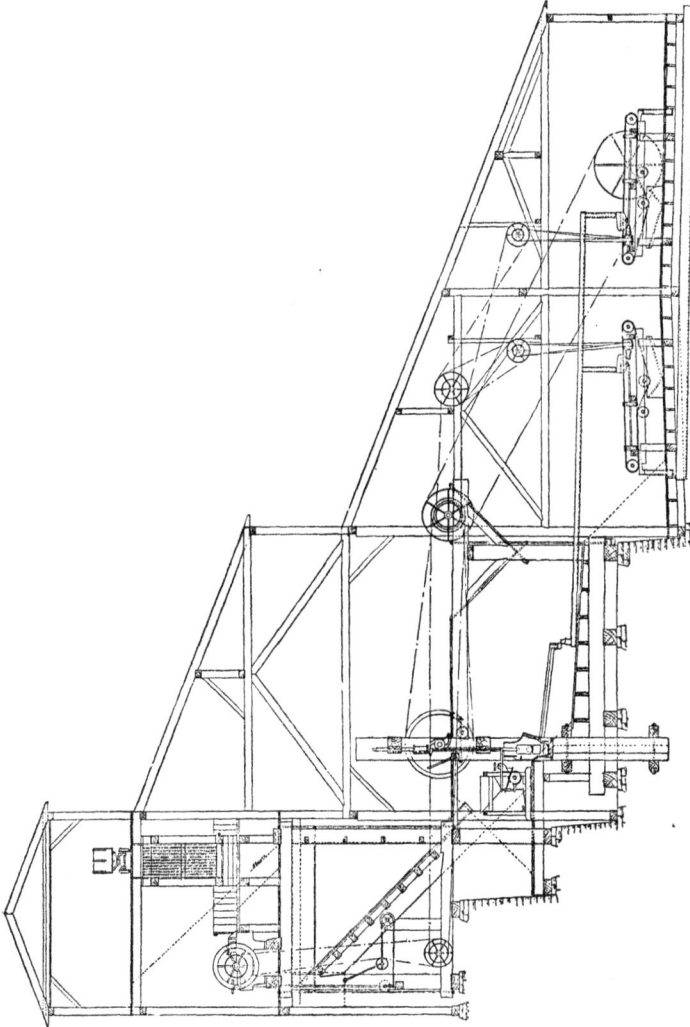

Fig. 78.—General Arrangement of Machinery in Stamp Mill.

men from the weather, and to prevent theft of amalgam. On the Rand, theft is further guarded against, under the compulsion of the law, by the provision of covers to the amalgamating tables. The covers are securely fastened and can be unlocked and raised when required. The interior of the mill should be lighted by as many windows as possible, in order to facilitate superintendence and repairs. In cold climates, the mill buildings are

12

sometimes warmed either by passing the flues from the boiler fires through them from end to end, before leading the products of combustion to the stack, or by steam pipes.

The tailing is discharged into a sluice by which it is carried into a river, or into the sea, or run into settling pits, or impounded behind dams, or elevated and piled in heaps. One of the latter courses is adopted, either if water is scarce, so that it is necessary to use it over again, or if the discharge of tailing is forbidden by law. When the tailing is rich enough to be subjected to further treatment, as is almost always the case in modern practice, it is passed to the cyanide or concentration plant.

For a general discussion of stamp milling, see p. 209.

CHAPTER VIII.

AMALGAMATION IN THE STAMP BATTERY.

Amalgamated Plates.—These plates are usually made of copper, and are as much as $\frac{3}{8}$ inch thick for the inside of the battery and $\frac{1}{16}$ to $\frac{3}{16}$ inch thick for the outside. The average weight used in California is 3 lbs. per square foot. It was formerly laid down as a general rule that the heaviest plates were the best, as they last longer and are not so easily dented, but comparatively light plates are now used. The copper should be of the best quality, and, if it is hard, it must be annealed before applying the mercury, so as to make it absorbent. This was formerly done by heating it from below as uniformly as possible until sawdust laid on the upper surface was ignited. The plate is then straightened by blows of a wooden mallet, striking a block of wood laid on it, and the surface is carefully cleaned by scouring with sand or fine emery paper until quite bright and free from all traces of the surface tarnish of oxide and carbonate of copper. It is then washed with strong soda to remove all traces of grease. Cleaning may also be effected by nitric acid diluted with 9 parts of water, or by a $2\frac{1}{2}$ per cent. solution of potassium cyanide, rubbed on with a woollen rag and carefully washed off with water. As soon as the plate is clean, it is rubbed with a mixture of fine sand, sal-ammoniac and mercury by means of a brush, the sal-ammoniac preventing the recommencement of oxidation. Potassium cyanide is now more often used than sal-ammoniac, and the sand is sometimes omitted. More mercury is sprinkled on and wiped over with the brush or a piece of rubber until the surface is pasty. Care is taken not to make the plate " wet " with mercury. If any drops of mercury appear on the plate losses occur through scouring.

·If the plates were used in this condition they would not catch the gold very well at first, but would continually improve until they had become coated with gold amalgam. In order to make them efficient from the start, they are usually coated with gold amalgam or silver amalgam before being laid down in the mill. Gold amalgam is most effective, but is seldom used, as it is so much more expensive than silver. The amalgam is rubbed on with a piece of india-rubber, the plate being wetted with a solution of sal-ammoniac or potassium cyanide to keep it bright. The whole operation is called *setting* the plates. According to Roskelley,[1] 16 ozs. of mercury and 6 ozs. of gold amalgam are enough for setting a plate 16 feet by 4 feet 9 inches.

Silver amalgam is generally used in the Transvaal. The simplest method of making it is to reduce clean precipitated silver chloride with hydrochloric acid and iron nails, and after washing and picking out the nails, to rub the finely-divided silver with mercury in a porcelain mortar.[2] The pure amalgam of buttery consistency containing the silver in the finest state of division,

---

[1] Roskelley, *J. Chem. Met. and Mng. Soc. of S. Africa*, Feb. 1904, p. 286.
[2] E. H. Croghan, *J. Chem. Met. and Mng. Soc. of S. Africa*, Aug. 1909, 10, 43.

is applied to the copper plates after the usual scouring, and some two or three
weeks before the plate is put into commission. The plate is frequently
dressed with the same amalgam during this period. About ¼ oz. amalgam
per square foot of plate is required. This method gives time for the plate to
absorb the amalgam, and tarnishing on exposure to the air is thus prevented.
After milling has begun, the coating of silver amalgam is gradually removed
from the plates as gold amalgam is scraped off, but by the time the former
is all gone its place will have been taken by a permanent and equally efficient
coating of gold amalgam.[1]

A more usual method of preparing the plates in California is to coat
them with *electro-deposited silver*. This plating is done by certain establishments in California for most of the mills, but it can be done on the spot
without much difficulty, the plant required being inexpensive. After being
silvered, the plates have the mercury applied to them. They absorb a large
amount of mercury, catch gold well, and are little trouble to keep clean.
The plates need not be re-silvered, except after scraping and sweating (see
under " The Clean-up," p. 193), as they become coated with gold amalgam
in the course of time. About 1 oz. of electro-deposited silver is required
per square foot of copper plate, but some mills use 2 ozs. or more.[2] Silvered
plates are not used inside the mortar.

The position of the battery plates is as follows :—The lower edges of the
inside plates are level with the upper surface of the pulp, when the battery
is working properly—*i.e.*, they are usually at 1½ or 2 inches above the surface
of the dies. The plate on the feed side is generally about 9 to 12 inches
wide, and is of the same length as the battery ; it is bolted to the mortar
itself, and its angle of inclination varies with the shape of the latter, so that
the angle of inclination is sometimes 40°, and it is sometimes nearly vertical.
The plate on the discharge side is inclined at an angle of 10° to 50° to the
vertical, and is as wide as the space below the screen permits, being usually
from 3 to 6 inches wide. Sometimes the plate is curved to fit the chuck-
block and sometimes it is bent. It is fixed to a wooden chuck-block, which
has its top bevelled off so as not to obstruct the screen opening. The block
is bolted to the mortar with some thickness of blanketing between, in order
to make a tight joint. Several sizes of these chuck-blocks, with their copper
plates attached, are kept in the mill, a wider block being substituted for a
narrower one when the wear of the dies has proceeded to a certain extent.

At the Alaska Treadwell Mill, and at one or two mills on the Rand, the
cast-steel linings of the mortars were furnished with several horizontal slots
or recesses for the collection of amalgam, and these took the place of the
back copper plates. The front or chuck-plate was, however, retained in
these mortars. This arrangement is shown in Fig. 66 (p. 161). Inside plates
are now generally discarded on the Rand.

The *outside plates* are fastened to a wooden table with copper nails, or
wooden clamps, or by wedges driven into the raised edges of the table. The
battery tables are heavy and are unconnected with the battery frame, in
order to avoid excessive vibration due to the stamps. Possibly some vibration
is advantageous. The table is as wide as the battery (4 feet 9 inches to 5 feet),
and usually from 6 to 8 feet long. On the Rand they are 12 to 20 feet long.
In California a length of 2 or 3 feet of plates of the same width, the *apron*

---

[1] W. A. Caldecott, *ibid.*, 1908, **9,** 142.
[2] Richards, *Ore Dressing*, 1903, p. 748.

plates, are interposed between the battery and the tables proper, on to which there is a drop of 2 or 3 inches. On the Rand, a steel plate 18 inches long, flush with the copper plates, receives the ore from the battery. This is to take up the wear caused by the falling pulp.[1] Figs. 80 and 81 show amalgamated tables and copper plates as used on the Rand in cases where the battery screens are of 600 mesh per square inch or finer.[2] The inclination of the plates varies from $\frac{1}{2}$ to $2\frac{1}{2}$ inches per foot (see below, p. 186).

Below the amalgamated tables in California there is often a succession of four or five *sluice plates*, each about 30 inches long, with *drops* of 1 to 3 inches in vertical height between them. They are usually made narrower than the others, and are frequently only 12 or 18 inches wide. This practice is doubtless due to the traditions as to the treatment of auriferous gravel

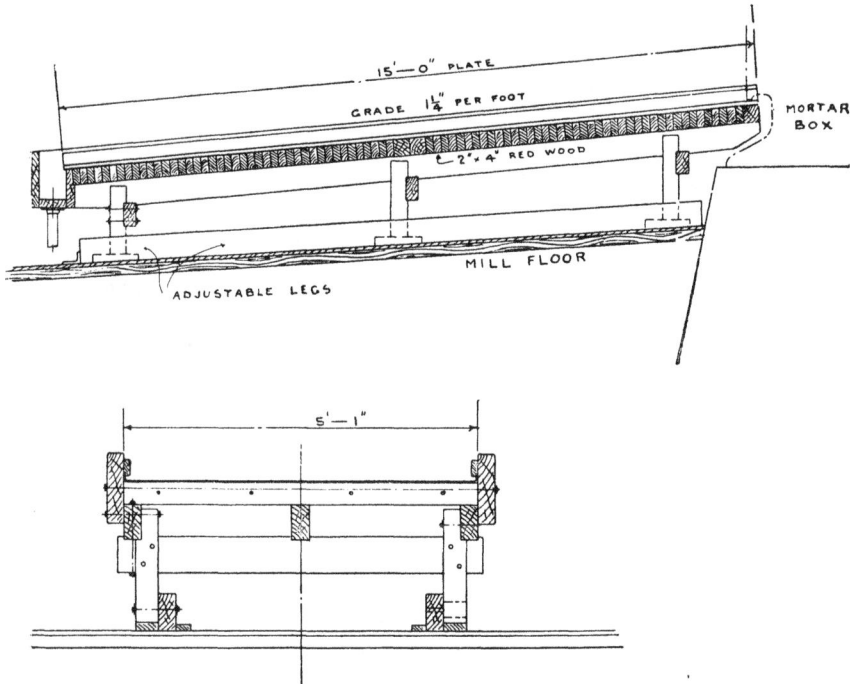

Figs. 80 and 81.—Amalgamating Plate for Stamp Batteries.

in sluice boxes existing when amalgamated plates were first devised and introduced. It is not to be commended, as the stream of ore and water, forced into a narrower channel, becomes deeper and flows more rapidly and tumultuously, with the result that the contact between the ore and amalgamated plates is much reduced, and very little gold is caught. The use of the drops is to assist in catching the float gold and to separate the amalgam which has become floured and mixed with the pulp.

Drops of 2 or 3 inches are also in use in some mills on the tables proper. At the Wildman Mill, Sutter Creek, California, for example, there are five

---

[1] Schmitt, *Rand Metallurgical Practice*, vol. ii., p. 171.
[2] Schmitt, *op. cit.*, p. 170.

drops in a total length of 23 feet 9 inches.[1] Drops may cause scouring, and, according to MacFarren, should not exceed $\frac{1}{2}$ to $\frac{3}{4}$ inch in height.

The use of outside battery plates is advantageous with fine crushing, but when the screens are of 8-mesh (64 holes per square inch) or coarser, the scouring action of the pulp prevents the gold from being caught. On the Rand, where coarse crushing (up to a maximum of $\frac{1}{2}$-inch screen apertures) is practised, battery plates have been discarded. The coarse battery pulp is passed through tube mills and then over stationary plates. In modern practice, suggested by Caldecott, the product of one tube mill is passed over three plates, each $4\frac{1}{2}$ feet by 12 feet. Two plates are considered to be enough,

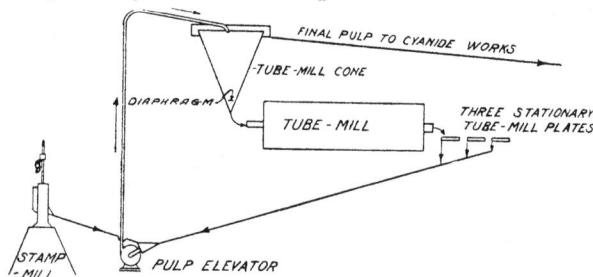

Fig. 82.—Arrangement at the Simmer and Jack Mill.

but the third plate is added as a measure of precaution and to enable two plates to be in operation whilst the third is being cleaned up or dressed. A rapid accumulation of black sand (q.v.) takes place. The arrangement at the Simmer and Jack Mill is shown in Fig. 82,[2] and another arrangement in Fig. 83.[3] The area of the plates is greatly reduced by these arrangements, and is given as only 2 square feet of amalgamated plate per stamp unit,[4] as against 15 square feet or more under the old arrangements with far less

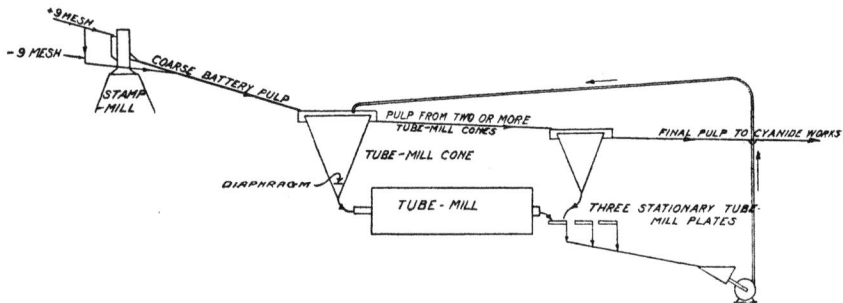

Fig. 83.—Another Rand arrangement of Amalgamated Plates.

output per stamp. This innovation gives less work in dressing and cleaning-up the plates, and reduces the amount of gold locked up in them. The amalgam is caught more evenly over the plates, and the daily scrape is over a greater area, not only over the top part.

Treatment of the Plates.—In order to keep the plates in proper condition so that successful amalgamation may be maintained, the closest watch

[1] Richards, *Ore Dressing*, 1903, p. 733.
[2] Dowling, *Rand Metallurgical Practice*, vol. i., p. 123.
[3] *Ibid.*, p. 129.   [4] Gowland, *Non-Ferrous Metals*, 1914, p. 232.

must be kept over them. The silver-plated copper table is preferred in California from the ease with which it is kept clean, but is not used in the Transvaal. It is not considered desirable to put on the plates as much mercury as they will hold, since, if the amalgam is too fluid, losses are sustained by scouring, but, on the other hand, if the amalgam becomes too hard and dry from absorption of gold and silver, further amalgamation is checked and fresh mercury must be added.

The condition of the inside plates is regulated by the amount of mercury supplied to the mortar. In Colorado there is an opening at the front of the battery and above the screen frame, ordinarily covered with canvas, which can be lifted up by the millman, who introduces his arm, and determines by passing his hand over the front plate, whether the right amount of mercury is being added by the feeder. The regulation of the addition of mercury is thus effected without removing the screen frame. Mercury is sometimes added direct to the outside plates, and sometimes their condition is regulated by the additions of mercury in the mortar-box. The amount added varies with the conditions of crushing and the richness of the ore, but in general from 1 to 2 ozs. of mercury are fed in for every ounce of gold contained in the ore. The finer the state of division of the gold and the more sulphides there are contained in the ore, the more mercury is required. It is fed into the battery at stated intervals of from half an hour to an hour. In some mills amalgamation in the battery is not attempted, no inside copper plates being provided, and under these circumstances it is not usual to feed mercury into the battery.

The practice of feeding mercury into the battery, although still frequently pursued, has been more generally discontinued. The objections urged against it are mainly that the mercury so introduced, and the amalgam formed through its agency, tend to become so excessively subdivided that a high percentage is lost through flouring; moreover the mercury is liable to sicken when the ores contain sulphides. These evils, no doubt, exist, and tend to increase with the percentage of sulphides present, while arsenic and antimony in particular cause heavy losses of both mercury and amalgam if battery amalgamation is attempted, but with ordinary free-milling ores such losses do not appear to be serious. When coarse free gold is present, inside amalgamation is probably advantageous, but otherwise outside amalgamation is now generally preferred.

The amalgamated plates are *dressed* as frequently as is necessary, the length of time allowed to elapse between two operations depending partly on the richness of the ore. To dress the plates, the battery is stopped by hanging up the stamps and a flow of clear water continued to wash off the loose sand. The "black sand" is swept off by brushing from the bottom to the top of the plate and kept separately for grinding with mercury. Black sand consists of so-called "mixed" grains composed of particles of gold adherent to particles of pyrite. The gold is amalgamated and adheres to the plate, which is rendered inoperative for further catching at that spot, owing to the covering of pyrite.[1]

The plates are then rubbed with a hard brush so as to soften and distribute the amalgam, and fresh mercury is added where there are hard spots, if any. The amount of mercury put on the plates should be enough to keep their surfaces in a pasty condition, but not enough to gather into liquid

---

[1] Smart, *Rand Metallurgical Practice*, vol. i., p. 81.

drops or to run off. Sometimes cyanide is added in dressing the plates, to assist in the removal of stains, but sal-ammoniac is now seldom used. As cyanide causes a loss of gold by dissolution, its use has been discontinued on the Rand, and if any chemical is considered necessary a 10 per cent. solution of hydrochloric acid suffices and is innocuous (Smart). The plates are dressed once or twice a day, or as often as every two hours when rich ores are being treated, or when stains tend to appear owing to the nature of the ore. Usually in dressing the plates no amalgam is removed, but sometimes a partial clean-up is effected at the same time, the surplus amalgam being wiped off with rubber squeegees, working from the bottom to the top.

On the Rand,[1] in 1904, mercury was sprayed on about every eight hours to the upper 3 feet of the plates, the part that is scraped daily. The mercury oozing downwards supplied the lower part. Then a solution of 0·08 per cent. cyanide was sprinkled over the plate, and the mercury was rubbed in with a hard brush. Afterwards the plate was brushed smooth with a soft brush, and then washed down.

Discoloration of the Copper Plates.—The plates often become stained by the formation on them of oxides, carbonates, or other compounds of copper through the corrosive action of the water and pulp. Ores containing decomposing sulphides acidify the water and thus cause the corrosion of the plates, a yellow film being formed on the surface of the metal. The presence of carbonic acid in the water is equally harmful, but Aaron pointed out many years ago that the addition of slaked lime to the water neutralises the acid substances and diminishes the tarnishing. On the Rand sufficient lime is added to the mill ore-bins or to the mortar-boxes to neutralise the acidity of the ore and give a distinctly alkaline reaction (Smart), and the discoloration of the plates is thereby prevented, and the use of cyanide in dressing rendered unnecessary. The yield of gold by amalgamation is increased in this way and the clarifying of the returned mill-water facilitated, but the yield of slime in the cyanide plant is diminished.[2] Where lime is not added, the yellow, brownish, or greenish discoloration, the so-called "verdigris," appears in spots and spreads quickly, especially on new plates, those which have been silver-plated being less liable to become dirty than the others; whilst when a plate has become covered with a thick layer of amalgam it is not readily discoloured. When these stains appear the plate must be at once cleaned, as the stained part catches little or no gold. The chemicals used for the purpose have been sal-ammoniac and potassium cyanide, the operation being conducted as follows :—The battery is stopped, the plates rinsed with clean water, and a solution of sal-ammoniac applied to the stained parts with a scrubbing brush, and left covering them for a few minutes in order to dissolve the oxides. It is then washed off, a solution of potassium cyanide rubbed on to brighten the plate, and almost instantly washed off, fresh mercury being then added if necessary. Janin states[3] that long brushing with potassium cyanide is necessary, as otherwise the spots reappear when the water is turned on.

The use of cyanide is now unusual, as it has been found to cause serious losses by dissolving gold. Soda is used to remove grease, and hydrochloric

[1] Roskelley, *J. Chem. Met. and Mng. Soc. of S. Africa*, Feb. 1904, p. 287.
[2] J. R. Williams, *J. Chem. Met. and Mng. Soc. of S. Africa*, 1899, **2**, 657.
[3] *Mineral Industry*, 1894, p. 332.

acid for spots discoloured by oxide, etc. These reagents are effective and harmless.

The merest trace of any kind of grease or oil is very prejudicial to successful amalgamation, forming a film over the plates and over the little globules of mercury, and thus preventing contact between them and the particles of gold. Grease may consist of the tallow dropped from the miners' candles, of the oil from the loose steam which is sometimes used to warm the battery feed-water, or from the bearings of the machinery.

Effect of Temperature on Amalgamation.—By an increase in temperature the wetting of the gold by the mercury and the " catching " of it by the plates is facilitated, as is the coalescing of the globules of mercury.[1] This is due mainly to the reduction in the surface tension of the mercury, which, contrary to the opinion of Read, is a property tending to prevent the sinking of gold in mercury. Increase of temperature also causes an increase in the rate of absorption of mercury by the gold and by the plates, owing to an increase in the rate of diffusion. On the other hand, Read points out that undesirable effects in raising the temperature are the increased solubility of harmful salts and a corresponding increase in the precipitation of base metals into the mercury. This effect both hinders the proper action of the mercury and leads to its loss. An increase of temperature softens the amalgam on the plates, and may cause loss by scouring. Hence Smart states[2] that the temperature of the feed-water should not exceed 80° F. When the water is too cold, the amalgam becomes hard and powdery and loss may occur, whilst its catching powers may be seriously reduced. For this reason, the feed water is warmed by steam in winter in many mills. It is generally admitted that sudden variations in the temperature of the water should be avoided.

A comparatively low temperature is often advantageous. Thus, Professor Le Neve Foster obtained the following results[3] at the Pestarena Mill in Italy, during the years 1869-70 :—The average temperature of the water supplied to the mill during the six summer months was 52° F., and the average temperature of the water supplied in the winter months was 39·4° F., and yet, in spite of that, and of the fact that the average temperature, for instance, of the month of January, 1870, was as low as 33·6° F., he extracted 3·1 per cent. more gold with the cold water than with the warm. These figures do not necessarily prove that cold water is better for amalgamation, as there were in this instance other matters to be taken into consideration, but they show that amalgamation is possible even when the temperature of the water is on an average only 39°. The difference in the results in this case might have been due to the turbidity of the water (which was derived from the glaciers at Monte Rosa) in the summer, and its clearness in winter, or to the fact that the pyrite was more liable to decompose in warm weather, and so additional sickening of the mercury was caused in summer.

Read concludes[4] that when the influence of soluble salts in the ore may be neglected, as high a temperature as can economically be maintained, without variation, is most favourable to successful amalgamation.

---

[1] T. T. Read, *Trans. Amer. Inst. Mng. Eng.*, 1906, **37,** 56; *J. Chem. Met. and Mng. Soc. of S. Africa*, July, 1906, p. 18.
[2] Smart, *Rand Metallurgical Practice*, vol. i., p. 72.
[3] Discussion on paper by A. H. Curtis on Gold Quartz Reduction, *Proc. Inst. Civil Eng.*, 1892, **108,** [ii.], 56.
[4] Read, *loc. cit.*

Inclination of Plates.—The grade or slope of the plates varies with the nature of the ore to be treated, heavy pyritic ores requiring a higher grade than light quartz, while the coarser the crushing the steeper must be the grade. In California the copper tables have an inclination of from $1\frac{1}{2}$ to 2 inches per foot, the apron plates from $\frac{1}{2}$ to $1\frac{3}{4}$ inches per foot, with an average of $1\frac{1}{2}$ inches, and the sluice plates from $1\frac{1}{4}$ to $1\frac{1}{2}$ inches. The narrower plates have a lower slope in order to avoid increased scouring action. With heavily sulphuretted ores a grade of from 2 to $2\frac{1}{2}$ inches per foot is used. On the Rand a slope of $1\frac{1}{4}$ to $1\frac{1}{2}$ inches per foot was generally used for battery plates, with six or seven parts of water to one of ore. It has been found possible, however, to amalgamate efficiently the tube mill product with only $1\frac{1}{2}$ parts of water to one of ore on stationary plates, but in this case an inclination of 18 per cent. ($2 \cdot 16$ inches per foot) is necessary to prevent the banking of the sand and the formation of channels on the plates. This is now the standard grade on the Rand for stationary tube mill plates. The steepness of the grade is of great importance, as on it, and on the amount of water supplied, the attainment of the necessary contact between the ore and the plate depends. When the pulp is flowing properly, it travels down in a series of little waves and ripples, and, in consequence of the friction between the plate and the film of water in contact with it, the upper portions of these little waves travel faster than the lower parts, so that the motion becomes one of tumbling over and over. As a result of this, if the plate is long enough, every particle of pulp comes in contact with the amalgamated surface, and the perfect extraction of amalgamable gold, mercury and amalgam is obtained.

Muntz Metal Plates.—The use of Muntz metal (which consists of copper 60 per cent., zinc 40 per cent.) for amalgamated plates at the Thames Gold-field, New Zealand, is described by Rickard.[1] Muntz metal differs from copper in catching gold well as soon as the plate is amalgamated, not requiring to be covered with gold- or silver-amalgam before it begins to do good work. Moreover, the amalgamated surface is very superficial, since the mercury does not sink in so far as it does into a plate composed of pure copper, so that only a small quantity of mercury is required to cover it. The result is that cleaning up is easy and rapid, no iron instrument being necessary, but rubber being always sufficient. These properties make it particularly valuable for custom mills, where it is desirable to catch as much as possible without mixing the amalgam obtained from two parcels of ore crushed in succession. On the other hand, as it holds little mercury, it cannot absorb much gold, and must be cleaned-up at frequent intervals.

The mercury on Muntz metal plates does not suffer so easily from " sickening " as that on copper plates ; it has been suggested that this is due to the electrolytic action of the copper-zinc couple, which sets free nascent hydrogen, and so reduces the compounds of mercury and other metals which have been formed. It follows that Muntz metal plates are preferable for ores containing large amounts of heavy sulphides or arsenides. The greenish-yellow stains (called " verdigris " by millmen) which are formed on copper plates when grease and other impurities are present in the battery water, do not appear when Muntz metal is used, and such discolorations as occur on these plates can be better removed by dilute sulphuric acid than by potassium cyanide. At the Saxon Mill, New Zealand, the copper plates formerly required 7 lbs. of cyanide, costing 23s., per month to keep them

---

[1] T. A. Rickard, *Stamp Milling of Gold Ores*, pp. 179-182.

in order, while the Muntz metal plates, by which they were replaced, could be kept clean by 5 lbs. of sulphuric acid per month, the cost being 3s. 4d. It is stated, however, that, in the treatment of highly acid ores, which have been weathered for some time so that they contain large quantities of soluble sulphates, or in cases where the battery water contains acids, copper plates are less affected than Muntz metal, over which a scum is rapidly formed. In the Thames Valley, N.Z., Muntz metal is preferred in spite of the extremely acid nature of the water and ore.

In dressing new Muntz metal plates the following method is adopted in New Zealand :—The surface of the plate is scoured with fine, clean sand ; then it is rinsed with water, and washed with a dilute (1 to 6) solution of sulphuric acid. Mercury is then applied and rubbed in with a flannel mop until it wets the surface of the plate (*i.e.*, amalgamates with it) in one or more places, after which the mop is given a circular movement, passing through these spots, until the amalgamation of the surface spreads from them over the whole plate.

The discoloration of the Muntz metal plates is prevented by the weak electric current produced by the Cu-Zn couple, as has been already stated. The same effect, according to Aaron, can be obtained when ordinary copper plates are in use, by placing them in contact with iron or some other metal which is positive to copper. Strips of iron bolted to the top and sides of the plate are said to be suffiicent for the purpose, the copper being in that case unaffected by the acidity of the water, which causes oxidation and dissolution of the iron only. Janin's experience does not support these views.

Shaking Copper Plates.—A shaking copper plate has been recommended to be used either below or in place of the ordinary amalgamating tables, especially in cases where these do not appear to give good results. An ordinary fixed copper plate requires an inclination of from 1 to 2 inches per foot, in treating battery pulp, in order to keep it clear of sand, when the plate is of the same width as the battery. If, however, the plate is subjected to a short rapid shake, the sand is kept from packing, and amalgamation is well performed with a grade of only $\frac{1}{4}$ to $\frac{1}{2}$ inch per foot, or the amount of water needed with the pulp may be greatly reduced and better contact thus obtained. For these plates, silver-plated copper is the material employed. They are affixed to a light wooden frame which is moved by a crank-shaft, revolving 180 to 200 times per minute, placed on one side, with a throw of 1 inch at right angles to the direction of the flow of the pulp. In some mills, a longitudinal shake is given to the plate instead of this side shake. The frame may be hung on rods from above, but is more conveniently supported on four short iron springs, forming rocking legs. The width of the tables is made as great as possible, while the length is of less importance, as, the thinner the current of pulp flowing over them, the better the chance of the gold particles coming in contact with the plates and being retained. These shaking plates were first used in Montana in 1878, and have since been employed at several Californian mills. It is advantageous to add to them an amalgam- and mercury-saver. A simple device for this purpose is to nail a strip of wood, half an inch thick, across the copper plate near the top, thus forming a shallow riffle, the angle of which is soon filled with sulphides and coarse sand, which are kept in agitation by the movement of the table. This is stated by W. McDermott and P. W. Duffield [1] to be the

---

[1] McDermott and Duffield, *Gold Amalgamation* (London and New York, 1890), p. 16.

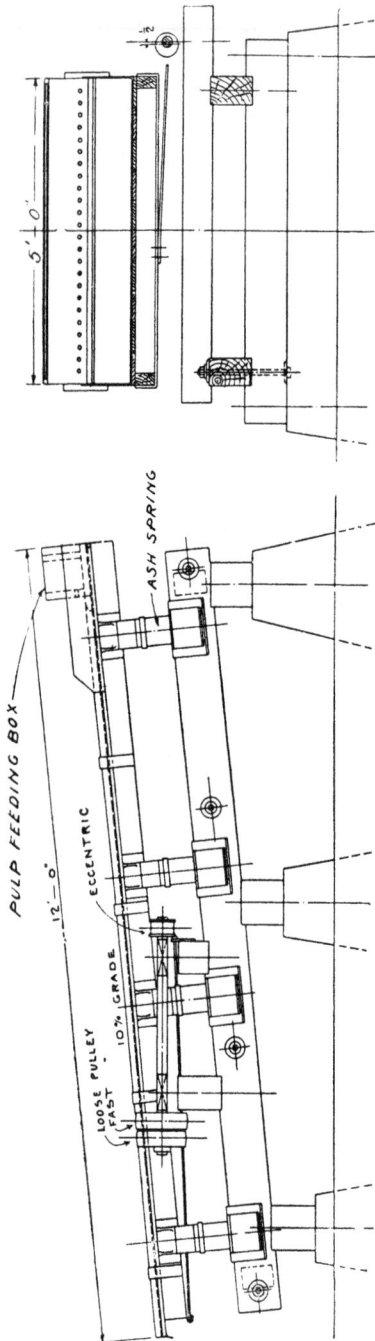

Fig. 84.—Shaking Table.

most effective contrivance yet devised for catching quicksilver and hard amalgam. If the inside copper plate should become hard by accident or neglect, chips of amalgam escaping through the screens are retained in this riffle, and, becoming spherical by rolling up and down under the effect of the shaking motion, increase in size just as a snowball does when rolled in snow.

In 1904, P. Carter[1] described the shaking plates used below the ordinary plates at the Ferreira Gold Mine. Each plate caught an average of 15·05 ozs. of fine gold per month. Each ordinary plate at the same time was catching nearly 400 ozs. per month. Two-thirds of the values caught on the shaking plates came from black sands. The plates were 11 feet long by 4 feet 6 inches wide. They were fixed on old Frue vanners, and had a fall of 4 inches in 12 feet.

Shaking plates were in general use for some years on the Rand to amalgamate the tube mill product, although by 1914 they had been generally displaced by stationary plates. In these tube-mill shaking plates, " the copper plate is usually $\frac{3}{16}$ inch thick, 5 feet wide and 12 feet long, and is mounted on a light frame, generally of timber, with a timber cover $\frac{1}{2}$ inch to 1 inch thick to form a bed for the copper plate. The table frame is carried on eight ash springs, and the shaking motion is obtained by an eccentric having a 1-inch stroke, and driven by a countershaft at 200 revolutions per minute. The power required is less than $\frac{1}{2}$ H.P. Fig. 84 shows a complete shaking table." [2] The

[1] Carter, *J. Chem. Met. and Mng. Soc. of S. Africa*, May, 1904, p. 405.
[2] Schmitt, *Rand Metallurgical Practice*, vol. ii., p. 173.

grade was usually 10 per cent., enough to assure the free flow of the diluted tube-mill pulp, which was in the ratio of 1 part of solids to $2\frac{1}{2}$ parts of water.

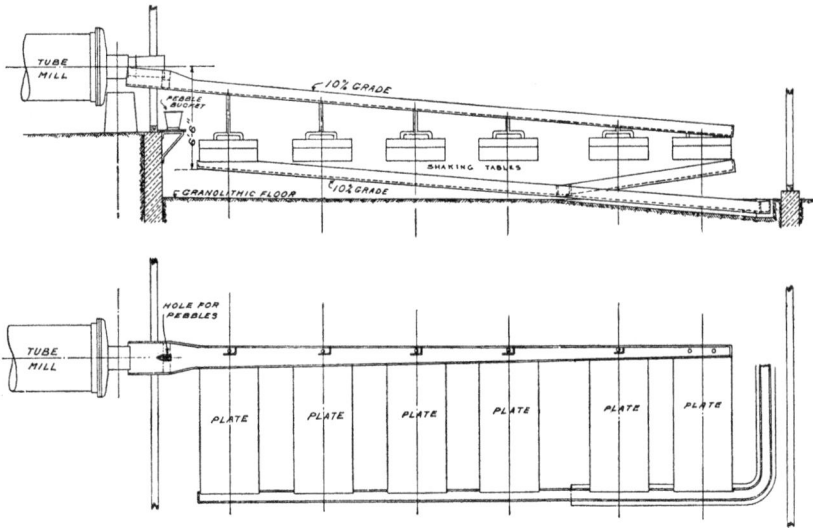

Fig. 85.—Arrangement of Shaking Plates with reference to the Tube-Mill. All Plates on one level.

The number of shaking plates was about five in parallel to each tube mill. The pulp was distributed evenly over the centre width of the plates by branched pipes and perforated boxes. The arrangement of the plates with

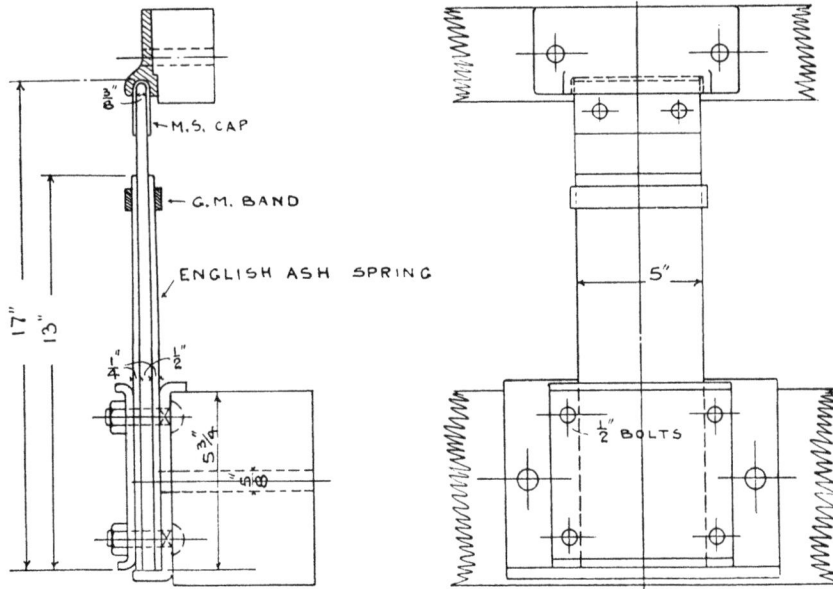

Fig. 86.—Ash Spring for Shaking Plates.

reference to the tube mill is shown in Fig. 85.[1] The ash springs were placed in a vertical position at the City Deep plant, an improvement on the arrangement shown in Fig. 86.[2] The area of shaking plates was about 1½ square feet per ton of ore milled per day.

Mercury Traps.—Another method of saving mercury and amalgam, which would otherwise be lost in the tailing, consists in the application of mercury wells or riffles. A mercury well consists of a shallow gutter filled with mercury, over the surface of which the pulp flows or through which

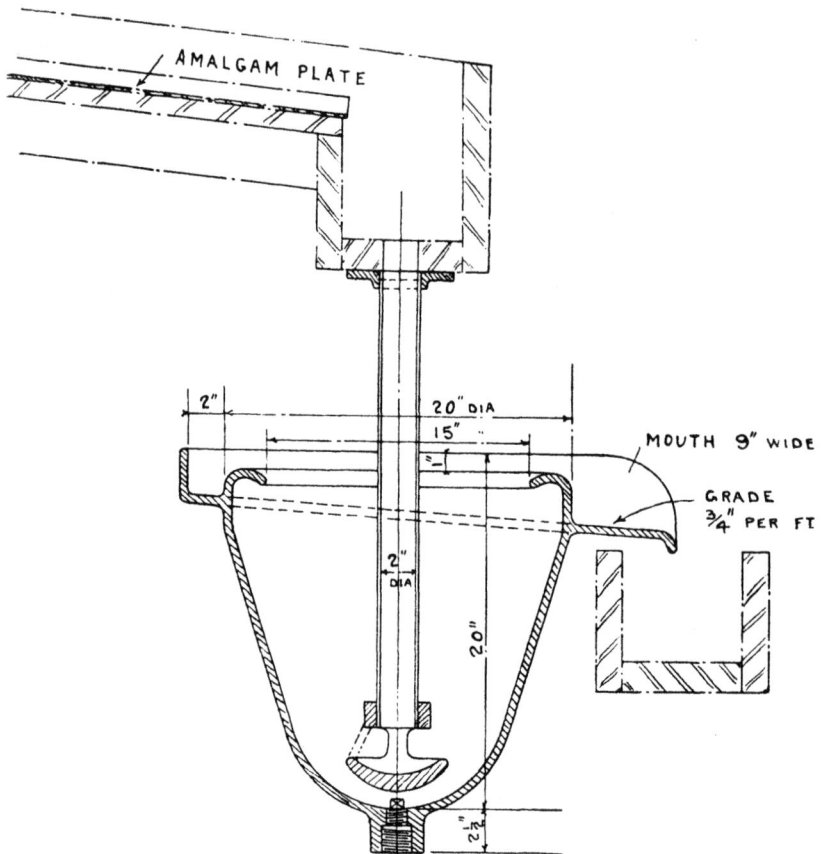

Fig. 87.—Circular Mercury Trap.

it is forced to pass by suitable machinery. Attwood's amalgamator, formerly much used in California, was a machine of the latter class. Such wells or traps are usually placed between the successive plates, the pulp dropping from the end of a plate on to the surface of the mercury in the well, and then passing on to the next plate. The practice of placing a well between the screens and the amalgamating tables has been condemned, as it prevents

---

[1] *Ibid.*, p. 175.   [2] Schmitt, *ibid.*, p. 174.

proper supervision being kept over the feeding of mercury into the battery, over-feeding being difficult to detect under these circumstances. Large particles of amalgam are caught in mercury traps, but finely-divided or floured mercury or amalgam are carried over and escape (Smart). For the work of classifiers as mercury traps, see Chap. XI.

On the Rand,[1] " the pulp overflowing the lower edges of the amalgamated plate falls into a short transverse launder secured to the table frame, and sloping slightly from either side to the centre. At this point the tailing pulp escapes through a short vertical pipe, the upper end of which projects about an inch above the bottom of this transverse launder, which constitutes the first amalgam and mercury trap. The vertical pipe usually delivers into another (circular) mercury trap (Fig. 87),[2] after which the overflow enters the main tailing launder."

Sodium amalgam is used to revivify sickened mercury, or to maintain it in good condition. It is prepared by heating a basin or iron flask of mercury to about 300° F., and dropping in little pieces of sodium not larger than a pea, one by one. Each addition causes a slight explosion and a bright flash of flame. The sodium may be added with less loss and less danger to the operator if the mercury is kept at a somewhat lower temperature and the sodium stirred into the mercury with an iron pestle or pressed below its surface with a spatula. When about 3 per cent. of sodium has been added to the mercury, the reaction becomes less active and the amalgam is then poured out upon a slab or shallow dish, allowed to cool and solidify, and then broken up and kept in stoppered bottles under naphtha. When it is necessary to revivify a quantity of mercury, a few small pieces of the amalgam are added to it and stirred in, or are previously dissolved in clean mercury before being added to the impure stuff. The strong affinity of sodium for oxygen, chlorine, etc., enables it to reduce the oxides and other compounds of the base metals which are coating the mercury globules. The sodium hydroxide passes into solution, neutralising part of the acidity of the water at the same time. The base metals are redissolved by the mercury which is then in good condition to take up the precious metals or to be caught on amalgamated surfaces or in riffles, but the mercury is not really purified and the base metals in it are soon oxidised again. Sodium amalgam is not much used except in amalgamating-pans or in mercury wells or riffles, or in cleaning retorted mercury —i.e., wherever large bodies of mercury can be directly acted on by it. It is of comparatively small value when added to the mortar of a stamp battery, although this use of it is not unknown. The use of electric currents, galvanic couples, etc., has an effect similar to that of sodium amalgam.

Galvanic Action in Amalgamation.—In amalgamation in the mortar, on plates, or in pans, not only are free metals absorbed, but the dissolved salts, and, to a less extent, the insoluble compounds of the heavy metals, are reduced and amalgamated, chiefly by galvanic action. The copper of the plates, or the iron of the mortar or pan, constitutes the positive element, and all metals less oxidisable than this reacting metal are reduced by it, and are then amalgamated by the mercury. In this way iron reduces both lead and copper, although, if these are present in the form of undecomposed sulphides, this action will be very slight. Now, if lead is introduced into the amalgam, the latter becomes pasty, and is subjected to considerable losses, and copper has an equally harmful effect. It is for this reason that the

---

[1] G. O. Smart, *Rand Metallurgical Practice*, vol. i., p. 76.   [2] *Ibid.*, p. 77.

arrastra is found to be better than the stamp battery or even the pan for certain plumbiferous ores.   This action of iron is of course enormously increased if the ores are subjected to a chloridising roast before being amalgamated, as in the Reese River process for the treatment of auriferous silver ores.

In some mills, this galvanic action has been increased by the passage of a weak electric current through the charge by means of a dynamo.   The amalgamated plates or the walls of the pan are connected with the negative pole, while the positive pole is formed of a plate of graphite, lead or iron dipping into the pulp.   Under these conditions the mercury is still further protected from attack, and remains bright and lively, but the deposition of base metals in it is favoured, and the stronger the current the more this action is induced.   Consequently, such methods are attended with the best results when dealing with ores containing little or no copper, lead, etc., since in these cases the strength of current can be increased, and the mercury kept clean, without any ill effects.   The principle is made use of in Bazin's centrifugal amalgamator, Molloy's hydrogen amalgamator, and other similar machines.

An electro-chemical amalgamator was described by G. Warnford Lock in 1892.[1]   Further experiments in California [2] showed that with a small current good results are obtained by adding a solution of mercuric chloride to the pulp, when mercury is electro-deposited on the plate.   If common salt is added sodium is deposited.   The amalgam formed is tenacious and bright, resisting scouring.   The method may be used in the sluices of placer workings.

Designolle Process of Amalgamation.[3]—In this process, which may be conveniently referred to here, as the principle is worth bearing in mind, a solution of mercuric chloride is used.   It was tried at the Haile Mine, South Carolina, the method being as follows :—Charges of 600 lbs. of roasted ore were placed in cast-iron barrels with 1,000 lbs. of cast-iron balls, or in pans. The barrel was partly filled with water, and 1 gallon was added of a solution containing 1·7 per cent. of mercuric chloride, the same amount of hydrochloric acid, and twice as much salt, if the ore contained less than 15 dwts. of gold per ton.   This would be equivalent to about 10 parts of mercury to 1 part of gold.   The barrel was rotated for twenty minutes and then discharged into a settler, and the suspended amalgam caught on copper plates. The mercury was supposed to be reduced by the iron thus—

$$HgCl_2 + Fe = FeCl_2 + Hg,$$

and the metallic mercury thus freed amalgamated with the gold.   If no common salt was present, some mercurous chloride was formed according to the equation—

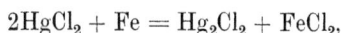

$$2HgCl_2 + Fe = Hg_2Cl_2 + FeCl_2,$$

and the subsequent reduction of the insoluble calomel by iron was not complete.   Hydrochloric acid was supposed to hasten the amalgamation by setting up some electrolytic action.

The total cost of the process at the Haile Mine was said to be only 35 cents, or 1s. 7½d., per ton, but it was abandoned when the percentage of iron in the

---

[1] Lock, *Trans. Inst. Min. and Met.*, 1892, **I**, 215.
[2] E. E. Carey, *Mining J.*, May 15, 1909, p. 617.
[3] This account is abridged from that given by Louis Janin, Junr., in *Mineral Industry*, 1894, p. 346.

material under treatment increased, owing to improved methods of concentration. Large quantities of oxide of iron were then amalgamated, and rendered the resulting mass harder to treat than the ore itself. By repeated washing, settling, and regrinding with fresh mercury, it could be partially purified, but not without a loss of gold. It is stated that 87 per cent. of the gold in the ore was extracted.

The Clean-up.—The amalgam, both on the inside and outside plates, does not accumulate evenly, but in ridges and knots which serve as nuclei for the collection of more. It is not advisable to allow the coating of amalgam to become very thick, since, although the plates catch better as the amalgam accumulates, losses may be experienced by scouring. The removal of amalgam takes place as often as necessary, once a day being the rule on the Rand, but both longer and shorter intervals are not uncommon in either districts, dependent on the richness of the ore.

The operation usually consists in the removal of the black sand, followed by a vigorous brushing over all parts of the plate with a stiff brush, mercury being sprinkled on at the same time and the amalgam thoroughly softened throughout. The amalgam is then removed either by a sharp-edged piece of hard rubber or by scraping with broad flexible steel scrapers, such as that shown in Fig. 88, care being taken not to scratch the plates.[1] In some

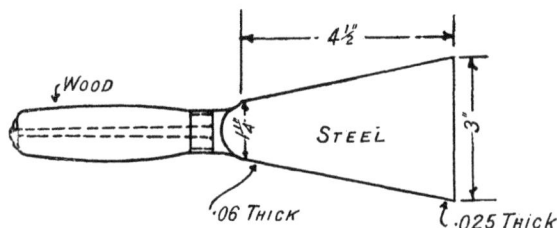

Fig. 88.—Scraper for removing Amalgam from Plates.

mills scraping is not thought advisable on the ground that it may lead to too close removal of the amalgam, leaving the plate liable to discoloration. Scraping is universal on the Rand, according to Smart.[2] The scraping of the outside plates usually takes from ten to fifteen minutes for each battery. The amalgam so obtained is ground with more mercury in a clean-up pan in order to soften it, the skimmings from the mercury wells, etc., being added to the charge. The inside plates are not scraped until the amalgam stands up in ridges on it; the operation may be necessary as often as twice a week, but it usually takes place twice a month, when a general clean-up is made.

In cleaning-up, the stamps are hung up, two batteries at a time; the screens, inside plates, and dies are all taken out and washed in tubs; and the "heading," or contents of the mortar, consisting of the pulp, mercury, sulphides, and pieces of iron and steel, amounting in all to a quantity sufficient to fill two or three buckets, is carefully scraped out and panned or fed into the mortar of one of the other batteries, which has not yet been cleaned up. In California, the heading from the last batteries is panned,

---

[1] *Rand Metallurgical Practice*, vol. i., p. 83.
[2] Smart, *loc. cit.*

the iron removed with a magnet, and the remainder ground with mercury in the clean-up pan.   G. O. Smith states [1] that mortar-box sand, as well as the sand from mercury traps and from washing the screens, is very rich, and must be ground with mercury in a clean-up pan.   Roskelley [2] considers that mortar-box sand from 10-dwt. rock would contain about 10 ozs. of gold to the ton after a three months' run.   Amalgam is found adhering to the inside of the mortar and to the dies, and is carefully detached and added to the clean-up pan.   Hard amalgam is removed with a chisel, care being taken not to lay bare the copper surface.   Fresh mercury is then added, and brushed over the plates, which are finally smoothed with a soft brush.   After the plates have been redressed, the batteries are restarted, and the next ones stopped and cleaned up.   Three men can clean up forty stamps in from five to seven hours, ten stamps being thus idle for the whole of this time.

The amalgam obtained from the batteries, outside plates, mercury wells, or sluices, is rarely clean enough for immediate retorting ; it is usually found to contain mixed with it grains of sand, pyrite, magnetite, and other minerals, together with fragments of iron and other foreign substances.   The skimmings from mercury wells are still more impure.   These materials must be purified by grinding with fresh mercury and washing before they can be passed to the retort.   The scraps of iron consist of fragments of shoes, dies, shovels, picks, hammers, and drills ; they are knocked about in the mortar until a quantity of gold and amalgam has been driven into their interstices.   At the Jefferson Mill, Yuba County, California, about $\frac{1}{2}$ ton of such scrap, picked out by hand or with a magnet, had accumulated in 1885.   It was attacked with warm dilute sulphuric acid until the surface had all been dissolved off, and the residue was then well washed, and gold to the value of $3,000 thus recovered.   The shoes, dies, etc., which were too large for this treatment, were boiled in water for half an hour, and then struck with a hammer, when the gold dropped out. [3]   On the Rand the die sand is passed through a screen of about four holes to the linear inch, which removes the larger pieces of iron and steel and ore.   The ore is returned to the mortar-box and the coarse pieces of iron and steel sold as *battery chips*.   The finer sand, etc., is treated in the clean-up barrel or tube mill. [4]

In small mills the dirty amalgam is ground in a mortar by hand with fresh mercury and hot water, until it is reduced to an even thin consistency, when the dirty water is poured off, and the mercury poured backwards and forwards from one clean porcelain basin to another until the pyrite, dirt, etc., have risen to the surface, when they are skimmed off.   The skimmings obtained are put back into the mortar, and re-ground by themselves with fresh mercury. The clean mercury is then squeezed through canvas or wash-leather, when the greater part of the gold and silver contained in it, together with about one and a-half times its weight in mercury, remains in the bag, the rest of the mercury, with a small quantity of the precious metals dissolved in it, passing through.   The amalgam is now often squeezed in an amalgam press, [5] the excess mercury being expelled through a perforated disc covered with duck, by means of a plunger worked by a screw.

---

[1] Smith, *J. Chem. Met. and Mng. Soc. of S. Africa*, June, 1904, p. 454.
[2] Roskelley, *op. cit.*, Aug. 1904, p. 53.
[3] *Sixth Report Cal. State Mineralogist*, 1886.
[4] Smart, *Rand Metallurgical Practice*, vol. i., p. 88.
[5] See Smart, *loc. cit.* for details.

In large mills a clean-up revolving **barrel** is often employed to mix the amalgam. Fig. 89 [1] is a section of a cast-iron barrel in use on the Rand. At the Plumas Eureka Mill, the barrel is 3 feet in diameter and 4 feet long, and revolves twenty times a minute ; the charge is 700 lbs. of amalgam and 20 lbs. of mercury, or more if the amalgam is very rich. A dozen or more iron balls or pieces of iron, such as worn-out battery shoe shanks, are put into the barrel, together with sufficient mercury to yield a fluid amalgam, when grinding is complete, and enough water to make a thin pulp. The use of the iron is to help to mix the amalgam and mercury, but it causes some loss by flouring, and is omitted in Australian mills. After being revolved for from two to twelve hours, or until the amalgam has run together and the sand ground to slime, the barrel is opened and washed out with water, the tailing being run over amalgamated plates and through a mercury well or some other form of amalgam-saver (a large power-driven batea being used on the Rand), after which the amalgam is scooped out of the barrel and squeezed in wash-leather.

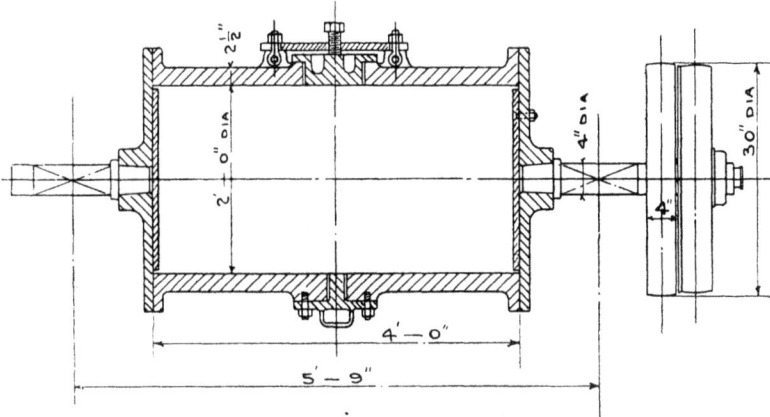

Fig. 89.—Amalgam Clean-up Barrel.

The Clean-up Pan is also extensively used. One of the oldest in use in the United States, the **Knox Pan**, is still a great favourite, especially for treating battery sand, skimmings, etc. It consists (Fig. 90) of an iron pan 5 feet in diameter and 14 inches deep. Wooden or iron shoes are attached to the arms, g, which make from twelve to fourteen revolutions per minute. Iron shoes are considered better for brightening or polishing the particles of gold contained in the pyrite, and so rendering them fit for amalgamation. The charge for this pan is about 300 lbs. of impure amalgam, mercury, concentrate, skimmings, etc. The charge is made into a pulp with water and ground for three or four hours, after which more mercury is added, and mixing is carried on for a few hours longer, before the pulp is diluted, settled, and discharged. The tailing suspended in the water is usually passed over amalgamated plates, and is then often caught in settling pits, and either sold or subjected to further treatment on the mill, as it is frequently of high value. The mercury is squeezed through canvas, and the amalgam retorted.

---

[1] Smart, *Rand Metallurgical Practice*, vol. i., p. 85.

At the Simmer and Jack Proprietary Mines, where the mill contains 280 stamps, there are five clean-up pans 36½ inches in diameter by 13½ inches deep.[1]

On the Rand, " the tailing from pans, barrel and batea, and the poorer sand is fed to the clean-up tube mill, which is the final clean-up grinding machine, and through which all tailing obtained during the clean-up should be finally passed." [2] The clean-up tube mill is a miniature tube mill,[3] about 6 feet long and 4 feet in diameter, fed by hand. The outflow passes over a small shaking amalgamated plate, and thence to a settling box. The overflow from the box is slime, which is treated with cyanide in a small air agitation vat (Fig. 91),[4] and thence after decantation to the ordinary cyanide slime plant.

The position from which the greater part of the gold is obtained in a clean-up varies according to the ore and the method of treatment. In California from 50 to 80 per cent. of the gold saved is caught on the single plate inside

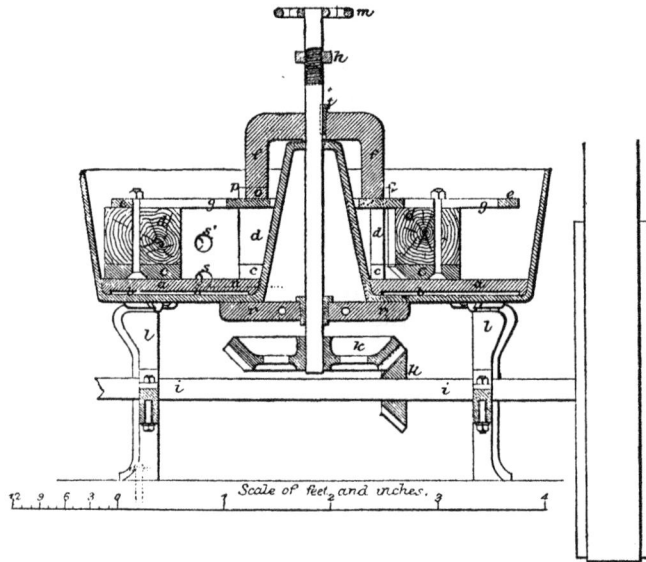

Fig. 90.—Knox Pan.

the battery, the remainder being caught on the outside plates and the sluice plates, or being contained in the concentrate. In the Grass Valley, at the Original Empire and the North Star Mills, from 70 to 85 per cent. was caught inside the battery. The amalgam from the battery plates is usually richer than that from the outside plates, especially if the gold is coarse. The reason for this is that coarse gold, being easily amalgamable, is almost all caught on the inside plates, while fine gold, even if amalgamated in the battery, forms a more fluid amalgam which passes through the screens and is caught outside. For the same reason, amalgam from near the top of the tables is richer

[1] John Yates, *Metallurgical Engineering on the Rand* (1898), p. 47.
[2] Smart, *Rand Metallurgical Practice*, vol. i., p. 88.
[3] See below, Chap. x.
[4] Smart, *loc. cit.*

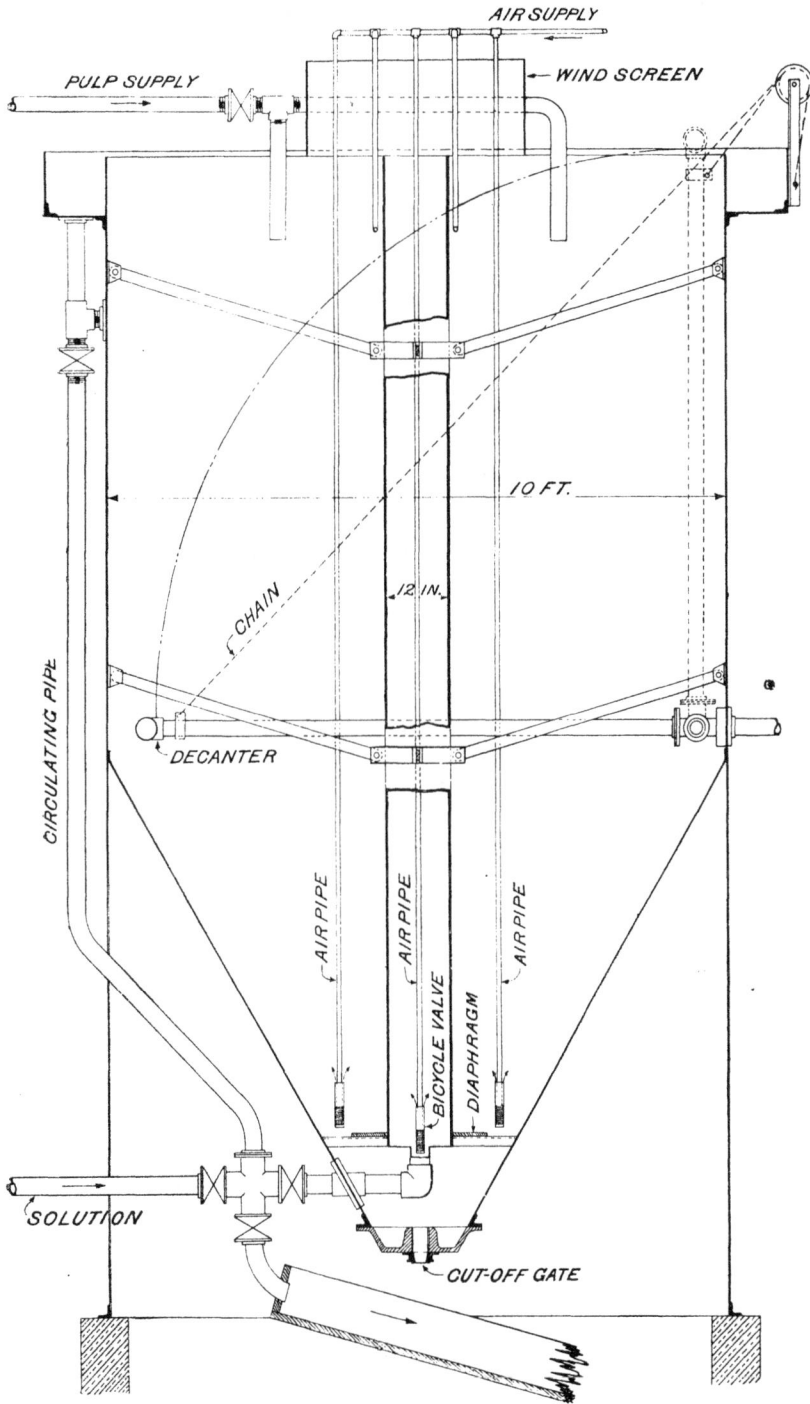

Fig. 91.—Air Agitation Vat for Treatment of Slime resulting from the Crushing of Battery Sand.

than that from the bottom. Coarse gold forms a richer and stiffer amalgam than fine gold, for the reason already given on p. 43. At the two mills last named the value of the plate amalgam was $4.50 per oz., and that of the battery amalgam $8.50 per oz. According to Peplar,[1] the amalgam caught on inside plates is of very low grade, on account of the large percentage of iron that amalgamates on them. Roskelley [2] found at the Robinson Deep Mine, where inside plates are not used, that 95 per cent. of the gold saved by amalgamation was caught within 3 feet of the top of the table.

Steaming and Scaling Amalgamated Plates.—In spite of ordinary cleaning-up, amalgam gradually accumulates as a hard scale on the plates, and is removed periodically by steaming or otherwise heating them before they are scraped. At the North Star Mill, California, the plates are occasionally immersed in boiling water so as to soften the amalgam before they are scraped. On the Rand the plates are generally steamed and scraped every three or four months, according to Smart. " Steaming is carried out by placing a wooden cover over the whole surface of the plate and introducing steam through a $\frac{3}{4}$-inch pipe in the middle for from 10 to 15 minutes. The cover is then removed and the amalgam scraped from the whole of the surface before the plate cools." [3] Mercury is then applied to the plate while it is still warm. The plates are sometimes " sweated " in California by heating them over a wood fire before they are scraped. At the Empire Mill, Grass Valley, California, the sweating of the outside and apron plates of four batteries produced bullion to the value of $19,000.[4] These plates had been in use for eighteen months, and the ore which had been run over them averaged 18 dwts. of free gold per ton. After scaling and sweating, the plates may require replating, especially if they are scoured with sand to help in the removal of the amalgam. In course of time they are worn out, the copper becoming brittle and worn into holes, but they usually contain enough gold when discarded to pay for a new set.

Several methods are in use for recovering the gold from old plates. For example, they may be dissolved in nitric acid, when the gold is left nearly pure. A more economical method of detaching the gold, much used in Australia, is described by W. M'Cutcheon,[5] as follows :—The plate is placed on the hearth of a reverberatory furnace, or on a fire made with logs in the open air, and the mercury expelled at a gentle heat. If the temperature is too high, the gold sinks into the copper at once, and the copper must then be dissolved. After the mercury has been driven off, the plate appears to be more or less coated with gold on one side. This surface is treated with hydrochloric acid for eight or ten hours, and the plate is then replaced on the hearth and exposed to a dull red heat until well blackened. On plunging it into cold water, the gold now scales off, and is collected and freed from copper by boiling in nitric acid.

The method in use on the Rand for scaling plates before putting them aside is described by I. Roskelley [6] as follows :—A mixture of $\frac{1}{2}$ lb. sal-ammoniac, $\frac{1}{2}$ lb. nitre, $\frac{1}{2}$ lb. hydrochloric acid, and a pint of water is applied to the plate with a soft brush, and allowed to stand for about fifteen minutes.

---

[1] Peplar, *J. Chem. Met. and Mng. Soc. of S. Africa*, May, 1904, p. 402.
[2] Roskelley, *op. cit.*, Feb. 1904, p. 287.
[3] Smart, *Rand Metallurgical Practice*, vol. i., p. 84.
[4] *Eighth Report Cal. State Mineralogist*, 1888, p. 714.
[5] Private communication.   [6] Roskelley, *loc. cit.*

Afterwards the plate is heated over a good fire. When it has become quite black, which will be in about half an hour, it is dipped into a bath of water, when the scaling can be washed off. Some millmen drive off the mercury before applying the above mixture ; others do so afterwards. Should all the scaling not come off, the parts needing it are treated over again. The scalings are afterwards collected and mixed as follows :—1 part scalings, 1 part sulphur, 1 part borax, and 1 part sand. This is melted in a crucible and the gold extracted. Smart[1] gives the charge for the crucible as, scales 100 parts, borax 50, sand 25, and manganese dioxide 8 parts, yielding gold bullion 800 to 900 fine.

The methods given in this section are instrumental in recovering the gold amalgam which has " accumulated " on the copper plates.[2] It is this accumulation which is in part responsible for the poor returns from new mills, usually ascribed to "absorption by copper plates." Thomae quotes figures showing that the usual absorption is about $\frac{1}{4}$ oz. of gold per square foot of plate, most of which is taken up in the first fortnight. Wilkinson[3] found that after a long run some plates in a Rand mill yielded by scaling 36·8 ozs. of fine gold, the total area of the plates being 191 square feet. Halse found in a mill in Columbia that from 3 to 5 ozs. of gold per square foot of electro-silvered plate accumulated as scale in about two years.

Old copper plates are usually melted down and sold to refineries, where they are useful for mixing with gold and silver bullion in making-up the alloy for parting. They always contain absorbed gold which cannot be removed without destroying the plates. The total amount is less than $\frac{1}{4}$ oz. per square foot of plate (Thomae).[4] Read has shown[5] that the rate of absorption is greater with higher temperature, and is greater in ordinary plates than in electro-silvered plates.

Retorting.—The solid amalgam, which is retained in the canvas or wash-leather filters, usually contains from 30 to 45 per cent. of gold and silver, according to the state of division of the gold present in the ore, and also to the degree of care exercised in squeezing out the excess of mercury. For separating the gold from the mercury there are two kinds of retorts in general use—the *pot-shaped* retort, which is sometimes cast with trunnions to swing on supports, in small mills ; and the cylindrical retort, shown in Fig. 92, in larger mills. In Figs. 93 and 94[6] the retort furnace in use on the Rand is shown ; the cylinder is 5 feet 3 inches long and 12 inches in diameter.

The pasty amalgam is rolled up into balls or kneaded into cakes, and squeezed into the pot-shaped retort, and often rammed down with a bolt-head, although this course is deprecated by some metallurgists, who prefer to leave the amalgam as spongy and open in texture as possible, believing that a uniform product is thus obtained more rapidly at a lower temperature and without so much loss. In the horizontal retort the amalgam is placed in iron trays divided into compartments by partitions. In either case, the retorted metal is prevented from adhering to the iron, either by laying it on three or four thicknesses of paper, the ashes of which remain beneath

[1] Smart, *loc. cit.*
[2] W. F. A. Thomae, *Trans. Inst. Mng. and Met.*, 1908, 17, 482 ; E. Halse, *ibid.*, p. 486.
[3] Wilkinson, *ibid.*, p. 493.
[4] See also R. T. Bayliss, *Eng. and Mng. J.*, Aug. 11, 1906 ; *Trans. Amer. Inst. Mng. and Met.*, 1896, 26, 33, 1048.
[5] Read, *Trans. Amer. Inst. Mng. Eng.*, 1906, 37, 56.
[6] Smart, *Rand Metallurgical Practice*, vol. i., p. 92.

the amalgam, or by covering the iron trays with a coating of lime white. The mercury is condensed in cooling tubes passed through water ; the loss by volatilisation is usually very small, and may be taken as being about one grain of gold per pound of mercury.

The charge is heated slowly until the boiling point of mercury is reached, when the fire is checked, and the retort kept at an even temperature for three or four hours, or until the bulk of the mercury has been driven off. The retort is then raised gradually to a bright red heat to expel the remainder ; after cooling, it is opened, the trays withdrawn, the retorted metal loosened by a chisel, if necessary, and turned out on a table.

In retorting amalgam containing considerable quantities of base materials, there is a danger of the vent being choked up by condensation of solid material. The retort should be so arranged that a rod can be passed through the condensing pipe so as to clear it of obstructions, if necessary. The front of the retort is luted on carefully with chalk or wood-ashes and salt, and firmly

Fig. 92.—Cylindrical Amalgam Retort.

clamped with an asbestos joint, so as to be quite tight, otherwise a loss of mercury is incurred. In all retorts the lid is turned and ground so as to fit on perfectly. The condensing pipe should not have an open end dipping freely into water, as in that case a sudden cooling of the retort would cause the water to be sucked in, and an explosion would occur. The open end of the pipe may be enclosed in a bag of sacking or rubber immersed in water, or the pipe may be continued by a piece of sacking which dips under water.

The pot-shaped retort requires no brick fittings, and can be heated over an assay furnace or forge fire, or in a fire built on the ground, when it is placed on a tripod stand. In the latter case the fire is lit at the top and burns slowly downwards. The pot-shaped retort is not filled to more than two-thirds its capacity, and must be heated very gradually at first.

The retorted metal is porous and from 500 to 950 fine in gold, the remainder being in general chiefly silver, with base metals and sulphides in

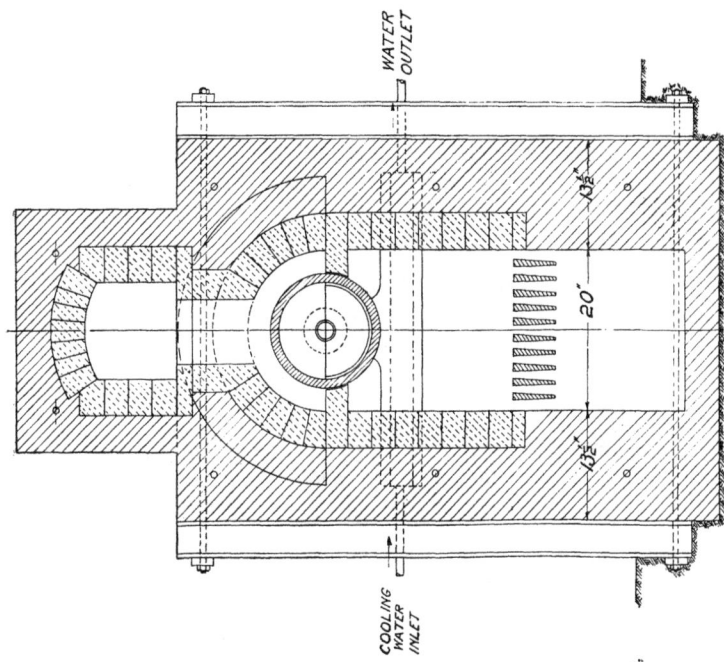

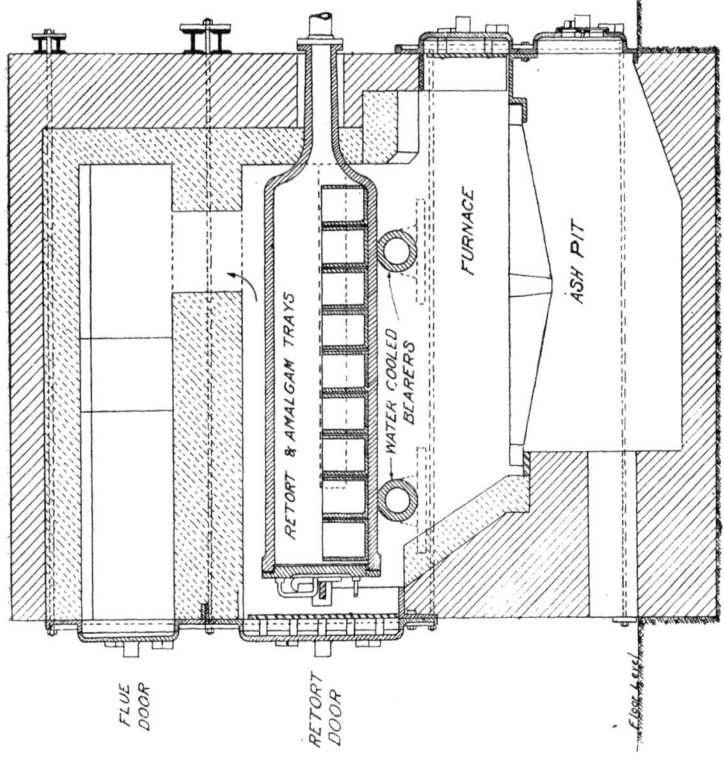

WATER OUTLET

COOLING WATER INLET

$13\frac{1}{2}'$

20'

$13\frac{1}{2}'$

B.

FLUE DOOR

RETORT DOOR

RETORT & AMALGAM TRAYS

WATER COOLED BEARERS

FURNACE

ASH PIT

A.

Figs. 93 and 94.—Retort Furnace.

smaller quantities. The gold is melted in crucibles with sand, carbonate of soda and borax, and suffers a further loss in weight, due to the slagging off of oxides, earthy impurities, etc., and to the volatilisation of a small quantity of mercury, which is obstinately retained until the melting takes place. Richards states [1] that if the amount of mercury is to be reduced below 1 or $1\frac{1}{2}$ per cent. in retorting, a white heat must be used, by which the retorts are damaged and soon worn-out. Additions of nitre, corrosive sublimate, etc., are not to be recommended. The melting of bullion is dealt with fully in Chapter XVIII. The loss of weight in melting is given by Richards [2] as 1·5 per cent. at the Homestake Mill and 7 per cent. in the Caledonia Mill. It is given by Smart as 0·5 to 1 per cent. on the Rand.[3] Part of this loss consists of gold and silver retained by the slag which is either remelted or passed to the stamp mill or clean-up barrel.

Loss of Mercury.—The loss of mercury in stamp milling is due to (1) "flouring," or minute mechanical subdivision, due to excessive stamping or grinding, and (2) "sickening," or extreme subdivision caused by chemical means.[4] In the latter case, a coating of some impurity is formed over the minute globules of mercury, which are thereby prevented from coalescing, from taking up gold and silver, or from being caught by the plates and wells, as the coating prevents all contact between the mercury and other bodies. The impurity may be an oxide, sulphate, sulphide, or arsenide of some base metal, either originally present in the mercury, or taken up from the ore by it; occasionally the mercury itself may be partly converted into a sulphate or other salt, although this latter condition is not common. The employment of pure mercury, containing no base metals dissolved in it, will reduce the loss due to sickening, but such pure mercury is not always obtainable (except by very careful distillation, in which the first and last portions condensed are rejected) and it soon takes up fresh impurities when used with sulphuretted ores. The base metals usually present in mercury are rapidly oxidised in the air, especially in contact with water; the oxidation is made much more rapid by the presence of any acid in the water, and this acidity (due to the presence of acid sulphates from decomposing pyrites) is rarely quite absent from battery and mine waters, although it is often neutralised by lime. The metallic oxides thus formed are not soluble in mercury, and they float on its surface in the form of little black scales, which soon form a coating. Impure mercury, when used to amalgamate the plates, causes their discoloration by oxidation of the dissolved metals. One of the impurities in mercury most to be feared is lead, as the amalgam of this metal tends to separate out of the bath of mercury in which it is dissolved. According to Prof. J. Cosmo Newbery, it rises to the surface by degrees, taking with it any gold amalgam that may have been formed, and floats as a frothy scum, coating the mercury and preventing any further action by it, whilst it is readily powdered and carried away in suspension by a current of water flowing over it, so that the gold contained in it is lost.

---

[1] Richards, *Ore Dressing*, 1903, vol. ii., p. 782.
[2] Richards, *op. cit.*, p. 784.
[3] Smart, *Rand Metallurgical Practice*, vol. i., p. 93.
[4] Richards points out (*Ore Dressing*, p. 751) that there is no advantage in trying to draw a distinction between these two terms, which was apparently first done by T. A. Rickard (*Stamp Milling of Gold Ores*, 1897, p. 216). Richards holds the view that the globules are in all cases prevented from re-uniting by a film of some foreign substance. The term "deadening" of mercury is also used.

Sickening of the mercury is also promoted by base minerals present in the ore. Most gangues, except heavy spar, hydrous silicates, etc., have no action on the quicksilver; even clean cubical iron pyrite, and other iron and copper pyrites, if they are undecomposed, are harmless, although the materials last named cause sickening when partly decomposed. The other sulphides are all more or less harmful, their action being, however, much less energetic than the compounds of arsenic and antimony. J. Cosmo Newbery conducted a number of experiments in Australia many years ago, to determine the action of some of the base metals on mercury, and found that compounds of arsenic and antimony are particularly harmful, and that if gold containing metallic arsenic is amalgamated, the resulting amalgam is black and powdery, and floats on the mercury, being coated with black metallic arsenic, which separates out and refuses to unite with mercury. Arsenical pyrite seems to act in the same way as metallic arsenic, a large amount of black sickened mercury being produced by it, the action being especially energetic if the pyrite is partly decomposed. The black coating is, in this case, a mixture of pyrite, arsenic, and mercury, in a very finely-divided state. Sodium amalgam acts beneficially when arsenic is causing loss of mercury.

Sulphide of antimony breaks the mercury into black powder even more quickly than arsenic, some sulphide of mercury being formed if there is any trituration, whilst the antimony forms an amalgam. The action of sodium amalgam on this mixture is of no avail, as sodium sulphide is formed, more antimony amalgam produced, and sulphuretted hydrogen set free, the results on the amalgamation of the gold being very disastrous. Bismuth sulphide acts similarly, but with less rapidity.

Floured mercury is perfectly white in appearance, like flour, sickened mercury, as already stated, being blackish. If this floured mercury is examined with a lens, it is seen to consist of a number of minute particles—many of them microscopic—each of which is perfectly bright and pure, shining like a mirror. They are prevented from coming into contact and coalescing by being surrounded by films either of air or of some transparent foreign substance. Floured mercury is readily carried away and lost in the tailings, but if passed through and agitated with a large body of clean mercury much of it is at once absorbed in the mass. The loss through flouring is experienced in the milling both of refractory and free-milling ores. The effects of grease and also of talc, serpentine, clay, and other hydrous silicates in subdividing mercury are doubtless due to mechanical action only.

In California the total loss of mercury varies from $\frac{1}{5}$ to 1 oz. of mercury per ton of ore crushed, the mean being about $\frac{1}{2}$ oz. per ton. Most of the mercury is lost as such and not in the form of amalgam, as is proved by the fact that where the largest proportion of mercury is fed into the battery the greatest loss takes place but the highest percentage of gold is recovered. Thus, at the North Star and Empire Mills the greatest loss in the State occurs, 1 oz. of mercury being lost per ton, but over 90 per cent. of the gold is extracted. In the Blackhawk Mills, Colorado, where base ores are crushed, containing from 12 to 20 per cent. of pyrite, the loss of mercury is from $\frac{1}{5}$ to $\frac{1}{2}$ oz. of mercury per ton. The mills in which the greatest loss of mercury occurs have the deepest discharge, the ore and mercury being in these cases pounded together for a greater length of time before being ejected from the mortar, so that more flouring takes place. At the Ferreira Deep Mill

the loss of mercury is given by D. J. Peplar [1] as 0·35 oz. per oz. of gold
or 0·16 oz. per ton of ore crushed. Caldecott gives it as 0·1 oz. per ton of
ore milled "in good modern practice." [2]

Many suggestions have been made at various times for the reduction
of the loss of mercury. The use of various methods of keeping it clean
and lively and of neutralising the bad effect of base minerals has already
been noticed (see pp. 184 and 191).

Properties and Purification of Mercury.—Mercury solidifies at − 39°
and boils at 357°, but is slightly volatile even at ordinary temperatures.
Its specific gravity is 13·59, its electrical conductivity is 1·6, and its thermal
conductivity 1·8, if that of silver be taken as 100. Pure mercury is unaffected
by the air at ordinary temperatures, but is slowly oxidised if heated to about
350° C. When mercury is impure from the presence of other (oxidisable)
metals, these are rapidly oxidised in air, forming powdery scales. It is not
acted on by hydrochloric acid, and is almost unaffected by dilute sulphuric
acid, but with hot concentrated sulphuric acid it forms $HgSO_4$. Mercury
is dissolved even by cold dilute nitric acid, and is rapidly dissolved by hot
nitric acid. It is dissolved by aqua regia with the formation of mercuric
chloride, $HgCl_2$. Pure mercury will roll down an inclined surface without
forming a pronounced "tail" and without leaving any streak behind it.
If a blackish film is left behind, the mercury requires purification.

When agitated with oil, fats, turpentine, many organic substances, sulphur, etc., mercury is split up into minute globules, not easily re-united.
This is known as the "flouring" of mercury (see also p. 202). Vegetable
or animal oils cause more flouring than mineral oils. Coalescence of floured
mercury is effected by the action of certain reducing agents, such as water
and sodium, the passage of an electric current, or with some loss by the
action of nitric acid.

The vapour of mercury has a poisonous effect (salivation) on the animal
system. Among the remedies are cleanliness, fresh air, acid foods, abstention
from alcohol, and potassium iodide as medicine.

Amalgams.—Mercury forms amalgams directly with gold, silver (more
readily if heated), copper, lead, zinc, bismuth, magnesium, tellurium, and
thorium. Amalgams of tin and cadmium are formed directly with great
ease. Mercury unites with antimony and arsenic only if heated. Antimony
gradually separates from its amalgam as a black powder. Iron amalgam
is formed directly only if the iron is finely divided. When iron amalgam is
retorted, pyrophoric iron is formed, yielding a somewhat troublesome mixture
with gold. Amalgams of nickel, cobalt, manganese, chromium, aluminium,
palladium, and platinum are not formed directly, but are formed indirectly
by electrolysis of their salts with mercury as the negative pole, or merely
by the presence of acid and a stick of zinc. Amalgams of sodium and potassium are formed directly with the aid of heat. The amalgam $Hg_{12}Na_2$ is
solid, containing about 2 per cent. of sodium.

Purification of Mercury.—It is very important to use pure mercury in
ordinary amalgamation processes, so as to reduce the losses as far as possible.
The purification may be effected by distillation with lime and iron filings.
The iron filings decompose sulphides and prevent bumping. An addition
of charcoal powder is mentioned by Richards, its use being to prevent the

[1] Peplar, *J. Chem. Met. and Mng. Soc. of S. Africa*, May, 1904, p. 403.
[2] Caldecott, *Rand Metallurgical Practice*, vol. i., p. 384.

formation of volatile oxides. Lead, cadmium and zinc pass over in part with the mercury. Hulett and Minchin[1] have shown that volatile metals such as zinc and cadmium are not removed by distillation unless it is carried out without bumping and in a current of air, which oxidises them. Zinc, tin, copper, and iron may be removed by shaking with dilute hydrochloric acid; shaking with mercurous nitrate, ferric chloride or potassium dichromate and strong sulphuric acid has also been recommended. Floating impurities are removed by running the mercury through a glass funnel, regulating the discharge by a finger placed over the stem-hole. If mercury is covered with dilute nitric acid (one part of acid to three parts of water) it is gradually purified, especially if stirred occasionally; mercurous nitrate is formed and acts on the base metals. A more rapid way of removing base metals is to pass a stream of air through mercury covered with dilute nitric or sulphuric acid. The base metals are rapidly oxidised by the air and dissolved by the nitric acid. This method has been found useful by T. C. Cloud. L. Meyer[2] lets the mercury fall through a long column of mercurous nitrate, for which nitric acid may be substituted. J. H. Hildebrand[3] recommends that the falling mercury should be broken up by passing it through muslin.

Loss of Gold.—The losses of gold in amalgamation may be ranged under the following heads :—

1. Loss of free gold contained in amalgam, due to flouring of mercury, scouring of plates, etc. This has been dealt with above under the heads " Treatment of the Plates " and " Loss of Mercury."

2. Loss of gold which " floats " in water and is carried away with the slimes. See section on " Float Gold," p. 206.

The remedy for the losses due to the above two causes is the use of drops and mercury traps.

3. Loss of gold which is not in a condition to be directly amalgamated.

The last heading may be subdivided into three, viz. :—

(a) Loss of gold contained in sulphides, tellurides, etc. See " Gold in Pyrite " and the treatment of tellurides described in Chapter XVII.

(b) Loss of free gold, which is prevented from being amalgamated by being coated with a film of some mineral (" rusty " gold), or with grease. See the section on " Rusty Gold," p. 207.

(c) Loss of " free " gold imbedded in particles of rock. The remedy is finer crushing in the battery or in re-grinding machines.

In the preparation of gold ores for amalgamation every care must be taken that the course best suited to each particular case is being pursued. In some instances, which are not common, the whole of the gold may be present in a form in which it can be directly amalgamated. In general, however, the gold is present in two or more forms, one capable and the others not capable of amalgamation. In such cases there is no reason to be dissatisfied with the action of an amalgamating machine if it extracts a high percentage of the free gold, even though the total extraction obtained by it is comparatively low.

---

[1] Hulett and Minchin, *Phys. Rev.*, 1905, **21**, 388; Hildebrand, *J. Chem. Met. and Mng. Soc. of S. Africa*, 1909, **10**, p. 224.
[2] L. Meyer, *Zeitsch. anal. Chem.*, 1863, **2**, 241.
[3] Hildebrand, *J. Amer. Chem. Soc.*, Aug. 1909, p. 934; *J. Chem. Met. and Mng. Soc. of S. Africa*, 1909, **10**, 224.

The following scheme of examining tailing with a view to determine the causes and amount of loss is given by McDermott and Duffield [1] :—

Small samples are taken at intervals from the waste outflow of the mill, until a bucketful is collected ; this is allowed to settle for several hours, the clear water is decanted, preferably through a filter, and the remainder evaporated to dryness. Care must be taken to avoid spilling anything out of the vessels containing the samples. The sample having been well mixed, portions are treated as follows :—

A. One part is panned and examined for free gold, amalgam, and quick-silver. If these are present, it is probably the fault of the millman, and nothing further need be done until this state of things is remedied.

B. The tailing is sized by screening, and the coarse, medium, and fine materials (the latter consisting, say, of that portion which passes a 100-mesh screen) are weighed and assayed separately, the coarser portions being reground and panned to find whether their values are in free gold or in sulphides.

C. The sulphides are separated from the tailing on a vanning shovel or batea, and are weighed and assayed. It may thus be determined whether they are worth saving, and the size of the mesh used for the screens will depend largely on this, and on the nature of the sulphides, which will in many cases be badly slimed and difficult to catch if the ore is finely crushed.

D. The loss due to fine or "float" gold may be determined by assaying the slimes after the sulphides have been carefully removed by concentration. This requires much skill and patience, but can in almost all cases be success-fully accomplished by the vanning shovel. The concentrate may be examined under the microscope for fine specks of gold, but these and the fine sulphides can be recovered by concentration on suitable machinery. The assay value of the tailing from the vanning shovel will give some idea of the amount of float gold which is being lost. It will usually be found to be smaller than may be expected. If it is large, the use of some system of amalgamation more perfect than that by copper plates (such as pan-amalgamation) or of a method of smelting, or of a wet method, may be considered, if the advantages appear sufficient to pay for the presumably increased cost.

"**Float**" **Gold.**—The loss of finely divided or "float" gold, particularly when it cannot be checked by the use of swinging plates, or of drops between the amalgamated plates, is often another name for the loss of slimed sulphide. Many examples have been adduced of the large percentage of the gold in the ores crushed in particular mills, which has been carried away suspended in water in a form not easily recoverable by settling. In the majority of these cases, however, no attempt seems to have been made to distinguish between the values contained in slimed sulphide and those existing as particles of free gold. Where this is not done there are no grounds for the assumption that any free gold is escaping at all. Thus G. M'Dougal, of Grass Valley, California, found [2] that a gallon of water in a stream, $\frac{3}{4}$ mile below two mills, contained on an average 1·18 cents worth of gold. He called this "float" gold, but did not try to find out its physical condition, and it was very likely contained in sulphide. Again at the Spring Gully Mine, in Queensland, the tailings from the battery, if settled in the ordinary way by running off the water, were found to contain 7 dwts. of gold per ton, but if carefully

[1] McDermott and Duffield, *Gold Amalgamation* (London and New York, 1890), p. 7.
[2] *Mines West of the Rocky Mountains*, by R. W. Raymond, 1873.

filtered, assayed 15 dwts. All such examples prove only that the slime is rich, not that "float" gold is being lost, and although it is of course likely that some finely divided gold is carried away in suspension in water during the treatment of many ores, nevertheless, if suffiicent care were taken in ascertaining this loss, it would probably prove to be less than is generally believed.

To aid in catching the float gold, swinging amalgamated plates have been introduced, and are in use in the sluices below the batteries of many Californian mills. They are also used in hydraulic mining. The swinging plate consists of a curved strip of silver-plated amalgamated plate about 3 inches deep, and of the same width as the sluice in which it is hung ; it is suspended on eyes through which wires pass. The plate thus hangs, half submerged, with its concave side up-stream, and is kept swinging by the current, so that all floating particles of gold must come in contact with it. It is found in practice that, immediately under each plate, across the sluice, a line of amalgam which has dropped from the plate accumulates. The plates are placed a few feet apart. They cost little and are very effective.

"Rusty" Gold.—The appearance in the tailing of free gold, which is not especially finely divided, but, nevertheless, is not in a condition to be amalgamated, may be regarded as a rare occurrence, but deserves some consideration. Amalgamation is in these cases prevented by the existence of a thin film of some neutral substance over the surface of the gold. The film may be so thin as to be transparent, but it is enough to prevent contact between the gold and the mercury. The disastrous effect of a film of grease covering gold particles has already been remarked upon. It is said to have been a fruitful source of loss in the treatment of certain ores in the Transvaal that they were impregnated with mineral oil. The effects of grease may be combated by the use of chemicals (caustic alkalies, potassium cyanide, etc.), but it is, of course, better to use every precaution to avoid the introduction into the pulp of candle grease from the mine or of oil from bearings, guides, etc., or contained in steam from the boiler. Losses in amalgamation are also caused by the greasy substances contained in some ores, such as the powdered hydrated silicates of magnesia and of alumina, which cause frothing, and coat the gold with a slime which prevents the action of the mercury.

Other films are formed of oxide of iron, compounds of sulphur, arsenic, etc., or of silica. Some years ago J. Hankey, of San Francisco, had a collection of particles of native gold which appeared as bright and lustrous as usual, but were coated by thin translucent films of red oxide of iron. These particles of "rusty" gold could not be wetted with mercury, but if a piece were snipped off one end, the mercury seized on the fractured surface at once. Such gold seems to be rare in nature.

In 1867, William Skey, of the Geological Survey of New Zealand, after a series of experiments on the ores and tailings of the Thames Valley, came to the conclusion that the bright gold particles which refused to amalgamate were always coated by some compound of sulphur. He found that gold takes up sulphur from sulphide of ammonium or sodium, or from sulphuretted hydrogen, when brought in contact with their solutions, and that after this the gold refuses to amalgamate. He supposed that these compounds of sulphur were often formed by the action of acidulated water on the minerals in ores, and that consequently " a large area of the natural surfaces of native gold is covered with a thin film of auriferous sulphide, and that the greater part of the gold which escapes amalgamation at the battery consists of this sulphurised gold."

Gold in Pyrite.—In the preceding pages no account has been taken of the loss of gold which is contained in pyrite, as it has been assumed that the latter is saved by concentration if it is valuable, and this subject is dealt with in Chapter XI. Nevertheless, as this gold comes under the head of non-amalgamable gold, its physical state and the causes of its disinclination to unite with mercury may conveniently be considered here. In general, pyrite yields only a moderate proportion of its gold contents if it is run over the amalgamated plates (see "Black sand," p. 183), and if it is ground very fine in a pan with mercury the percentage extraction is better. Among the old processes used for the amalgamation of the gold in pyrite may be mentioned the treatment in revolving wooden barrels with mercury, as practised at the St. John del Rey Mine, and the practice of leaving the pyrite to be decomposed by weathering before grinding it with mercury. This method of oxidation seems to be decidedly inferior to the alternative plan of roasting the sulphide, by which the oxidation is rendered more complete and the particles of gold agglomerated to some extent. However, the amalgamation of pyrite, even when roasted, is far from perfect, part of the gold still remaining in a condition unfit for extraction in this way. The ores, which have been met with in various parts of the world, consisting mainly of limonite or hydrated oxide of iron, and in most cases believed to be the result of decomposition of pyritic ores by atmospheric agencies, are also extremely refractory, causing the mercury to sicken rapidly, and yielding only about the same percentage of gold as can be obtained from unoxidised pyrite.

The most celebrated case of this kind is that of the surface ore at Mount Morgan, in Queensland, which was an ironstone gossan consisting of siliceous brown iron ore, derived according to one view from the decomposition of pyrite. Although the gold appeared to be free, it could not be amalgamated, yielding only about 30 per cent. when crushed in batteries and subjected to prolonged grinding in pans with mercury. When the ore was dehydrated by roasting in reverberatory furnaces the extremely fine particles of gold were agglomerated, and between 80 and 90 per cent. could then be extracted by amalgamation, the remainder being presumably coated with oxides of iron. The richness of the ore, however, made even this result unsatisfactory, and a process of chlorination was adopted in practice. It was subsequently found that the ore contained tellurium, which may have been the cause of the difficulties in its treatment.

A similar case was noticed by Mactear [1] in South America, where a limonite ore which only yielded from 35 to 40 per cent. of its gold when treated in Huntington pans, was made to yield between 85 and 90 per cent. by merely subjecting it to a dehydrating calcination before amalgamating it. Louis Janin, Jr., mentions another case [2] in the ores of the Southern Cross Mine, Deer Lodge County, Montana, which consist of limonite derived from the alteration of pyrite. In panning large samples, only one or two specks of gold could be seen, although the ore contained from 1 to 2 ozs. per ton. This ore yielded only about 40 per cent. on being amalgamated, but over 90 per cent. was dissolved out by leaching the raw ore with cyanide of potassium, and similar results were obtained by chlorination. Here the ore was thoroughly

---

[1] Mactear, *Mining Journal*, Jan. 24, 1893, p. 70.
[2] Janin, Junr., *Mineral Industry*, 1892, p. 249.

decomposed, but yet the gold would not amalgamate to a much greater extent than if it were still contained in the original pyrite, whilst the chemicals at once dissolved it. For the condition of gold in pyrite, see p. 79.

Discussion of Stamp-milling.—The stamp battery must be regarded from two different points of view—viz., (a) as a crushing machine, (b) as an amalgamating machine, and it should be remembered that the modifications designed to make it a more efficient crusher often reduce its power as an amalgamator, and *vice versâ*.

Stamps were originally designed as crushing machines, and the tendency has been lately towards a return to this view of their proper function. A practice which seems to be growing in popularity consists briefly in crushing the ore and effecting its discharge from the battery as rapidly as possible. With this object in view heavy stamps are used, running very fast, with a moderate drop ; the screens are coarse, the screen area is large and placed as low down as possible, and the mortar is made narrow, with nearly vertical sides. These arrangements all increase the output of the battery. Amalgamated copper plates are either not placed inside the mortar, or there is a plate on the discharge side only. In the extreme case, no mercury is fed into the battery, and the coarsely crushed ore is re-ground in tube mills before being passed over amalgamated plates. Under these conditions, the output is very large, and may even be as high as 20 or more tons of ore per stamp per day. The percentage of gold extracted by amalgamation is low.

When the stamp battery is used as an amalgamator, the height of discharge (height of screen above dies) is great, amounting even to 12 inches or more, the screens are finer, the supply of water is reduced, and all efforts are directed to fine crushing in the battery. Mercury is fed into the mortar, and amalgamated plates are put inside the mortar on both feed and discharge sides. The object of these arrangements is to keep the ore in the mortar for a long time, so as to increase the chance of catching the gold on the inside plates. The duty of the stamps is, of course, greatly diminished, and may be as little as 1 ton of ore per stamp per day. Light stamps must be used to avoid excessive scouring of the inside plates. Under such circumstances a high percentage of gold is amalgamated. Between these two extremes there are many gradations.

It is probable that no two ores, between which there are considerable physical or chemical differences, can be treated to the best advantage under exactly the same conditions. A millman experienced in the treatment of the ores of one district may be quite at fault when attempting to amalgamate an ore unlike those to which he has been accustomed. A silver mill, in particular, has been pronounced to be the worst possible school for a gold amalgamator, whose work must be closer in proportion as his amalgam is richer than that obtained from silver ores.

It is obvious that the stamps and screens must be such as are calculated to produce the largest possible output, without rendering the pulp unsuitable for the processes of amalgamation or of concentration, or both, which are to follow. The ideal crushing has been often stated to be to " crack the nut and leave the kernel entire," or in other words, to liberate the particles of gold without breaking them. This suggestion, however, is not a helpful one, inasmuch as it has been found in many cases that amalgamable gold remains locked up in particles of ore of all sizes except the very finest slimes. The tendency is now in the direction of finer crushing, if not in the stamp battery, then in re-grinding machines (see below, Chap. X.).

14

The subject of delivery is closely connected with that of crushing and must be considered at the same time. The screens are not usually placed quite close to the level of the dies in the mortar on account of the rapid wear caused by the violent projection of pulp against them when in that position. Their height above the dies is varied according to the ore, the delivery being slower in proportion to this depth of discharge when it exceeds 2 or 3 inches or, according to Peplar, 5 inches. The banking of sand against the screens checks discharge through their lower part if the depth of discharge is small.

The best size for the mesh of the screens must be determined by direct experiment. It has often been contended that, as the crushing must be fine enough to liberate the particles of free gold from their matrix, therefore the size of the screen mesh depends on the state of division of the precious metal in the ore. Even if this be so, however, it does not follow that the apertures in the screen need be small. When coarse screens are used, it is found that, in the course of the crushing, much of the ore has been reduced to a comparatively fine state of division, and usually this portion is found after amalgamation to contain but little gold ; from this, the coarser material, in which the gold is still locked up, may be separated by sizing in suitable machines (see Chap. XI.) and reground in an amalgamating pan or tube mill. If, on the other hand, the slime is found to be as rich as the coarse sand, it may be that no finer crushing is required, as the output would be thereby diminished without any corresponding increase in the yield per ton. If the slime, after separation of the sulphides, is found to contain more free gold than the coarse sand does, this fact points to the conclusion that a coarser screen might be used without detriment, and experiments in this direction should be made, and the limit of economy thus found by trial.

The evils of overstamping, due to slowness of discharge, have been often dwelt on, and probably are frequently exaggerated. It is true that the excessive production of slime thus caused may sometimes be disadvantageous, but, setting this aside, it has been frequently asserted that particles of gold are reduced in size by overstamping, so that they will float off in suspension in water, whilst, even if not reduced in size, they are hammered and flattened so as to be rendered incapable of amalgamation. " Float " gold has been considered above, p. 206, and as regards hammered gold, it does not seem to be beyond doubt that flattening and hardening alone will prevent gold from being amalgamated. Prof. T. Egleston has described a number of experiments [1] which tend to show that amalgamation is retarded by this treatment of gold, but, on attempting to repeat these experiments at the Royal Mint, the author could not obtain results similar to those of Professor Egleston. Pieces of pure gold when subjected to repeated blows with a clean 7-lb. hammer on a clean anvil occasionally showed a disinclination to amalgamate, but if these pieces were washed with dilute ammonia, so as to remove any grease that might be adhering to them, they were instantly wetted by mercury and were dissolved by it at about the same rate as clean annealed gold. It thus appeared that, in these cases at least, grease formed a more potent preventive of amalgamation than the hardness of the gold. Moreover, gold-leaf which has been subjected to an extended course of hammering

---

[1] Egleston, *Metallurgy of Gold, Silver and Mercury in the United States* (1890), vol. ii., p. 586.

is readily amalgamable. It seems probable that the supposed difficulty of amalgamating " hammered " gold has no real existence.

There is still much difference of opinion as to the desirability of attempting to amalgamate the gold inside the battery. The view is now widely held, following the views of the Rand metallurgists, that the addition of mercury to the mortar is a mistake, and that no copper plates should be put inside. It is considered by increasing numbers of millmen that no machine can be successful at once as a crusher and an amalgamator. Nevertheless, the opposite view has not yet been completely abandoned. The practice of adding mercury to the mortar when no inside plates are used is certainly not now much favoured, although it is still adhered to in Ballarat and some other districts in Australia. In treating rich ores, however, when the gold is coarse grained and nearly pure, there does not seem to be any valid objection to be raised against catching the gold on inside plates in a concentrated form, instead of letting it all go to the outside. In this case it is better to add mercury to the mortar, and this is probably not so important a cause of loss of mercury by flouring and sickening as is often assumed. If a decomposing ore is mixed with lime beforehand, and the acidity of the water used is corrected, the conditions do not appear to be favourable to the production of salts injurious to the mercury, and the latter when charged in is probably almost instantly washed through the screens or else dashed against and retained by the plates. The mercury does not remain on the die, subjected to repeated blows, which would no doubt cause much flouring. There is also the point of view that the distribution of mercury through the ore favours amalgamation, and that amalgam is more easily caught on stationary or shaking plates than very finely divided gold. This last point remains a matter of some doubt. It is often denied on the ground that amalgam is of less density than gold. Finely divided gold, however, particularly if it is very impure, containing much silver, requires more mercury for its amalgamation than coarse gold, and in this case it is difficult to keep the plates in good order, so that it is usually advantageous to save the extra trouble and labour, caused by looking after and cleaning-up inside plates, by putting all the plates outside. This has been the experience in a number of mills, including that of the Montana Company, where the inside plates have been entirely discarded.

Amalgamation outside the battery has also been the subject of much discussion and some careful investigation. Numerous amalgamating machines have been patented, the inventors in every case praising their own contrivances and decrying the copper plate, but the latter has not as yet been superseded, and its principle is applied to almost all its more promising rivals.

To secure successful amalgamation it is necessary that the particles of gold should be brought into absolute contact with the mercury. This contact is obtained in one of three ways, viz. :—

1. The mercury and ore are ground together in pans, arrastras, and similar machines, contact being secured by pressing the gold and mercury together.

2. The ore is allowed to flow over or even through a bath of liquid mercury, or the endeavour is made to ensure contact by letting the pulp fall from a height upon the mercury.

3. The ore is allowed to flow over either stationary or shaking amalgamated copper plates, drops being sometimes introduced between the

plates to break up the pulp and to assist in catching the amalgam and " float " gold.

The first method is undoubtedly the best for ensuring contact, but the operation is tedious and, in most cases, unnecessary. As to the relative merits of the last two methods, the great majority of metallurgists are advocates of the superiority of the plates. They point out that in a mercury bath, in spite of the first impression to the contrary felt by every one on approaching the subject, contact with the ore is very difficult to obtain. If wet pulp is introduced at the bottom of a bath of mercury, it rises to the surface in lumps, surrounded by films of water, and dry pulp is still more effectively protected from contact with the mercury by films of air. If a thin stream of pulp is run over the surface of a bath of mercury of sufficient size, the chances of the particles of gold coming in contact with the quicksilver (by settling through the stream) are greatly improved, but in this case the more convenient copper plate could be substituted for the bath. Moreover, it is well known that rich gold or silver amalgam catches gold more readily than pure mercury, and, whilst the surface of the plates can be readily covered with such pasty amalgam, it would involve a larger and unnecessary sinking of capital to keep any considerable percentage of precious metals in the baths. For these reasons, and for the practical reason that plates are found to work better than baths, the use of the latter has been gradually more and more restricted, until they are now only to be found in narrow wells and riffles for the purpose of catching hard amalgam and floured mercury, and as purely supplementary aids to the plates, whilst they are sometimes dispensed with altogether. When a proper disposition of the plates is made, it is rare to find amalgamable gold escaping into the tailing. Even if the latter contain several pennyweights of gold to the ton, this does not show that the amalgamation effected on the plates is unsatisfactory, and, where tailing has to be reground or roasted, or treated in any way that may alter the condition of the gold, before a further extraction by mercury is obtained, no proof is afforded that the plates are not doing their work.

CHAPTER IX.

OTHER FORMS OF CRUSHING AND AMALGAMATING MACHINERY.

Special Forms of Stamps.—Since within certain limits and under certain conditions the capacity of a stamp battery depends on the number of blows given per minute and on the momentum of the fall, various contrivances have been suggested with a view to increase both of these.

In *pneumatic stamps,* such as the Husband and Phoenix stamps, a crank shaft raises the stem, which is attached to an air cylinder. The air in this is compressed, and the stamp head attached to a piston is thus raised. It is forced down to deliver the blow by similar means.

In *spring stamps,* such as the Elephant stamp, a crank shaft raises the stamps against the action of powerful springs, which increase the power of the blow. The rate of wear and cost of repairs have been found to be high in all these stamps, and they have never been much used.

Morison's high-speed stamp [1] is lifted like an ordinary pneumatic stamp, except that the compression cylinder contains water instead of air.

Steam Stamps.—The ordinary form of steam stamp consists of a direct-acting vertical engine, having a steam cylinder and slide valve at the top, the piston-rod being rigidly connected with the stamp. Each stamp head works in a separate round or rectangular mortar, with screens both at the front and back, and sometimes all round. The screens are of Russia sheet-iron with punched holes of about $\frac{1}{5}$ to $\frac{5}{8}$ inch in diameter, the steam stamp being best adapted for coarse crushing. The speed of working is from 90 to 130 blows per minute, and the output is from 100 tons to as much as 400 tons of ore per head in twenty-four hours. It is obvious that gold could not be economically saved on plates inside the mortar of one of these stamps, and as a matter of fact, until recently they were only employed in coarsely crushing the copper ores of the Lake Superior region. Nevertheless, the curious fact seems to be well established that these stamps, with their heavy blow, do not make so much slime as the ordinary gravitation stamp. They have been tried in crushing silver ore in Montana, and gold ore at the Homestake Mine, in the Black Hills, through a 30-mesh screen. As their capacity is so great their use is limited to cases in which large quantities of ore are available, one Ball or Nordberg steam stamp, such as is used in the Lake Superior district, being equal to from twenty to fifty head of gravitation stamps. They may be useful in preparing ores for re-grinding.

The chief advantages of the steam stamp are economy of space and labour. The advantage of subdividing the work among a number of batteries is that stoppages for repairs and breakages affect only a small part of the crushing capacity at one time.

---

[1] D. B. Morison, *Proc. N.E. Coast Inst. of Engineers and Shipbuilders,* 1896-7. Session xiii.

The Tremain Steam Stamp [1] works in sets of two in a mortar, which has front and side screens. The upper ends of the stems are two pistons moving in cylinders actuated by steam at a pressure of about 110 lbs. per square inch. The stem and shoe weigh about 300 lbs. The whole plant is very

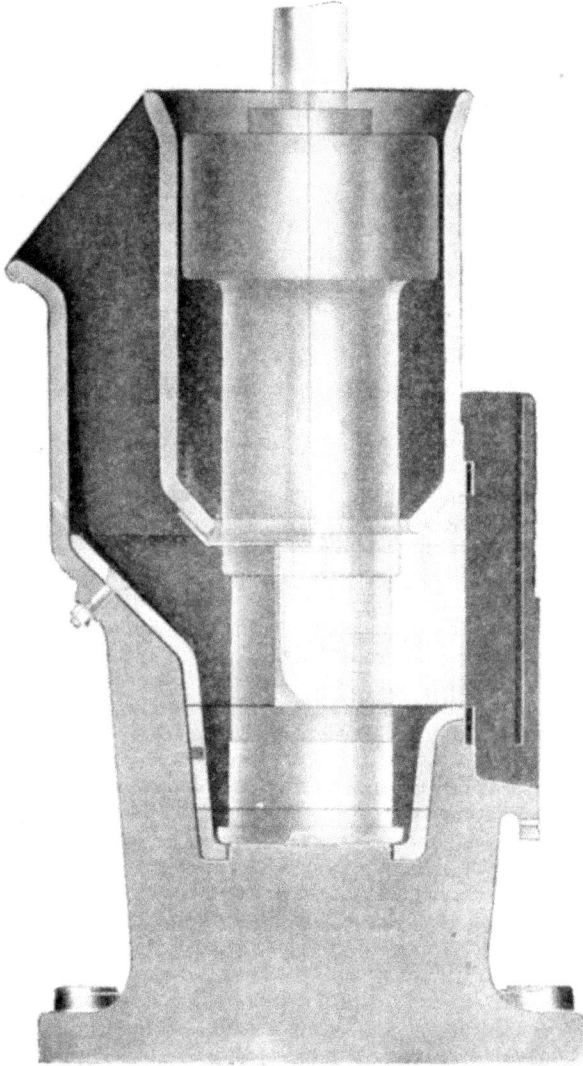

Fig. 95.—Nissen Mortar-Box.

cheap and light, so that it is suitable for work during development and on small mines. Over 40 mills were in use in Southern Rhodesia in 1908, and

---

[1] *Trans. Amer. Inst. Mng. Eng.*, 1896, **26**, 545; S. H. Loran, *ibid.*, Feb. 1904; C. E. Parsons, *Mng. and Sci. Press*, Sept. 19, 1908, p. 386.

they are also at work in South America. The usual output of the two stamps is about 10 or 12 tons per day using 20-mesh screens. They require an excessive amount of power and consume much oil which tends to leak into the mortar and interfere with amalgamation. Inside plates are often used and mercury fed into the mortar-box. These stamps are generally regarded as far inferior to gravitation stamps for treating gold ores, and their use is limited to special cases.

The *Holman Steam Stamp* is also well spoken of.

The Nissen Stamp.—The Nissen stamp [1] (see Fig. 95),[2] introduced in 1903, is a gravity stamp mill with heavy stamps, each stamp having its own cylindrical mortar-box. The latter is made of special steel with easily removable fronts to facilitate removals and repairs. Manganese steel liners, introduced into the mortar in three sections, are used to protect it from wear. Since the blow of the stamp is always given in the direction of the axis of the box, the latter is not subjected to oscillating blows which would tend to weaken the fastenings and the foundations. The mortar usually rests on concrete foundations, to which it is fastened by bolts, as in ordinary stamp mill practice.

The screen has the large area of $3\frac{3}{4}$ to 4 square feet, and passes half-way round the mortar, so that discharge openings on the same horizontal plane are equidistant from the die, and the pulp strikes the screen everywhere at right angles, thus facilitating discharge. A steel frame drawn up tight by two heavy gib keys fixes the screen to the mortar.

The shoes and dies are made of chrome steel and are 10 inches in diameter for the 2,000-lb. stamp. The dies have round bases fitting into recesses in the mortar-box, and are thus kept central. Evenness of wear is claimed as a special feature of the dies. The boss head is enlarged at its upper end, providing a shoulder which counteracts any upward splash of pulp between the boss and the contracted lower portion of the neck. Double-faggotted iron is used for the stems, which have a diameter of 5 inches and a length of 10 feet against a minimum length of 12 or 13 feet in the 2,000-lb. ordinary stamp. There is a greater rotation per lift to the stamp than with ordinary stamps.

Table XXI. appended gives the weights of the falling parts for one stamp. For batteries with more than one stamp the weights are in proportion. The centre of gravity of the Nissen stamp is low, partly in consequence of the great diameter of the head.

TABLE XXI.

Part.	Weight for 2,000-lb. Stamp. Lbs.	
Stem,	652 = 32·6	per cent.
Shoe,	248 = 12·4	,,
Boss head,	798 = 39·9	,,
Head collar,	31 = 1·55	,,
Tappet,	271 = 13·55	,,
Total,	2,000 100·00	,,
Weight of stamp and all accessories,	19,470	
For eight-stamp battery,	95,750	

[1] Nissen, *J. Chem. Met. and Mng. Soc. of S. Africa*, 1911, **12**, 111.
[2] Schmitt, *Rand Metallurgical Practice*, vol. ii., p. 141.

The 1,750-lb. stamp has a boss head weighing 250 lbs. less than that for the 2,000-lb. stamp. The head and shoe together make up 53·8 per cent. of the total weight, against 58·1 per cent. in the case of the City Deep gravity stamp. The cams are made of chrome steel and are self-tightening. The

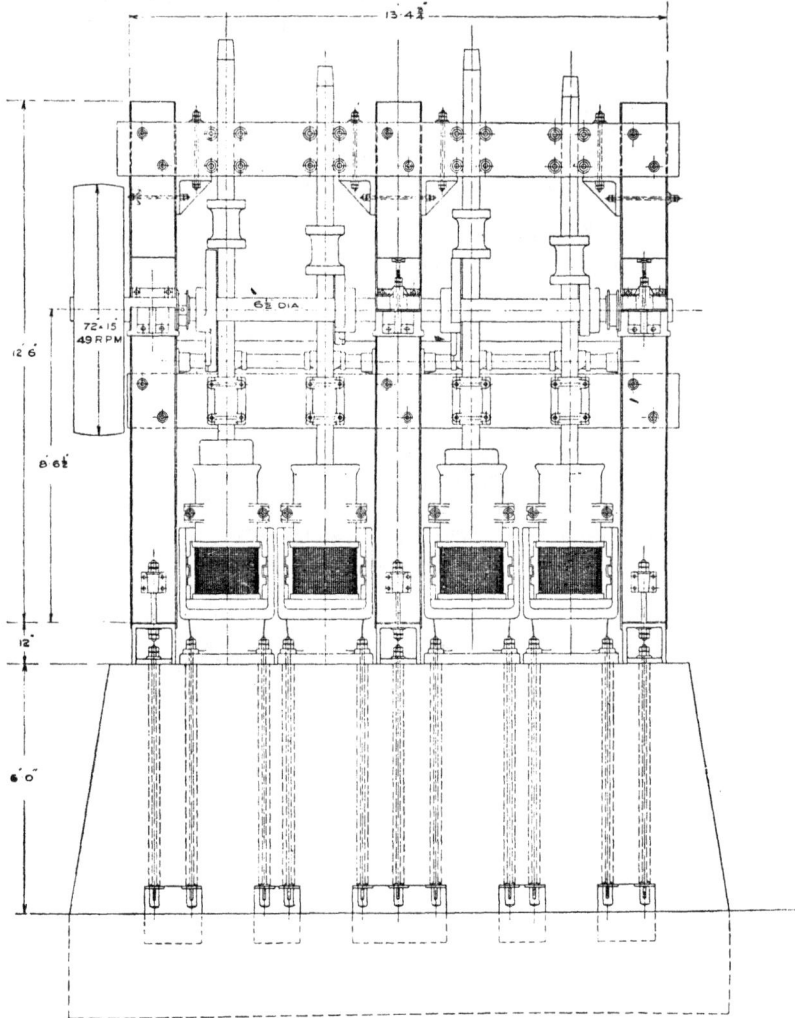

Fig. 96.—Nissen Stamp Battery (Elevation).

cam shaft is $6\frac{1}{2}$ inches in diameter. One feature is that excessive vibration in the cam-shaft is prevented by providing caps for the bearings, as shown in Fig. 96.[1] The design of the king-post (Fig. 97)[2] is also a special feature which is expected to prevent the timber from warping. The water supply

---

[1] Schmitt, *Rand Metallurgical Practice*, vol. ii., p. 143.
[2] Schmitt, *ibid.*, p. 146.

enters through the chute leading from the ore feeders, ore and water thus passing into the mortar together.

Where it is desired to have inside amalgamation a circular conical-shaped copper plate is provided to fit into the mortar. This plate may be removed when loaded with amalgam and replaced by a spare one.

The total height of the battery frame above the foundations is 13 feet, and this admits of buildings being erected of less height than in the old mills. The centre of the cam-shaft is usually 9 to 10 feet above the foundation.

It is apparent that with a single stamp in each mortar, any single unit can be stopped either on account of damage to itself, or because of a breakdown in a part of the mill which it feeds, without interfering in any way with the work of other units.

Several objections have been formulated [1] against the single stamp, and they may be summarised as follows :—

1. The quick drop of a heavy stamp on soft friable ores tends to loosen the battery frame and the foundations.

2. An individual stamp is said to demand as much attention as a five-stamp mill.

3. The mortars tend to fill up and the height of discharge varies, so that there is considerable wear on screens with new dies.

A series of tests [2] which were carried out by the Central Mining and Investment Corporation with four Nissen stamps and a battery of ten ordinary stamps working on similar material at

Fig. 97.—Nissen Battery ; Design of King-Post.

the City Deep Mine, showed a distinct superiority in favour of the former. A summary of figures from these tests is given in Table XXII.

TABLE XXII.

	NISSEN. Four Stamps.	CITY DEEP. Ten Stamps.
Running weight, . . .	1,927 to 2,245 lbs.	1,775 to 1,863 lbs.
Drops per minute, . . .	103	100
Height of drop, . . .	8½ inches.	8½ inches.
Screen, ...	9 holes per sq. in. ⅜ inch aperture.	9 holes per sq. in. ⅜ inch aperture.
Height of discharge, . .	2 to 2⅝ inches.	2¼ to 2½ inches.
Stamp duty (9-mesh), . .	24·47 to 30·85 tons.	18·26 to 20·95 tons.
,, (⅜ inch aperture), .	36·69 to 37·74 tons.	22·72 to 24·34 tons.
Power consumption per ton of ore,	2·7 to 4 H.P.-hours.	4·45 to 5·5 H.P.-hours.
Quantity of ore crushed per pound of falling weight, .	25·35 to 37·91 lbs.	19·6 to 27·45 lbs.

The superiority of the Nissen stamp in this test is considered to be due

---

[1] *Mng. and Sci. Press*, 1907, **94,** 115, 147, 303.
[2] P. N. Nissen, *J. Chem. Met. and Mng. Soc. of S. Africa*, 1911, **12,** 111.

to the greater facility with which the feed can be adjusted to the capacity of each individual stamp, in conjunction with the greater screen area.

As a result of the above tests Nissen stamps have since been installed in the City Deep Mill and in some new mills in Rhodesia, and so far as published results show are working satisfactorily.

It is claimed (1) that on the average the Nissen stamp requires 30 per cent. less power than the ordinary stamp for the same tonnage, representing a saving of 8d. to 10d. per ton of ore milled ; (2) the maintenance costs are considerably reduced ; and (3) the plant requires less capital expenditure.

Del Mar calculates [1] from a study of the crushing efficiencies—according to the law that the work done is proportional to the area exposed and to the power required per unit of falling weight—that the mechanical effects of a one-stamp Nissen mill and an ordinary five-stamp mill are in the ratio of 100 : 59.

Stadler,[2] in commenting on the Nissen stamp, observes that in the ordinary stamp battery the particles issuing from the mortar are more or less rounded, showing that the reduction has been by attrition and abrasion. He considers that in the Nissen mill the particles are less subject to this action and less energy is lost by the churning of the water. Stadler also expresses his belief that this stamp will prove very efficient as a fine grinder.

The Huntington Mill.—The Huntington Roller Mill, here described as a type of the many good roller mills now in use, is best suited for the fine crushing of ores which are not too hard. It consists of an iron pan, at the top of which a ring, B (Fig. 98), is set, and attached to this are three stems, D, each of which has a steel shoe, E, fastened to it. The stems are suspended from the ring and are free to swing in a radial direction, as well as to rotate round their own axes, whilst the whole ring, B, with the stems and shoes, revolves round the central shaft, G. The shoes or rollers, as they are called, are thus driven outwards by centrifugal force and press against the replaceable ring die, C. In front of each roller is a scraper, F, which keeps the ore from packing. The rollers are suspended with their bases at the distance of 1 inch from the bottom of the pan, which can also be replaced when worn. The lowest part of the screen is situated a little above the top of the rollers, and outside it there is a deep gutter into which the ore is discharged, and from which a passage leads to the amalgamated plates. The ore and water being fed into the mill through the hopper, A, generally by an automatic feeder, the rotating rollers and the scrapers throw the ore against the sides, where it is crushed to any required degree of fineness by the centrifugal force of the rollers acting against the ring die. From 17 to 25 lbs. of mercury are placed in the bottom of the pan, the clearance below the rollers permitting them to pass freely over the mercury without coming in contact with it, so that it is not stirred up and " floured," but the motion is such as to bring the pulp in contact with the quicksilver. The speed of the mill is from 45 to 75 revolutions per minute. The ore should be broken in rock-breakers to a maximum size equal to that of a walnut, or, better still, of a cobnut, before being fed in. The action of the rollers is one of impact rather than of grinding, the ore being granulated without the production of much slimes. The free gold, as soon as it is liberated from its matrix, is in great part amalgamated and retained by the mercury at the bottom of the pan, the remainder being

---

[1] Del Mar, *Eng. and Mng. J.*, Dec. 14, 1912, p. 1129.
[2] Stadler, *South African Mng. J.*, Dec. 2, 1911.

caught on the plates outside the mill. Coarse gold is caught inside and fine gold outside the mill, but the yield inside is comparatively small when ores with high percentages of sulphides are in course of treatment.

The mill is particularly adapted for the treatment of ores containing brittle sulphides, which, if pulverised by stamps, are liable to become slimed, and so to be in an unsuitable condition for concentration. It is also suitable for argillaceous quartz, which yields its gold more readily under the " puddling " action of the rollers than when pounded by stamps. Moreover, the Huntington mill does much more satisfactory work than stamps on soft ores or in regrinding coarse tailing. The reason for this lies, of course, in the relatively large amount of screen area in the mill and its consequent high efficiency of discharge, a point in which stamps are decidedly inferior to it. As the splash is heavy against the sides the wear of the screens is somewhat rapid, but they can be very quickly replaced.

Fig. 98. Huntington Mill.

The capacity of a 5-foot mill, the one which is most commonly in use, is from 10 to 20 tons of rock per day through a 30-mesh screen, the power required being from 10 to 12 H.P. The weight of each of the roller shells, which are replaceable, is about 170 lbs. The wear and tear on the replaceable parts is very great, amounting to about 14 ozs. per ton of rock crushed when soft ores, previously broken small, are being treated. If large pieces of hard quartz are fed into the mill, or if the mill is overfed, the mercury is splashed against the screens and passes through with the pulp, and when by accident pieces of iron or steel are introduced, the ring die is occasionally broken. Another source of disaster in Huntington mills lies in the use of acidulated water, such as that derived from mines or encountered when decomposing pyritic ores are treated ; the mill is rapidly corroded and rendered unfit for work by such water.

The chief advantages supposed to be gained by the use of Huntington mills instead of stamps may be thus epitomised :—

1. *Reduced First Cost.*—The cost for the same capacity is not more than two-thirds that of stamps, even at the manufacturers' shops, while the difference in favour of the mill is even more in outlying districts from its light weight, and corresponding low freight, and from the cheapness of its erection.

2. *Saving of Power.*—The mill is said to run with about one-half the power per ton of ore crushed.

3. *The wear and cost of renewals is less* for the mill than for stamps, the cost being from twopence to threepence per ton of ore for the former, against about fivepence or sixpence for the latter.

4. *There is less loss by flouring* of amalgam and quicksilver, while the good discharge and absence of grinding leaves the pulp in a better condition for concentration.

The first three advantages appear to refer only to such soft and brittle ores as are especially suited to the Huntington mill. The mill requires to be set to work in an intelligent manner by experienced and skilful hands, and watched carefully. The dangers of over-feeding have been already alluded to. One difficulty in automatic feeding is that self-feeding, such as is carried on by stamps, is impossible. The automatic feeder must work separately and be set to feed a certain weight of ore per hour, this weight having been determined by trial. If, after this, there is any change in the hardness of the rock, no automatic change in the rate of feeding takes place, and the machine may be choked up or run at below its maximum capacity unless watched and the feeder regulated. Another difficulty is in the quantity of water to be added. An excess of water, making thin pulp, does not favour internal amalgamation, and it may be stated that, in general, the pulp should be kept as thick as possible, consistent with its prompt discharge through the screens when sufficiently fine. If the pulp is too thick to run easily over the copper plates outside the mill, water may be added there by means of a perforated pipe. The rate of running should be as high as possible, since, if the other conditions are the same, the crushing power varies as the cube of the number of revolutions per minute.

A few examples are appended of the results obtained in actual practice by this excellent machine. At the Spanish Mine, Nevada County, California, there are [1] four Huntington mills, three of 5 feet diameter and one of 4 feet diameter. The ore is free-milling and is passed through a Blake stone-breaker and thence to the mills. The four mills run at 58 revolutions per minute, and pulverise 35 tons of ore each in twenty-four hours, to pass through a slot screen equal to 20 mesh. The pulp is passed over the usual amalgamated plates after leaving the mills, and $\frac{1}{4}$ oz. of mercury is added with each ton of ore. Forty-five per cent. of the gold recovered comes from the inside of the mill, where the amalgam obtained is much richer than that from the plates. The loss of quicksilver is from $\frac{1}{15}$ to $\frac{1}{30}$ oz. per ton of ore, and the total cost of milling is about one shilling per ton, while for the month of November, 1887, it was only tenpence per ton. The ore is a soft talcose slate, containing streaks and veins of ferruginous quartz, carrying gold. The chief trouble in working lies in the frequent re-adjustment of feed which is found necessary. In a special test run of one month 42·4 per cent.

---

[1] This account refers to the practice in the year 1887.

of the gold contents was extracted, and the remainder lost in the tailings. This poor result was probably due to over-feeding, but profits were made. nevertheless, although the ore yielded only a little over 1 dwt. of gold per ton in 1887 and 1888. Twenty-two horse-power were used by the Huntington mills in crushing from 120 to 140 tons per day.

At the Shaw Mine, El Dorado County, a 5-foot mill, making 50 revolutions per minute, pulverised 10 to 12 tons per day, so as to pass through a 25- to 30-mesh screen. At the Mathines-Creek Mine, in the same county, a 5-foot mill pulverised 9 to 10 tons per twenty-four hours, so as to pass a screen equal to a 40 mesh. At the Monto Cristo Mine, Mono County, two 5-foot mills, running at from 65 to 75 revolutions per minute, pulverised 2 tons of ore per hour to pass a screen equal to a 40 mesh ; 25 lbs. of quicksilver were charged into each mill at the commencement of the run, and about ½ oz. more added each half hour.

In an account of Huntington mill practice at Kalgoorlie,[1] von Bernewitz gives a list of ten 5-feet mills in use there in 1912. In these mills, the percentage of gold recovered by amalgamation, in the cases where it was stated, varied from 20 to 60 per cent. The ores consisted of mixtures of oxide of iron, ferruginous clay and quartz. Punched screens equal to about 30-mesh wire screens were used in all cases. The speed of the mills was from 68 to 74 revolutions per minute, and the daily capacity was from 30 to 80 tons of ore per mill, according to the hardness of the ore. The cost of milling in four cases cited was from 1s. 8d. to 3s. 4d. per ton, including the cost of water. The pulp from the Huntington mills was collected and the sand cyanided.

Although cases have been adduced in which 90 per cent. of the gold contents of an ore crushed in the Huntington mill was retained inside the machine, this is decidedly exceptional, and the mill is probably inferior to the stamp mill as an amalgamator on many ores.

The high cost of repairs in the Huntington mill limits its use. The objection to the mill is not so much the actual wear, but that certain parts of the mill wear out in one spot, so that a large casting, little worn in other places, has to be scrapped (Semple). Suggestions are made by C. C. Semple to reduce the excessive wear and to make replacement of parts cheaper.[2] The absence of sliming effect is in favour of the use of the mill for re-grinding tailing for concentration.

Among other roller mills, the *Bryan Mill* is one of the most successful.[3] It resembles the Chilian mill and consists of an annular mortar in which are fixed segmental annular steel dies. The ore is crushed by three vertical rollers with fixed horizontal axles, which rotate in journals fixed to an annular horizontal rotating plate. This plate revolves round the central axis of the machine, and the rollers are made to run round on the dies. A scraper follows each roller to keep the dies clean, and to discharge the pulp through the screens which form the wall of the trough all round. The pulp runs round at the rate of 300 feet per minute or more at the periphery, the plate moving the rollers making 30 revolutions per minute. The mill is suitable for hard as well as for soft ores. The capacity of a 4-foot mill is about 15 or 20 tons

---

[1] Von Bernewitz, *Mng. and Sci. Press*, Nov. 16, 1912 ; *Cyanide Practice*, 1910 to 1913, p. 177.
[2] Semple, *Trans. Amer. Inst. Mng. Eng.*, 1911, **42**, 602.
[3] See also article by F. Furman in *Metal Miner*, 1896, p. 266 : and E. A. Tays, *Trans. Amer. Inst. Mng. Eng.*, 1899, **29**, 776.

of quartz ore per day through a 40-mesh screen. The Bryan Mill has been introduced on the Pacific Coast, in Mexico, and in Australia for crushing gold ores.

The *Chilian Mill* and other fine grinders are described in Chap. X.

The Tyrolean Mill.—In Hungary, in Transylvania, and in the Tyrol, the principle of separation of the operations of crushing and amalgamation is still successfully used, the mills employed having a strong resemblance to the amalgamation pans described below (pp. 223-227). In some districts of Hungary and Transylvania, Californian stamps have been recently introduced, displacing to some extent the square German stamps, but the amalgamation is still effected in bowls or pans, called in Germany " *Quickmills*," which are the modern equivalents of the old Tyrolean mills still to be found at work in certain retired valleys in the Eastern Alps (see p. 152). The best known of these modern machines are the *Schemnitz Mill* and the *Lazzlo Amalgamator*.

The *Schemnitz Mill*, shown in Fig. 99, closely resembles the ancient form

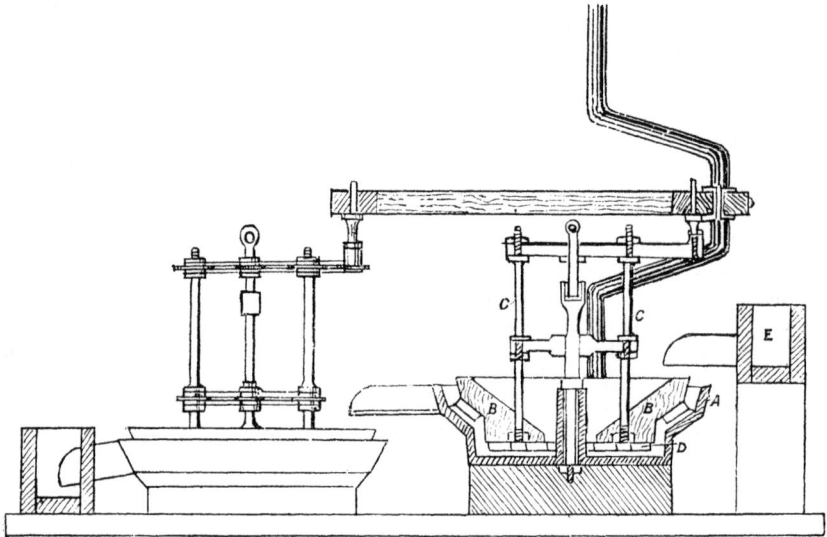

Fig. 99.—Schemnitz Mill.

and is always used in pairs, one overflowing into the other. It consists of a cast-iron bowl, about 2 feet in diameter at the top and 18 inches at the bottom, with an internal depth of 7 inches. A massive wooden muller, B, hollowed out inside in the form of a cone, is suspended in the bowl by the iron rods, C, and is revolved at the rate of about thirty turns per minute. About 26 lbs. of mercury are charged into the bowl, and the lower face of the muller is set with twenty iron teeth, D, placed radially, which almost touch the mercury and force the pulp (delivered into the mill from the trough, E) outwards towards the periphery of the bowl, whence it overflows into the second mill.[1]

In the neighbourhood of Schemnitz, in Hungary, lead ores, containing

---

[1] Schnabel and Louis, *Metallurgy*, 1905, vol. i., p. 961.

2 to 4 dwts. of gold per ton are crushed by stamps and are treated in these mills, the tailing being concentrated for smelting. At Nagybanya, in Hungary, the loss of mercury in these mills amounts to from $\frac{1}{2}$ to 1 part for each part of bullion recovered. Thus at the Kreuzberg works, near Nagybanya, in one month 155 metric tons of ore were treated, and 1·957 kilogrammes (62·9 ozs.) of gold were recovered with the loss of 1·9 kilogrammes of mercury, and in another month 754·5 metric tons of ore yielded 5·756 kilogrammes (185·8 ozs.) of gold, with the loss of 2·9 kilogrammes of mercury.[1] The clean-up usually takes place once or twice a month, and the capacity of each pair of mills is from $\frac{3}{4}$ to 1 ton of ore per day.

The *Lazzlo Amalgamator* [2] differs from the Schemnitz mill mainly in having a flat-bottomed bowl, which is furnished with two circular iron partitions dividing the bowl into three concentric compartments. The ore is fed into the centre one and overflows into the others in succession, and thence into a second smaller bowl. The muller is of iron, and dips into each compartment of the bowl compelling the pulp to pass down and come in contact with the mercury three times before it escapes, and the gold, owing to its density, does not readily pass upwards and over the partitions.

At the Füzesd Dreifaltigkeit Mine, near Boicza, in Hungary, the ore is crushed by Californian stamps, the capacity of which is 0·8 ton per head in 24 hours, and then passed through Lazzlo amalgamators. The amalgam is collected in settling pans, which resemble the amalgamators, but have no iron teeth, and the tailing is classified in spitzkasten, and concentrated by passing over buddles and then over canvas tables. The auriferous pyrite is caught on the buddles and sold to smelters, but the product of the canvas tables is very rich in free gold, and is ground with mercury in iron mortars by hand with pestles. The Lazzlo amalgamators are about 25 inches in diameter, and each pair treats from 1·7 to 2 tons of ore in twenty-four hours, 75 to 80 per cent. of the gold being saved. The loss of mercury is about 1 oz. per ton of ore, and the power required for twenty-four pairs of amalgamators and eight settlers is 4 H.P.

At a mine near Brad, in Transylvania, each pair of amalgamators treats from 3 to $3\frac{1}{2}$ tons of ore per day, but only 55 per cent. of the gold is extracted, the tailing being treated in American pans, and concentrated on Bilharz tables for smelting. Specimens of the ore containing visible gold are amalgamated by hand in mortars, and the rest contains about 8 dwts. per ton.

Amalgamation Pans.—An amalgamation pan consists of a circular cast-iron pan, provided on the inside with a renewable false bottom of cast iron—constituting the lower grinding surface—and a " muller," or upper grinding surface (d, Fig. 100), attached to a vertical revolving spindle, g, which is set in motion by bevel wheels, t, placed below the pan. The muller grinds to impalpable pulp ore which has been already reduced to a coarse powder by stamps, and also mixes the ore with mercury, introduced into the bottom of the pan, and so amalgamates the gold and silver. The origin of the pan is probably to be traced to the Mexican arrastra, and some of the varieties of the pan are merely slightly modified arrastras. One variety consists of a sectional wrought-iron pan fitted with a granite grinding bottom and with granite mullers, which are attached to a vertical spindle rotated by hand or by animal power.

In work on gold ores the use of amalgamating pans was formerly mainly

---

[1] *Loc. cit.*   [2] *Ibid.*, p. 963.

limited to regrinding skimmings, blanket sand, and concentrate obtained
in working a stamp mill. Pans are now largely used at Kalgoorlie in grinding
roasted sulpho-telluride ore (see below, p. 227, and also Chap. XVII). In
this case some gold is amalgamated, but the object of the pans is in part to
prepare the ore for cyaniding. They are also used, at Kalgoorlie, in the wet
crushing mills, merely as fine grinders of unroasted ore, as an alternative
to tube mills (q.v.). In this case, amalgamation is not practised, and the
ore is merely prepared for cyaniding or concentration.

Silver ores were formerly often crushed in a battery, roasted with salt
if necessary, and then amalgamated in pans. Silver ores containing considerable quantities of gold were often similarly treated, but with gold
ores proper it was seldom necessary to resort to this process, which has
now become obsolete, and only a brief account is given below.[1]

Gold ores which do not yield a fair percentage of their values when run
over amalgamated plates are occasionally treated in pans. In such cases the
ore may be roasted or treated raw. As already stated (p. 208) it is seldom
advantageous to roast a gold ore before amalgamation, since, although in
a roasted pyritic ore specks of free gold may often be detected where none
were visible in the raw ore, a part of the precious metal usually appears
after roasting to be difficult to bring in contact with mercury. The cause
of this is not always easy to discover, but it may sometimes be due to the
coating of gold by thin films of iron oxide or other minerals. Moreover,
the addition of salt to a gold ore in the roasting furnace, as is pointed out in
the chapter on chlorination, is often attended by appreciable losses by
volatilisation. These two causes are sufficient to account for the low percentage of gold usually extracted when an auriferous silver ore is treated
by roasting with salt and pan-amalgamation. Under exceptional circumstances a gold ore may prove to be satisfactorily handled by roasting and
amalgamation.

Pan-amalgamation, whether the ores treated are raw or roasted, may
be conducted in one of two ways. The older system is to crush wet in the
stamp mill, and collect the ore in large shallow settling pits or pointed boxes
(see p. 256). A sufficiently dry pulp having been obtained by draining, it
is dug out by hand and charged into the pans. A newer method, the continuous system, is briefly described on p. 226.

Old System.—The amalgamating pans in use are very numerous, and vary
greatly in form. The shape of the bottom was formerly much in dispute,
flat, cone-shaped, and hemispherical bottoms each having its advocates,
but it is now generally believed that flat-bottomed pans are the best, wearing
more evenly and doing more work. The pans are often heated, so as to
increase the rate of amalgamation, by means of steam led through a chamber
below a false bottom in the pan, but the more economical device of introducing steam into the pulp itself has also at all times been in use. The
objections to the latter course are that the pulp may be so much diluted
that amalgamation is checked, and that oil is liable to be introduced with
the steam with equally disastrous results. When the ore is roasted before
being treated in the pan, it is in some mills charged in hot, hot water being
added also, and as the pan is covered up and is still warm from the previous
charge, it remains at a sufficiently high temperature throughout the operation

---

[1] See H. F. Collins, *Metallurgy of Silver*, pp. 80-111, for a detailed description of the
Pan-amalgamation Process.

without further treatment. The grinding of the ore by the muller is an additional source of heat.

One of the common forms of amalgamating pans is shown in Fig. 100. This pan is 5 feet in diameter, with cast-iron bottom, *a*, and wooden sides, *h*. The mullers are shown resting on the cast-iron dies, *c*, which protect the bottom from wear, whilst replaceable shoes attached to the lower surface of the mullers are also shown. The shoes and dies can be kept in contact while the spindle, *g*, is rotated, so that the ore can be ground, or the muller can be raised by rotating the hand-wheel and centre screw, *j*, on the top of the spindle, so that only circulation and mixing of the charge take place. In some pans copper plates, *l*, are introduced, being attached to the side walls and projecting into the interior. These plates are intended both to mix the pulp and to catch the amalgam, much of which is retained on them.

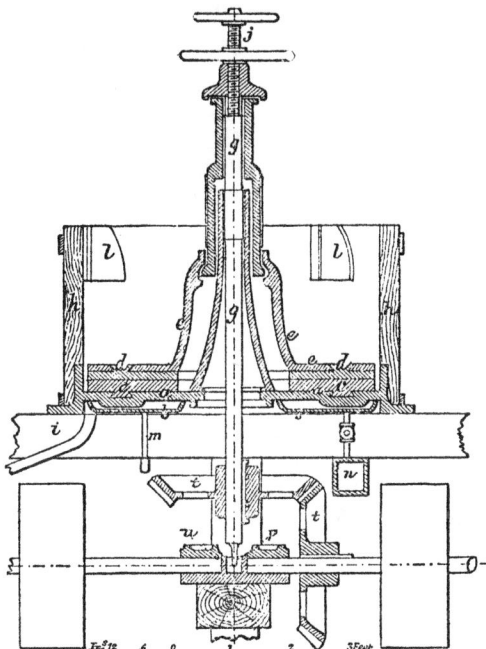

Fig. 100.—Amalgamation Pan.

The more usual system is to employ separate vessels called settlers for the collection of the quicksilver and amalgam, after the pans are discharged. The speed of the muller is usually from 65 to 75 revolutions per minute. Below the muller the pulp is continually worked from the centre of the pan to the circumference, being returned towards the centre above the muller and passing down through the latter by inclined slots which terminate near the centre. In Fig. 100, which represents the form known as the Patton pan, *n* is the main through which steam is passed into the chamber, *b*, to heat the pulp, and *m* is the outlet pipe.

The method of operation is as follows :—The charge of ore is introduced with the mullers raised slightly and kept revolving, water being added at the same time in quantities sufficient to make the pulp of a pasty consistency,

15

so that globules of mercury remain suspended in it without subsiding. The mullers are then lowered and the ore ground for from two to four hours, after which the mullers are raised and the mercury added gradually, and thoroughly mixed with the pulp for six to eight hours longer. The object in raising the mullers is to prevent the sulphides from being ground up with mercury, which would cause considerable losses by flouring and sickening. Nevertheless this raising of the mullers is not an invariable practice. When the amalgamation is thought to be complete, water is introduced to dilute the pulp, and the whole is discharged into a settler; or else the diluted pulp is stirred by the raised muller at a reduced rate of speed until the globules of mercury have re-united and sunk to the bottom, when the pulp is gradually run off, beginning at the top, usually by pulling out in succession plugs set in the side of the pan at different levels. The discharge takes place into a bucket or tub, where some of the mercury accidentally carried over is caught. The bulk of the mercury in some mills is drawn off from the bottom of the pan before the pulp is discharged.

In order to facilitate amalgamation various chemicals have been recommended as desirable additions to the charge. In later practice, however, only salt, sulphate of copper, nitre, cyanide of potassium, lime and sodium amalgam were used (see pp. 184 and 191). In treating gold ores, cyanide of potassium and sodium amalgam were added to keep the mercury clean and lively, but both chemicals are now comparatively rarely resorted to. Salt and sulphate of copper are chiefly added to silver ores, their use having been suggested by the Patio process. They are believed to decompose certain base minerals, and so to prevent the sickening of mercury, which would otherwise be caused by their presence, and also to liberate silver from some of its compounds and thus render it capable of amalgamation. The use of lime is of course to neutralise any acid sulphates of iron, etc., which may be formed by the partial decomposition of the ore, and so prevent the sickening of the mercury. If added when the pulp is diluted, lime is said to be efficacious in assisting the mercury to collect together and settle.[1]

The results of a number of careful laboratory experiments with a small amalgamating pan are given by H. O. Hofman and C. R. Hayward.[2] The influence of time, and amounts of salt and blue vitriol were examined.

The Boss Continuous System.—In this system of pan-amalgamation the pulp is continuously run direct from the stamp battery through a series of pans arranged so that each overflows into the next one, which is placed at a slightly lower level. The first two or three pans are arranged as grinders, the battery pulp not being fine enough for complete amalgamation, and the pulp is then passed through a series of amalgamating pans supplied with mercury. After this the mercury and amalgam are separated from the ore in settlers, which are larger pans in which the pulp is diluted and stirred less vigorously. The tailing overflows from the settlers, and is run to waste or led over concentrators. The number of pans arranged in series through which the pulp must pass, in order to yield a fair percentage of its precious metals, is determined by experiment for each particular ore. It is obvious that the consistency of the pulp must be thinner than has usually been considered desirable for successful amalgamation, and, as a matter of fact,

---

[1] For a full account of the chemical reactions involved in pan-amalgamation, see the *Report of the United States Survey of the Fortieth Parallel*, vol. iii., chap. v.
[2] Hofman and Hayward, *Trans. Amer. Inst. Mng. Eng.*, 1909, **40,** 382.

its volume is usually doubled by the introduction of the continuous process, but in spite of this the percentage of extraction is not lower than by the old method. By the Boss system there is a large saving in labour, fuel, and in wear and tear ; the settling pits or pointed boxes are dispensed with, and no movement of the pulp by hand is needed. The mercury is collected in wells and pumped up into tanks, whence it is fed automatically into the amalgamating pans.

Modern Practice with Pans.—The pans used at Kalgoorlie, Western Australia, are of the Wheeler type.[1] The Wheeler pan differs from the combination pan chiefly in the shape of the shoes and dies, which have curved radial and bevelled sides instead of straight radial sides (Gowland). The Kalgoorlie Wheeler pans are usually 5 feet in diameter, revolving at 45 to 60 turns per minute and requiring about 5 horse-power. They range up to 8 feet in diameter.[2] They have heavy mullers, which can be raised or lowered as usual. " A set of shoes and dies lasts from four to six months on roasted ore, and when half-worn down a compensating weight of some 600 lbs. is put on the top of the plate. They grind about 10 tons per day, at a cost of 1s. 7·8d. per ton milled. The side feed is satisfactory and is almost universal," [3] the alternative being central feed. The Cobbe-Middleton pan [4] was introduced at Kalgoorlie in 1907. It differs from the Wheeler pan in having the pressure between the shoes and dies maintained constant by means of weighted levers acting below the body of the pan and pressing it upwards. The body of the pan is held by guides, up and down which it can slide, and when it is desired to release the pressure between the shoes and dies, the body of the pan can be depressed by rotating a hand wheel. The feed is central. The ore from 40 stamps at the Hainault Mine without previously passing over amalgamated plates was sent to five of these pans, which amalgamated the pulp and reduced it from 8-mesh to 40-mesh actual. About 7 horse-power was required per pan. The tailing was classified and concentrated, and the concentrate roasted and cyanided.

At Kalgoorlie pans are used for grinding and amalgamating wherever the roasting process is in vogue, and at the wet crushing mills both pans and tube mills are used for fine grinding. Pans take the ball-mill product, of which 40 per cent. will pass a 150-mesh screen, and reduce it so that 90 per cent. will pass the same screen (von Bernewitz).

Wheeler pans have been found advantageous at the Ivanhoe Mill [5] for regrinding, and Broadbridge suggests that the pan should be used as an intermediate grinding medium, afterwards passing the product through tube mills.

At the Ivanhoe Mill it was found by Nicholson [6] that the product of 10 stamps could be dealt with by two pans, using battery screens of 10 mesh or 100 holes to the square inch. The screen tests gave the results in Table XXIII.

Nicholson introduced the use of compensating weights to fit on the muller when it had become partly worn. Even when these are used, the product

---

[1] For description of the original Wheeler pan, introduced in 1862, see J. A. Phillips, *Gold and Silver*, 1867, p. 397 ; Schnabel and Louis, *Metallurgy*, 1905, vol. i., p. 793.
[2] Gowland, *Non-Ferrous Metals*, 1914, p. 276.
[3] Von Bernewitz, *J. Chem. Met. and Mng. Soc. of S. Africa*, 1909, **10**, 222.
[4] *Mining Mag.*, 1909, **1**, 213.
[5] W. Broadbridge, *Trans. Inst. Mng. and Met.*, 1904, **14**, 104.
[6] Broadbridge, *ibid.*, p. 149.

of pans is not quite so fine as that given by tube mills, but there are compensating advantages. Nicholson considered that five 5-foot pans would do the same work as two 16-foot tube mills (so that the first cost of pans would be less than half that of the tube mills), and that the cost of running would be about the same. The power required per pan was about 5·3 H.P. at the Ivanhoe Mill.

TABLE XXIII.

	Pulp Leaving Mortar Boxes.	Product of Spitzkasten Entering Pans.	Product Leaving Pans.
	Per Cent.	Per Cent.	Per Cent.
On 20 mesh, . . .	15·6	26·3	Nil.
,, 30 ,, . . .	9·7	15·7	,,
,, 40 ,, . . .	4·2	6·4	,,
,, 60 ,, . . .	16·4	16·4	10·6
,, 90 ,, . . .	8·0	8·7	18·6
Through 90 ,, . . .	46·1	26·5	70·8

M. G. F. Söhnlein [1] found a 5-foot pan efficient for fine grinding small quantities of sand concentrate in Bolivia. The pulp was de-watered to contain 48 per cent. of solid material and fed centrally through the cylinder around the central column, so that it all passed under the mullers, and was discharged after the single passage. At one passage about 30 per cent. of the feed was sufficiently ground. The oversize was returned from a Dorr classifier. The speed of the mullers was 60 revolutions per minute. The feed was 38 tons of sand from Overstrom tables, the quantity of sand actually ground to pass 200-mesh being 30 tons per day. The material was not hard (consisting of 50 per cent. of quartz, felspar and slate and 50 per cent. of iron oxide obtained by roasting pyrite), and was graded as follows, before going to the pan :—

On 40 mesh, ... 20·0 per cent.
,,   60 ,, ... 20·5 ,,
,,   80 ,, ... 16·5 ,,
,, 100 ,, ... 10·5 ,,
,, 150 ,, ... 8·5 ,,
,, 200 ,, ... 5·5 ,,
Through 200 ,, ... 18·5 ,,

The power required was 6·26 H.P., so that 4·76 tons were slimed per H.P. per hour. The total cost was 30 cents per ton.

In regard to Söhnlein's results, A. James remarks [2] that at Kalgoorlie, where the central feed has also been tried, the practice was to grind the coarse sand to an intermediate size in two pans in parallel and to slime their product in a third pan. The usual results were not so good as those claimed by Söhnlein.

---

[1] Söhnlein, *Eng. and Mng. J.*, 1913, **96,** 581 ; see also E. E. Wann, *ibid.*, p. 1183, and Söhnlein, *ibid.*, 1914, **97,** 822.
[2] James, *Mng. and Sci. Press*, 1914, **108,** 73.

G. A. and H. S. Denny [1] give some of the advantages of using pans for fine grinding instead of tube mills, as follows :—

(1) Their accessibility.
(2) The possibility of internal amalgamation.
(3) The smaller capacity of the unit.
(4) The saving of time in the renewals of working parts.

Nevertheless, tube mills seem to be generally preferred to pans, partly because the former yield a finer product.

Amalgamation of Concentrate in Pans.—The treatment of concentrate by pan-amalgamation is almost if not entirely obsolete, and does not represent modern practice, in which concentrate is either smelted or treated by wet methods. Whether concentrate has been previously roasted or not, its treatment in pans is seldom attended by the successful extraction of a high percentage of the gold. A stone arrastra usually gives better results in treating roasted concentrate than an iron pan. In Australia the method was often adopted of employing a large excess of mercury and little water, and of keeping the roasted material from contact with iron, and in some experiments conducted in Mexico, C. A. Stetefeldt found that by the use of gold amalgam instead of mercury, and by grinding in stone vessels, a high percentage of gold was extracted from low-grade ores.

Among special forms of pans designed to treat concentrate are the Berdan pan and the Britten pan, which were both introduced many years ago.

The *Berdan pan* [2] is a shallow annular basin about 4 feet in diameter, surrounding a cone which is attached to a spindle set at an angle of about 15° to the vertical. The spindle is rotated by bevel gearing at 20 to 30 revolutions per minute, and carries the basin round with it. In the annular basin are one or two loose iron balls which remain at the lower side of the cone when the pan revolves. The pulp is fed in with mercury at the higher side, and is ground by the balls and discharged over the lower edge of the pan. The capacity of the Berdan pan is from 1 to $2\frac{1}{2}$ tons per day. It is used chiefly in Australia.

The *Britten pan* is a deeper cast-iron stationary basin almost hemispherical in shape, in which a pear-shaped muller rolls round. It was formerly used for grinding and amalgamating rich specimen ore and concentrate in Wales. It is of small capacity, and is not a continuous machine like the Berdan pan, but grinds the charge put into it as long as is desired.

---

[1] G. A. and H. S. Denny, *Mining Mag.* (New York), Sept. 1905, p. 180.
[2] For full description and illustrations, see Louis, *Gold Milling*, 1894, pp. 343-346.

CHAPTER X.

FINE GRINDING.

Discussion of Coarse and Fine Crushing.—The ordinary system of treating "free-milling" gold ores consists in crushing in the stamp battery, amalgamating either in or outside the battery, and treating the tailing by the cyanide process. The free or amalgamable gold is caught on amalgamated plates, and the gold contained in pyrite, tellurides, etc., is left for the attack of cyanide. Ores are of infinite variation, however, and the percentage of gold recoverable by amalgamation can hardly be fixed even for a single ore. It depends on many factors, the main one being in most cases the degree of fineness to which the ore is crushed. In general some of the gold is distributed through the ore in an extremely fine state of division, and the finer the crushing the more gold is laid open to the attack either of mercury or of cyanide. This consideration points to the desirability of "sliming" all the ore, if it can be done cheaply enough.

On the other hand, the gold is generally in part contained in sulphides, tellurides, etc. These form the richest part of most ores, and retain their gold with greater obstinacy than the other constituents of the ore. They are also the most brittle, so that if the whole ore is reduced to a fine state of division the sulphides, etc., are converted into an impalpable slime. In this condition they cannot readily be saved by concentration. At one time another objection to fine crushing was that the crushed material could not be treated by cyanide or chlorine on account of the mechanical difficulties of leaching. It was necessary in chlorination to use pressure in shallow leaching vats, and in cyaniding to separate the slime from the sand, and to leave the former untreated. Now that slime can be treated by decantation or filter pressing, or with the use of vacuum filters, the objection to its formation has less force.

An exact definition of slime is desirable. Perhaps the most useful definition is "that portion of the crushed ore which, owing to its physical condition, cannot be leached by percolation under the action of gravity." The "physical condition" is defined by H. A. White[1] as "minutely subdivided condition and the presence of colloidal substances." Such material does not subside readily in water. It has been proposed that finely crushed material which settles readily in water and can be leached should be called "grit," but this word has not passed into general use.

Before the introduction of the cyanide process, when the general methods of treatment were crushing and amalgamation, followed by concentration and the treatment of the concentrate by roasting and chlorination, the aim was to avoid the formation of slime. At that time the value of crushers giving a uniform product seemed especially great, and rolls were strongly

---

[1] White, *Rand Metallurgical Practice*, vol. i., p. 189.

advocated as against stamps, which produced a greater proportion of very finely divided material. Later, when the Kalgoorlie ores began to be treated by dry crushing and roasting, the uniform product of ball mills was found to be better than the partially slimed product of Griffin mills, because of the greater difficulty of roasting the latter.[1] When ores are subsequently to be roasted there is little advantage in fine crushing, because in the course of roasting the particles usually become porous, so that chlorine or cyanide can penetrate into their interior and dissolve gold which has not been exposed on the surface of the grains.

Now, however, it is fully realised that, in the case of ore which is treated without roasting, the finer the state of division the higher the percentage of gold that can be extracted. The method of regrinding the product of the stamp battery or of dry crushers in other machines has been developed, resulting in a higher percentage of extraction in many cases. This has been followed by attempts to increase output by using coarser battery screens, and also to use a method of gradual reduction in successive machines, rather than one of great reduction in size in a single machine.

It has already been pointed out that if ore is retained for a long time in the mortar of a stamp battery by increasing the height of discharge and the fineness of the screens, and by diminishing the screen area, a higher percentage of gold can be amalgamated in the battery and the fineness of the product increased. The result is that only 1 or 2 tons of ore are crushed per stamp per diem, but most of the gold is caught by the inside copper plates. When the screens are coarse and set low down, the crushed ore is rapidly discharged, a smaller proportion of fine material is produced, and the percentage of gold which could be amalgamated in the mortar is small, so that the advantage of adding mercury with the ore disappears and all gold-catching by amalgamation is left to outside plates. The coarse battery-pulp, however, yields little to amalgamation and scours the plates. There is, therefore, a tendency to omit this step also, and to pass the ore through fine grinders before amalgamation. So far, this is a method similar to the old German practice (see p. 152), and it appears to be well established that the reversion to ancient methods is desirable, and that the stamp battery is neither the most economical fine crusher nor the most efficient amalgamator.

For example, it has been established that for such hard, brittle ores as those of the Rand it is most advantageous to reduce the ore to $1\frac{3}{4}$ inches in rock-breakers before feeding it to the stamps, to crush to about $\frac{1}{4}$ inch through screens (9-mesh screens, aperture 0·272 inch) in the stamp batteries ($\frac{1}{2}$ inch is also used), and to grind to about 90-mesh in tube mills, after which the pulp is amalgamated by being passed over plates.

The influence of the screen on the output of ore is illustrated by the following results obtained in 1904 at the Glen Deep Mill on the Witwatersrand :—

Screen, Holes per Square Inch.	Output per Stamp in Twenty-four Hours.	
800	4·9 tons.	
200	6·9 ,,	
150	7·8 ,,	
100	8·04 ,,	
64	9·4 ,,	{ 10·68 tons under the most favourable circumstances.

[1] W. Evan Simpson, *Trans. Inst. Min. and Met.*, 1904, **13**, 22.

With screens of four holes to the square inch the output per stamp is 20 tons and upwards per day, as at the Consolidated Langlaagte and Van Ryn Deep Mills.

As the Rand ore must finally be treated with cyanide in any case, the question naturally arises, "Cannot amalgamation be dispensed with altogether?" One answer is that it is well to remove coarse gold by amalgamation, and to leave to the cyanide plant any gold which is finely divided enough to be dissolved. This method is maintained on the Rand. The alternative method is to pass the coarse particles of gold through the tube mill again and again, returning them with the oversize, until they are sufficiently comminuted to be readily soluble in cyanide. This method has been adopted at many mills in America.

Energy consumed in Crushing.—Consideration has lately been given by H. Stadler,[1] A. O. Gates,[2] A. F. Taggert,[3] S. J. Speak,[4] and others to the amount of energy consumed in crushing. Attempts have been made to make use of the "laws" of Rittinger and Kick, and to adapt them to ore crushing, but neither of them provides a perfectly satisfactory basis for calculation. Rittinger's law is to the effect that the energy absorbed in crushing is proportional to the surface produced (Gates) or to the reduction in diameter (Speak); and Kick's law, as stated by Stadler, is that "the energy required for producing analogous changes of configuration of geometrically similar bodies of equal technological state varies as the volumes or weights of these bodies."

When the energy required for crushing an ore has been estimated, data are obtained for approaching from the side of theory the problem of determining the relative efficiency of various crushing machines, such as stamps, rolls, pans, and tube mills; and some help may be given in the selection of the machine to be adopted for a definite purpose in a particular case. At present this selection is difficult, and mistakes are sometimes made. Up to the present, however, theory has given no certain guidance.

Tube Mills.—Fine grinding of coarsely crushed ore is effected in pans, tube mills, conical pebble mills, such as the Hardinge mill, and Chilian mills. In pans, amalgamation is simultaneously carried on, and they are dealt with above, p. 223. The other machines are considered in this chapter.

Tube mills are *par excellence* the machines for fine grinding. They were first used in grinding tin ore in Cornwall about the year 1879,[5] and in grinding gold ore at Butte, Montana, in 1894,[6] and are also used in the cement industry for dry grinding. They were applied to gold ores by Dr. Diehl and by Mr. Sutherland in West Australia in 1899, and are used for wet grinding. They consist of revolving cylinders rather more than half-filled with pebbles, by the impact of which, in falling, coarse particles of sand are crushed fine. They differ from ball mills (*q.v.*) essentially in having the inlet for ore at one end and the outlet at the other, and in the absence of provision in the machine itself for the return of uncrushed material (oversize). It is, therefore, necessary for the cylinders to be of greater length than in ball mills. The number of pebbles used in tube mills is also much larger than that of the steel balls in ball mills.

[1] Stadler, *Trans. Inst. Mng. and Met.*, 1910, **19**, 478.
[2] Gates, *Eng. and Mng. J.*, May 24, 1913, p. 1039 ; April 18, 1914, p. 795.
[3] Taggert, "The Work of Crushing," *Bull. Amer. Inst. Mng. Eng.*, Jan. 1914, p. 143.
[4] Speak, *Trans. Inst. Mng. and Met.*, 1914, **23**, 482.
[5] *Eng. and Mng. J.*, June 16 and July 28, 1906.
[6] Abbé, *Eng. and Mng. J.*, May 26, 1906, p. 1010.

The *cylinders* consist of ⅜ to ¾ inch steel plate with cast-iron or steel heads, which are strengthened by radial ribs. Fig. 101 [1] shows a tube mill with cast-iron ends.

The variation in size of tube mills used in different parts of the world is considerable.[2] At Kalgoorlie some of the earlier tube mills were 3¼ feet by 13 feet, but they are now generally 4 feet by 16 feet. At El Oro, in Mexico, some mills 5 feet by 24 feet are in use, but generally in North America the 5 feet by 18 feet mills are most popular. On the Rand the standard size is 5½ feet by 22 feet, and mills of 6 feet by 16½ feet, installed at the new Consolidated Langlaagte Mill, are not considered by Caldecott to do better work. However, there is at present a tendency in favour of shorter mills, with the hope of increased efficiency by removing the slimed portion of the ore sooner. In 1914, short tube mills only 6 feet long and 7 or 8 feet in diameter were introduced, but there are no available data as to their success in practice.

The cylinders of tube mills have recently been made in sections for transportation on mule back (see Fig. 102).[3] The sectional cylinder is sometimes made of great size, up to 5 feet in diameter and 16 feet long.

The cylinder is carried on two hollow *trunnions*, one at each end. The

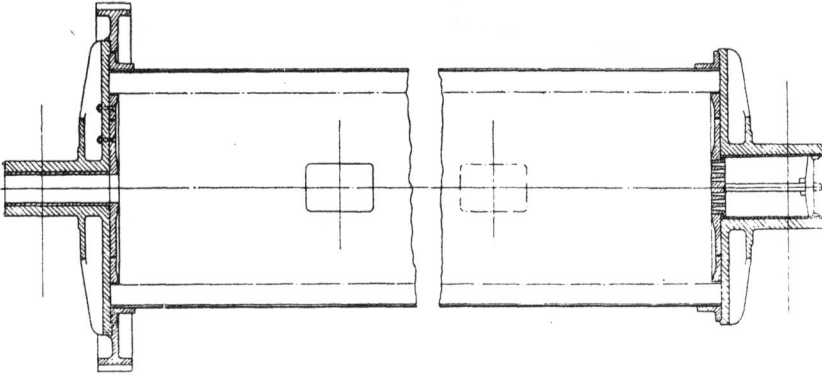

Fig. 101.—Section of Tube Mill.

pulp enters through one of these (left-hand end in Fig. 101), and is discharged through the other. The Cooper roller bearings for the trunnions have recently been introduced, with the object of saving power.

The *lining* consists of hard wood, chilled cast iron, hard homogeneous iron, steel, manganese steel, quartzite, or flint (silex). Of these, wooden blocks were stated to be worn out in three days in South Africa, and cast-iron plates in three weeks, but in this case coarse ore was said to have been treated. Manganese steel linings lasted fifteen months at Kalgoorlie, and chilled iron liners from seven to ten months.[4] The great cost of manganese steel made it less economical than hard iron.[5]

The usual lining has been silex or flint blocks about 6 inches thick cemented in. On the Rand, where local chert is used for lining, the life of the lining

[1] Dowling. *Rand Metallurgical Practice*, vol. i., p. 106.
[2] See Gieser, *Eng. and Mng. J.*, 1914, **97**, 463, for complete data.
[3] Reproduced with the permission of the Cyanide Plant Supply Co.
[4] A. James, *Trans. Inst. Mng. and Met.*, 1904, **14**, 98.
[5] Trewartha James, *ibid.*, p. 121.

is about 80 days; at the expiration of this period it is usually found that it has been worn down to 1½ inches thick at the end near the inlet. The wear of flint linings is equal in weight to the wear of the flint pebbles, according to Dr. Diehl, but is far less according to S. Robinson.[1] The wear of liners is more rapid at the inflow end than near the outflow. The thickness of the flint lining was at first 2½ inches, and was afterwards increased to 4 inches, and then to 6 or 7 inches.

Iron-ribbed liners were introduced first at El Oro (Gieser), and are in wide use. The ribbed form is designed to reduce the slip of the pebbles, which was very great with smooth iron liners, and not inconsiderable with new silex linings. When the pebbles slip, they are not carried high enough for effective impact in falling (see below, p. 242), and the output of the mill falls off. Nevertheless, smooth iron liners are preferred at Tonopah, Nevada,[2] for very fine grinding.

S. H. Pearce found [3] at the Glen Deep Mill that the introduction of a new lining, made of rings of manganese steel, was attended with a cessation of the usual rumbling noise. There was a tendency for the pebbles to wear

Fig. 102.—Palmarejo Sectionalised Tube Mills (showing Pebble Gratings after Neal's Discharge).

flat, and the crushing efficiency dropped. After a time the rumbling began again, and the crushing efficiency increased. It was due to the fact that the pebbles slipped on the new smooth surface, but that the lining acquired a rough surface by wear, and the pebbles were then raised higher, causing crushing (by impact) instead of reduction by grinding.

W. R. Dowling [4] confirms this view from the experience at the Robinson Deep Mill. He found that a silex lining takes a larger feed and gives a finer ground product than manganese steel, owing to the absence of sliding action when the former is used. Ribbed, corrugated or honey-combed steel liners, however, prevent the slip of the pebbles better than silex blocks, and are

[1] Robinson, *ibid.*, p. 135.
[2] Megraw, *Eng. and Mng. J.*, 1913, 95, 414.
[3] Pearce, *J. Chem. Met. and Mng. Soc. of S. Africa*, 1905, 5, 304.
[4] Dowling, *ibid.*

much favoured. They absorb more power, but probably increase the efficiency of the mill. The speed of rotation may be reduced and power saved without loss of efficiency in the case of some of these liners.

The Osborn liner (Fig. 103)[1] consists of a series of slightly wedge-shaped iron bars spaced at from $2\frac{1}{2}$ to 5 inches apart at the base round the periphery of the tube mill, held in place by iron wedges. The liner is used on the Rand, and its life is found to be much greater than that of silex blocks, and is said to be 300 days.[2] It can be replaced more rapidly than silex, and the unworn ribs near the outflow can be used again. At one mine the cost per tube mill per day was 38s. 6d. for silex liners and 21s. 10d. for Osborn

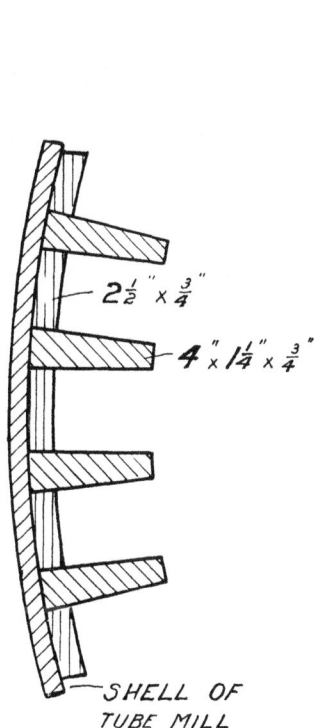

Fig. 103.—Osborn Liner.

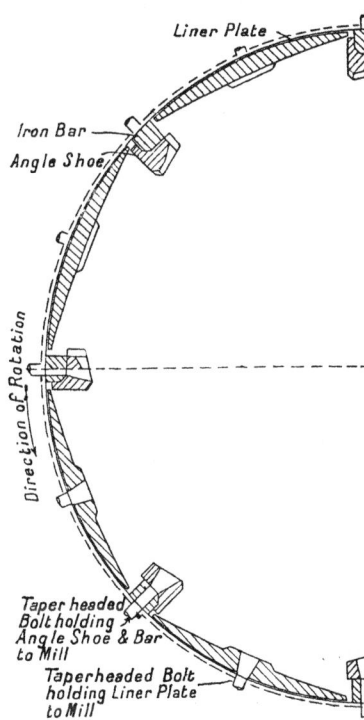

Fig. 104.—Komata Liner.

liners.[3] The pebbles lodge between the ribs, take up the wear and fall out as the tube revolves.

The El Oro liner (cast-iron segments), the honey-comb liner and the Gibson liner (short pegs of steel set in cement) have also been used on the Rand. The Komata liner,[2] consisting of longitudinal ribs about 18 inches apart with plates between, was introduced at Komata in New Zealand by F. C. Brown, and is now in wide use, especially in New Zealand and Nevada. It is shown in section in Fig. 104.[4] This liner is thin, and occupies little space,

---

[1] Schmitt, *Rand Metallurgical Practice*, vol. ii., p. 153.
[2] Gieser, *Eng. and Mng. J.*, 1914, **97,** 465, 466.
[3] Schmitt, *Rand Metallurgical Practice*, vol. ii., p. 153.
[4] Reproduced with the permission of the Cyanide Plant Supply Co.

so that the capacity of the mill is increased, but although thin, the liner has a long life. The ribs are much further apart than in other ribbed liners.

The *pebbles* usually consist of flint, although the use of steel balls has been advocated. The size varies up to 3 or 4 inches in diameter. According to M. Davidsen,[1] in wet crushing, large pebbles wear less rapidly than small pebbles and crush as finely. The larger the pebbles the coarser may be the ore fed in. For dry grinding, small pebbles of 1 to 2 inches in diameter are preferable. The Greenland flints, found on the Danish shores, are more durable than the chalk flints of the French coast. The mill is kept about half filled with pebbles, the upper level of which varies from 3 inches above the centre line of the mill (as in S. Africa) to as much as 7 inches below. On the Rand pieces of banket ore about 4 inches in diameter are fed in instead of pebbles. The amount required may be as much as $2\frac{1}{2}$ per cent. of the total tonnage milled.[2] Dowling[3] found that the load of pebbles in a 22 feet by 5 feet mill was 12·8 tons, if maintained 3 inches above the axis. Hence he calculated that the 40,000 pebbles in the mill delivered 1,200,000 blows per minute. The weight of pebble load for a 22 feet mill at all heights and diameters and the method of calculation is given by Caldecott.[4]

There are two *manholes* in the shell for use when the mill has to be relined or repaired.

The *feeding* of ore and of pebbles into a tube mill is effected through one of the trunnions, which is about 8 inches in internal diameter. Several different methods of feed have been proposed. One of these, designed by L. Pryce, is shown in Fig 105.[5] It consists of two parallel discs 22 inches in diameter and 7 inches apart. A spiral between the discs connects an opening at the periphery with the centre. The pebbles fed from a hopper, designed by J. E. Thomas, into the peripheral opening are carried to the centre and pushed into the hollow trunnion by an Archimedean screw. The pulp coming from a classifier or dewaterer enters as shown.

The *discharge* takes place through the other trunnion, which is also hollow and is of greater diameter, so that the outflow is about 5 inches lower than the inflow. The outflow is through a perforated plate (right hand of Fig. 101), flush with the end liner, with holes of $\frac{1}{2}$ inch diameter inside the mill expanding to $\frac{3}{4}$ inch towards the outflow to prevent choking (Dowling).

Neal's discharge, introduced at El Oro in Mexico, is an alternative to the perforated plate. It consists of a baffle placed inside the tube mill, and a reverse screw which returns to the tube mill any pebbles which might escape the baffle, although it permits of the free escape of the water-borne pulverised pulp. Neal's discharge is said by Alfred James to be in general use in America and in wide use in Asia. The internal scoop discharge is mentioned by Dr. Caldecott (see Chap. XVII.). Peripheral discharge has been tried, but though suitable in dry crushing, is not much used in wet crushing.

Fragments of pebbles and small worn-out pebbles too small to assist in the crushing pass out with the pulp, and are separated from it by a circular screen or pebble-catcher of $\frac{1}{4}$ inch mesh, through which the pulp passes into a launder, while the pebbles are delivered at the end of the screen.

[1] Davidsen, *Trans. Inst. Min. and Met.*, 1904, **14**, 155.
[2] W. R. Dowling, *Rand Metallurgical Practice*, vol. i., p. 116.
[3] Dowling, *loc. cit.*
[4] Caldecott, *J. Chem. Met. and Min. Soc. of S. Africa*, 1913, **13**, 363.
[5] Dowling, *Rand Metallurgical Practice*, vol. i., p. 114.

The *speed of revolution* of tube mills varies with the diameter. Krupp uses the formula $\dfrac{32}{\sqrt{D}}$ for the correct number of revolutions per minute, where D = internal diameter in metres. The Davidsen formula is $\dfrac{200}{\sqrt{D}}$, where D is the diameter in inches. These two formulæ are almost identical and would give about 26 revolutions per minute for a mill of 5 feet in diameter. White's formula is $\dfrac{34\cdot22}{\sqrt{D}}$, where D is in metres. The speed in Rand tube mills varies from 28 to 34 revolutions per minute with a diameter inside the shell of 5 feet 6 inches.[1] The peripheral speed inside the lining at 32 revolutions with new linings 6 inches thick would be 452 feet per minute, and with worn linings only 1 inch thick it would be 536 feet per minute. As the lining wears, the number of revolutions per minute is reduced to keep the peripheral speed approximately unchanged. These are higher speeds than were formerly used, and somewhat higher than that generally used elsewhere, but the speed is dependent to some extent on the other conditions. There is a

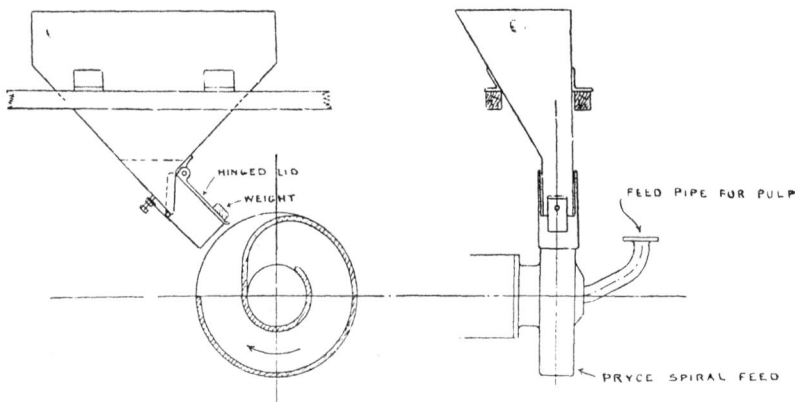

Fig. 105.—Pryce Pulp and Pebble Feeder, with Thomas's Hopper.

certain peripheral rate of speed which is most advantageous with a particular ore. The *power* required varies with the rate of feed (load) and speed. A standard Rand tube mill of 22 feet long by $5\frac{1}{2}$ feet diameter requires from 90 to 120 H.P., increasing as the liners wear, owing to increased size and capacity of the mill. Motors of 125 H.P. are used, but these are to be increased to 175 H.P. in new mills. According to Davidsen the formula P (in H.P.) $= 0\cdot15$ N, where N = capacity of mill in cubic feet, gives the power required, but this is too low for Rand practice.

The *capacity* (output) of tube mills depends on the hardness and size of the material supplied to them, and on the fineness of the product. The amount of grinding done and the fineness of the product depend on the rate of feed and the length of the tube, as well as on the hardness of the ore. According to Gruessner,[2] at the Hannan's Star Mill, Kalgoorlie, the ore from

---

[1] Dowling, *Rand Metallurgical Practice*, vol. i., p. 117. See also Ball, *Trans. Inst. Mng. and Met.*, 1912, **21**, 3.

[2] Gruessner, *Trans. Inst. Mng. and Met.*, 1904, **14**, 87.

two No. 5 dry ball mills was treated wet in a tube mill 16 feet 5 inches long
and 3 feet 11 inches in diameter. The ball mills crushed 78 tons per day
and, after classification, the coarser part of this, amounting to 38 tons, was
ground to about 250 mesh in the tube mill. The pulp fed into the mill
contained : coarser than 40 mesh, 20·5 per cent. ; 40 to 60 mesh, 43·8 per
cent. ; 60 to 100 mesh, 28·1 per cent. ; 100 to 150 mesh, 7·6 per cent.
The product was not examined,[1] but returned to the spitzkasten and
re-classified, the sand returning to the mill. The overflow (final product)
contained : coarser than 100 mesh, 0·7 per cent. ; 100 to 150 mesh, 4 per
cent. ; finer than 150 mesh, 95·3 per cent. The power consumption of the
16-foot mill was 30 H.P. The cost of fine grinding per ton of crude ore
was as follows :—

Power, . . . 11·53d.
Flints and liners, . 1·85d.
Labour, . . . 3·50d.
Repairs, . . 0·99d.

Total, . 1s. 5·87d., or 3s. 0·68d. per ton of material actually slimed
in the tube mill.

At the Oroya Brownhill Plant, Kalgoorlie, according to W. Broadbridge,[2]
about 7,500 tons per month were treated in six flint mills (five 13 feet 7 inches
by 3 feet 8 inches, and one 12 feet 11 inches by 4 feet 1 inch), during the
period February to July, 1904. The average amount of sand slimed per
mill per hour was 1·77 tons, and the cost 1s. 9·33d. per ton treated.

Much greater outputs are now obtained on the Rand, where, however,
a product only fine enough to pass through a 90-mesh screen is aimed at.
The usual feed is about 400 tons of underflow from the classifiers, to a
standard tube mill of 22 feet by 5½ feet. If a product passing a 200-mesh
screen is required, the capacity is, of course, less.

The effect of refitting a tube mill at the Ivanhoe Mine was given by
W. Broadbridge as follows [3] :—

TABLE XXIV.

	Sand before Grinding.	Sand after Grinding.	
		16-foot Mill fitted with Old Liners and Small Flints.	Same Mill fitted with New Liners and Large Flints.
	Per Cent.	Per Cent.	Per Cent.
On 40 mesh, . .	38·4	0·5	0·10
„ 60 „ . .	41·6	14·0	6·28
„ 100 „ . .	15·4	43·0	36·26
„ 150 „ . .	1·6	9·0	10·27
Through 150 „ . .	2·8	33·5	47·09

The *size* of particles in the inflow has been the subject of much investigation. On the Rand, with 4-inch pebbles, it has been found that the best

---

[1] According to W. H. Trewartha-James (*ibid.*, p. 122) the amount of oversize left after
one passage through the mill was so great that some 268 tons of sands were passed through
the mill per day for a product of 38 tons of slimes.

[2] Broadbridge, *ibid.*, p. 101.   [3] Broadbridge, *loc. cit.*

size is the material passed by the stamps through wire screens containing nine holes to the square inch, the diameter of the aperture being 0·27 inch.

With coarser particles the weight of the falling pebbles is insufficient to crush them; with finer particles the energy is partly wasted. The greater the diameter of the mill, and the larger the pebbles, the coarser the feed may be. F. C. Brown [1] gives a diameter of 4 feet inside the liners as suitable for a tube mill intended to treat battery pulp which has been passed through a 6-mesh screen. He recommends smaller diameters for finer feeds. The fineness of the product depends on the length of the mill, a short mill giving a granular product and a long mill giving slimed ore. At Broken Hill a 4 × 13 feet mill grinds tailing which has passed through 10-mesh, reducing the proportion of + 40-mesh material from 60 to 10 per cent. Brown considers that two short mills in tandem with a classifier between them will do better work than one mill equal in length to the short mills combined. Short mills, however, are not considered suitable to the conditions on the Rand.

The amount of *moisture* in the pulp varies with the nature of the ore and other conditions, and usually amounts to about 40 per cent. (see below, p. 243). In dry tube milling a small percentage of moisture is very injurious, 1 per cent. of moisture reducing the capacity in certain cases by one-half.[2]

A section of the *Smidth-Davidsen Mill* is shown in Fig. 106, which is from a drawing made by the Cyanide Plant Supply Company. It may be seen that the feed is through the hollow trunnion, but the discharge is through a grating near the periphery. In other mills the discharge, like the feed, is central. Tube mills of the same type at Waihi are shown in Fig. 107.

The Krupp Tube Mill [3] shown in Fig. 108 is a longitudinal section of a mill for dry crushing; Fig. 109 is a plan of the mill, and Fig. 110 a cross-section of the tube or drum, showing the discharge apertures at *c*. The drum is 1·2 metres (3 feet 11¼ inches) in diameter, and 5 metres (16 feet 5 inches) long, inside measurement, and revolves at the rate of 29 revolutions per minute. The drum consists of sheet iron 12 mm. ($\frac{15}{32}$ inch) thick, has cast-steel ends, and is lined with hard cast-iron plates. The grinding balls are introduced

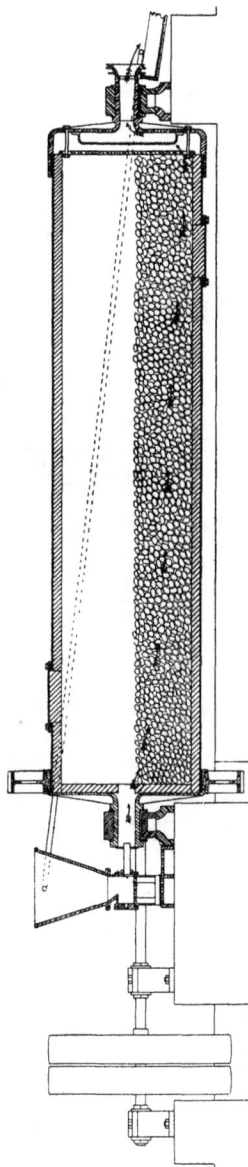

Fig. 106.—Smidth-Davidsen Tube Mill.

---

[1] Brown, *Mng. and Sci. Press*, 1912, **104**, 206; *Mineral Industry*, 1912, p. 938.
[2] Durant, *Eng. and Mng. J.*, 1912, **94**, 111. *Mineral Industry*, 1912, p. 939.
[3] From description by H. Fischer in *Zeitsch. Ver. deut. Ing.*, March 26, 1904.

through a manhole, whilst the material to be crushed is fed and
discharged through the hollow trunnions cast together with the end walls.
The material passes first of all into the hopper, *a*, in which a studded shaft

Fig. 107.—Tube Mill at Waihi.   Reproduced by permission of Mr. Alfred James.

rotates. A longitudinally-grooved drum, *b*, regulates the supply of material.
This drum is rotated by toothed-wheel gearing and a pair of five-step pulleys
at different speeds, and contains a few balls which serve to shake the drum,

and thereby secure the emptying of its grooves. The material supplied is conveyed by a screw into the hollow trunnion, which is conical inside, enlarging towards the drum. Here it is scooped up by two helical blades which

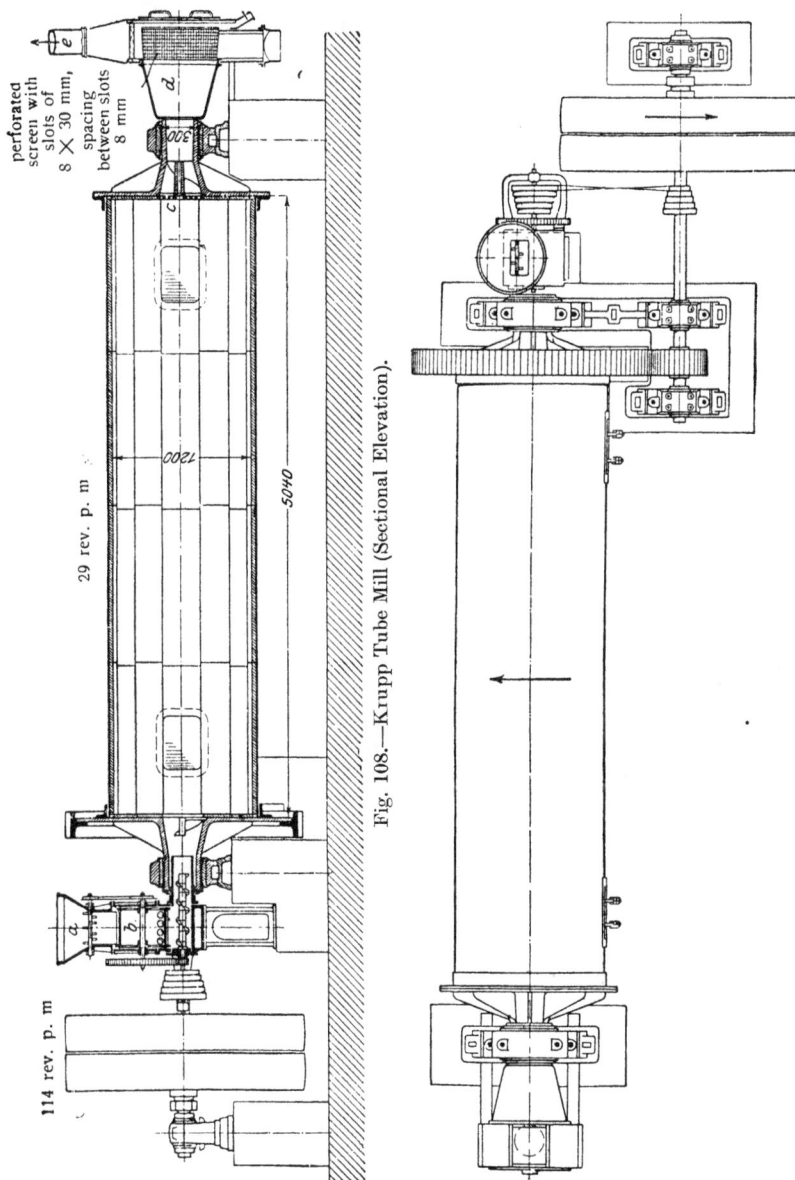

Fig. 108.—Krupp Tube Mill (Sectional Elevation).

perforated screen with slots of 8 × 30 mm, spacing between slots 8 mm

29 rev. p. m

114 rev. p. m

Fig. 109.—Krupp Tube Mill (Plan).

pass it into the drum, and prevent the balls from being thrown out of the mill. The ore then passes through the drum and through the grating, c (Figs. 108 and 110), into the outlet trunnion.

16

The grating, c, is provided with curved slots of about 25 millimetres (1 inch) in width. The material passes therefrom into a hopper, d, attached to the trunnion. A perforated screen is connected to the hopper. The slots of the screen are 8 millimetres ($\frac{5}{16}$ inch) in width and 30 millimetres ($1\frac{3}{16}$ inches) in length; they permit the sufficiently crushed material to drop through, whilst very hard particles which are not crushed, and splinters of the flintstone balls, are retained and subsequently drop out on the right side (Fig. 108). The sieve is enclosed by a casing, out of which the air is drawn by means of the pipe, e, so that air enters the drum at all other apertures. This is to avoid loss by dusting.

Theory of Tube Mills.—According to H. Fischer,[1] who made experiments for the Krupp firm, the tube mill does its work mainly by impact, not by grinding. Glass drums and drums with gratings at the ends were constructed, and the action in the interior observed. One such drum 1 metre (39·3 inches) in diameter was filled with flint balls to a height of 450 mm. (17·7 inches) and rotated. Prof. Fischer found that, at the slow speed of rotation of 21 to 23 revolutions

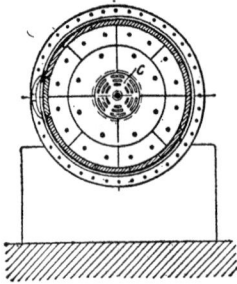

Fig. 110.—Krupp Tube Mill (Cross-section showing Discharge Apertures).

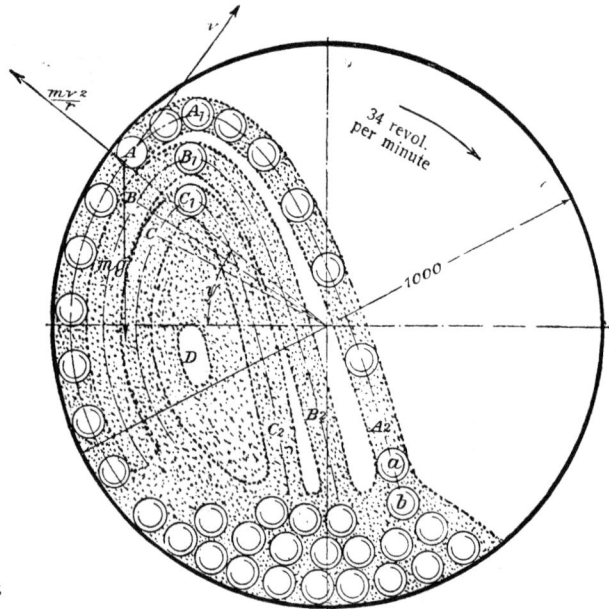

Fig. 111.—Diagram showing Action of Tube Mill.

per minute, the balls rolled slowly down the slope. At 32 revolutions per minute (the correct speed for a tube of this diameter, according to the formulæ given above) the charge of balls had become looser, and their bulk was considerably more than half the capacity of the drum. At 34 revolutions (Fig. 111) the balls (A A_1, etc.), next the drum were carried up without sliding or rolling on the drum, until at a certain height they separated from it and were projected outwards in a curve, falling near the other side of the charge of pebbles. Prof. Fischer concluded that at this speed each ball fell separately, crushing and spattering the cushion of ore between it and the balls that had previously fallen. The other layers of pebbles described

---

[1] Fischer, *Zeitsch. Ver. deut. Ing.*, Mar. 26, 1904.

similar but shorter courses, and a hollow space, **D**, continually varying in size and shape, was always recognisable. The blow struck by each ball of the outer layer was a sliding one, the relative velocity of two balls at the moment of impact being 3·5 metres per second in the line joining their centres, and 1·2 metres per second in the direction at right angles to this. These velocities give some measure of the respective effects of the impact and grinding actions.

There is no doubt that some effect is produced by grinding, especially at low velocities of revolution, but probably the mill becomes more effective when the impact of the pebbles is at its maximum. The wear of the pebbles is doubtless greater in proportion as the grinding action is allowed to increase. Soft ores would be more amenable to grinding than the hard ores of the Rand, Waihi, etc.

H. A. White agrees [1] with Fischer that the pulverising action is due almost entirely to actual impact of the falling balls in dry crushing, but points out that when the mill is half full of water the effect of impact is of less importance. A ball falling into 2 or 3 feet of water will not strike the bottom with enough force to do much crushing, and in this case it is probable that the grinding action between the balls will do a great proportion of the work.

However, the best results have been obtained with very thick pulp, containing about 40 per cent. of water. The correct proportion of water to ore depends on the specific gravity of the ore, on its coarseness, on its composition or hardness, and on the amount of the feed. The correct proportion of water is 39 per cent. with Rand ore, if 400 tons are fed per day into a standard tube mill of 22 by 5½ feet. In a tube mill containing such pulp there is no " free " water for the balls to fall through. It is considered that the amount of moisture present should be enough to make the particles of ore adhere to the pebbles and to the lining, so that when impact occurs between the pebbles there are some particles between the pebbles ready to be crushed. If only 200 tons of solids are fed in 24 hours to a standard mill, the best percentage of moisture in Rand ore is said to be 27 per cent. (Dowling). The amount of water giving the best results for each ore must be accurately determined correct to 0·1 per cent. and adhered to rigidly.

White finds that the average fall of the balls is at a maximum, if the mill is half-filled with balls, when the number of revolutions per minute, N, is equal to $\dfrac{34 \cdot 22}{\sqrt{D}}$, where D is the diameter of the mill expressed in metres. White's result, obtained by calculation, gives the same speed as that at which Fischer found by experiment that each ball fell separately. A somewhat faster rate is now used on the Rand, and, consequently, it appears to be probable that the theoretical investigations were not made without error.

Hardinge Conical Pebble Mill.—It has been urged that, in the case of the ordinary cylindrical form of tube mill, particles which are crushed to the required fineness near the feed end of the mill go gradually forward to the discharge end, using up needlessly the energy of the falling pebbles, which could otherwise be employed in crushing the larger fragments still remaining. Moreover, both ends of the tube are loaded with the same quantity and size of pebbles. When these become worn a considerable amount of energy is expended by the larger pieces in acting upon the smaller ones. Hardinge suggests that in the perfect machine the crushing should theoretically be done

---

[1] White, *J. Chem. Met. and Mng. Soc. of S. Africa*, 1905, **5**. 290.

in stages,[1] the larger pebbles crushing the larger particles of ore, leaving the smaller pebbles to finish the pulverising to the mesh for which the mill is designed. Also, as it is reduced, the fine material should be removed, leaving the space available for larger particles to come under the influence of the pebbles.

The Hardinge conical mill[2] was designed to avoid these possible defects in the tube mill, and is said to be adjustable for producing an all-slime product with a minimum of coarse grains, or, on the other hand, for granulating without the formation of an excess of slime. In practice it has been found to be better suited to the latter purpose than to the former one.

The conical mill consists essentially of two hollow cones (see Fig. 112) the rims of whose bases are attached by a short cylindrical section, and whose apices are formed into hollow trunnions providing for the support of the mill and for the feed and discharge of the ore. The shell is made of heavy steel plate, the joints, both longitudinal and circumferential, being

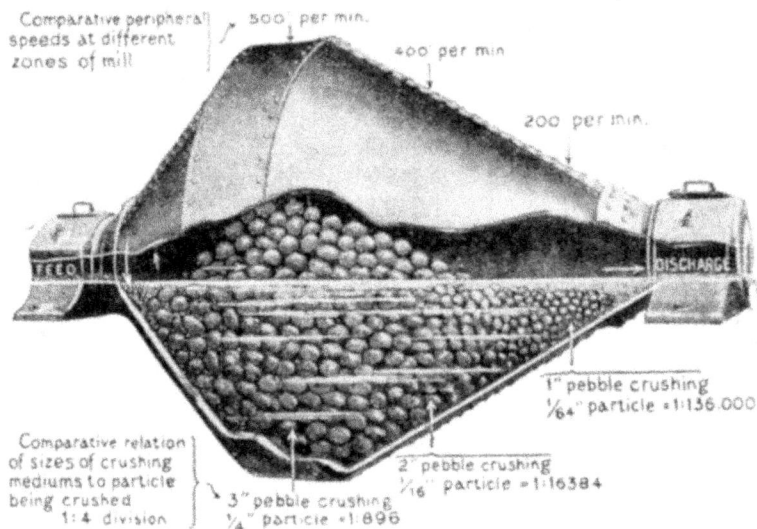

Comparative peripheral speeds at different zones of mill

500 per min.
400 per min.
200 per min.

FEED   DISCHARGE

1" pebble crushing
1/64" particle = 1:136.000

2" pebble crushing
1/16" particle = 1:16384

Comparative relation of sizes of crushing mediums to particle being crushed
1:4 division

3" pebble crushing
1/4" particle = 1:896

Fig. 112.—Hardinge Conical Pebble Mill

butted, strapped and rivetted. The diameter of the cylindrical portion is usually from $4\frac{1}{2}$ to 8 feet, and its corresponding length from 13 to 30 inches. These are the usual dimensions by which the mills are designated. For producing a maximum of slimes the cylindrical portion is made of much greater length, up to 72 inches. A shell of $4\frac{1}{2}$ feet × 13 inches carries a silex lining $2\frac{1}{2}$ inches thick and a charge of pebbles of 1,500 lbs. The capacity is 24 to 36 tons per 24 hours. A mill 8 feet in diameter, and having a cylindrical portion 30 inches long, occupies a space of 11 feet × 13 feet. The weights of the mill, lining and pebbles are 11,500 lbs., 7,500 lbs., and 12,000 lbs. respectively, and 35 to 45 horse-power is necessary for driving purposes.

---

[1] H. W. Hardinge, *Trans. Amer. Inst. Min. Eng.*, Feb. 1913.
[2] HW. H. ardinge, *Trans. Amer. Inst. Min. Eng.*, 1908, **39**, 336. *Electrochem. and Metallurgical Industry*, Jan. 1909, p. 47. *Eng. and Min. J.*, Nov. 16, 1907, p. 925.

The capacity is rated at 48 to 170 tons per twenty-four hours, and the peripheral speed is maintained at about 750 feet per min. (29·8 revolutions per minute). The capacity and power depend on the weight of the pebble charge, and vary also according to the hardness and size of the material to be ground.

The ore is fed in by a scoop or spiral feed, attached by an extension piece to the trunnion casting at the apex of the cone with greatest slope. This device picks up a charge of ore at each revolution and feeds it into the mill through the trunnion. No screens are used at the discharge end. The trunnions are turned on the inside after the parts have been fitted together, and are run in bearings of cylindrical sleeve and ball and socket construction. The axis is inclined slightly to the horizontal and according as the obliquity is large or small the discharge is quicker or slower and the crushing coarser or finer. The foundations on which the supports rest are usually now made of concrete with foundation bolts as in ordinary stamp mill practice. The lining of the mill and the character of the pebbles used are much the same as those ordinarily employed in tube mills. Silex bricks are most common as a lining, but some metallurgists prefer a metal plate lining. This possesses the distinct advantages that it can be made in convenient sections for passing into the mill and that these can be easily removed for repairs, necessitating but little delay in the actual grinding process. Danish pebbles are mostly used and recommended, but mine-ore has replaced them in a few instances, although not with complete success. Steel balls are also used, and in this case the lining is of steel.

David Cole[1] cites a case in which the silex lining was dispensed with and a pebble lining substituted. After 425 days the lining was in good condition, whereas the silex lining they had previously used had an average life of 75 days. The pebble lining, too, being rough, requires no " lifters."

The work of crushing progresses as follows :—As the ore enters it comes into contact with the larger pebbles, and, when crushed to a certain degree passes on between slightly smaller pebbles, which by a sizing action have found positions on the incline of the outlet end. The charge is lifted at right angles to the axis, which, as before mentioned, is slightly inclined, either by pure friction or by specially arranged lifters attached to the lining, and then tends to fall vertically owing to the action of gravity. Thus a gradual progression of the ore towards the exit is maintained. The ore is crushed partly by the impact and partly by the grinding action of the falling pebbles. The peculiar feature of the machine is the gradual sizing of both the pebbles which are crushing and the material which is being crushed. The great difference in the size of the particles is comparatively equalised by the diminished fall and the reduced speed at the periphery. Gradation in size is dependent on the amount of feed, the inclination of the mill, and the rapidity of the discharge. By having the mill only slightly inclined and carrying a large load of pebbles, it is stated by Hardinge that the material may be crushed to a very fine state of division.

Hardinge mills may be run singly, in tandem or in series. In some cases[2] the Hardinge mill has replaced stamps altogether, the ore passing straight from the rock breakers to a ball mill, after which it is classified and the oversize further reduced in a pebble mill whose dimensions are determined by the degree of fineness required. The oversize from the pebble mill is classified

[1] Cole, *Eng. and Mng. J.*, 1913, **96**, 927.
[2] Hardinge, *Eng. and Mng. J.*, 1913, **95**, 663.

and returned to the latter for re-grinding. At the MacIntyre plant, Porcupine, Ontario, the material from the rock crusher passing a 1½-inch ring is fed to a 4½ feet Hardinge ball mill and then slimed in a 72 × 72 inch pebble mill. The following table shows the grading analyses at various stages :—

TABLE XXV.

Mesh.	+ 10	+ 20	+ 40	+ 80	+ 100	+ 200	– 200
Feed to 4½ feet Hardinge mill. Crusher product through 1·5 in.,	88·56	4·17	7·0	...	...	...	...
Product from 4·5 feet Hardinge ball mill,	8·50	18·4	20·78	15·0	34·0	34·0	...
Product from 72 × 72 pebble mill,	...	...	...	0·65	4·92	3·56	90·87

In order to obtain a more granular product the 72 × 72 inch pebble mill may be replaced by one of 72 × 22 inches.

The following figures, supplied by the Hardinge Company, show the work of a Hardinge ball mill used in place of stamps and a Hardinge pebble mill for fine grinding at the Vipond Gold Mining Company's Mill, Porcupine, Ontario (see Fig. 113) :—

TABLE XXVI.

4½′ BALL MILL.			72″ × 72″ PEBBLE MILL.		
Material.--From Mill Bin.			Material.—Classifier Heads.		
Mesh.	Feed to Mill.	Product from Mill.	Mesh.	Feed to Mill.	Product from Mill.
	Per cent.	Per cent.		Per cent.	Per cent.
On 1 mesh	4·50	...	On 10 mesh	16·64	...
,, ¾ ,,	17·00	...	,, 20 ,,	27·42	...
,, ½ ,,	31·00	...	,, 40 ,,	27·04	...
,, ¼ ,,	24·70	...	,, 60 ,,	8·88	0·0
,, 10 ,,	11·86	8·50	,, 80 ,,	7·04	0·15
,, 20 ,,	4·17	18·40	,, 100 ,,	5·00	1·95
,, –20 ,,	7·00	...	,, 150 ,,	5·05	...
,, 40 ,,	...	20·78	,, 200 ,,	1·95	29·10
,, 60 ,,	...	8·30	Through 200 ,,	...	68·80
,, 80 ,,	...	6·70			
,, 100 ,,	...	3·30			
,, 150 ,,	...	· ·			
Through 100 ,,	...	34·0			

Capacity, . . 50 tons per 24 hours.		Capacity, . . 90 tons per 24 hours.
Charge, . . 9,000 lbs. balls.		Charge, . . 9,000 lbs. pebbles.
H.P., . . 15.		H.P., . . 35.
Revolutions, . 33½ per minute.		Revolutions, . 27 per minute.
Water, . . 66 per cent.		Water, . . 50 per cent.
No inclination of feed end.		No inclination of feed end.

R. Franke[1] found that Hardinge mills were superior to Chilian mills in steadiness of operation, labour charge, consumption of power, cost and depreciation. A. O. Gates[2] does not agree that in the Hardinge mill each

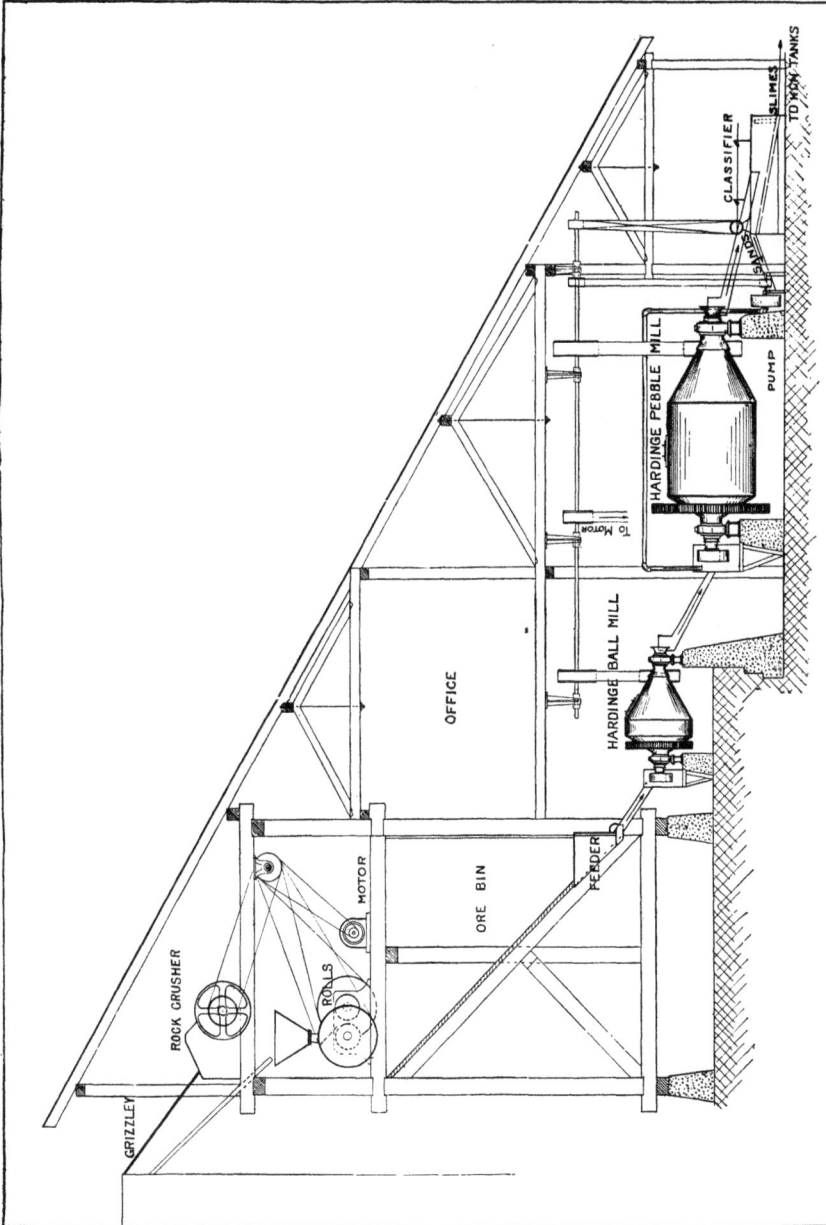

Fig. 113.—Vipond Gold Mining Company's Mill, Porcupine, Ontario, showing Hardinge Ball Mill and Hardinge Pebble Mill.

[1] Franke, *Mng. and Sci. Press*, 1913, **107**, 223.
[2] Gates, *Eng. and Mng. J.*, 1914, **97,** 12 ; *Bull. Amer. Inst. Mng. Eng.*, Nov. 1913.

particle receives the blow which it requires at any particular moment to reduce it further. He gives a diagram (Fig. 114) on which are drawn curves showing (1) the rate of pulp flow, (2) energy per lb. of pulp, (3) total pebble weight per foot length, (4) probable effective pebble weight per foot and (5) energy per foot length.

From the curves it is seen that for about a third of the length of the cone the effective pebble weight is considerably less than the total weight, but that afterwards the curves merge into one another and slope gently to zero along the axis. The rate of pulp flow is constant in the cylindrical part of the mill, and then it quickly increases as the outlet apex is approached. The energy per foot length is at a maximum at the cylindrical portion of the shell, but very soon diminishes at a rapid rate until finally it becomes zero at the exit.

Gates considers that the active zone in a Hardinge mill is really equivalent to a short tube mill. Thus an 8-foot size conical mill produces about the same effect as a 5 × 5 feet or 6 × 6 feet tube mill, and Gates further expresses

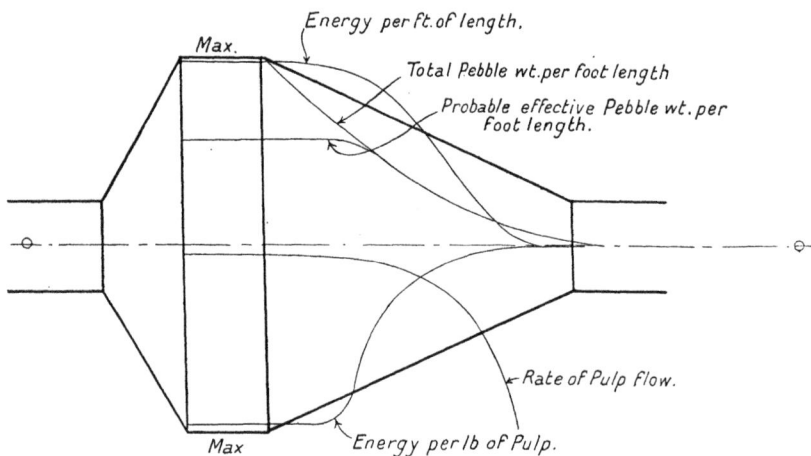

Fig. 114.—Hardinge Mill Diagram.

his belief that " the fine crushing machine of the future concentrating mill will be a short tube mill, followed by an efficient sizer to remove more of the fine material than is done at present and followed by another short tube mill."

In practice generally the Hardinge mill has proved to be better suited for producing a granular material of intermediate size than for fine grinding, in which it is inferior to tube mills.

The Chilian Mill.—This machine probably originated as an improvement on its prototype, the *trapiche*, see p. 146, and has also been known as the " edge runner." It was formerly used in the Patio process for the extraction of silver from silver ores, chiefly to prepare ores for the arrastra (*q.v.*), and is now used in the fine crushing of gold ores. Its re-introduction in the United States dates from about 1904.

The Chilian mill consists of a circular cast-iron pan, to which are bolted wrought-iron sides. Runners or crushing rollers travel round a cone in the centre of the pan. The track or die-ring on which they run consists of sections of cast steel, and the tires of the crushing wheels are also of cast steel.

Chilian mills are of two kinds [1] :—

1. High-speed—with die-rings of small diameter (4 feet to 6 feet) and crushing wheels also small (2 or 3 feet, with 4 to 6 inches face). The rate of speed is 30 to 40 revolutions per minute.

2. Low-speed—with die-rings of large diameter (6 to 10 feet) and large, heavy crushing wheels (7 to 8 feet, with 20 inches face). They run at 8 to 12 revolutions per minute.

High-speed mills are commonly used for regrinding, following rolls, breakers and stamps. Low-speed mills are better adapted to amalgamation.[2]

Two forms of the *low-speed Chilian mills* are in use. The old form had rollers which turned on separate horizontal axles in a vertical sliding hub. This type has many mechanical defects. In the modern form the rollers are

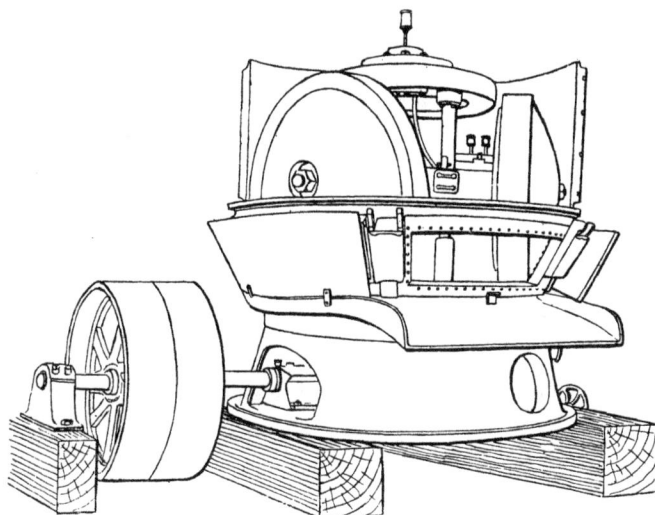

Fig. 115.—Chilian Mill, Gaika Mine, Rhodesia.

carried on an off-set axle, which slides, vertically without turning, in a vertical spindle driven by the gearing. The rollers retain their vertical position. The sliming capacity of the mill is increased by the drag due to the off-set axle.

In many cases low-speed mills have been proved to be superior to high-speed machines, and Urbiter,[3] has shown by a comparison with other grinding machines that, although they are not really " slimers," they are capable of producing efficiently a large percentage of very fine material. There is a limit, however, to their usefulness, and they cannot be considered as regrinding machines. They are best fed with coarse material, using a large volume of water in the process of crushing. According to Urbiter it only requires one

---

[1] H. A. Megraw, *Eng. and Min. J.*, Nov. 12, 1910, p. 967 ; 1913, **96**, 18, 821.
[2] A. Maclaren, *Eng. and Min. J.*, Aug. 12, 1911, p. 305.
Urbiter, *Eng. and Min. J.*, Aug. 5, 1911, p. 257.

passage of the ore through a mill to produce a high proportion of material which is finer than 200 mesh. An instance is given by Webster,[1] in which ore crushed size to $1\frac{1}{2}$-inch was fed into a Chilian mill and in one operation 80 per cent. of — 200-mesh material was produced with a mechanical efficiency of 46·13 per cent. By classifying the mill product and returning the oversize, no appreciable increase in the amount of slimes produced was noticed.

J. B. Empson[2] gives a number of figures showing the work of low-speed Chilian mills as fine grinders, and compares them favourably with stamps and stamp-tube mill combinations. He mentions the following advantages of Chilian mills over stamps :—

1. Less height required.
2. Cheaper foundations.
3. Fewer repairs.
4. Withstand rough handling.
5. Overfeeding or underfeeding does not hurt mill.
6. In some cases absence of all screens.
7. Repairs can be made on the ground.
8. Minimum vibration.

At the Goldfield Consolidated Mill[3] six Chilian mills were installed between the existing stamps and the tubes, instead of adding forty extra stamps. The capacity of the plant was thereby increased 40 per cent.

The *Akron mill* is described by Eaton as a typical instance of a Chilian mill of *high speed*[4] (see Fig. 116).

The base of the mill is a circular casting, and is designed so that the mill can run in either a right-hand or left-hand direction.

The mortar consists of one casting, or may be made with the upper, lighter portion separate from the lower, heavier portion. The total screen area is 1,800 square inches, and is divided into five sections round the mill. A semicircular launder, placed round the mortar and inside the splash plates serves to carry the pulp towards the apron.

The pan should be as narrow as possible without allowing any coarse material to pass through the screens.

When new, the die-ring, which fits into the mortar, is 1 inch below the screen. It is 5 feet in diameter, and has a cross-section 7 × 4 inches.

The rollers, three in number, are heavy solid castings, weighing 3,000 lbs. each, and are carried on trunnions which fit into boxes on the drive head over the central spindle. The rollers can assume any angle when passing over uneven ore. The tyre is made (like the die-ring) of rolled steel, and is fixed on to the casting by wooden wedges. The centre of the roller shaft is placed a little higher than the centre of the trunnion, thus allowing the centrifugal force to increase the crushing capacity. The correct position of the rollers is to have them inclined slightly inwards at the top. It is recommended[5] that the tyres should have a groove along their centre, and that compensating weights should be added as the tyres wear.

---

[1] Webster, *Eng. and Mng. J.*, 1912, **93**, 393.
[2] Empson, *Eng. and Mng. J.*, Feb. 3, 1912, p. 259 ; *Mexican Inst. Mng. Met.*, Dec. 1911.
[3] Hutchinson, *Mng. and Sci. Press*, May 6, 1911, p. 616.
[4] J. M. Eaton, *Trans. Inst. Mng. and Met.*, 1911, **20**, 161.
[5] R. H. Richards and C. E. Locke, *Mineral Industry*, 1912, p. 936.

Ore is fed in from an annular casting fixed to the cover, through three $3\frac{1}{2}$-inch pipes, one directly in front of each roller and after the corresponding "plough."

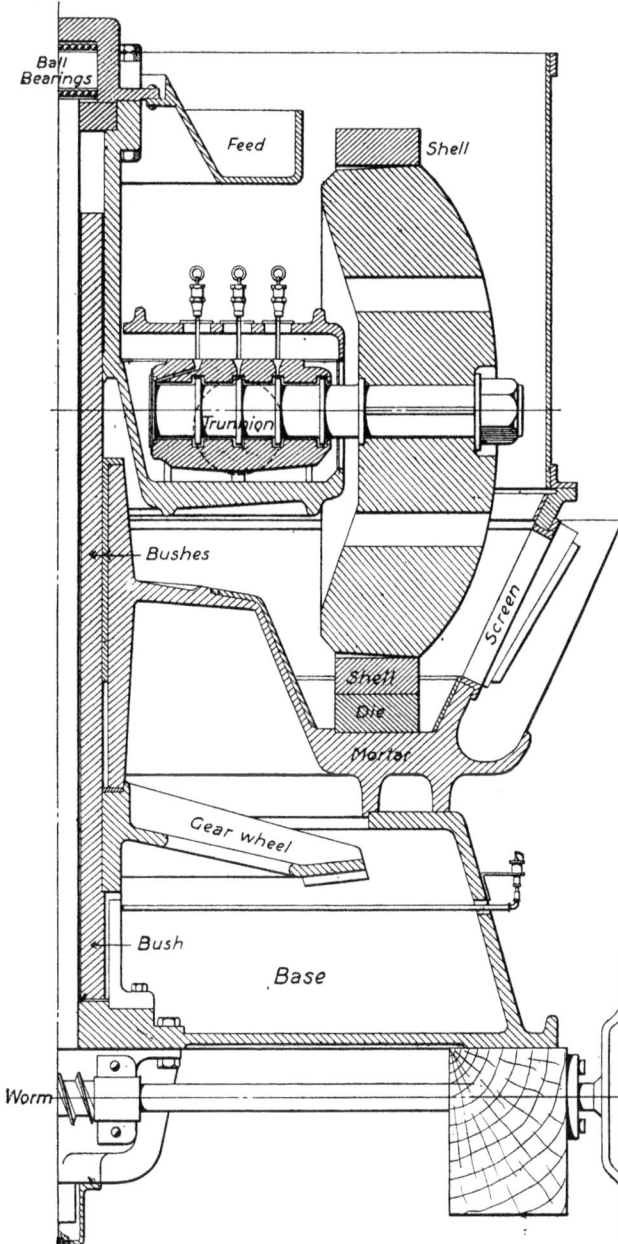

Fig. 116.—Section of Chilian Mill (Akron Type).

The driving gear is underneath the mill, and is actuated from a shaft outside.

There is considerable variance of opinion as to the depth of feed to be maintained in the mortar, but with the mill in question this is adjusted to be twice the height at which the ploughs are set above the die.

It is found that Chilian mills, whether of high or low speed, do more efficient work when fed with coarse material of an average size of $\frac{1}{4}$ to $\frac{1}{2}$ inch. Such a product may be obtained from roll crushing, and in some cases stamps have been entirely replaced by this combination, as, for example, at Stratton's Independence Mill, Colorado.[1] At this mill the material from the breakers goes to rolls and then to Akron Chilian mills, which treat 100 to 130 tons each per day, according to the feed, the screens used and the character of the ore. These particular mills revolve at 33 revolutions per minute. Argall considers that though the Chilian mill is a slimer it is capable of adjustment to give a fine but granular product. At Stratton's Independence the best results were obtained with a 0·046-inch screen aperture.

Megraw found that by feeding into a low-speed Chilian mill material of 80 to 40 mesh there was considerable packing on the die-rings, thus preventing the true grinding action. Webster suggests the adoption of a Dorr classifier to overcome this.

The Lane mill is a typical low-speed machine, and is employed for fine crushing and amalgamation. There are six rollers, each 42 inches in diameter, with a 5-inch face, and having tyres $2\frac{1}{2}$ inches thick. The mill revolves eight times per minute on a track 10 feet in diameter.

Argall, in discussing [2] the work of Akron mills, says that the reduction of the ore results from two causes :—

1. The direct crushing effect due to the weight and speed of the rollers.

2. The grinding or abrading effect due to the rollers being constrained to travel in a circular path.

The smaller the diameter of the die the greater is the twisting effect on the ore particles ; with a die of greater diameter, the action is similar to that of small thin cylinders rolling between the faces of the rollers and dies. The abrasion thus becomes less, and the action of low-speed mills accordingly approximates to that of rolls, and the ore is not finely ground. It is noticed that the capacity increases as the steel die and rollers wear. This is explained by the fact that wear introduces hollows into the wearing parts, and that all ore enclosed within these is subjected to the twisting effect of the rollers.

G. A. Denny [3] thinks that there is a critical speed for each mill, at which it works most advantageously. This critical speed is affected by the size of the feed.

Chilian mills were introduced in Russia in 1870, and have been in use ever since, improvements in construction being made from time to time. The type of mill most used in Russia [4] in recent years, in many instances replacing stamps, is one of 7 to $10\frac{1}{2}$ feet in diameter, revolving eleven times

[1] Argall, *Mng. Mag.*, 1911, **5,** 365.
[2] Argall, *ibid.*, p. 366.
[3] Denny, *Can. Mng. J.*, 1912, **23,** 832.
[4] H. C. Bayldon, *Trans. Inst. Mng. and Met.*, 1910, **20,** 125.

per minute. It has a capacity of 16 to 26 tons per twenty-four hours. Bayldon[1] describes the latest type of mill used in Russia. It runs at 16 revolutions per minute, and crushes 39 tons of ore per day.

E. E. Carter[2] quotes the case of an antimonial gold ore in Idaho which is amalgamated in a low-speed Chilian mill after being crushed by stamps to which no plates are attached.

Pans.—For fine grinding in pans, see p. 227.

---

[1] Bayldon, *ibid.*, p. 133.
[2] Carter, *Mng. and Sci. Press*, Mar. 11, 1911, p. 370.

CHAPTER XI.

CONCENTRATION IN GOLD MILLS.

Concentration.—The object of concentration is the separation of the heavy valuable mineral from the light worthless gangue. Complications are often introduced by the fact that various base minerals must be separated from one another, an ore being subdivided into several products. Many gold ores, however, only require separation into two parts—the " concentrate " in which the precious metal is contained, and the " tailing," which is thrown away. The German system of coarse crushing, sizing by means of screens, and concentrating on jigs, is not, as a rule, applicable to gold ores proper, although much used on auriferous lead, zinc, and copper ores. This system will not be described here, where only the methods in use for the treatment of pulp from the stamp battery and some other machines will be considered.

All the concentrating machines depend for their action on the effect of a difference of densities on the fall of bodies in a fluid. The fluid employed in almost every instance is water, although several machines have been devised in which air is used as the concentrating medium. It has often been proposed to use some solution which shall have a lower density than the valuable mineral to be saved, but a higher density than the worthless gangue ; the mineral would then sink, while the gangue would remain floating. The high cost of any such solution is sufficient to put this method out of the question, without discussing any further disadvantages. For the various flotation processes see p. 271.

The fall in still water of solid materials takes place according to two laws, one applicable to very shallow water, through which the particles fall with increasing velocity, while the other is true when the depth of the water is considerable, so that the particles for the greater part of their course proceed at their maximum velocity. In shallow water the fall is almost entirely according to density, so that those machines which utilise only the first instants of the fall will have great efficacy in concentrating.[1] It is this fact which has necessitated the use of shallow currents in concentrating tables, sluices, etc. In almost all these machines the fine sand and slime is brought into suspension in water, and the liquid is then run over an inclined surface. The deposit of sand, which is thus formed on the table, tends to become enriched in heavy minerals, because the stream moves faster at the surface of the water, where the lighter particles remain, than it does next the bed, where the heavy particles have settled. The deposit is continually worked up and brought again into suspension by a rake or broom, or by a series of shakes or blows imparted to the apparatus, so that the effect mentioned above is repeated frequently. If the stirring up is

---

[1] The matter is complicated by the quantity of solid matter present, as free settling differs from hindered settling. See, among other papers, G. G. Bring, *Jern-Kontorets Annaler*, 1906, p. 321 ; Richards and Locke, *Mineral Industry*, 1907, p. 970.

violently performed all the slime and very fine particles are kept in suspension in the water and carried away and lost, a slow stream of water and very slight agitation being favourable to their retention in the deposit which is formed. When the stream of water is rapid and voluminous, fine material, whether heavy or light, is swept away and lost in the tailing, whilst if too small a stream of water is used, much worthless sand is deposited with the concentrate. It follows that the amount of water used must be regulated according to the work which it is proposed to do. Frequently, it happens that clear water must be added to the pulp to dilute it sufficiently. On the other hand, it often happens that the pulp is too thin, and water is then removed by means of pyramidal boxes or cones (de-waterers), described below.[1]

Other operations, which it is often of the utmost importance to perform before concentration, are *classifying* and *sizing*. The necessity of sizing is obvious, when it is remembered that a shallow stream of water swift enough to carry down fine sulphides, might be powerless to move a pebble of quartz. The usual method of classifying is based on the varying rates of fall of particles through a deep column of water, or on the various movements of particles when in an upward-moving column of water, which are dependent on the same properties. In this way *equal-falling* particles are obtained together, and since a sphere of galena is equal-falling with a sphere of quartz of from 1·5 to nearly 4 times the diameter, according to the absolute sizes,[2] it follows that classifying has very different results from sizing. Nevertheless, classifying, when efficiently performed, is of great assistance as a preparation for treatment by shallow-stream concentrators, and especially to separate sand from slime before treatment with cyanide, and to prepare material for feeding into tube mills. The *Stanley classifier* [3] is a combination of screen and hydraulic classifier, and is designed to size the materials in Rand dumps irrespective of their specific gravity.

Sizing by Screens.—Screening to equal sized particles is a desirable preparation for concentration on most machines, but in the past has not usually been employed on material finer than about 8 mesh, owing to difficulty in working, and to wear and tear of fine screens if used wet in the usual way. The two ordinary types of screens are the cylindrical revolving and flat shaking forms. For the finer sizes it has been usual to substitute hydraulic classifiers—i.e., boxes with ascending currents of water, described below— for screens ; but the one is not an exact equivalent of the other. Screens produce classes of particles of nearly equal size regardless of their respective specific gravities. Classifiers produce classes of equal falling particles, the sizes of which depend on their respective specific gravities.

Several new devices have been brought forward for using much finer screens than were formerly considered practicable. Steeply inclined and rapidly jarred flat screens, with sprays of water, are reported to give good results ; and very fine slightly inclined screens submerged in water (with jarring motion, and a special construction for delivering the coarser over-size into clear water) have been recommended. The advantages of this last

---

[1] See also Caldecott's sand filter table, and Dorr's thickener, Chap. xvi.

[2] R. H. Richards, *Ore Dressing*, 1st edition, 1903, p. 471. Richards states (*Mineral Industry*, 1896, p. 705) that in practice, with particles between 10- and 60-mesh screens, the diameter of the quartz is from $2\frac{1}{4}$ to $3\frac{3}{4}$ times that of the galena, varying with the absolute size. See also *Bull. Amer. Inst. Mng. Eng.*, May, 1907, p. 435 ; *Mineral Industry*, 1897, p. 694 ; 1907, p. 969.

[3] Stanley, *Mng. Mag.*, 1912, **6,** 141.

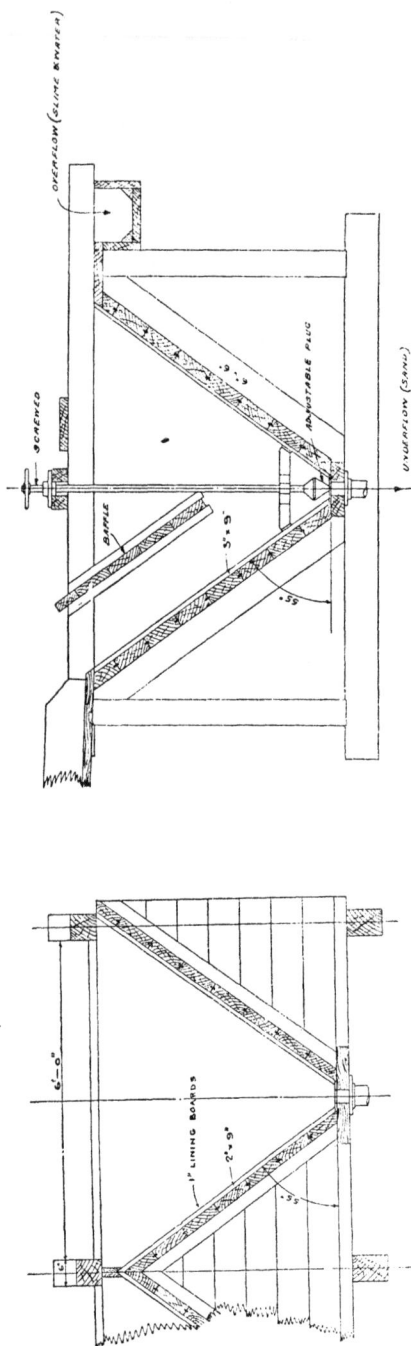

Figs. 117 and 118.—Slime Spitzkasten made of Wood.

type are stated to be that no extra wash water is used, screening is more perfect under water than with washing sprays, and the wear of screens is reduced to a minimum. Such screens have been tried as fine as 60 mesh—that is, with 3,600 holes to the square inch. There is also the Callow endless travelling belt type of screen.[1]

Hydraulic Classifiers were introduced by Prof. P. R. von Rittinger in the middle of the last century for use in the Hartz,[2] and from their shape were known as *spitzkasten* or *pointed boxes*. These boxes have the shape of inverted pyramids, the stream of unsized pulp entering at one side and flowing out at the other, whilst there is also a small discharge at the apex. The current slackening on entering the box, the heavier and larger particles in suspension at once begin to settle, and, escaping the influence of the current, fall quietly to the bottom of the box, where they are discharged. Rittinger recommended the use of four spitzkasten ranged in series, each box being twice as wide as the preceding one, with lengths increasing in arithmetical progression—*e.g.*, lengths of boxes, 6, 9, 12, and 15 feet respectively. In each successive box a number of approximately equal-falling particles are removed, the closeness with which the subdivision is made varying with the size of the box, and the corresponding extent to which the current carrying the pulp is checked on entering it. The larger the box the more material

[1] The Callow Screen, *J. Chem. Met. and Mng. Soc. of S. Africa*, 1909, **9**, 313.
[2] *Aufbereitungskunde* (1867) ; Richards, *Ore Dressing*, vol. i., p. 439.

is collected by it, and, therefore, the more heterogeneous the particles caught. In a small box the material collected consists more nearly of truly equal-falling particles.

Figs. 117 and 118 [1] show slime spitzkasten made of wood as used on the Rand. In spitzkasten the sides are at an angle of at least 50° to the horizontal, to insure the uninterrupted descent of the slimes. The pulp is delivered evenly across the end of the box. Surface currents are prevented and the incoming pulp thoroughly mixed with the mass of the water in the box by the baffle plate (Fig. 118), which extends across the box near the inflow end. The discharge is made by cutting the other end of the box from 2 to 3 inches lower than the sides ; this overflow is made perfectly level so that the water flows out in an even sheet. The discharge pipe for the pulp is at the bottom of the box. A contracted orifice is liable to be choked up, and

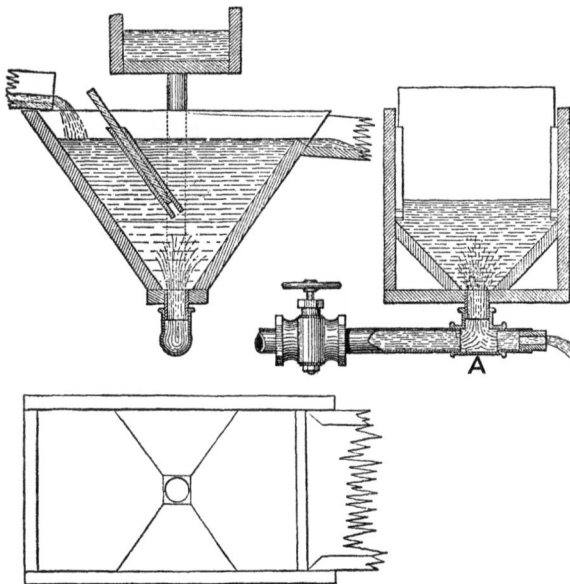

Fig. 119.—Spitzlutte with Ascending Current.

a smaller diameter than $1\frac{1}{2}$ to 2 inches is to be deprecated. A large orifice, however, delivers too much pulp.

These boxes have the advantage over the ordinary settling-pits used to retain tailings and to catch pulp for pan-amalgamation, that they do not require to be dug out, while the settling, owing to the elimination of surface currents, is more perfect.

Spitzlutte.—Equal-falling particles, in any box, whatever its size may be, are carried away through the aperture in the apex by muddy water containing material of all sizes down to the finest slime, and it was to eliminate this material that the *ascending current* was introduced. This is a current

---

[1] Schmitt, *Rand Metallurgical Practice*, vol. ii., p. 194.

of clear water, which enters at the apex of the box (*spitzlutte* [1]) in greater quantity than can be discharged by the outflow near the same spot, so that there is an upward current of water into the box. The result is that no muddy water is discharged below, but only the particles of ore which have weight

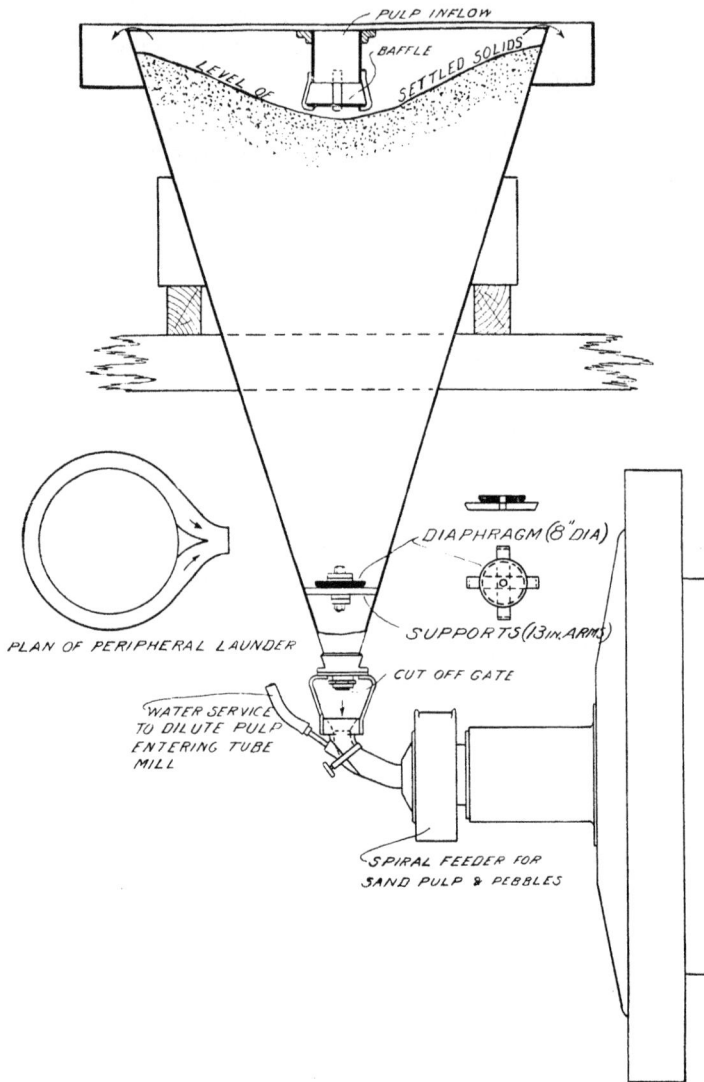

PULP INFLOW

BAFFLE

LEVEL OF SETTLED SOLIDS

DIAPHRAGM (8″ DIA)

SUPPORTS (13 in. ARMS)

CUT OFF GATE

PLAN OF PERIPHERAL LAUNDER

WATER SERVICE TO DILUTE PULP ENTERING TUBE MILL

SPIRAL FEEDER FOR SAND PULP & PEBBLES

Fig. 120.—Caldecott's Hydraulic Cone Classifier, with Diaphragm.

[1] There seems to be some difficulty in the definition of spitzlutten. Richards (*Ore Dressing*, p. 421) states that the word signifies a pointed tube, and that Rittinger used it in this sense. "Lutte," however, appears to mean a "gutter in which gold is washed." It is certainly convenient to limit the use of the word spitzlutten to pointed boxes with an ascending current of clear water, and spitzkasten to boxes without such a current.

enough to drop through the ascending current, so that by regulating the strength of this, any desired class of ore can be obtained. In Fig. 119, a form is shown in which the sliding partition assists the settling, by causing the pulp to pass downwards, rapid surface currents across the box to the overflow being thus prevented. The discharge of the heavy particles is effected through A, the clear water pipe itself, by the arrangement shown. The launder above the box supplies the clear water current, and shows the head of water used, which must be kept constant to ensure uniformity of results.

The chief defect of the early forms of pyramidal boxes was that, as the area of the vat became larger and larger towards the top, the velocity of the ascending water naturally became less and less, so that many particles were able to settle down below the level of the overflow, but were stopped by the increasing force of the current, so that an accumulation of the ore took place half way up the box, and ultimately became so great as to interfere with the classification. Several remedies have been devised for this defect, of which one of the simplest and most effectual is to make the box pyramidal below, but with vertical sides in the upper part. This construction is partly carried out in the box described above. A slime-pit is added to catch the stuff which is too light to settle in the boxes, in all cases in which these slimes are of sufficient value to pay for treatment. The number of boxes used depends on the tonnage to be treated and on the number of classes of material which it is deemed advisable to make. Usually two or three classes are sufficient.

On the Rand, the spitzlutte is about 2 feet by 3 feet at the top, with the upper sides vertical for about 18 inches. The inverted pyramid below has an angle of about 60°. The baffle plate is in the middle and movable vertically. Spitzlutten are usually arranged in a series of three.[1] Choking, and the difficulty of regulating the classification with constantly varying conditions are the chief defects of these appliances.

Cone Classifiers.[2]—The most recent form of hydraulic classifiers are large cones, 5 to 8 feet in diameter at the top and from 7 to 10 feet deep (see Fig. 120).[3] The cones are built of $\frac{3}{16}$-inch sheet steel with the lower 18 inches of the cone of cast iron, to resist the hard wear of that part and to facilitate renewals. The pulp enters from a launder through a large central pipe, which delivers about 12 inches below the level of the pulp, and is supplied with a circular baffle plate placed horizontally about 3 inches below the open end of the pipe. The fine pulp overflows all round the cone into an annular launder. The underflow passes through a nozzle at the apex of the cone, and passes vertically downwards into the inlet of the tube mill, or into launders, as horizontal or inclined pipes tend to choke. The nozzle is regulated with a cut-off gate (see Fig. 121).[4]

The cone classifier is kept nearly full of sand by means of a circular diaphragm, due to Caldecott, about 8 to 10 inches in diameter, placed in the axis of the cone near the apex. The diaphragm prevents the sand from settling and forming a channel in the middle, but allows it free passage in the annular space round the diaphragm, and ensures a steady flow of coarse material through the nozzle. The level of the sand is kept constant under varying inflow by opening or closing a sliding shutter or gate at the nozzle. Coming from below so great a depth of sand, the underflow contains

[1] Dowling, *Rand Metallurgical Practice*, vol. i., p. 99.

[2] W. A. Caldecott, *J. Chem. Met. and Mng. Soc. of S. Africa*, 1909, **9**, 312 ; Dowling, *Rand Metallurgical Practice*, vol i., p. 99.

[3] Dowling, *Rand Metallurgical Practice*, vol. i., p. 100.   [4] *Ibid.*, p. 101.

little water, say 25 to 28 per cent., and issues very slowly even with a free vertical discharge. A large outlet is required, the diameter of the nozzle being from $1\frac{3}{4}$ to $2\frac{1}{4}$ inches. Separate de-watering appliances are not required. In working, coarse grains settle near the centre and fine grains near the periphery.

The capacity of these cone classifiers is very great. A cone 8 feet in diameter at the top and 10 feet deep is capable of delivering from 400 to 600 tons of sand per twenty-four hours at its underflow, from pulp composed of 44 per cent. slime and 56 per cent. sand,[1] the product thus obtained carrying rather less than 30 per cent. moisture, against about twenty times

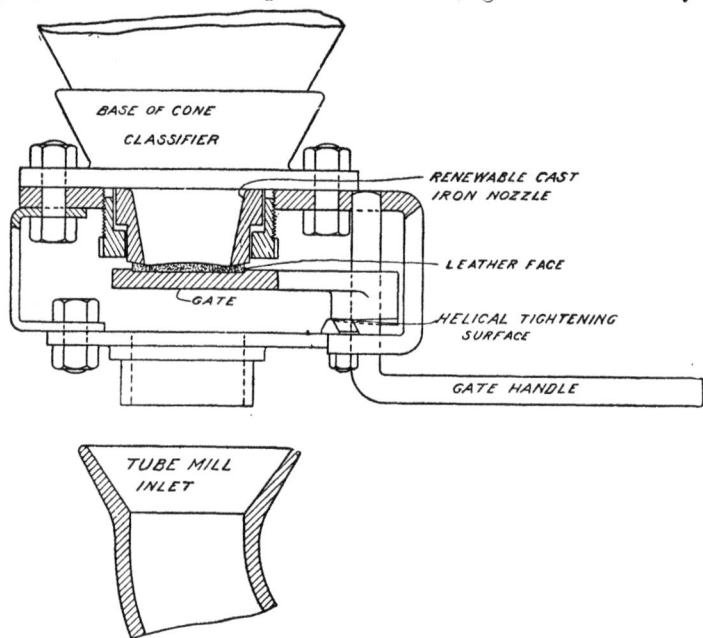

Fig. 121.—Adams's Cut-off Gate.

that amount in the original pulp. The cone is more efficient than pyramidal spitzkasten, as is shown in the following table :—

TABLE XXVII.—PERCENTAGES OF PRODUCT.[2]

	+ 60 Mesh, Size of Product over 0·01 Inch.	− 60 + 90 Mesh, 0·01 to 0·006 Inch.	− 90 Mesh, Less than 0·006 Inch.	Total.
Tube mill feed (underflow)—				
Pyramidal spitzkasten,	61·8	22·3	15·9	100
Cone classifiers,	67·1	20·7	12·2	100
Pulp entering cyanide works (overflow)—				
Pyramidal spitzkasten,	10·9	16·0	73·1	100
Cone classifiers,	7·4	15·9	76·7	100

[1] J. E. Thomas, *Rand Metallurgical Practice*, vol. i., p. 152.
[2] G. O. Smart, *J. Chem. Met. and Mng. Soc. of S. Africa*, 1910, 10, 287.

The conditions were similar, so that a true comparison was given by the results of the tests given above.

The objection has been made to the Caldecott cone classifier, just described, that bubbles of air are carried down by the inflow and are still rising through the liquid at the outflow, thus interfering with efficiency. There is also the difficulty that the thick spigot product becomes coarser if the amount of feed is increased.[1] This entails close watching of the underflow. A jet of water below the diaphragm has been tried and considered an improvement.

In the Waterman settling cone,[2] adopted at the Butters Co.'s plants, the cone is fed at the periphery and discharged through the central pipe (see Fig. 122), which is made larger. The slime follows the course shown by the arrows. The annular space between the side of the cylinder and the cone is $\frac{1}{2}$ to $\frac{5}{8}$ inch wide. The cylinder is divided by a vertical partition shown in the figure, extending to near the point of the cone, which prevents a vortex from being formed by the discharge at the spigot. The rising current for discharge being through a cylinder, the difficulty of the formation of suspended sand banks is avoided (see p. 259). This can be put right,

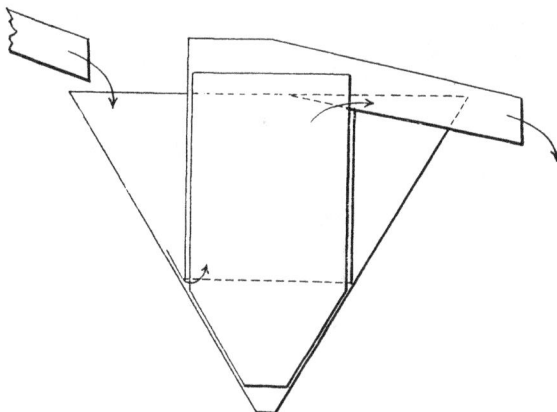

Fig. 122.—Waterman Settling Cone.

however, in the ordinary cone classifier by making the sides of the cone vertical near the top, or even inclined upwards towards the central line, so that the classifier consists of two superposed cones, the lower one inverted, the other truncated but not inverted, with their common base at the level of the bottom of the feed pipe.

The Dorr Classifier is shown in Fig. 123. It consists of a settling box in the form of an inclined trough open at the upper end, in which mechanically operated rakes are placed to remove the heavy material as fast as it settles, the liquid and slime overflowing at the closed end.

The rakes are suspended by suitable hangers from bell cranks connected by rods to levers which terminate in rollers. The latter press against cams attached to the crank shaft. The rakes are lifted and lowered at opposite ends of the stroke by the action of the cams transmitted through the levers and rods to the bell cranks. The horizontal motion is produced by cranks.

[1] Robertson, *Mineral Industry*, 1912, p. 944.
[2] D. Waterman, *Mng. and Sci. Press*, April 20, 1912, p. 567 ; *Mng. Mag.*, 1912, **6**, 457.

The pulp is fed across the centre of the trough and the sand settles to the bottom, after which it is advanced up the inclined bottom of the trough by the scrapers reciprocating with a slow raking motion. After emerging from the liquid the sand passes across a washing device and is discharged from the open end of the machine with about 26 per cent. moisture. If the machine is stopped while full of sand, the rakes can be raised from the bottom and, after it has been started, they can be gradually lowered while running until they assume their normal position.

The slime is prevented from settling by the flow of the liquid as well as by the agitation near the bottom caused by the reciprocating motion of the scrapers, and overflows at the lower end of the machine. The agitation produced, while ample to prevent the slime from settling, is not sufficient to cause the sand to overflow with the slime.

The height of fall in this machine is small. The capacity is $3\frac{1}{2}$ to 8 tons of solids per hour in four to six parts of water. The overflow contains little sand, and the machine is highly efficient. The repairs are small.

Fig. 123.—Duplex Dorr Classifier.

Dorr classifiers were introduced in 1904, and are extensively used, especially in America. Besides the duplex machine shown in the figure, there is a simplex machine with only one set of rakes instead of two.

Early Concentrating Machinery.—One of the oldest and most primitive machines employed in the concentration of fine sand by means of a shallow stream of water was the German *buddle*, which has a distinct but imperfect resemblance to the Long Tom described on p. 103.[1] Canvas tables were used below these buddles in Germany, and probably suggested the use of *blanket-strakes* or *tables*, which were adopted in the early days of the goldfields of the United States and Australia, and are still retained in some places. The rough surface of the blanketing seems to be particularly efficacious in catching

---

[1] For a description of the similar buddle formerly used in Colorado for the treatment of battery sand, see Raymond, *Mines, Mills, and Furnaces of the Pacific States* (New York, 1871), p. 357.

and holding thin plates and spangles of free gold or of sulphides, which are readily washed off smooth surfaces by a current of water, and these rough appliances, although almost useless for catching slime, still find favour where considerations of economy prevent the purchase of modern high-priced concentrators. The blanketing is usually in strips of 16 or 18 inches wide, and several feet long, and is nailed or stretched on wooden frames, which have an inclination of about ½ inch per foot.

At intervals of about half an hour, when a quantity of mineral has already been collected on the rough surface, the blankets are taken off and washed in tubs of water, where a deposit collects which is afterwards dug out. At the St. John del Rey Mine, the framework supporting the blankets was hung on pivots above a shallow tank. When it was necessary to clean them, the framework was turned so that the upper surface of the blanket was inclined downwards, and the mineral washed off its surface by a hose, much time and labour being thus saved.

At this mine the trays supporting the blankets were 18 inches wide and 30 feet long, with a fall of 1 inch per foot. The upper 16 feet were covered with bullocks' skins, tanned with the hair on them, and in lengths of 26 inches ; below these was a series of blankets or baize cloths of the same length, made of coarse wool with a long nap. The fall from the battery box upon the tray was 4 inches, a screen being placed across the end to break the fall of the water, and cause it to strike the tray nearly at right angles. About 90 per cent. of the gold contained in the ore was caught on these blankets. The blanket sand contained 95 per cent. of sulphides, and was so fine that 90 per cent. of it passed through a 100-mesh sieve. It was amalgamated in revolving wooden barrels, yielding 96 per cent. of its assay value, but this was due to the fact that very little gold was contained in the pyrite, most of it being present in the form of free particles.

Blanket sluices have been declared unsuitable for catching fine sulphides, and their concentrate is usually contaminated by admixture with much sand. If set at a proper inclination they will save fine amalgam and free gold, but even in this respect they are less satisfactory than shaking copper plates and riffles.

Riffled sluices were employed at the same time as blankets for effecting rough concentration. The riffles were formed of half-inch strips of wood nailed across the sluice box, the grade of which was about three-quarters of an inch to the foot. As soon as the concentrate had accumulated until it reached the top of the riffle, another strip was nailed on, and the process was repeated until the bed of concentrate was several inches thick, when it was scraped out and a fresh start made. Similar to this device was the *raising-gate concentrator*, which was practically a riffled sluice in which the riffle was raised continuously by machinery, instead of being adjusted at intervals by hand.

The Round Buddle was invented in Cornwall, where until recently it was used in dressing the tin ores to the exclusion of almost every other concentrator. There are two varieties.

1. The convex round buddle, in which the ore and water are added at the centre of the machine, and flow down over the surface to the periphery.

2. The concave buddle, in which the pulp is added at the periphery of the machine and flows down to the centre.

In both cases revolving arms, carrying brushes, pass over the surface and stir the deposit as it is being formed, and the spouts distributing the ore also rotate so as to deliver the pulp evenly. The buddles are from 12 to 18 feet in diameter.

These machines are not continuous in action, and after the deposit has accumulated to a depth of a few inches the operation is suspended and the deposit dug out. The " headings " or material on the upper 12 to 18 inches of the inclined surface are kept separate, and the stuff near the bottom of the slope is called the " tailings." Round buddles are not adapted to obtain a finished product in one operation. The headings and tailings must, as a rule, be subjected to further treatment, and between them is a large quantity of material which differs little from the original ore. Thus handling and re-handling of the stuff is necessitated, and it is on this account that these machines are not now much used on gold ores. The principle on which they depend is favourable to the collection of slime, and the modern improved buddles are perhaps better adapted for fine ores than any other machines, except those employing a travelling belt. Modifications have been proposed to adapt these buddles to the treatment of gold ores by adding riffles containing mercury, and by other devices, but have not found much favour. One of the chief changes, which was proposed in the United States, is to keep the brushes and ore-spouts stationary, and to rotate the inclined bed.

Among concentrators with circular revolving beds, which were apparently suggested by the revolving buddle, may be mentioned Hendy's and Duncan's machines. These were formerly employed in California, but have given place to the vanners which are better adapted for saving slime.

Percussion Tables.—In these machines the work of keeping the pulp in a state of agitation, done by the rakes or brushes in the German and Cornish buddles described above, is effected by sudden blows or bumps imparted sideways or endways to the table. The table is made of wood or sheet metal, the surface being either smooth or riffled.

End-bump tables are hung by chains or in some similar manner, so as to be capable of limited movement, and receive a number of blows delivered on the upper end. These blows are given by cams acting through rods, or else the table is pushed forward against the action of strong springs by cams on a revolving shaft, and then being suddenly released is thrown back violently by the springs against a fixed horizontal beam. The movement of the pulp depends on the inertia of the particles, which are thrown backward up the inclined table by the blow given to the table, the amount of movement varying with their mass, and depending, therefore, both on their size and density. The vibrations produced by the percussion also perform the work of the rakes in destroying the cohesion between the particles, and a stream of water washes them down. The result is that the larger and heavier particles may be made to travel up the table in the direction in which they are thrown by the blow, by regulating the quantity of water, while the smaller and lighter particles are carried down. These machines yield only two classes of material, heading and tailing. One such machine, the *Gilpin County* "*Gilt-Edge Concentrator*," was devised in Colorado, and soon displaced the blanket sluices at almost all the mills at Blackhawk. It consists (Fig. 124) essentially of a cast-iron or copper table, 7 feet long and 3 feet wide, divided into two equal sections by a 4-inch square bumping-beam. The table has raised edges, and its inclination is about 4 inches in $5\frac{1}{2}$ feet at its lower end, the remaining $1\frac{1}{2}$ feet at the head having a somewhat steeper grade. The table is hung by iron rods to an iron frame, the length of the rods being altered by screw_threads, so as to regulate the inclination to the required amount. A shaft with double cams, A, making 65 revolutions per minute, enables 130 blows per minute to be given to the table in the following manner ; on being released by the cam, the table is forced forward by the strong spring, B, so that its head strikes against the solid beam, C, which is firmly united to the rest of the frame. The pulp coming from the copper plates is fed on to the table near its upper end by a distributing box, D, and is spread out and kept in agitation by the rapid blows. The sulphide settles to the bottom of the pulp, and is thrown forward by the shock, and eventually discharged over the head of the table at the left hand of the figure, while the gangue is carried down by the water and discharged at the other end. One machine is enough to concentrate the pulp from five stamps, treating about 8 cwts. of ore per hour. If the table consists of amalgamated copper plates, it is of some use for catching free gold also. This machine is not so effective in saving slimed pyrite as the Wilfley table or the vanners.

The Frue Vanner.—This machine is described in detail as being typical of the shaking travelling-belt concentrators. Machines of this class are especially adapted for treating finely-crushed battery sand which does not contain a large percentage of "mineral" (that is, sulphide and other heavy

Fig. 124. — End-bump Table.— " Gilt-Edge Concentrator."

materials). They are frequently set to concentrate unsized pulp coming straight from the amalgamating tables.

The Frue vanner (Figs. 125 to 127) consists of an endless rubber belt, mounted on a frame, with its upper surface slightly inclined to the horizontal, and subjected to two movements, a slow constant longitudinal movement and a slight and rapid side shake. The belt forms the bed or plane on which the dressing of the ore is effected, being an inclined plane, 12 feet long, and bounded down the two sides by projecting rubber flanges, which prevent the water and sand from dropping over the sides. An arrangement of rollers permits of the belt being slowly revolved in the direction of its length and *up* the incline ; thus, though the dimensions of the working plane remain always the same, its surface is constantly travelling. The crushed rock in a stream of water is delivered near the upper end of the belt by means of the sand distributor, No. 1, Figs. 125 to 127, and flows down the belt towards its lower end. Now, as the inclination at which the belt is set is only from 3 to 6 inches on the 12 feet, and as the stream of water is not large and spreads over the whole width of 4 feet, it is obvious that, if it were not for the movements of the belt, much of the crushed rock contained in the water would settle on the belt, while the water and the finer and lighter particles of sand would alone reach the foot of the table and drop over into a waste launder.

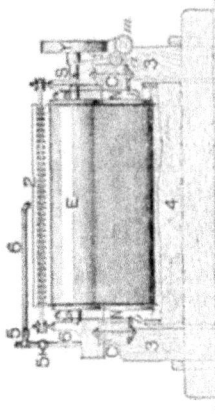

Fig. 127.

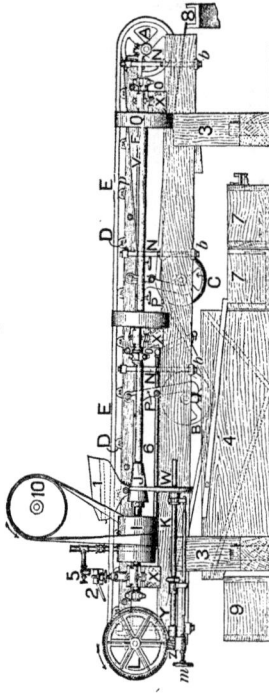

Fig. 126.

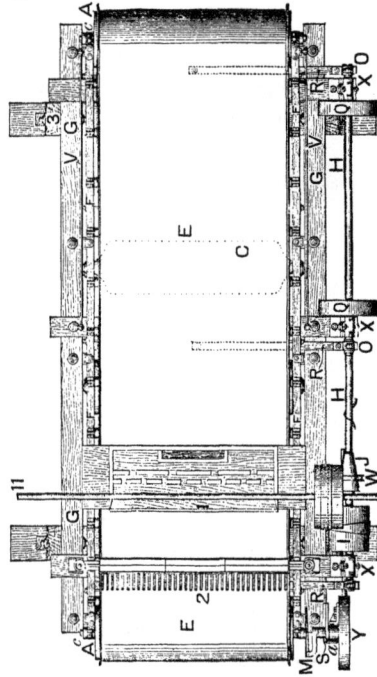

Fig 125.

Figs. 125 to 127.—Frue Vanner.

In order to separate the heavy metallic minerals from the accompanying gangue or rock, a second stream of water is applied, whilst a gentle side shake is given to the belt, to keep the sand in a state of agitation, which prevents it from " packing," and facilitates the sorting process. The water distributor, 2, is placed about a foot above the pulp distributor, and delivers small jets of water, 3 inches apart, over the entire width of the belt. The side shake thoroughly mixes this water with the pulp, spreads the whole uniformly on the belt, and enables the heavy particles of mineral to settle through the sand and cling to the belt, when they are carried up by it past the small jets of water and deposited in the collecting tank, while the lighter gangue is carried down by the stream and delivered into the tailings launder.

The Frue vanner is shown in plan in Fig. 125, in side elevation in Fig. 126, and in end elevation in Fig. 127. The following is a brief description of the machine :—

A A are the main rollers that carry the belt and form the ends of the table, B and C being other rollers. The belt E, which is 4 feet wide and $27\frac{1}{2}$ feet in entire length, passes through water underneath B, depositing its concentrates in the box, No. 4 ; and then passes over C, the tightening roller. D D are small galvanised iron rollers, and their support causes the belt E to form the surface of the evenly inclined plane table.

The cranks attached to the crank shaft, H, are $\frac{1}{2}$ inch out of centre, thus giving a throw of 1 inch, which is the amount of the lateral throw. I is the driving pulley that forms with its belt the entire connection with the power. J is a cone pulley on the crank shaft, H. By shifting the small leather belt connecting J and W, the uphill travel of the main belt, E, is increased or diminished at will ; the pulley, W, is moved by the hand-screw, m. R, R, R are three flat-steel spring connections bolted underneath the cross-pieces of the frame, F, and attached to the cranks of the shaft, H. These springs give the quick lateral motion—about 200 per minute.

No. 2 is the clear water distributor, and is a wooden trough which is supplied with water by a pipe, and the water discharges on the belt in drops through grooves 3 inches apart.

No. 1 is the ore-spreader, which moves with F, and delivers the ore and water evenly on the belt. In some cases, where the pulp from the stamps contains too much water, a box is placed between the stamps and the concentrator. From the bottom of this box the sand can be drawn with the proper amount of water, the superfluous water passing away over the top. A depth of three-eighths to one-half inch of sand and water is maintained on the table.

There is also a copper well that fits in (and shakes with) the ore-spreader as shown in the drawing. This is used in concentrating gold ores, for saving amalgam and quicksilver escaping from the silvered plates above, and can be taken out and emptied at any time. Into this well falls all the pulp from the battery. Its ends are lower than the wooden blocks of the spreader, so that the pulp passes over the ends of the well and is evenly distributed.

For some gold ores it is desirable to use on the ore-spreader a silvered copper plate the size of the spreader, and, when this is used, the wooden blocks of the spreader are fastened to a movable frame on top, so that they can be removed when the plate is cleaned-up, once or twice a month. No. 8 is a section of the launder to carry off the tailings.

The speed of the uphill travel of the belt varies from 2 to 12 feet per minute. If the ore treated be poor in pyrite, the upward motion of the belt does not exceed 20 inches per minute ; if richer, the speed is increased accordingly, and in agreement with the inclination of the belt, being greater as this inclination increases, but usually not exceeding $3\frac{1}{2}$ feet per minute. The inclination can be changed at will by wedges at the foot of the machine. The motion, the water used, the grade, and the uphill travel is regulated for every ore individually, and must be adjusted with every change in the pulp, if good work is to be maintained.

The amount of water used on the machine is from 1 to $1\frac{1}{2}$ gallons per minute of clear water at the head, and from $1\frac{1}{2}$ to 3 gallons per minute with the pulp.

The capacity per day of twenty-four hours for a 4-foot vanner is usually put down as about 6 tons of material, fine enough to pass a 40-mesh screen. In California, in general, two Frue vanners treat the product of each battery of five stamps, though in some cases three 4-foot vanners to ten stamps yield good results. In the Empire Mill of the Plymouth Consolidated Gold Mining Company, California, where the gangue is light, the ratio of stamps to vanners is 5 to 1.

Sizing of the pulp is usually omitted if the material is fine enough to pass a 30- or 40-mesh screen. If particles coarser than this are present they are removed by classifiers.

A riffled or corrugated belt is sometimes used on the Frue vanner in cases where large quantities of comparatively coarse sulphide is present.

The treatment of slime on Frue vanners was studied by R. Gahl in 1909,[1] and in particular the relative advantages of corrugated and smooth belts were examined, with inconclusive results. A high belt speed was found to be advantageous, and the correct amount of load, slope, etc., was determined for both kinds of belt.

In the *Lührig vanner* the endless travelling india-rubber band is not flanged at the sides, and has a slight side inclination, which is, therefore, at right angles to the direction of travel ; the latter is horizontal, not inclined ; by an arrangement of cams and springs, end-blows are given to the framework carrying the belt. The pulp from the batteries is collected in settling boxes placed overhead and delivered on to the belt through a small distributing box situated above its right-hand upper corner, the belt being driven from right to left. Clear water is supplied through a perforated pipe fixed diagonally across the belt. The pulp moves across the belt from the higher side to the lower, this motion being assisted by the clear water ; the light particles of gangue are washed down in this direction at a faster rate than the heavy particles of pyrite. On the other hand, the travel of the belt and the end-blows move the ore in the direction of the length of the belt. The results of the combined motions are as follows :—

1. Tailing passes off the table nearly opposite the distributing box at the right-hand end.

2. A middle product, containing both sulphide and gangue, is delivered near the middle of the side of the belt.

3. Clean concentrate is delivered near the left-hand end of the belt, having travelled the greatest distance before being washed off at the side.

Each of these three products is delivered into a separate hopper, and by a simple arrangement of sliding plates the exact points on the belt at which the delivery of the middle product is divided from those of the heading and tailing can be altered, so that the percentage of each product can be regulated.

The amount of ore treated in twenty-four hours by one of these machines

---

[1] Gahl, *Trans. Amer. Inst. Mng. Eng.*, 1909, **40**, 517.

is stated to be 4 to 5 tons if the concentrate is iron pyrite, and 3 to 4 tons if it consists of galena and blende.

The Wilfley Table is shown in Fig. 128. It consists of a smooth shaking

Fig. 128.—The Wilfley Table.

surface, of which Fig. 129 is a diagrammatic plan. The light, worthless material is washed down sideways towards F G by a flowing sheet of water, while the heavy concentrate is thrown towards the upper end, C G, against gravity by a longitudinal jerking action. The pulp is supplied from the wooden feed box, A B, 5 feet 1½ inches long, through a number of small holes. Clear water is supplied from the trough, B C, which is 9 feet 10½ inches.

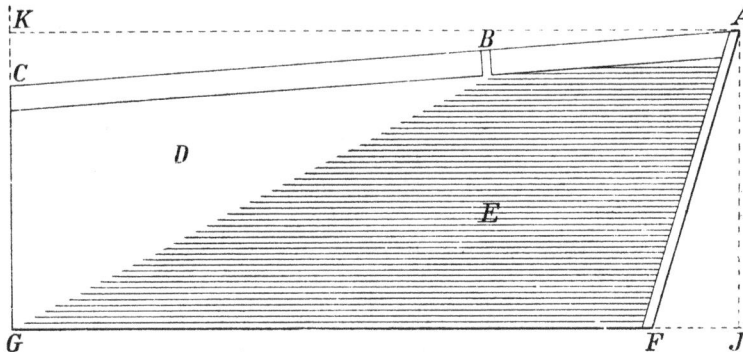

Fig. 129.—Diagrammatic Plan of Wilfley Table.

long, also through small holes. A sheet of linoleum is tacked down over the whole table and is fully exposed at D. A number (usually 46) of wooden riffles shown at E are placed on the top of the linoleum and are nailed down.

The longest riffle runs the whole length of the table, and the shortest riffle is about as long as the feed box. The riffles are $\frac{1}{4}$ inch high at the " head end," A F, and taper down to nothing at the " concentrate end," C G. They are about $\frac{1}{2}$ inch wide, the shorter ones being narrower, and are separated by $\frac{7}{8}$-inch spaces, so that the distance between centres is about $1\frac{3}{8}$ inches. The fall from the back, A B C, to the front, F G, can be adjusted by wedges, and is usually about 1 inch in 13 inches. The rise from the " head end," A F, to the concentrate end, C G, is usually set at about $\frac{1}{2}$ inch. The length of the side F G is 13 feet 9 inches, and the width at C G is 5 feet $\frac{1}{4}$ inch. The length of the perpendicular from A on F G is 6 feet 1 inch, and from A on C G is 15 feet $9\frac{1}{2}$ inches. The angle at G is a right angle and that at F obtuse. The shaking mechanism is situated at the end A F, and consists of a movement towards C G, beginning slowly and uniformly accelerated by means of a toggle joint, and a return movement towards A F effected by a spring, beginning suddenly and uniformly retarded. The effect of this vanning motion is to a certain extent similar to that of a bump delivered at the end C G, the concentrate moving uphill towards G. The length of the stroke is about $\frac{3}{4}$ inch and the number about 240 per minute.

The tailing passes over the front F G, and the concentrate over the end C G. The middle product passes over near G, in the last 16 inches of the front, falling into a special launder, whence it is returned to the feed box or sent to other machines.

The size of the pulp treated on this table is usually from 16 to 30 mesh. The table does not save the finest slime, and the Wilfley slime table, which consists of a series of moving troughs with canvas bottoms, is recommended in some cases for the treatment of the tailing from the ordinary Wilfley tables. The capacity of the Wilfley table is given as varying from $12\frac{1}{2}$ to 30 tons per day, the amount of wash water required being also variable. The power required is about 1 H.P. per machine.

Experiments were made by R. H. Richards,[1] with the view of determining how far it is desirable to feed sized products or classified products instead of natural products to a Wilfley table. He tried the separation of galena (specific gravity 7·5) from quartz and also the separation of cupriferous pyrite (specific gravity 4·68) from quartz, and concluded that a classified feed gave better results than either a sized feed or a natural feed, and that " there remains only to design classifiers sufficiently perfect to realise this natural advantage of classified feed over sized feed as applied to the Wilfley table."

Many other tables of the riffle type are in use—e.g., Deister concentrators. Of late there has been a tendency towards multi-deck riffle tables.

Hartz Jigs.—These machines differ in principle from all those previously described, inasmuch as the particles are separated by their fall through a somewhat deeper column of water than is the case on inclined tables, while a series of blows from below, causing waves moving upwards, continually brings the particles into suspension, and allows them to drop again. The initial period of the fall in water, during which the motion depends chiefly on density, is thus continually reproduced, and the result is a perfect separation of heavy from light particles of ore when working on any materials except the finest pulp. Jigs consist of sieves supporting beds of ore, which are completely immersed in water ; the ore is raised and allowed to fall

---

[1] Richards, *Trans. Amer. Inst. Mng. Eng.*, 1907, **38**, 556 ; 1908, **39**, 303.

by a quick succession of currents of water caused by the sudden action of a piston below, which is so worked that the upward movement or pulsation resembles that produced by a blow, while the downward movement is gradual. Under these conditions the heavy particles work downwards and pass through the sieve, while the lighter gangue is carried away horizontally by a stream of water introduced either from below or from above. Such machines are especially suitable for coarse ores.

In the Hartz jig a layer of coarse heavy particles is spread on the sieve to prevent too much of the ore from passing through. The stuff is fed in regularly at the head of the jig, and the strokes of the piston raise both the bed of heavy particles and the ore. The heaviest grains of ore find their way during the downstroke into the interstices of the bed, gradually pass through it, and coming to the screen, fall through into the tank below. The lighter particles cannot descend, and are gradually washed over the end partition by the continuous supply of water. Two products are therefore given, neither requiring further treatment if the conditions are favourable, and the machine properly adjusted. The wire meshes of the screen are always much larger than the ore treated, and the bed is composed of material of as nearly as possible the same density as the concentrate to be obtained, and is usually from $\frac{1}{2}$ to 1 inch in depth. The number of strokes of the piston per minute is from 60 to 80 with coarse sand of $\frac{1}{12}$ inch in diameter, and 200, 300, or even 400 with very fine sand, approaching slime. The length of stroke varies under the same conditions from $\frac{2}{5}$ to $\frac{1}{5}$ inch, and in the case of very fine, almost impalpable sand, the stroke may be diminished till it becomes a mere tremor. Callon and Le Neve Foster expressed the view that for enriching even very fine sand, the Hartz jig is the simplest and most economical machine yet invented, and requires the least amount of labour.[1]

A complete investigation on jigging was made by R. P. Jarvis [2] with useful results.

Flotation Processes.—It was found in 1899 by Elmore that in the case of certain ores, residuum oil separates sulphides and metallic particles from earthy or stony materials without reference to their respective densities. The oil after being mixed with pulp floats to the top, carrying any sulphide with it, and can then be separated from the tailing and from the concentrate. Other flotation processes are based on the action of acids on carbonates, or on the creation of a vacuum, the sulphide being raised by bubbles of air or other gas, which are formed in the liquid and entangled in the valuable particles. These processes are not, as a rule, applicable to ordinary gold ores.

Concentration by air instead of water is effected in a number of machines, which are used in certain cases. In hot dry climates, where water is scarce, where the ore is readily made quite dry, or in the event of water exercising a harmful chemical action on the ore, air concentration may be sometimes usefully employed. The jigging principle is used in Krom's pneumatic jig, centrifugal force in Clark & Stansfield's machine, and a steady current of air in Edison's dry blower (see p. 112).

[1] Callon's *Lectures on Mining*, English edition, vol. iii., p. 103.
[2] Jarvis, *Trans. Amer. Inst. Mng. Eng.*, 1908, **39**, 451-521.

CHAPTER XII.

DRY CRUSHING.

Introduction.—In those cases in which gold ores are treated by crushing and amalgamation, and the whole of the tailing, or only the concentrate obtained from it, is subsequently treated by the cyanide process or by chlorination, the method of crushing will be determined by the considerations discussed in the chapters on Amalgamation, Concentration, and Fine Grinding, and will depend partly on the state of aggregation of the free gold. If, on the other hand, ores are to be roasted before treatment by solvents, as is invariably the case in chlorination and sometimes in cyaniding, dry crushing is always resorted to. It is also sometimes used when the ore is not roasted. The method is especially valuable if water is scarce.

In dry crushing before roasting, it is usually not necessary to reduce the ore to a very fine state of division. It is usually passed through screens with only 8 to 30 holes to the linear inch. One of the points to be attended to is the production of a granular product, as if it is minutely subdivided the quantity of ore lost by dusting both in dry crushing and the subsequent roasting is greatly increased. Dry crushing is usually effected by stamps, by rolls, by ball-mills, or by roller-mills. The ore is previously passed through rock-breakers and then dried before it is finely crushed.

Drying.—Most ores must be dried before they can be finely crushed by means of rolls or other dry crushing plant. The screens soon become clogged by damp ore, especially if it is argillaceous. The moisture must be reduced to about 1 per cent., and Argall points out [1] that the removal of water of hydration by heating to about 250° makes it much easier to screen hydrated ores.

The oldest method of drying was to spread the ore, after the large lumps had been removed by a grizzly and crushed to 1½-inch size, on large flat areas heated from below by flues from the roasting furnace. The floor was usually covered with iron plates. After being dried, the ore was shovelled up and passed to the crushing mill. This plan involved much additional handling of the ore, and was a source of ill-health among labourers, besides requiring great floor space. It has been superseded by the adoption of inclined continuous-discharge, revolving iron cylinders, similar to the Howell-White furnace (q.v.), but not lined with bricks. The ore is passed through these, and is dried by the products of combustion of a fire, which are also passed through it. One such cylinder, 3 feet in diameter and 18 feet long, will dry from 30 to 40 tons of ore per day, at a small cost for fuel and power.

Argall [2] uses a tubular drier, consisting of a nest of four or six tubes similar to his roasting furnace (q.v.), but of lighter construction. He gives the capacity at from 80 to 300 tons of ore per day, the consumption of coal

---

[1] Argall, *Trans. Inst. Mng. and Met.*, 1902, **10**, 241.   [2] Argall, *loc. cit.*

at about 30 lbs. per ton of ore, and the cost of drying ore containing 6 per cent. of moisture at 5 cents per ton in Colorado.

An alternative furnace—viz., Stetefeldt's shelf drying kiln—was described in a paper read by the inventor at the meeting of the American Institute of Mining Engineers, held at Roanoke, Virginia, in June, 1883. In principle it resembles the Hasenclever furnace, a number of shelves being arranged zig-zag above each other in a vertical shaft, down which the ore slides, falling from shelf to shelf, while the products of combustion from a furnace rise through it. It is 21 feet high, and dries from 30 to 50 tons of ore per day. Although the first cost of this furnace is considerable, its working expenses are said to be low.

Stamps.—When stamps are used for dry crushing, double discharge mortars are provided (see Fig. 130), and the screens are put low down, giving a small height of discharge. These arrangements are necessitated by the difficulty of discharging the crushed ore from the mortar, the only means of doing this being the dashing of the dry ore against the screens, due to the fall of the stamps. Expediting the delivery of the ore by a blast of air forced into the mortar, or by suction from outside, is not often attempted. The result of the slowness of discharge is that with a given fineness of the mesh of the screens, dry crushing gives a finer product than wet crushing. This is an advantage with many ores, in respect that the gold is more readily accessible to the solutions. The output in dry crushing is much lower than in wet crushing by stamps, and the cost per ton of ore is usually stated to be from 50 to 100 per cent. more.

One of the difficulties in dry crushing is the production and loss of dust, which is generally richer than the bulk of the ore, for the reason that the sulphides are more brittle than the gangue. This difficulty is partly met by surrounding the crushing machine with an outer casing, and drawing off the escaping dust with air suction. Nevertheless, some dust escapes, involving a loss of gold and also some injury to the machinery and to the health of the workmen.

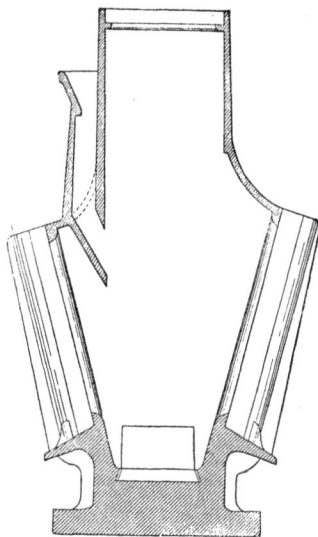

Fig. 130.—Double Discharge Mortar for Dry Crushing.

Dry crushing by stamps was formerly largely used in the treatment of silver ores by pan-amalgamation or by hyposulphite (thiosulphate) leaching before the abandonment of these processes. It was introduced on a large scale in 1894 for crushing the oxidised quartzose ores from the upper levels of the Waihi and other Ohinemuri (Upper Thames) Mines, New Zealand, and was finally abandoned in favour of wet crushing in 1904, partly owing to the fact that the character of the ore had changed with the working-out of the upper levels.

At the Waihi Mill [1] the ore was dried and partly roasted in kilns, with

---

[1] J. McConnell, *Trans. Inst. Mng. and Met.*, 1899, **7**, 26; F. M. Merricks, *ibid.*, p. 35; P. Morgan, *Trans. Australasian Inst. of Mng. Eng.*, 1904, **9**, [ii.], 161; *Eng. and Mng. J.*, Aug. 4, 1904, p. 183.

wood as fuel, and then broken down in rock-breakers before being fed to the stamps. The stamps weighed 1,000 lbs. each, and gave 94 blows a minute with a 6½-inch drop. The screens were 30-mesh steel wire (900 holes to the square inch), and the output was about 1¼ tons per stamp per day. Over 77 per cent. of the product was fine enough to pass a 90-mesh sieve. The ore was then charged into shallow leaching vats and cyanided, the solutions being run in from below. The depth of the charge was from 20 to 30 inches, and vacuum leaching was used. A depth of 3 feet of ungraded pulp was found to be unmanageable, and no attempts were made to separate the slime. According to James,[1] the cost at the Waihi Mill was stated to be 9·58d. per ton less for dry stamping than for wet stamping on the same ores.

Rolls.—It is generally considered that rolls are specially adapted for intermediate crushing, taking a product with a maximum diameter of from ¾ inch to 1½ inches from rock-breakers, and reducing it to a size passing through a screen of 12 or 16 meshes to the square inch. Finer crushing, however, has often been carried out. The exact limit at which rolls cease to be economical machines is still a matter of doubt. Like other dry crushing machines, they are better suited for soft friable material than for either hard or caking (clayey) ores.

Rolls are cylinders, between two of which pieces of ore are drawn and crushed by compression. They are placed with their faces a short distance apart, this distance varying with the degree of fineness to which the ore must be reduced. The main driving power is applied to the shaft of only one roll, by means of belt pulleys, the other roll being driven only with sufficient force to ensure that the rollers will always take hold of the ore, and also to keep them in motion when no ore is passing between them.

In the older forms, tooth-gearing was used instead of belt pulleys for the application of the power, and the two rolls were compelled to revolve at an equal rate of speed by gear-wheels, connecting together, placed on the axles. The advantages of belted rolls are, that a higher speed can be easily attained, and also that if the rolls were to become jammed from any cause, the belts would slip or be thrown off, while the tooth-gearing would be broken. Geared rolls, however, are still largely used for coarse crushing. The rolls have crushing tires made of steel or of chilled iron. Chilled iron is much cheaper, but the wearing of the faces is more rapid and less uniform than in the case of steel rolls. Emery wheels are used for levelling the unevenly worn faces of the rolls. The crushing tires can be taken off and replaced when they are worn out. In some rolls the crushing strain is taken up by powerful springs, which press the rolls towards one another ; when particularly hard fragments are passing through the rolls, they are forced apart against the action of the springs. It is desirable, in order to keep the wear of the faces even, that the rolls should always be kept parallel, and special appliances are used to ensure this. The hopper is designed to spread the ore evenly across the crushing face, and the rolls, screens, elevators, etc., are all securely boxed in with a wooden housing. This last precaution is necessary in order to prevent loss by floating dust, which otherwise may be large, the richest part of the ore thus passing off, and not only making the atmosphere of the mill insupportable, but having a disastrous effect on the bearings of the machinery. Rolls for moderately fine crushing are usually from 12 to 16 inches across the face, and from 22 to 36 inches in diameter, but for coarse crushing larger

---

[1] James, *Cyanide Practice* (1901, 1st edition), p. 104.

rolls, 36 to 54 inches in diameter, with faces 15 to 28 inches wide, are now used.

Richards points out [1] that ore may be crushed by rolls in two ways, according to the rate of feed, speed of rolls, etc. If the speed is high and the feed light, each particle of ore is crushed separately between the rolls. In this case, which he calls " free " crushing, there will be a maximum of coarse and a minimum of fine material produced. In " choke " crushing, the ore is fed in a thick stream between the rolls, so that the particles crush each other and a maximum of fine material is produced. Part of the power will be used in compressing the loosely-packed stream. According to Richards, free crushing is the more advantageous course, provided that a very fine product is not required.

The prevailing opinion seems to be that rolls are not economical fine crushers. A certain percentage of the material passing through rolls is crushed very fine, and if the whole product is to be fine, the remainder is sieved out and returned to the rolls. A further quantity, smaller than before, will be sufficiently reduced by the second passage through the rolls. According to Fischer Wilkinson,[2] if fresh coarse ore, equal in quantity to the separated fines, is added to the intermediate product, it is probable that the output will remain constant, and that the power required will be less than in the stamp battery.

Rolls are not generally used to crush finer than about 20 mesh, but Edison proposes to crush to 200 mesh, using choke crushing and corrugated rolls.[3] Corrugated rolls were formerly tried at Mount Morgan and elsewhere, but were abandoned as being of little value, wearing unevenly, and soon getting out of order. Edison's process appears to necessitate a reduction of comparatively coarse material to excessively fine particles in a single pair of rolls. In general, however, it is considered that gradual reduction in successive pairs of rolls is more economical, the product of one pair of rolls passing to another set slightly closer together. An application of this principle is given in the following description [4] of the Mount Morgan plant, Queensland, formerly in use :—

At Mount Morgan the ore was soft and friable (the " Gossan Cap "), with about 10 per cent. of hard boulders of quartz. It was broken in Krom jaw-breakers (hinge at bottom) to $\frac{3}{4}$-inch gauge, dried, and then passed successively through four pairs of belted rolls, all running at 112 revolutions per minute. The distances between the faces were $\frac{3}{8}$ inch, $\frac{3}{16}$ inch, $\frac{1}{16}$ inch, and 0 respectively, the first two rolls being 26 inches in diameter and 15 inches wide, and the last two 36 inches in diameter and 16 inches wide. Revolving hexagonal screens or trommels were used after each pair of rolls, the brass screens having 20 holes to the linear inch. The first trommel was protected inside by a steel screen with 16 holes to the square inch. The coarse product was in each case sent to the next pair of rolls. The roll tires were of cast steel, 3 inches thick, and required turning up again in from one to three months. They lasted twelve months, and were discarded when worn down to $\frac{1}{2}$ to 1 inch thick. The wear was 0·108 lb. of steel per ton of ore crushed. The power required was 0·8 I.H.P. per ton crushed in 24 hours. The capacity was $62\frac{1}{2}$ tons per day for each set of eight rolls, and the total cost, exclusive

[1] Richards, *Ore Dressing*, p. 98.
[2] Fischer Wilkinson, *Trans. Inst. Min. and Met.*, 1905, **14,** 159.
[3] Simpkin & Ballantyne, *ibid.*, p. 62.
[4] N. F. White, *Trans. Australasian Inst. Min. Eng.*, 1900, **6,** 37.

of lighting, breaking, and drying, was 3s. 10d. per ton. The crushed ore was roasted and chlorinated (see Chap. XIV.). The rolls have now been superseded by Krupp ball mills (*q.v.*).

"At the Bertrand Mining Company's Lixiviation Mill, Nevada, where Krom's rolls are used for pulverising silver ores with a quartz gangue, 15,000 tons of ore were crushed before it was found necessary to put new tires on the finishing rolls ; while after a further crushing of 5,000 tons of ore, the tires of the roughing rolls were still expected to be good for two or three month's work. In neither case were the tires found to be at all grooved, the reason for their renewal being that they had become worn too thin for further work. The rolls in the Bertrand Mill are stated to crush 50 tons of hard quartzose ore in twelve hours to pass through a 16-mesh screen ; while in the Mount Cory Mill, Nevada, 50 tons are reduced to 30 mesh in the same time." [1] In these mills silver ores were crushed for treatment by roasting and lixiviation with a hyposulphite solution.

The Frazee rolls [2] have been designed for fine crushing, better control being obtained over the wear of the roll-shells by new devices.

Comparison between Rolls and Stamps.—As the subsequent treatment of an ore determines its method of crushing, no accurate general comparison of stamps and rolls can be made. A comparison is only possible in the special cases where both methods of crushing are applicable. Wet crushing by rolls need not be considered, as it is not practised ; even where the advocates of rolls wish to replace stamps by rolls for wet crushing and amalgamation they propose that the ore should be crushed dry and then wetted down. Dry crushing by stamps is usually about one-third more expensive than wet crushing, as the capacity falls off to that extent, in spite of double discharge, assisted occasionally by currents of air given by blowers, which are designed to carry the crushed ore against the screens. The amount of slime made is also large. Advantages in the use of rolls for dry crushing have been stated to be " the fewness of the wearing parts, and consequent small cost of repairs ; the great efficiency of the process, in that the ore escapes from the rolls immediately it is crushed, so that overcrushing is unlikely to occur, and the great capacity of rolls, effective work being constantly done, and the amount of crushing surface brought into contact with the ore per minute being very large. The prime cost of the rolls is considerably less than that of stamps of the same capacity." [3]

The capacity of rolls has perhaps been frequently over-estimated owing to the assumption having been made, that the product of equal crushing surfaces must be the same. Thus, Curtis observes in the paper already cited :—" As an index of the capacity of Krom's rolls, it may be stated that two sets of 26-inch (diameter) rolls, with faces 15 inches long, give rather more effective crushing surface than fifty gravitation stamps, each 8 inches in diameter, falling at the rate of ninety drops per minute. In making this calculation, the average diameter of the rolls is taken as only 24 inches, so as to allow for their gradual wearing, while their speed is taken at 100 revolutions per minute."

Although the calculated crushing surface is as stated, it does not follow that the capacity of the two sets of machinery compared with one another

---

[1] Curtis on "Gold Quartz Reduction," *Proc. Inst. Civil Eng.*, 1892, **108**, 97.
[2] C. Q. Payne, *Trans. Amer. Inst. Mng. Eng.*, 1912, **43**, 327.
[3] Curtis, *loc. cit.*

is the same, since it will depend on the pressure as well as the crushing surface. No doubt the pressure at the moment of impact of a 900-lb. stamp is enormously greater than that exercised by the faces of the rolls.

On this point J. Richards, M.E., writes:[1] " The coefficient of (crushing) effect in such machines (as revolving rollers and reciprocating machines) is as the area of the acting surfaces, and the speed with which they approach each other. The mistake in respect of the crushing power of rollers comes from confounding their circumferential velocity with the working one— that is, with the parallel velocity at which the surfaces approach and leave each other," that is, the velocity at right angles to the crushing surface. Richards elsewhere shows that a Blake machine in this way actually outruns a revolving roller, although its crushing face travels more slowly than the periphery of the roll. He proceeds :—" The same principle holds good in respect to stamps, the crushing surfaces having a parallel approach, while with rotary machinery the approach is not parallel, except on an imaginary line at the centre."

A point to be noticed is that, in machines which act by pressure applied on the principle of the lever, such as reciprocating-jaw crushers or rolls, the whole force necessary to crush the lumps of quartz is transmitted to the fulcrum, this fulcrum being represented in the case of rolls by the bearing surface at the axle. The consequence is that the frame must be made very strong and heavy, and the axle bearings attended to with great care, if rapid wear is to be avoided.

The question whether gravity stamps can be advantageously replaced by rolls or other machines for crushing material from about $1\frac{3}{4}$ inches to $\frac{1}{4}$ inch diameter has again been revived recently. In certain cases, especially for soft ores, the stamps may be the less efficient machine. For hard ores, such as those of the Rand, there is little prospect of stamps being given up.

Ball Mills.—Mills in which ore is crushed by iron balls have long been in use. In the earlier forms the balls travelled round in a slot which they fitted loosely, the crushing being effected partly by centrifugal force. The wear of the balls soon interfered with the efficiency of some of these machines, and none proved very successful. In later machines, a number of small balls enclosed in a rotating drum crush the ore by falling from little shelves, and this type has passed into extensive use for dry crushing in Australia and America.

The Krupp or *Krupp Grüsonwerk* Ball Mill [2] has been highly successful. It consists of a cylindrical drum rotated by belting, but geared down by toothed wheels (see Fig. 131). The interior of the drum consists of from 5 to 10 overlapping grinding steel plates, a, b, so arranged that the balls drop a few inches in passing from each plate to the next. The half, a, of the grinding plates is perforated with $\frac{1}{2}$-inch tapered holes, and the other or main grinding half, b, is bent inwards towards the axis to raise the balls above the next succeeding plate. The crushed ore passes through the perforations in a, and falls upon a coarse sieve, J, constructed of finely perforated steel plates, which protect the outer fine sieves, K, of phosphor bronze or steel wire gauze. The coarse particles retained on both inner and outer sieves, which enclose the crushing drum and rotate with it, are returned by elevator plates, l, m, to the interior of the drum through perforated plates, n, guarding apertures between the grinding plates. The

---

[1] J. Richards, *Production of Gold and Silver in the United States*, 1880, p. 369.
[2] See "Ball Mill Practice at Kalgoorlie," von Bernewitz, *Mng. Mag.*, 1911, 5, 139.

material, which need not be smaller than 2 or 3 inches in diameter and is sometimes larger, is fed into the drum from a hopper at the side. The balls are of chilled cast iron or cast steel and of various sizes, the largest being about 5 inches in diameter. As they wear down, others are added. The

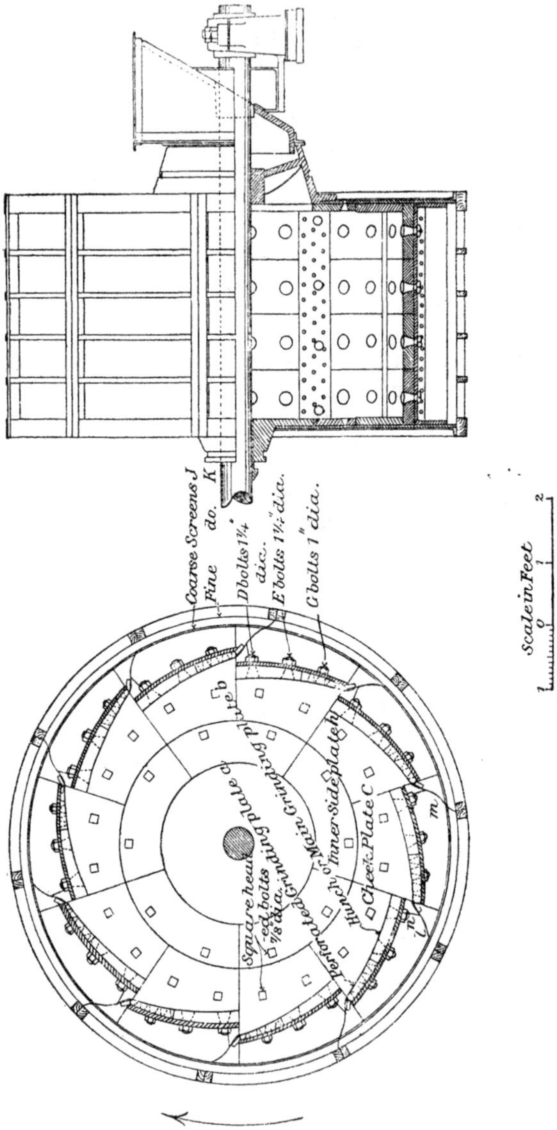

Fig. 131.—Krupp Ball Mill.

material which passes through the fine sieves falls into a hopper which is continuous with the dust-proof casing of sheet iron. The interior of the sides of the drum is lined with chilled iron or steel plates. The mills are made in eleven sizes, the largest having a drum 8 feet 10 inches in diameter and

4 feet 6 inches in width. This size rotates 21 to 24 times per minute and contains about 2 tons of steel balls. The power required is about 50 to 60 H.P., and the amount crushed is about 3 tons of quartzose ore per hour through a 40-mesh sieve. Smaller mills, however, are usually employed, especially the No. 5 mill (see below) which contains 20 to 22 cwts. of steel balls and has an output of about 1½ tons of Kalgoorlie ore per hour, using a 20-mesh screen. One steel ball weighing 18 lbs. is added each day to compensate for wear.[1]

The feed for the Krupp ball is regulated by sound, a correctly fed mill giving a rumbling sound, and a mill running light giving a clear metallic sound due to the balls striking one another.[2] It is usually stated that the mill crushes by impact, but Von Bernewitz is satisfied that the main action is by attrition, the balls rolling over and rubbing and grinding the ore. The size of the ore feed is about 3-inch cubes. The wear of steel is about 0·55 lb. per ton of ore at Kalgoorlie, and the total cost varies from 17·8d. to 32·3d. per ton (Von Bernewitz). The fine screens are brushed once a day.

The same mill is made for wet crushing with an increase of output of about 30 per cent. Sprays of water are introduced inside the drum and are directed against the sieves, and the lower part of the dust casing is converted into a spitzkasten, which separates slime from sand. The wet crushing mill, however, has not yet passed into regular use.

The Krupp ball mills installed at Mount Morgan,[3] where they were for the first time used for crushing gold ores, are shown in Fig. 131, which is reproduced by the kind permission of the Australasian Institute of Mining Engineers. They are No. 5 mills, 7 feet 6 inches in diameter and 3 feet 9 inches wide, and containing about a ton of steel balls. There are 10 cast-steel grinding plates, and the fine screen has 400 holes to the square inch. The mills were driven at from 21 to 22 revolutions per minute, and crushed soft friable ores containing 10 per cent. of hard-quartz boulders (see above, p. 275). The ores were delivered from rock breakers. Sixteen mills crushed about 123 tons in eight hours, so that the product of each mill was 23 tons per day of 24 hours. The power required was 13 I.H.P. per mill. The wear amounted to 0·919 lb. of steel plates, bolts, etc., and 0·725 lb. of steel balls per ton of ore on a run of 12 months, ending 31st May, 1899. The total cost for breaking, drying, and crushing was 3s. 11d. per ton against 6s. 9d. per ton when Krom rolls were used. Exclusive of power, the cost of crushing in the ball mills only was 1s. 3d. per ton.

These mills have also been largely used in crushing at Kalgoorlie, where the experience is similar to that at Mount Morgan. The screens used at the Great Boulder Main Reef [4] were 900 to the square inch, and 60 per cent. of the product passed a 90-mesh screen. The cost of crushing, exclusive of power, was 1s. 6d. per ton.

Von Bernewitz [5] gives another instance of a West Australian mill using No. 5 Krupp ball mills, which ran at 26 revolutions per minute, taking from 16 to 20 H.P. each. A steady feed of ore was maintained from a shaking

---

[1] W. E. Simpson, *Trans. Inst. Mng. and Met.*, 1903, **13**, 25.
[2] Von Bernewitz, *loc. cit.*
[3] "Some Notes on Dry Crushing," by N. F. White, *Trans. Australasian Inst. Mng. Eng.*, 1900, **6**, 37.
[4] A. G. Charleton, *Gold Mining and Milling in Western Australia* (1903), p. 247.
[5] Von Bernewitz, *Mining J.*, Sept. 4, 1909, p. 299 ; *J. Chem. Met. and Mng. Soc. of S. Africa*, 1909, **10**, 222.

feeder. Each mill was loaded with 2,350 lbs. of steel balls and crushed 40 tons daily through a 27-mesh screen. Using Krupp steel liners, the mill ran for eight to ten months without renewals. Three balls were added per week, the consumption in balls being $3\frac{1}{2}$ ozs. steel per ton crushed. The balls were worn to $1\frac{1}{2}$ inches diameter. A fan drew off the dust and helped to keep the fine screens clear. The cost of crushing was 1s. $8\frac{1}{2}$d. per ton. Von Bernewitz considered these mills better for dry crushing than stamps, rolls, or Griffin mills.

The Griffin Mill is also used at Kalgoorlie and elsewhere. It consists of a single roller about 18 inches in diameter with a vertical axis. It is suspended from above somewhat in the manner of the Huntington mill, and made to rotate about 200 times a minute. At the same time it rolls round on the inside of a die-ring 30 inches in diameter, so that it has a gyratory motion. The screens are placed above the die ring. The machine is also used for wet crushing. It produces a finer product than ball mills when using an equivalent screen, and is thus more suitable when the ore is to be slimed than when the product is to be treated by leaching without separation of the slime. James gives the cost of crushing through a 15-mesh screen at Kalgoorlie as 1s. 10d. per ton, exclusive of power.[1] In December, 1904. 34 Griffin mills and 14 Huntington mills were in use in Western Australia for preparing ores for cyaniding.[2]

---

[1] James, *Cyanide Practice* (1901), p. 111.
[2] *Govt. Gazette of Western Australia*, Dec. 1904, pp. 530-554.

CHAPTER XIII.

ROASTING.

Introduction.—The roasting of gold ores as a preliminary to chlorination or cyanidation was formerly largely practised. With the discontinuance of the chlorination process and the progress of other methods of treating telluride ores by cyanide (see Cnap. XVII.), it is becoming an unusual course to roast gold ores, except at Kalgoorlie. Brief descriptions of the usual methods are given below.

The operation of roasting, as a preliminary to chlorination, has for its object the expulsion of the sulphur, arsenic, antimony and other volatile substances existing in the ore, and the oxidation of the metals left behind, so as to leave nothing (except metallic gold) which can combine with chlorine when the ore is subsequently treated with it in aqueous solution. For this purpose the ore is heated in a furnace, through which a current of air is passed, salt being added if oxide of copper, lime, magnesia, etc., are present.

As a preliminary to cyanidation, roasting is used to remove the sulphur or tellurium and to prevent the waste of cyanide by cyanicides. The ore is roasted " dead " because the compounds formed in partly oxidised pyrite ($FeSO_4$, etc.), act as cyanicides, whereas the insoluble ferric oxide, Fe_2O_3, finally formed, is not attacked by cyanide. Telluride ores are roasted with the object of rendering the gold soluble, telluride of gold being only very slowly attacked by cyanide. As sulphur is always present in telluride ores, the roasting is continued until the whole of the sulphur is expelled. At one time ores were also often merely dried before cyanidation, where they were clayey and plastic, in order to open the pores to the attack of cyanide, and to make the ores more granular and readily leached. With the progress in the cyanide treatment of slime, the preliminary drying of ore has become less common (MacFarren).

The ordinary reverberatory furnace, worked by hand labour, was in wide use as a preliminary to chlorination, especially where only a few tons, or less, of concentrate are to be treated per day. Various mechanical furnaces, capable of handling large quantities of ore, were devised to supersede the old-fashioned contrivance, and some of these will be described in the sequel.

Hand-worked Reverberatory Furnace.—The construction of the ordinary reverberatory furnace is too well known to need detailed description here.[1] It consists of a vaulted chamber, containing the ore ; through this chamber, the flames and products of combustion from a furnace and a current of air are made to pass in a horizontal direction above the ore, which is thus heated. The ore is also stirred by hand with iron rakes, which are passed through small working doors. The hearth of the vault (also called the " laboratory " of the furnace) is formed of bricks placed on edge, as close together as possible.

---

[1] See *An Introduction to the Study of Metallurgy*, Roberts-Austen and Harbord, 1910, p. 338.

No mortar is used, but a little clay is plastered between the bricks. The height of the furnace hearth is about 3½ feet above the floor of the building, on which the labourers stand, and the space underneath the hearth is either occupied by vaults or filled with well tamped rubble. The arch is usually one course of bricks (8 inches) thick ; the height between it and the hearth is, in long furnaces, about 24 inches near the bridge, and gradually diminishes towards the other end. This height is less in short furnaces. The best firebricks are used for the fire-box and bridge, and for the hearth and arch of the first few feet of the " laboratory." The remainder is made of common brick. It is necessary to have a damper in the flue to regulate the draught ; the aperture of the flue should not be on a level with the hearth, as in that case the loss by dusting is increased. The brickwork of the furnace is supported by longitudinal and transverse iron braces which may be eased as the furnace becomes hot, in order to avoid any buckling of the walls. The working doors have cast-iron frames, and are about 15 inches wide and 9 inches high. The fuel used must of course be a long-flame coal, or wood.

For the particular purpose of roasting pyritic ores before chlorination, the temperature on the working floor of the furnace is kept low when the ore is first charged in, and high in the later stages. If a single small floor is used the fire must be alternately checked and urged to secure these conditions. Moreover, when the roasting is nearly complete, the high temperature required renders the gases passing into the flue very hot, and so a corresponding waste of fuel results. To prevent this waste, it is customary for roasting furnaces to be built with a very long hearth, or to have several successive hearths, so as to utilise the waste heat from the portion of the working floor next the fire. Furnaces with three floors at slightly different levels are much favoured ; in these a charge remains for a few hours on each of the floors in succession. It is first placed on the floor farthest removed from the fire, and, after a time, is raked down on to the middle hearth, and thence to that nearest the fire, fresh charges being put on the spaces just cleared, so that there are always three charges in the furnace in various stages of oxidation.

The most usual form in the Western States of America is the " 4-hearth," in which the length of the hearth is four times its width, so that the dimensions are, say, width 15 feet, length 60 feet. In this case there should be eight working doors on each side. Instead of three floors at different levels, a single continuous floor, sloping down from the flue towards the fire, has been used at many works in Australia, Mexico, and the United States. At Suter Creek a continuous-hearth furnace was 12 feet wide and 80 feet long, and the mineral was worked in three distinct parts, as though there were three floors. The angle of slope is made large in some Australian furnaces, so that in the course of the "rabbling" or stirring, the ore continually travels towards the fire-box. Furnaces with two or three superposed floors are also used to a limited extent ; the lowest floor is next the fire-box, and communicates by a vertical flue with the floor above, and so on. The ore is charged in on the top floor, and after a time is raked down through the vertical flue on to the next floor. In this case the floors are heated by the gases passing below them as well as above them, and fuel is economised. Merton's furnace, subsequently described, is one of this kind with automatic rabbling.

The operation of roasting pyritic ore in an ordinary furnace with three floors may be described as follows :—The furnace being hot, and the flame from the fire-box reaching completely across the first floor, the ore is charged in on the third floor and spread out by the rabbling tool. The weight of the charge may be taken as from 12 to 18 lbs. per square foot of floor space, varying according to the nature of the ore, a high percentage of sulphur necessitating small charges. The layer of ore is 2 or 3 inches deep. It is not spread quite evenly, but made to form a series of parallel ridges by means of the rabbling tool, so as to increase the surface exposed to the air. The working doors may be closed at first to heat the ore quickly. Moisture is at once given off in great quantities, and the sulphur soon begins to burn with a blue flame. When this is seen to take place, all the working doors are opened, and the charge is energetically rabbled, with little intermission, until the sulphur flame disappears. If this is not done, clots are formed which are afterwards difficult to break up. The air for the combustion of the sulphur is supplied by holes in the fire-bridge and from the working doors. The flames and heated products of combustion from the fire tend to rise above the colder air, and move along next to the arch of the furnace, while the air forms a sheet between these gases and the ore. In practice, although the air is introduced below them, nevertheless the reducing gases from the fire partially mix with the air, and greatly reduce its oxidising power. Moreover, the combustion of the sulphur in the ore on the first two floors further reduces the amount of free oxygen present in the current of air, and roasting on the third floor is, therefore, largely dependent on air derived from the working doors.

When the sulphur flames have abated, and the charge has been heated almost to redness, it is transferred to the middle floor, where it is raised to a dull red heat and most of the sulphating takes place. During this stage the ore swells considerably, so as to occupy much more than its original bulk. All the lumps previously formed should be broken up on this floor. Rabbling is continued until the ore is uniformly dull throughout, so that, on turning it over, the fresh surfaces appear but little brighter than that which has been exposed for some time. The charge is then transferred to the floor next the fire. There is now little risk of the formation of lumps, and the charge may be allowed to reach a bright red heat. Rabbling is of less importance than before, as little oxidation takes place, the chief reaction which occurs being the decomposition of the sulphates already formed. As long as this is still going on, the ore emits the odour of sulphur dioxide. When no further odour can be detected, and the ore can be piled up so as to maintain a vertical face, shows no bright specks on its glowing surface, emits no sparks if some of it is tossed up by the working tool, and is inclined to become black very readily from cooling, the charge is said to be " dead " or " sweet," and is ready to be withdrawn. It should be observed that, when the ore contains much sulphur, its particles at a low red heat appear less coherent than when cold, and flow almost like water, so that the charge cannot be made to form a heap with steep sides. Care must therefore be taken when the ore is on the middle floor to prevent any part of the charge from flowing out of the working doors, which it is very liable to do when being rabbled.

The best means of determining whether a charge is completely roasted is to throw some of the ore into water and test the solution for soluble sulphates with barium chloride.

Certain insoluble sulphides may be detected by boiling the roasted ore with caustic potash, filtering, and adding a solution of lead carbonate. A brown colour or black precipitate denotes the presence of sulphides. This test is used at the Great Boulder, according to D. Clark.[1]

Although roasting is carried on until the ore is said to be " dead," it is not practicable to eliminate the whole of the sulphur, and a small percentage both of insoluble and of soluble sulphur can be found in roasted ores. The insoluble sulphur is, of course, more detrimental to gold extraction by chlorine than is soluble sulphur. It is not usual in roasting for chlorination to leave more than about 0·1 to 0·15 per cent. of insoluble sulphur in the ore, but much larger percentages of soluble sulphates are often left undecomposed in roasted ore.

The charge is withdrawn by a scraper, and falls by gravity through a hole in the floor of the furnace near the working door into a pit below. This hole is covered by a plate while roasting is being performed.

The time of roasting depends chiefly on the ore, but may be shortened by more continuous rabbling than most workmen can perform, the work being somewhat exhausting. In a three-floor furnace, concentrate with 15 per cent. of sulphur usually remains eight hours on each of the three floors. The consumption of coal for fuel is usually from 10 to 20 per cent. of the weight of the ore.

Chemistry of Oxidising Roasting.—Sir Wm. Roberts-Austen discusses as follows [2] the roasting of a " mixture consisting of sulphides mainly of iron and copper, with some sulphide of lead, small quantities of arsenic and antimony as arsenides, antimonides, and sulpho-salts, usually with copper as a base. The temperature of the furnace in which the operation is to be performed is gradually raised, the atmosphere being an oxidising one. The first effect of the elevation of the temperature is to distil off sulphur, reducing the sulphides to a lower stage of sulphurisation. This sulphur burns in the furnace atmosphere to sulphurous anhydride (SO_2), and coming in contact with the material undergoing oxidation is converted into sulphuric anhydride (SO_3). It should be noted that the material of the brickwork does not intervene in the reactions, except by its presence as a hot porous mass, but its influence is, nevertheless, considerable. The roasting of these sulphides presents a good case for the study of chemical equilibrium. As soon as the sulphurous anhydride reaches a certain tension, the oxidation of the sulphide is arrested, even though an excess of oxygen be present, and the oxidation is not resumed until the actions of the draught change the conditions of the atmosphere of the furnace, when the lower sulphides remaining are slowly oxidised, the copper sulphide being converted into copper sulphate, mainly by the intervention of the sulphuric anhydride, formed as indicated. Probably by far the greater part of the iron sulphide only becomes sulphate for a very brief period, being decomposed into the oxides of iron, mainly ferric oxide, the sulphur passing off. Any silver sulphide that is present would have been converted into metallic silver at the outset were it not for the simultaneous presence of other sulphides, notably those of copper and of iron, which enables the silver sulphide to become converted into sulphate. The lead sulphide is also converted into sulphate at this low temperature (viz., about 500°). The heat is now raised still further with a view to split up the sulphate of copper, the decomposition of which leaves oxide of copper. If, as in this case, the bases are weak, the sulphuric anhydride

[1] Clark, *Australian Mining and Metallurgy*, p. 24.

[2] Roberts-Austen, Presidential address to the Chemical Section, British Association, Cardiff Meeting, 1891 ; Roberts-Austen and Harbord, *op. cit.*, p. 7.

escapes mainly as such ; but when the sulphates of stronger bases are decomposed the sulphuric anhydride is to a great extent decomposed into a mixture of sulphurous anhydride and oxygen. The sulphuric anhydride, resulting from the decomposition of this copper sulphate, converts the silver into sulphate, and maintains it as such, just as, in turn, at a lower temperature, the copper itself had been maintained in the form of sulphate by the sulphuric anhydride eliminated from the iron sulphide. When only a little of the copper sulphate remains undecomposed, the silver sulphate begins to split up (viz., at about 700°) . . . partly by the direct action of heat alone, and partly by reactions such as those shown in the following equations :—

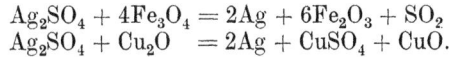

$$Ag_2SO_4 + 4Fe_3O_4 = 2Ag + 6Fe_2O_3 + SO_2$$
$$Ag_2SO_4 + Cu_2O = 2Ag + CuSO_4 + CuO.$$

The charge still contains lead sulphate, which cannot be completely decomposed at any temperature attainable in the roasting furnace except in the presence of silica. . . . The elimination of arsenic and antimony gives rise to problems of much interest, and again confronts the smelter with a case of chemical equilibrium. For the sake of brevity it will be well for the present to limit the consideration to the removal of antimony, which may be supposed to be present as sulphide. Some sulphide of antimony is distilled off, but this is not its only mode of escape. An attempt to remove antimony by rapid oxidation would be attended with the danger of converting it into insoluble antimoniates of the metals present in the charge. In the early stages of the roasting it is, therefore, necessary to employ a very low temperature, and the presence of steam is found to be useful as a source of hydrogen, which removes sulphur as hydrogen sulphide, the gas being freely evolved. The reaction

$$Sb_2S_3 + 3H_2 = 3H_2S + 2Sb$$

between hydrogen and sulphide of antimony is, however, endothermic, and could not, therefore, take place without the aid which is afforded by external heat. The facts appear to be as follows :—Sulphide of antimony, when heated, dissociates, and the tension of the sulphur vapour would produce a state of equilibrium if the sulphur thus liberated were not seized by the hydrogen, and removed from the system. The equilibrium is thus destroyed, and fresh sulphide is dissociated. The general result being that the equilibrium is continually restored and destroyed until the sulphide is decomposed. The antimony combines with oxygen and escapes as volatile oxide, as does also the arsenic, a portion of which is volatilised as sulphide.

" The main object of the process which has been considered is the formation of soluble sulphate of silver." The reactions, however, are precisely similar in an ordinary oxidising roast.

The following remarks on the decomposition of the various minerals present in complex ores may be of use in assisting the student to understand the reactions which proceed in the roasting furnace :—

1. *Pyrite*, FeS_2.—On heating this compound sulphur is volatilised, the reactions being probably expressed thus :—

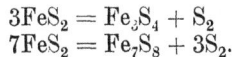

$$3FeS_2 = Fe_3S_4 + S_2$$
$$7FeS_2 = Fe_7S_8 + 3S_2.$$

The sulphur burns to SO_2, which is partly converted by the heated quartz, etc.,[1] into SO_3, uniting with the free oxygen present. The ferrous sulphate formed by this sulphuric acid is split up by the heat and the ferrous oxide (FeO) converted into ferric oxide (Fe_2O_3) which gives the ore a red colour when cold. Some basic sulphates always remain undecomposed. If the temperature of the part of the charge next the fire-bridge has been too high, or if the charge is kept too long in the furnace, especially when not freely exposed to the air, some magnetic oxide is formed, thus :—

$$3Fe_2O_3 = 2Fe_3O_4 + O.$$

The presence of magnetic oxide makes the ore darker in colour. This is an undesirable change, as the magnetic oxide is acted on by chlorine far more readily than the sesquioxide.

W. E. Greenawalt[2] states that the dark magnetic oxide can be reconverted to the red sesquioxide by subjection to a lower temperature and an abundant supply of air. For this reason he advocates finishing at a lower temperature than is used for the previous stages of roasting. In four successive tests on 100-ton lots he found that a higher percentage of extraction by chlorination was obtainable by roasting at a high initial heat and a low finishing heat than by roasting at a lower initial heat and a higher finishing heat, although, in the latter case, the elimination of the sulphur was more complete.

2. *Chalcopyrite.*—The decomposition of the copper sulphate formed in the furnace leaves a mixture of cuprous and cupric oxides, both soluble in chlorine.

3. *Galena*, PbS.—The presence of this mineral in any but small quantities is very detrimental, as both lead sulphate and lead silicate (formed by its decomposition in the presence of silica) are very fusible, and, at the temperature required to split up copper sulphate, cause the ore to become pasty and form lumps. Roasting must be performed very slowly and cautiously to avoid this effect.

4. *Arsenopyrite*, FeAsS.—Arsenates of iron, copper, lead, etc., when formed are not easily decomposed, as they resist a high temperature, and are only slowly converted into sulphates by sulphuric acid at a red heat. It is, therefore, desirable to avoid their formation, and with this end in view the precautions which have been already mentioned above are taken.

5. *Antimonial Sulphides* are still more difficult to deal with, the antimonates formed being less easily decomposed than arsenates. Their formation is avoided in the manner already described.

6. *Blende*, ZnS, forms oxide and sulphate of zinc, of which the latter can only be split up by a very high temperature. At a bright red heat a basic sulphate is formed which is converted to oxide at a white heat. If blende is roasted at a high temperature and with a plentiful supply of air, sulphate of zinc is not formed to a large extent.

7. *Calcium Carbonate*, $CaCO_3$, is decomposed at a red heat, CO_2 being given off and caustic lime, CaO, left in the charge. The change is slow at 600° (low red heat) and rapid at 800° (full red heat). Magnesium carbonate is similarly decomposed.

8. *Tellurides* containing gold are fusible far below a red heat, and heavy losses of gold may occur through absorption by the furnace bottom. The

[1] Plattner, *Metallurgische Röstprozesse* (Freiburg, 1856).
[2] Greenawalt, *Eng. and Mng. J.*, July 29, 1905, p. 145.

melted telluride of gold may remain in the ore, and in that case at a red heat the tellurium is partly volatilised and partly oxidised to the oxide TeO_2, which also sublimes. Spherical beads of gold remain behind and are difficult to dissolve. *Selenides* behave similarly.

9. Metallic *Gold, Silver,* etc., are fused at high temperatures, forming spherical beads which are difficult to dissolve. The temperature of the ore should never be allowed to exceed 1,000°, for this reason.

Elimination of Arsenic and Antimony.—H. M. Howe, in explaining how this is effected, distinguishes three horizontal zones in the ore: [1] (1) the upper surface, where oxidation is only slightly hindered by sulphurous and sulphuric acids by the products of combustion of the fuel; (2) the middle layers where oxidation proceeds to a very limited extent; (3) the lowest layers, where " a pellet of ore is simply exposed to the action of the other pellets with which it is in contact, of volatilised sulphur, and of sulphurous and sulphuric anhydrides generated by the action of sulphur on previously formed metallic oxides." He proceeds—" The expulsion of arsenic and antimony as sulphides is favoured in the middle and lower zones by the presence of volatilised sulphur, mixed with sulphurous acid and at most a very limited supply of free oxygen and sulphuric acid. In the upper part of the middle layer, to which a small amount of free oxygen penetrates, we have the gently oxidising conditions favourable to the formation of arsenious acid and trioxide of antimony. In the upper zone the stronger oxidising conditions rather favour the formation of fixed arseniates and antimoniates, though, even here, part of the arsenic and antimony may volatilise and escape while passing through their intermediate volatile condition of arsenious acid and trioxide of antimony." On stirring the mass, these arsenates and antimonates, being exposed to the reducing action of volatilised sulphur and undecomposed sulphides in the lower zones, may again be converted into volatile oxides. Protoxide of iron, suboxide of copper, and sulphurous acid are also efficacious in reducing arsenic acid, higher oxides of iron and copper, and sulphuric acid being formed. " Thus, every individual atom of arsenic may travel forth and back many times through the volatile condition, being oxidised at the surface and reduced below the surface, . . . and every time it arrives at this volatile condition an opportunity is offered it to volatilise and escape." If a small quantity of coal or coke dust is mixed with the ore, after it has been completely oxidised, and the air excluded, the arsenates and antimonates are again reduced to the lower oxides, and, if they are " carried past the volatile state "—*i.e.,* reduced to metals—they may again be passed through it by an oxidising atmosphere. " Of course the expulsion of arsenic and antimony is favoured by the presence of a large proportion of pyrites, both because the sulphur distilled from the pyrites tends to drag them off as sulphides, and because the presence of the pyrites prolongs the roasting, and thus increases the number of times which the arsenic and antimony pass back and forth past their volatile conditions; hence, it is sometimes desirable to mix pyrites with impure ores to further the expulsion of their impurities."

The Use of Salt in Roasting.—Certain ores require the addition of salt in roasting in order to chloridise material which would otherwise absorb chlorine when the ore came to be " gassed," and so cause additional expense as well as inconvenience. If silver as well as gold is to be extracted from

[1] Howe, Copper Smelting, *Bull.* No. 26 *U.S.A. Geol. Survey,* Washington, 1885.

the ore, the addition of salt is necessary in order to form chloride of silver in the furnace, since metallic silver is not attacked by chlorine at the highest temperature ever employed in the leaching vat. The silver chloride is then dissolved out by sodium thiosulphate or some other solvent either before or after the extraction of the gold.

Even if no silver is present, an ore must be roasted with salt if it contains much copper (as sulphide, or as an oxidised salt), lime, magnesia, or other substance which, after being subjected to an oxidising roasting, is rapidly attacked by chlorine at ordinary temperatures. The salt is usually added towards the end of the operation, when no sulphides and only a small percentage of sulphates are left undecomposed; sometimes, however, the ore and salt are mixed before charging in. To some sulphides only 5 lbs. of salt per ton of ore are added, but others require as much as 90 lbs. per ton. The weight of salt added must be at least six to eight times that of the silver present in the ore. If a large amount of salt is used, it is desirable to leach the roasted ore with water, before treating it with chlorine gas, in order to remove the coating of soluble sulphates and chlorides remaining on the surface of the granules of ore.

The chemical action of the salt is due to a double decomposition between it and the sulphates of the heavy metals, by which sulphate of soda and the chlorides of the heavy metals are produced. The following general equation approximately represents the reaction :—

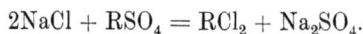

$$2NaCl + RSO_4 = RCl_2 + Na_2SO_4.$$

Chlorine is also set free by the action of sulphuric anhydride on salt, and the presence of water vapour induces the formation of much hydrochloric acid. These gases act directly on the several constituents in the ore, forming chlorides and oxychlorides. The metallic chlorides and oxychlorides formed are in many cases volatile (e.g., the compounds of copper, iron, lead, arsenic, antimony, etc.), and, in passing off, the volatile compounds carry away with them varying proportions of gold and silver, which, as a rule, are not recoverable in the dust chambers. The chloride of copper is especially active in causing these losses.

Other reactions which probably take place are as follows :—

1. Ferrous sulphate, acted on by salt at a red heat in the presence of moist air, yields hydrochloric acid and chlorine, which act on the gold and silver, while ferric sesquioxide and sodium sulphate are produced.

2. Ferric chloride, Fe_2Cl_6, is also produced at the same time. This is volatile, and chloridises silver with great energy at a red heat, sesquioxide of iron being produced.

3. Cupric chloride, $CuCl_2$, is easily decomposed into cuprous chloride, Cu_2Cl_2, and free chlorine, or into the oxychloride, $Cu_2O.Cl_2$, and free chlorine. The vapours of $CuCl_2$ thus give rise to further supplies of nascent chlorine available for the chlorination of the silver.

4. Arsenic and antimony form volatile chlorides which are decomposed by means of oxygen and water vapour, yielding arsenious and antimonious acids and nascent chlorine or hydrochloric acid.

It is thus obvious that the presence of base minerals is advantageous in that they may cause nascent chlorine to be set free in the presence of silver in all parts of the furnace. On the other hand, the loss of gold is increased by any increase in the quantities either of silver or of the base metals, since in the former case the time of the roasting is prolonged. The best chloridising effect is obtained in a highly oxidising atmosphere, so that very little sulphur is required in the ore, and, if much is present, the practice of eliminating the greater part before adding the salt is not likely to be attended with any diminution in the percentage of silver chloride formed. Moreover, the water vapour in the furnace gases promotes the formation of hydrochloric acid. When salt is used in roasting, the ore is often allowed to cool slowly in heaps after being withdrawn from the furnace. If treated in this way, a higher percentage of the silver, etc., is found to be chloridised than if the ore is wetted down at once, or even spread out to cool in a thin layer. On the other hand, the losses of gold by volatilisation are increased by slow cooling. Chlorine continues to be evolved for a long time after the withdrawal of the charge has taken place, the heaps smelling strongly of the gas.

Losses of Gold in Roasting.—Plattner proved in 1856 [1] that in the oxidising roasting of ordinary auriferous pyrite, a loss of gold can take place only when the operation is carried on so rapidly that fine particles are carried off mechanically by the draught. This conclusion, as far as sulphides and arsenides are concerned, has been confirmed by Küstel,[2] and by Prof. S. B. Christy,[3] but the latter adds that it is extremely difficult to prevent all mechanical loss by dusting, which is caused by even a moderate draught. Küstel records the loss of 20 per cent. of the gold present during the oxidising roasting of certain tellurides of gold and silver, and states that this is not a mechanical loss, but is due to volatilisation. This seems to be a mistake, see p. 10.

The losses of gold which are sustained when salt is added to the furnace charge may be very great. Küstel found that a telluride ore, on being roasted with 4 per cent. of salt, lost 8 per cent. of its gold before the ore was red hot. Aaron [4] found that certain ores, containing only gangue and pyrite, suffered great loss of gold in roasting with salt which had been added at the commencement of the operation; only a small part of this gold was condensed in the flue, in which was found a yellowish fluffy precipitate, consisting largely of chlorides of copper and iron, and containing nearly 30 ozs. of gold to the ton. He found that the loss was greatly reduced by diminishing the quantity of salt, and by reserving it until the dead roasting was nearly complete.

In the chloridising roasting of a Mexican ore, consisting mainly of magnetite and pyrite with 3·5 to 7 per cent. of chalcopyrite, C. A. Stetefeldt found the losses of gold to be from 42·8 to 93 per cent. of the total gold contained. He states [5] that " there is no doubt that the volatilisation of the gold takes place with that of the copper chlorides. The loss increased with the quantity of these chlorides formed and volatilised." He further shows, however, that the presence of copper chloride is not the only possible cause of loss, since an ore consisting of hard white quartz, intimately mixed with about 7 per cent. of calcite and a little pyrite, lost 70 to 80 per cent. of its silver, and 68 to 85 per cent. of its gold, when roasted with 5 per cent. of salt. When subjected to an oxidising roast, no loss of gold took place. The reason for the extraordinary behaviour of this ore was not discovered.

Prof. Christy [6] found that, in the ores on which he experimented on a

[1] Plattner, *Metallurgische Röstprozesse* (Freiburg), p. 128.
[2] Küstel, *Roasting of Gold and Silver Ores* (1880), p. 56.
[3] Christy, *Trans. Amer. Inst. Mng. Eng.*, 1888, **17,** 3.
[4] Aaron, *Leaching of Gold and Silver Ores* (1881), p. 121.
[5] Stetefeldt, *Trans. Amer. Inst. Mng. Eng.*, 1885, **14,** 339.
[6] Christy, *Trans. Amer. Inst. Mng. Eng.*, May, 1888, **17,** 14.

19

small scale in a muffle furnace, a greater loss was sustained by adding the salt near the end of the roasting operation than by mixing the same weight of salt with the ore at the start. He explained that this is due to the fact that the amount of gold volatilised varies with the amount of chlorine which comes in contact with it. When the salt is added at the start, the chlorine is at first removed by the sulphur as fast as it is formed, escaping as chloride of sulphur, and thus the gold is protected from attack. When the salt is added after a long oxidising roast, the chlorine is rapidly generated (the ore being red hot and containing large quantities of sulphates), and the gold is no longer protected from attack by the sulphur. The loss of gold is also in all cases increased by working at a higher temperature, owing to the large amount of chlorine generated, and to the increase in the volatility of the gold. It is apparent from the results given on p. 61 that the temperature used in chloridising roasting must be very carefully regulated, the loss of gold being increased far more by high temperature than by a lengthening of the time in the furnace. Moreover, the salt must be reduced to the least possible quantity. It must, however, be remembered that the maximum volatility of gold chloride is at about 250°, or far below a red heat.

The advantage found to be gained in practice by adding salt near the end of the operation is due to the fact that, in the continuous roasting of ores in long-bedded furnaces, the gases given off from the finishing floor pass over a great length of comparatively cold, unsalted, and unoxidised ore before reaching the flue. The quantity of gold chloride mixed with the chlorine, which is evolved from the red-hot ore as soon as the salt is added, is no doubt large; but the SO_2 from the colder ore, and the steam from the fuel, " offer excellent means for the reduction of the chloride of gold right within the furnace, while the most efficient means probably is the pyrites themselves," which have been proved to be readily capable of condensing gold on their surface. If all the salt is added at the start, there is a continued volatilisation of chloride of gold throughout the furnace, and a less favourable opportunity for it to condense. The difference between the results in the muffle and in the reverberatory furnace is thus explained.

At Nevada City, at the Merrifield Mine, and in other works in the neighbourhood, the old-fashioned long furnace, with a single step separating the finishing hearth from the rest of the furnace, was still used in 1888.[1] These furnaces are from 55 to 65 feet long, holding from 6 to 9 tons, and producing about 3 tons of roasted ore per day, so that the ore remains in the furnace from two to three days. The custom there was to give the ore a long oxidising roast at a low red heat, ending in a low cherry-red heat, and then, when the ore reached the finishing floor, the temperature was slightly lowered, and the salt added. The salt was stirred thoroughly into the ore, and as soon as it was " dissolved " by the roasted ore—i.e., in about half an hour—the charge was drawn into the cooling pit. This lowering of the temperature is evidently of great importance in reducing the loss, but the time occupied in roasting is regarded as less material if no salt is present. These mills were on custom work, charging $15 to $20 per ton of ore for treatment, and guaranteeing a yield of 90 per cent. of the gold and 60 per cent. of the silver. Their method of roasting seems to be considered in California as that best suited to concentrate containing a high percentage of sulphur, but their loss in roasting has not been ascertained.

---

[1] Christy, *op. cit.*, p. 42.

At one of the Californian chlorination mills it was found by experiment in 1882 that nearly 50 per cent. of the gold and 28 per cent. of the silver were being lost by volatilisation. In this case the pyrite was roasted on two hearths for thirty-six hours, 1 per cent. of salt being added four hours before the charge was drawn. The reason for the great loss was thought by Professor Christy to be the high temperature of roasting, particularly on the charging-in floor.

The variation of the loss in different ores which are treated precisely alike is doubtless due partly to the presence or absence of metals forming volatile chlorides which carry off the gold, and partly to the physical condition of the latter, the volatilisation being greater if it is in a state of minute subdivision.

Volatilisation Process for Treating Gold Ore.—S. Croasdale found,[1] from experiments on a number of ores, that from 80 to 99 per cent. of the gold contained in them could be volatilised by roasting with salt. This formed the basis of a process for the extraction of gold and silver from pyritic ore. If more than 3 per cent. of sulphur was present in the ore it was partly removed by roasting after dry crushing. From 5 to 20 lbs. of salt per ton of ore was then added and oxidising roasting continued at a temperature of about 1000°. The gold and silver were volatilised and condensed in scrubbing towers or in filters made of burlap sprayed with water. The process was tried at Mayer, Yavapai County, Arizona, in 1906,[2] and the same method was applied in 1913 by B. Howe to antimonial ore at the Gwalia Consolidated Mine, West Australia.[3] In both cases the process was soon abandoned.

MECHANICAL FURNACES.

The furnaces which have been designed with the object of saving the labour necessary to work the reverberatory furnaces may be divided into four classes, viz. :—

1. Stationary hearth furnaces, supplied with iron hoes moved by machinery by which the ore is rabbled. The O'Hara, Pearce Turret, Brown, Ropp, Edwards, Merton, MacDougall, Evans-Klepetko and Herreshoff furnaces are examples of this class.

2. Rotating-bed furnaces, in which the hoes or stirrers are stationary, while the bed supporting the ore revolves. The Godfrey calciner [4] and the old Bunker Hill furnace are examples of this class.

3. Rotating cylindrical furnaces, which consist of brick-lined iron cylinders capable of being rotated, so that the ore is tumbled over and over by their motion while it is being roasted. Examples are the Brückner, the White-Howell, the Hofman, and the Argall furnaces.

4. Shaft furnaces, in which the powdered ore falls by gravity, in a shower, through an ascending column of hot air, the oxidation being effected in the course of the fall. The Stetefeldt furnace, which is the only one based on this principle, is not used for dead roasting, as it is not adapted for the purpose. It is used for the chloridising roasting of silver ores, and will not be described in this volume.

---

[1] Croasdale, *Eng. and Mng. J.*, Aug. 29, 1903, p. 312 ; *Mng. Mag.*, 1914, 10, 200.
[2] O. H. Fairchild, *Mng. and Sci. Press*, Sept. 1, 1906.
[3] Howe, *Mng. Mag.*, 1913, 9, 437.
[4] *Trans. Inst. Mng. and Met.*, 1899, 7, 323.

1. **Furnaces with Mechanical Stirrers**—*O'Hara Furnace*.—This is the oldest mechanical furnace, and it bears a great resemblance to the old-fashioned reverberatory furnace. It has two superposed hearths, in each of which the arch is very low, so as to confine the heat close to the ore. An endless chain, set in motion by suitable machinery, passes through the furnace, resting on the upper hearth, and returns along the lower hearth. Attached to the chain at proper intervals are iron frames of a triangular shape ; on these frames are a number of ploughs or hoes set at an angle, so that one set of hoes turns the ore to the centre, and the next set turns it in an opposite direction towards the walls. The ploughs thus stir the ore thoroughly, and at the same time move it gradually towards the fire. The ore falls from the upper to the lower hearth by gravity, and similarly falls from the lower hearth into a pit when it arrives at the hottest place in the furnace.

One of the chief causes of difficulty and expense in working furnaces with mechanical stirring apparatus is that the iron hoes gradually become heated, and they are then rapidly corroded by the sulphur in the ore. In hand stirring the rabbling tool is withdrawn as soon as it is hot, and allowed to cool, while another is substituted for it meanwhile, thus prolonging its life. In furnaces with mechanical rabbles the trouble and expense caused by the wearing of the hoes were very great until the device was adopted of placing the moving parts of the mechanism outside the ore-hearth with the stirring hoes passing through a slot in the furnace wall. This device is used in some of the furnaces described below. In the Edwards and Merton furnaces, described below, pp. 294 and 297, which have displaced all other furnaces at Kalgoorlie, the rabbles are water-cooled.

MacDougall, Herreshoff, Wedge, and Evans-Klepetko Furnaces.[1]—These furnaces have from 3 to 8 circular superimposed hearths, the ore being dried on the top floor and roasted on the lower floors. The ore is stirred by radial arms carrying teeth, the arms being moved by a rotating central shaft to which they are attached. The furnaces are made hot and in many cases subsequently work without fuel other than the sulphur in the ore, reducing the sulphur in concentrate at Butte to about 4 to 6 per cent. They are not used in roasting gold ores.

Pearce Turret Furnace.—This consists of an ordinary reverberatory hearth built in an annular form (see Fig. 132). In the centre of the circular space surrounded by the hearth is a vertical iron column carrying four hollow horizontal arms projecting through a slot into the reverberatory hearth which they cross transversely. The column revolves and the arms carry rabble blades which traverse the hearth, stirring the ore and moving it round the circle by degrees. " Air is forced through the hollow arms and is discharged against the rabble blades, performing the double duty of cooling the iron work and of furnishing heated air for the oxidation of the ore." The ore is discharged automatically after passing once round the furnace. Two or more fireplaces are used. The hearth is 6, 7, or 8 feet wide, and is now sometimes built in two and even six floors, one above the other. Moreover, the top of the arch is sometimes used as a drying hearth. An improvement on the air blast for cooling the rabble arms consists in supplying them with water-jackets, by which their life is lengthened, the air being let in through

---

[1] "Notes on the Metallurgy of Copper of Montana," by H. O. Hofman, *Trans. Amer. Inst. Mng. Eng.*, 1903, **34**, 270-282 ; Gowland, *Non-Ferrous Metals*, pp. 30-33.

the fire-boxes and the apertures in the outer wall. This furnace is not used in roasting gold ores.

The Brown Horse-shoe Furnace.—In this furnace the ore-hearth (L, Fig. 133, which is a cross-section of the furnace through the ore-hearth and fire-box), has a narrow chamber, M, on each side of it and separated from it by tiles projecting downwards from the arch and upwards from the hearth, so that only a narrow slot is left between the two series. In the lateral chambers, M, are laid rails on which the carriages supporting the stirrers

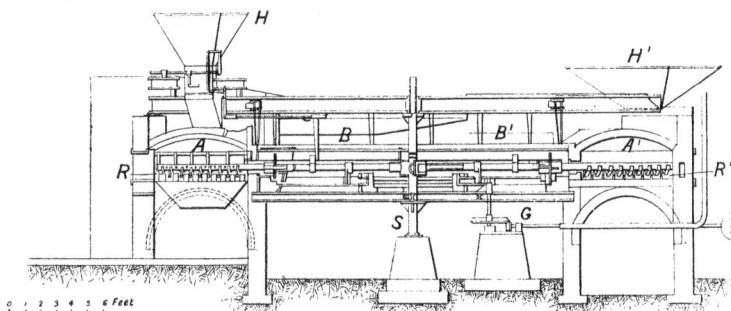

Fig. 132.—Pearce Turret Furnace. The tunnel is shown at A A′, and the rabbles at R R′. These rabbles are moved from the central support S by the gearing G. Hoppers are indicated at H H′; B B′ are the supporting girders of the tunnel. (From Roberts-Austen and Harbord, *Introduction to the Study of Metallurgy*, 1910, p. 349.)

are moved by a cable running continuously on grooved wheels. These are placed in doorways in the wall of the furnace to keep them cool. The carriage spans the ore-hearth, passing through the slots connecting L and M, and is arranged to grip the cable or release it. The carriage supports steel stirrers which just reach the ore-hearth and stir and move on the ore. The carriages run on a continuous elliptical track, only part of which is inside the furnace. In operation half the carriages are traversing the furnace while half are resting in a cooling space outside the furnace. When two carriages are used,

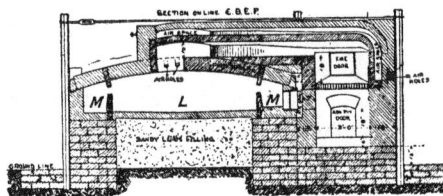

Fig. 133.—Brown Horse-shoe Furnace—Transverse Section through Ore-hearth and Fire-box.

the one moving out of the furnace strikes the other, which is at rest in the cooling space, and pushes it forward until it grips the cable automatically and starts off to take up the stirring, and the hot moving carriage is at the same time automatically released and comes to rest. The stirrers are thus prevented from becoming overheated.

There are one, two, or more fire-boxes, according to requirements, and there is generally a prolongation of the ore-hearth outside the furnace walls, used as a cooling hearth, over which the ore is pushed by the stirrers, until it falls into a hopper ready for chlorination. Fig. 133 is a section of the furnace erected in 1896 at the Golden Reward Chlorination Mill, at Deadwood, Dakota. In this case the roasting hearth was 180 feet long and 8 feet wide, and the cooling hearth 78 feet long. Three rabbling carriages were used, following one another at intervals of 90 seconds, and moving the ore forward at the rate of 22 feet per hour. The ore contained from $2\frac{1}{2}$ to 8 per cent. of sulphur, and 65 tons of ore crushed through 30-mesh sieves were roasted per day, the product containing 0·3 per cent. of sulphur. The capacity was thus one ton to every 22 square feet of hearth per day, and the total cost less than 50 cents per ton for roasting, cooling, and conveying the ore to the barrels,[1] the cost per day of twenty-four hours being made up as follows :—

Fuel, 6 cords of wood, ...	$19.50
Labour, two men, one each shift, ...	5.00
Power, 5 H.P., ...	1.25
Oil, lights, repairs, etc., ...	1.50
	$27.25

The Ropp Straight-line Furnace differs from the Brown furnace in having the cable placed underneath the hearth, the slot being in the hearth itself. Through the slot an arm, connected below with a four-wheeled truck, extends upwards, and across the upper end of this arm is the supporting bar from which the stirring teeth project. The chamber beneath the hearth is easily accessible for repairs, and the trucks and rabbles are cooled outside the furnace while traversing the track leading back to the starting point. The hearth of the standard furnace is 105 feet long and 11 feet wide. It was tried at Kalgoorlie, but, like the Brown furnace and Holthoff-Wethey, is no longer used in roasting gold ores.

The Holthoff-Wethey Furnace[2] is of similar design. Six of these were used at the Great Boulder Perseverance Sulphide Mill in 1901. The fixed horizontal hearths were 120 feet long and 12 feet wide, the capacity of each furnace being 60 to 70 tons per day. The ore, which apparently contains about 3 or 4 per cent. of sulphur, is stirred and carried forward by means of eight travelling rabbles in each furnace, and less than 0·02 per cent. of sulphur in undecomposed sulphides is left in the ore after roasting. These have now been replaced by the Edwards Duplex furnace.

The Richards Furnace[3] was introduced at Mount Morgan and subsequently installed at the Great Boulder Main Reef Sulphide Works. It consists of eleven superimposed hearths, the whole furnace being 30 feet long, 12 feet wide, and 65 feet high. It is heated by a single fire, and the issuing gases are stated to be at a temperature of 300°. The ore is not stirred, but banks up on each hearth until it forms a natural slope when it slides down, falling from floor to floor through vertical flues. At Mount Morgan the cost of roasting in this furnace was put at 5s. per ton.

The Edwards Furnace[4] (see Figs. 134 to 136).—This furnace and the

---

[1] *Eng. and Mng. J.*, July 4, 1896, p. 8.

[2] Charleton, *Gold Mining and Milling in Western Australia*, p. 348 ; *Mineral Industry*, 1897, p. 448 ; and "Notes on the Metallurgy of Copper of Montana," by Hofman, *Trans. Amer. Inst. Mng. Eng.*, Feb. 1903, **34**, 272.

[3] Charleton, *op. cit.*, p. 328.

[4] W. E. Simpson, *Trans Inst. Mng. and Met.*, 1903, **13**, 28 ; F. D. Power, *Eng. and Mng. J.*, Feb. 11, 1904, p. 242 ; Gowland, *Non-Ferrous Metals*, p. 25.

Merton furnace are the only ones now in use in roasting gold ores at Kalgoorlie. The Edwards furnace is made in two patterns, the stationary and

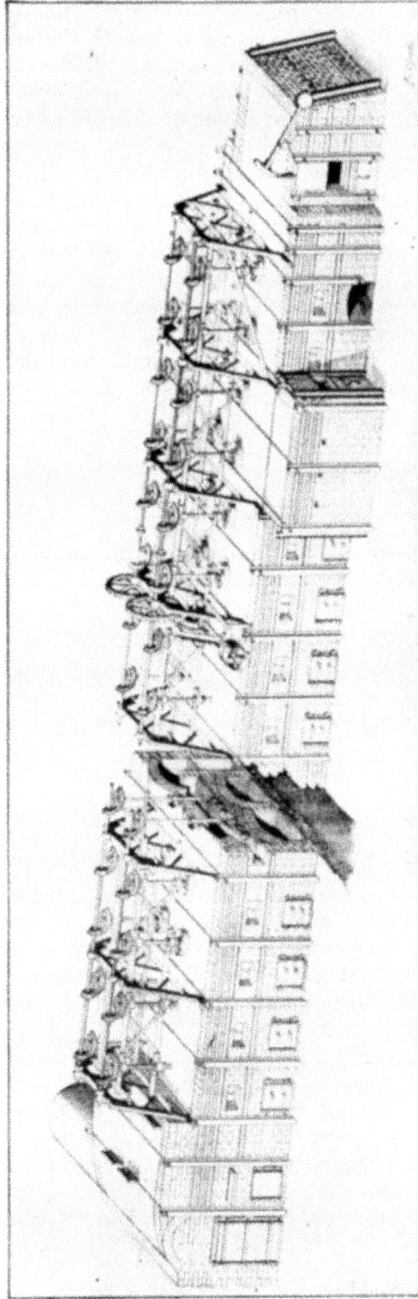

Fig. 134.—Edwards' Furnace.

the tilting. The stationary " Duplex " furnace is built of brick with a flue under the hearth. It has one or two side fire-boxes and one at the end. The length varies, but is usually from 100 to 120 feet, and the width from 9 to 12 feet. There is a number of rotating rabbles, which are hollow and water-cooled. The spindles of the rabbles pass through the arch of the furnace, and above the arch are fitted with crown cog-wheels which gear with bevel pinion cog-wheels, carried on two lines of shafting running longitudinally along the top of the furnace. The rabbles are geared right and left alternately so that two adjacent rabbles rotate in opposite directions. The circular paths of the feet of the rabbles overlap, and there are little or no dead spaces on the hearth. The ore moves down the inclined hearth in a zig-zag fashion under the influence of the rabbles. The shoes of the rabbles are cast iron, and when worn out can be easily replaced on the feet without cooling the furnace. In the older furnaces there was only one line of rabbles, instead of two. The width was 6 feet, and the amount of ore roasted at Kalgoorlie (containing about 6 per cent. of sulphur) was 12 to 18 tons per day. The duplex furnaces roast from 60 to 100 tons of Kalgoorlie ore per day in preparing it for treatment by the cyanide process.

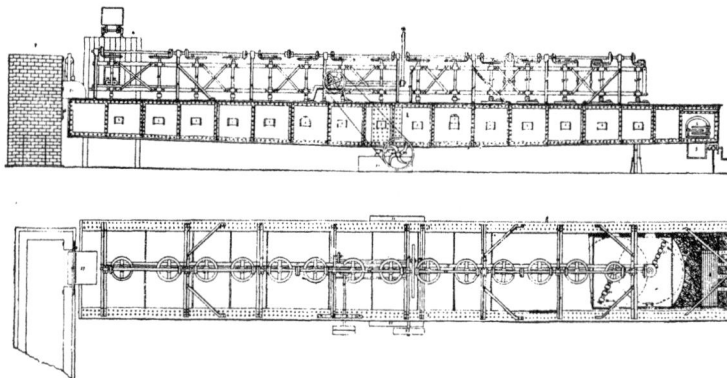

Figs. 135 and 136.—Edwards' Tilting Furnace.

The hearths at the Associated Mine furnaces [1] are $11\frac{3}{4} \times 107$ feet, the fall 0·31 inch per foot, and the 52 rabbles move at 2·62 revolutions per minute. There are three fire-boxes, and fuel consumption averages 11 per cent. of the ore roasted, which amounts to 95 tons per day in each furnace. The sulphur in the ore amounted to 5·5 per cent., and the cost of roasting was 2s. 6d. per ton in 1911. If bad work was detected the furnace was stopped for a time to heat up.

The tilting furnace (Figs. 135 and 136) is made of iron externally and lined with brick. It consists of two iron girders 63 feet long, resting on pivots at the centre, and the movement of ore can be regulated by tilting the furnace from the horizontal to any desired angle to suit the ore, thereby increasing or decreasing the rate at which the ore passes through the furnace. The tilting furnace is also made in duplex form. The output for the simplex pattern is from 8 to 20 tons per day, and for the duplex about double.

---

[1] M. W. von Bernewitz, Mng. and Sci. Press, May 13, 1911, p. 661.

The Merton Furnace[1] (Fig. 137) consists of three superimposed hearths and a fourth hearth in front of them and at a lower level. The main hearths

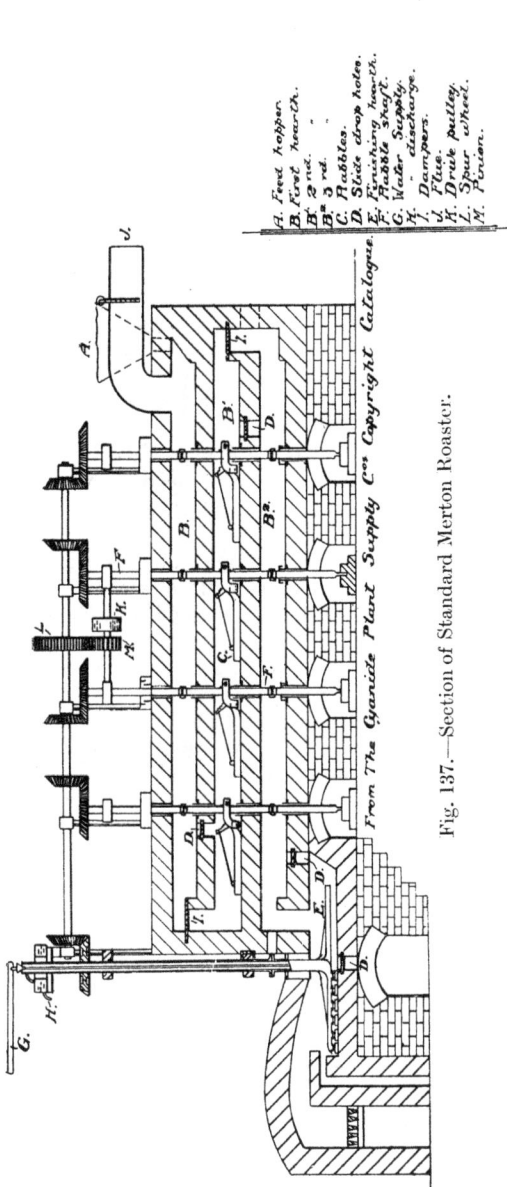

A. Feed hopper.
B. First hearth.
B¹. 2 nd.
B². 3 rd.
C. Rabbles.
D. Slide drop holes.
E. Finishing hearth.
F. Rabble shaft.
G. Water Supply.
H. — discharge.
J. Dampers.
J. Flue.
H. Drive pulley.
L. Spur wheel.
M. Pinion.

From The Cyanide Plant Supply Cᵒˢ Copyright Catalogue.

Fig. 137.—Section of Standard Merton Roaster.

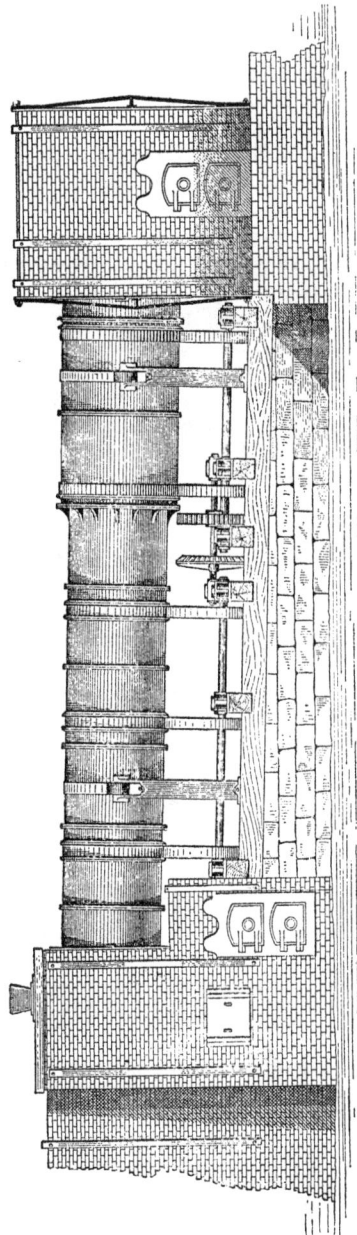

Fig. 138.—White-Howell Furnace.

---

[1] Clark, *Australian Mining and Metallurgy*, p. 21 ; *Eng. and Mng. J.*, Oct. 20, 1904, p. 634 ; W. E. Simpson, *loc. cit.* ; Gowland, *op. cit.*, p. 28.

are 23 feet 9 inches long and 8 feet wide over all. The height of each arch
is 16½ inches in the middle and 9 inches at the side. The ore is charged in
on the upper hearth, stirred and moved forward by rotating rabbles attached
to vertical shafts, and delivered in succession to the middle and lower hearths
and finally to the finishing hearth. There are four shafts, with a rabble
attached to each shaft on each of the three floors. The hearths are horizontal, having no fall. The capacity of the furnace when treating Kalgoorlie
sulpho-telluride ore is said to be from 18 to 25 tons of ore per day.

In the most recent type of Merton furnace, the shafts on the primary
hearths are hollow and water-cooled, like that on the finishing hearth, and
the finishing hearth is made of greater area, being now 7 feet in diameter.
The furnace is said to treat from 20 to 33 tons of telluride gold ore per day
at Kalgoorlie, bringing the sulphur from 3 to 5 per cent. down to 0·01 per
cent. as sulphide, with a fuel consumption of 14 per cent.

At the Associated Northern Blocks, Kalgoorlie,[1] six Merton furnaces were
treating daily 120 tons of ore containing 3 per cent. sulphur. The hearths
were 30 feet by 6½ feet, and most of the elimination of the sulphur was on
the third or lowest hearth. The consumption of fuel was 14 per cent. by
weight of the ore, and the cost of roasting 2s. 9d. per ton.

The Associated Furnace[2] at work at Kalgoorlie at the same time had two
superimposed hearths, each 46 feet long, of which the upper one only was
heated by a fire, although about half the sulphur was burnt off on the lower
hearth.

2. **Revolving Cylindrical Furnaces.**—In the fixed-charge intermittent
furnaces, such as the Brückner cylinder and the Hofman furnace, a charge
of ore was introduced through apertures, which were then closed and the
furnace rotated until the roasting was complete.

The *White-Howell* furnace (Fig. 138) is an old example of the continuous
working furnaces of this type. The ore fed in at the upper end travels down
through the inclined cylinder towards the fire, and is discharged automatically
at the lower end. There is a considerable loss of ore by dusting. In this
furnace only the enlarged part next the fire-box is lined with firebrick, the
remainder being left unlined. Cast-iron spirally-arranged shelves assist
in raising and showering the pulp through the flames. John E. Rothwell[3]
added a hopper-shaped dust chamber with its bottom consisting of an inclined
cast-iron plate projecting about 8 inches into the upper end of the cylinder.
The dust carried out of the cylinder settled in this chamber, and, as it accumulated, slid down the sides and mixed with the fresh ore. Rothwell used a
cylinder 36 feet long and 5 feet in diameter with an inclination of 14 inches
only, rotating once per minute. The lining was of firebrick, 6 inches thick.

The Argall Roasting Furnace.—This is a modification of the older forms
of rotating cylindrical furnaces, and consists of four small parallel tubes
instead of one large one. The tubes are about 28 feet long, 2 feet in diameter,
and lined with firebrick, which is thicker at the lower end nearest the fire.
The ore is equally distributed between the tubes, so that the furnace is
balanced and requires but little power to rotate it. The furnace is rotated
by friction drivers, and makes about one revolution in four minutes, the
slowness of movement of the ore diminishing the loss by dusting. The
working drawings reproduced in Figs. 139 and 140 were supplied for the

[1] Von Bernewitz, *loc. cit.*   [2] *Ibid.*
[3] Rothwell, *Mineral Industry*, 1892, p. 234.

Air Conduct

Fire Brick

Stoker

Discharge Hopper

Fire Brick

Rabble Brick

Hard Burnt Tile

Porous Tile

10' T Beam

8' T Beam

8' T Beam

Scale

13 Ft

10

5

Damper

Speed 1 rev" in 4 min."

8' T Beam

10' T Beam

33 rev. per min.

Ore Hopper

Damper

9" Rods 3' 0" C. to C.

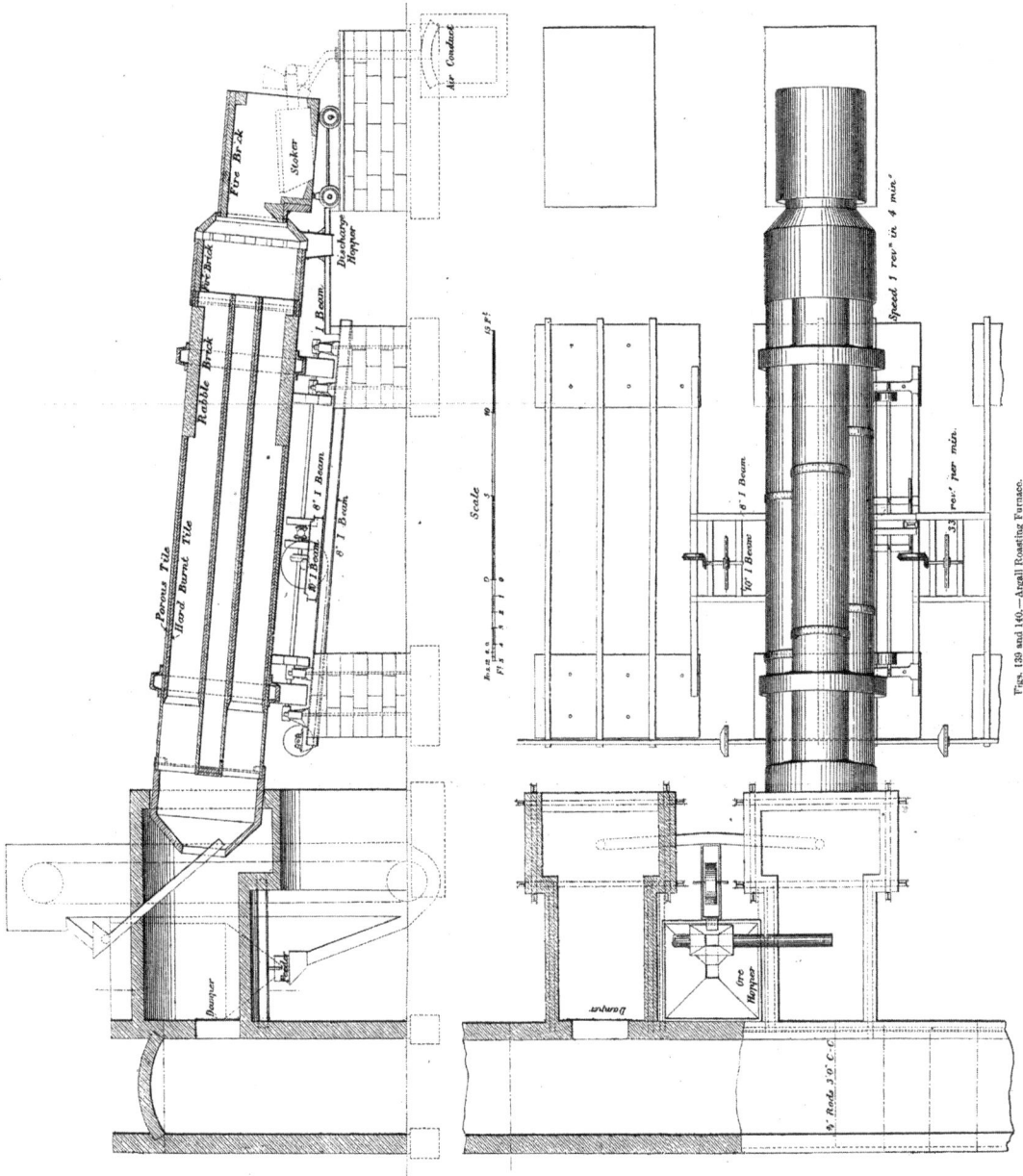

Figs. 139 and 140.—Argall Roasting Furnace.

purpose by Philip Argall. The furnace as shown is arranged for burning coal, but some have been erected and were used for some time with oil as fuel. The capacity of the furnace when roasting ores containing 2 per cent. of sulphur is from 45 to 50 tons per day, the sulphur being reduced to 0·1 per cent. ; the ore is from two to three hours in the furnace, and the consumption of slack coal for this ore is 150 to 160 lbs. per ton. On pyritic ores containing 12 per cent. of sulphur, the capacity is said to be from 25 to 30 tons per day. The cost of roasting is very small, hardly any repairs and supervision being necessary. The furnace was used continuously for some years at Cyanide, Colorado, in roasting Cripple Creek ores for cyanide treatment. Similar furnaces with modifications, are used for drying ore and also for cooling roasted ore. In the cooler, water tubes and jackets reduce the temperature of 300 tons of ore per day from a red heat to the ordinary temperature. The drying furnaces were in use in 1914 at the Cam and Motor Mill, Rhodesia.

Use of Producer Gas in Roasting.—The use of producer gas in roasting may be mentioned. It was introduced at the Holden Mill, Aspen, Colorado, in 1891, and soon completely displaced other fuel there for both drying and roasting.[1] Producer gas was introduced at the Great Boulder Sulphide Mill, Kalgoorlie, in 1901 [2] for use in the Edwards furnaces. It is produced from a mixture of coal and sawdust. There are three Dowson producers, 15 feet high and 6 feet in diameter. The gas is delivered to each furnace by a 6-inch pipe, from a 20-inch main, and distributed to various parts of the furnace by smaller pipes. The use of wood for firing was apparently not entirely dispensed with at first.

Cooling Roasted Ore.—This was formerly effected by spreading the roasted ore on sheet-iron floors, and either leaving it for twenty-four hours, or damping it down with a hose-pipe. More recently [3] it has been passed through plain or multi-tubular cylinders cooled with water, or passed back by automatic rabbles above or below the stationary hearth furnaces, or allowed to fall from shelf to shelf through air in towers built on the Hasenclever principle.

---

[1] W. S. Morse, *Trans. Amer. Inst. Mng. Eng.*, 1893, **21**, 919.
[2] A. G Charleton, *Gold Mining and Milling in Western Australia*, p. 336.
[3] Rothwell, *Mineral Industry*, 1900, **9**, 364.

CHAPTER XIV.

CHLORINATION.

Introduction.—The chlorination process depends on the fact that chlorine readily attacks gold (see above, p. 61), and forms soluble gold chloride. The solution is separated from the ore and the gold precipitated by suitable reagents. The process was widely used for the treatment of concentrate from about 1850 downwards, and for the treatment of suitable ore from about 1888 to 1912. It has now been superseded, mainly by the cyanide process, except for the treatment of small quantities of concentrate in a few scattered mills, and will probably soon become entirely obsolete. A brief account of it is given in this chapter.

The use of chlorine as an agent for the extraction of gold from ores was first suggested by Dr. John Percy, F.R.S., at the Swansea meeting of the British Association, held in August, 1848, in a paper [1] embodying the results of experiments carried out in the year 1846. Simultaneously, in 1848, Prof. C. F. Plattner, Assay Master at the Royal Freiberg Smelting Works, applied chlorine gas to the assay of the Reichenstein residues, and proposed that a similar method of treatment should be adopted on a large scale. These residues were the result of treating the Reichenstein ore with the object of extracting the arsenic. They consisted chiefly of oxides of iron and oxidised arsenical compounds, and had been roasted in the course of the process for the extraction of the arsenic. The residues had been accumulating for more than a century, and contained from 15 dwts. to 1 oz. of gold per ton ; they were considered too poor to smelt, while they could not be made to yield the gold contained in them by amalgamation.

Prof. Plattner's suggestion was followed up by investigations made by Dr. Duflos in 1848,[2] and by Lange in 1849.[3] Dr. Duflos compared the results obtained by treating the residues with chlorine water by percolation in a stationary vat, and by agitation in a revolving barrel ; and as these results were the same, he recommended the stationary vat as being more economical. He also obtained identical results with chlorine-water and with dilute solutions of chloride of lime and hydrochloric acid mixed together. On the other hand, Lange believed that gaseous chlorine, applied to the ore in the same manner as had been used by Plattner in assaying, was a more efficient agent than a solution of chlorine in water, and it seems to have been in accordance with his advice that the first chlorination works, that at Reichenstein, was established in 1849. The chlorinating vessels were small earthenware pots, and the precipitant employed was sulphuretted hydrogen. Plattner subsequently recommended wooden vats coated with pitch, and ferrous sulphate as a precipitant, and although these were not at first used at Reichenstein

---

[1] Percy, *Phil. Mag.*, 1853, **36**, 1-8.
[2] Duflos, *J. prakt. Chem.*, 1849, **48**, 65-70.
[3] Lange, *Karsten's Archiv*, 1851, **24**, 395-429.

they were adopted by G. F. Deetken in 1857, when he introduced the system into California.

The following three systems of chlorination were developed, and used in different parts of the world :—

1. The *Plattner process*, in which the gold is dissolved by means of gaseous chlorine acting on moist ore. The actual solvent is really a saturated solution of chlorine in water. The process is almost obsolete.

2. The *Barrel process*, in which a supersaturated solution of chlorine in water is used, the chlorine being kept in solution by an atmosphere of chlorine of a few pounds pressure inside the barrel. The process is obsolete.

3. The *Vat-solution process*, in which weaker, unsaturated solutions of chlorine are used. The process is obsolete.

These processes are described separately below, in the order of their introduction on a working scale. For the volatilisation (chlorine roasting) process, see above, p. 291.

THE PLATTNER PROCESS.

Plattner Process at Reichenstein.[1]—The material treated consisted of the residues obtained by roasting arsenical iron pyrites for the production of white arsenic, which was volatilised and condensed in brick chambers. There were forty-eight earthen chlorination pots, each holding 150 lbs. of ore. These pots were strengthened with iron hoops, and suspended on two journals, so that they could be discharged by inverting them.

The lower part of the pots was of a conical shape, and this part was filled with pebbles and sand covered with a perforated earthen plate, the function of which was to prevent the ore from mixing with the filter bed. The ore filled the cylindrical part of the pot above the earthen plate. The chlorine was generated by the action of hydrochloric and sulphuric acids on manganese dioxide in earthenware vessels, and was conveyed thence to the ore-pots through leaden pipes. The gas was introduced below the filter bed, and passed upwards through the ore for an hour ; a wooden cover was then fitted on, but not luted down until chlorine had been passed for from six to seven hours longer, after which all joints were luted down with dough, and the vat left until the next day. The cover was then removed and water, at a temperature of from 64° to 77° F., poured on, and allowed to percolate through the ore and filter bed by gravity. The liquid coming from twenty-four pots was conveyed to four vats, the first one being filled with solution before the second was used, and so on ; the contents of the fourth vat, being too poor for precipitation, were used over again for leaching. The leaching was stopped when 90 cubic feet of water had passed through the total charge of 3,600 lbs., this being at the rate of 312 gallons per ton of 2,000 lbs. The liquid from the first three vats was drawn off into twenty glass globes, which were heated on a sand bath so as to raise the temperature to 77° F. Sulphuretted hydrogen, obtained from fused sulphide of lead and sulphuric acid, was passed through until the saturation point was reached, when the liquid was left to settle until the next day ; after this the clear supernatant liquid was passed through sawdust filters to catch all the sulphide of gold, which might still have been in suspension. The sulphides were refined by dissolving them in acids, precipitating the metallic gold

---

[1] B. Kerl, *Hüttenkunde*, 1865, **4**, 372.

with ferrous sulphate, and melting it in clay crucibles with nitre and borax. The amount of roasted ore treated daily was 3,600 lbs., containing about $\frac{5}{9}$ oz. of gold per ton, so that only about 250 ozs. of gold were extracted yearly, and it is difficult to believe that the enterprise could have been a commercial success.

Later Practice in the Plattner Process.—The system ascribed to Plattner consists of the following operations :—

1. The concentrate or residue is subjected to a "dead" roast in a reverberatory furnace.

2. The roasted ore is slightly damped with water and charged into wooden vats, holding from 1 ton upwards. The vats have false bottoms consisting of filter beds of gravel or of cloth. Chlorine gas, generated in another vessel, is introduced at the bottom of the vat, and rises through the ore, permeating every part of it. The vat is then closed up and left undisturbed for twenty-four hours or more, by which time all the gold is converted into soluble chloride of gold.

3. The soluble salts are then washed out with water, which is allowed to flow on to the surface of the ore, and, passing through it, drains through the filter bed. When all the gold has thus been removed in solution, the tailing is thrown away.

4. The solution of gold chloride is acted on by ferrous sulphate, or some other suitable reagent, by which the gold is precipitated ; the particles of the precious metal are allowed to settle, and then are collected and melted down.

This process may be used for the treatment of small quantities of roasted concentrate. The principal modifications which were made in it were introduced mainly in order to deal with larger quantities of material.

The Vats.—The vats used for impregnating the ore with chlorine are usually, in California, about 7 feet in diameter, and are made of staves 3 feet long and 2 inches thick, which consist of the best split sugar-pine. They are coated with a mixture of pitch and tar to protect them from the corrosive action of the chlorine. Before being used for the first time, the vats are thoroughly soaked with water to diminish their absorptive action on the chloride of gold solution, but all wood brought in contact with this solution is nevertheless invariably impregnated with a certain amount of gold. This may be recovered after the vats, etc., are worn out, by burning them and fusing the ashes with suitable fluxes. The false bottom generally consists of quartz pebbles, the lowest layer being of the size of hazel nuts, and each successive layer consisting of finer material until, at the top, a thin layer of fine sand (passing a 20-mesh sieve, but retained on a 60-mesh sieve) is spread evenly over the surface. The thickness of the filter bed (which is not shown in the figure) is usually from 6 to 12 inches. It is supported on boards (A, Fig. 141), 1 inch thick, in which numerous $\frac{1}{2}$-inch auger holes are drilled ; these boards rest on wooden strips (not shown in the figure), 3 inches wide and 1 inch thick, which do not reach the edge of the vat, and so keep a clear space 1 inch deep, just above the true bottom of the vat, in which the solution can accumulate. The solution is drawn off by a leaden pipe fitted with a stopcock, preferably of stoneware. Deetken states [1] that fine sea-shells (consisting of carbonate of lime) have been used instead of quartz pebbles for the filter beds without any prejudicial result. Talcose rocks, and

---

[1] Deetken, *Mineral Resources West of the Rocky Mountains*, 1873, p. 342.

particularly silicates of alumina, must not be used on account of their power of absorbing the chlorine. For the same reason sulphides, magnetic iron oxide, metallic iron, fragments of wood or other organic matter, or, briefly, any substance capable of being acted on by chlorine or of reducing the chloride of gold must be carefully excluded from the filter bed.

The surface of the filter bed is covered with boards, not fitted closely together, but made into a framework by cross-pieces and pierced with many auger holes. This cover is useful when the tailing is being cleared out, otherwise, in shovelling away the ore, the surface of the filter bed is partly removed also. Filter cloths of canvas, burlap, or cocoa-nut fibre matting are also frequently used above the filter-bed, stretched tightly over a framework of wood which accurately fits the inside of the vat. The space between the canvas and the wall of the vat is packed with hemp or other material closely tamped down. Filter-cloths of every material, except asbestos, are soon rotted by the action of the chlorine, and their use is frequently dispensed with. Wool lasts longer than cotton.

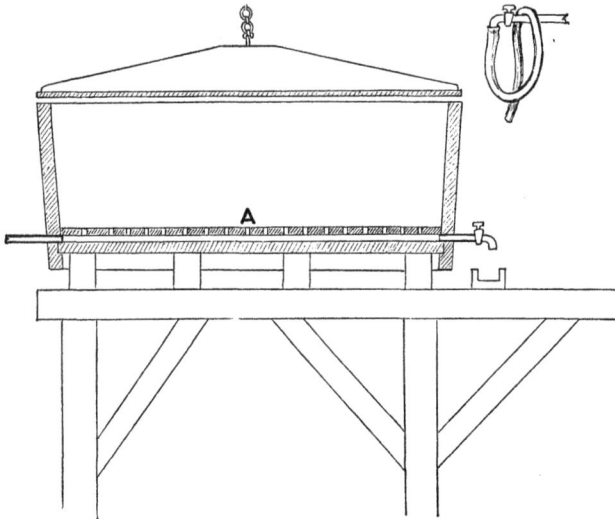

Fig. 141.—Plattner Chlorination Vat.

Charging in the Ore.—When the vats are ready to be charged, a layer of dry ore is spread over the false bottom, and time given for the water from the filter bed to be drawn up into this layer by capillary attraction. If attention is not paid to this point, the lowest layer of ore becomes too wet from the combined effect of the water added to it before charging in and that absorbed from the false bottom. The result is that the passage of the chlorine through the mass is resisted and there is a great increase in the consumption of the gas. Deetken states that the whole of the usual charge of gas may be thus consumed, not rising more than a few inches above the bottom. The greater part of the charge is damped by sprinkling with water and thoroughly mixing. The amount of the water added varies with the nature of the ore, but the usual amount is from 6 to 12 per cent. for roasted ores. If it is made too wet, dry ore to the required amount is mixed with it. A good rough method of ascertaining when it is of the proper degree of dampness is to compress some in the hand ; balls of ore should be readily formed in this way, but should be just dry enough to crumble up again. The reason for the addition of water is that perfectly dry chlorine has scarcely any action on metallic gold at ordinary temperatures, and up to a certain point an increase in the amount of water present raises the rate of dissolution of the gold. The limit of the amount of water that can be added is, however, determined by physical conditions, as the mass must be of loose porous texture in order to permit the gas to permeate readily through every portion of it. The ore is passed through a 4-mesh sieve, and falls into the vat in a light shower. Although left undisturbed as far as possible, the charge must be levelled off with a rake occasionally

When the vat is filled to within 6 inches of the top, the surface of the ore is made concave or saucer-shaped, higher at the sides than in the middle. The cover, usually of wood, is then lowered on to the vat by means of a chain and pulley, and the rim luted with a mixture of wet clay and sand, or, more usually in former times, with dough. These joints are kept moist during the " gassing " process by wet rags. The gas is introduced through a lead pipe, which is shown on the left-hand side of Fig. 141, passing into the vat below the false bottom. A small hole is left in the cover through which the

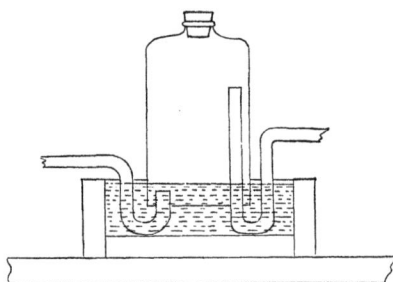

Fig. 142.—Wash-Bottle for Chlorine Gas.

displaced air may escape, and the issuing gases are tested from time to time by means of a rag tied to a stick and moistened with dilute ammonia. As soon as chlorine is found to be coming off freely, the hole is plugged, but the current of gas is not stopped until after the lapse of one or two hours more, when the charge is supposed to be saturated with the gas, the total time required for the impregnation being usually from five to eight hours.

Generation of the Chlorine.[1]—The chlorine is generated in air-tight vessels of lead heated by a water bath and fitted with a stirring apparatus passing through the lid and worked from the outside. The gas necessary for a 3-ton charge of roasted concentrate may be generated in a leaden vessel of 20 inches in diameter and 12 inches deep. The charge for 3 tons of ore consists of 20 to 24 lbs. of rock salt, 15 to 20 lbs. of manganese dioxide, containing 70 per cent. of available material, and 35 lbs. of oil of vitriol of 66° B., diluted with half its weight of water.

The outlet tube is of lead, but connections in pipes are often made by short pieces of indiarubber-tubing, well greased on the inside. These resist the action of chlorine fairly well. The gas is passed through about $\frac{1}{2}$ inch of water in the wash-bottle shown in Fig. 142. The use of the wash-bottle is partly to free the chlorine from hydrochloric acid, or other impurities with which it is contaminated, but mainly to give an indication of the rate of flow of the gas. A little hydrochloric acid does no great harm, and in any case some hydrochloric acid is formed in the charge by the

[1] See description and figure of the Edwards generator in the section on Miller's chlorine process, Chap. XVIII.

action of chlorine on traces of sulphides left undecomposed in the ore, thus :—

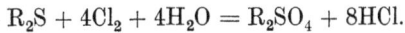

$$R_2S + 4Cl_2 + 4H_2O = R_2SO_4 + 8HCl.$$

The only disadvantage due to the presence of much hydrochloric acid in the gas lies in the fact that certain metallic oxides (oxides of iron, copper, etc.) are much more readily soluble in the acid than in chlorine, chlorides and water being formed, and the resulting solution will be contaminated with these chlorides, so that special precautions are necessitated to prevent the bullion from becoming base.

Impregnation of the Ore.—The ore is allowed to remain impregnated with the gas for from twenty-four to forty-eight hours, the continued presence of a strong excess of gas being ascertained at intervals by removing the plug from the cover and applying the ammonia test. When, as is usually the case, there is a large excess of gas when the impregnation is at an end, it may be disposed of in one of several ways. It may be dissolved by adding water before raising the cover (the usual method of procedure), or it may be withdrawn by aspiration and discharged outside the building, or stored in a gasometer for use in a subsequent charge.

The time of impregnation varies according to the size of the particles of gold, the fineness of the metal, and the temperature employed. Chlorine has a very slow action on pure gold, the rate increasing gradually with the temperature up to 100°, see pp. 22 and 61.

Fine gold is acted on more slowly than that of less fineness. The alloys containing base metals (copper, etc.) are dissolved very rapidly, and small quantities even of silver appear to increase the rate of solution, but if the percentage of silver amounts to 10 per cent. or over,[1] an insoluble coating of chloride of silver is formed over the granule, and further action is checked or completely stopped.

Reactions in the Impregnation Vat.—The amount of chlorine to be used depends mainly on the substances present, other than gold, by which chlorine is absorbed. If any sulphides are present they are oxidised by the chlorine in presence of water ; sulphates, chlorides, and sulphuric and hydrochloric acids being formed.

Protosulphates or any other protosalts present are converted almost instantaneously to persalts by the chlorine, as follows :—

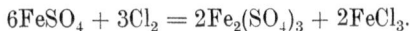

$$6FeSO_4 + 3Cl_2 = 2Fe_2(SO_4)_3 + 2FeCl_3.$$

It is obvious from these reactions that great waste of chlorine in the impregnation vat is caused by imperfect roasting, 1 per cent. of unoxidised sulphur present in pyrite, if open to the attack of the chlorine, being enough to convert 8·9 per cent. of chlorine (or about 200 lbs. per ton of ore) into hydrochloric acid. It is, therefore, desirable that the ore should be roasted dead.

Sulphate of copper ($CuSO_4$) does not appear to be acted on by chlorine, but, nevertheless, when much of it is present in a roasted ore, chlorination generally seems to be rendered impracticable. This is possibly due to the fact that some sulphate of iron accompanies it. Whenever sulphates of these metals are left in the roasted ore by accident or design it is necessary to remove them by a preliminary leaching with water before the chlorine is introduced.

---

[1] Hofman and Magnuson, *Trans. Amer. Inst. Mng. Eng.*, 1904, **35**, 948.

Of course if the ore is chlorinated in tubs by gas, it must be partially dried and sieved back into the tub before impregnation can be attempted.[1]

Organic matter is also oxidised by chlorine, although much more slowly. At ordinary temperatures, pitch and tar are almost unaffected, and the fibres of matting, canvas, etc., are acted on very gradually. Pieces of decaying wood or dried leaves must not be introduced with the water into the leaching vat, and if surface water is used it should always be carefully strained before being run in. A rough analysis of the water employed will often be serviceable, as it is frequently strongly alkaline in dry countries, and may be softened with advantage.

The absorption of chlorine by metallic oxides is the most frequent cause of waste, and, in the vat process, there are usually no efforts made to prevent this. Well roasted sesquioxide of iron (Fe_2O_3) is scarcely attacked by chlorine, especially if the temperature attained in the furnace has been high. If any magnetic oxide (Fe_3O_4), however, has been formed from over-heating or has been originally present in the ore, the absorption of chlorine is considerably greater, ferric chloride being formed and dissolved. Ferrous oxide is instantly converted into a mixture of chloride and sesquioxide of iron. Hydrochloric acid acts more rapidly than chlorine on all these oxides, but is nevertheless very slow in dissolving the well roasted sesquioxide. Metallic iron, which is sometimes accidentally introduced, is dissolved at once by both chlorine and HCl. The oxides of copper and zinc are quickly dissolved by chlorine, and still more readily by HCl. Lime and magnesia also readily absorb chlorine, forming hypochlorites, chlorates and chlorides, but hypochlorites are decomposed by any acid which may be present.

If any appreciable quantity of oxides capable of absorbing chlorine are present, it is cheaper to dissolve them by adding dilute sulphuric acid to the ore, and then, if possible, to leach out the soluble sulphates formed, before subjecting the ore to the action of the gas.

Amount of Chlorine required.—The amount of chlorine required varies greatly, both with the nature of the ore and the manner in which it is roasted. In order to roast pyrite dead, a long time in the furnace terminating at a high temperature is necessary, and the addition of salt may be desirable in order to chloridise oxides which would otherwise absorb the more expensive chlorine in the impregnation vat. These conditions in the furnace, however, may cause enormous losses by volatilisation, the endeavour to save a few pounds of chlorine in the vat causing the loss of 30 or 40 per cent. of the gold in the furnace. In ores where the percentage of copper, etc., is not large, and where, in consequence, salt need not be used in the furnace, the roasting may be finished at a high temperature without any disadvantage, and the consumption of chlorine may be thus reduced to a very low point. Butters states [2] that at his mill in Kennel, California, where all descriptions of concentrate and pyritic ores were treated by the Plattner process, the average consumption of chlorine was 12 lbs. per 2,000 lbs. of ore.

Leaching the Charge.—When it is judged that the impregnation has lasted long enough for all the gold to be dissolved, the excess of chlorine gas is removed, the lid is taken off, and water is added to the charge to wash

---

[1] Ore roasted "dead" or "sweet" usually contains from 0·15 to 0·4 per cent. of sulphur chiefly in the form of sulphates, or locked up in uncrushed ore The lower the amount of sulphur which is left in the roasted ore the better is the extraction of gold and the smaller the amount of chlorine required.

[2] Butters, *Eng. and Mng. J.*, Dec. 20, 1890.

out the soluble chloride of gold. The water may be added from below, but it is more usual to pour on water at the top, and let it flow out at the bottom.

The water must be added carefully, as otherwise the ore may pack unevenly, and channels may be formed through the mass, and the leaching thus rendered imperfect. Water is usually run from a tap on to a layer of gunny-sacking placed over the ore, or through a coil of lead pipe, pierced with small holes, underneath the cover. The stopcock below the false bottom is then opened, and the yellow- or blue-coloured solution (coloured by salts of iron, gold, and copper), which should have a strong odour of chlorine, is run into a settling tub and thence to the precipitation vat. The leaching is continued as long as any trace of gold can be detected in the issuing liquid, after filtering, by ferrous sulphate or stannous chloride (see above, p. 65). When large quantities of copper salts are present in the solution, their strong bluish tints mask the slight discoloration due to a precipitate of a small quantity of metallic gold, and, also they appear to interfere with the precipitation by ferrous sulphate, in some cases at least preventing it from taking place.[1]

The last charges of wash-water are not mixed with the strong solution, but stored in other vats and used again for the first washings of other charges. Re-precipitation of dissolved gold in storage vats or impregnation vats is not to be feared, so long as there is an excess of chlorine present in the liquid, and this can easily be ensured by adding a small quantity to any solutions not smelling strongly of the gas.

The amount of water used in leaching is usually about 2 tons of water to 1 ton of ore, but in most cases part of the water is used again in the next charge.

Precipitation of the Gold.—The precipitating vat is of the same material as the leaching tubs, and may be from 5 to 7 feet in diameter and 3 feet deep. There is no false bottom, and the vat is often made wider at the bottom than at the top to prevent any adherence of the gold to the sides. The wood is protected by a coating of pitch or paraffin-paint, or is left without paint of any kind. The vat receives a smooth finish inside to facilitate perfect cleaning, and is set perfectly level to avoid loss of gold while the waste liquor is being drawn off. The precipitating solution of ferrous sulphate, which is usually prepared at the mill by dissolving iron in sulphuric acid, and contains some free acid, is introduced into the precipitating vat before the gold solution. The mixture is well stirred with wooden paddles, in order to make the precipitate settle better.

Precipitation takes place according to the equation—

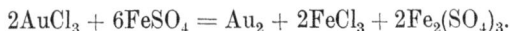

$$2AuCl_3 + 6FeSO_4 = Au_2 + 2FeCl_3 + 2Fe_2(SO_4)_3.$$

The oxidation of the ferrous salt is also effected in other ways, notably by the excess of free chlorine present in the solution, so that much more sulphate of iron is required than is indicated by the equation. The difficulty of collecting and saving the precipitated gold is due to the fine state of division of the precipitate, which settles slowly and tends to pass through filters. The gold settles better if it is well stirred, and Aaron recommends an addition of more sulphuric acid and vigorous stirring, two hours after the precipitation is complete, as a means of assisting the settling. Besides gold, the only other metals precipitated by ferrous sulphate are those which form insoluble

---

[1] Butters, *loc. cit.*

sulphates—viz., lead, calcium, strontium, and barium. The last two of these are rarely present, and the others are dealt with in the manner described below. Basic iron salts are not precipitated if enough free sulphuric acid is present, and when precipitated, they may be removed from the gold by treatment with acids, or by slagging them off in the furnace.

If lead or calcium is present, dissolved in the solution, it will be precipitated as an insoluble sulphate on the addition of the ferrous sulphate, and thus render the gold-precipitate impure and less easy to treat. The amount of lead in the solution is usually small, unless hot water has been used for leaching, and most of the lead chloride is, in any case, separated by the canvas filter. The usual method of removing the calcium is to add sulphuric acid to the gold solution, and to let it stand for a few hours, when calcium sulphate crystallises out, forming a crust on the sides and bottom of the vat. The liquid is then drawn off and transferred to another vat for the precipitation of the gold.

When precipitation is complete the liquid is allowed to remain at rest for some time, in order to allow the gold to settle to the bottom. Butters [1] states that forty-eight hours is usually sufficient, but that sixty hours is better, and the determination of the extent to which the settling has progressed may be made by tapping the solution at various heights and filtering the liquid thus obtained. When a quart of liquid, drawn from a point 2 inches above the bottom of the vat, gives only a slight dark stain to filter paper, on being passed through it, the settling may be regarded as complete. C. H. Aaron quotes instances [2] where, after forty-eight hours settling, as much gold remained in suspension in the liquid which was drawn off as was equivalent to 50 cents per ton of the ore treated.

When the waste liquid has been drawn off by a floating siphon, more ferrous sulphate and fresh solutions from the leaching vat are poured into the vat, and the process repeated until enough gold has accumulated at the bottom to warrant a clean-up. This may take place at intervals of from a fortnight to three months. The clear liquid is drawn off as closely as possible, and the slime scooped out and filtered through paper, or, by means of a press, through canvas. The gold precipitate is then dried carefully and fused in graphite pots, with salt, sand, nitre, borax, etc., as fluxes, according to the requirements of the case. If the precipitate contains any considerable amount of impurities (such as oxides and basic salts of iron), which is usually the case, it may be treated with hydrochloric acid before fusion. The bullion produced varies from 920 to 990 fine, the alloying metals consisting chiefly of iron and lead.

Cost of Working.—The cost of treating concentrate or ore by the Plattner process depends chiefly on the cost of roasting. In 1867, the total cost in California was stated by Küstel to be $14.55 per ton, but in 1872 it had been reduced to $11, the expense of roasting being in each case about two-thirds of the whole. At the works of the Plymouth Consolidated Mining Company, California, in 1886,[3] the cost of treating 100 tons per month was $9.40 per ton (the roasting accounting for $4.60 per ton, or nearly one-half), and at the Providence Mine, in the same State, it was about $6.30, without including the expenses of general supervision, interest on first cost, and depreciation of plant.

---

[1] Butters, *loc. cit.*
[2] Aaron, *Eighth Report Cal. State Mineralogist*, 1888, p. 836.
[3] Small, *Trans. Amer. Inst. Mng. Eng.*, 1886, **15**, 305.

It was estimated in 1888 [1] that, generally, throughout California, a plant, capable of treating 6 tons of concentrate daily, cost from $6,000 to $7,000 for its erection, while the cost of extraction was about $10 per ton, and the proportion of the gold extracted was from 90 to 92 per cent.

Use of Liquid Chlorine.—At the Utica Mine, California,[2] the concentrate is roasted in long reverberatory furnaces, allowed to cool, moistened with sulphuric acid and water and sieved into wooden vats. Chlorine gas derived from a drum of liquid chlorine is then introduced into the charge, and after forty-eight hours the ore is leached in the usual way. The cost of treatment was $7.80 per ton, when the chlorine was generated from manganese dioxide, salt and sulphuric acid, and only $6.90 per ton after the introduction of liquid chlorine. One drum of liquid chlorine weighs 300 lbs., holds 115 lbs. of chlorine, and takes the place of 972 lbs. of manganese dioxide, 1,080 lbs. of salt, and 2,160 lbs. of sulphuric acid. The drum is 10 inches in diameter and 5 feet long, and costs $40 at the Utica Mine. The time of introducing the gas into a charge is reduced by its use from six or eight hours to twenty minutes.

The Plattner Process at the Plymouth Consolidated Gold Mining Company, Amador County, California.[3]—The ore contains 11 dwts. of gold, mostly in the free state. It is crushed in a stamp battery, No. 8 screens (40 mesh) being used, passed over a length of 20 feet of amalgamated plates (the upper one of which is copper and the rest silver-plated) and then concentrated on Frue vanners. The concentrate amounts to from $1\frac{1}{4}$ to $1\frac{1}{2}$ per cent. of the weight of the ore, and contains from 5 to 10 ozs. of gold per ton. It is treated at the rate of 100 tons per month, being kept damp until charged into the roasting furnace, to prevent the formation of lumps. A "Fort-schaufelungsofen" is used for roasting, 80 feet long by 12 feet wide, its hearth consisting of a long continuous plane, holding three charges at one time, which are kept separate. The three stages are called the "drying," "burning," and "cooking" stages. In the (middle) burning stage, the bed of ore is kept thin and occupies double the space of each of the other charges. The furnace is worked by three eight-hour shifts of one man each. The charges weigh 2,400 lbs., including 10 per cent. of moisture, and on an average contain 20 per cent. of sulphur. Just before the sulphur ceases to flame, $\frac{3}{4}$ per cent.—i.e., 18 lbs. —of salt is added.

The chlorinating vat is 9 feet in diameter and 3 feet deep; it holds 4 tons of ore. The filter-bed is 6 inches deep, and consists of, (a) at the bottom, wooden strips, $\frac{3}{4}$ inch wide, placed 1 inch apart; (b) above this, 6-inch boards placed 1 inch apart and laid across the strips; (c) coarse lumps of quartz, diminishing upwards to fine stuff; and (d) at the top, a cover of 6-inch boards placed similarly to the lower layer, but crosswise to them. Chlorine is introduced on both sides of the vat, which is left luted-up for two days, and then leached for four or five hours, the tank being kept full of water during the operation. A gunny-sack protects the surface of the ore from the direct impact of the water from the hose. The ore in the impregnation vat contains about 6 per cent. of water (crumbling after it has been sieved), and is sieved into the vat through a screen of $\frac{1}{2}$-inch mesh.

The gold solution is acidulated with sulphuric acid to precipitate the lead,

[1] *Eighth Report Cal. State Mineralogist*, 1888.
[2] T. N. Smith, *Eng. and Mng. J.*, April 22, 1899, p. 467.
[3] For a more complete description, see that given in the *Eighth Report Cal. State Mineralogist*, 1888, of which the account appended is an abstract.

and ferrous sulphate is then added to it in another tank, after which it is allowed to settle for two days before the supernatant liquor is siphoned off. The gold is left to accumulate for fourteen days and the wet gold is then filtered and fused in graphite pots. The average extraction is from 95 to 96 per cent. of the gold, and 7 dwts. per ton are left in the tailing. All wood is protected from the action of the acids by paraffin-paint. The cost of milling is said to be 39 cents per ton of ore, and the cost of concentration, roasting, and chlorination, $13.40 per ton of concentrate.

The Plattner Process at the Alaska Treadwell Mine.[1]—The material treated consists of the sulphides collected by Frue vanners from a stamp battery, and contains 40 per cent. of sulphur, mostly pyrite. The gangue is quartzose, containing from 2 to 5 per cent. of calcite, which necessitates the addition of salt in the roasting furnace.

The roasting was first effected in Brückner cylinders, which were abandoned owing to the large amount of fuel consumed and the enormous losses by dusting and volatilisation (which are said to have amounted to 30 per cent. of the gold). The automatic Spence furnace was then tried and proved to be useless until it was used as a reverberatory. Six were erected, and the cost of roasting was reduced by one-half, but the capacity was small (10 tons per day in six furnaces) and the consumption of fuel great, and a reverberatory furnace being erected, was found to be more satisfactory. The Spence furnaces were accordingly discarded, and 25 tons of concentrate were roasted per day in five reverberatory furnaces of 13 by 65 feet, inside measurement, at a cost of $3½ per ton. The ore is not roasted quite dead, owing to the idea that loss of gold by volatilisation would be thereby diminished.

The roasted ore is spread on the cooling floor, wetted down and sifted carefully into vats, each of which holds 4½ tons, and is then impregnated with the gas. This operation occupies four hours, after which the vat is left untouched for thirty hours, fresh gas being forced in a few hours before leaching. The leaching usually requires twelve hours. The tailing is sampled and assayed, and, if found sufficiently poor, is sluiced into the sea.

The solution is run into collecting tanks and thence to the precipitating vats, which already contain the necessary amount of ferrous sulphate in solution. The precipitation is complete when all the solution has been run in, or when the vat is full. It is then stirred briskly for a few minutes and left to settle for from eighteen to twenty-four hours, when the supernatant liquor is siphoned off and passed through a large filter. The supernatant liquor usually contains from 23 to 25 cents of gold per ton of material treated, and this is all saved in the filter.

The clean-up is made twice a month, the drying and melting of about 750 ozs. of gold being done by one man in one day. The gold is washed into a small tub, allowed to settle, and the supernatant liquor returned to the precipitating vat. The gold is dried in an iron pan without filtering, and melted with a little borax.

Each one of the chlorination vats, holding 4½ tons of ore, cost $50, and lasts for three years without any repairs. The filter in it costs only the price of a few gunny-sacks, and lasts for six months without any attention. The ferrous sulphate is prepared on the works from sulphuric acid and scrap-

---

[1] This description applies to the practice in the year 1891.

iron. In 1890, the cost was at the rate of about $10 per ton, the yield in gold being over $40 per ton. The results for the period June, 1890, to May, 1892, were as follows :—

TABLE XXVIII.

Year.	Tons Chlorinated.	Yield per Ton.	Cost per Ton.		
			Labour.	Supplies.	Total.
June, 1890, to May, 1891, .	5,869	2·02 ozs.	$5.03	$3.99	$9.02
June, 1891, to May, 1892, .	6,177	1·52 ,,	4.80	2.81	7.61

THE BARREL PROCESS.

Historical.—It was mentioned on p. 300 that Dr. Duflos used a revolving barrel in some of his experiments at Breslau, in 1848, and obtained results almost identical with those given by the vat percolation method. He, therefore, preferred the latter as being cheaper. In 1877, Dr. Howell Mears, of Philadelphia, patented a process which was gradually improved in practice, after having been adopted by several mines in the United States, and in particular the improvements introduced by Adolph Thies, in 1881, caused the name of the Thies process to be applied to the amended method of procedure. In 1887 the barrel process was applied by Cosmo Newberry and C. J. T. Vautin to the ore of the Mount Morgan Mine, Queensland, where it was successfully worked for some time before it was replaced by the vat solution process. In 1889 and succeeding years the barrel process was developed in the United States, and appeared likely to become of great importance. It reached its zenith about 1898, and has now been superseded by the cyanide process, the last mills at Cripple Creek and Colorado City having closed down in 1912.

The Mears Process.—In this process, the roasted ore was charged into lead-lined iron cylindrical barrels, together with enough water to make an easily flowing pulp ; chlorine was then forced in under pressure through the hollow trunnion of the barrel, which was revolved until the gold had been dissolved. The pressure of chlorine was stated to be as much as 40 or 50 lbs. to the square inch. The manhole of the barrel was then opened and the ore discharged by gravity into a leaching vat below, where the soluble gold was washed out and precipitated by any of the known methods.

The process was formerly in use at the Phœnix and Haile Mines in Carolina, and at the Bunker Hill Mine, California, but was superseded by the Thies process.

The Thies Process.—In this process the chlorine gas was generated inside the barrel itself by means of bleaching powder and sulphuric acid. The method was simplified in this way, and a number of leaky joints dispensed with. Thies found that a moderate pressure of chlorine, of a few pounds to the square inch, was enough. His barrel held from 2,000 to 2,500 lbs. of ore.

Practice in Dakota and Colorado.[1]—Improvements were introduced in

---

[1] Rothwell, *Mineral Industry*, 1892, **1**, 233 ; 1896, **5**, 261 ; 1900, **9**, 359.

Western America by J. E. Rothwell and others to enable larger quantities of ore to be dealt with, and to overcome the mechanical difficulties of leaching. These were due to the endeavour to leach mixtures of sand and slime, which are always separated from each other before being treated in the cyanide process.

The barrels were increased in size and varied from 12 feet long and 6 feet in diameter, with a capacity of 8 or 9 tons of ore, to 16 feet long and $5\frac{1}{2}$ feet in diameter, with a capacity of 12 tons. Barrels holding 18 tons were also in use at Cripple Creek. The shell of the barrel was of $\frac{5}{8}$-inch mild steel, and the ends of cast iron heavily ribbed. They were designed to withstand a pressure of 100 lbs. per square inch. The lining was of 24-lb. lead—*i.e.*, $\frac{1}{2}$ inch in thickness. The barrel was rotated on trunnions or on friction rollers.

The filter was placed inside the barrel. It consisted of a diaphragm parallel to the axis of the barrel, and forming (in transverse section) the chord of an arc of the circle of the barrel. The filter was at first made of asbestos cloth, resting on a framework consisting of oaken planks, each 11 inches wide and as long as the barrel, and 2 inches thick. The area of the filter in the $12\frac{1}{2}$-ton barrel was 4 by 16 feet, or 64 square feet. The planks were grooved, and the filter-cloth rested on the sharp ridges between the grooves, the surface being almost entirely available for filtration. Above the asbestos cloth was placed an open wooden grating, and the whole was held in place by cross-pieces, the ends of which rested under straps bolted to the inside of the shell. The cloth lasted for from 15 to 18 charges, or from $2\frac{1}{2}$ to 3 days. In later practice the asbestos cloth was replaced by 4 lbs. sheet-lead, with fine perforations, supported by sheet-lead $\frac{3}{8}$-inch thick with $\frac{3}{8}$-inch perforations, and enclosed between wooden gratings. The thin sheet-lead lasted from 6 to 60 charges. A sand filter has also been used inside the barrel.

There were two manholes in the shell above the diaphragm for charging and discharging the ore, and a third opening below the diaphragm for the discharge of the solution and washings.

The barrel is [1] charged by first filling the space under the filter with water, which at the same time is allowed to pass through the filtering medium, and wash it ; then the quantity of water required to make the ore free flowing (usually from 40 to 60 per cent. of the weight of the ore) is put in above the filter. The sulphuric acid is then cautiously poured into the water, through which it sinks in a mass to the bottom, without mixing with it ; the ore is then charged in, as follows :—The hoppers are furnished with two shoots, one for each charging-hole. The ore is let fall through these shoots alternately, the hole through which ore is not being passed serving as an air-vent. Meanwhile the bleaching powder has been weighed-out and placed in two small kegs. When the ore has all been introduced into the barrel, a workman, stationed at each charging-hole, hollows out a space in the surface of the dry ore with his hands, and, emptying one of the kegs into the barrel, closes up the charging-hole as quickly as possible. If all these operations have been conducted rapidly without a hitch, there is no immediate evolution of chlorine, but, if some time is suffered to elapse after charging-in the ore, the acid liquid, thoroughly stirred-up and mixed by the fall of the ore into it, gradually rises through and wets the charge, and the bleaching powder,

---

[1] This description applies to the practice in Dakota between the years 1890 and 1900.

falling on ore which has been wetted with acid, gives off copious fumes of chlorine before the cover-plate can be screwed on.

The amounts of bleaching powder and sulphuric acid added depends on the composition of the roasted ore. There must be a considerable excess of chlorine at the end of the treatment. The chlorine required varies from 3 to 10 lbs. per ton of ore. At the Valley Custom Mills, Cripple Creek, 100 to 200 lbs. of bleaching powder (yielding 30 to 33 per cent. of chlorine) and 300 to 400 lbs. of sulphuric acid of 60° B. were used in 1905 per barrel charge.[1] The equation representing the action of the acid on the bleaching powder is usually given as follows :—

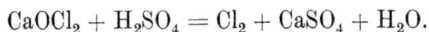

$$CaOCl_2 + H_2SO_4 = Cl_2 + CaSO_4 + H_2O.$$

Less chlorine is given off than is shown in the equation, as bleaching powder usually contains only from 25 to 35 per cent. of " available " chlorine.

The barrel is revolved at the rate of 4 or 5 up to 12 revolutions per minute for 3, 4, or sometimes 6 hours. The chlorine must be present in excess, and this is ascertained from time to time by holding a rag wetted with ammonia solution opposite a small valve which is opened momentarily.

It was generally supposed that the gold would be cleaned by attrition, and in particular that silver chloride, formed by the action of chlorine, would be rubbed off the surface of particles of native gold-silver alloys, and clean surfaces of gold would thus be continually offered to the attack of the chlorine.

Hofman and Magnuson,[2] however, have shown that there is a limit to this action, if it exists. They found that with pure gold and quartz grains an excess of chlorine in a rotating bottle dissolved pure gold a little faster than gold containing 10 per cent. of silver. The alloy gold 80, silver 20 was almost as readily dissolved by a supersaturated solution of chlorine as pure gold, but the alloy gold 70, silver 30 was acted on much less rapidly. The solvent power of somewhat weaker solutions was far more seriously interfered with by the presence of 20 per cent. of silver in the alloy.

After the chlorination is complete the barrel is stopped, so that the filter assumes a horizontal position ; the hose is attached to one of the outlet pipes, and after waiting for a few minutes to allow the pulp to settle, as recommended by Rothwell, the discharge valve is opened and the solution allowed to run out, the pressure of gas left in the barrel being enough to start the leaching. A hose is also attached to the inlet pipe, and water is pumped in under a pressure seldom exceeding 40 lbs. per square inch, the air in the top part of the barrel being compressed and forming an elastic cushion. By washing in this manner, no chlorine is allowed to escape into the building, as it is all absorbed by the water. If necessary the leaching is suspended at intervals, and the barrel is again revolved for a few minutes, so that its contents are thoroughly mixed-up together again. In this way the formation of permanent channels in the ore is prevented, and perfect leaching is ensured. The wash-water coming from the barrel is tested for gold with sulphuretted hydrogen. The full charge of ore is said by Rothwell to occupy a depth of 38 inches on the filter in his $12\frac{1}{2}$-ton barrel, and the average time of leaching to be $2\frac{1}{2}$ hours. The amount of water required for leaching is about 120 gallons per ton, besides the 100 gallons per ton contained in the charge.

---

[1] Gowland, *Non-Ferrous Metals*, p. 214.
[2] Hofman and Magnuson, *Trans. Amer. Inst. Mng. Eng.*, 1904, **35**, 948.

The tailing is discharged into a car which will hold the whole charge of ore and water, and is then run out of the building ; or, if water is abundant, the tailing is discharged into a sluice, and washed away. In some cases, as at Cripple Creek, it was concentrated on Wilfley tables, and the concentrate was sent to a cyanide plant for the further extraction of gold (Gowland). The filter-cloth is washed clean by a jet of water under pressure directed successively to all parts of it. This water is discharged by revolving the barrel.

The solution coming from the barrel, passes first to a tank with a gravel filter and then to the settling tanks, which are of sufficient capacity to allow several hours for settling. The clear liquid from these is forced up by air pressure into the precipitating vats.

The gold is precipitated with sulphuretted hydrogen. In Dakota, the practice prevailed for a time of destroying the free chlorine with sulphurous acid gas before passing sulphuretted hydrogen, but later, at Cripple Creek, the use of SO_2 was given up. The sulphuretted hydrogen, generated by the action of sulphuric acid on iron matte, is forced into the solution, usually diluted with air, for stirring purposes and for the expulsion of chlorine. It destroys the last traces of chlorine and precipitates the gold, thus :—

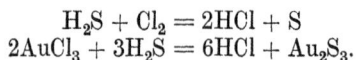

$$H_2S + Cl_2 = 2HCl + S$$
$$2AuCl_3 + 3H_2S = 6HCl + Au_2S_3.$$

Metallic gold and free sulphur are also precipitated.

It is said to take less than an hour at the Golden Reward Works to precipitate the gold from 5,000 gallons of solution (resulting from the lixiviation of from 25 to 50 tons of ore). The liquid is quite cold, but the precipitate is in a collected, voluminous and flocculent form which settles quickly. It is left undisturbed for two hours, and the liquid is then drawn off to within 4 inches of the bottom of the vat, and passed through a filter-press, provided with a set of heavy, canton-flannel filter-cloths. The head of liquid used for filtering is 25 feet, and the filtration is said to occupy from three to four hours, according to the amount of sulphides already contained in the filter. When the latter is full, a small air-pump is connected with it and a current of air passed through it for an hour to dry the mass of sulphides into hard cakes, which are easily handled and removed. The precipitate is then roasted in iron trays, which are placed in cast-iron muffles and heated only from the top, no stirring being necessary, so that the loss from dusting is small. The filter-cloths are burnt with the precipitate, when they have become either clogged with sulphides, or untrustworthy owing to the action of the acid liquors. The precipitate is then melted down with sodium carbonate, borax, nitre and sand, the total loss in handling being very small. The bullion is about 900 to 950 fine in gold, the remainder consisting chiefly of silver, copper, lead and arsenic. The bulk of the precipitate remains at the bottom of the vat. It is allowed to accumulate for a fortnight, and is then filtered and treated as above. The slag resulting from the fusion of the gold is crushed and melted down with litharge, and the reduced lead cupelled (Rothwell).

It has usually been assumed that all the lead, copper, and silver contained in the liquids is precipitated with the gold, and that if much copper is present, the bullion will be very base, and Langguth suggested the removal of the copper from the precipitate by dilute nitric acid. Rothwell, however, has stated [1] that, with careful work in the presence of acid, the gold can be precipitated, forming a bullion from 820 to 960 fine, whilst the copper, arsenic, antimony, etc., remain in solution together with barely a trace of gold. If much base metal is present in the solution, ferrous sulphate is sometimes used.

The amount of chemicals used in the chlorination and precipitation departments at the Golden Reward Mill, Dakota, in 1891 is given below :—[2]

TABLE XXIX.

	JULY, 1891.		AUGUST, 1891.		SEPTEMBER, 1891.	
	Lbs. per Ton of Ore.	Cost in Cents.	Lbs. per Ton of Ore.	Cost in Cents.	Lbs. per Ton of Ore.	Cost in Cents.
Sulphuric acid, . .	26·82	45·6	25·14	44·0	24·77	43·3
Chloride of lime, . .	11·04	35·8	10·17	33·1	10·09	32·8
Crude sulphur, . .	0·37	1·14	0·31	0·9	0·29	1·2
Iron sulphide, . .	0·77	4·54	0·79	4·0	0·84	4·2
Total cost, . .	..	87·1	..	82·05	..	81·5

The total working costs at Cripple Creek in 1900 were given by J. E. Rothwell as $3.53 per ton.[3]

Bromine was used at Rapid City, Dokota, in place of chlorine in 1892, but with poor results, the tailing being rich. The amount of bromine used seems to have been insufficient.

THE VAT-SOLUTION PROCESS.

Practice in Europe.—The process was devised by W. Munktell at Fahlun, Sweden, where it was worked from 1885 to 1888 on the tailing of copper ore. The material was leached in vats with dilute solutions of bleaching powder and hydrochloric acid, which were mixed just before they were run on to the ore. The method was afterwards used on concentrate in Hungary.

Somewhat similar treatment is followed at Bovisa, in N. Italy, where about 25 tons per day of pyritic ore, containing 34 per cent. of sulphur, 10 to 12 per cent. of arsenic, and 0·6 to 0·7 oz. of gold per ton, are roasted dead (the arsenic being recovered) and chlorinated in lead-lined wooden vats holding 10 tons each. Weak solutions of bleaching powder and sulphuric acid are allowed to pass slowly through the ore, and the total time of treatment is three days. From 85 to 87 per cent. of the gold is recovered. [4]

The Permanganate Chlorination Process.—This process differs from the ordinary treatment of ore in vats by chlorine water only in the use of a different solution. M. Etard originally proposed a solution containing

---

[1] Rothwell, *Mineral Industry*, 1896, p. 278.
[2] Rothwell, *Eng. and Mng. J.*, Mar. 25, 1893, p. 269.
[3] Rothwell, *Mineral Industry*, 1900, p. 370.
[4] F. Clerici, *Eng. and Mng. J.*, Aug. 29, 1896, p. 201.

45 lbs. of strong commercial muriatic acid, and 12 ozs. of crystals of permanganate of potash. Professor Black, of Otago University, as an alternative suggested 12 lbs. of common salt, 14 lbs. of sulphuric acid, and 6 or
7 ozs. of permanganate crystals per 100 gallons of water. These solutions
smell slightly of chlorine, but it is claimed that in the absence of gold or other
material attacked by chlorine, practically no free chlorine is produced. Both
solutions dissolve gold at less than one-tenth the rate of a saturated solution
of chlorine in water (see p. 22).

The method was tested by the treatment of 2,000 tons of pyritic ore
at the Bethanga Mill, Victoria, in 1900.[1] It was crushed dry by a ball mill
to 20 mesh and roasted in hand-worked reverberatory furnaces, 40 lbs. of
salt per ton being added a few minutes before withdrawal. Wooden vats,
5 feet deep, holding 17 tons of roasted ore were used for chlorination. The
solution contained 110 lbs. of commercial sulphuric acid, 90 lbs. of common
salt, and 12 lbs. of potassium permanganate (per 1,000 gallons ?). The average
time of treatment was 140 hours, the tailing being tested by panning, and
finally leached with hot water. The ruby colour of the solution was used
to gauge its solvent strength. The average contents of the roasted ore during
one month were gold 2·313 ozs., silver 1·63 ozs. per ton, and the tailing
contained an average of 0·147 oz. gold per ton, or nearly 3 dwts. This corresponds to an extraction of 93·7 per cent. of gold. According to D. D. Rose-
warne,[2] the cost of chlorination, using permanganate, during four months
averaged 11s. 5d. per ton, and using chlorine at the same time the average
cost was only 7s. 4d. per ton. On this exhaustive test the permanganate
process was abandoned.

Chlorination at the Mount Morgan Mine, Queensland.[3]—This was the
largest and most important chlorination mill in the world, but the method
of treatment has now been given up. It consists in subjecting the ore in
open vats to the action of a solution of chlorine in water. It was adopted
in 1888 after chlorination in 1-ton barrels had been used for about a year.

The ore from the upper levels consisted of sinter, kaolin, quartz, and
ironstone, with a small percentage of sulphur. The ore from the lower levels
consists of hard quartz, heavily charged with pyrite. The gold in the latter
is believed to be mainly in the form of telluride, and after roasting, all the
ores contain only finely divided gold. The ores are treated in different
plants by different methods.

At the " West Works " plant, a plan of which is shown in Fig. 143, over
100,000 tons of low-grade ore from the upper levels were treated annually.
The ore is crushed dry in Krupp ball-mills (see above, p. 279), to pass a
20-mesh screen, roasted in revolving cylindrical furnaces and stored in main
hoppers higher than the vats. The roasting is mainly to dehydrate the ore,
so as to improve the speed of leaching, which is otherwise excessively slow.
A small percentage of sulphur is also removed. The vats are 60 feet long,
12 feet 6 inches wide, and 5 feet deep, and hold 100 tons of ore. They are
made of concrete lined with neat cement, which in turn is covered with
pitch and tar applied hot. There are sixteen of these vats. The filter bed

    [1] " The Permanganate Chlorination Process at Bethanga," by E. Harris. *Inst. Mng.
and Met.*, report of meeting on Dec. 15, 1904.
    [2] Rosewarne, *Inst. Mng. and Met.*, report of meeting on Dec. 15, 1904, p. 25.
    [3] See E. W. Nardin, *Proc. Inst. Civil Eng.*, 1900, **142**, 297-307 ; and *Australian Mining
and Metallurgy*, by Donald Clark, pp. 265-294. Figs. 143-146 are reproduced by kind permission of the Institution of Civil Engineers.

is 20 inches thick, consisting of 16 inches of sand and gravel resting on perforated planks supported by wooden bearers. The vat has a fall of 3 inches from one end to the other for draining off the solution. The depth of the bed of ore is little more than 3 feet. The ore is tipped in from trucks and roughly levelled.

The chlorine is generated in ordinary flagstone stills from sulphuric acid, manganese dioxide and common salt. The acid is made on the spot from imported sulphur. The manganese and salt are crushed and thoroughly mixed before charging into the still. The chlorine gas is passed into scrubbing towers (see Fig. 144), each 2 feet 3 inches in diameter and 20 feet high, filled with glass bottles and old assay crucibles, always kept wet by trickling water. The solution of chlorine contains about 600 grains of chlorine to the cubic foot, or 0·14 per cent. of chlorine. The weight of chlorine used is about 2½ lbs. per ton of ore, which corresponds to about 30 cubic feet of solution

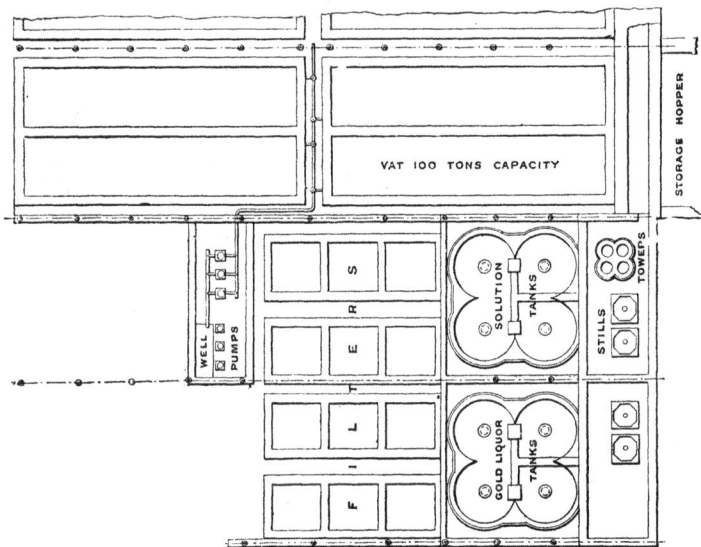

Scale, 1 inch = 40 feet.

Fig. 143.—Plan of Chlorination Plant, Mount Morgan.

per ton. The solution is stored in solution tanks (see Fig. 145), closed chambers with concrete domes, whence it flows through a 6-inch earthenware main to the vats. There are four stills, four towers, and four solution tanks, the latter having a capacity of about 950 cubic feet each.

The solution is allowed to run on the charge, from 6-inch earthenware mains (see Fig. 146), until the vat is full, when the valve below the false bottom is opened and the vacuum of about 5 lbs. per square inch in the pump main draws the liquor through. Fresh solution is run on to keep the ore covered, and this is continued until the gold is dissolved, which appears to take about thirty-six hours. During the first few hours all the chlorine is absorbed by the ore which becomes hot, doubtless from the attack on ferrous and basic salts as well as sulphides left in the ore after roasting. After about fifteen hours, the liquor issuing from the vat is found to contain gold, which gradually increases in amount.

When the liquor gives a good black precipitate with ferrous sulphate, and smells strongly of chlorine gas, it is assumed that all the gold is dissolved, and the vat is then said to be " chlorinated," and wash-water is turned on and passed through the ore until no gold can be detected in it by means of ferrous sulphate. The ore is then discharged by means of shovelling into trucks, though it is intended to replace this by a steam dredge. The solution of gold is passed through charcoal filters as described below.

At the " Top Works," the ore treated is hard compact quartz heavily charged with iron pyrite ; it contains an average of 4 ozs. gold per ton and 11 per cent. sulphur, which after roasting is reduced to 0·15 per cent. The ore is roasted in a Richards' shaft furnace (see p. 294) at a cost of about 5s. per ton. The amount of solution required is larger than that at

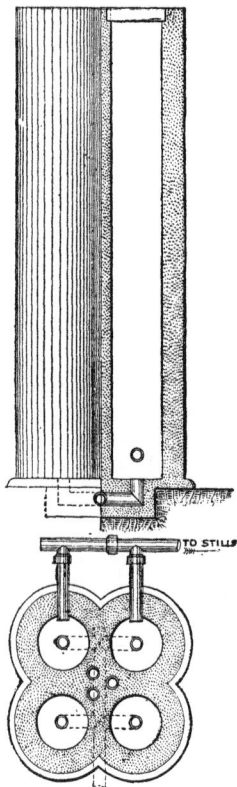

Scale, $\frac{1}{16}$ inch = 1 foot.

Fig. 145.—Chlorine Solution Tanks, Mount Morgan.

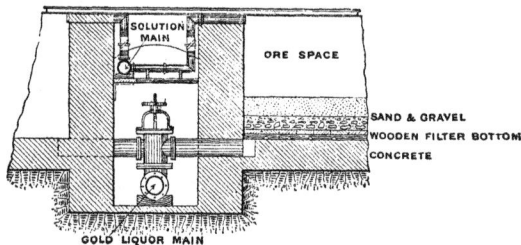

Scale, $\frac{1}{8}$ inch = 1 foot.

Fig. 144.—Scrubbing Towers, Mount Morgan.

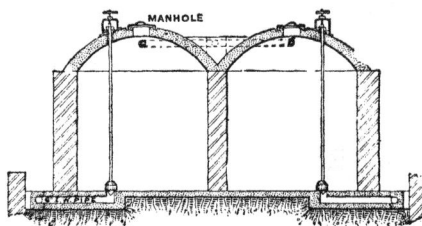

Scale, $\frac{1}{8}$ inch = 1 foot.

Fig. 146.—Solution Mains and Filter Beds, Mount Morgan.

the West Works, but the time of treatment is about the same, as the ore is more porous and the washing rapid.

Precipitation.—The gold is precipitated by charcoal. In 1801, Henry referred to the reducing action of charcoal on chloride of gold in solution, and observed that the gold will be precipitated on the charcoal, if the solution is either exposed to the direct light of the sun or heated to 212° F.[1] In 1869,

[1] Wm. Henry, *An Epitome of Chemistry* (London, 1801), p. 95.

in Percy's laboratory, some sticks of wood-charcoal were immersed in water, and 32·50 grains of gold in the form of chloride were added on August 7. Eighty-five more grains of gold, in the form of chloride, were added on November 3, 1869, and the bottle was left to stand. This bottle is in the Percy collection, at South Kensington, and the surface of the charcoal is now coated over with metallic gold, which shows on its surface the fibres and vessels of the stem.

Vegetable charcoal was adopted as a precipitant at the Mount Morgan Mine, Queensland, in 1887. The method adopted there is as follows :— The solution is heated, the free chlorine being thus expelled ; the liquid is then made to run slowly through large shallow tanks, each 10 feet by 11 feet and 4 feet 6 inches deep (see Fig. 147).[1] They are built of brick on a concrete foundation, and lined with Portland cement and tar. The layer of charcoal in them is 2 feet deep, rests on a filter-bottom, and is covered with thick per-forated sheets of lead. The liquor flows downwards through three charcoal filters in succession, 660 cubic feet of charcoal being sufficient for the treat-ment of the liquor from about 100 tons per day of 11 dwts. ore. In preparing the charcoal for the filters,[2] it is crushed, and all that passes a 20-mesh and remains on a 30-mesh screen is called " coarse," whilst that which passes the 30-mesh and remains on a 40-mesh screen is called " fine." All that passes 40-mesh is thrown away. Two cubic feet of fine are used to 1 cubic

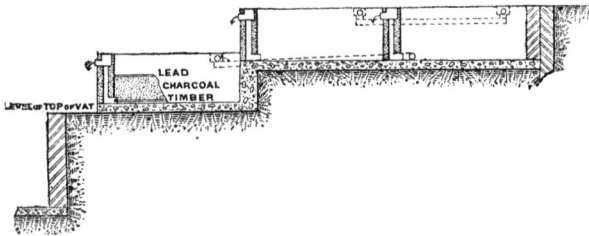

Scale, ₁⁄₁₀ inch = 1 foot.

Fig. 147.—Precipitation Vats, Mount Morgan.

foot of coarse, the coarse being at the top and bottom. The charcoal is well rammed in with the foot, especially round the walls. For seven months the liquor running away from these filters assayed in one mill 0·76 grain gold per ton, and in the other 1·89 grains per ton. After passing the charcoal, all liquors are run through a concrete tank with a bed of sawdust 1 foot thick and more gold caught in this way. When the charcoal is coated suffi-ciently with gold, it is burnt in small furnaces furnished with dust chambers. The ashes were formerly amalgamated, and the rich slimes sold to the smelters. Afterwards the ashes were fluxed in crucibles and eventually in a small reverberatory furnace, the fluxes used being soda ash and borax, which were mixed with the ashes and then moistened with water. D. Clark states that the charcoal ash contains from 20 to 40 per cent. of gold.[3] The slag con-tained from 20 to 30 ozs. per ton. Animal charcoal cannot be used owing to the difficulty of burning it afterwards.

---

[1] Nardin, *Proc. Inst. Civil Eng.*, 1900, **142**, 302. Fig. 147 is reproduced by kind per-mission of the Institution of Civil Engineers.
[2] *Eng. and Mng. J.*, Dec. 17, 1898, p. 724.
[3] Clark, *Australian Mining and Metallurgy*, p. 292.

The exact action of the charcoal has not been fully demonstrated. It acts slowly on cold solutions, and its action is not rapid even at boiling point. It is under the disadvantage that it does not destroy free chlorine, which must therefore be expelled by boiling or by passing a current of air through the liquid before the precipitation of the gold is begun. W. M. Davis states that 240 parts of charcoal are required for the precipitation of 19¼ parts of gold. The prevailing opinion is that the hydrogen and hydrocarbons remaining in the charcoal are the active agents in the precipitation, hydrochloric acid and free gold being formed.

The following are the chief items in connection with the chlorination at Mount Morgan during May, 1898 :—[1]

TABLE XXX.

	West Works.	Top Works.
Tons chlorinated, ...	9,852	1,534
Average assay of ore, ...	11 dwts. 13 grs.	4 ozs. 4 dwts. 10 grs.
„ „ residue, . . .	22 grs.	4 dwts. 6 grs.
„ extraction, ...	92·06 per cent.	95 per cent.
„ time under " solution," . .	33 hours.	64 hours.
„ „ of washing, . . .	35 „	6 „
„ weight of chlorine per ton of ore,	2·51 lbs.	11·47 lbs.
Cost per lb. of chlorine, . . .	5·85d.	5·53d.
Weight of chlorine per cubic foot of " solution,"	620 grains.	599 grains.
Total cost per ton of ore for chlorinating and precipitating, ...	4s. 4·58d.	..

[1] Nardin, *op. cit.*, pp. 304, 305.

CHAPTER XV.

THE CYANIDE PROCESS. CHEMICAL REACTIONS.

Introduction.—The cyanide process will always be indissolubly connected with the names of its discoverers, J. S. MacArthur and the Forrests. The solvent action of potassium cyanide on metallic gold and silver has long been known, as is pointed out below, p. 323, but it was believed that the use of an electric current in conjunction was needed to quicken the action, which was otherwise too slow to be of any practical value. Until the surprising results of the MacArthur-Forrest process, as it was called for some years, were made public, metallurgists were all disposed to condemn the use of cyanide as a solvent for gold in ores as being chimerical, on account of the instability of potassium cyanide and the slowness of its action on gold, combined with its high cost and poisonous character.

The first attempt at the direct extraction of gold by the use of cyanide was made by J. H. Rae, who took out a patent in the United States in 1867 for a process dependent on the removal of gold and silver from their ores by the combined action of a " current of electricity and of suitable solvents or chemicals," such, for instance, as cyanide of potassium, the gold being simultaneously precipitated on copper plates by the electric current.

In 1885, J. W. Simpson, of New Jersey, proposed to treat ores containing gold, silver, and copper with a solution containing 3·0 per cent. of potassium cyanide and 0·19 per cent. of ammonium carbonate. Copper was to be dissolved at the same time as the gold ; if silver was present also, an addition of common salt was made. The inventor appears to have believed that, by using carbonate of ammonia, the necessity was obviated of employing an electric current, in conjunction with cyanide of potassium, in order to dissolve the gold. The precipitation of gold was effected by " a piece or plate of zinc." The process was not tried on a large scale.

In 1886, however, a series of experiments on wet processes of treating gold ores was begun by J. S. MacArthur and R. W. and Wm. Forrest in Glasgow,[1] and it was entirely owing to their energy and skill that cyanide of potassium was successfully applied in practice to the treatment of gold ores. Their process consists essentially in attacking gold and silver ores with dilute solutions containing less than 1 per cent. of KCy, caustic soda or lime being added to ores rendered acid by the oxidation of pyrite, and then in precipitating the precious metals by zinc shavings. This process was introduced at Karangahake, New Zealand, in 1889, and near Johannesburg, in the Transvaal, in 1890, and has now passed into use in all parts of the world. Its success is complete on many ores, and its extension has become very great, partly at the expense of the chlorination process, and partly in treating the tailing of ore which has been crushed and amalgamated. Such tailing was

---

[1] MacArthur and R. W. and Wm. Forrest, *English Patent*, 14,174, Oct. 19, 1887; 10,223 of 1888.

formerly run to waste after concentration. In a large number of cases, ores which were formerly regarded as too poor to treat at a profit by any known process, are treated by cyanide with or without previous amalgamation. The chief advantage of the cyanide process over the chlorination process is that roasting is not necessary, even if sulphides are present; this is a most important point in the treatment of low-grade ores, especially in places where fuel and labour are costly. Moreover, silver as well as gold is extracted by cyanide solutions, and the process has been extensively applied to the treatment of silver ores, superseding the patio, pan-amalgamation and thiosulphate processes. If tellurides are present, ordinary cyanide is inefficient, and the ores must either be roasted or treated with bromo-cyanogen.

Messrs. MacArthur and Forrest, in the course of their investigation on wet methods, became dissatisfied with chlorine as a solvent, owing to its energetic and preferential action on sulphides of base metals and other bodies, and its inapplicability to ores containing silver. They desired to find a solvent which would exercise a selective action in favour of the precious metals, instead of other substances. With this object in view, they experimented with a number of solvents (such as ferric bromide, ferric chloride, etc.), and finally decided that potassium cyanide possessed advantages over all other substances. They showed that it was essential to use very dilute solutions, which do not energetically attack pyrite and many other base minerals, but readily dissolve gold and silver. Certain compounds, especially those of copper, destroy cyanide and interfere with extraction. These compounds are called " cyanicides."

" The potassium cyanide supplied to the gold mines in the early days of the gold-extracting process was a black mass containing some 75 per cent. of KCN and all the iron and carbon liberated by the decomposition of the ferrocyanide. This was replaced by the pure white crystalline product of the Erlenmeyer process, which contained about 2 molecules of KCN to 1 molecule of NaCN, the cyanogen content being equivalent to 98 to 100 per cent. of KCN. Shortly after the introduction of the Castner process in 1900, solid cakes of pure white crystalline sodium cyanide containing 97 to 98 per cent. of NaCN (equivalent to 129 to 130 per cent. of KCN), became an article of commerce, and this is at present the form in which the greater part of the cyanide employed in gold-extracting is used."[1] Sodium cyanide has been objected to in some cyanide mills.[2]

The process was originally applied to the residue after amalgamation of gold ore, but is now in many cases used for the direct treatment of ore, especially in America, as described in Chap. XVII. Much attention was devoted to the study of the chemical side of cyaniding in the early days, but of late years the sum of knowledge of the chemistry involved has increased less rapidly. " The physical and mechanical side of cyanidation has been in a state of continuous development since the first introduction of the process. . . . The field of cyanidation has been and is constantly widening through its encroachment upon amalgamation, concentration and smelting."[3]

It is desirable to understand the chemical foundations of the cyanide process before considering the details of practice, and accordingly a study of the chief chemical reactions involved is appended here.

[1] Ewan, *Thorpe's Dict. of Applied Chemistry*, 1912, vol. ii., p. 182.
[2] Magenau, *Eng. and Mng. J.*, Aug. 25, 1906, p. 363.
[3] MacFarren, *Cyanide Practice*, p. 6.

Action of Potassium Cyanide [1] on Gold and Silver.—Dr. Wright, of Birmingham, discovered in 1840 that metallic gold is soluble in potassium cyanide, when a current of electricity is passing, and Elkington's patent, taken out in the same year, was partly based on this discovery. Bagration, in 1843,[2] studied the action of cyanides on plates of gold in the absence of a current of electricity, and announced that they were slowly dissolved. Faraday, in 1857,[3] pointed out that gold-leaf is dissolved by a dilute solution of the salt, and also showed that if the gold floats on the surface of the liquid, so that the side of the leaf is in contact with the air, while the other is bathed by the solvent, the action is much more rapid than if the metal is completely submerged. Elsner had previously, in 1846, furnished some evidence [4] that the presence of oxygen is required for the solution of the gold.

On evaporating the solution, colourless octahedral crystals of auropotassium cyanide, $KAuCy_2$, are formed, which may be viewed as being a double cyanide, produced as follows :—

$$4Au + 8KCy + O_2 + 2H_2O = 4KAuCy_2 + 4KOH.$$

The equation for the solution of silver is similar.

W. A. Dixon found, in 1877,[5] that although cyanide by itself was slow in dissolving gold, its action was hastened by the addition of alkaline oxidising agents. He also mentions calcium hypochlorite, potassium ferrocyanide and manganese dioxide as hastening this action.

According to the equation given above, 130 parts by weight of KCy in the presence of 8 parts of oxygen suffice for the solution of 197 parts of gold. This has been proved by J. S. Maclaurin [6] to be the case in all carefully conducted experiments. The amount of oxygen dissolved in liquids not specially prepared, to say nothing of that contained in a porous mass of pulverised ore, is consequently enough for the solution of great quantities of gold.

According to G. Bodländer,[7] the chemical action in the dissolution of the gold is as follows :—

$$2Au + 4KCy + 2H_2O + O_2 = 2KAuCy_2 + 2KOH + H_2O_2$$

and subsequently

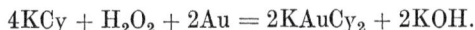

$$4KCy + H_2O_2 + 2Au = 2KAuCy_2 + 2KOH.$$

Bettel suggests the following in place of this :—[8]

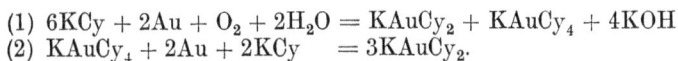

(1) $6KCy + 2Au + O_2 + 2H_2O = KAuCy_2 + KAuCy_4 + 4KOH$
(2) $KAuCy_4 + 2Au + 2KCy \quad = 3KAuCy_2.$

Evidence has been adduced that a substance is formed reacting like H_2O_2, but it does not follow that the actions represented by the equations

---

[1] In general, when potassium cyanide is named, it may be taken to include sodium cyanide and mixtures of the two salts. Also in the equations given in this chapter, K and Na are interchangeable, except from the point of view of ionic dissociation.

[2] Bagration, *Bull. Acad. Sci. St. Petersbourg*, 1843, **2**, 136.

[3] Faraday, *Proc. Roy. Inst.*, 1857, **2**, 308.

[4] Elsner, *J. prakt. Chem.*, 1846, **37**, 441-446.

[5] Dixon, *Proc. Roy. Soc. of N.S.W.*, Aug. 1, 1877; *Chem. News*, Dec. 20, 1878, p. 293.

[6] Maclaurin, *J. Chem. Soc.*, 1893, **63**, 724.

[7] Bodländer, *Zeitsch. angew. Chem.*, 1896, p. 583; see also *J. Chem. Met. and Mng. Soc. of S. Africa*, 1903, **4**, 273.

[8] Bettel, *South African Mining Journal*, May 8, 1897.

given by Bodländer and Bettel are not limited to an insignificant part of the whole mass.

It is clear that oxygen or some other substance acting similarly is necessary to assist cyanide of potassium in dissolving gold. " The decomposition of the potassium cyanide is facilitated by the affinity of the potassium for the free oxygen present, so that nascent cyanogen is liberated to combine with the metallic gold " (Caldecott). It has, however, been frequently pointed out that in the interior of a mass of ore undergoing treatment the conditions are not favourable for the maintenance of a sufficient quantity of oxygen in a free state. Certain constituents of the ore tend to unite with it, and further absorption of free oxygen from the air is extremely slow. Hence the time required for the treatment of a charge is many hours, or even days, although under favourable conditions the gold could be dissolved in a few minutes, or at most in two or three hours. To supply the oxygen, various oxidising substances have been tried, such as the passage of a current of air through the solution, and the addition of various materials. The charges are also sometimes drained and turned over or transferred to other vats, in order to aerate the damp ore. In cyanide solution oxygen is dissolved to the extent of about one-fifth of an ounce per ton.[1]

Bettel[2] found that gold dissolves in the absence of oxygen if the crushed ore contains basic ferric sulphate, by which potassium ferricyanide is formed, the reactions being expressed thus—

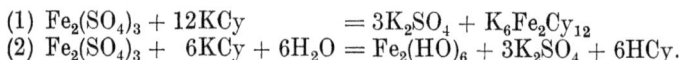

(1) $Fe_2(SO_4)_3 + 12KCy \qquad = 3K_2SO_4 + K_6Fe_2Cy_{12}$
(2) $Fe_2(SO_4)_3 + 6KCy + 6H_2O = Fe_2(HO)_6 + 3K_2SO_4 + 6HCy.$

Bettel and Marais also found in 1894 that gold leaf would not dissolve in a solution of potassium cyanide from which the air had been expelled by the passage of a current of hydrogen, and that the addition, under these circumstances, of either potassium dichromate, chromate, chlorate, perchlorate, nitrate, or nitrite or of ferric hydrate or bleaching powder, did not enable the gold to dissolve. The addition of pyrolusite gave a doubtful result, and lead dioxide caused very slow dissolution of the gold. On the other hand, gold dissolved slowly if chlorine, iodine dissolved in potassium iodide, or ferric chloride were added ; rapidly, if bromine were added ; and decidedly, if potassium ferricyanide or permanganate, sodium dioxide, hydrogen peroxide or barium dioxide were added. Michailenko and Meshtscherjakoff[3] confirmed most of these conclusions, and found that sodium dioxide exercised its maximum influence at a concentration of 0·02 of an equivalent (or about 0·2 per cent.). Morris Green[4] finds that potassium permanganate as such does not aid dissolution. It is clear, therefore, that certain oxidisers are ineffective, and in practice, when air is not used, potassium permanganate, sodium dioxide, bromine (Sulman-Teed process), and umber (Adair process)[5] have been mainly employed to assist in the dissolution of the gold. In general, artificial oxidation is not resorted to unless there is some reducing agent present in the ore or the water which absorbs the oxygen. The oxidising substances generally act as " cyanicides," destroying some of the solvent.

[1] A. F. Cross, *J. Chem. Met. and Mng. Soc. of S. Africa*, 1898, **2,** 402.
[2] Bettel, *S. African Mng. J.*, *loc. cit.*
[3] Michailenko and Meshtscherjakoff, *J. Russ. Phys. Chem. Soc.*, **44,** 567 ; *Mng. and Sci. Press*, Oct. 19, 1912, p. 500.
[4] Green, *J. Chem. Met. and Mng. Soc. of S. Africa*, 1913, **13,** 355.
[5] Adair, *J. Chem. Met. and Mng. Soc. of S. Africa*, 1908, **8,** 331.

" The effect of adding to a pyritic ore the necessary amount of powerful oxidisers is really the formation of active absorbents of oxygen " (Caldecott), so that they defeat their own object. It was formerly supposed that any oxygen coming in contact with the solution would be at once taken up by the cyanide, forming cyanates, but it is now well known that oxygen and cyanide remain side by side without rapid union between them taking place. Cyanates are without action on metallic gold.

When bromine is used, it is in the form of cyanogen bromide (bromocyanogen, bromocyanide), as proposed by H. L. Sulman,[1] and used in the Diehl process, see p. 394. When this substance is added to cyanide solutions, the presence of oxygen is not required, and the rate of dissolution is markedly increased. The cyanogen bromide is set free as follows :—[2]

$$2KBr + KBrO_3 + 3KCy + 3H_2SO_4 = 3BrCy + 3K_2SO_4 + 3H_2O.$$

A mixture of the bromide and bromate salts in the proper proportions is manufactured and sold. When required for use the bromocyanogen is made by agitation of the mixture with cyanide and sulphuric acid in a closed vessel. Its action in the treatment vat is supposed to be as follows :—

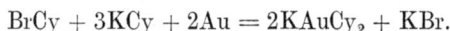

$$BrCy + 3KCy + 2Au = 2KAuCy_2 + KBr.$$

Cyanogen chloride and iodide behave similarly, but are very unstable. For the effect on tellurides, see p. 394.

The addition of a small quantity of potassium mercuric cyanide, $HgCy_2.2KCy$ to ordinary cyanide solutions to accelerate their action has also been proposed by J. S. MacArthur and by N. S. Keith.[3]

From the fact that mercury is electro-negative to gold in cyanide solutions, Skey in 1876[4] concluded that metallic gold in contact with a solution of mercury cyanide would rapidly dissolve and mercury be reduced. He found this to be the case alike with gold and silver, which dissolved with almost equal readiness. The mercury is precipitated on the surface of the particles of gold and forms amalgam, the equation being probably—

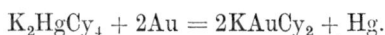

$$K_2HgCy_4 + 2Au = 2KAuCy_2 + Hg.$$

The accelerating action in practice is, however, inappreciable. Mercury dissolves in excess of cyanide, but less readily than gold ; thus, according to Julian and Smart[5]—

$$Hg + 4KCy + 2H_2O = HgCy_2.2KCy + 2KOH + H_2.$$

It has been suggested by Christy,[6] and Julian and Smart,[7] that the dissolution of gold and other metals in cyanide depends primarily on the electromotive force exerted. H. F. Julian[8] attached great weight to the formation of local voltaic circuits.

---

[1] Sulman, *Trans. Inst. Mng. and Met.*, 1895, **3,** 202.
[2] E. W. Nardin, *Trans. Australasian Inst. Mng. Eng.*, 1907, **12** ; *Mng. and Sci. Press,* Oct. 24, 1908, p. 562.
[3] Keith, *Engineering*, 1895, **59,** 379.
[4] Skey, *Trans. New Zealand Inst.*, 1876, **8,** 334.
[5] Julian and Smart, *Cyaniding Gold and Silver Ores* (1904), p. 107.
[6] Christy, "The E.M.F. of Metals in Cyanide Solutions," *Trans. Amer. Inst. Mng. Eng.*, 1900, **30,** 864.
[7] Julian and Smart, *op. cit.*, pp. 62-94.
[8] Julian, Report Brit. Assoc., 1905, p. 369.

Christy contends [1] that the rate of dissolution of gold in cyanide depends on two factors, the electromotive force of gold in cyanide solutions, and the number of hydroxyl ions in the solution. Michailenko,[2] however, denies the latter contention, and states that OH ions do not accelerate the velocity of solution, but that an excess of them has a retarding influence. The E.M.F. of gold continually rises as the solutions become stronger, but is never large enough to enable gold to dissolve in the absence of oxygen. In pure water the E.M.F. of gold is -- 0·72 volt; in solutions containing about 0·043 per cent. KCy the E.M.F. becomes zero, and it then rises somewhat rapidly to + 0·42 volt in a solution of 6·5 per cent. KCy, and thereafter more slowly to 0·468 volt in a solution containing 41·7 per cent. KCy. It follows from Nernst's theory of solution pressure that solutions of gold in cyanide free from dissolved oxygen would remain undecomposed if they contained more than 0·043 per cent. KCy, but that gold would tend to precipitate spontaneously in the absence of oxygen in solutions containing less than 0·043 per cent. KCy. This requires experimental proof. The author, in conjunction with L. L. Fermor, made a series of experiments in 1903, in the attempt to decide the matter, but obtained no evidence of spontaneous precipitation in cyanide solutions of any strength. On the contrary, when metallic gold was present a little of it always dissolved, although the amount was excessively small in the absence of oxygen, amounting to about one-tenth of that observed by Maclaurin (see below). It appeared probable, therefore, that in these experiments the exclusion of free oxygen was not quite perfect, in spite of the addition of pyrogallol and other deoxidisers, among other precautions.

The presence of dissolved oxygen completely alters these results, and by direct experiments Christy found that solutions containing no more than 0·00065 per cent. KCy did not dissolve gold, and that for all practical purposes the cyanide of potassium solution ceases to act when its strength falls below 0·001 per cent. Above that strength, however, the solubility of gold increases rapidly when free oxygen is present.

J. S. Maclaurin found [3] that the rate of dissolution of pure gold, in the form of plates, in potassium cyanide solution passes through a maximum when proceeding from dilute to concentrated solutions. The maximum is reached when the solution contains 0·25 per cent. of KCy. The solubility of gold is very slight in solutions containing less than 0·005 per cent., but increases rapidly as the strength rises to 0·01 per cent., when the rate of dissolution is ten times as great as in the 0·005 per cent. solution, and about half as great as that in the 0·25 per cent. The rate increases slowly as the strength rises to 0·25 per cent., and thereafter decreases much more slowly, until in 15 per cent. solutions the rate of dissolution is about equal to that in 0·01 per cent. solutions. Higher strengths show a gradual diminution in the rate of dissolution up to saturation point. Silver is also dissolved at a maximum rate in solutions containing 0·25 per cent. of cyanide, and the changes in the rate are similar to those noted above in the case of gold, the rates for silver being always about two-thirds of the corresponding rates for gold, or, roughly, in the same ratio as the atomic weights of the two metals. In both cases there is hardly any change in the rate of solubility as the strength rises from 0·1 to 0·25 per cent. It is remarkable that the solubility of oxygen in cyanide solutions continually decreases as the

---

[1] Christy, *loc. cit.*   [2] Michailenko and Meshtscherjakoff, *loc. cit.*
[3] Maclaurin, *J. Chem. Soc.*, 1893, **63**, 724; 1895, **67**, 199.

concentration increases, and it is to this fact that Maclaurin is disposed to attribute the variations in the rate of dissolution of the gold.

Bagration [1] states that the solution of metallic gold in cyanide of potassium is facilitated by a rising temperature. Little attention was paid to this point in the early days of the cyanide process, but in 1899 Loewy stated [2] that experiments at the May Consolidated Gold Mine in 1893 yielded 10 to 14 per cent. higher extraction when heat was applied. The ore treated was pyritic containing 50 to 60 per cent. of its gold in the free state, which was more affected by the heating of the solutions than the gold in the pyrite. Loewy considered that solutions should be heated to above 35° C. G. A. Darling in the discussion [3] pointed out that in the six "cold" months of the year in the Transvaal, from April to September, the average extraction during seven years, between 1891 and 1898, at the Robinson Gold Mine was 69·5 per cent., and in the "hot months," October to March, the extraction was 72·2 per cent. Von Gernet at the same time stated [4] that at the Worcester Cyanide Works during September, 1898, exhaust steam was passed through pipes to heat the solution, and that the residues had contained only 8 grains per ton, although no comparative figures were available. On the other hand, it has been pointed out that in the "hot months" in South Africa there is more rainfall and the organic matter in the water is reduced, and the extraction might be improved from that cause, since the presence of organic matter reduces the free oxygen in the solutions. For present practice in the Transvaal, see p. 407.

An improvement in extraction of about \$1 per ton on Tonopah ore, due to raising the temperature of the solution from 60° F. to 90° F., at a cost of 16 cents, is recorded by A. H. Jones,[5] and further details are given by von Bernewitz.[6] The increase in the consumption of cyanide if any is not given. In practice, the solutions are rarely heated. It is generally believed that an increase of temperature is accompanied by an increase in the consumption of cyanide, but exact data are lacking.

A. F. Crosse has shown that gold is also dissolved in HCy (or acidulated KCy solution) in the presence of oxygen. In this case auricyanides are formed instead of the usual aurocyanide.

Dissolution of Base Metals.—The dissolution of mercury has already been mentioned on p. 325. Iron dissolves slowly in cyanide solutions, but is much less soluble than gold or silver.[7] The equation is probably as follows :—

$$Fe + 6KCy + H_2O + O = K_4FeCy_6 + 2KOH.$$

Dr. Gore has shown that the rate of solution of gold is increased by 50 per cent. if it is placed in contact with iron, and that it is then five times more soluble than the iron.

Zinc and copper dissolve readily in cyanide solution without the presence of free oxygen but with evolution of hydrogen, thus—

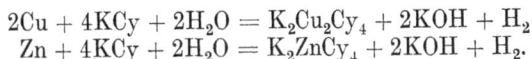

$$2Cu + 4KCy + 2H_2O = K_2Cu_2Cy_4 + 2KOH + H_2$$
$$Zn + 4KCy + 2H_2O = K_2ZnCy_4 + 2KOH + H_2.$$

[1] Bagration, *J. prakt. Chem.*, 1844, **46**, 367.
[2] Loewy, *J. Chem. Met. and Mng. Soc. of S. Africa*, 1898, **2**, 390.
[3] Darling, *J. Chem. Met. and Mng. Soc. of S. Africa*, 1898, **2**, 458.
[4] Von Gernet, *loc. cit.*
[5] Jones, *Mng. and Sci. Press*, Jan. 27, 1912, p. 176.
[6] Von Bernewitz, *ibid.*, Dec. 28, 1912, p. 828.
[7] Julian and Smart, *Cyaniding Gold and Silver Ores*, p. 103.

Lead and tellurium dissolve very slowly, and aluminium apparently not at all, but the latter is readily soluble in alkali; arsenic (according to Julian), platinum, and carbon do not dissolve in cyanide. Tellurium dissolves in alkali in presence of oxygen.

Owing to the use of the cyanide process, and owing also to the practice (now generally condemned) of using potassium cyanide to clean the amalgamated plates, there is some danger of gold being dissolved in the stamp mill and lost. Care is taken (when water is used over again in stamp mills) that the water is not contaminated by mixture with waste cyanide solutions. Von Gernet found that the water in the dams on the Witwatersrand contained gold in amounts varying from traces up to 12 grains per ton.[1] Crushing in cyanide solution is discussed subsequently.

Action of Cyanide on Tellurides of Gold.—Cyanide solutions act very slowly on tellurides of gold, but dissolve both elements, although it is possible that tellurium is attacked only by the free alkali in the solution. The following is a description of an experiment made by the author on fused gold telluride, $AuTe_2$:—Gold and tellurium were fused together in the correct proportions, and the result, a blackish-grey substance, found to contain 45 per cent. of gold. It was crushed through a 160-mesh sieve and afterwards separated into two parts, coarse and fine, by elutriation. Each sample was treated on an asbestos filter with a solution maintained at a strength of 0·4 per cent. KCy, continually drawn through the filter and aerated before it was returned.

The results were as follows :—

A. *Fine Sample.*—Time of treatment, 327 hours.

	Before Treatment.	After Treatment.	Loss.
Au, Te,	Grammes. 0·888 1·086	Grammes. 0·008 0·552	Grammes. 0·880 or 99·1 per cent. 0·534 or 49·2 per cent.
	1·974	0·560	1·414

B. *Coarse Sample.*—Time of treatment, 1,120 hours.

	Before Treatment.	After Treatment.	Loss.
Au, Te,	Grammes. 1·129 1·379	Grammes. 0·004 0·457	Grammes. 1·125 or 99·7 per cent. 0·922 or 66·8 per cent.
	2·508	0·461	2·047

It is evident that the gold and tellurium in a fused mixture are both soluble, and native tellurides probably behave similarly, some specimens at least dissolving more readily than fused mixtures.

---

[1] Von Gernet, *J. Chem. Met. and Mng. Soc. of S. Africa*, Jan. 1899, p. 3.

" It is evident that there is a wide difference in the solubility of the gold in different telluride minerals, and even in the same minerals from different localities. Thus in Western Australia, where the gold is in a sulpho-telluride, the ore can be dealt with by fine grinding and long treatment, whereas the same treatment applied to certain Mexican and United States tellurides extracts little or no gold." [1]

In a cyanide solution from the Goldfield Consolidated Mill, von Schultz and Low found 0·00008 per cent. of tellurium and 0·0006 per cent. of gold.[2]

The rate of dissolution is greatly increased by the addition of cyanogen bromide, by which it is made possible to treat telluride ores with cyanide without previous roasting (see *Diehl process*, p. 394).

Action of Cyanide on Compounds of Silver.—A number of experiments made by Louis Janin, Jun.,[3] on various salts of silver point to the following conclusions :—Silver chloride is readily soluble in cyanide, and the arsenate is also rapidly dissolved. Silver sulphide and antimonide are less easily acted on, but are not so refractory as metallic (cement) silver. The presence of copper salts appears to exercise a detrimental action on the solubility of silver sulphide.

Silver sulphide is more slowly dissolved than metallic gold, and strong solutions, containing up to 1 per cent. KCy or more are required. Caldecott has shown [4] that its dissolution by the primary reaction

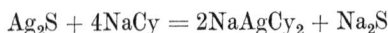

$$Ag_2S + 4NaCy = 2NaAgCy_2 + Na_2S$$

is stopped by the sulphide of sodium, which reprecipitates silver if present in excess and soon establishes an equilibrium. The soluble sulphides may be removed by secondary reactions, the principal one being

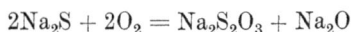

$$2Na_2S + 2O_2 = Na_2S_2O_3 + Na_2O$$

followed by

$$Na_2S_2O_3 + Na_2O + 2O_2 = 2Na_2SO_4.$$

The other secondary reactions also involve the presence of oxygen, so that aeration is as important in the treatment of silver ores as it is in that of gold ores. In the treatment either of gold or silver ores soluble sulphides may be removed or rendered harmless by the addition of lead acetate, lead nitrate, litharge, or mercury salts.

The arsenical and antimonial sulphides of silver (ruby silver, stephanite, etc.), are not readily dissolved by cyanide, and silver contained in native galena, tetrahedrite and zinc blende is not amenable to treatment.

Action of Potassium Cyanide on Metallic Salts and Minerals Occurring in Ores.—The ordinary gangue of most ores (viz., silica and silicates of the alkalies and alkaline earths) exercises no direct influence on the cyanide solution. The carbonates of the alkaline earths are also probably without influence. The decomposing effects of sulphides of the heavy metals vary with the physical state of the sulphides ; hard undecomposed sulphides are very slowly attacked, but soft partly decomposed sulphides are more soluble, and in many cases act as cyanicides. Arsenical ores behave similarly.

---

[1] Julian and Smart, *Cyaniding Gold and Silver Ores*, p. 107.
[2] Von Schultz and Low, *Cyanide Practice*, 1910 to 1913, p. 566.
[3] Janin, Junr., *Eng. and Mng. J.*, Dec. 29, 1888.
[4] Caldecott, *J. Chem. Met. and Mng. Soc. of S. Africa*, 1908, **8**, 266.

Messrs. MacArthur and Forrest found that dilute cyanide solutions exercise a "selective action" in dissolving gold and silver in whatever form they may be present, in preference to sulphides or other salts of the base metals. There are exceptions to this rule, some of which are noted in the sequel. "For instance, cyanide of potassium solution has a strong tendency to dissolve precipitated sulphide of zinc, but its action on the natural sulphide of zinc, blende, is almost *nil*. The same holds for compounds of iron, and thus we prove selective action by the average result of a series of experiments on ores. Let us suppose a pyritic ore containing about 7 per cent. of iron and 8 per cent. of sulphur with about 1 oz. of gold to the ton. After grinding, this ore is treated with a solution containing about 1 per cent. of cyanide of potassium. The most of the gold will be dissolved and the rest of the ore left practically untouched. It is obvious that the amount of cyanogen contained in the solution is insufficient to combine with the iron present in the ore, yet, notwithstanding the much greater mass of iron sulphide present and open to attack, it is the gold that is selected for action by the cyanide solution. Taking the average result of our work we find that a higher percentage of gold than of silver is extracted, which justifies us in concluding that the selective action is greater on the former than on the latter. One of the ores on which our early investigations were done was composed as under :—

Copper,	. . . . .	0·15 per cent.
Arsenic,	. . . .	15·09 ,,
Antimony,	. . . .	Traces.
Sulphur,	. . . .	14·65 per cent.
Iron,	. . . . .	18·77 ,,
Silica,	. . . . .	36·20 ,,
Lead,	. . . . .	2·66 ,,
Zinc,	. . . . .	4·00 ,,
Alumina,	. . . .	4·20 ,,
Gold, per ton,	. . . .	2 ozs. 2 dwts. 16 grs.
Silver, per ton,	. . . .	2 ozs. 13 dwts. 8 grs.

"In this ore we had an extraction of gold 85 per cent., silver 50 per cent., for a consumption of cyanide of about 0·45 per cent., and investigations showed that the action was directed in the order, gold, silver, iron, zinc, copper. For the amount of cyanide consumed it is obvious that the amount of base metals dissolved must have been very slight.

"The consumption of cyanide on fresh concentrates varies naturally with the composition of the concentrates. In many cases it is less than 0·2 per cent. of their weight. When the concentrates contain marcasite there is a greater consumption of cyanide than when the pyrites is entirely of the ordinary yellow cubical description. The presence of compounds of copper, physically soft, also tends to increase the consumption." [1]

It has been laid down as a general rule that oxides, hydroxides, carbonates, sulphates and sulphides of those metals which are electro-positive to gold in cyanide solutions are dissolved more rapidly than the last-named metal, whether it is present in the metallic form or contained in its commonly occurring salts.

This rule certainly applies to the precipitated salts commonly occurring in the laboratory, but J. S. MacArthur has shown that the case may be quite different when the naturally occurring minerals are concerned. He states

---

[1] Private communication from J. S. MacArthur.

that " precipitated sulphide of copper is rapidly dissolved, and also a sooty form of the same substance occasionally met with as a mineral occurring in ores. On the other hand, fused copper matte is scarcely acted on at all, and in a great many cases the same may be said of the hard dense sulphides of copper usually found in nature. Sulphide of zinc exhibits the same differences of behaviour : the ' black-jack ' concentrates of the Ravenswood Mine, Queensland, can be treated with good results, little zinc being dissolved. Again, oxide of copper, if freshly precipitated, is strongly acted on by the cyanide, but if it is heated to dull redness in a muffle it becomes insoluble, and a large excess of this material added to a gold ore makes no difference in the percentage of extraction, while the consumption of cyanide is not increased by its presence. The action of cyanide solutions on sulphide of silver is similarly dependent on its physical state."

Prof. Christy's experiments on the electromotive force of metals and minerals in cyanide solutions, compared with Ostwald's normal electrode, show [1] that many minerals tend to be less rapidly dissolved than pure metallic gold. The following table is abridged from the one given by him :—

TABLE XXXI.

	OSTWALD'S NORMAL ELECTRODE = − 0·56 Volt.		
	$\frac{M}{1}$ KCy or 6·5 %.	$\frac{M}{10}$ KCy or 0·65 %.	$\frac{M}{100}$ KCy or 0·065 %.
	E.M.F. in Volts.	E.M.F. in Volts.	E.M.F. in Volts.
Zinc, commercial,	Not given.	+0·77	+0·59
Bornite,	+0·45	+0·25	−0·16
Gold, pure,	+0·37	+0·23	+0·09
Silver, pure,	+0·33	+0·15	−0·05
Copper glance,	+0·29 (?)	+0·25	+0·05
Lead,	+0·13	+0·05	+0·01
Mercury, pure,	−0·09	+0·01	−0·11
Gold, amalgamated,	..	..	−0·13
Niccolite,	−0·11	−0·17	−0·44
Iron,	−0·17	−0·24	−0·24
Chalcopyrite,	−0·20	−0·34	−0·44
Pyrite,	−0·28	−0·42	−0·48
Galena,	−0·28	−0·48	−0·52
Argentite,	−0·28	−0·56	−0·55 (?)
Berthierite,	−0·30	−0·52	−0·52
Magnetic pyrites,	−0·30	−0·40	−0·54
Fahlore,	−0·36	0·52	−0·52
Mispickel,	−0·40	0·45	−0·54
Cuprite,	−0·43	−0·55	−0·57
Electric light carbon,	−0·46	−0·52 (?)	−0·57
Blende,	−0·48	−0·52	−0·55
Boulangerite,	−0·50	−0·55	−0·55
Bournonite,	−0·50	−0·55	−0·56
Ruby silver ore,	−0·54	−0·53 (?)	−0·54
Stephanite,	−0·54	−0·55	−0·52 (?)
Stibnite,	−0·56	−0·56	−0·56

[1] Christy, *Trans. Amer. Inst. Mng. Eng.*, 1900, **30,** 864.

Prof. Christy observed that in testing the minerals it was difficult to get good electrical contact between the conducting platinum wires and the rough surface of the mineral fragment, so that the results are only provisional, especially as the resistance in some cases, such as zinc blende, stibnite, etc., was very high. This would tend to make the results for the minerals too low. Nevertheless, the table is interesting, and shows, for example, that chalcopyrite has hardly more tendency to go into solution than pure pyrite, while bornite and copper glance have a strong tendency to set up electric currents and to dissolve. It is clear from the table that, in the case of the samples which Christy used, pure chalcopyrite, galena, argentite, magnetic pyrites, fahlore, mispickel, blende, boulangerite, bournonite, ruby silver ore, stephanite and stibnite, when free from their oxidation products, are apparently little acted on by cyanide solutions, but he gives no information as to the physical condition of the minerals tested.

Action of Potassium Cyanide on Oxidised Sulphides.—When pyrite occurs in tailing which has been subjected to the action of the weather for some time before treatment, compounds are formed which are more prejudicial to the solution than the sulphides. Sulphide of iron, FeS_2, is oxidised by air and water, ferrous sulphate and free sulphuric acid being formed, thus—

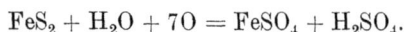

$$FeS_2 + H_2O + 7O = FeSO_4 + H_2SO_4.$$

Ferrous sulphate suffers further oxidation, and normal ferric sulphate, $Fe_2(SO_4)_3$, is produced, which eventually loses acid and becomes a basic sulphate, $2Fe_2O_3 . SO_3$. Other basic salts of complex and unknown compositions appear to be formed also.

W. A. Caldecott has investigated the products of decomposition of pyrite contained in slime accumulated in dams or pits in the Transvaal, with the following results.[1] He finds that the main stages in the oxidation of pyrite or marcasite are as follows :—

(1) FeS_2,	. .	Pyrite.
(2) $FeS + S$,	. .	Ferrous sulphide and sulphur.
(3) $FeSO_4 + H_2SO_4$,		Ferrous sulphate and sulphuric acid.
(4) $Fe_2(SO_4)_3$,	.	Ferric sulphate.
(5) $2Fe_2O_3.SO_3$,	.	Insoluble basic ferric sulphate.
(6) $Fe(OH)_3$,	.	Ferric hydroxide.

All the products named are found in weathered tailing and $Fe(OH)_2$ in addition.

Oxidation of pyrite also occurs to some slight extent during treatment, on the surface layers of sand charges and during prolonged aeration and agitation of slime charges. Ferrous sulphide is produced by grinding pyrite between iron surfaces,

$$FeS_2 + Fe = 2FeS,$$

and the reaction occurs in the stamp battery and steel-lined tube mills, etc., to some extent (Caldecott). The sulphide yields deleterious products as in weathered tailings.

---

[1] Caldecott, *J. Chem. Met. and Mng. Soc. of S. Africa*, 1897, **2**, 98, 122; 1907, **7**, 315. *Proc. Chem. Soc.*, April 29, 1897.

In the presence of such oxidised copper and iron pyrites, the following reactions take place :—

(1) The free sulphuric acid liberates hydrocyanic acid,

$$H_2SO_4 + 2KCy = K_2SO_4 + 2HCy.$$

The sulphuric acid also attacks silicate of alumina or magnesia present in the ore, thus—

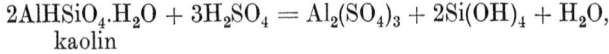

$$2AlHSiO_4.H_2O + 3H_2SO_4 = Al_2(SO_4)_3 + 2Si(OH)_4 + H_2O,$$
kaolin

producing gelatinous colloidal silicic acid which retards the settlement of slime, necessitating the heating of solutions.[1] The soluble sulphates can be neutralised by lime.

(2) Ferrous sulphate reacts with the cyanide, forming ferrous cyanide, which dissolves in the excess of potassium cyanide, so that it does not appear in the free state.

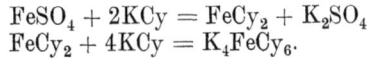

$$FeSO_4 + 2KCy = FeCy_2 + K_2SO_4$$
$$FeCy_2 + 4KCy = K_4FeCy_6.$$

The potassium ferrocyanide, if sufficient acid be present, reacts with fresh ferrous sulphate forming a bluish-white precipitate.

$$FeSO_4 + K_4FeCy_6 = K_2Fe_2Cy_6 + K_2SO_4.$$

This precipitate oxidises in the air to Prussian blue if free acid is present—

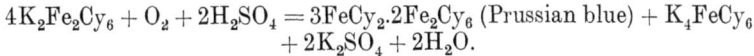

$$4K_2Fe_2Cy_6 + O_2 + 2H_2SO_4 = 3FeCy_2.2Fe_2Cy_6 \text{ (Prussian blue) } + K_4FeCy_6$$
$$+ 2K_2SO_4 + 2H_2O.$$

Both these precipitates are decomposed by potash or soda and, therefore, cannot be formed in their presence. The reactions may be represented as follows :—

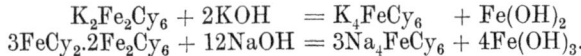

$$K_2Fe_2Cy_6 + 2KOH = K_4FeCy_6 + Fe(OH)_2$$
$$3FeCy_2.2Fe_2Cy_6 + 12NaOH = 3Na_4FeCy_6 + 4Fe(OH)_3.$$

Consequently, if free acid is not present Prussian blue is hardly formed at all, as the solution becomes alkaline, and the precipitate is decomposed as fast as it is formed.

It follows from these reactions that if the blue colour of Prussian blue is visible in the vats or on the surface of the tailing heaps, waste of cyanide must have taken place.

(3) Ferric sulphates are decomposed by potassium cyanide, hydrocyanic acid being evolved and ferric hydroxide precipitated.

(4) A mixture of ferrous and ferric sulphates produce Prussian blue by reacting with potassium cyanide, ferrocyanide of potassium being formed at first as above ; the equation is—

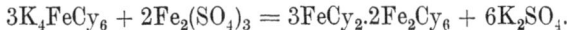

$$3K_4FeCy_6 + 2Fe_2(SO_4)_3 = 3FeCy_2.2Fe_2Cy_6 + 6K_2SO_4.$$

Here again the waste of cyanide is prevented by keeping the solutions alkaline.

---

[1] W. A. Caldecott, *J. Chem. Met. and Mng. Soc. of S. Africa*, Jan. 1907, p. 217 ; *Rand Metallurgical Practice*, vol. i., p. 383.

(5) Sulphate of copper, $CuSO_4$, acts differently from $FeSO_4$, cuprous cyanide, Cu_2Cy_2, being formed, soluble in excess of KCy to $K_2Cu_2Cy_4$, a compound very prone to decomposition. Copper sulphate also gives a precipitate with potassium ferrocyanide, thus—

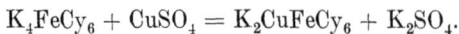

$$K_4FeCy_6 + CuSO_4 = K_2CuFeCy_6 + K_2SO_4.$$

Sulphate of copper acting on KCy or HCy appears to produce nascent cyanogen, and so to promote dissolution of gold, thus :—

$$CuSO_4 + 2HCy = CuCy + H_2SO_4 + Cy.$$

Copper carbonate is also a troublesome cyanicide. According to Clennell, the reaction which occurs is as follows :—

$$2CuCO_3 + 7KCy + 2KOH = K_4Cu_2Cy_6 + KCyO + 2K_2CO_3 + H_2O.$$

(6) Ferrous hydroxide, when formed as above, is instantly dissolved in KCy, thus—

$$Fe(OH)_2 + 6KCy = K_4FeCy_6 + 2KOH$$

and also

$$2Fe(OH)_2 + O + OH_2 = 2Fe(OH)_3.$$

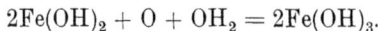

Ferric hydroxide, however formed, does not act on potassium cyanide.

(7) Ferrous sulphide is acted on as follows :—[1]

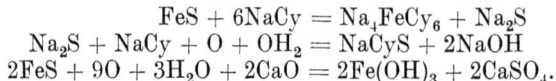

$$FeS + 6NaCy = Na_4FeCy_6 + Na_2S$$
$$Na_2S + NaCy + O + OH_2 = NaCyS + 2NaOH$$
$$2FeS + 9O + 3H_2O + 2CaO = 2Fe(OH)_3 + 2CaSO_4.$$

Copper and zinc in the condition of hydroxides or carbonates are quickly dissolved in preference to the precious metals. If sulphates of these metals are formed in an ore containing limestone or clay, double decomposition occurs with the production of sulphate of lime or alumina, and oxides or carbonates of the heavy metals, which are dissolved by the cyanide, thus—

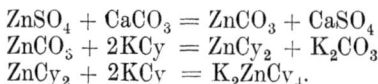

$$ZnSO_4 + CaCO_3 = ZnCO_3 + CaSO_4$$
$$ZnCO_3 + 2KCy = ZnCy_2 + K_2CO_3$$
$$ZnCy_2 + 2KCy = K_2ZnCy_4.$$

Arsenopyrite similarly causes a waste of cyanide when decomposing. Alkaline sulphides are formed which absorb oxygen and necessitate the aeration of the solution. Stibnite is worse in these respects, and ores containing it are generally refractory.

Speaking generally, decomposing sulphides act as cyanicides, destroying the cyanide, but copper ores are probably the most fatal in this respect. They also act as deoxidisers, necessitating much aeration, the removal of soluble sulphides by soluble solutions of lead, mercury, etc. For the effect of lead sulphide, see below, p. 336.

Effect of Alkalies.—Since acidity of the ore causes decomposition of the cyanide, an obvious method of reducing the loss is to add alkali in some form. Before doing this, the free sulphuric acid and soluble salts may be removed by leaching with water, and then a solution of caustic soda or lime is run on to the ore, and after standing for some time is drained off and

---

[1] Caldecott, *Rand Metallurgical Practice*, vol. i., p. 387.

followed by the cyanide solution. Lime is usually added either to the ore or to the mill service water; the amount required is carefully estimated. Water and alkali washes have thus been superseded. The insoluble basic salt is thus converted into ferric hydroxide and soluble sulphates—

$$2Fe_2O_3.SO_3 + Ca(OH)_2 + 5H_2O = 4Fe(OH)_3 + CaSO_4$$

Also,

$$H_2SO_4 + Ca(OH)_2 = CaSO_4 + 2H_2O$$
$$FeSO_4 + Ca(OH)_2 = Fe(OH)_2 + CaSO_4.$$

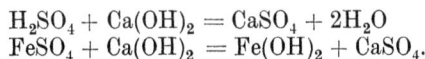

The calcium sulphate forms a saturated solution, and is deposited in the mill pipes, necessitating cleaning (Caldecott). The ferric hydroxide does no harm. Ferrous hydroxide dissolves in cyanide, forming ferrocyanides, and absorbs oxygen, as above, p. 334. Material containing ferrous sulphate and ferrous sulphide are, therefore, not rendered innocuous by lime. During the ore treatment they are gradually converted into ferric salts by the action of air or cyanide or both (Caldecott). Alkali does not protect cyanide from copper salts.

"Caustic soda was at one time commonly used as a protective alkali, but besides being dearer than lime, it tends to introduce sulphides into solution from the ore and to retard flocculation and settlement or percolation, possibly by converting the colloids present into the 'sol' form of turbid suspension,[1] and likewise hinders precipitation, possibly by inducing more rapid formation of $Zn(OH)_2$ coating on the zinc shavings.[2] When much organic matter is present in tailings under treatment, the use of caustic soda assists the introduction into the solution of soluble organic compounds, which resemble a solution of soap in preventing the ready escape of hydrogen given off by the zinc sponge, so that the latter may rise with the froth above the solution level. Lime, in this respect as in others, displays a marked clarifying effect."[3]

Decomposition of Potassium Cyanide.—Hydrocyanic acid is one of the weakest acids known, and is expelled from its salts by all mineral acids and many organic acids. Carbonic acid decomposes potassium cyanide in presence of water thus:—

$$2KCy + CO_2 + H_2O = 2HCy + K_2CO_3.$$

The smell of hydrocyanic acid, noticeable whenever KCy or its solutions are exposed to air, is accounted for by this reaction. The cyanide is protected by the presence of a slight excess of alkali.

In the presence of air, potassium cyanide takes up oxygen, and is converted first to cyanate and then to carbonate—

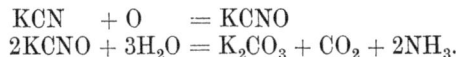

$$KCN + O = KCNO$$
$$2KCNO + 3H_2O = K_2CO_3 + CO_2 + 2NH_3.$$

These reactions are very slow, but become much more rapid if heat is applied. Strong solutions turn brown in the air. In dilute solutions, potassium cyanide suffers hydrolytic dissociation, and is partly changed into HCy and KOH.

---

[1] For a discussion of "gel" and "sol" forms of colloids and the influence of reagents on them as affecting slime settlement, see H. E. Ashley, *Mng. and Sci. Press*, June 12, 1909, p. 831, and *Trans. Amer. Inst. Mng. Eng.*, 1910, **41**, 380.

[2] *J. Chem. Met. and Mng. Soc. of S. Africa*, April, 1898, p. 319; Park, *Cyanide Process*, p. 319.

[3] W. A. Caldecott, *Rand Metallurgical Practice*, vol. i., p. 388.

It results from this that the passage of a stream of any neutral gas such as nitrogen through the solution causes an evolution of hydrocyanic acid, while the solution becomes alkaline.　The equation is—

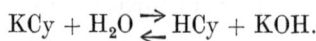

$$KCy + H_2O \rightleftarrows HCy + KOH.$$

The extent of the hydrolytic dissociation is 1·12 per cent. of the salt in a solution containing 0·65 per cent. KCy and 2·34 per cent. of the salt in a solution containing 0·16 per cent. KCy.[1]　In weaker solutions, the extent of the hydrolytic dissociation is greater.　If the solution is boiled with acids or alkalies, hydrolysis of the cyanide occurs rapidly, ammonia and formates being formed, thus—

$$KCN + 2H_2O = NH_3 + HCO_2K.$$

This equation does not represent the whole effect, as acetates and other organic substances are also formed.　The reactions also proceed, although very slowly, if the solution is cold and neither acids nor alkalies are present.

　　The destruction of cyanide by cyanicides, as described in the previous section, is far more serious than this spontaneous decomposition in water.

　　Soluble Sulphides in Cyanide Solutions.—A small percentage of a soluble sulphide present in the cyanide solution greatly delays the dissolution of gold.　Doubtless this is partly owing to the abstraction of oxygen from the solution by the sulphide, for gold sulphide is freely soluble in KCy, so that the surface of the metal should remain free from sulphide.　Bettel, however, points out [2] that silver sulphide is far less soluble than gold sulphide, and that if native gold containing 20 per cent. of silver is treated, a film almost insoluble in (dilute) cyanide solutions may be formed.　It is certain that some specimens of pure gold leaf dissolve with great difficulty if they have been previously dipped in sulphide solutions, or if traces of soluble sulphides or sulpho-cyanides are present in the solution.　The difficulty disappears if the sulphides are removed, either by being precipitated with lead salts, or by the action of certain oxidisers.

　　Caldecott observes that the soluble sulphides are derived from that present in commercial cyanide, and from the action of cyanide on sulphides of iron in the ore.　They become converted into sulpho-cyanides at the expense of cyanide and oxygen, thus :—

$$Na_2S + NaCy + O + OH_2 = NaCyS + 2NaOH.[3]$$

　　The alkaline sulphides are usually removed by dissolved salts of lead or mercury.

　　Dissolved lead salts precipitate soluble sulphides.　Lead acetate reacts with cyanide solutions yielding hydrate or oxycyanide of lead, and this is dissolved as an alkaline plumbite, $PbO.xNa_2O$.　The reaction proceeds thus —[4]

$$PbO.xNa_2O + Na_2S = PbS + (x + 1)Na_2O$$
$$PbS + NaCy + O = PbO + NaCyS.$$

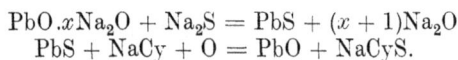

[1] J. Shields, *Phil. Mag.*, 1893, **35**, 387.
[2] Bettel, *South African Mining J.*, May 8, 1897.
[3] Caldecott, *Rand Metallurgical Practice*, vol. i., p. 388.
[4] *Op. cit.*, p. 387.

so that the insoluble lead sulphide is also an absorbent of oxygen and cyanide. Any soluble lead remaining " passes into the zinc boxes, where it is deposited as metallic lead, thus constantly renewing the activity of the zinc-lead couple by forming fresh uncoated precipitating surfaces " (Caldecott).

G. H. Clevenger observes[1] that soluble sulphides are readily oxidised and seldom occur in mill solutions, but that lead salts may be advantageous,

(1) By removing sulphur ions and so accelerating the dissolution of silver occurring as sulphide.

(2) By retarding the formation of sulphocyanide and the separation of free sulphur, and so saving cyanide.

(3) By checking the fouling tendency of copper on the solution.

He also finds that—

(4) An excess of lead is disadvantageous, but this is less observable when litharge is used.

(5) Lead is more advantageous with silver or silver-gold ores than with gold ores.

Re-Precipitation of Gold and Silver in the Leaching Vats.—If the solution is acid there is said to be some danger of a precipitation of gold previously dissolved, insoluble aurous cyanide being thrown down, according to the equation—

$$KAuCy_2 + HCl = KCl + HCy + AuCy.$$

This, however, need not be feared as long as there is an excess of KCy, which must all be destroyed by the acid before the aurous cyanide can be precipitated. Moreover, aurocyanhydric acid, $HAuCy_2$, is formed and remains in solution, AuCy being precipitated only as the HCy leaves the solution by evaporation.[2] There is danger in transferring a solution containing gold to a vat containing pyritic material. If the latter should contain any soluble salts of the heavy metals, insoluble salts are thrown down—e.g. :—

$$2KAuCy_2 + ZnSO_4 = K_2SO_4 + ZnAu_2Cy_4.$$

A. N. Mackay attributes re-precipitation of gold to the action of free alkali on fine sulphides,[3] by which soluble sulphides are produced. In one case in which the protective alkali amounted to 0·04 per cent. and the cyanide to 0·02 per cent., complete re-precipitation of dissolved gold took place in 18 hours. The slime in contact with the solution contained 0·5 per cent. of FeS_2. By reducing the alkalinity to 0·01 per cent., the re-precipitation was prevented from occurring.

"The presence of carbonaceous or decomposing organic matter in the sand or slime undergoing cyanide treatment is distinctly detrimental on account of its oxygen-absorbing properties and tendency to precipitate gold already in solution. Semi-burnt coal in commercial lime," dirt of various kinds, decomposing products of vegetation, or carbon in certain forms may be " the cause in certain cases of re-precipitation of gold from cyanide solution when kept in contact with sand for a few days in a closed vessel ; and further, its reducing capacity being stimulated by heat, serves to explain

[1] Clevenger, *Mng. and Sci. Press*, Oct. 24, 1914, p. 635.

[2] Lindbom, *Bull. Soc. chim.*, 1878, **29**, 422. See also treatment by acid solutions, p. 397.

[3] Mackay, Re-precipitation of Gold from Cyanide Solutions, *Trans. Inst. Mng. and Met.*, May, 1905, **14**, 541.

why in practice warming of sand solutions may result in less rapid dissolution of the gold content than with solutions at ordinary temperatures." [1]

Reactions in the Zinc Boxes.[2]—It was believed at one time that the precipitation of gold and silver by zinc was effected by simple displacement, according to the equation—

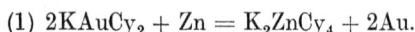

$$(1) \quad 2KAuCy_2 + Zn = K_2ZnCy_4 + 2Au.$$

This simple view is now generally discredited. The principal observed facts are as follows :—

(1) An excess of free cyanide favours the precipitation of gold.[3] Sometimes it is necessary to add cyanide direct to the solution in the zinc boxes.

(2) An evolution of hydrogen takes place during precipitation, but free gaseous hydrogen does not precipitate gold.

(3) Zinc dissolves in the solution, the amount depending on its strength. The waste of zinc is such that it requires, say, from 5 to 20 parts of zinc to precipitate 1 part of gold, or from 15 to 60 atoms of zinc to 1 atom of gold. Part of the zinc (usually at least half) is disintegrated and remains with the precipitated gold.

(4) The precipitation is aided by the presence of metals, such as lead, which are electro-negative to zinc in cyanide solutions. The lead increases the rate of dissolution of the zinc. It must, of course, be in contact with it, as in the lead-coated zinc shavings usually employed.

(5) It is necessary that the zinc should be of very great area for efficient precipitation, especially in solutions weak in cyanide. Zinc plates are generally found unsuitable.[4] Zinc shavings (or "thread") or in the alternative zinc dust ("fume") are suitable.

The large area is required because "no doubt but a fraction of the zinc area usually does useful work, the remainder being rapidly coated with lime salts and zinc compounds (see below, p. 342), and thus put out of action. . . . The lead-zinc couple was first introduced to reduce the tendency of precipitated copper to form a smooth coherent metallic coating upon the zinc shavings." [5]

"Hydrogen to some extent polarises the zinc by coating it with a thin film of gas, which prevents its actual contact with the gold in the solution ; and no doubt the efficacy of a lead coating is in part due to the roughened surface, consequent on its use, assisting bubbles of hydrogen to form and escape." [6] Zinc shavings, by reason of their ragged edges, have a similar advantage over zinc plates.

(6) "In addition to gold, any silver, mercury, copper, cobalt and nickel present in solution as double cyanides are precipitated by zinc in the same way as gold, the copper being occasionally visible in the red coating produced on the shavings, whilst the zinc shavings are rendered brittle should much mercury be present." [7] Other elements, such as dissolved lead, tellurium and selenium, are also precipitated, the precipitate sometimes containing appreciable amounts of metals which previously existed in the ore and the solutions in such small proportions as to escape detection.

[1] Caldecott, *Rand Metallurgical Practice*, vol. i., p. 387.
[2] See W. A. Caldecott and E. H. Johnson, *J. Chem. Met. and Mng. Soc. of S. Africa*, 1903, **4,** 263.
[3] W. R. Feldtmann, *Eng. and Mng. J.*, Aug. 11, 1894, p. 126.
[4] See, however, MacArthur, *J. Chem. Met. and Mng. Soc. of S. Africa*, 1913, **13,** 310.
[5] Caldecott, *Rand Metallurgical Practice*, vol. i., p. 391.
[6] *Op. cit.*, p. 390.   [7] *Op. cit.*, p. 391.

(7) "During the passage of the solution through the boxes some cyanide is consumed, and the free alkali in solution is increased, though in neither case is the amount very great in practice."[1]

(8) "Heat assists precipitation, but increases the waste of cyanide by decomposition." Ehrmann has shown[2] by experiments on solutions of various strengths, that about as much gold is precipitated by zinc in twenty-four hours at 20° C. as in two hours at 80° C. In the latter case there is, according to MacArthur, "an enormous waste of cyanide by the formation of urea, which manifests itself by its strong unpleasant odour."

(9) The proportion of dissolved gold which is precipitated in the zinc boxes diminishes as the solutions become weaker in gold. Hence some gold, say 0·02 dwt. per ton, is always left in the solution.

(10) The gold is redissolved if it becomes detached from contact with zinc, provided that oxygen is still present, and "air blown through gold-coated shavings immersed in cyanide solution raises the gold value of such solution, and a similar effect can likewise be produced by exposing such shavings to the air in a shallow vessel partly filled with solution."[3] The solutions leaving the boxes are practically free from oxygen.[4]

Theories as to the precipitation must fit the observed facts, and are of value only so far as they enable the operator to form such a mental picture of what is happening as to help him in his work.

Precipitation is a process of reduction, or at least is favoured by reducing conditions and prevented by oxidising conditions. It is thus the exact opposite of dissolution. Caldecott states the main reactions, thus :—[5]

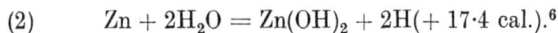

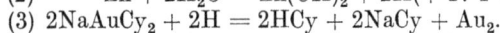

(2) $Zn + 2H_2O = Zn(OH)_2 + 2H(+ 17·4 cal.).$[6]
(3) $2NaAuCy_2 + 2H = 2HCy + 2NaCy + Au_2.$
(4) $HCy + NaOH = NaCy + H_2O.$

He points out that the oxidation of zinc raises the temperature of the solution by amounts up to 1° F. in passing through the boxes. This oxidation can hardly take place except owing to contact with an element, such as lead, gold, mercury, etc., which is electro-negative to zinc, and it may be argued that the hydrogen is in all probability mainly produced according to the reaction—

(5) $Zn + 4NaCy + 2H_2O = Na_2ZnCy_4 + 2NaOH + 2H$

which must take place in the absence of metal to be precipitated, and, therefore, can hardly fail to be generally proceeding. This view of the method of production of hydrogen by zinc brings zinc precipitation into line with precipitation by aluminium, the essential fact in each case being the liberation of hydrogen during the dissolution of the metal. In the case of aluminium, Moldenhauer found[7] that an excess of alkali was necessary, and this dissolves aluminium with the formation of an aluminate, hydrogen being given off.

---

[1] *Loc. cit.*
[2] Ehrmann, *J. Chem. Met. and Mng. Soc. of S. Africa*, July, 1899, p. 147.
[3] Caldecott, *Rand Metallurgical Practice*, vol. i., p. 393.
[4] A. F. Crosse, *J. Chem. Met. and Mng. Soc. of S. Africa*, Aug. 1898, **2**, 402.
[5] Caldecott, *Rand Metallurgical Practice*, vol. i., 389.
[6] There is also the reaction $Zn + 2NaOH = Na_2ZnO_2 + 2H$ to be considered, especially on the application of heat.
[7] Julian and Smart, *Cyaniding Gold and Silver Ores*, 1st ed., p. 158.

It follows that free cyanide should not be required in precipitation by aluminium.[1] However, the equations given by E. W. Hamilton [2] are the following :—

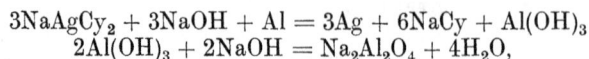

$$3NaAgCy_2 + 3NaOH + Al = 3Ag + 6NaCy + Al(OH)_3$$
$$2Al(OH)_3 + 2NaOH = Na_2Al_2O_4 + 4H_2O,$$

but for low-grade solutions, when caustic and aluminium are both present in excess, he prefers—

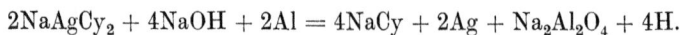

$$2NaAgCy_2 + 4NaOH + 2Al = 4NaCy + 2Ag + Na_2Al_2O_4 + 4H.$$

Aluminium does not react directly with cyanide.

Reaction No. (5) occurs the more readily the more cyanide is present, and if it is regarded as the essential reaction, the desirability of the presence of free cyanide is at once apparent. In this respect Dr. Caldecott's explanation is less obvious, and it even calls for a gradual increase in the free cyanide present during precipitation, which is the reverse of the observed facts. Moreover, it does not account for the well-known production of free alkali. If the zinc hydrate formed in equation (2) is dissolved in cyanide at the moment of formation, thus—

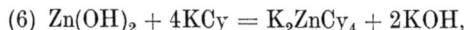

$$(6) \quad Zn(OH)_2 + 4KCy = K_2ZnCy_4 + 2KOH,$$

then the loss of cyanide and increase of alkali are accounted for and the necessity of the presence of free cyanide is explained as the need of keeping the metallic zinc clean. This explanation, however, makes equations (2) and (6) identical with equation (5), which states the facts more simply, and gives the same evolution of heat.

MacFarren,[3] following J. S. Clennell, gives the precipitation of gold in the presence of cyanide as—

$$(7) \quad KAuCy_2 + 2KCy + Zn + H_2O = K_2ZnCy_4 + Au + H + KOH$$

and in the absence of free cyanide as—

$$(8) \quad KAuCy_2 + Zn + H_2O = ZnCy_2 + Au + H + KOH$$

making the precipitation an electro-chemical displacement of K and Au by Zn, followed by the displacement of hydrogen in water by potassium. Equation (8) seems the more probable when it is remembered that the potassium ions in KCy itself are displaced by zinc in the presence of water. Christy also gives equation (8),[4] but instead of (7) he gives—

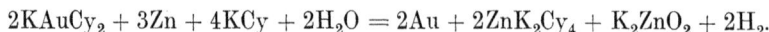

$$2KAuCy_2 + 3Zn + 4KCy + 2H_2O = 2Au + 2ZnK_2Cy_4 + K_2ZnO_2 + 2H_2.$$

The caustic alkali produced by equation (5) or (7) is partly neutralised by carbonic acid from the air, and calcium carbonate is thus thrown down.

Further experimental data may decide the matter, but in its absence, always bearing in mind the nature of the help that theory can give to practice,

---

[1] See Julian and Smart, *Cyaniding of Gold and Silver Ores*, p. 158.
[2] Hamilton, *Eng. and Mng. J.*, May 10, 1913, p. 936.
[3] MacFarren, *Cyanide Practice*, p. 159.
[4] Christy, "On the Solution and Precipitation of Gold." *Trans. Amer. Inst. Mng. Eng.*, 1896, **26**, 735.

Caldecott's main contention that nascent hydrogen is the agent of precipitation of gold may be accepted. The hydrogen which becomes gaseous has escaped its true function and become inert. It would also tend to form an imperceptible layer on the surface of the zinc, which would thereby be protected from further action, and consequently the sooner it is got rid of the better. The ragged edges of thread zinc and the rough surface of lead-coated zinc are both beneficial in this respect. Circulation of the solution should enable a greater proportion of the nascent hydrogen to be used in precipitation. The evolution of hydrogen in zinc dust precipitation has not been described, and it is probable that in this case the hydrogen is practically all used in precipitating gold, thus reducing the waste of zinc.

Caldecott has observed that, if there is a rapid flow of solution in the boxes, particles of lead and gold and of mercury and gold may be detached from the zinc and float to the surface, buoyed up by bubbles of hydrogen gas. Such particles may be carried away and the gold redissolved.

With regard to the effect of electro-negative metals, Caldecott observes that the ordinary impurities of commercial zinc, such as lead, carbon, iron and arsenic, do not dissolve, but like deposited gold and other metals assist the zinc-lead couple to promote efficient precipitation. It has been suggested that iron being electro-negative to zinc, gold might be deposited on the inside of steel extractor boxes, and on the trays of wire screening used to support the zinc shavings in the boxes. In practice, however, Caldecott finds that very little gold is deposited in this way, " probably owing to the fact that a protective coating of rust and calcium sulphate speedily forms on such iron surfaces and prevents actual contact of the metal with the gold-bearing solution."

The presence of an excess of alkali and a deficiency of cyanide apparently favours the production of sodium zincate, thus—

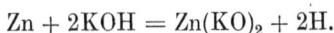

$$Zn + 2KOH = Zn(KO)_2 + 2H.$$

The presence of large quantities of the double cyanide of zinc and potassium in the solutions is not prejudicial to the solvent action of the simple cyanides. At the Mercur Mine the stock solution was apparently as efficacious after nine months' use as at the start, although it must have contained large quantities of zinc cyanide. Feldtmann showed that gold in ores can be dissolved by zinc-potassium cyanide, but J. S. C. Wells points out that the double cyanide remains undecomposed by gold so long as any simple cyanide is present. The simultaneous presence of zinc and some other constituent (possibly arsenic) was found by Hamilton to be prejudicial to the dissolution of silver at Cobalt.[1] An accumulation of base heavy metals in the solution can be got rid of by the addition of soluble sulphides.

The " white precipitate " formed on the zinc shavings in weak and " medium " cyanide solutions, and tending to prevent good precipitation, has been the subject of several investigations (G. W. Williams, A. Whitby, B. Bay and A. Prister).[2] According to an analysis by Prister and

---

[1] Hamilton, *Eng. and Mng. J.*, May, 10, 1913, p. 936.
[2] Williams, Whitby, Bay and Prister, *J. Chem. Met. and Mng. Soc. of S. Africa*, 1904-6, 4 and 5,

Bay, the composition of a sample from the Ferreira Deep Mill was as follows :—

Ferrocyanide of zinc and potash, $K_2Zn_3(FeCy_6)_2$, 10·45 per cent.
Zinc cyanide, $ZnCy_2$, ... 22·73 ,,
Zinc hydroxide, $Zn(OH)_2$. . . . . 54·79 ,,
Copper oxide, ... 0·40 ,,
Iron oxide, Fe_2O_3, ... 1·00 ,,
Silica, ... 1·03 ,,
Moisture, ... 8·07 ,,

98·47 ,,

There seems to be some doubt as to the cause of the formation of insoluble zinc cyanide (see, however, MacFarren's *Cyanide Practice*, p. 160), although there is no doubt that in a strong solution of KCy it would redissolve, thus—

$$ZnCy_2 + 2KCy = K_2ZnCy_4$$

or would be prevented from forming. Similarly $Zn(OH)_2$ is redissolved by KCy, thus—

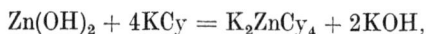

$$Zn(OH)_2 + 4KCy = K_2ZnCy_4 + 2KOH,$$

or

$$2Zn(OH)_2 + 4KCy = K_2ZnCy_4 + Zn(OK)_2 + 2H_2O,$$

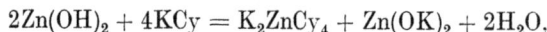

followed, according to N. Anderson,[1] by

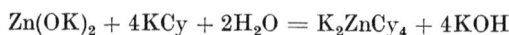

$$Zn(OK)_2 + 4KCy + 2H_2O = K_2ZnCy_4 + 4KOH$$

the potassium zincate being decomposed by cyanide. Strong cyanide solutions accordingly do not permit the formation of visible zinc hydroxide.

According to Caldecott,[2] the absence of free alkali in solution and the presence of hydrocyanic acid instead, as produced by the preliminary addition of sodium bicarbonate, prevents the formation of any white zinc hydroxide coating on zinc shavings, thus—

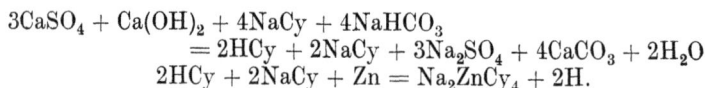

$$3CaSO_4 + Ca(OH)_2 + 4NaCy + 4NaHCO_3$$
$$= 2HCy + 2NaCy + 3Na_2SO_4 + 4CaCO_3 + 2H_2O$$
$$2HCy + 2NaCy + Zn = Na_2ZnCy_4 + 2H.$$

MacFarren, however,[3] assuming that the white precipitate is mainly $ZnCy_2$ and $Zn(OH)_2$, states that the white precipitate appears only to a slight extent in strongly alkaline solutions, even though weak in cyanide. Caldecott proceeds that the white precipitate is " undesirably abundant in cold weather or when the treatment of weathered ore or accumulated slime has introduced an abnormal amount of potassium ferrocyanide into the working solutions." An assumption is made here that the white precipitate is partly zinc ferrocyanide.

Among other substances mentioned by Caldecott as being deposited in the zinc boxes are zinc sulphide; silica and alumina dissolved from the

[1] Anderson, *Mineral Industry*, 1895, p. 330.
[2] Caldecott, *Rand Metallurgical Practice*, vol. i., p. 390; *J. Chem. Met. and Mng. Soc. of S. Africa*, Dec. 1903, p. 639.
[3] MacFarren, *op. cit.*, p. 160.

ore and " presumably precipitated as zinc aluminate and silicate " ; calcium sulphate, especially when partly weathered ore is treated ; calcium carbonate formed by the action of the carbonic acid of the air on dissolved calcium hydroxide ; and minute particles of ore and especially slime mechanically deposited.

One more point must not be forgotten, and that is that the soluble compounds are all more or less ionised, and consequently that all of them will be present together in the solution in amounts depending on their degree of ionisation, mass action having free play.

The Precipitation of Gold by Charcoal.—It has been shown that graphite from Natal and from Barberton does not precipitate gold from aurocyanide solutions,[1] but that graphite in West African schist may do so.[2] The power of charcoal to precipitate gold from cyanide was proved by Morris Green [3] to depend on carbon monoxide gas occluded in the charcoal. Free carbon monoxide is without effect, and its condition in charcoal is unknown. The reaction involved in precipitation is presumably a reduction analogous to that effected by nascent hydrogen.[4] It was found by N. S. Keith [5] that finely divided carbon when agitated with ore in cyanide solutions caused the gold to be more rapidly dissolved, presumably by its electronegative action, just as iron in contact with gold increases its solubility, according to Gore (see p. 327).

The Precipitation of Insoluble Cyanides of Gold.—In the absence of free cyanide, the sodium in sodium aurocyanide may be replaced by certain heavy metals, such as silver or copper, in which case the gold comes down as an insoluble double metallic cyanide. As the latter is soluble in alkaline cyanide, it is necessary to destroy the cyanide in the solution before the precipitation takes place, and consequently the method is not used in practice except in assaying.

P. de Wilde proposed [6] to acidify the solution with sulphur dioxide and to precipitate aurous and cuprous cyanides with sulphate of copper, leaving the solution to settle for twelve hours. This method involves the destruction of the cyanide.

S. B. Christy proposed to acidify the solution with sulphuric acid and to stir in freshly precipitated sulphide of copper or any cuprous salt, when the whole of the aurous cyanide is precipitated in a few hours. Wilde subsequently proposed carefully to neutralise auriferous cyanide solutions with dilute sulphuric acid, and then to add a solution of cuprous chloride in common salt, assisting the settlement of the precipitate by adding potassium chlorate or blowing air through the solution. The precipitate is then dried and calcined, and the oxide of copper removed by dissolving it in sulphuric acid, leaving the gold nearly pure.

Effect of Thiocyanates on Metallic Gold.—This has been investigated by H. A. White,[7] who found that potassium cyanate, though inactive by itself, would dissolve gold if mixed with various oxidising agents, such as potassium permanganate, ferricyanide of potassium, dilute nitric acid, etc. Ferric thiocyanate also dissolved gold, but in this case it was supposed that the presence of dissolved oxygen was necessary.

[1] Green, *J. Chem. Met. and Mng. Soc. of S. Africa*, Sept. 1912, **13**, 84.
[2] Brühl, *Trans. Inst. Mng. and Met.*, 1914, **23**, 82.
[3] Green, *ibid.*, p. 65.   [4] See, however, Feldtmann, *ibid.*, April, 1915.
[5] Keith, *Engineering*, 1895, **59**, 379.
[6] De Wilde, *Revue Universelle des Mines*, etc., Brussels, Oct. 1, 1895.
[7] White, *J. Chem. Met. and Mng. Soc. of S. Africa*, October, 1905, **6**, 109.

The Stark process of treating dumps [1] was said to be explained by these results.

TESTING OF CYANIDE SOLUTIONS. [2]

Free Cyanide.—The ordinary method of estimating the amount of free cyanide present in a liquid is by titration with a standard solution of silver nitrate. Silver cyanide is formed, and redissolves in the excess of potassium cyanide until one-half of the latter has been decomposed. The equations are as follows :—

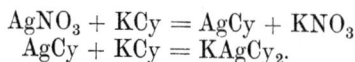

$$AgNO_3 + KCy = AgCy + KNO_3$$
$$AgCy + KCy = KAgCy_2.$$

When one-half of the KCy present has been converted to AgCy, an additional drop of $AgNO_3$ solution causes the formation of a permanent white precipitate of AgCy. The amount of silver solution added is then read off, and the percentage of cyanide calculated. The equation of the end reaction is—

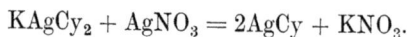

$$KAgCy_2 + AgNO_3 = 2AgCy + KNO_3.$$

A few drops of a 10 per cent. solution of potassium iodide are often added, in accordance with a suggestion made by J. S. MacArthur, to make the end reaction sharper, and to prevent inaccuracy through the presence of ammonia or other substances in which silver cyanide is soluble. The results are, however, slightly too high, owing to the presence in mill solution of alkali and the double zinc cyanide. A. M'A. Johnston dissolves [3] 13·046 grains of triple crystallised silver nitrate in distilled water and makes up to a litre. He titrates by running this into 100 c.c. of cyanide solution (previously filtered if turbid) containing one or two drops of potassium iodide solution. Then each c.c. of $AgNO_3$ solution used represents 0·01 per cent. KCy in the cyanide solution.

Total Cyanide.—This includes the free cyanide (NaCy or KCy), hydrocyanic acid and the cyanide in the double cyanide of zinc, K_2ZnCy_4. An excess of caustic soda (say 5 c.c. of a 10 per cent. solution) is added to the solution, together with a few drops of the potassium iodide indicator and a drop or two of ammonia (Johnston). The solution is titrated with silver nitrate as before.

Protective Alkali. [4]—The solutions required are silver nitrate (as above), phenolphthalein, prepared by dissolving 1 gramme of the powder in 100 c.c. of 60 per cent. alcohol or methylated spirits; decinormal oxalic acid, $N/10$ $C_2H_2O_4$, prepared by dissolving 6·3 grammes of oxalic acid crystals, $C_2H_2O_4.2H_2O$, in distilled water and diluting to 1,000 c.c. ; potassium ferrocyanide, K_4FeCy_6, prepared by dissolving 10 grammes of K_4FeCy_6 in 100 c.c. of distilled water.

The test is as follows :—Measure into a conical flask 100 c.c. of cyanide solution. Add just sufficient silver nitrate solution to give a permanent

---

[1] *Rand Metallurgical Practice*, vol. i., p. 394.
[2] For full details of analysis and the testing of ores, see J. E. Clennell, *Chemistry of Cyanide Solutions;* A. M'A. Johnston, *Rand Metallurgical Practice*, vol. i., pp. 322-379 ; MacFarren, *Cyanide Practice*, pp. 28-86.
[3] Johnston, *Rand Metallurgical Practice*, vol. i., p. 323.
[4] For this account of the determination the author is indebted to Mr. A. M'A. Johnston and to Dr. Caldecott.

precipitate, and then 5 c.c. of the ferrocyanide solution and two drops of phenolphthalein. Titrate with decinormal oxalic acid till the phenolphthalein colour is discharged. Then each c.c. of oxalic acid solution used multiplied by 0·004 gives the percentage of alkalinity in terms of caustic soda. The portion of solution used to determine free cyanide may be afterwards used to estimate the protective alkali.

The use of potassium ferrocyanide in this determination is desirable, so as neither to include K_2ZnCy_4 as free alkali nor to neutralise free alkali present by excess of $AgNO_3$.

For the determination of gold and silver contained in cyanide solutions, see Chap. XIX., and for the numerous other tests occasionally required in a mill laboratory, see the references cited in the footnote on p. 344.

CYANIDE POISONING.

Treatment in Case of Cyanide Poisoning.—A coating of oil or kerosene (or rubber gloves) protects the hands from the action of the cyanide solution, by which a rash is sometimes developed.

If cyanide solution is swallowed, anything that causes vomiting may be administered, but the action of cyanide is extremely rapid and every second is of value. Freshly precipitated ferrous hydroxide, made by mixing oxide of magnesium, caustic potash, and ferrous sulphate, is a useful antidote. In the reaction which occurs ferrocyanide of potassium or sodium is formed, which is innocuous. The antidote recommended in S. Africa[1] consists of 30 c.c. of a 23 per cent. solution of ferrous sulphate and 30 c.c. of a 5 per cent. solution of caustic potash in separate phials. These are mixed in a mug and 2 grammes of powdered oxide of magnesium stirred in and the dose administered to the patient. The use of nitrate of cobalt and of sodium thiosulphate, both internally and as injections, have also been suggested. Twenty minims (1·2 c.c.) or more of a 0·5 to 1 per cent. solution of cobalt nitrate, or of a 5 to 10 per cent. solution of sodium thiosulphate, is repeatedly injected under the skin. Comparatively harmless cobalt cyanide or sodium sulphocyanide are formed.[2] As an alternative copious subcutaneous injections of hydrogen peroxide are recommended. The patient may also be rubbed with camphor and alcohol, or cold water may be dashed upon the skin.

If hydrocyanic acid has been inhaled, an inhalation of ammonia, or chlorine, or ether, should be administered, but respirators should be worn in cases, such as in the acid treatment of gold slime, where evolution of poisonous gases are probable. Park records the warning that in the acid treatment of gold slime from arsenical ores, an evolution of arsenuretted hydrogen may occur, and the inhalation of this gas is very deadly, no antidote being known.

Potassium sulphocyanide is also exceedingly poisonous, and ammonium thiocyanate in a somewhat less degree.[3] The fumes of molten cyanide appear to be more or less harmless.

---

[1] *J. Chem. Met. and Mng. Soc. of S. Africa*, 1904, **4,** 679.
[2] Dixon Mann and Brend, *Forensic Medicine and Toxicology*, 5th ed., p. 578.
[3] For further particulars on cyanide poisoning, see *Rand Metallurgical Practice*, vol i., p. 331; *Eng. and Mng. J.*, Nov. 3, 1906, p. 835.

CHAPTER XVI.

THE CYANIDE PROCESS. GENERAL METHODS.

THE method of treatment may be conveniently considered as being divided into five distinct operations, viz. :—

(1) Preparation of the ore for treatment.
(2) The cyaniding of sand.
(3) The cyaniding of slime.
(4) Precipitation of the dissolved gold.
(5) Conversion of the precipitated gold into bullion.

1. PREPARATION OF THE ORE.

In the treatment of gold ores by cyanide, the same mechanical difficulties which had been met with in the chlorination process and in other leaching processes soon presented themselves. It is necessary to crush finely, for otherwise too great a proportion of the gold remains locked up in the larger grains and escapes dissolution. Crushed ore and especially finely crushed ore, however, is always found to contain a proportion of slime, and when this is allowed to remain intermixed with the rest of the ore, the mixture resists the passage of liquids through it. Many ores contain so much slime that a bed only 12 or 15 inches thick cannot be leached at a reasonable rate, even with a vacuum below the filter bed. Moreover, the gold is more readily dissolved from slime than from less finely divided ore, and this constitutes another reason for separating sand from slime. Such a separation has, therefore, become a part of the usual practice in cyaniding. Sand and slime are treated separately with different appliances.

Primary crushing is generally effected in the stamp battery and secondary reduction in tube mills (*q.v.*) or other machines. Sand is crushed more finely than formerly as a preliminary to treatment with cyanide. "All-sliming" methods have also been applied.

The separation of sand from slime by means of classifiers has been dealt with in Chap. XI. Some other methods in use are described below.

In places where the separate treatment of slime is not desirable, either from its small quantity or from other reasons, the slime and sand are thoroughly mixed before and at the time of charging them into the leaching vat, but the results are seldom quite satisfactory. In India, according to Alfred James,[1] siliceous slime was, as a temporary expedient, dried in the sun and then treated in the ordinary way by leaching.

Dry crushing was formerly much in vogue before the methods of treating slime had been evolved. In dry crushing the sand and slime are kept mixed. The ore must usually be thoroughly dried before being crushed, especially

---

[1] James, *Cyanide Practice*, 1901, p. 82.

if it is of a clayey or talcose nature. For drying furnaces, see above, p. 272. It is necessary to dry at a low temperature to avoid partial oxidation of sulphides by which cyanicides are formed. The ore is crushed by means of rolls or similar machines (in which the formation of slime is reduced to a minimum) or by stamps. Water and alkali washes are often necessary to remove cyanicides.

In 1895, at the George and May Mill in South Africa, the ore was coarsely crushed in a Gates crusher and dumped at once into leaching vats (Hatch and Chalmers). About 75 per cent. of the gold was extracted.

Down to 1898, at the Waihi Mine, New Zealand,[1] ore was dried in brick-lined kilns, which were charged with alternate layers of wood and ore, crushed in Gates crushers and stamps, and conveyed to the leaching-vats. The vats were 4 feet deep, and the depth of the charge about 2 feet, as a greater depth would have entailed difficulties in leaching. About 90 per cent. of the gold was recovered, but the tailing was rich (5 or 6 dwts.), and the method was superseded by wet crushing.

Dry crushing is employed before *roasting*, which has been mainly used on sulpho-telluride ores at Kalgoorlie and Cripple Creek (see p. 393). Before leaf filters were introduced, oxidised clayey ores were sometimes roasted for purposes of dehydration, by which they were rendered more granular and

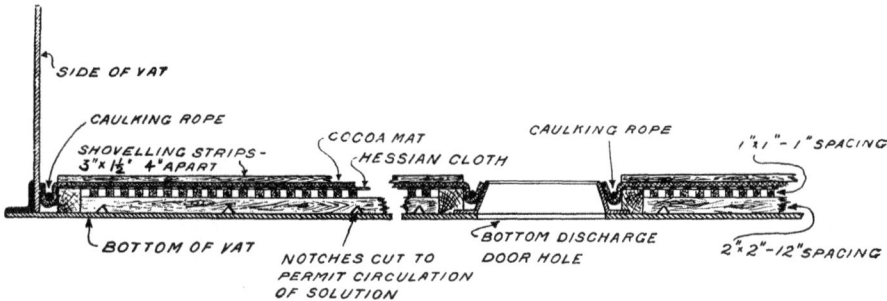

Fig. 148.—Filter Bottom in Sand Vats.

leachable. Roasting before cyaniding must be complete, as ores containing sulphides, when partially roasted (not quite " dead ") contain ferrous salts and other cyanicides. It is noteworthy that unoxidised pyrite and sesqui-oxide of iron, Fe_2O_3, the final product of roasting, are alike without effect on cyanide solutions, but all intermediate compounds are cyanicides. The efficiency of roasting is tested by adding cyanide to a clear solution obtained by shaking the roasted ore with water. If a discoloration appears the ore still contains soluble salts, which will destroy cyanide and make foul solutions.

In ordinary practice ores are not roasted, but if they contain a fair pro-portion of amalgamable gold, especially if some of the gold is not very finely divided, they are treated by amalgamation before being cyanided. There is now a strong movement in the direction of the omission of amalgamation, especially in America. In this case a cyanide solution can be used in the battery, and examples of the practice are given in the sequel. When ore is crushed and amalgamated, the tailing is separated into sand and slime, the sand is collected for treatment by percolation (p. 352), and the slime dealt with as described in Section 3, p. 356.

---

[1] Jas. Park, *Trans. Amer. Inst. Min. Eng.*, 1899, **29**, 666.

In the Transvaal sand is collected for treatment in one of three ways :—

　　(*a*) By hose filling.
　　(*b*) By the " Butters and Mein " pulp distributor.
　　(*c*) By the " Caldecott " continuous collecting plant.

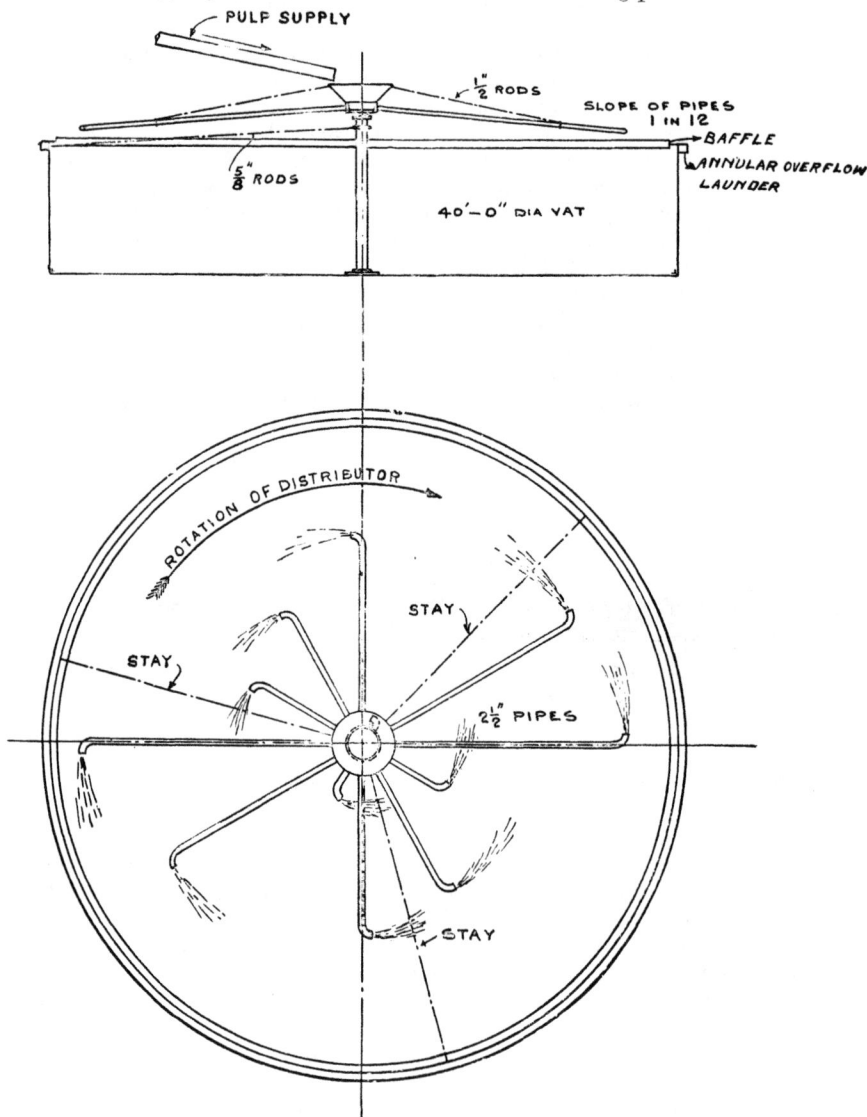

Figs. 149 and 150.—" Butters and Mein " Pulp Distributor for Filling
Sand-collecting Vats.

　　(*a*) In **hose-filling** the underflow from the classifiers passes through pipes and is fed into the collecting vats by one or more hose pipes, the position of which is changed from time to time, in order to distribute the pulp evenly.

The vats were formerly wooden, but are now constructed of steel, and are from 25 to 50 feet in diameter and 6 to 9 feet deep. The sand settles to the bottom and the slime and water overflow through discharge holes in the side of the vat. The level of discharge is raised from time to time as the level of the collected sand rises. When the vat is full, the residual water is drained off through the filter bed at the bottom of the vat. The construction of the

Fig. 151.—Collecting Tanks, Champion Reef Cyanide Plant.

filter bed is shown in Fig. 148.[1] The collected sand is usually transferred to other vats for treatment with cyanide.

(b) **The Butters and Mein distributor** is shown in Figs. 149 and 150.[2] It is an automatic revolving apparatus working on the principle of the garden

---

[1] Schmitt, *Rand Metallurgical Practice*, vol. ii., p. 203.   [2] *Op. cit.*, p. 205.

sprinkler. Pipes of several different lengths conveying the pulp revolve round a central spindle, and spread the pulp evenly over the surface of the water with which the vat is filled. The sand settles to the bottom and accumulates there, while the slime overflows at the top. The vat is filled with water before sand collection is begun. The collecting tanks of the Champion Reef cyanide plant are shown in Fig. 151.[1] When the accumulation of sand in the vat reaches nearly to the top, the water is drained off and the ore discharged by shovelling, or by means of machinery, such as the Blaisdell vat excavator (see p. 355), in which steel discs attached to revolving arms push the sand towards the centre of the vat. In either case the sand falls through apertures in the bottom into ore-cars, or on to a travelling belt below. The sand is now in a suitable condition for leaching, containing only small quantities of slime. Sometimes the pulp collected in this way is treated directly without removal. In other mills, where the method of "double treatment" of the tailing is employed, the first solution of cyanide is directly applied to these vats, and, after draining, the pulp, wetted with cyanide solution, is transferred to the second treatment vats. This has the advantage of an additional aëration of the charge during treatment. Double treatment is made cheaper by using two super-imposed vats. As already stated, the usual method is to remove the collected sand to the cyanide vats for treatment.

The slime overflowing from the collecting vats is treated with the rest of the slime which has previously been separated from the pulp by cone or other classifiers.

(c) **Caldecott's Continuous Collection.**—In this method collecting vats are not used and time is saved. One arrangement is shown in Fig. 152.[2] In this case sand from which the surplus water has been drained on Caldecott's sand filter table is charged into the treatment vat. In other arrangements now generally

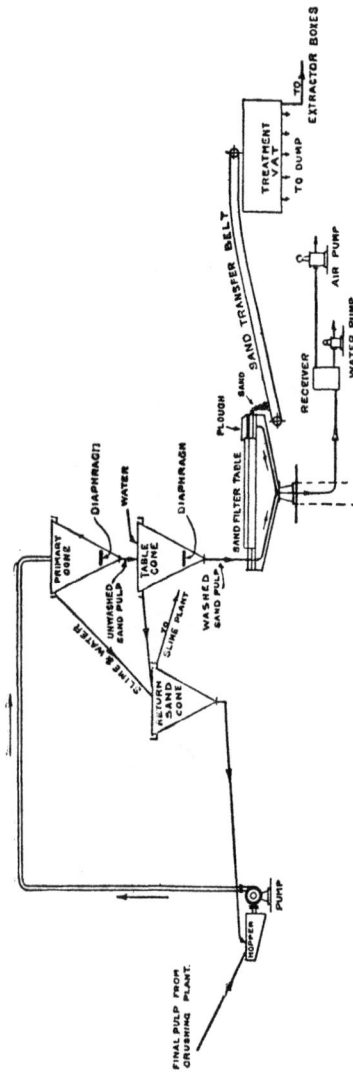

Fig. 152.—"Caldecott" Continuous Sand-collecting Plant.

---

[1] Reproduced with the permission of the Cyanide Plant Supply Co.
[2] *Rand Metallurgical Practice*, vol. ii., p. 208.

used the sand is mixed with cyanide solution in a hopper and forced through pipes to the vat by a centrifugal pump. The sand filter table is merely a de-waterer, and consists of a rotary filter bed with a vacuum pump, and a stationary plough to remove the comparatively dry sand. The

Fig. 153.—Sand Filter Tables at Simmer and Jack Mine, 1913.

pulp is distributed at a point about 3 feet behind the plough, so that the sand makes almost a complete revolution on the table before being removed.

The sand filter tables at the Simmer and Jack Proprietary Mines in 1913 are shown in Fig. 153, which is from a photograph kindly sent by Dr.

W. A. Caldecott. The tables are each 25 feet in diameter, with 3 feet breadth of filter, and a daily capacity of 750 tons of sand. They receive the thick sand pulp underflow from the diaphragm cone classifiers. The sand can be collected, aërated and leached in the same vat when these tables are employed and the dissolving of the gold by cyanide begins within half an hour of the ore being crushed in the battery. For the collection of slime, see pp. 356, 360.

2. THE CYANIDING OF SAND.

This is effected in round vats, constructed of wood, concrete, or mild steel. If wood is used, it is covered with a coating of paraffin paint or a mixture of asphaltum and coal-tar. Concrete and brick vats are not now advocated owing to their great cost and less convenient working, and wooden vats suffer more than iron and steel ones from exposure and from being alternately emptied and filled.[1] Steel vats are painted inside and out to prevent rusting. The steel vats at the Jumpers Deep Mine are 40 feet in diameter and 7 feet deep inside. The sides are of $\frac{1}{4}$-inch steel plate, and the bottoms are $\frac{5}{16}$ inch thick. The plates are riveted together and strengthened with angle iron at the top and bottom of the vat.

Wooden vats are made of staves, 4 to 6 inches wide and 3 inches or more thick, held together by round iron hoops, with bottom planks 3 inches thick, fitting into a slot in the staves. Sometimes in new mining districts square tanks are built, as being cheaper and more easily constructed, but they do not last long and are more difficult to keep tight.[2]

The dimensions of vats vary with the work to be done. In calculating the capacity of a leaching vat, the volume of a ton of collected sand on the Rand is taken as 21·5 cubic feet (C. O. Schmitt), and that of transferred sand as 26 cubic feet. When settled, clean Rand sand occupies about 23 cubic feet per ton (Caldecott).[3] On the Rand vats are from 25 to 65 feet in diameter, and from 7 to 10 feet in depth. If the material to be treated is such that percolation is difficult, the depth of the vat is kept small, and the diameter made as large as possible. In the direct treatment of dry-crushed ores (see p. 347), this is usually necessary, as they always contain some slime. A depth of 2 or 3 feet of dry-crushed ore is usually as much as can be conveniently leached, and vacuum pumps are often added to expedite the work.

The vacuum may be obtained by a direct-acting pump, or by the use of a large boiler, in which a vacuum is created by a Westinghouse or other pump. As soon as the pressure in the boiler falls to about half an atmosphere, it is connected with the aperture of the vat below the filter-bed. The rate of leaching is often doubled by the diminution of the pressure, below the filter-bed, to half an atmosphere, and in some cases it is increased from $\frac{1}{2}$ or 1 inch to 7 or 8 inches of liquid (in the leaching vat) per hour.

The false bottom is usually a wooden framework, constructed of boards pierced with numerous auger holes, or in larger vats of wooden slats crossing each other. The framework is covered by cotton twill, canvas or cocoa-nut matting, which are not so rapidly destroyed by the solution as they are in the chlorination process. Thick canvas duck, resting on matting, forms a

---

[1] J. Yates, *Metallurgical Engineering on the Rand*, p. 97.
[2] A. James, *op. cit.*, p. 19.
[3] For the capacity of vats of various dimensions, see *Rand Metallurgical Practice*, vol. ii., pp. 198 and 216.

trustworthy filter-cloth. The filter-cloth is protected by wooden slats, which prevent it from being injured by workmen's shovels in clearing out the charge. The construction of a filter-bed is shown in Fig. 148.

An iron pipe communicates with the space below the false bottom, and conveys the liquid to the pumps or to the zinc boxes. The solution does not attack wood or iron ; brass and bronzes are attacked and corroded rapidly.

The vats are filled to within a few inches of the top, and the charge is levelled by means of hoes. The amount of ore charged in is such that, after the solutions have been applied, the surface of the charge may stand at about 12 inches below the rim. In levelling the ore, the labourer must not step into the vat or forcibly press down the ore, as irregular filtering is produced by this cause. The shrinkage of the charge on the addition of liquid is from 10 to 18 per cent. The ore is charged in as dry as possible, but a few per cent. of moisture (up to say 15 per cent.) makes very little difference to the subsequent leaching.

The use of wash-water to remove soluble salts is not required with fresh tailing, and lime being now added in the battery, it is not often required at this stage to neutralise acidity. A solution of lead acetate is sprinkled on to the charge before it is transferred to the treatment vat. " From 15 to 25 lbs. dissolved in water will be found sufficient to precipitate the soluble sulphides from a charge of 750 tons of current sand " on the Rand.[1]

The " strong " solution of cyanide is then run on. " From 25 to 30 per cent. by weight of the charge of strong solution of 0·12 per cent. KCy will be found sufficient to treat the majority of Rand sand, and also to keep the solutions in circuit up to strength. The practice of closing the leaching cock and allowing the solution to saturate the charge thoroughly has the drawback of expelling the air. . . . A better method is to pump sufficient solution on to the charge to cover it to about a foot in depth, and as fast as this solution leaches down, to follow with more until the desired quantity has been added."[2] The charge is then drained and air drawn through by the vacuum pump.

The strong solution is usually conveyed at once to the zinc boxes, but it was formerly preferred either to raise it and pass it through the charge again (circulation method), or to transfer it to a second or even a third charged vat before precipitating the gold. The advantage of these " circulation " and " transference " methods is that the solutions become much richer in gold than if they were only allowed to percolate through a single charge of ore, and consequently they give a cleaner deposit on the zinc with much less consumption of cyanide, the volume of solution passing through the precipitation boxes being less. At the Mercur Mine the circulation was formerly kept up for from 24 to 240 hours, according to the speed of leaching, the usual time being about 60 hours. At the Robinson Mine 20 tons of solution covered the ore in a 75-ton vat, and were continually pumped back into the same vat for 36 hours, and then passed to the zinc boxes. Such methods are now abandoned, partly from fears of re-precipitation of gold in the leaching-vat.

When the strong solution has drained away and aëration is complete, the weak solution containing 0·02 or 0·03 per cent. KCy is run on. The amount used is four or five times as much as that of the strong solution. The weak

[1] J. E. Thomas, *Rand Metallurgical Practice*, vol. i., p. 161.
[2] J. E. Thomas, *ibid.*, p. 165.

solution is drained off, and followed by a final wash with water or by an extremely weak solution such as waste liquors from the slime plant, in order to remove cyanide and soluble gold as completely as possible.

The whole time of treatment in the sand vats on the Rand is seven or eight days, and the ratio of solution to sand about two to one by weight (Thomas).

Strength of Solutions.—Although the strong solution contains only 0·12 per cent. KCy on the Rand, a strength of 0·5 per cent. or more is necessary in the treatment of some ores, especially silver ores.

Stronger solutions were used in the early days of the process, but were found to be unnecessary. In 1895, Caldecott[1] made a series of experiments on a working scale to test the dissolving powers of solutions of different strengths. Five similar vats were charged with 34·03 tons of fresh slightly-pyritic tailing of uniform composition from the Jumpers battery, containing 100 grains of gold per ton. The treatment that each vat received was exactly the same except that the strength of solution employed was different in every case. To each charge, 15·31 tons of solution were added, and circulated for three days, after which the solution was drained off and displaced by 8·4 tons of wash-water. When the charges had drained dry, they were sampled, the result being as follows :—

TABLE XXXII.

	I.	II.	III.	IV.	V.
Strength of cyanide, per cent.—					
(1) Before treatment,	0·041	0·110	0·373	1·021	3·333
(2) After treatment,	0·010	0·068	0·277	0·860	3·020
Loss of cyanide (lbs. per ton),	0·34	0·48	1·1	1·9	3·6
Extraction (per cent.),	83	84	83	85	83

The stronger solutions may have dissolved the gold much more rapidly than the weaker ones did although the final results were the same. The conditions under which the experiments were carried out prevented that point from being determined. Since, however, it takes two or three days in practice to wash out soluble salts from Transvaal tailing, there is little or no real advantage to be gained by dissolving the gold quickly. Crosse also found[2] that almost equally good results were obtained by treating rich Bonanza tailing with solutions containing KCy varying in amount from 0·1 to 0·5 per cent., the time of treatment being twenty-four hours.

For the composition of ordinary working cyanide solutions used on the Rand, see p. 407.

Sumps.—The solutions, after passing the zinc boxes, are stored in sumps, which are now usually large steel tanks. These may be placed either at a lower or a higher level than the leaching vats. The former is the more convenient plan. A single storage tank to hold enough solution for one day's work is then placed above the leaching tanks, and is filled by pumping from the sumps once a day. In the alternative, a single sump of moderate size

[1] Caldecott, *J. Chem. Met. and Mng. Soc. of S. Africa*, 1896, **I**, 293.
[2] A. F. Crosse, *ibid.*, p. 326.

receives the liquid from the boxes, and as soon as it is nearly full its contents are pumped to a large vat above the leaching tubs.

Disposal of the Tailing.—This is sometimes sluiced out of the tanks by

Fig. 154.—Disposal of Sand Tailing in Underground Workings by Truck Transport.   Robinson Deep Mine.

water under pressure. Where the supply of water is not large, or the fall of the ground is insufficient for sluicing out the tailing—*e.g.*, on the Rand— it is removed by shovelling out or by the Blaisdell excavator. The vats are furnished with doors in the bottom, which are usually circular, constructed of steel, and about 16 inches in diameter. The sand falls into trucks, which are hauled to the dump, or on to a conveyor belt. On the Rand the sand residue from several mills is now returned underground into old workings to prevent subsidences.[1] It may be transported by pumping through pipes or on trucks, as at the Robinson Deep (see Fig. 154, which is from a photograph kindly supplied by Dr. Caldecott). Here the sand is dumped from the trucks into an inclined (20°) tunnel and sluiced down with water and a solution of permanganate of potash as cyanicide supplied from the dissolving box in the left foreground.

Samples of the issuing solutions are taken for assay during treatment, and samples for assay of the residue are taken from the ore-trucks, usually by means of a long iron semi-circular probe, shaped like a cheese-taster, which is thrust to the bottom of the vat, then revolved by means of the handle, and withdrawn with the tailing adhering to it.

Figs. 155 and 156[2] show a simple form of a cyanide plant for the treatment of sand.

3. THE CYANIDING OF SLIME.

The most useful definition of slime is probably that given in Chap. X. on p. 230. A good definition given by H. A. White[3] describes slime as "that portion of crushed ore which, owing to its minutely subdivided condition and the presence of colloidal substances, settles very slowly in water and cannot be leached without extra pressure." The slime separated from sand by the methods given in Chap. XI. and on p. 348 is suspended in water, and to promote its settlement a solution of an electrolyte, in practice always lime, is added. This causes agglomeration or flocculation of the particles, "clouds of large and indefinite diameter" are formed, and these quickly subside in the liquid. The largest flocks, according to White, are produced with an alkalinity of about 0·03 per cent. In practice with an alkalinity of not more than 0·005 per cent., Rand slime settles at the rate of 2 to 4 feet per hour with a clear overflow. Heat assists settlement.[4]

There are two principal methods of slime treatment, viz. :—

A. The decantation method, used in South Africa, but in course of displacement by B. A modified method, counter-current decantation, is now being used in America.

B. Agitation and slime filtration used elsewhere in many forms.

These are described in succession below.

A. The Decantation Method.—Lime is added to the stream of slime on entering the slime plant. Among the devices for adding the lime may be mentioned the use of a grinding pan, into which lime is fed, and out of which it is carried by a small stream of water as milk of lime. Ordinary Rand pyritic banket requires about 6 lbs. of lime per ton of slime, in addition to that fed into the stamp mill. The slime enters the *collecting vat*, which has a conical bottom, and may be 50 feet in diameter, through a vertical pipe

---

[1] Caldecott and Power, *J. Chem. Met. and Mng. Soc. of S. Africa*, Sept. 1913.
[2] Gowland, *Non-Ferrous Metals*, p. 218.
[3] White, *Rand Metallurgical Practice*, vol. i., p. 189, where a full discussion on slime is given.
[4] A. Salkinson, *J. Chem. Met. and Mng. Soc. of S. Africa*, 1907, **7**, 403 ; also 1908, **8**, 142.

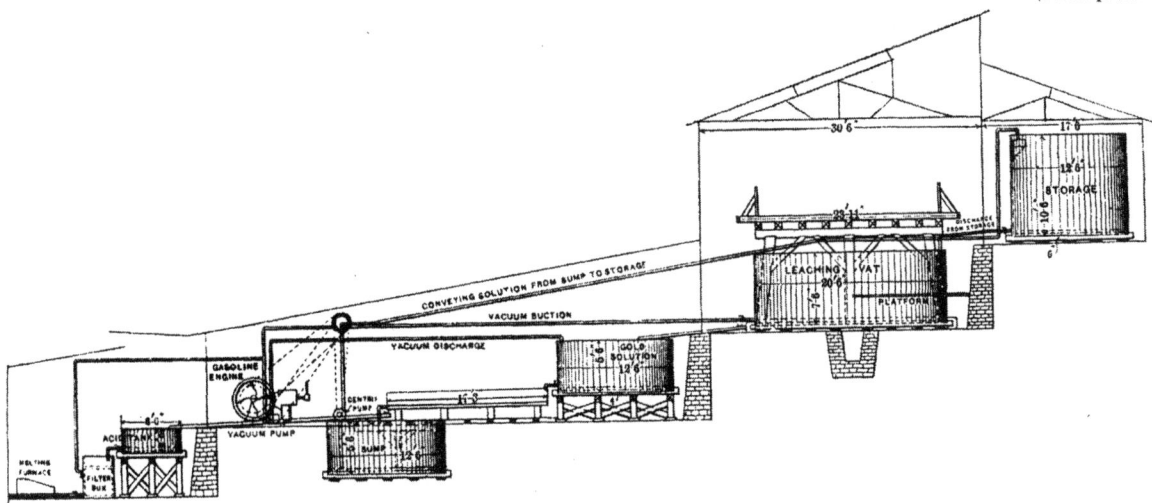

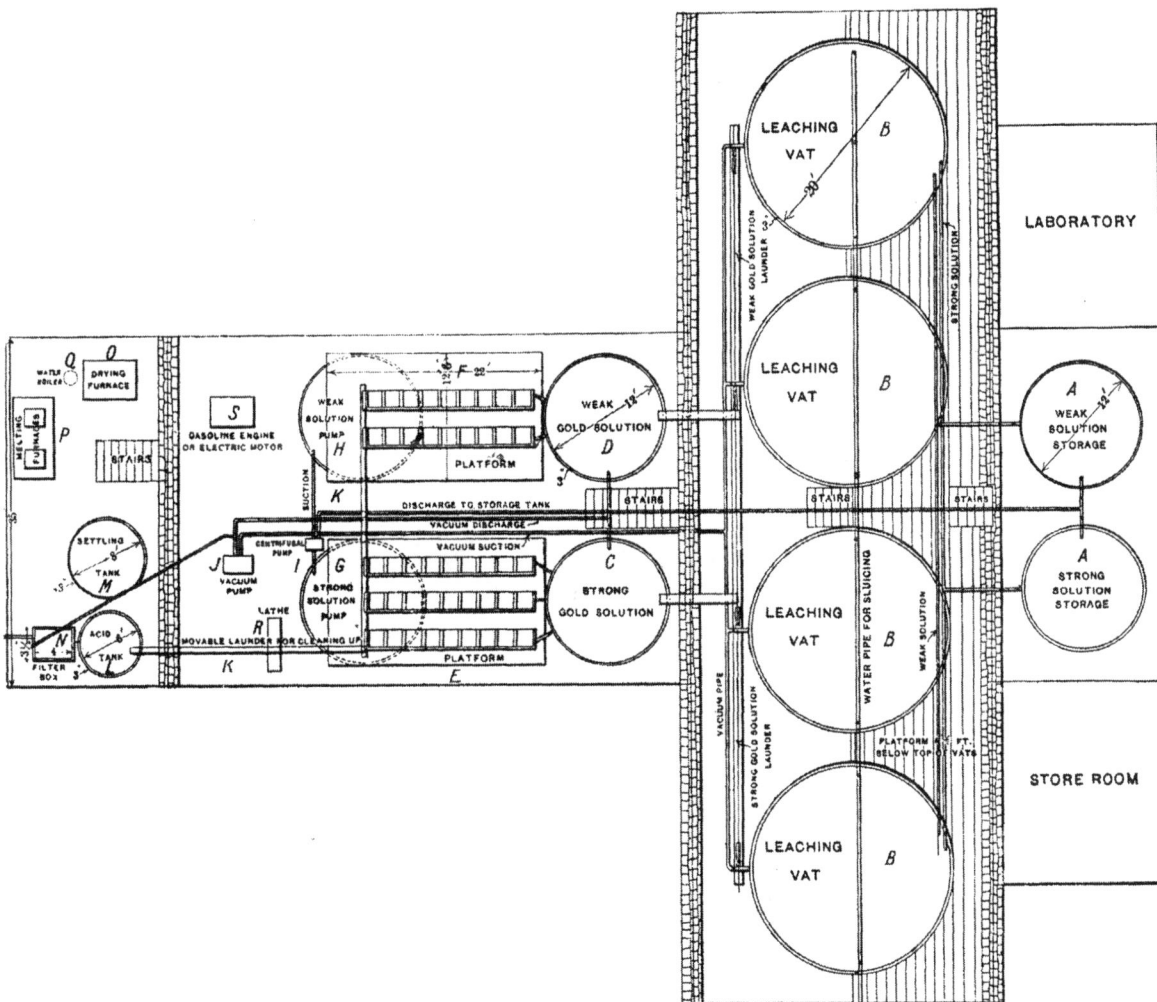

Figs. 155 and 156.—Simple Cyanide Plant.

entering from above in the centre of the vat. The stream is delivered below the surface, and strikes a baffle plate, which spreads it out horizontally. Air bubbles pass up through a second larger pipe or sleeve, which surrounds the feed pipe. The slime settles and clear water overflows the rim of the vat all round, into a launder, at the same time that thin pulp is entering through the feed pipe.

When the collecting vat has received its charge of settled slime, the supernatant water is drawn off through a *decanter* (see Fig. 157),[1] which consists of a hinged pipe so arranged that the height of its intake can be adjusted to a point near the surface of the settled slime. The slime (in which the ratio of solid to liquid is about 1 to 1) is then sluiced out through the apex

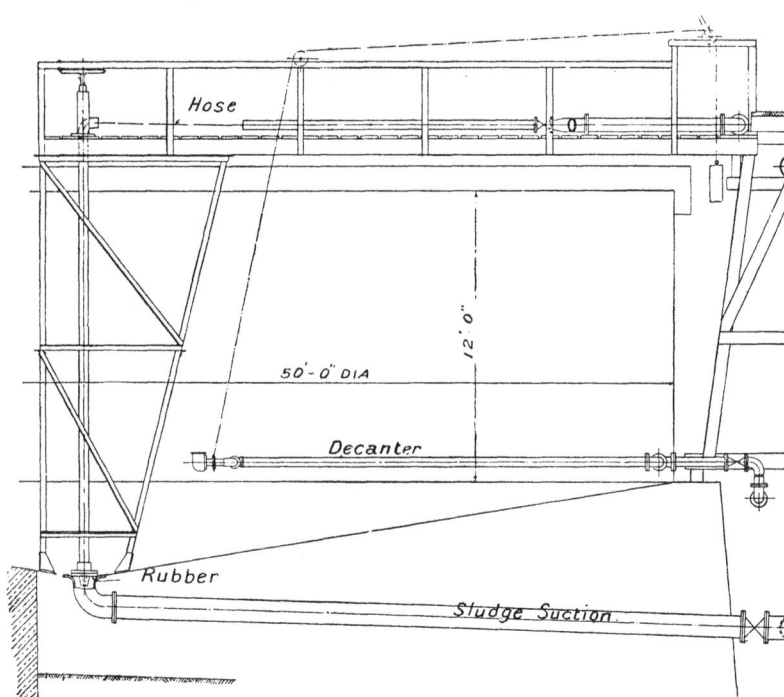

Fig. 157.—Vat with Decanter.

of the conical bottom by means of a jet of cyanide solution, and carried to a pump, by which it is delivered tangentially into other vats, giving the charge a rotary motion, so as to mix it thoroughly and keep it in suspension. The ratio of solution to solid is now about $3\frac{1}{2}$ or 4 to 1. The strength of the solution is 0·01 or 0·02 per cent. KCy and 0·004 to 0·02 per cent. free alkali. About 80 per cent. of the gold is dissolved in the passage through the pumps, but agitation is continued for a few hours by withdrawing the solution at the bottom and discharging it in oblique jets at the top. In the alternative, the charge may be transferred to an *intermediate vat*, and thence to the *first settlement vat*. Lead acetate solution is added to the charge

[1] Dowling, *Rand Metallurgical Practice*, vol. i., p. 217.

in the collecting vat in order to precipitate soluble sulphides. Aëration is effected by air drawn into the pumps and by air bubbles carried down by the sludge entering the vats. The solution also becomes aërated in passing through pipes and pumps.

After the gold has been dissolved, the slime is allowed to settle, the solution removed by decantation, and the pulp sluiced out into the *second settlement vat*, with " precipitated solution " which has been through the zinc boxes and still contains some gold and cyanide. This second wash is used again for the first treatment of the next charge. The settled slime, still containing about one-sixteenth of the dissolved gold, is then diluted and discharged to the dam. Settlement of slime is facilitated by the use of warmed working solutions, for which purpose waste steam is commonly employed (Caldecott).

The first solution is clarified by being passed through a sand filter vat or Johnson filter press (*q.v.*), and then goes to the zinc boxes. Filter presses are being superseded by sand filter vats for this purpose. The solutions are generally heated (say to 86° F.) by waste steam. Centrifugal pumps are in general use.

Although the gold in slime fresh from the battery is very readily soluble in cyanide solution, it is quite otherwise with slime which has accumulated in dams and settling pits and has been exposed to the weather for some time. The presence in these materials of finely divided ferrous sulphide, ferrous hydroxide, and other ferrous salts and decomposing organic matter gives rise to the rapid abstraction of oxygen from the cyanide solutions used in their treatment, and the solution of the gold is in this way absolutely prevented. Even in fresh slime some ferrous sulphide, produced during the crushing in the battery, may be present, as Caldecott has shown that this substance is produced by triturating pyrite in a mortar. The obvious remedy for this difficulty is to supply oxygen artificially, either in the form of air delivered into the agitation pump, or in the form of an oxidising agent, such as potassium permanganate, a method now abandoned. In 1897 Caldecott [1] applied air passed through perforated pipes in the bottom of the vat to the oxidation of accumulated slime, and experimented in the use of the air-lift for slime treatment as a means of agitation.

The *Usher process* [2] or Adair-Usher process is a modification of the decantation process, by which it is made more continuous. The charge is collected and the gold dissolved as usual, but after settlement has proceeded so far that there are 3 to 6 inches of clear solution above the slime, decantation is begun and at the same time precipitated solution is added through horizontal perforated pipes at the bottom of the charge. The barren solution displaces the gold solution which rises upwards through the slime, and is drawn off at the top. Theoretically more dissolved gold should be extracted with less solution by this method than by ordinary decantation.

The decantation method described above has several disadvantages. It requires a number of large vats and much space, and the slime is discharged still containing 6 or 7 per cent. of dissolved gold. The residue is, therefore, unduly rich, and the filtration processes described in the next section are preferable in this respect, as they separate 98 per cent. or more of the soluble gold. Besides this, the amount of gold solution requiring precipitation

[1] Caldecott, *J. Chem. Met. and Mng. Soc. of S. Africa*, July, 1897, **2**, 100.
[2] Adair, *J. Chem. Met. and Mng. Soc. of S. Africa*, 1908, **8**, 331; Dowling, *Rand Metallurgical Practice*, vol. i., p. 225.

yielded by the decantation method is about 3 tons per ton of slime, whereas by filtration 2 tons or less of gold solution are produced.

Dr. W. A. Caldecott observes that for Rand slime—" in general the advantages of vacuum filtration increase with the value of the slime treated, but assuming with vacuum filtration the same capital expenditure on plant as for a decantation plant of equal capacity, 4 per cent. higher recovery, and 3d. per ton higher working cost, it would follow that for slime above 1·5 dwts. in value vacuum filtration is preferable to decantation, and has the advantage of being independent of fluctuations in the rate of settlement due to variations in temperature between summer and winter. In either case the ordinary slime collecting vats are employed and the residue discharged by pump as pulp."

Eleven vacuum filter plants were in operation on the Rand in June, 1914.

A later method of continuous decantation, reminiscent of the Usher process, has been made possible by the success of the Dorr thickener (q.v.), and has been developed in North America. The method is usually called *counter-current decantation.* The pulp is washed in a series of thickeners, being diluted after thickening and transferred to another thickener, and lastly goes to a vacuum filter, which, however, has been omitted in some mills where the solution itself is low in cyanide and contains small metal values.[1] The method is in use at some mills at Porcupine, Ontario, and elsewhere.

The advantages and limitations of the method are discussed by H. St. J. Brooks,[2] who considers it inapplicable to ores in which the greater part of the gold and silver is locked up in sulphides, so that long-continued agitation with cyanide is required to dissolve the metals. High-grade ores requiring strong solutions are also unsuitable. He would apply the system to low-grade siliceous ores requiring solutions of only moderate strength, and finish with filtration.

B. **Treatment of Slime by Agitation with Cyanide and Filtration.**—The difficulty in the filtration of slime is that even a thin layer of slime packs down and offers great resistance to the passage of liquids. A pressure much greater than that of the atmosphere is required, and in practice layers of ½ inch to 1 inch are used with vacuum leaching or layers of 2 to 3 inches with pressures of 40 to 100 lbs. per square inch given by pumps. An enormous area of filtering surface is required for operations on a large scale, and this is obtained by using a number of parallel plates or leaves placed side by side. The dissolution of the gold is easy and rapid, and is usually effected by agitation with cyanide, although some exceptions are noted below.

When slime is forced against a filtering surface either by the pressure of the atmosphere (vacuum filtration) or by direct and higher pressure, the slime forms a coherent and approximately homogeneous layer or cake, through which liquids pass almost evenly. There is little tendency for the formation of channels, and so washing is complete and satisfactory with little water.

Warwick discusses the matter as follows :—[3] " In removing the soluble values from slime cakes, the principle of displacement is used rather than the laws of continuous or repeated dilution." In cakes of uniform permeability the wash-water would pass through like a wall pushing the gold-bearing or " pregnant " solution before it. In practice a wash equal in volume

---

[1] H. A. Megraw, *Eng. and Mng. J.*, May 23, 1914, p. 1061; also Oct. 17, 1914, p. 683.
[2] Brooks, *Mng. and Sci. Press*, April 26, 1913, p. 624.
[3] A. W. Warwick, *Mng. Eng. World*, 1913, **38**, 665, 797, 1135; *Mineral Industry*, 1913, p. 347.

to the moisture in the cake and filter leaves will remove 80 to 85 per cent. of the dissolved gold from the slime cake. In order to recover 98 per cent. of the dissolved gold, a volume of wash equal to 1·5 to twice that of the retained solution must be used. To avoid increasing the volume of mill solution unduly, the larger part of the wash consists not of water but of barren solution, low in cyanide. This results in a residue low in gold but containing some cyanide, which is thrown away with the residue.

The filter surface is usually vertical, but the slime cake forms and gradually thickens or builds up on the canvas, adhering well so long as the vacuum is maintained. Some material builds up a slime cake readily and quickly—e.g., 1 inch thick in a few minutes—but with argillaceous or talcose slime the building-up is slower. If cracks form in the cakes, washing becomes impossible, as the liquid follows the line of least resistance.

In filter-pressing, the pulp to be filtered is thickened, in order to make settlement inside the presses less likely to occur. Such settlement would imply the formation of non-homogeneous cakes, with the result that the washing would not be uniform.

There are many different slime filters in use, and sufficient time has not yet elapsed for the best types to drive the others from the field. Several well-known varieties are described below.

Thickeners.—It is necessary that slime should be de-watered or thickened before it is treated in agitators. One method of doing this is by settlement in large vats, as described above on p. 357. The appliances for de-watering sand (cones, filter table, etc.), are obviously not suitable for slime. One successful slime de-waterer in wide use is the *Dorr continuous thickener* (see Fig. 158).[1] This consists of a vat from 20 to 35 feet in diameter and from 8 to 12 feet deep, in which a central vertical shaft, reaching nearly to the bottom, and carrying four radial arms, revolves slowly. Pieces of angle iron attached to the arms are so placed that they move the settled thickened pulp to the centre of the bottom of the vat, where it is discharged through a pipe. The thin pulp flows into the tank at the centre, just below the surface, and the clear liquor overflows continuously at the periphery over a lip or through perforations. The shaft can be raised and lowered while it is running.

At the Liberty Mill in 1908, where the size of the vat was 33 feet by 10 feet, the speed was 4·8 revolutions per hour, the feed was 120 tons solid and 660 tons solution, and the discharge at the bottom contained the 120 tons solids and 288 tons of liquid, or in the ratio of 1 to 2·4. It is claimed that a ratio of 1 of solid to 10 of water can be reduced to a ratio of 1 to 1½ in continuous work or of 1 to 1 in intermittent work. The theoretical advantages of the method of thickening by removing slime at the bottom as fast as it accumulates there are discussed by H. G. Nichols.[2]

Agitators.—The method of agitation in the decantation method is described above, p. 357. It consists in withdrawal from the bottom of the vat and delivery at the top by centrifugal pumps. In Western Australia, revolving-arm or paddle agitators were used. "In these it is not possible to use a thick or very sandy pulp, and in the majority of them air is introduced into the pulp, generally by an air jet at the bottom of the vat, but in some cases in small jets along the agitator paddles, it having been delivered through the hollow shafting of the agitator gear. The smaller sizes of these mechanical

---

[1] Gowland, *Non-Ferrous Metals*, p. 245.
[2] Nichols, *Trans. Inst. Min. and Met.*, 1908, **17,** 293.

agitators are often arranged so that the stirring gear can be raised out of the pulp, when a stoppage is imperative." [1]

Modern agitators may be classified as follows :—

      1. Air agitation—
          The Brown agitator.
          The Parral vat.
          The Just agitator.
      2. Solution jet agitation—
          The Trent vat.
      3. Screw propeller agitation—
          The Hendryx agitator.

Other forms, such as the Dorr agitator, are in use.

Fig. 158.—Dorr Continuous Thickener.

The Brown Agitator or Pachuca Tank.—This agitator was introduced in New Zealand in 1902, and afterwards adopted at Pachuca, in Mexico. Such vats are also known as air-lift vats. " In this type of vat the agitation is effected by lessening the specific gravity of a central column of pulp, by the introduction of air at just such a pressure as will overcome the pressure of the column at the point of introduction " (Gowland). In Fig. 159 [2] is shown a vertical section of a Brown agitator. It consists of a tall cylinder of steel of great height in comparison with its breadth, and conical at the bottom. Inside the cylinder is a smaller central pipe, B, extending to a point near the

[1] R. Allen, *J. Chem. Met. and Mng. Soc. of S. Africa*, 1911, **11**, 428.
[2] Gowland, *Non-Ferrous Metals*, p. 246 ; F. C. Brown, *Mng. and Sci. Press*, Sept. 26, 1908.

bottom. Air is introduced into this central pipe through the internal pipe, D, which discharges at G through a rubber valve. C is a second air pipe outside the tube. It serves to keep the pulp in motion during the filling and discharge of the vat. The pipe, E, is used for supplying solution, water

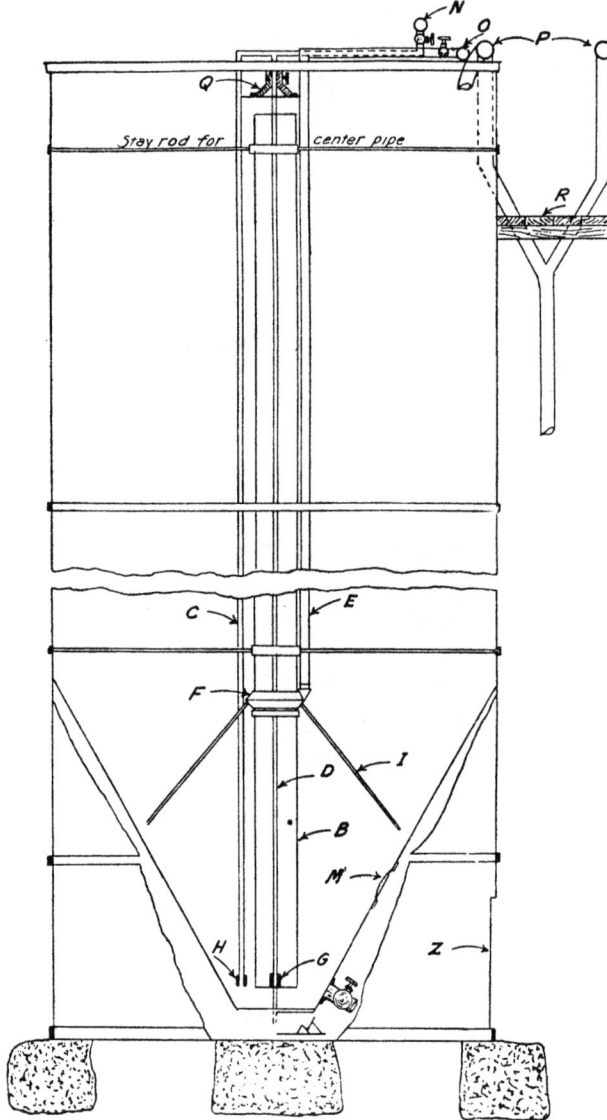

Fig. 159.—Brown Agitator.

or air to the distributor, F, which discharges through the pipes, I, to prevent packing before agitation is started and when it is stopped. Air is supplied to pipes C, D and E from pipe O, and water or solution from pipe N. An adjustable splash plate, Q, is fitted near the top of the central tube. The slime pulp is fed through the pipe P. The discharge after agitation takes place through a pipe with a stopcock shown close to the apex of the cone. The vats are of various sizes, such as 45 feet in height and 15 feet in diameter.

The method of working is as follows :—As soon as the tank is filled with

Fig. 160.—" Brown " Agitators at the East Rand Proprietary Mines, Johannesburg.

slime and solution, air is turned on in the pipe D, and as it bubbles up through the central tube it lightens the column of pulp inside and rapidly lifts it, causing it to overflow. Fresh pulp is drawn in at the bottom, and a perfect circulation is given.

In starting up old charges, or those that have been allowed to pack, the material at the bottom of the cone is softened by the introduction of water or solution through the wash-ring F, so that it may be readily lifted through the tube B or upcast column ; or a suspended pipe with compressed air jet may be left to bubble in the packed centre tube until it works its way down to the cone, when agitation quickly becomes general.[1] The use at the Treadwell Mine of a " Spider " or adjustable hollow annular casting with radiating fingers is described by W. P. Lass.[2]

In beginning agitation, the air pressure required is considerable, but when circulation is fully established, the air pressure falls off, and to avoid excessive circulation the quantity of air is kept moderate in amount. As originally used, the Brown agitator was intermittent in its action, but more recently four to six vats have been placed side by side, and worked continuously as a series.[3] In this case pulp is continuously fed into the first one near the entrance to the air-lift, and is drawn off into a pipe placed about midway between the air-lift and the periphery of the vat, and conveyed to the second vat, and thence in succession to the others. After passing through the series of four vats, the pulp is discharged to the filter plant.

Fig. 160 is from a photograph of the Brown agitators at the East Rand Proprietary Mines, Johannesburg.[4] These four agitators are stated to be continuously agitating 1,600 tons of slime a day for an extraction of 97 per cent.

The Parral Vat (Fig. 161)[5] has a flat bottom and two to four air-lifts near the periphery of the tank instead of one at the centre, as in the Pachuca tank. In this way vats of greater diameter and capacity are made possible. An elbow or turn is fixed to the top of the air-lift pipe, and the pulp is delivered circumferentially so that the contents of the vat acquire a rotary motion, which is designed to avoid the settling of solids, and assists agitation. The Parral vat is generally made of less height than the Pachuca tank.

The Just Agitator.[6]—In this agitator air from a blower is introduced through a special porous brick or " silica sponge " floor in the bottom of the vat, agitating 9 feet of pulp above it. The bricks are not so porous that the slime can pass through them. The agitator was introduced at Guanajuato, Mexico, about 1910. The silica sponge soon became clogged with lime salts.

The Trent Vat [7] is also flat-bottomed. Circulation is effected by drawing off the thin slime from the top of the tank and forcing the liquor by means of a centrifugal pump through a central pipe (see Fig. 162)[8] to near the bottom of the tank. The central pipe has four arms or pipes radiating from it, which are fitted with short jet pipes or nozzles inclined downwards toward the bottom of the tank. The pressure of the solution and slime discharging through these restricted nozzles causes the arms to revolve at a speed sufficient for thorough agitation. Air is admitted to the suction part of the pump

---

[1] J. E. Alley, *Mng. and Sci. Press*, July 27, 1912, p. 118.
[2] Lass, *ibid.*, Oct. 21, 1911, p. 517.
[3] Mennell and Grothe, *Mineral Industry*, 1909, p. 348 ; Kuryla, Grothe and Lamb, *Mineral Industry*, 1910, p. 302.
[4] By courtesy of the Cyanide Plant Supply Co.
[5] Gowland, *Non-Ferrous Metals*, p. 248. See also *Eng. and Mng. J.*, Feb. 21, 1914, p. 422.
[6] *Trans. Amer. Inst. Mng. Eng.*, 1910, **41**, 371.
[7] *Eng. and Mng. J.*, May 21, 1910. p. 1066.
[8] Gowland, *op. cit.*, p. 249.

through a small valve, and " the top of the tank presents a seething effer-
vescent surface." The arms revolve within a few minutes after starting

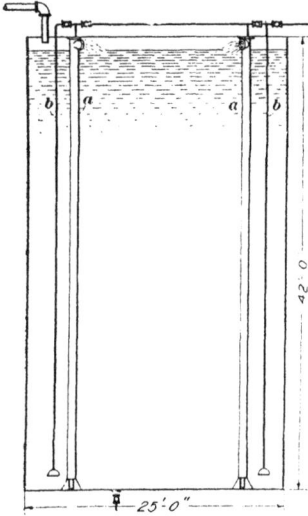

Fig. 161.—Parral Vat.

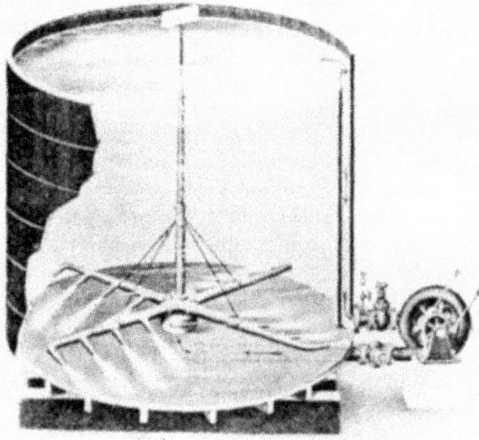

Fig. 162.—Trent Agitation Vat.

the pump, even after the slimes have been allowed to settle for several hours.
The Trent vat was introduced at Tonopah, Nevada, in 1910. Continuous

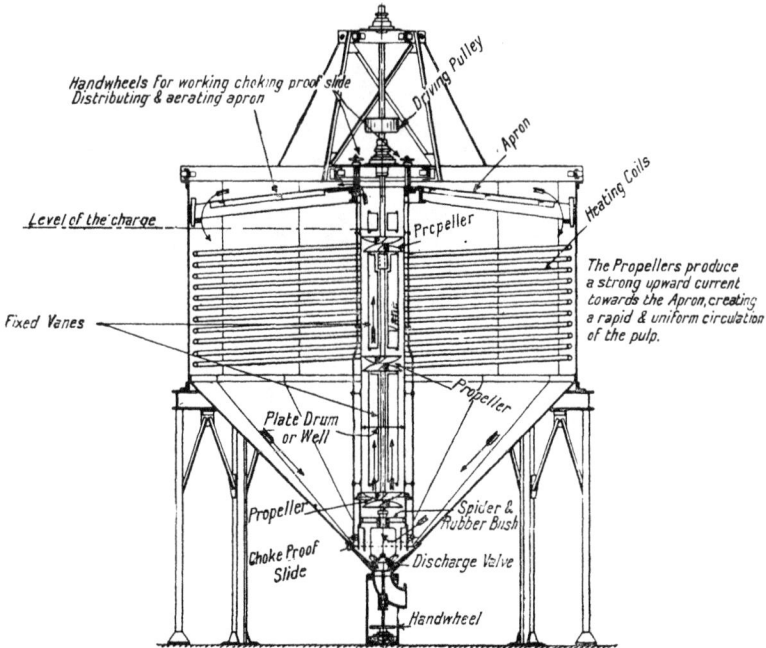

Fig. 163.—Hendryx Agitator.

treatment in Trent agitators at the West-End Mill at Tonopah is described by J. A. Carpenter.[1]

The Hendryx Agitator (Fig. 163)[2] is intended for the treatment with cyanide of sand and slime or of the pulp from tube mills without classification. It consists of a cylindrical tank with a conical bottom. In the centre of the tank is a well or tube, as in the Pachuca tank. In the well is a hollow shaft carrying a number of screw propellers, and driven from the top by a driving pulley. These raise the pulp in the well and so circulate the contents of the vat. The well has a circular apron at the top, which slopes gently towards the circumference, thus spreading out the overflowing pulp in a thin sheet, so that it is aërated by contact with air. A coil of steam pipe in the vat serves to raise the temperature of the charge if that is required.[3] The agitation requires more power than other machines, but clogging owing to stoppages does not take place.[4]

The Dorr Agitator (Fig. 164) can be affixed to any flat-bottomed vat.

Fig. 164.—Dorr Agitator.

It consists of a central vertical cylinder carried by a shaft supported from the top of the tank and equipped with two arms carrying ploughs, as in the Dorr thickener, which travel round near the bottom of the vat and draw the pulp to the centre. The pulp is raised through the cylinder by an air-lift and distributed evenly over the surface of the charge in the vat by revolving launders. The air pipe for the lift is inserted through the bottom of the vat.

The arms are hinged so that they can be raised and stand close to the cylinder during a shut-down. On beginning again, they are lowered by

[1] Carpenter, *Mng. and Sci. Press*, May 3, 1913, p. 646.
[2] Gowland, *Non-Ferrous Metals*, p. 250.
[3] L. D. Bishop, *West. Chem. and Met.*, vol. iii., p. 187 ; *Mineral Industry*, 1907, p. 544.
[4] *Mng. Mag.*, Sept. 1912, p. 228.

degrees until the settled pulp has been brought into suspension, so that there is no danger of the arms being broken. The Dorr agitator is said to be well adapted for continuous agitation, as the vat cannot become partly filled with settled sand during agitation. The tanks used (*e.g.*, 30 feet diameter by 12 feet deep) are of small depth, so that the air pressure required is small. The arms revolve at from 1 to 4 revolutions per minute.

Filtration of Slime. — Slime filters are classified by G. J. Young as follows :—[1]

I. Vacuum or suction filters.

> A. Appliances using a thin slime-cake and practically continuous in their action (Oliver and Ridgway filters).
> B. Appliances using a thick slime-cake and intermittent in their action (Moore and Butters filters).

II. Pressure filters, in which greater pressure is used than is possible with vacuum filters.

> C. Ordinary filter presses (Johnson and Dehne presses).
> D. Sluicing filter presses (the Merrill filter press).
> E. Filtering chambers or cylinders ; filters in which the filtering basket is enclosed in a cylinder (Burt, Kelly, and Sweetland filter presses).

III. Centrifugal filters. (No examples of these are given.)

Brief descriptions of most of these filters are given below.

The Moore Vacuum Filter.[2]—This is the oldest of the vacuum slime filters. It consists of a series of parallel plates or leaves. Each leaf is simply a light framework with canvas on both sides, and is of great dimensions, 20 feet by 4 feet, or later 16 feet by 5 feet. A suction pipe and also two pressure pipes communicate with the interior. When a vacuum is formed the canvas sides are prevented from collapsing by wooden strips and wire netting. A number of these leaves (49 in later practice) are hung (4 inches apart) from a steel frame, forming a " basket " or unit. The basket hangs by cables from an overhead traveller, and can be raised or lowered. The suction pipes communicate with one main pipe.

In operation the filter basket is lowered into the vat containing the thin slime, so as to be submerged, and the vacuum pump started. The slime is agitated to prevent settling. After one or two hours a cake of slime of $\frac{3}{4}$ inch to 1 inch thick is formed on both the outer surfaces of each leaf, and during this time the pump is continuously discharging clear gold solution. The basket is then lifted out of the vat and transferred to a vat containing weak cyanide solution, which is drawn through the cakes for washing purposes for twenty minutes. The washing is finished by ten minutes' immersion in a vat of wash-water with continuous pumping. The basket is then lifted out and run over the discharge hopper, and after the cakes are dry the suction is for the first time discontinued and a blast of air forced into the leaves,

---

[1] Young, *Mng. and Sci. Press*, Oct. 28, 1911, p. 552 ; *Trans. Amer. Inst. Mng. Eng.*, 1911, **42**, 752.
[2] *Mining Reporter*, Denver, Nov. 12, 1903 ; *Eng. and Mng. J.*, Dec. 5, 1903, p. 855 ; *J. Chem. Met. and Mng. Soc. of S. Africa*, 1903, **4**, 758.

by which the waste slime-cakes are dislodged and fall into ore trucks. The arrangement of the vats is shown in Fig. 165.[1]

The Butters Filter[2] also consists of a number of parallel plates or leaves, but they remain stationary in a filtering vat. Thin pulp after agitation is run into the vat, and as liquid is withdrawn by the vacuum pump more pulp is added. When cakes of slime $\frac{3}{4}$ to 1 inch thick have been formed on the canvas leaves (inside which cocoa matting is placed), the pulp is run off and succeeded by barren solution or water for washing as in the Moore filter. When washing is complete, the cakes are detached by water pressure from within and, disintegrating in the water, are carried away when it is run off.

The canvas filtering surfaces in both these filters become clogged with calcium salts in course of time, and this is removed with dilute (2 per cent.) hydrochloric acid. Both these filters are of great efficiency. Sometimes a lower vacuum gives better results than a higher one.

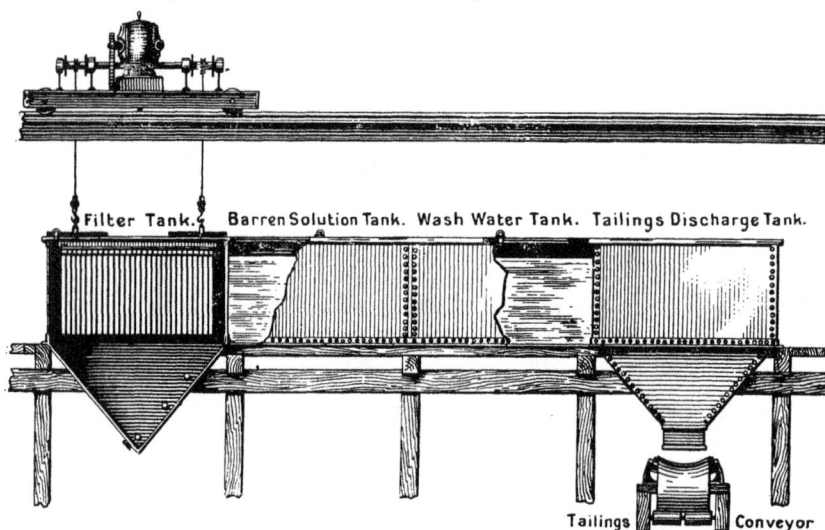

Fig. 165.—Moore Filter Vats.

The Butters filter installation at Brakpan Mines, Transvaal, is shown in Fig. 166.[3] This plant contains 336 filter leaves, and has a capacity of 1,200 tons of dry slime per twenty-four hours. The cost of operation is about 3d. per ton, the slime assays 2·6 dwts., and the average extraction is 95·5 per cent. Air-lift vats are in common use with vacuum filters on the Rand.

The Ridgway Filter.[4]—This machine consists of 12 or 14 cast-iron horizontal filtering frames, which are in the form of sectors of a circle. They are suspended from hollow arms which radiate from a central hollow revolving column, and also run on rollers at the periphery. The frames revolve in a shallow annular trough divided by partitions into three compartments, one filled with slime pulp and the second with wash-water, whilst the third is the discharge chamber. The under side of the frames is the filter surface, and is

[1] Gowland, *Non-Ferrous Metals*, p. 252.
[2] *Mineral Industry*, 1906, p. 411 ; *Mng. and Sci. Press*, June 22 and 29, 1907.
[3] From a photograph kindly supplied by Mr. F. L. Bosqui.
[4] *Mineral Industry*, 1907, p. 540 ; Gowland, *Non-Ferrous Metals*, p. 255.

covered with filter cloth. The frames revolve with suction on, dip into the slime pulp and take on the filter cake of $\frac{1}{8}$ to $\frac{3}{8}$ inch in thickness, rise over the elevated portion and pass to the solution part of the trough, the valve being

Fig. 166.—" Butters " Slime Filter at Brakpan Mines.

automatically changed; here they suck solution through the slime cakes, pass over the second elevated portion (the valve changing automatically to compressed air) and discharge the slime cakes into the third part of the trough,

24

then again making the cycle as described. The total cycle requires 60 seconds, of which about 13 are in pulp, 30 in wash solution, and 17 on the raised portions of the track and in discharging. Each frame has a filter surface of 4 square feet. The machine has a capacity of 30 to 60 tons per day ; it is shown in Fig. 167.[1] In practice its use is limited by the small filtering surface to the treatment of slime which can be rapidly filtered, as, if longer than fifteen seconds are taken for cake formation, the output of the machine is too small. There is also no opportunity of increasing the wash period with the grade of material to be treated. In these respects it is inferior to the Moore filter and to the reciprocating Ridgway filter, an improved form described below.

The Reciprocating Ridgway Filter.[2]—In this machine a set of twenty vertical filtering frames are suspended together, and are lowered into a slime pulp vat and kept there until the frames have received a cake of about 1 inch in thickness. The frames are then lifted out and rapidly immersed in a washing vat for the necessary time, after which the cakes are discharged.

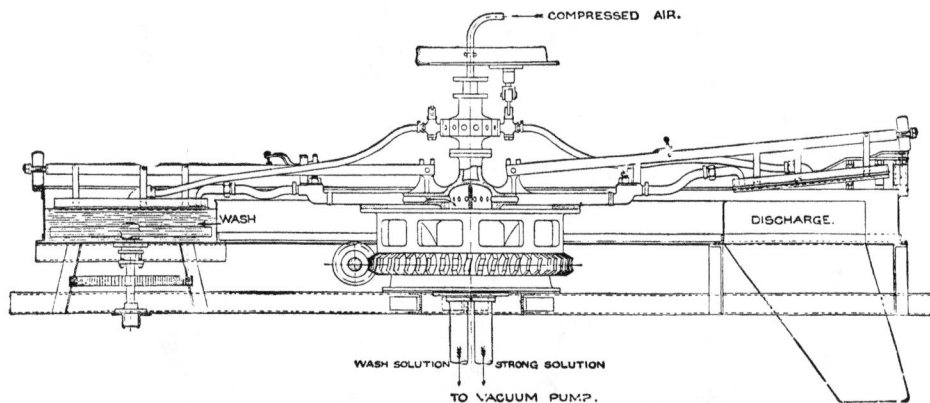

Fig. 167.—Ogle-Ridgway Filter.

This machine is shown in Fig. 168,[3] as installed at the Great Boulder Mine, West Australia. The cakes are being lifted from the pulp vat to the washing tank.

It is claimed that it is the rapid means of transferring the cake to the solution that constitutes the advantage of the filter, " as it is practically impossible to hold a heavy roasted sulphide cake on for more than sixty seconds without the appearance of cracks."

The Oliver Filter [4] consists of a drum containing the filtering area, which is situated on the peripheral surface or cylinder. The drum is revolved once every five or six minutes on horizontal trunnions, and is partly submerged in pulp, the level of which reaches above the centre, so that the filter surface is submerged for three-fifths of a revolution. The filter surface is divided into 24 compartments, and under the influence of a vacuum a slime cake of $\frac{1}{4}$ to

---

[1] By courtesy of the Cyanide Plant Supply Co.

[2] *Mng. and Sci. Press*, June 28, 1913, p. 993.

[3] By courtesy of the Cyanide Plant Supply Co.

[4] A. H. Martin, *Mng. and Sci. Press*, Nov. 27, 1909, p. 715 ; G. A. Tweedy and R. T. Beals, *Trans. Amer. Inst. Mng. Eng.*, 1910, **41**, 324.

½ inch thick accumulates on the canvas. On emerging from the tank the cake is partly dried and then washed by sprays. Finally, the vacuum is cut off and air pressure applied inside the filter, the cake is detached, and assisted by a water spray slides down over a scraper into a launder. The filter surface then immediately enters the slime pulp again.

The Hunt Continuous Sand Filter [1] is a stationary horizontal annular filter bed partly consisting of sand. The filtration, aided by a vacuum, takes

Fig. 168.—Reciprocating Ridgway at the Great Boulder Mine lifting Cakes from Pulp to Wash.

place downwards, and a mixture of sand and slime is treated. After being washed, the mixture is scraped off and discharged.

Ordinary Filter Presses.—An example of one of these is shown in Fig. 169, in use on sulpho-telluride slimed ore. In these presses the action is intermittent. A number of parallel filter leaves are clamped together, and the

---

[1] *Mineral Industry*, 1908, p. 454.

slime pulp is forced by pressure into the spaces between the filter cloths, when the liquid passes through the cloths and the slime forms a layer on their surface. The slime cakes accumulate until they are 2 or 3 inches thick.

Fig. 169.—Filter Press, Kalgoorlie, Western Australia. From a Photograph kindly supplied by Mr. A. G. Charleton.

When washing is complete, the cakes are partly dried by compressed air, and the press is then opened and the cakes discharged. Pressures of 45 to 100 lbs. per square inch have been used. The output per press is about

8 or 10 charges in 24 hours. A press of standard size has 50 chambers, each 40 inches square, and will hold 9,000 lbs. of dry slime. Two types are known as the "Chamber" and "Frame" presses.[1] In the former the plates are kept apart by flanges, and in the latter by separate frames. The *Dehne press*, belonging to the frame type, has been largely used in Western Australia. In this press (Fig. 170)[2] there is a number of cast-iron plates, with a hollow frame between each two plates. Filter cloths are hung over the corrugated surface of each plate. Two kinds of plates are used, marked A and B respectively in Fig. 170. The pulp is passed into the spaces within the frames, *f*, between the two filter cloths, through which the solution passes. When the spaces are filled with pulp, wash liquor enters from the channel, *b*, and is forced horizontally through both filter cloths and the pulp between them and passes out through *d*.

The *Merrill press*[3] has corrugated plates and hollow frames, placed alternately as in the Dehne press. The slime pulp enters each frame through a feed channel at the top, B (Fig. 171). The solution or wash-water is drawn off through the channels, C. The distinguishing feature of the Merrill press is the removal of the slime cake by sluicing instead of by opening the press. The sluicing water under a pressure of 60 to 90 lbs. per sq. inch is introduced through the pipe, H, which has a sluicing nozzle, I, opposite the central plane of each frame. In Fig. 171, D is the partially sluiced slime cake, E is the filter cloth, partially removed to show the corrugated filter plate, F, and G is a horse-shoe clamp for holding the filter cloth against the filter plate. The pipe, H, slowly rotates round its axis through an arc of nearly 180°, backwards and forwards, so as to direct the jet of water in all directions inside the pulp chamber. As the cake is washed away by the jet, the mixture of slime-residue and water flows into the channel underneath the sluicing pipe and through the outlet cocks to the waste conduit below.

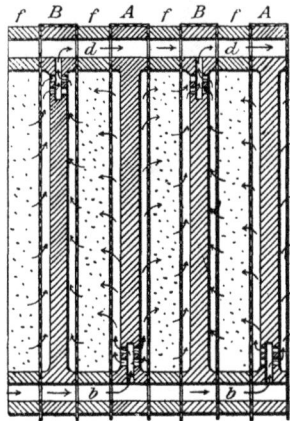

Fig. 170.—Dehne Filter Press (Section).

There are two varieties of Merrill filters—(1) the "solid filling" press, in which the compartment is filled with slime cake and the wash-water passes through the whole cake, as in the Dehne press, and (2) the "partial filling" or "centre washing" press, in which filling is stopped when the cake consists of two equal portions; one of these adheres to the filter cloth on each side, with a vertical space between them extending from the top to the bottom of the frames. The wash solution is then forced into the central space, and passes through the cakes from the centre outwards.

The Merrill press is sometimes used as a treatment press, see below, p. 412.

[1] Julian and Smart, *Cyaniding Gold and Silver Ores*, p. 236.
[2] Julian and Smart, *ibid.*, p. 239.
[3] *Mineral Industry*, 1905, p. 284; M. Ehle, *Mines and Minerals*, Mar. 1907.

The Burt press [1] consists of a pressure cylinder inclined at 45°, with a discharge door at the lower end. Inside the cylinder 28 filter mats or leaves are loosely suspended, each of 8 square feet area, with a pipe leading from their interior compartments to the outside of the cylinder. Slime pulp is forced into the cylinder under 40 to 60 lbs. pressure, and slime cakes are formed on the filter leaves. The surplus slime is then expelled from the cylinder by compressed air and wash-water is pumped into the cylinder,

Fig. 171.—Merrill Press.

and passes through the cakes to the solution pipe. The cakes are dried by air, and the discharge door is then opened and air admitted into the interior of the filter leaves, with the result that the slime cakes drop off and slide out of the cylinder by gravity.

---

[1] *Mineral Industry*, 1907, p. 541.

Examples of the use in practice of some of the agitators and filters described above are given in Chap. XVII.

4. PRECIPITATION OF THE GOLD.

This is effected by means of zinc either in the form of shavings or dust. Electrical precipitation, the charcoal method, and other methods have now been discontinued, but are briefly described below.

Precipitation by Zinc Shavings.—These are freshly turned on a lathe just before use, and kept dry to avoid oxidation. The material is sometimes called " filiform zinc " or " zinc thread." The shavings form a light spongy mass easily traversed by the solution and presenting a large surface for precipitation. The shavings are $\frac{1}{500}$ to $\frac{1}{1500}$ inch in thickness and of no great width, but are sometimes 2 or 3 feet long. They weigh from 6 to 15 lbs. per cubic foot in the boxes, and are so fine that they can be ignited with a match, burning to zinc oxide. One pound of zinc $\frac{1}{500}$ inch thick exposes an area of 28 square feet (E. H. Johnson).

Since 1898, on the Rand and in many mills elsewhere, the new zinc going to the extractor boxes is immersed in a solution containing about 10 per cent. of acetate of lead. The zinc soon acquires a dark-coloured hue, due to a coating of precipitated lead, and it is then transferred to the extractor boxes and at once covered with solution to prevent surface oxidation, which occurs very rapidly on exposing the couple to the air. The *lead-zinc couple* is more active in precipitating gold than zinc alone, and is especially required for weak and low-grade solutions. It also rapidly removes copper from the solution, and so prevents the " plating " action of copper on zinc shavings, which is a source of trouble in weak solutions if ordinary zinc is used. Any excess of the lead acetate added to sand and slime charges to precipitate soluble sulphides tends to keep the zinc active by forming fresh coatings. Lead is also sometimes contained in zinc dust, and lead acetate solution is added to the mixing device in zinc dust precipitation. An objection to the use of lead is that it contaminates the bullion and requires removal.

The shavings are placed in steel troughs, the " zinc-boxes," which are divided into compartments by partitions. They are supplied with baffle boards, which cause the solution to flow upward through the shavings. Argall uses boxes in which the solution flows alternately downward and upward through the zinc, entering the first compartment at the top. Fig. 172 shows a steel extractor box stated to be one with ten compartments.[1]

Fig. 173, from a photograph kindly supplied by Dr. W. A. Caldecott, gives a general view of the interior of the extractor house at the Knight's Deep and Simmer East Joint Plant, Transvaal.

The shavings are supported on wire-screen trays, formed of $\frac{1}{8}$-inch or $\frac{1}{4}$-inch iron wire gauze, so that the finely divided, precipitated gold falls through to the bottom of the trough, while the unaltered zinc remains on the sieve. The trays are removable and furnished with handles so that they can be lifted out. The bottom of each compartment slopes towards one corner, where there is a plug hole through which the gold slime is washed in cleaning up. The zinc boxes are fitted with lids and kept locked.

There are usually about ten compartments, of which the first and last

---

[1] Schmitt, *Rand Metallurgical Practice*, vol. ii., p. 265. Schmitt's illustration, reproduced in Fig. 172, appears to show twelve compartments.

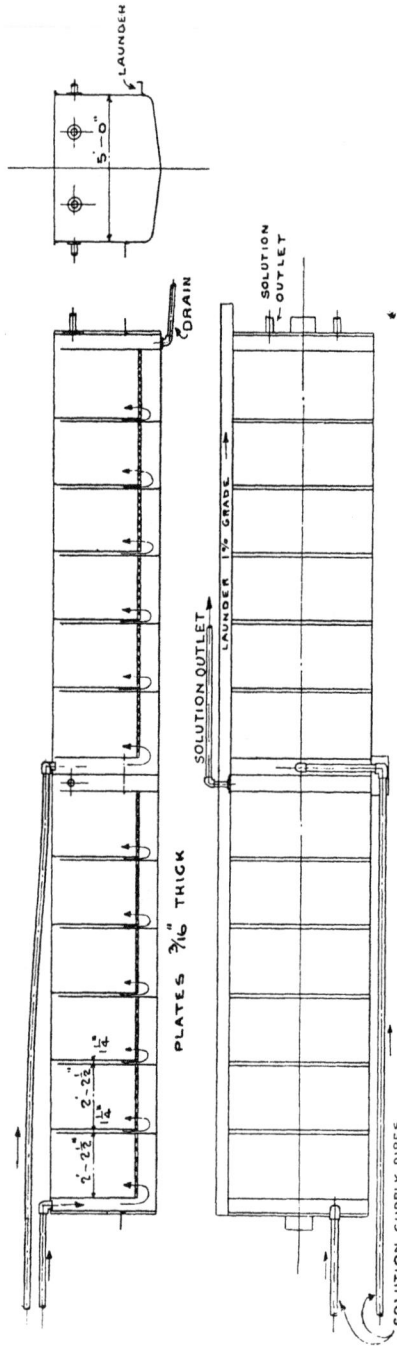

Fig. 172.—Steel Extractor (Twelve Compartments).

Fig. 173.—Interior of Extractor House, Knight's Deep and Simmer East Joint Works.

are left empty or are supplied with a sand filter to clarify the solution by settling or filtering on entering and leaving the box. The solutions are, however, often filtered in a separate box, as at the Glencairn Mine where cocoa-nut matting is used.[1] Other filtering arrangements are in use. The use of iron in a box in contact with the zinc increases the loss of zinc by dissolution, but iron wire trays form the most convenient support. Wooden supports have been used.

The dimensions of the boxes depend on the amount of work to be done. It is usual to allow from 1 to $1\frac{1}{2}$ cubic feet of zinc-box space per ton of solution to be precipitated in 24 hours. This gives contact for 30 to 45 minutes. The coarser the zinc the more time is required. On the Rand about $2\frac{1}{2}$ tons of solution are precipitated per ton of ore treated.

The slime solution from the decantation process is clarified before precipitation by means of filter presses of the frame and distance type (q.v.), or by ordinary sand filter beds.

It is usual on the Rand to have three sets of boxes in a cyanide works, one for the " strong " solution, one for the " weak " solution, and one for very dilute washings. In the last two boxes less zinc is consumed, but the gold slime is poorer, less gold being contained in the solution.

When the solution comes in contact with bright zinc, the latter turns black at once, owing to the deposition on it of finely divided gold. The zinc is gradually dissolved, and the shavings fall to pieces, those in the first compartment being consumed most rapidly. As the precipitation proceeds, the zinc is transferred from the lower compartments to the upper ones, and fresh zinc is added at the foot of the box. In thus " dressing " the zinc boxes, it is essential to distribute the shavings evenly to avoid too close packing in places and the formation of free channels. The corners are carefully filled in. Short and rotten zinc is placed on the top. The boxes are dressed at the clean-up (say once in 7 to 10 days), or in some mills more frequently.

The consumption of zinc is generally from 4 to 20 ozs. of zinc for 1 oz. of gold recovered. The consumption is partly chemical (by dissolution) and partly mechanical (short zinc removed with the gold in cleaning up). Virgoe found [2] that in a sand plant in Mexico using a solution of 0·32 per cent. KCy, the chemical loss was 86·1 per cent. and the mechanical loss 13·9 per cent. of the total loss of zinc. In the slime plant during the same period, using a solution of 0·04 per cent. KCy, the corresponding losses were 36·8 and 63·2 per cent. The total losses per ton of solution were, sand plant 0·296 kg., slime plant 0·096 kg. The average loss on the Rand is about 0·4 lb. zinc per ton of ore milled, or, say, 0·15 lb. per ton of solution. For the chemical reactions occurring during precipitation, see p. 338.

Precipitated solution from the zinc boxes should contain no more than 0·02 dwt. of gold per ton (Caldecott).

Changes in composition of the solution in passing through the zinc boxes have been investigated by P. Argall.[3] In Fig. 174, which is kindly supplied by him, some of his results obtained in studying the conditions at the Metallic Extraction Co.'s Mill at Florence are shown. Each double compartment contains a depth of 35 inches of zinc, and the solution was analysed when

---

[1] John Yates, *Metallurgical Engineering on the Rand*, 1898, p. 31.
[2] W. H. Virgoe, *J. Chem. Met. and Mng. Soc. of S. Africa*, 1903, **4**, 615.
[3] Private Communication, 1904.

passing from each compartment to the next. The gold is seen to be precipitated mainly in the first compartment, and the amounts of alkali and zinc in solution to increase most rapidly at the same time. No reason is assigned for the shape of the curve of alkalinity, and the existence of subsidiary maxima and minima may perhaps be doubted. The cause of the precipitation of the lime may be explained by assuming that it is in existence

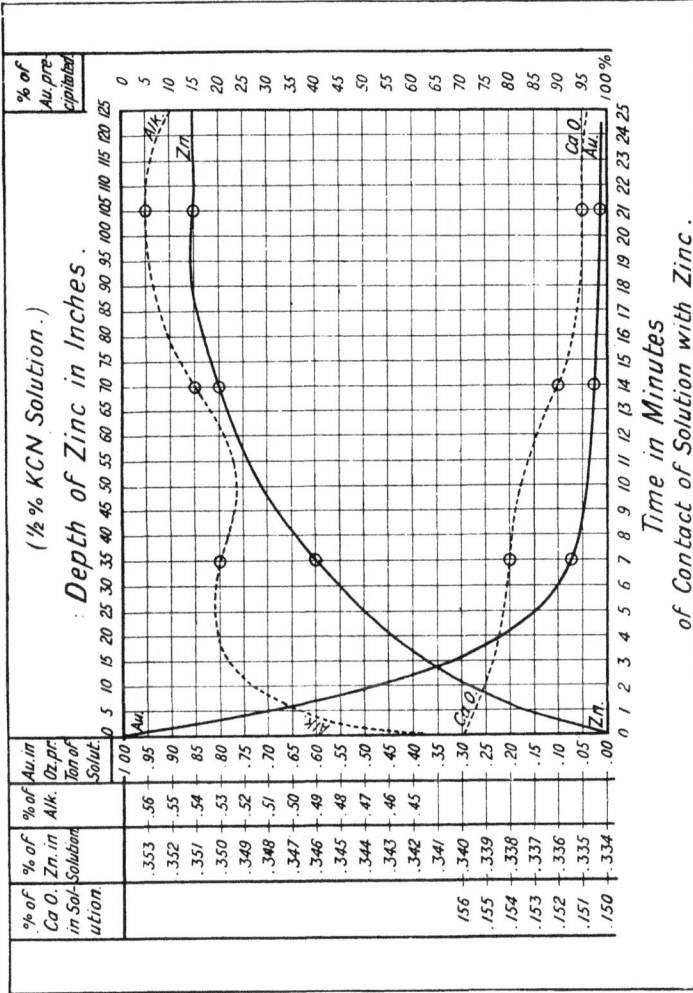

Fig. 174.—Chemical Changes during Passage of Cyanide Solution through Zinc Boxes.

in the solution largely as cyanide of calcium, $CaCy_2$. Exposure to air causes this to oxidise and the hydroxide is then precipitated, a saturated solution in water containing only 0·13 per cent. CaO at 15°, and less at higher temperatures. Carbonic acid from the air would also precipitate lime as calcium carbonate.

Clean up of Zinc Boxes.—The clean up usually takes place two or three times a month, and is often coincident with dressing. As a rule, only the first three compartments are cleaned up. The solution is displaced by clean water, and the latter drawn off down to the level of the wire gratings. The screens containing the undissolved shavings are lifted out and the zinc-gold slime, remaining at the bottom of the boxes, allowed to run out by withdrawing a plug in the bottom of the box, and drained through a 20- or 30-mesh screen, which retains a small part only. After the residue has been gently rubbed on the screen and well washed, it is put back again into the first compartment of the box, on the top of fresh shavings, as it consists mainly of small pieces of unconsumed zinc. The shavings proper are thoroughly washed and rinsed on a sieve to separate the gold slime as closely as possible, and are returned to the zinc boxes. Short rotten zinc containing gold is not now usually returned to the zinc boxes.

In large plants, the zinc shavings may be washed in Thomas's trommel,[1] a cylinder with screening which can be fitted in a compartment of the zinc box. Zinc shavings are put into the trommel, which is immersed in water to a depth of one-third of its diameter and rotated.

The gold slime is run into a small filtering vat, or more usually into a press of the Johnson type, where it is washed and sent to the acid tanks, or partly dried for fluxing.

The richness of the dried precipitate depends on the strength of the cyanide solution, and on the time of contact as well as on the quantities of base metals which are present in the solution. By the prolonged action of the cyanide, the zinc shavings become partly corroded and disintegrated. so that the precipitated gold is mixed with zinc debris. Ordinary commercial zinc contains a considerable proportion, generally over 1 per cent., of lead, and a small quantity of carbon, besides other impurities, such as arsenic and antimony. Since the introduction of the lead-zinc couple, the proportion of lead has risen to about 5 per cent. All these impurities accumulate and are collected with the gold.

According to C. Butters and J. E. Clennel,[2] the pans in use at the Robinson Works contained about 5 or 6 gallons of dried precipitate or gold slime, which might contain as much as 150 ozs., or as little as 20 ozs., of gold. A little silver was also contained in it, the remainder being chiefly zinc and lead, with smaller quantities of tin, antimony, organic matter, etc. The average composition of dried gold slime in South Africa was at that time approximately as follows :—

Gold and silver, ...	10 to 20 per cent.
Zinc, ...	30 to 60 ,,
Base metals, silica, alumina, oxides of iron,	
etc., ...	20 to 40 ,,

The Hanauer Smelting Company found the precipitate from the Mercur Mine to contain the following substances :—[3]

---

[1] J. E. Thomas, *J. Chem. Met. and Mng. Soc. of S. Africa*, 1903, **4,** 315.
[2] Butters and Clennel, *Eng. and Mng. J.*, Oct. 15, 1892, p. 365.
[3] C. W. Merrill, *ibid.*, Nov. 5, 1892, p. 440.

Zinc,	.	.	.	.	.	.	.	39·1 per cent.
Calcium carbonate,	.	.	.	.	.	.	36·7	,,
Gold,	.	.	.	.	.	.	.	4·4	,,
Cyanogen,	.	.	.	.	.	.	3·5	,,
Sulphur,	.	.	.	.	.	.	2·6	,,
Iron,	.	.	.	.	.	.	.	2·4	,,
Residue,	.	.	.	.	.	.	6·0	,,

$$94·7 \quad ,,$$

The calcium carbonate was probably deposited in the boxes chiefly from suspension in the solution, the gangue of the ore mainly consisting of limestone.

In cleaning up the precipitate from gold solutions it is profitable to separate as much gold from the zinc as possible, to avoid locking up values in the zinc boxes. Zinc containing gold is accordingly not returned to the boxes. It is otherwise with silver solutions. In the treatment of silver ores, the precipitate often contains from 60 to 75 per cent. of silver, as it is more profitable to return as much zinc as possible to the boxes, even at the cost of locking up silver in them. The saving of zinc more than compensates for the loss of interest in this case.

Precipitation by Zinc Dust.—In this process zinc dust (" zinc fume " or " blue powder ") is agitated with gold solution and the precipitate separated from the impoverished solution by filter pressing. It was introduced by H. L. Sulman at Deloro, Canada, in 1894,[1] and at the Mercur Mine, Utah, in 1896.[2] It was developed at the Homestake Mine in 1906-1908,[3] and has since spread widely.

Analyses of a number of samples of zinc dust are given by W. J. Sharwood.[4] Those used in gold precipitation contain from 85 to 95 per cent. of metallic zinc, 3 to 10 per cent. of zinc oxide, which is harmless but inoperative, and usually 2 or 3 per cent. of lead, which is advantageous. Zinc dust is rapidly oxidised in moist air. It should be so fine that about 95 per cent. will pass a 200-mesh sieve and contain no lumps. Full details of testing zinc dust for use in the cyanide process are given by Sharwood.

The earlier method of using zinc dust was to run the solution into large vats of considerable depth, agitate with compressed air and sprinkle dust into the charge. Precipitation was complete in about fifteen minutes, and the solution was then passed through filter presses. In the filter press, the solution continued to act on zinc accumulated from previous charges, and this feature is retained in the Merrill plant described below. The air agitation, however, tended to defeat its own object, one effect being to redissolve the gold.

Later practice is typified by the Merrill plant. One form of this as originally used at the Homestake Mill is shown in diagrammatic form in Fig. 175,[5]

[1] Sulman, *J. Soc. Chem. Ind.*, 1897, **16,** 961 ; *Trans. Fed. Inst. Mng. Eng.*, 1897, **15,** 417.
[2] G. A. Packard, *J. Chem. Met. and Mng. Soc. of S. Africa*, 1899, **2,** 720.
[3] A. J. Clark, *ibid.*, 1909, **9,** 222.
[4] Sharwood, *Mng. and Sci. Press,* May 11, 1912, p. 659 ; *Cyanide Practice*, 1910-1913, p. 403.
[5] Gowland, *Non-Ferrous Metals*, p. 262.

and is described by Clark and Sharwood.[1] The dry zinc dust was fed continuously by a travelling belt into a mixing cone, where it was agitated with a small stream of barren solution, by means of compressed air forced into the cone, and an emulsion formed. The emulsion overflowing by a side pipe near the rim of the cone was delivered at the base of the suction column of a pump, the contact of the zinc and the gold-bearing or " pregnant " solution taking place during the elevation of the solution from the sump to the presses. It was found, however, that a deleterious precipitation of calcium carbonate took place in the mixing cone, owing to the introduction of air.

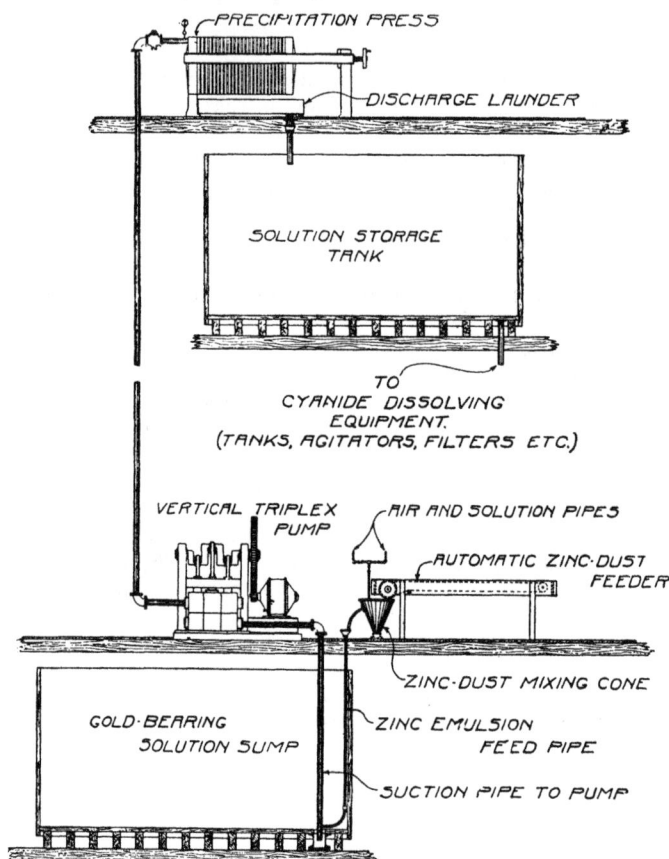

Fig. 175.—Merrill Precipitation Plant.

Agitation with air is now omitted, and a continuous feeder shown in Fig. 176[2] is used instead of the mixing cone. In this, the zinc dust is conveyed from a hopper, A, by means of a revolving feeder to a mixing cylinder, E, where it is carried by a small stream of solution, entering at C, to the

[1] Clark and Sharwood, Trans. Inst. Min. and Met., 1913, 22, 126-132; A. J. Clark, Mng. Mag., 1911, 4, 289.
[2] A. J. Clark, Mng. Mag., loc. cit., or J. Chem. Met. and Mng. Soc. of S. Africa, Sept. 1911, 12, 103.

pump-suction, no air being used to assist in the mixing. It is generally the practice to maintain a drip of a solution of lead nitrate or acetate to the feed pipe carrying the zinc emulsion, to obtain the advantage of the lead-zinc couple. Merrill filter presses of triangular section (see Fig. 177) [1] are used, the solution entering at the lowest point or apex of the triangle. The mixing takes place partly in the upcast pipe and partly in the press.

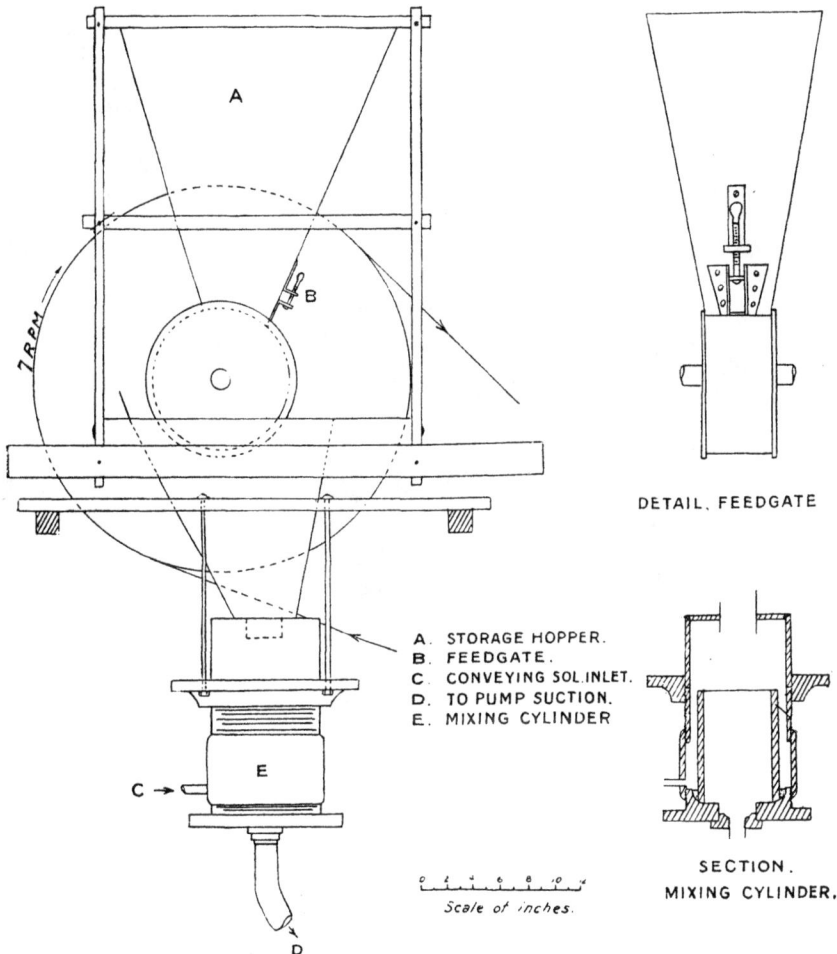

A. STORAGE HOPPER.
B. FEEDGATE.
C. CONVEYING SOL. INLET.
D. TO PUMP SUCTION.
E. MIXING CYLINDER

DETAIL. FEEDGATE

SECTION.
MIXING CYLINDER.

Scale of inches.

Fig. 176.—Todd Zinc-Dust Feeder, Homestake.

The filter cloth consists of two thicknesses of twilled cotton. The clean up takes place once a month, when the press is opened and the filter cloths scraped and burnt or washed and replaced.

The details of the results at the Homestake Mill are summarised as follows :—[2]

---

[1] Clark, *Mng. Mag.*, April, 1911, **4,** 291.   [2] Clark and Sharwood, *loc. cit.*

In September, 1911, taking sand and slime plants together, 0·535 ton of solution per ton of ore was precipitated ; the consumption of zinc dust was 0·165 lb. per ton of solution, or 0·089 lb. per ton of ore, or 1·84 lbs. per oz. of gold precipitated ; the assay of the pregnant solution was $1.867 per ton, and that of the barren solution was $0.026 per ton. (The value of 1 dwt. of gold is $1.033.)

It has not yet been shown whether zinc in the form of dust or of shavings is more advantageous for precipitation generally. The consumption of zinc dust is about equal to that of shavings, but it costs less per lb. The precipitation is about equally efficient in the two processes, but with the exercise of great care and skill better results and a higher grade precipitate can be obtained with dust than with shavings. The clearest advantage of the use of zinc dust is that all the precipitated gold is cleaned up and brought to account, whereas with shavings part always remains in the boxes for a future

Fig. 177.—Triangular Merrill Filters for Gold Precipitate.

clean up. The difficulty of the " white precipitate " is avoided, and copper in the solution is not prejudicial when zinc dust is used. By the Merrill system, too, the zinc is always kept from the air when wet, and surface oxidation is thereby prevented. On the other hand, the short time of contact and the incompleteness of the zinc-lead couple formed in zinc-dust precipitation are against its adoption in dealing with very dilute or poor gold solutions.

Precipitation by Zinc Wafers.[1]—This was introduced at the Caveira Mine, Portugal, in 1907. Some advantages are claimed.

Precipitation by Aluminium.—The first practical application of aluminium (originally proposed by Moldenhauer in 1893, see p. 339) was in the form of dust, and was made at Deloro in 1908, and by Kirkpatrick[2] and E. M.

---

[1] J. S. MacArthur, *J. Chem. Met. and Mng. Soc. of S. Africa*, Jan. 1913, **13**, 310.
[2] S. F. Kirkpatrick, *Eng. and Mng. J.*, June 28, 1913, p. 1277.

Hamilton,[1] working on silver solutions in 1913. At Nipissing, Ontario, Hamilton used Merrrill's zinc dust machinery with modifications necessitated by the fact that it is difficult to wet aluminium dust, and that even after it has been wetted it tends to rise to the surface and float as a thick scum. The emulsifier is omitted, and the aluminium is agitated with the solution in tanks by means of paddles or screw propellers. The precipitation occupies from 10 to 15 minutes, and the solution then passes to filter presses. The consumption of aluminium is about 0·02 lb. per oz. of silver (or about 1 part aluminium to 3·5 parts of silver), and cyanide is regenerated. The presence of caustic alkali is necessary (see also p. 393). Some special constituents in the solutions at Cobalt (possibly due to the presence of arsenic) prevented the use of zinc there. A difficulty in the use of aluminium shavings caused by the formation of quantities of insoluble alumina is at least in part due to the action of the air, and does not seem to occur in the aluminium dust method described above.

Electrical Precipitation.—The *Siemens-Halske process*,[2] using iron anodes and thin lead sheets for cathodes, was used on the Rand in certain mills from 1894 to 1899, but was abandoned after the introduction of the lead-zinc couple. Its disadvantages were its failure to effect complete precipitation, from 0·3 to 0·5 dwt. gold being left in solution, and the large amount of by-product, containing variable amounts of gold, formed by the decomposition of the iron anodes. (It may be recalled that the ions of $KAuCy_2$ are K and $AuCy_2$, so that insoluble cyanide of gold tends to appear at the anodes.) " It is said to serve as an excellent means of removing copper from solution prior to the precipitation of the gold on zinc shavings " (Caldecott).

In a later modification of electrical precipitation used at Minas Prietas and elsewhere,[3] a heavy current density of 0·3 to 0·8 ampère per sq. ft. was used or 10 to 100 times as much as that used in the Siemens-Halske process. The gold is deposited as a black slime, which can be wiped off the cathodes, which are of tinned iron. The anodes are peroxidised lead plates or graphite. Perfect precipitation is not obtained, and Lamb suggests that the method should be supplemented by zinc precipitation. The advantages are the avoidance of foul solutions (because base metals are precipitated by the current) and the regeneration of cyanide. " The process is still in use in some plants." [4]

The Molloy,[5] Pfleger,[6] and Pelatan-Clerici [7] processes were also electrical, but were not found to be satisfactory, and are only of historical interest. In the last-named process the precipitation took place on amalgamated copper plates simultaneously with dissolution, without separation of ore and solution, and in that respect is similar to the Gilmour-Young process,[8] in which, however, the precipitation on mercury was chemical, and was effected by the use of a mixture of amalgams of zinc and copper, without the aid of a current of electricity.

Precipitation by Charcoal.—Charcoal was formerly used [9] in a number of small plants in Victoria. The charcoal is placed in small tubs, or in boxes

[1] Hamilton, *Eng. and Mng. J.*, May 10, 1913, p. 935.
[2] A. von Gernet, *J. Chem. Met. and Mng. Soc. of S. Africa*, 1894, **I**, 28.
[3] M. R. Lamb, *Eng. and Mng. J.*, April 3, 1909, p. 705.
[4] H. A. Megraw, *Eng. and Mng. J.*, June 20, 1914, p. 1232.
[5] Julian and Smart, *Cyaniding Gold and Silver Ores*, 1st ed., p. 158. [6] *Ibid.*, p. 159.
[7] Bosqui, *The Cyanide Process*, 1899, p. 136.
[8] A. James, *Trans. Inst. Mng. and Met.*, 1898, **7**, 63.
[9] J. I. Lowles, *Trans. Inst. Mng. and Met.*, 1899, **7**, 190.

resembling ordinary zinc boxes, and the solution passed in succession through a number of these. The precipitation of gold is said to bear no relation to the strength of cyanide, but no mention has been made of the precipitation of gold from solutions containing less than 0·04 per cent. of KCy. The sump liquors may contain as little as 3½ grains of gold per ton. The clean-up methods are similar to those employed in connection with the chlorination process. The cost is supposed to be about double that of precipitation by zinc.

When the charcoal becomes ineffective with use, it is re-burned (see above. p. 343). " The use of charcoal may be reasonably recommended in cases where a limited quantity of tailings have to be treated which contain base metals, such as copper, zinc, etc." [1]

5. Production of Bullion from the Precipitate.

Various methods have been suggested for effecting the elimination of the zinc and other base metals. The chief ones are—

(a) Direct fusion with fluxes.
(b) Roasting, followed by fusion.
(c) Treatment with acid, followed by fusion with or without roasting.
(d) Reverberatory furnace lead fusion and cupellation (Tavener process), often preceded by acid treatment.
(e) Blast furnace lead fusion and cupellation.

(a) **Direct Fusion.**—This method was used in the early days of the cyanide process, but is now superseded, except perhaps in certain cases where high-grade precipitate is produced. The slime is dried in iron pans or merely by air-blowing in a filter press. If the zinc has been imperfectly separated from the slime, nitre must be added. Other fluxes, such as borax, carbonate of soda, sand and fluorspar are always added.

The slag, which consists of silicates of zinc, soda, etc., corrodes the pots rapidly. Large quantities of zinc oxide are given off as fumes, forming thick crusts in the flues, and evil-smelling products of decomposition of the cyanides are also evolved. The bullion produced by this method varies in colour from iron grey to pale yellow, and cannot be obtained uniform in composition, so that accurate assays are difficult to obtain.

The results of analyses made on three ingots of bullion produced in this way in South Africa, and shipped to London, are appended :—

TABLE XXXIII.

	I.	II.	III.
Gold,	60·3	61·7	72·6
Silver,	7·3	8·1	9·2
Zinc,	15·0	9·5	7·1
Lead,	7·0	16·4	4·9
Copper,	6·5	4·0	4·8
Iron,	2·2	0·3	1·4
Nickel,	2·0	..	..
	100·0	100·0	100·0

[1] H. A. Megraw, *Eng. and Mng. J.*, June 20, 1914, p. 1234. See also M. Green, *Trans. Inst. Mng. and Met.*, 1913, **23**, 65.

The slags obtained in this way are always rich in gold, part of which is sometimes in the form of shots, and may be recovered by crushing and panning. The crushed slag is then fused again with the addition of granulated lead, or of litharge, when all the gold is concentrated in the lead. If the lead thus obtained is granulated, it can be used again until rich enough to be worth cupelling.

When by the careful use of a 40-mesh screen, followed by the squeezing of the slime in a filter bag, the zinc has been almost all eliminated by mechanical means, very little nitre is required. According to James,[1] the zinc can be reduced to about 1 per cent., and bullion obtained 960 fine in gold and silver without the use of nitre, roasting, or acid. The dry slime is fused in graphite pots with about half its weight of borax, a little sodium carbonate, and sand or fluorspar. if necessary. The proportion of the fluxes must, however, be determined in each case by experiment. It is necessary to form a fluid slag, but it is difficult to avoid pasty slags in the presence of much zinc, and hence acid treatment is preferred.

(b) **Roasting followed by Fusion.**—The dried slime is spread in a thin layer on iron trays, and heated to a barely-perceptible red heat in the flue of a furnace, or above a special grate under a hood. The carbon, zinc, arsenic, etc., ignite readily, being in a very fine state of division, and roasting proceeds regularly without much stirring, which would tend to cause loss through dusting. Dense fumes of zinc oxide are given off, and the residue consists chiefly of the oxides of lead and zinc with gold and silver and a variable quantity of sand. A little nitre is sometimes mixed with the wet slime to aid in the oxidation of the base metals. Fusion with fluxes is carried out as in the previous case. The bullion produced is said to be about 800 fine on the Rand. The method, like the one previously described, appears to have been discarded, owing to the loss by dusting, which was always an unknown quantity.

(c) **Acid Treatment.**—In this process, which is of almost universal application, the zinc is dissolved in sulphuric acid and the residue dried and smelted. Hydrochloric acid is occasionally used, on account of its solvent power on lime and lead. The wet slime from the filter press is charged into a large vat and sulphuric acid is then added a little at a time, the strength used being from 10 to 15 per cent. On the Rand the slime was added little by little to the acid. Violent action ensues, and the fumes are highly poisonous, sometimes containing arsenuretted hydrogen, and must be carried off by a good draught with the aid of a hood. Cold water may be added to moderate excessive action. The effervescence is at first great and the vat must be deep to prevent frothing over.

Where white precipitate (q.v.) is present in the boxes in considerable quantity, special precautions are taken on the Rand,[2] including (a) emulsification of the white product and avoidance of the formation of greasy lumps, and (b) long contact, say 12 hours, with hot acid in excess. Some of the coarser zinc is then reserved to neutralise the acid. The treatment improves the value of the calcined slime. Care is taken that the solution shall be sufficiently dilute to prevent the separation and crystallisation of $ZnSO_4.7H_2O$. To aid in this the charge is sometimes heated by steam, and must be occasionally stirred. For this purpose a two- or a four-armed

[1] James, *Trans. Inst. Mng. and Met.*, 1895, **3,** 404.
[2] H. A. White, *J. Chem. Met. and Mng. Soc. of S. Africa*, Sept. 1914, **15,** 50.

stirrer is provided. The vat is made of wood, which may be lined with lead, and is usually from 6 to 10 feet in diameter and 5 or 6 feet deep on the Rand. The vat is closed with a lid in which is a charging hole and a pipe for the escape of the gases.

J. E. Thomas and G. W. Williams have shown [1] that sodium bisulphate is a cheaper and more convenient solvent than sulphuric acid for zinc. The use of bisulphate of soda is now general on the Rand. There is less tendency to boil over, but a larger plant and an extra vat for dissolving the bisulphate are required.

After treatment with acid the slime is settled, washed with hot water, and separated by decantation or by a vacuum filter or a filter press. It is then dried, calcined and fused in the usual way.

The following analyses of slime, after treatment with acid at the Brodie Works, Cripple Creek, Colorado, are given by W. R. Ingalls :—[2]

TABLE XXXIV.

	I.	II.	III.
Gold, ...	49·85	26·18	23·60
Silver, ...	9·54	6·84	6·00
Silica, ...	15·60	10·00	6·60
Lead, ...	3·00	Trace	Trace
Copper oxide, . . .	5·58	8·11	6·40
Zinc oxide, . . .	0·14	31·80	41·50
Lime, ...	0·51	5·32	0·03
Ferric sesquioxide, } Alumina, . . } . .	4·00	6·80	9·26
Sulphur trioxide, . .	11·32	4·80	6·45
	99·54	99·85	99·84

In Nos. II. and III. the treatment with sulphuric acid was very incomplete. Ingalls states that in fusing these slimes No. I. would require about 50 per cent. of sodium carbonate and 25 per cent. of borax, but silica in addition would be required by the others.

The analyses in Table XXXV. show the composition of acid-treated and calcined precipitate obtained on the Rand.

In No. IV., made by A. Whitby,[3] the organic matter, including insoluble cyanide compounds, was not determined. The high percentage of silica was said to be due to blown sand. In No. V.,[4] given by W. A. Caldecott, some lead existed as sulphate. The slime was from the Robinson Deep Mill. Nos. VI. and VII., made by G. W. Williams,[5] are of slimes from the East Rand Proprietary. No. VI. is before and No. VII. is after calcination. No. VI. also contained from 7 to 10 per cent. of ferrocyanides (K_4FeCy_6). About half the sulphur was present as PbS, which was oxidised in the ensuing calcination.

---

[1] Thomas and Williams, *J. Chem. Met. and Mng. Soc. of S. Africa*, June, 1905, **5**, 334.
[2] Ingalls, *Mineral Industry*, 1895, p. 336.
[3] Johnson and Caldecott, *J. Chem. Met. and Mng. Soc. of S. Africa*, July, 1902, **3**, 47.
[4] T. K. Rose, *Trans. Inst. Mng. and Met.*, 1905, **14**, 400.   [5] *Loc. cit.*

TABLE XXXV.

	IV.	V.	VI.	VII.
Au, ...	34·50	34·0	36·07	37·3
Ag, ...	4·75	4·3	3·5	3·7
Zn, ...	..	..	7·50	6·16
ZnO, ...	7·00	22·5	..	..
Pb, ...	12·50	..	20·81	20·00
PbO, ...	..	15·6	..	..
Cu, ...	2·55	..	2·46	2·5
CuO, ...	..	0·7	..	..
Fe, ...	..	..	0·02	Trace
Fe_2O_3, ...	3·65	2·9	..	..
Ni, ...	..	..	Trace	Trace
NiO, ...	1·00	0·7	..	..
SO_3, ...	6·95	6·8	6·3	13·7
SiO_2, ...	21·00	7·5	12·6	16·2
	93·90	95·0	89·26	99·56

The prepared gold precipitate is melted in graphite crucibles with clay liners, and poured into conical moulds (see pp. 427, 433). Even after roasting, the precipitate appears to contain reducing substances, and an oxidiser is added, which is still often nitre. This does not readily oxidise lead, and the use of manganese dioxide was introduced in 1902 by Johnson and Caldecott.[1] It readily oxidises lead, but carries more silver into the slag than nitre does, and an excess is therefore avoided. If the slags pass to the Tavener furnace the silver is saved there. Other fluxes are sand, borax and sodium carbonate, the latter being especially useful if much zinc is present. Fluorspar is sometimes added. On some mines, fluxes and manganese dioxide are not necessary, but a cover of borax and sand is added.[2] The precipitate and fluxes are mixed, sometimes in a rotating machine, and are charged into the crucible by degrees, more being added as the charge melts and sinks down. There is some loss by dusting due to making an intimate mixture, and the practice is now obsolete (H. A. White).

A flux used on the Rand is given by E. E. Meyer as follows :—[3]

Acid-treated dried slime, ...	100	parts.
Borax, ...	45	,,
Sodium carbonate, ...	1	,,
Fluorspar, ...	7	,,
Manganese dioxide, ...	2	,,
Sand, ...	15	,,

This gave bullion from 875 to 900 fine.

[1] Johnson and Caldecott, *J. Chem. Met. and Mng. Soc. of S. Africa*, 1902, **3**, 46.
[2] H. A. White, *ibid.*, Sept. 1914, **15**, 51.
[3] E. E. Meyer, *J. Chem. Met. and Mng. Soc. of S. Africa*, 1905, **5**, 169.

Flux charges given by MacFarren are as follows :—[1]

Precipitate,	. . . . . .	100 parts.
Borax glass,	. . . . .	12 to 30 ,,
Sodium carbonate,	. . . .	6 to 15 ,.
Silica,	. . . .	3 to 8 ,.

If the fluxes are in suitable proportions the slag is fluid and glassy and contains, after panning, about 20 ozs. of gold per ton, which is recovered by fusion with litharge. The buttons from the conical moulds, which are sometimes as much as 100 ozs. in weight, are remelted in plumbago crucibles and cast into ingot moulds. The precipitate is sometimes melted in Faber du Four tilting furnaces,[2] fired with coke, oil or gas, and remelted in crucibles. The bullion prepared as above may be as high as 985 fine in gold and silver, and should not be lower than 850 fine. It is sometimes brittle from the presence of zinc, etc. The loss in roasting and subsequent handling is estimated by Bettel to be from 0·01 to 0·025 per cent., or less than $\frac{1}{4}$d. per oz.

The method of converting precipitate into bullion in use at the Princess Works, Transvaal, in 1897, has been described by E. H. Johnson, as follows :—[3] The slime is transferred direct in buckets from the zinc boxes to a filter vat, 6 feet in diameter, and filtered by a vacuum-pump through closely-woven canvas, the clear liquid being returned to the boxes. Water is then added and pumped through until all soluble cyanides have been removed, and the slime is then baled out (being weighed in the buckets during the process) into a sheet-iron tray. A closed vat, fitted with a stirring apparatus, is then charged with dilute (1 in 10) sulphuric acid, 1 lb. of concentrated acid being added for each pound of moist slime, the agitator is started, and the slime added little by little through a hopper, the fumes being carried away by a 3-inch pipe. After all the slime has been charged in, the agitation is continued for half an hour, and the hopper and sides are then washed down, the agitator washed and removed, and the vat filled with water and allowed to settle. No heat is used, except that caused by the mixing of strong sulphuric acid with water in the vat and by the action on the slime. The clear liquid is siphoned off, and washing by decantation repeated four or five times, the washings containing about 13 grains of gold per ton of solution. The gold residue is then dried in enamelled iron dishes, roasted gently in sheet-iron trays, and then ground, mixed with fluxes, and transferred to the crucible. The charge fuses quietly with little fume, yielding bullion amounting to 50 to 60 per cent. of the weight of the slime. The average fineness in 1896 was 821·9, and the slags assay 23 ozs. per ton, but only 0·2 per cent. of the gold is locked up in this way.

In cleaning-up on the Rand by modern methods the gold contained in the by-products, such as pots, liners and slag, should not exceed 0·5 per cent. of the gold produced, and in some cases is less than 0·25 per cent.[4]

Crucible tilting furnaces are coming into use for melting precipitate. They have the advantage that the crucible remains in the furnace during pouring, whereas in the stationary furnace the crucible is usually lifted out for pouring, although the charge could be ladled out. The result is that the crucible is

---

[1] MacFarren, *Cyanide Practice*, p. 187.
[2] For description, see Gowland, *Non-Ferrous Metals*, p. 181.
[3] Johnson, *J. Chem. Met. and Mng. Soc. of S. Africa*, June 19, 1897, **2**, 73.
[4] H. A. White, *J. Chem. Met. and Mng. Soc. of S. Africa*, Sept. 1914, **15**, 51.

cooled down before being replaced in the furnace, and does not last so long as in the tilting furnace. The melting is also more quickly done in the tilting furnace, in which oil or gas is used as fuel; on the other hand, tilting furnaces are more expensive to install.

Oil-fired tilting reverberatory furnaces without crucibles are used at the Tonopah-Belmont Mill [1] for melting precipitate without previous acid treatment. The precipitate contains 74 per cent. of gold and silver.[2] The cost of treatment is very low (0·1d. per oz.), and for a large plant such furnaces seem to be more suitable than crucible tilting furnaces, the cost of furnace liners being less than that of crucibles, and the labour costs lower.

Electric furnaces are also sometimes used for smelting precipitate.[3]

(d) **Tavener Process.**—In this process the precipitate is smelted with litharge, fluxes and coal in a small reverberatory (or " pan ") furnace, and the auriferous lead cupelled. It was originally introduced [4] with the intention of avoiding acid treatment, but is now generally used for acid-treated slime.

The charge for the pan furnace is approximately as follows :—[5]

Gold slime, . . .	100 parts	
Litharge, ...	100 ,,	{ (Varied to produce bullion containing 7 to 10 per cent. gold).
Assay slag, . . .	55 ,,	
Coal dust, . . .	10 ,,	
Silica, . . .	25 ,,	
Iron (any scrap), . .	13 ,,	

The damp charge is put into the warm furnace and covered with some of the litharge. The fire is then lighted and the charge slowly melted, and the iron scrap afterwards added to clean the slag. Washes of litharge and coal are also used. When action ceases, the slag is run off through the slag door into slag pots and the lead tapped into moulds.

The lead pigs are cupelled in an English cupellation furnace until a cake of gold-silver is obtained on the cupel. The cake is broken up while hot, and the pieces melted in crucibles and cast into ingots. The fineness of the bullion is 960 to 980 in gold and silver, and the losses small.

The slags from the pan furnace are smelted for lead with other by-products, such as assay slag, old cupellation " tests," old crucibles, filter cloths, extractor house sweepings, etc. This is done either in the pan furnace itself or in a small blast furnace,[6] and the lead bullion so produced is cupelled.

(e) **Blast Furnace and Cupellation.**—Instead of a reverberatory furnace a small blast furnace is used for smelting gold slime in most of the plants installed by C. W. Merrill in America.[7] The method at the Homestake Mill is typical of this practice.[8] Here the precipitate is partly dried, mixed with fluxes and briquetted. The briquettes are dried and the rich ones added to the lead bath in the cupel furnace. The lower grade material is smelted with by-products in an ordinary water-jacketed lead blast furnace with three tuyers, and the lead passed to the cupel. At the Goldfield Consolidated Mill, direct blast-furnace smelting of briquetted precipitate is practised.

[1] A. H. Jones, *Eng. and Mng. J.*, June 14, 1913, p. 1197.
[2] *Ibid.*, May 9, 1914, p. 967.
[3] H. R. Conklin, *ibid.*, June 15, 1912, p. 1191.
[4] P. S. Tavener, *J. Chem. Met. and Mng. Soc. of S. Africa*, Oct. 1902, p. 112.
[5] L. A. E. Swinney, *Trans. Inst. Mng. and Met.*, 1907, **16,** 115.
[6] E. H. Johnson, *Rand Metallurgical Practice*, vol. i., pp. 277-285.
[7] H. A. Megraw, *Eng. and Mng J.*, Mar. 21, 1914, p. 606.
[8] Clark and Sharwood, *Trans. Inst. Mng. and Met.*, 1912, **22,** 137.

CHAPTER XVII.

CYANIDE PROCESS.

Special Methods and Examples of Practice.

Treatment of Concentrate.—When gold and silver in an ore are largely contained in constituents such as sulphides, it is often expedient to separate them by concentration and to treat the products separately. The removal of the concentrate enables the residue to be cyanided with comparative ease and simplicity. The concentrate was formerly in general treated by roasting and chlorination, or shipped to smelters. Chlorination has now become obsolete, and concentrate is cyanided in many mills. The methods include " roasting, leaching, agitation, filtration, decantation, oxidation and fine grinding as with ordinary ores " (MacFarren). Free gold dissolves slowly, being as a rule comparatively coarse-grained, and can be in part recovered by amalgamation. " No standard method for the treatment of concentrate has yet been or can be decided on, as it must vary with the nature of the concentrate and the mode of occurrence of gold and silver in it. Hence the practice differs at almost every mill " (Gowland).

Roasting has been generally discontinued except for sulpho-telluride ores (*q.v.*). *Percolation* may require as much as three or four weeks to obtain a good extraction. The concentrate tends to pack down and become impermeable, and is sometimes mixed with sand to facilitate leaching. Shallow charges are usual. Strong solutions, containing as much as 0·5 per cent. KCy or more, are used. The work is begun and ended with weak solutions and water washes. It is necessary to apply special methods of aëration, such as the forcing of air through the charge, or turning the material over by shovels, or transferring it to another vat. Acidity and dissolved sulphides are dealt with as usual. Cyanicides, especially iron and copper salts, are sometimes removed by means of dilute sulphuric or hydrochloric acid followed by water washes (MacFarren).

Fine grinding and agitation saves much time. Air agitation is preferred to the use of mechanical agitators, owing to the tendency of the charge to pack, on account of its rapid settlement in water. Concentrate is often ground in cyanide solution if the removal of cyanicides is not required. The addition of fresh cyanide solution to the charge during treatment is sometimes found to be necessary.

Decantation is facilitated in the treatment of slimed concentrate by the fact that it settles rapidly. For the same reason *filtration* by some of the leaf filters is difficult. The Kelly filter press [1] is mentioned by MacFarren as giving satisfactory results, on the ground that in this machine cake formation by pressure is rapid. With such filters as the Moore or the Butters the

---

[1] For description see MacFarren, *Cyanide Practice*, pp. 148 and 201.

concentrate settles in the vat before the cake can be formed. With such a filter press as the Dehne, settlement takes place inside the press, and channels are formed preventing efficient washing.

An interesting observation made by J. J. Denny,[1] that sulphide, sulph-antimonide and sulpharsenide of silver are decomposed by the action of nascent hydrogen, produced by the dissolution of aluminium in caustic soda, may prove to be of value in the treatment of gold concentrate and sulpho-telluride ores. The reactions are given as follows :—

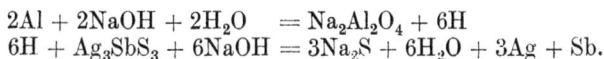

$$2Al + 2NaOH + 2H_2O = Na_2Al_2O_4 + 6H$$
$$6H + Ag_3SbS_3 + 6NaOH = 3Na_2S + 6H_2O + 3Ag + Sb.$$

The silver is set free and rendered accessible to the action of cyanide. The method has been used for a short time in the treatment of silver ore at Nipissing, Ontario, but has not yet been applied to gold ores.

The Treatment of Sulpho-telluride Ores.—These ores occur in large quantities at Kalgoorlie, West Australia, and at Cripple Creek, Colorado. The gold is contained almost entirely in the sulpho-tellurides, and probably exists mainly in the form of telluride of gold (*q.v.*). The gold cannot readily be extracted from these by cyanide solutions, owing to the slight solubility of tellurium in cyanide. The successful methods of treatment have been (1) dead roasting with the expulsion of tellurium, sulphur, etc., followed by cyaniding : used at Kalgoorlie and formerly at Cripple Creek ; (2) wet crushing and concentration, followed by treatment of concentrate and various classes of the tailing with cyanide to which cyanogen bromide is added to assist in the dissolution of the gold tellurides : used at Kalgoorlie, and with modifications at Cripple Creek. These methods are described in succession below.

(1) *Roasting Processes.*—The *Marriner process,*[2] in work at certain mines at Kalgoorlie, consists in (1) dry crushing in Gates' rock breakers, followed by ball mills or roller mills ; (2) roasting dead ; (3) grinding wet in pans with a large quantity of mercury and hot alkaline cyanide solutions ; (4) agitating by means of paddles in large vats with cyanide solutions containing from 0·01 to 0·08 per cent. KCy ; (5) filter-pressing ; and (6) precipitation of the gold by zinc.

In roasting, a large amount of soluble sulphates is formed, and the losses by dusting must be carefully attended to. Edwards or Merton furnaces (*q.v.*) had replaced all others by 1911. The pans are used primarily to grind the ore to fine slime, which is necessary to enable the cyanide to dissolve the gold, but any coarse gold is extracted in them by amalgamation. The roasted ore is carried by a stream of weak cyanide solution to spitzkasten, where it is separated into sand and slime, the sand going to the pans and the slime to the agitation vats. The pans are of the Wheeler type (see p. 227). The continuous overflow from the pans is again passed through spitzkasten and the sand returned to the pan. The agitation vats are 15 to 22 feet in diameter and 4½ to 9 feet deep, and are provided with radial arms or stirrers revolved by bevel gearing. The pulp is agitated for about eight hours. It is then drawn off and forced by pumps into filter presses of the Johnson type (see p. 372), where the gold solution is separated and the residue washed.

---

[1] Denny, *Mng. and Sci. Press*, Sept. 27, 1913, p. 484.
[2] James, *Trans. Inst. Mng. and Met.*, 1900, **8**, 490 ; Donald Clark, *Australian Mining and Metallurgy*, pp. 15-44.

According to von Bernewitz,[1] filter presses are being discarded in favour of vacuum filters, which are already engaged in filtering the current mill slime at the Associated Northern, Boulder and other mills. The reason is that vacuum filtration is cheaper, especially in labour, although the press washing is very efficient. The total cost of the process at Kalgoorlie is given by von Bernewitz at about 10s. per ton, with an extraction of 93 per cent. on 8-dwt. ore.[2]

(2) *Treatment with Cyanogen Bromide* (*Bromo-cyanogen*, *Bromo-cyanide*), BrCy.—In the *Diehl process*,[3] which was introduced at the Hannan's Star Mine, and subsequently used at the Lake View Consols and at Hannan's Brownhill, Kalgoorlie, the ore is wet crushed, amalgamated, concentrated, separated into sand and slime, the sand reground in tube mills, and the whole agitated with cyanide (to which cyanogen bromide is added) and filter-pressed. The concentrate, which contains 30 to 40 per cent. of the gold, is roasted and sent back to the wet crushing mill.

According to Knutsen, the ore at Hannan's Brownhill is crushed in a 20-stamp battery fitted with inside and outside amalgamated plates. An average of 75 tons of ore is crushed per diem through 30-mesh screens. There are 4 Wilfley concentrators, and from 3 to 5 per cent. of concentrate is produced. This is roasted in an Edward's mechanical furnace and sent back to the stamp mill. The tailing passes through two classifiers and the separated sand is reground in two slime or tube mills, each containing $2\frac{1}{2}$ tons of flint balls. The product of the tube mills goes back to the classifiers. The slime is concentrated by the removal of water in spitzkasten, and the sludge, containing 40 to 45 per cent. of solid slime, flows to one of five agitators, each $20\frac{1}{2}$ feet in diameter and $7\frac{1}{2}$ feet deep. When the agitator is full, it is charged with strong KCy solution, and two hours later with strong BrCy solution. The amounts are KCy 0·20 per cent. and BrCy 0·05 per cent. of the solid material. After about 24 hours' agitation, lime is added, the quantity being 3 to 4 lbs. per ton of solid slime, and about two hours later the charge is filter-pressed. Each of the two presses holds 5 tons, and the time of treatment is about two hours. The gold solution is again filtered before flowing to the zinc boxes. The filter-press cakes, after being washed and then dried by air-blowing, go to the dump.

At the Hannan's Star Mill, Kalgoorlie,[4] the ore from the rock-breaker is dry-crushed in Krupp ball mills, mixed with water, passed over amalgamated plates and then ground fine in tube mills. The fine product is agitated in vats with cyanogen bromide. Nardin observes[5]—" It is useless to add bromo-cyanogen at the beginning of the agitation, as it is comparatively rapidly destroyed. The proper time to add it is when no further extraction can be gained by plain cyanide solution. Before adding the bromo-cyanogen it is essential to neutralise some of the alkalinity of the pulp solution by the addition of H_2SO_4, in order to prevent the rapid destruction of bromo-cyanogen." The alkalinity is reduced from 0·02 or 0·03 per cent. lime to 0·01 per cent. The method of preparation of cyanogen bromide from bromide and bromate of potassium, sulphuric acid and cyanide is given by Nardin.[6] It is found that finely divided iron and also fine pyrite are destructive of cyanogen bromide.

[1] Von Bernewitz, *Mng. and Sci. Press*, Mar. 15, 1913, p. 409.
[2] Von Bernewitz, *loc. cit.*
[3] H. Knutsen, *Trans. Inst. Mng. and Met.*, 1902, **12**, 1.
[4] E. W. Nardin, *Mineral Industry*, 1908, p. 444 ; G. W. Williams, *ibid.*, 1907, p. 537.
[5] Nardin, *loc. cit.*   [6] Nardin, *loc. cit.*

Cyanogen bromide is less used than formerly (von Bernewitz). The average cost of treatment at Kalgoorlie is about 9s. per ton, and the average extraction is 88 per cent. on 6⅔ dwts. ore.[1]

(3) *Cripple Creek Procedure.*—Ore treatment at Cripple Creek has always been influenced by the fact that neighbouring smelters are available to which ore can be sold. Consequently, rich ore and rich concentrate have always been shipped to them. Poorer ores and concentrate were formerly roasted and chlorinated, or in the alternative roasted and cyanided. Roasting of the original ore has now been given up, and the usual procedure is to crush in cyanide solution, concentrate, and treat the tailing by cyaniding with the aid of cyanogen bromide. Concentrate is shipped to a smelter or roasted and cyanided.

At the Stratton's Independence Mill,[2] where the method was worked out by Philip Argall, the practice is as follows :—The ore is broken to about 1½-inch cubes in Gates' gyratory crushers. It is then slightly moistened with cyanide solution, lime is also added, and the ore is reduced to about ¼-inch size by means of rolls. From the rolls it passes to 6-foot Akron Chilian mills (see p. 250) fitted with square wire screens having an aperture of 0·046 inch (or about 10 mesh). The mills give a fine granular product suitable for concentration. The crushing is done in cyanide solution containing ½ lb. KCy per ton, and from that point the ore is constantly in solution.

The pulp is led to Ovoca classifiers (double spiral screw machines), whence the nearly dry sand goes to Card concentrators. Here two products are obtained, a high-grade concentrate (containing 5 to 7 ozs. gold per ton), which is shipped to the smelters (although it could be roasted and cyanided) and a middling, which is reground and reconcentrated. The slime from the classifiers is thickened in cones and closely concentrated on Deister tables or on vanners. The sand tailing and the slime tailing are cyanided separately.

The sand is treated in leaching vats for four days with solution containing ½ lb. KCy per ton, and is discharged by sluicing. It contains only 1 dwt. of gold per ton before treatment. The slime contains 2 dwts. per ton, and after thickening it is agitated with solution containing ¾ lb. KCy per ton for six hours in a modified Pachuca tank, afterwards with cyanogen bromide for four hours in another similar tank. The cyanogen bromide treatment differs from that generally adopted, in that it is found necessary to maintain a comparatively high alkalinity during the treatment. The slime pulp is fed from a storage tank into a vacuum filter of the stationary or Butters type, and the solution is clarified in a filter press. Precipitation is effected by means of zinc shavings, and the precipitate is shipped to smelters.

The ore to be milled comes from the dump, and contains only $3.50 in gold and silver, and the extraction totals 71·5 per cent., of which 43·65 per cent. is obtained by concentration and 27·85 per cent. by cyaniding. The consumption of chemicals is about 0·45 lb. NaCy, 0·3 lb. zinc, 2·2 lbs. lime, and 0·4 lb. of the bromine salt per ton. The cost of treatment is given as $1.239 per ton.

Antimonial Ores.—Gold ores containing stibnite are difficult to treat with cyanide. "Antimony sulphide is very soluble in caustic alkali and decomposes cyanide, combining with the alkali and forming antimonite

---

[1] Von Bernewitz, *Mng. and Sci. Press*, Mar. 15, 1913, p. 409.
[2] H. A. Megraw, *Eng. and Mng. J.*, Feb. 8, 1913, p. 313.

and thio-antimonite; also some KCyS is formed and HCy is evolved. The antimony compounds act as strong deoxidisers " (Julian and Smart). Hence much cyanide is destroyed and the gold is not dissolved.

J. S. MacArthur [1] proposed to remove the antimony by dissolution in caustic soda solution containing 2 to 4 per cent. NaOH, and to treat the residue by roasting, amalgamating and cyaniding.

F. H. Mason [2] proposed to remove the antimony by alternately roasting the ore in an oxidising atmosphere and with the addition of coal. The antimonates at first formed are thus reduced. The residue was then treated by amalgamation and cyaniding, but the percentage of extraction in experiments on a Canadian ore [3] was only from 50 to 80 per cent. However, roasting and fine grinding represent the most successful treatment of antimony ores.

At the Globe and Phœnix Mill, Rhodesia, [4] the presence of 1 per cent. of stibnite reduced the extraction of gold by ordinary cyaniding to about 20 per cent. It was found that by exposing the slime (containing up to 6 dwts. gold per ton) to oxidation by weathering on the dam for six months, the extraction by cyanide became satisfactory. The sand was not equally amenable, but by fine grinding most of the stibnite passed into the slime owing to its brittleness. In the method adopted, the stibnite was partly removed by hand picking, and the tailing from the stamps after amalgamation was separated into sand and slime. The sand was concentrated and then ground and amalgamated in pans, and again concentrated. The sand is cyanided by percolation with fair results if the stibnite in it does not exceed 0·2 per cent., and the slime is passed to the dam for weathering. The concentrate is roasted in Merton furnaces, by which the antimony is reduced from 5 per cent. (as stibnite) to 3 per cent. (chiefly in the form of antimonate), and the roasted product is ground and amalgamated and then treated with cyanide in agitation vats and filter-pressed. The extraction is from 85 to 90 per cent., but the tailing is rich, and a further amount can be extracted after the residue has been weathered for some time.

At Hillgrove, New South Wales, [5] antimonial tailing which has been subjected to long weathering is treated by percolation. Care is taken to make the strongly acid tailing almost exactly neutral or very slightly alkaline, to avoid dissolution of antimony, which is thrown down in the zinc boxes and gives trouble in smelting the precipitate. Ore containing stibnite is also cyanided at Bidi, in Upper Sarawak, by direct treatment of coarsely crushed ore with 0·05 per cent. KCy solution. About 75 per cent. of the gold is extracted from 5 dwt. ore.[6]

Cupriferous Ores.—Unoxidised chalcopyrite is little acted on by cyanide, and does not occasion trouble, but " in some cases, as with oxides and carbonates, a few pounds of copper per ton may prohibit cyaniding by ordinary methods " (MacFarren), the soluble compounds acting as cyanicides. Sometimes the copper is removed by a dilute sulphuric acid wash before cyaniding.[7] A solution of ammonia has also been proposed [8] for the same purpose.

[1] A. Selwyn-Brown, *J. Chem. Met. and Mng. Soc. of S. Africa*, July, 1906, **7**, 20.
[2] Mason, *Mng. and Sci. Press*, April 28, 1906.   [3] Selwyn-Brown, *loc. cit.*
[4] H. T. Brett, *Mng. Mag.*, July, 1911, p. 51.
[5] W. A. Longbottom, *Mng. and Sci. Press*, June 29, 1912, p. 884.
[6] R. Pawle, *Trans. Inst. Mng. and Met.*, 1905, **15**, 79.
[7] W. S. Brown, *Trans. Inst. Mng. and Met.*, 1906, **15**, 445.
[8] A. Jarman and E. Le Gay Brereton, *ibid.*, 1905, **14**, 288; H. L. Sulman, *ibid.*, **14**, 363; *Electrochem. and Met. Ind.*, Mar. 1908, p. 128; *Pacific Miner*, Mar. 1910.

Apart from the action of ammonia on copper compounds, the addition of ammonia to cyanide solutions increases their power of dissolving gold and also their selective action on gold. Acid solutions of cyanide have also been used on cupriferous and antimonial tailing (Gitsham process).[1] In this process a little sulphuric acid is added to the cyanide solution, with which the ores are treated. Gitsham suggests the following equations :—

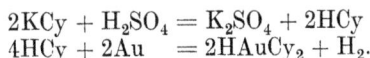

$$2KCy + H_2SO_4 = K_2SO_4 + 2HCy$$
$$4HCy + 2Au = 2HAuCy_2 + H_2.$$

Sulman and Picard point out that the dissolving action of the HCy depends on the presence in solution of $CuSO_4$ or some similarly acting substance, by which nascent cyanogen is produced. It is stated that very little copper is dissolved by acid cyanide solutions and consequently that ores containing copper carbonate can be treated. The solution is regenerated by alkali before precipitation. Wheelock points out [2] that cupriferous cyanide solutions can be regenerated by the addition of sulphuric acid—

$$K_2Cu_2Cy_4 + H_2SO_4 = K_2SO_4 + Cu_2Cy_2 + 2HCy$$

and the insoluble cuprous cyanide may be treated with hydrogen sulphide [3]—

$$Cu_2Cy_2 + H_2S = Cu_2S + 2HCy.$$

Cyanide solutions containing copper give trouble in precipitation on zinc shavings, the copper plating the zinc and preventing its further action. This difficulty is not experienced in zinc dust precipitation. Copper is not precipitated by zinc in solutions containing much KCy, but it is readily brought down by the lead-zinc couple (see p. 338). It is also precipitated by soluble sulphides.

Crushing in Cyanide Solution.—The use of cyanide solution instead of plain water in the stamp battery was tried as early as 1891 in the United States, and in 1892 at the May Consolidated, Transvaal.[4] It was successfully developed at the Crown Mines, New Zealand, in 1897,[5] and applied in South Dakota in 1903.[6] It has not been much practised in conjunction with amalgamation where coarse gold exists, because, although the gold is cleaned and brightened by the cyanide, and thus kept in excellent condition for amalgamation, the corrosive effect on the copper plates and the hardening of the amalgam by cyanide more than neutralises this advantage. Crushing in cyanide is used in many mills, however, in connection with " all sliming " methods, and the use of slime filters, amalgamation being omitted. " Where the gold or silver in the ore is very finely disseminated, all sliming may be required to expose a fair percentage of the precious metals to the action of the solvent cyanide. This feature is frequently associated with a very small percentage of recovery by amalgamation being practicable, as with gold telluride or silver sulphide ores. Under these conditions crushing with cyanide solution instead of water is often practised, if only in order that dissolution of the gold or silver, in any case usually a lengthy process, may

[1] Von Bernewitz, *Cyanide Practice*, 1910-13, p. 102.
[2] R. P. Wheelock, *Mng. and Sci. Press*, Dec. 18, 1909, p. 814.
[3] Von Bernewitz, *Cyanide Practice, loc. cit.*
[4] Butters and Clennell, *Eng. and Mng. J.*, Oct. 8, 1892, p. 342.
[5] J. McConnell, *Trans. Inst. Mng. and Met.*, 1898-9, 7, 26 ; *Trans. Amer. Inst. Mng. Eng.*, Sept. 1899.
[6] See Papers by C. H. Fulton and J. Gross, *Trans. Amer. Inst. Mng. Eng.*, Sept. 1904.

begin at as early a stage as possible ; air-lift vats preceded by mechanical pulp thickeners are employed to expedite this dissolution ; and vacuum or pressure filtration of the slime pulp, which frequently settles very slowly on account of its colloidal nature, is used to separate the residual solid slime from the gold and silver bearing solution " (Caldecott).

It is generally admitted that crushing in cyanide is desirable where the percentage of recovery by amalgamation is trivial, but it is otherwise where a high percentage of the gold can be thus recovered. It is true that in many cases amalgamation may be omitted and almost all the amalgamable gold recovered by crushing in cyanide. In the latter case the secondary crushing in the tube mill circuit is largely instrumental in dissolving the particles of free gold. These are freed and in part flattened or broken up by the pebbles, and thus rendered more readily soluble. It is, of course, obvious that the rate of dissolution of large particles is less than that of small ones, as their ratio of area to weight is lower.

On leaving the tube mill, the attenuated particles of gold are returned to the mill in the underflow of the classifiers, and circulate until they are completely dissolved or become so small as to be included in the overflow of slime. Nevertheless, this method of extracting amalgamable gold has not yet been generally accepted as the most advantageous method of procedure in ordinary cases, and remains a subject of controversy.

On the Rand, where about 64 per cent. of gold is recovered by amalgamation and 36 per cent. by cyanide, crushing in cyanide solution instead of water has not been accepted (although it has been tried on a large scale) because of trouble from the following causes :—[1]

(1) Corrosion of amalgamated copper plates.

(2) Baseness of the bullion, due to deposition of dissolved copper on zinc shavings.

(3) Need of more prolonged and more expensive cyanide treatment necessitated by richer tailing pulp.

(4) Liability to loss by leakage and overflow of gold-bearing mill service solution.

(5) Difficulty of obtaining accurate screen values (by assay) owing to presence of gold in solution in screen pulp.

It may be observed that the first two objections would disappear if amalgamation were entirely abandoned, but in that case long and costly treatment in tube mills would be required, involving all-sliming. " It is unlikely that any all-sliming process will ever be applied to banket ore, since the cost of converting into slime the residual hard granular sand particles, constituting nearly half the weight of the present crushed product, would far exceed the value of the small additional amount of gold recovered in that way " (Caldecott).

Another point is that in stamp battery practice from 5 to 10 tons of solution are required per ton of ore, and although most of this can be retained in the mill circuit, at least part of the solution in the mill circuit must be passed through the precipitating plant, thus increasing the loss of zinc and cyanide. There is also the fouling of the mill circuit solution to be considered, necessitating the use of fresh solution occasionally and the throwing away of the old solution. With soft ores other machines, such as the Huntington

---

[1] E. L. Bateman, *S. African Mining J.*, Jan. 10, 1914, p. 469. Endorsed in letter from Dr. Caldecott.

mill, requiring less solution (say 4 tons per ton of ore) can be used in milling, and the difficulty caused by the large amount of solution in the mill circuit is diminished.

On the other hand, among the advantages of crushing in cyanide, it is pointed out that—

(1) Dissolution of gold is promoted by the close mixing of ore and solution involved in crushing.

(2) Float gold is rapidly dissolved during crushing without a chance to escape.

(3) If water is used, the de-watered pulp carries some water into the cyanide plant. This amounts only to from 25 to 40 per cent. of water in the case of sand, but in slime it is considerably more. With all-sliming, followed by de-watering in Dorr's thickener, the ratio of water to dry slime is 1 or $1\frac{1}{2}$ to 1. The addition of this quantity of water to the cyanide solution in agitation vats seriously dilutes it. More cyanide must be added, and to avoid a continual increase in the quantity of solution in the cyanide plant, it is necessary to throw some solution away. When ore is crushed in cyanide, an amount equivalent to the solution in the thickened pulp can be returned to the mill circuit, and so the consumption of cyanide is *ipso facto* reduced.

(4) There is no risk of the theft of amalgam.

Comparisons made by N. Cunningham[1] between the practice at the Hollinger and Dome Mills, Porcupine, Ontario, are as follows, for the first few months of 1914 :—

	HOLLINGER MILL. Crushing in Cyanide Solution.	DOME MILL. Crushing in Water.
Solution precipitated per ton of ore,	2·67 tons.	1·70 tons.
Zinc consumption per ton of ore, .	0·59 lb.	0·34 lb.
Cyanide consumption per ton of ore,	0·46 lb.	0·75 lb.
Total recovery, ...	96·0 per cent.	95·0 per cent.

The Hollinger mill does not use amalgamation, but the Dome mill amalgamates the ore. The ores differ only in grade, the Hollinger ore being three times as rich as that of the Dome, so that the tailing at the Dome mill is poorer than that at the Hollinger. At the Dome mill, Dorr thickeners are used, giving an underflow of pulp in which the ratio of liquid to solid is about 1 to 1. The extra consumption of cyanide is instructive. The comparison is chiefly interesting as an indication of the excellent results which can be obtained on similar ore by both methods.[2] The respective costs of treatment are not given.

Examples of crushing in cyanide are given on pp. 415, 416, and 417.

EXAMPLES OF PRACTICE.

Besides the following examples, others have been already given, notably in the section on the treatment of sulpho-tellurides.

---

[1] Cunningham, *Mng. and Sci. Press*, July 4, 1914, p. 19.
[2] For a discussion on crushing in cyanide see A. W. Allen, *Mng. and Sci. Press*, Aug. 16, 1913 ; May 30, 1914 ; Sept. 5, 1914 ; J. B. Stewart, Sept. 20, 1913 ; N. Cunningham, July 4 and Oct. 17, 1914.

Witwatersrand.

The treatment is fully described in *Rand Metallurgical Practice*, vols. i. and ii. Briefly, it consists in sorting out waste rock ; breaking the ore in rock-breakers ; crushing in alkaline water through coarse screens by stamps ; de-watering in cones ; crushing in tube mills ; diluting and passing the pulp over amalgamated plates ; classifying the pulp in cones into sand and slime, with return of a classified underflow to the tube mill ; cyaniding of collected sand by percolation ; coagulation of slime with lime, and treatment with cyanide by decantation (p. 356) ; precipitation of gold by lead-zinc shavings ; dissolving the zinc in bisulphate of sodium, and fusing the residue with fluxes or with litharge. Owing to the flat nature of the ground, elevation of the pulp by pumping or tailing wheels is required during treatment and in the disposal of tailing. The methods are continuously undergoing revision, and new plants exhibit new features, while old machines, partly discredited, still continue at work in old plants, to be discarded when it is profitable to do so. There is accordingly considerable divergence between the methods in use side by side.

The following section is slightly abridged from a summary of recent progress, for which the author is indebted to Dr. W. A. Caldecott, the consulting metallurgist of the Consolidated Goldfields group. The account is supplementary to the information contained in the foregoing pages, and applies alike to crushing, amalgamation and cyaniding.

The frontispiece gives a general view of the Knight's Deep and Simmer East joint reduction works. The photograph was taken in 1907, and kindly sent by Dr. Caldecott for reproduction. It shows the 400-stamp mill and the elevator wheels on the right, and the dump at the back.

The practice on the Rand is the result of many years' evolution, with the co-ordinated efforts of many men, freely interchanging their experience, and applying to their special problem, wherever possible, the results obtained elsewhere. The main features which broadly characterise Rand practice are the large number of producing mines in close proximity to one another ; the large scale of operations ; the absence of chemical difficulties in the simple low-grade ore ; the use of as few and as large units as possible for each stage of treatment, those fulfilling similar functions being grouped together to facilitate supervision ; simplicity of plant arrangement and operation ; cheap power and unskilled labour ; and the group system of supervision, guidance and control of several subsidiary companies. As regards future tendencies, plants already exist such as the Randfontein Central and Knight's Deep, capable of treating 3,500 to 5,000 tons of ore daily, and such large installations may become more general.

The chief physical characteristic of banket ore is its hard abrasive nature, which involves the necessity of crushing by impact by means of stamps and tube mills, and precludes the economic use of all rolling, rubbing, or grinding machines for pulverising ore, on account of the heavy operating cost of such appliances through high steel consumption by abrasion. This is the reason for the failure of all attempts to introduce fine crushing rolls, ball mills, edge roller or Chilian mills and grinding pans. In general, it is not desirable to crush banket so fine as to obtain a residue containing, on the average, much less than 0·25 dwt. gold per ton after cyaniding. This necessitates a final tailing pulp containing from about 75 per cent. of − 90 product (*i.e.*, passing 90-mesh sieve, with 0·006 inch aperture) with 4-dwt. ore, to 85 per cent. of — 90 product with 7-dwt. ore.   About 70 per cent. of the — 90 grade will pass a 200-mesh (0·003 inch aperture) screen.

Sorting has been largely abandoned, as trials show that the percentage

Fig. 178.—The 40-Stamp Battery of the New Heriot Company, showing the Stamps, Water Supply and Copper Tables

26

extraction from sorted waste (say under 0·5 dwt. in value) is upwards of 90 per cent. or practically as high as from ordinary banket ore, while

Fig. 179.—Stamp Batteries without Amalgamated Plates. Robinson Deep Mine.

the additional tonnage crushed costs very little, when ample reduction plant is available. The elimination of sorting reduces the capital expenditure and the operating costs.

Breaking.—Underground breakers, where used, are set to make a product of 6 to 9 inches, and surface breakers one of $1\frac{1}{2}$ or 2 inches. Grizzlies screen off about 50 per cent. of the ore, which does not pass through the breakers. Both jaw and gyratory breakers are used, and efficient and careful breaking is regarded as essential. Tube mill pebbles of 4-inch cube are sorted out in the breaker station, and in order to minimise the breathing of dusty air by operators, sprays or exhaust fans with suctions to each breaker, to withdraw the dust produced, are now commonly installed.

Stamp Milling.—The use of heavy stamps up to 2,000 lbs. weight is general in all new mills. The Nissen stamps are especially adapted for small installations. With heavy stamps not more than 97 drops per minute, with 8 inches set height of drop, should be given to avoid undue breakages of cam shafts owing to shock of cam impact on the tappet.

The ratio of 10 heavy stamps per standard tube mill ($5\frac{1}{2}$ by 22 feet) is the usual modern practice, with coarse screens up to $\frac{1}{2}$ inch aperture and stamp duties up to 20 tons per stamp of 2,000 lbs., and without plates in the stamp mill. Figs. 178 and 179 illustrate the former practice with amalgamated plates at the New Heriot mill, and the practice in Jan. 1914 at the Robinson Deep installation. In Fig. 179 the tube mill motors are shown on the left.

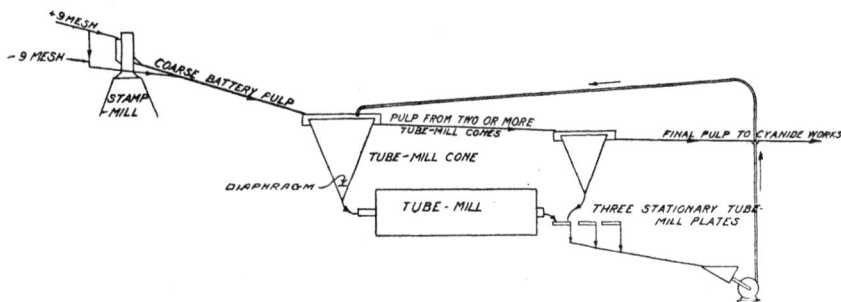

Fig. 180.

The ratio of water to solids in coarse battery screen pulp is very low, and varies between $4\frac{1}{2}$ and 5 to 1 ; tube-mill pulp contains about 39 per cent. of water, and the water ratio in the pulp flowing over the tube-mill amalgamated plates is $1\frac{1}{2}$ to 1. In general, launders in a modern crushing plant are steep, 10 per cent. being a frequent grade, and the proportion of water in the pulp in various stages is correspondingly reduced. The advantage of this procedure is obvious in reducing the cost of pulp elevation, the number of classifiers, and the size of the pumps returning the alkaline mill service water to the crushing plant.

To avoid having to pump coarse pulp from battery screens, the best practice, as at the Robinson Deep, is to gravitate from batteries into diaphragm tube-mill cones (6 by 9 feet), the underflow from which, containing up to 400 tons solids per 24 hours, enters the tube mills. This is illustrated in Fig. 180.[1] This diagram now represents accepted modern practice, except that it is not worth while to separate the mill ore feed into plus and minus 9-mesh products before the ore enters the mortar box.

[1] W. R. Dowling, *Rand Metallurgical Practice*, vol. i., p. 129.

The high stamp duty, up to 100 tons per 5-stamp battery of heavy stamps during 24 hours, and coarse screening at low cost have prevented the introduction of rolls in place of stamps for crushing to ½-inch mesh, although it would be quite practicable. This is due to the fact that the abrasive properties of banket ore make roll maintenance high, in addition to which there is the difficulty caused by the dust raised by dry-crushing, trommelling and return of oversize.

The use for pulp elevation of centrifugal pumps with renewable steel or hard cast-iron liners has become general on account of their relatively low capital cost, limited area required for installation, and flexibility in allowing the height of pulp elevation to be changed when required.

Tube-Milling.—In a modern Rand crushing plant about two-thirds of the crushing, as measured in tons of fine (— 90-mesh) product, is done by tube mills. The tube mills are large and are fed with the coarse pulp underflow of classifiers (see Fig. 181, which is from a photograph of the Robinson Deep installation taken in January, 1914). Each tube mill receives the coarse underflow from its diaphragm cone classifier on the right, into which the pulp gravitates from the stamp batteries. Slight variations from the dimensions of the standard mills (5½ by 22 feet) have lately been installed in the shape of mills of 6 by 16½ feet, but without much evident practical gain. Except for all-sliming, short tube mills merely serve to conform to limited sand feed or to limited power. For standard tube mills the rules should be observed of 31 revolutions per minute, high percentage of running time, adequate pebble load, and large (350 to 400 tons per 24 hours) sand feed as 40 per cent. moisture pulp, so as to provide sufficient coarse particles in the latter part of the tube mill to utilise fully the crushing effect of the falling pebbles. The use of an internal scoop discharge, lowering the level of the liquid pulp in the mill and equivalent to a large outlet trunnion, has enabled the crushing capacity to be increased by 25 per cent. and upwards with a corresponding increase from 7 tons to 15 tons per 24 hours of banket ore pebble consumption, more rapid liner wear, and higher power consumption. No new 5½ by 22 feet mills are likely to be installed with less than 150 H.P. motors and possibly 175 H.P. motors. Individual watt-meters for each tube mill are used to regulate the pebble feed so as to maintain a constant power consumption, and mechanical pebble feeders driven from the tube mill are employed in order to maintain continuously adequate pebble loads in each mill. The longitudinal hard steel bar liner has come into general use, and has a life of six months or more, owing to the wear being taken by pebbles wedged in the 2·5 inch spaces between the bars on edge.

Amalgamation.—The use of battery screening as coarse as 64 holes per square inch or coarser prevents plate amalgamation owing to the scouring of amalgam, and hence in modern stamp mills the use of amalgamated plates has been discontinued. They are retained, however, in the tube-mill circuit in the form of three stationary plates of the usual dimensions (4½ by 12 feet) per tube mill, but with a grade of 18 per cent. to prevent any banking of the thick pulp which flows over them. The percentage recovery by amalgamation under these conditions depends upon the fineness of crushing and the area of classifier surface in proportion to the tonnage milled, but recoveries of upwards of 60 per cent. are not unusual, and upwards of 70 per cent. has been regularly obtained with pyritic banket ore at the Princess Estate and Gold Mining Company. The advantages of the foregoing system are (1) more running time for stamps, which do not require to be stopped for plate dressing

Fig. 181.—Tube-mill Installation. Robinson Deep Mine.

or scraping, (2) less gold locked up in plant as amalgam used in the setting of plates, (3) greater security against theft, since amalgamation is confined to one room, (4) less labour for dressing plates, and (5) lower capital cost for plant equipment.

Classification.—During the last few years this has been greatly developed on the Rand. It is usually effected by a limited number of large diaphragm cones without additional water, as opposed to the use of numerous small cones with ascending water currents employed elsewhere. As pointed out in " The Finer Crushing of Banket Ore," [1] the classification in water of the battery screen pulp and of the tube-mill return pulp ensures the richer heavier pyrite being more finely crushed than the poorer lighter quartz, and renders exactness of battery screening of little importance, since the classifier instead of the screen determines what particles shall leave the crushing plant. Further, since all gold now recovered by amalgamation must be first classified out, the importance of classification is much increased, and its efficiency is evidenced by the high amalgamation recoveries referred to above. (These are higher than was formerly the case.)

As regards classification of sand and slime, this has now reached a very high degree of efficiency, with the result that sand can be treated in the vat where collected as a sand solution pulp from sand filter tables, without the cost of transfer and with a much reduced plant, whilst the clean sand residue is perfectly fitted for sand filling of old mine workings. Incidentally, the foregoing classification permits of lime being added to the ore entering the mill bins, instead of being crushed separately in a ball-mill for subsequent addition to the slime pulp. This is due to the fact that in cones no danger occurs of undue settlement of the slime with the sand, as was the case when the tailing pulp entered large sand collecting vats.

The use of safety or return sand cones receiving pulp from the common overflow launder of the main classifiers is useful in offsetting fluctuating delivery and uneven distribution of pulp to the main classifiers, and this system is followed both in regard to tube-mill classification and the classification of sand and slime.

Automatic regulators, for the purpose of preserving the thick underflow of diaphragm cones which are engaged in classifying sand and slime, are now in satisfactory operation on several mines. Owing to the action of the regulators the underflow ceases when the level of settled sand in the cone sinks and begins again when the level has risen slightly. The action of the regulator depends upon the pressure of the settled sand in the cone upon a large double-coned solid wooden float in the cone directly above the diaphragm, and attached to the lower end of a vertical rod whose upward motion is assisted by a spiral spring near its upper end. The vertical rod is connected above the cone with one end of a horizontal lever, whose other end is connected with a vertical rod outside the cone. The lower end of the vertical rod is connected with another horizontal lever at the level of the cone outlet, and terminates in a conical plug just below the outlet. As the pressure of the rising settled sand in the cone on the float increases, the latter is depressed, and the plug descends and opens the outlet, whilst the reverse action occurs as the sand in the cone falls.

In an efficient modern plant the proportion of slime is about 50 per cent.

---

[1] W. A. Caldecott, *Trans. Inst. Min. and Met.*, 1904-5, **14**, 48.

by weight, of which only 1·5 per cent. is + 200 mesh, and the percentage of colloidal slime in the sand is trivial.

Cyaniding.—The sand now frequently constitutes only 50 per cent. by weight of the ore milled. Owing to the absence of slime in the interstitial spaces, it occupies when settled about 23 cubic feet per ton, and can be satisfactorily collected and aërated and leached in the same vat when sand filter tables are employed (see Fig. 153, p. 351). The clean sand residue after cyaniding by percolation should not contain more than 0·33 dwt. gold per ton. For the methods of surface transport of pulp, see p. 356. The method of transport of sand, as pulp pumped through pipes by centrifugal pumps, is likewise largely employed on the Rand for tailing pulp elevation and solution pulp transfer from sand filter tables.

TABLE XXXVI.—Average Working Cyanide Solutions during December, 1913, in Plants of Simmer and Jack, Robinson Deep, Knight's Deep and Simmer Deep Companies.

	Per Cent. Free Cyanide (in Terms of KCy).	Per Cent. Protective Alkali (in Terms of NaOH).	Assay Value (Dwt. per Ton).	Temperature, ° F.
Strong solution applied to sand charges, . . .	0·095	0·021	0·015	65
Strong solution entering zinc boxes, . . .	0·028	0·012	2·312	70
Strong solution leaving zinc boxes, ...	0·025	0·021	0·015	65
Weak wash solution applied to sand charges, . .	0·011	0·013	0·013	67
Weak wash solution entering zinc boxes, . .	0·015	0·010	0·334	69
Weak wash solution leaving zinc boxes, . . .	0·010	0·012	0·013	67
Last drainings from sand charges, ...	0·008	0·008	0·044	...
First wash slime solution entering zinc boxes, .	0·007	0·008	0·399	76
First wash slime solution leaving zinc boxes, . .	0·006	0·009	0·015	72
Second wash slime solution,	0·008	0·007	0·129	80
Mill service water, . .	...	0·011	0·010	79

Slime treatment by decantation (see above, pp. 356 and 359) is still practised, but slime is now being treated in some plants by vacuum filtration. Air-lift vats are commonly used in Butters' vacuum filter plants, though for the dissolution of gold in current slime containing little reducing matter their use is not so essential as for accumulated slime, or for slimed rich battery sand. In both settlement of slime and precipitation, warmed working solutions are used, for which purpose the waste steam heat from the mill engine power plant is commonly employed. Lead-coated zinc shavings are still commonly retained in use for precipitation. Zinc dust precipitation is employed on two recent plants, but requires somewhat strong solution to induce sufficient chemical activity for complete precipitation during the limited period of contact before leaving the press, and is likewise dependent on the mechanical regularity of zinc dust feed.

As about $2\frac{1}{2}$ tons of solution are precipitated per ton of ore treated, and some reduction works treat 4,000 tons of ore per day, the enormous volumes involved are obvious. Of late the tendency has been to filter slime solution before precipitation through ordinary sand leaching vats specifically devoted to this purpose. These can be readily supplied with fresh sand and the old clogged sand discharged. Imperfect clarification, power consumption, and labour and cost involved in cleansing of cloths have caused clarification filter presses to be discarded in several plants.

Table XXXVI. (p. 407) shows the composition of ordinary working cyanide solutions.

The following table illustrates the consumption of the chief metallurgical stores in modern practice :—

TABLE XXXVII.—AVERAGE CONSUMPTION OF METALLURGICAL STORES PER TON OF ORE MILLED BY CONSOLIDATED GOLDFIELDS' COMPANIES DURING 1913.

Company.	Mercury.	Lime (75 Per Cent.).	Sodium Cyanide (130 Per Cent.).	Zinc.	Lead Acetate.	Bisulphate of Soda.
	Troy Ozs	Lbs.	Lbs.	Lbs.	Lbs.	Lbs.
Simmer & Jack,	0·0309	2·732	0·343	0·382	0·069	1·561
Robinson Deep,	0·0830	1·999	0·271	0 417	0·043	1·516
Knights Deep,	0·0864	1·965	0·300	0·376	0·039	1·216
Simmer Deep, .	0 0653	2·033	0·315	0·411	0·044	1 629
Averages, .	0·0664	2·1822	0·3072	0·3965	0·0487	1·4803

Actual results of present practice are illustrated in the tabular statement for 1913 (Table XXXVIII.), which represents the treatment of 3,387,230 tons of ore by four crushing companies on the Witwatersrand of the Consolidated Goldfields of South Africa, Limited. The following are the average working costs in pence per ton of ore for the various operations involved in the treatment of the ore from the headgear to the dump. The costs are for the same four companies and for the same period as those in Table XXXVIII.

Operation.					Cost per Ton (in Pence).
Transport of ore,	.	.	.	.	2·773
Breaking and sorting,	.	.	.	.	3·258
Stamp milling,	.	.	.	.	10·512
Tube milling, .	.	.	.	.	8·340
Sand treatment,	.	.	.	.	9·113
Slime treatment,	.	.	.	.	5·988
Total,	.	.	.		3s. 3·984

TABLE XXXVIII.—RESULT OF BANKET ORE TREATMENT FOR 1913.

Company.	Average Tonnage of Ore Crushed Monthly.	Per cent - 90 (·006 in.) Product in Tailing Pulp leaving Crushing Plant.	Percentage by Weight.		Screen Assay Value (Dwts. per 2000-lb. Ton).	Assay Value before Cyaniding (Dwts. per Ton.)			Assay Value after Cyaniding (Dwt. per Ton.)			Percentage Theoretical Extraction.			Percentage Actual Recovery (on Screen Assay Value.)	Total Working Cost per Ton Crushed. (Headgear to Dump.)
			Sand.	Slime.		Sand.	Slime.	Total.	Sand.	Slime.	Total.	Am'lga-mation.	Cyanid-ing.	Total.		
Simmer and Jack,	71,333	73·0	59·36	40·64	5·150	3·085	1·904	2·605	0·365	0·157	0·280	49·42	45·14	94·56	94·29	3s. 4·134d.
Robinson Deep, .	54,883	83·7	52·11	47·89	6·737	2·961	1·768	2·389	·303	·182	·245	64·54	31·82	96·36	96·17	3s. 6·880d.
Knights Deep, .	96,894	71·0	56·32	43·68	3·709	2·040	1·163	1·657	·366	·176	·283	55·32	37·05	92·37	93·69	3s. 2·924d.
Simmer Deep, .	59,158	77·8	47·96	52·04	3·986	2·561	1·280	1·895	·366	·169	·264	52·46	40·92	93·38	97·14	3s. 2·850d.
Averages (pro rata to Tonnage),	70,567	75·4	54·52	45·48	4·720	2·595	1·482	2·089	0·354	0·171	0·271	55·74	38·52	94·26	95·17	3s. 3·984d.

Homestake Mills, S. Dakota, 1911.[1]

Amalgamation followed by Cyanide—Merrill Filter Presses—Zinc Dust Precipitation.—The ore at this mine is of uniformly low grade, containing about 0·2 oz. gold per ton. When derived from deep levels it consists of chlorite or ferruginous hornblende, quartz, pyrite, pyrrhotite, and the carbonates of calcium, magnesium and iron. In some parts arseno-pyrite, garnet and mica are present. The mining is done by shafts, although open cuts are still being worked.

The ore as it comes from the mine is dumped over grizzlies, the rock passing through going to the storage bins, and the larger lumps travelling on to Gates' gyratory breakers. From these the broken ore is fed by means of a Challenge feeder into the battery mortar.

In all, there are six mills at Homestake, employing 1,000 stamps. Each set of ten stamps is driven by a 25 H.P. motor set on the level of the cam shaft. The mortars used are taller and narrower than those used in many other mills, and are not provided with back plates. The mortar blocks are formed of 2-inch creosoted pine planks. The falling weight of the stamps is about 850 to 900 lbs., and the initial drop with new shoes and dies is 10 inches. Water is fed into the top of the mortars through pipes of small diameter, and at some of the mills special nozzles have been fitted to these pipes, in order to preserve a maximum ratio of water to ore of 11 to 1. This ratio is large, according to general practice. Inside amalgamation is practised, a copper plate $\frac{1}{4}$ inch thick and 5 to 7 inches wide being attached to the chuck block under the screen.

The discharged ore passes through diagonal needle slot screens (approximately equivalent to 30 to 35-mesh screens) on to a series of outside amalgamation plates. The first of these is of pure copper, 4 feet 6 inches wide, 12 feet long, and $\frac{1}{8}$ inch thick. The remaining three sets of plates are made increasingly wider, in order to give a thin stream of pulp, and are silver-plated. The final row of plates is situated in a plate house, where the whole of the pulp from the mill is re-divided and spread uniformly over the tables.

Mercury is fed into the mortar box hourly. The amalgam which gathers in the mortar and on the outside plates is collected every day, while the inside amalgamation plates are scraped once in fifteen days. The loss of mercury is found to be about 0·13 oz. per ton of ore crushed.

The accumulated amalgam is retorted every ten days in cast-iron trays fixed within horizontal cylindrical retorts whose covers are luted on. Each retort holds 7,000 ozs. The whole of the mercury is volatilised in about seventeen hours, and the residual gold is then melted under borax in graphite crucibles.

Concentration is not now practised at this mill, as direct cyaniding has been found to give better results.

The tailing from the stamp mill is carried by launders to a sump in the regrinding mill, whence it is distributed to fourteen gravity cones, each 4 feet in diameter, and with sides sloping at 70°.

A large proportion (88 per cent.) of the pulp passes from these cones as overflow and goes to the classifiers. The spigot discharge from each gravity cone passes to a second concentrating cone (16 inches diameter), which is

[1] A. J. Clark and W. J. Sharwood, *Trans. Inst. Min. and Met.*, 1913, **22**, 68.

fed with an upward current of water at about 30 lbs. pressure. The overflow from this joins the previous overflow, while the sand is discharged to the regrinding machines through pipe launders.

The regrinding machinery consists of seven 5 feet pans and one tube-mill

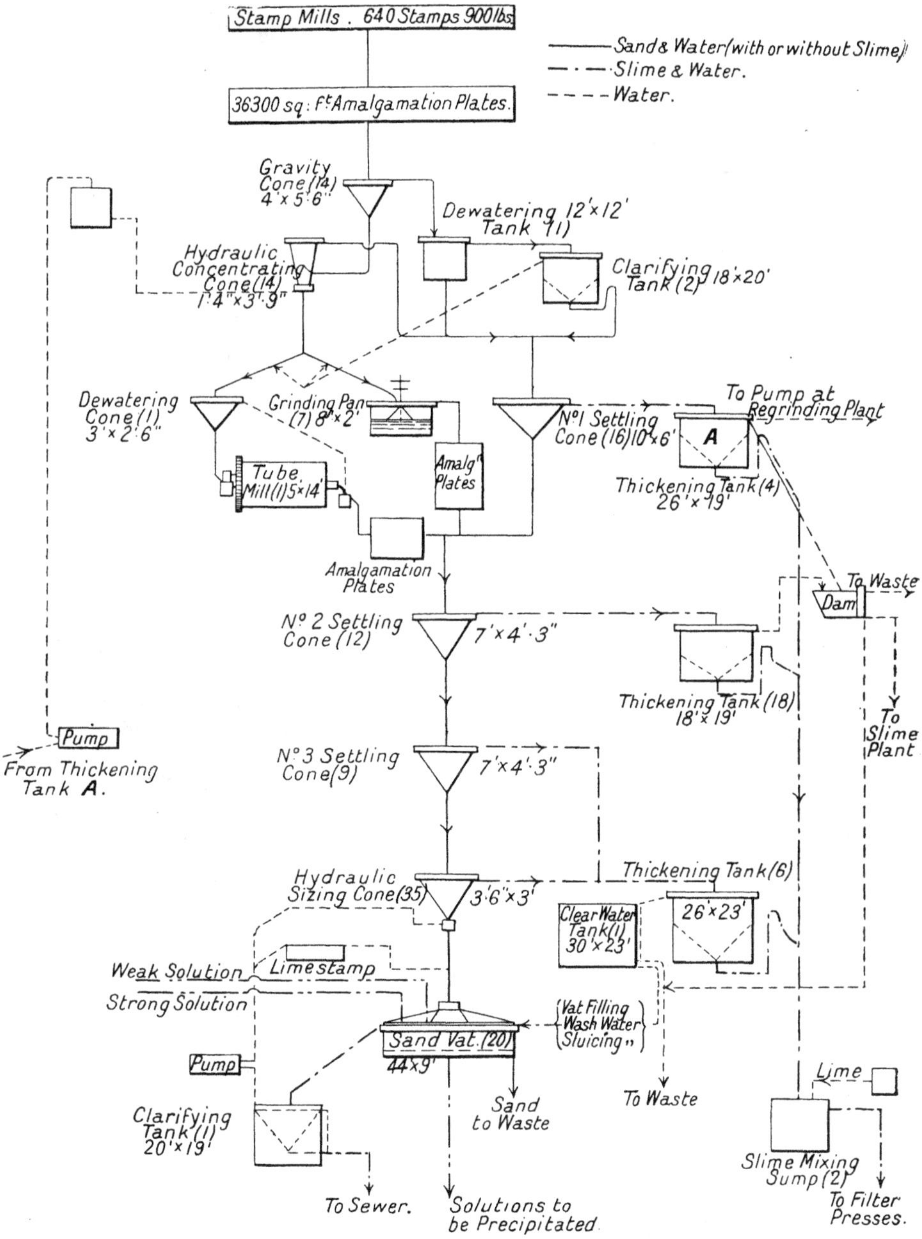

Stamp Mills . 640 Stamps 900 lbs

——— Sand & Water (with or without Slime)
— · — Slime & Water.
— — — Water.

36300 sq. Ft Amalgamation Plates.

Gravity Cone (14) 4' x 5'·6"

Dewatering 12'x 12' Tank (1)

Hydraulic Concentrating Cone (14) 1'·4"x 3'·9"

Clarifying Tank (2) 18'x 20'

Dewatering Cone (1) 3' x 2'·6"

Grinding Pan (7) 8'x 2'

Tube Mill (1) 5'x 14'

Amalg Plates

Nº I Settling Cone (16) 10'x 6'

To Pump at Regrinding Plant

A

Thickening Tank (4) 26' x 19'

Amalgamation Plates

Nº 2 Settling Cone (12)

7' x 4'·3"

To Waste

Dam

Thickening Tank (18) 18'x 19'

To Slime Plant

Pump

From Thickening Tank A.

Nº 3 Settling Cone (9)

7'x 4'·3"

Thickening Tank (6)

Hydraulic Sizing Cone (35)

3'·6"x 3'

Clear Water Tank (1) 30'x 23'

26'x 23'

Weak Solution

Lime stamp

Strong Solution

Pump

Clarifying Tank (1) 20'x 19'

Sand Vat (20) 44 x 9'

(Vat Filling
Wash Water
Sluicing „

To Waste

Lime

Slime Mixing Sump (2)

To Sewer.

Sand to Waste

Solutions to be Precipitated

To Filter Presses.

Fig. 182.—Flow Sheet of the Homestake Mill, South Dakota.

(5 × 14 feet). The pans are of the Wheeler type, and are designed to give a continuous peripheral discharge. They revolve 58 times per minute and treat 20 tons per day ; mercury is fed in at intervals. The tube-mill is fitted with a 5-inch Silex lining and employs flint pebbles as grinders. A de-watering cone is placed immediately before the tube-mill, in order to reduce the water content of the pulp to about 38 per cent. The mill grinds 73 tons per day. Silvered amalgamation plates are fixed after the pans and the tube mill, in order to recover any gold which may have been freed in the process of crushing. From a comparison of the two types of machine working on similar material, the tube mill is considered to be the more suitable for this particular ore.

The classification into sand and slime is carried out in sheet-iron cones fitted with replaceable cast-iron outlet nozzles. In the main system there are three series of cones, the underflow from the first and second passing to the succeeding series ; the overflow from each series is considered as slime. Only the last cone is fitted with arrangements which supply a current of water under pressure.

Practically the whole of the overflow from the classifying cones will pass a 200-mesh screen, and, before passing to the slime plant, is thickened in large tanks with conical bottoms, made of redwood. The feed into these tanks is central, and the discharge opening is at the apex of the cone.

The cyanide treatment of the sand is effected at one plant in a series of vats, twenty in number ; each is 44 feet in diameter and 9 feet deep, and holds 600 tons of dry sand. At another plant only five vats are used ; each is 54 × 13 feet, and holds 1,250 tons of dry sand.

A Butters and Mein distributor feeds each row of vats, and around each vat there is a series of syphons, which are used to facilitate the withdrawal of the surface water immediately after the solution has been run on. The weak and strong solutions are carried overhead in separate pipes. To every ton of pulp containing 43 per cent. solids 4 to 5 lbs. of lime, which has been crushed in a separate mortar, are added. A further 0·4 lb. of lime per ton is added on the top of the charge, and about 0·1 lb. per ton is added to the solution. Table **XXXIX.** is a summary of the sand treatment.

The slime is treated in Merrill presses (see p. 373), of which 28 are used, each containing 92 4-inch frames and 91 plates, and holding 25 tons of slime. The daily capacity is 70 tons. The slime contains 35 per cent. of solid matter, and is fed into two sludge tanks, 26 feet diameter and 24 feet deep, into which is also run a thick cream of slaked lime. From the sludge tanks the pulp passes to the presses through an 11-inch main. Filling is complete in about 70 minutes, the end being determined by measuring the quantity of water passing through in a given time. The treatment time is altered according as the rate of leaching is fast or slow.

		Hours.
Air,	. . . . . .	1 to 1·5
Weak solution,	. . . .	0·4 to 0·6
Air,	. . . . . .	1 to 1·3
Strong solution,	. . . .	1 to 1·5
Weak solution,	. . . .	0·8 to 1·7
Water,	. . . . . .	12 to 13 tons.

TABLE XXXIX.

	Approx. Time. Hours.	Strength of Solution, Pressure of Air, etc.	Effluent.
Filling, .	9	Overflow containing slimes to clarifying tank.	...
Draining,	16	...	To sewer.
Aërating,	16	4 lbs. per square inch.	
Leaching,	16	{ 0·10 per cent. NaCN. Au, $0.80.	{ Mainly to sewer. At end to precipitation boxes.
Draining,	16	...	Weak solution sump. } Precipitated in zinc boxes.
Aërating,	13	5 lbs. per square inch.	
Leaching,	14	{ 0·10 per cent. NaCN and some from weak solution sump.	Weak solution sump. }
Draining,	11	...	
Aërating,	9	5 lbs. per square inch.	...
Leaching,	12	{ 0·055 per cent. NaCN. Au, $0.02.	Strong solution sump. } NaCN and lime added and pumped to strong solution storage.
Washing,	16	{ Water tank. Clarifying tank overflow.	
Sampling,	3	...	{ Weak solution sump. } Precipitated and run to sewer.
Discharge, Filling, .	} 9	... ...	Sewer. ...

Table XL. gives the various operations of slime treatment in brief outline, and indicates the source and destination of all the solutions used.

TABLE XL.

	Approx. Time. Hours.	Source, Strength of Solutions, Pressure, etc.	Approx. Time. Hours.	Effluent.
Filling, -	1	{ Sludge pressure tank. 30 per cent. solids. Sp. gr. 1·25. Au., $0.87. 26 lbs. per square inch.	} 1	Part to clarifying tank overflow to wash water tank. Part to sewer.
Aërating, -	1	25 lbs. per square inch.	2	Sewer.
Leaching, -	0·3 – 0·4	{ 0·027 per cent. NaCN. Alk. 12. 18-20 lbs. per square inch. Au, $0.02.	} 1·3	Low solution sump. } Clarified, pptd. and to sewer.
Aërating, -	1	25 lbs. per square inch.		
Leaching, -	1	{ NaCN, 0·05 per cent. Au, $0.52. 18-20 lbs. per square inch.	} 1·7	Weak solution sump. } Clarified, pptd., and to weak soln. storage.
Leaching, -	1	{ NaCN, 0·027 per cent. Au, $0.02. 18-20 lbs. per square inch.		
Washing, -	1	{ Wash water tank. Through cloths and across cakes. 30 lbs. per sq. in.	0·7	Strong solution sump. } NaCN and lime added. Then to strong soln. tank.
Discharging,	1	{ Upper water tank. Through nozzles in sluicing bar. 60 lbs. per sq. in.	0·4	Low solution sump. Sewer. } Pptd. and run to sewer.

The precipitation of the gold in the enriched cyanide solutions is effected by means of zinc dust. At each plant there are two sumps containing weak solution, one of which is filling while the precious metals in the solution

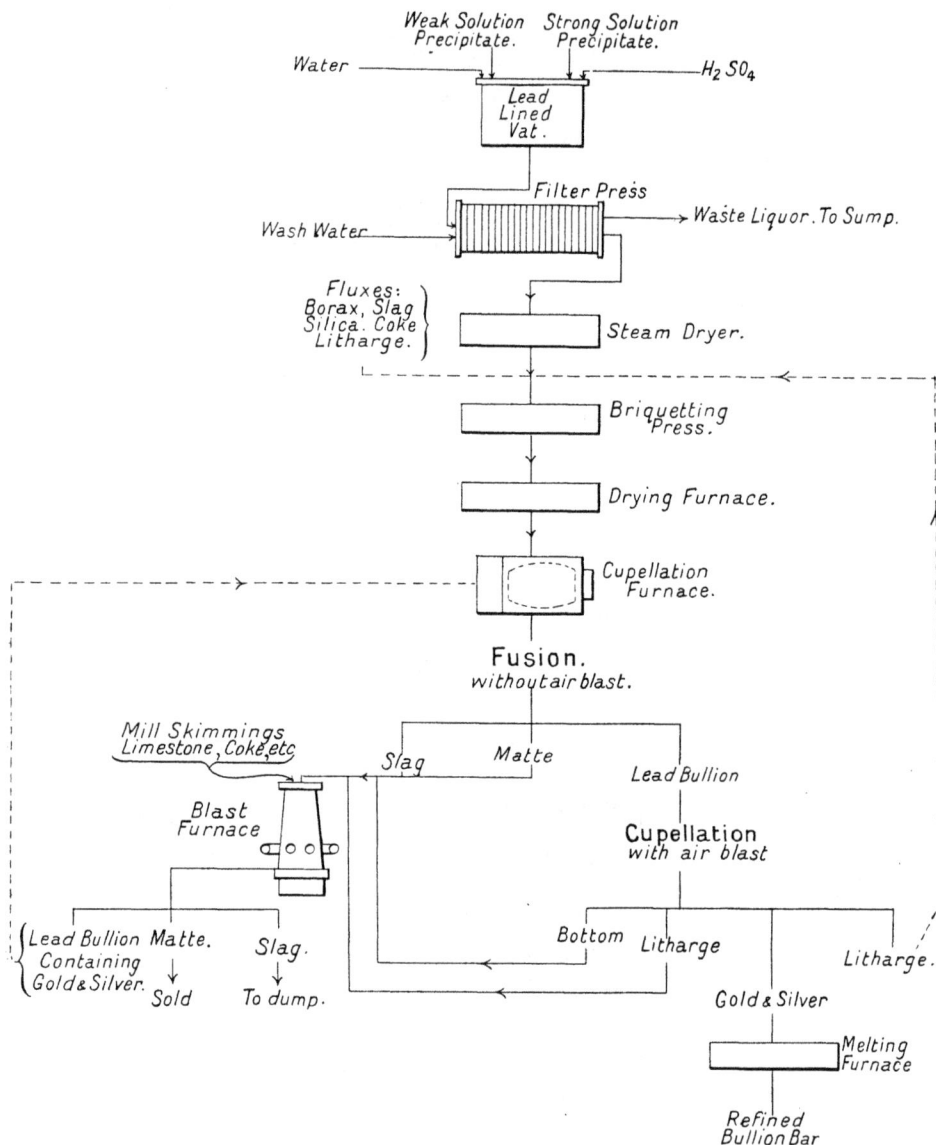

Fig. 183.—Treatment of Precipitate at the Homestake Mill, South Dakota.

from the other are being precipitated. The zinc dust is fed to a small stream of solution, which carries it forward to join the main stream from the sump at a point just beyond the discharge. Both zinc and solution are then pumped up to the Merrill presses, where the fine gold is caught on the filter cloths and the barren liquid is discharged to the weak solution storage to be used again. In the case of the " low " solution, the liquor is carried by gravity to a press situated below the level of the tank, and the zinc is added as before. The solution issuing from this press is run to waste.

Filter cloths are burnt after two months' use and the ash added to the bulk of the precipitate to be smelted. Sixty-two tons of the weak solution pass through the presses per hour when the process is regular and the machines are at their normal capacity. The following figures of the consumption of zinc are given as typical results from an average month's run :—

TABLE XLI.

| | All Solution Precipitated. | |
	Lbs. of Zinc per Ton of Solution.	Lbs. of Zinc per Oz. of Gold Precipitated.
No. 1 sand plant, . . .	0·173	1·22
No. 2 sand plant, . . .	0·174	1·31
Slime, ...	0·160	2·84
All, ...	0·165	1·84

At the end of each month a strong current of air is blown through the presses, in order to displace all remaining solution, and then the precipitate is discharged into a shallow tray placed beneath. The zinc is dissolved out of the precipitate by means of sulphuric acid (66° B.) in a lead-lined tank, and after the precipitate has been washed thoroughly and allowed to settle, the supernatant liquor is drawn off through a small filter press. Finally the precipitate itself is forced through the press, dried somewhat by a current of air, and then placed in trays to be dried over a steam chamber. After drying, the precipitate is mixed with suitable fluxes of litharge, borax and old slags, and moulded in a special machine into briquettes of 7 lbs. each. The briquettes are further dried in an enclosed furnace, and then charged on to the test of an English cupellation furnace. The test is made of 75 per cent. cement and 25 per cent. limestone. Lead bullion is first added to the furnace which has been heated already for some hours, and when the lead is molten the briquettes are charged in. The test soon becomes full of molten material, and at this stage the slag is drawn off and more briquettes added. Additional lead is now introduced and an air blast applied across the surface of the metal. Molten litharge is run off at intervals until none is left, when the heat is increased for a few minutes, and the bullion allowed to cool. This is then cut up, melted in a wind furnace, and cast into bars of about 980 fine. Most of the by-products, such as slag, spent litharge, matte, stained cupels, etc., are treated in a small water-jacketed blast furnace, in order to recover any values which they may contain.

Waihi Gold Mine, New Zealand,[1] 1906.

Crushing in Cyanide—Cyaniding of Concentrate.—The rock from the mine is crushed in two series of Gates' crushers until it passes a 1-inch to 1¼-inch ring. The crushed material is elevated to the stamps, of which

---

[1] F. N. Rhodes, *Mng. Mag.* (New York), Jan. 1906, p. 15.

there are 330 in use, ranging from 850 to 1,250 lbs. in weight. Cyanide solution is fed into the mortar box. The pulp issuing through the screens is sized in pyramidal shaped boxes, whose sides are arranged at an angle greater than 55° to the horizontal. The slime passes to the slime plant, and the sand to the tube mills.

The latter are 20 by 4½ feet, and one mill handles the product from thirty stamps. Muntz metal amalgamation plates are arranged in a separate building after the tube mills, and over these the slime from the classifiers is run. About 15 per cent. of the gold and silver in the ore is thus caught. The pulp is again sized after amalgamation, the slime going on to the agitators and the intermediate and coarser sand being treated on Union vanners and Wilfley tables. The sand from the concentrators is treated in spitz-luten and spitzkasten yielding 40 per cent. of slime and the rest sand. The latter is treated by leaching in large steel or concentrate vats for four to six days, the solution used containing 0·2 to 0·5 per cent. of cyanide. About 1 ton of sump solution and water washes is required for each ton of ore in the vat to remove the gold and silver in solution.

Lime is added to the slime to facilitate settlement, and it is then thickened and settled in large tanks. From these it passes to agitation vats, where it is stirred vigorously with 0·07 to 0·12 per cent. cyanide for about 45 hours. The slime next falls into a Monteju, or pressure tank of egg shape, and is forced into the filter presses. After washing with weak solution and then with water, the slime cakes are sluiced into the river.

The concentrate [1] from the vanners, consisting chiefly of the sulphides of iron, zinc, copper and lead, is kept under water to prevent oxidation, and is first treated in Krupp tube mills. The product is sized, the sand returned to the tube mills and the slime sent forward to be agitated with 0·12 to 0·3 per cent. cyanide solution in a series of ten connected cylinders (16 × 6 feet) with conical bottoms. The concentrate is finally filter-pressed to remove the rich solution. About 500 tons of concentrate are produced each month, assaying about 5·5 ozs. of gold and 65 ozs. silver per ton.

The enriched solutions are precipitated by zinc thread, the precipitate being fed on to the hearth of a cupellation furnace, and the resulting bullion refined by the Gutzkow process.

The capacity of the plant is about 1,000 tons of ore per day, and the extraction is given as 96 per cent.

The Argo Cyanide Mill, Idaho Springs, Colorado, 1913.[2]

Concentration—Counter-current Decantation.—This is a customs mill, and treats the ores from the mines in the surrounding district. The material is first crushed in gyratory breakers, and is then sampled, afterwards being sent to the stamp mill.

The stamps are twenty in number and weigh 1,000 lbs. each. They are electrically driven, and crush 135 tons per day, using a 12-mesh screen and a 4-inch height of discharge. Cyanide solution, to which a definite amount of lime has been added, is fed into the mortar box. Short (5 feet) amalgamation plates are used, as it was found that longer ones were more difficult

[1] E. G. Banks, *Mng. and Sci. Press*, Jan. 21, 1911, p. 142 ; W. Gowland, *Non-Ferrous Metals*, p. 279.
[2] S. L. Goodale, *Eng. and Mng. J.*, 1913, **96**, 385 ; J. V. N. Dorr, *ibid.*, p. 1036.

to keep in condition owing to the use of cyanide in the mortar. The amalgam traps used are electrolytic in action, the cathode being formed by a copper plate floating in a bath of mercury contained in a cone-shaped vessel, and the anode by the discharge pipe which is placed 3 inches above the copper plate.

The pulp from the stamps passes to a duplex Dorr classifier, and is then distributed to five Card concentrators, which give four products :—(1) Lead concentrate ; (2) iron concentrate (both these go to a Dorr classifier and then to separate bins) ; (3) middling, which is again concentrated on another Card table ; (4) tailing which goes through a three-unit Dorr classifier and then to waste, containing probably 50 cents per ton of value. The concentrate from the middling is shipped, while the overflow unites with that of the spitzkasten and finally goes to a small classifier. The slime from the latter goes to two 10-feet agitators, and the sand passes to a small tube mill, the discharge from which returns to the classifier. The underflow from the air agitators is treated again in one 12 by 20 feet tank and one Parral agitator, which are followed by four large Dorr thickeners. The last of these is higher than the first, and the others are placed intermediately so that the solution travels back by the action of gravity. The overflow from the first vat goes to the clarifying filter and then to the zinc boxes. The underflow from each of the other tanks passes to the next one in the series.

Three five-compartment zinc boxes are used, the head compartment of each box carrying a frame of 20 anodes and cathodes (5 × 4 inches), as an electrolytic oxidiser and an antidote to cyanicides in the solution.

Lluvia de Oro Mill, Chihuahua, Mexico, 1913.[1]

Crushing by Nissen Stamps—Counter-current Decantation.—The ore from this mine is very hard, and requires to be so crushed that at least 95 per cent. will pass through a 200-mesh sieve, in order to expose the fine particles of precious metal.

From the mine the ore passes to a grizzly. The oversize from this is crushed in a rock breaker, and joins the undersize on its way to the storage bin which feeds the stamp battery.

Twelve Nissen stamps, each weighing 1,500 lbs., and having individual mortars, are in use. The mortars are cast in two pieces, to facilitate repairs and removals. A 10 H.P. motor drives each set of four stamps, the height and frequency of drop being maintained at 8 inches and 105 times per minute respectively. The daily capacity of the mill is about 80 tons. Cyanide solution is run into the mortar, and amalgamation is not practised, as it was the opinion of the operators that the material which could be amalgamated could conveniently be concentrated, thus saving time and money in retorting, etc.

From the battery the pulp passes direct to two Baylis classifiers, the fine material from which goes to the distributing box for the concentrators, while the coarser portion is reground in two $3\frac{1}{2}$ by 16 feet tube mills, and eventually travels to the concentrators. The tube mills have an El Oro lining (see p. 235), and use pieces of ore from the mine instead of pebbles for crushing the pulp. The concentrators consist of five standard table machines. They deliver their concentrate to the appropriate bin

---

[1] H. R. Conklin, *Eng. and Min. J.*, Mar. 15, 1913, p. 551.

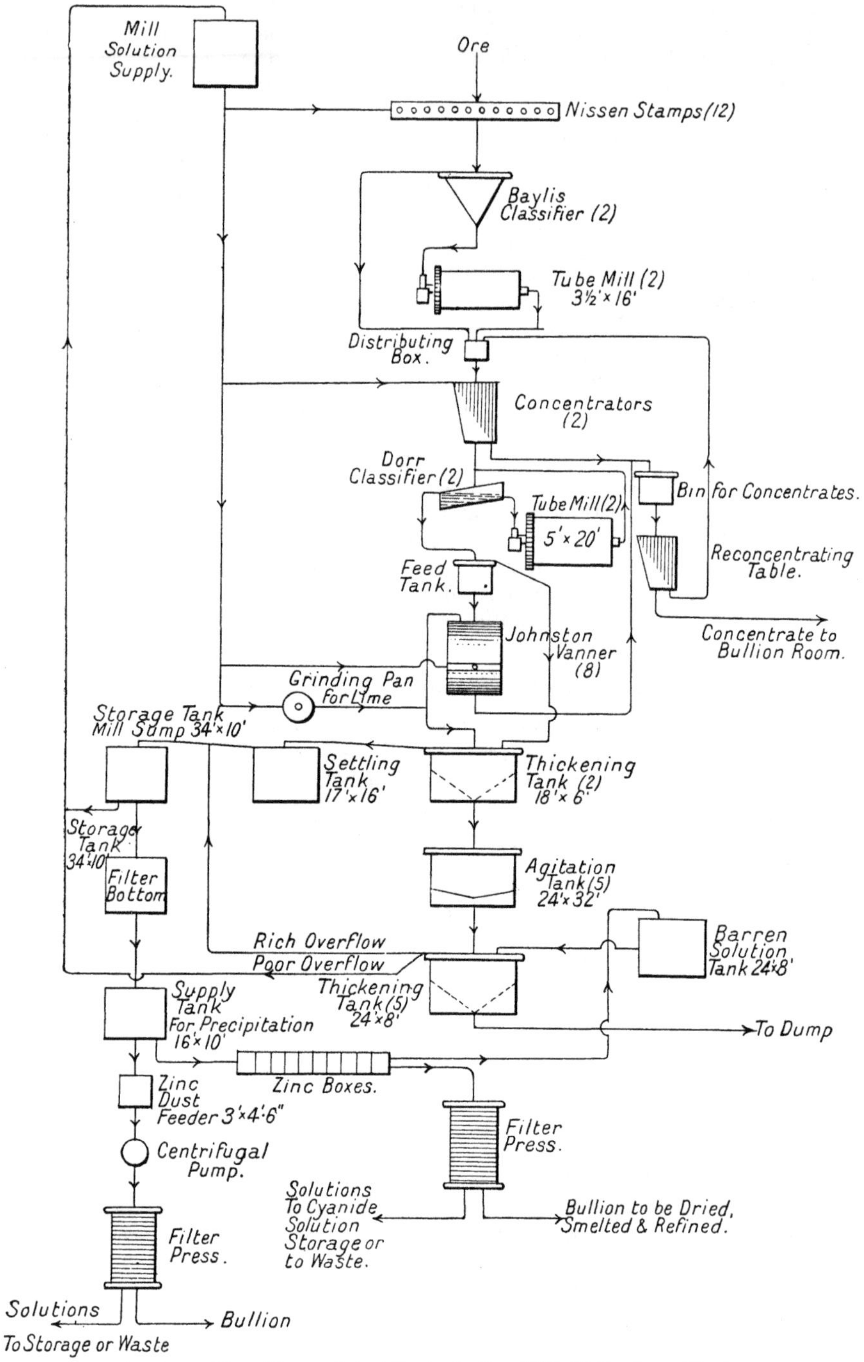

Fig. 184.—Flow Sheet of the Lluvia de Oro Mill, Chihuahua, Mexico.

and their tailing to two Dorr classifiers. From the bin the concentrate is re-treated on another table, and then sent to the bullion room. The sand from the Dorr classifiers is reduced in two tube mills in a closed circuit with the Dorr machines, the fine pulp being carried to a series of eight Johnston-Frue vanners. The concentrate from these joins that from the tables, while the tailing is thickened and rid of most of its contained mill solution in two tanks, 18×6 feet, and then enters the continuous agitation and thickening system, which is a feature of the mill. " The present method is continuous agitation in four of the five tanks in use, and washing in five thickening tanks. The piping used is automatic. Nearly all the gold and silver dissolved during agitation goes into solution in the first three tanks" (Conklin). Cyanide is added to the first agitation tank in a small stream of strong solution. The first thickening tank takes its supply from the last agitation tank ; its clear overflow goes to the precipitation tank, while the thickened material, together with barren solution, forms the charge for the second thickener. The overflow from the remaining tanks is used as mill solution, and the under-flow from each successive tank passes to the next one in the series, mixed with barren solution (or, in the case of the last one, with water), until finally it is discharged to the dump. It is claimed that " the most important fact developed in the operation of the mill is that the changing of the solution in contact with the pulp increases the extraction much more than either a longer time of agitation or the use of a stronger solution." The adoption of the continuous agitation principle has increased the extraction and enabled the use of a weaker cyanide solution, thus decreasing the loss of cyanide in the tailing moisture.

The metals in the solution issuing from the agitation tanks are precipitated by zinc in the form of shavings, and also as dust. The former is used in the ordinary type of zinc boxes, while the latter is added as a thick emulsion to the liquid in the precipitation tank. The precipitate is acid-treated, to rid it of zinc, and washed thoroughly. Finally, it is collected in a filter press under a pressure of 20 lbs., again washed and dried by a current of air.

The accumulated precipitate is smelted with fluxes in a battery of electric furnaces,[1] each 4 feet deep and 16 inches along both sides. The furnaces have a fire-brick lining, and are of the resistance type, the graphite electrodes being placed so that one just penetrates through the bottom of the furnace and the other enters the slag floating on the top of the molten metal. Alternating current is supplied at 60 cycles per second and 110 volts from a three-phase generator.

Yamagano Gold Mine, Satsuma, Japan, 1914.

This mine, situated in the locality of Kagoshima, is the fifth largest mine in Japan in the order of tons of ore treated, and the second in the order of gold produced.[2] Gold was discovered in this region early in the seventeenth century by ancestors of the present owners. Up to 1912 the district had yielded 1,200,000 ozs. of fine gold.

The ore is found in volcanic fissure veins, the country rock being composed principally of augite-andesite, tuff and volcanic conglomerate.

---

[1] H. R. Conklin, *Eng. and Mng. J.*, June 15, 1912, p. 1189.
[2] "14th Financial and Economic Annual of Japan," 1914, pp. 58, 59.

The ore is broken in a Blake machine and sized in a trommel, pieces larger than $\frac{1}{2}$ inch being returned to the breaker. Barren ore is largely picked out by hand on a sorting table, and the residual material passed to the ore bins.

There are two mills in operation,[1] one at Yamagano and the other at Nagano, the latter being the more important.

At Yamagano 20 stamps, each weighing 900 lbs., crush 660 tons (2,240 lbs.) monthly, the pulp running over four sets of fixed and shaking amalgamation plates, and then to spitzkasten. The slime from these goes direct to the cyanide plant; the sand is concentrated on three Wilfley tables, the concentrate being subjected to pan-amalgamation. The remaining sand passes to the cyanide plant to be treated by percolation.

At Nagano the ore is carried to a grizzly, 7 × 15 feet, and the oversize from this to a Gates' gyratory breaker, to be reduced to something less than 2-inch size.

The mill contains eighty 900-lb. stamps, set to an 8-inch drop and operating 80 times per minute. The monthly output is 2,200 tons (2,240 lbs.). Sixteen pairs of fixed and shaking amalgamation plates are fixed in front of the battery, and from these the pulp goes to four hydraulic cone classifiers, which discharge the sand on to a series of Wilfley tables and the overflow to the slime plant. The concentrate is smelted, the tailing being treated in the cyanide plant. The amalgam obtained from the plates has an average content of 19 per cent. of gold and 13 per cent. of silver.

The sand is fed into four collecting vats (30 × 6 feet 4 inches), by means of a Butters' distributor, and the water drained off with the aid of a vacuum pump. The leaching vats, into which the sand is then removed by shovelling and tipping, are 30 feet in diameter and 7 feet 4 inches deep, and are fitted with false bottoms, from beneath which all solutions enter. The acidity of the ore is neutralised with 0·5 per cent. of lime, and then a solution of 0·3 per cent. cyanide is allowed to remain in the tank for 24 hours. After this time, the solution is drained away from the bottom, the operation being completed by means of the pump. Dilute (0·08 to 0·1 per cent.) cyanide solution is next added, the volume being twice that of the sand to be treated, and this remains in contact for from 40 to 72 hours. Thorough washing follows, the whole treatment occupying about 160 hours. All solutions pass to the gold precipitation house.

The slime from the classifiers is settled in 14 ponds and the clear water run off. The thickened material (ratio $1\frac{1}{2}$ to 1) passes to three agitation tanks, 15 feet in diameter and 11 feet 6 inches deep, fed with 60 lbs. of potassium cyanide and 7 tons of weak cyanide solution for every 15 tons of slime. It is agitated for 10 hours by compressed air, with the aid of an outside centrifugal pump. From these tanks the slime passes to two Dehne presses, whence the solution and washings are carried to the precipitation house to be clarified and afterwards precipitated.

Solutions from the slime and sand plants are treated in separate zinc boxes, eight of which are provided, each having 14 compartments, and containing, in all, 700 lbs. of zinc shavings. Twice a month these are taken out, washed thoroughly to detach as great a proportion as possible of the precipitate, and the remainder dissolved in dilute sulphuric acid. The residue is washed and smelted in plumbago crucibles.

---

[1] T. M. Yoshida, *Eng. and Mng. J.*, Jan. 1914, **97**, 161, 217.

Uwarro Mill, Candor, North Carolina, 1914.[1]

All-Sliming by Rolls and Tube Mills.—The ore is a hard, tough quartz. It passes through a 1-inch grizzly, and the oversize is fed to a breaker of the Blake type, with chrome-steel jaws set for $1\frac{3}{4}$-inch product. A belt conveyor carries the broken ore to a bin, whence it passes with mill solution through two sets of rolls successively. The product from the coarse rolls passes through a screen with $\frac{1}{2}$-inch round holes, and that from the fine rolls through $\frac{3}{16}$-inch round holes, or $\frac{1}{4}$-inch woven wire screen. The crushed ore is classified and the coarse sand with some solution added passes to a tube mill and thence back to the classifier, a closed circuit being formed. The overflow goes to a Dorr thickener, and the overflow from this goes back to the mill solution tank. The thickened pulp goes to three Hendryx agitators arranged for continuous agitation. The agitation period is about eight hours, and an extraction varying from 94 to 97 per cent. is obtained. The agitators have steam coils inside the pulp, which is heated to about 90°. The filter is a Kelly press, which works only by day. The solution is expelled from the cake until it contains only 8 per cent. of moisture, and the cake is then discharged as tailing without any washing whatever. The tailing averages about 36 cents (0.35 dwt.), including both dissolved and unextracted gold and silver, and is sometimes as low as 20 cents. The pregnant solution is clarified in a press, and precipitated in zinc boxes. The precipitate is dried in a pan by steam and melted in a tilting furnace fired by kerosene.

---

[1] P. E. Barbour, *Eng. and Mng. J.*, Oct. 24, 1914, p. 729.

CHAPTER XVIII.

THE REFINING AND PARTING OF GOLD BULLION.

General Considerations.—By whatever process gold may have been extracted from its ores, it is necessary to melt the crude bullion and cast it into bars so that its value may be ascertained, and that it may be put into a form convenient for transportation and sale. The name " bullion " may be conveniently restricted to the precious metals, refined or unrefined, in bars, ingots, or any other uncoined condition, whether contaminated by admixture with base metals or not. It is, however, often applied to coin, and the appellation " base-bullion " is given to the pig-lead or to copper bottoms or pig-copper, which have been obtained in smelting operations, or as the result of melting worn-out amalgamated copper plates. Such materials may contain only a few parts per thousand of gold and silver, the main portion consisting of base metals. The treatment of base-bullion, however, properly belongs to the metallurgy of argentiferous lead and copper, and the descriptions given in this chapter apply to bullion which is valuable almost entirely on account of the gold and silver contained in it.

The operations to which the retorted metal, gold from the smelting operations in cyanide mills, etc., are subjected may be summarised as follows :—

1. The bullion is melted in crucibles (a rough refining operation being usually effected at the same time) and cast in ingot-moulds.

2. Assay-pieces are cut from the cast ingots or dipped from the molten metal before pouring, and assays are made on these, by which the value and composition of the bars are ascertained.

3. The bars are then usually sold to the refineries, where the base metals are eliminated and the gold and silver separated by " parting," and cast into bars separately. Both before and after the parting it is sometimes necessary to subject the bullion to further refining operations. The bars of gold and silver thus obtained, being of a high degree of purity, are in a condition to be used for minting, or for the various industrial purposes to which they are applied.

Composition of Bullion.—Bullion varies greatly in composition, and gold may be present in any proportion up to nearly 100 per cent. The gold obtained in some chlorination mills was of a high degree of purity and rarely contained much silver. This precipitated gold, however, generally made brittle bars owing to the presence of a few parts per thousand of lead, bismuth, antimony, etc. From some chlorination mills the gold was far from pure, owing to various causes. When ferrous sulphate was used as the precipitant, the precipitate sometimes contained ferric hydrate from which some iron was reduced in the crucible, and if sulphuretted hydrogen was used and the gold precipitated as sulphide, it was contaminated with all the heavy metals contained in the solution, copper, iron, and lead being most often encountered.

Retorted metal is of very different degrees of fineness, according to the nature of the ore and the course of treatment. It is usually about 990 fine in gold and silver taken together. Placer gold is usually finer than that derived from lodes, containing a smaller percentage of silver, while the nature of the material treated and the methods used in placer operations are not favourable to the contamination of the bullion with base metals, which vary in amount only from 0 to 20 parts per thousand, and seldom approach the latter figure.

Gold from battery plates is usually much less pure than placer gold, the percentage of base metals being sometimes much higher, a state of things due in great part to the difference in the method of treatment. The bullion from pan-amalgamation is less fine than battery gold, containing less gold, more silver, and more base metals. Retorted gold sometimes contains large quantities of iron, and copper is also a common impurity.

Bullion from the cyanide process is sometimes of very low standard, containing as little as 10 per cent. of gold. The predominating impurities are zinc and lead. Gold coming from the cupellation furnace (Tavener process) and from the crucible smelting of gold slime with manganese dioxide is usually from 960 to 980 fine in gold and silver.

The Melting Furnace.—The furnace used for melting the bullion is of simple construction. It may be round or square, with walls consisting of an outer layer of ordinary brick and an inner layer, at least 4 inches thick, of the best firebrick. There is often a complete outer casing of iron, which is useful in keeping the furnace from falling to pieces, but radiates more heat than the bricks. The fire-box in a small coke furnace may be about 1 foot square and about 2 feet deep. (The largest coke-fired furnaces at the Royal Mint were round, $21\frac{1}{2}$ inches in diameter and $31\frac{1}{2}$ inches deep.) Below is an ashpit, usually lined with a cast-iron tray and provided with a working iron door, through which the air-supply of the furnace passes, and by which it is regulated. The fire-bars are movable, and their ends rest loosely on iron supports. The top of the furnace may be made flat or sloping up towards the back at an angle of about 30°. In this case a wide flat ledge should be provided at the front, on which crucibles and moulds can rest. The top is always made of a cast-iron flanged plate, with an opening of the same area as the fire-box. This opening is closed by a cast-iron sliding door made in one or two pieces, and preferably lined with firebrick and running on rollers. The flue is placed at the back of the fire-box near the top; in a small 12-inch square furnace the cross-section of the flue should have an area of about 16 or 18 square inches—e.g., 4 inches square—varying, however, with the height of the stack, a higher stack going with a smaller flue. The flue communicates with a stack, which must be of brick for a distance of 2 or 3 feet from the furnace, but may be of wrought-iron tubing in its upper part. The height of the stack will depend on the position of the furnace, and should be as great as possible, 60 feet giving better results than any less amount. It has been stated that a height of 30 feet is the minimum that can be allowed in order to ensure a good draught, but very satisfactory results can be obtained with a stack only 16 feet high. The furnace can be built by any bricklayer acting under directions. No mortar is used in its construction, clay, mixed with an equal bulk of sand, being substituted for it. A sliding damper in the flue at a convenient height above the ground is necessary, so as to regulate the draught. The fuel used in such a furnace may be anthracite, charcoal or good coke, made in coke ovens, and broken into pieces of moderate size. If the coke is of high quality, it is the most satisfactory solid fuel, making a hot fire and lasting for a long time, so that it does not require very frequent replenishing. Neither dust, nor very small, nor very large pieces must be used. Charcoal is preferred in the United States Mints for small charges, and anthracite for large ones.

Gas furnaces are also used, as at the Royal Mint, the crucible being heated by one or more burners, each with a blast of air, placed at regular intervals in the circumference. Gasolene or naphtha furnaces are also in use.

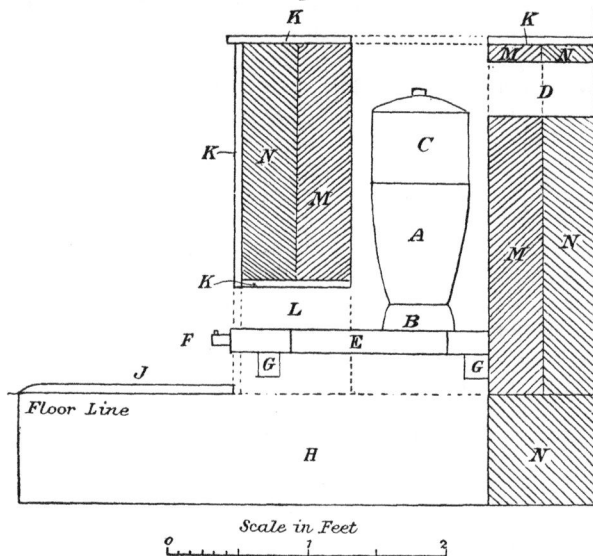

Fig. 185.—Coke Melting Furnace, Royal Mint.

The flues should pass into dust chambers. At the San Francisco Mint the dust chamber is 12 feet high, 1 foot wide, and of the same length as the row of furnaces, and at the Melbourne Mint the dust chamber is 8 feet high and 5 feet wide. Baffle plates should be inserted in the dust chambers, which are built of brick.

Royal Mint Furnace.—One of the gold melting furnaces formerly in use at the Royal Mint, London, is shown in sec-

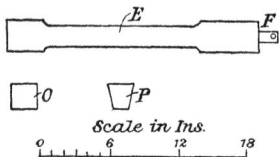

Fig. 186.—Fire-bar of Coke Melting Furnace, Royal Mint.

tional elevation in Fig. 185, and a fire-bar belonging to the furnace is shown in Fig. 186. The crucible A stands on B, which consists of the lower part of an old plumbago pot cut off about 2 inches from the bottom. C is the " muffle," a plumbago cylinder 6 inches high, resting on the crucible. It enables a deeper bed of coke to be used, and also gives space for charging-in bulky material. The furnace is 12 inches square and 2 feet deep above the fire-bars. The pot is $8\frac{5}{8}$ inches diameter at the widest part, and its charge is 1,200 ozs. of standard gold. The flue D is $4\frac{1}{2}$ inches deep and 5 inches wide, and communicates with a stack 45 feet high. The fire-bars, E, are $22\frac{1}{2}$ inches long and 2 inches deep. They are tapered downwards, especially in the middle portion. The section of the thick part near the ends is shown at O, Fig. 186, and that of the middle part at P, Fig. 186. The projection F is used for withdrawing the bars. G G, Fig. 185, are the supports for the bars ; they are of 2-inch square iron, and let into the brickwork. H is the ashpit, 12 inches by 3 feet 4 inches and 9 inches deep below the floor. It is lined with $\frac{3}{4}$-inch cast-iron plates. J is a sliding iron plate covering the exposed part of H. K K are cast-iron plates $\frac{3}{4}$ inch thick covering the furnace. There are four fire-bars and their thickened ends nearly touch, occupying 11 inches in the width of 12 inches. The spaces between the middle parts of the bars, however, are over $\frac{1}{2}$ inch wide, and some air passes in between the bars, but most of the air enters

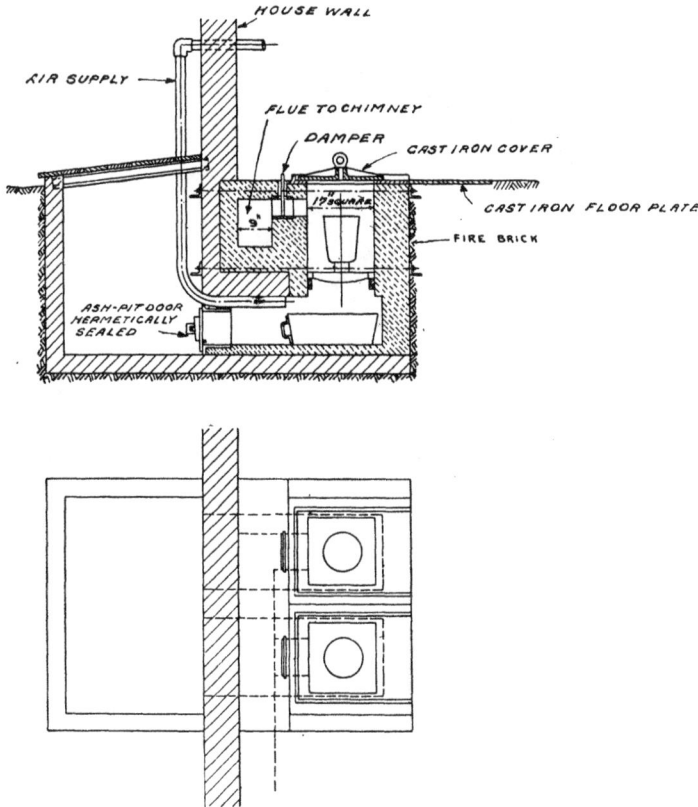

Figs. 187 and 188.—Bullion or Cornish Furnace, Two Pots.

the furnace through the aperture L which is $3\frac{1}{2}$ inches high above the fire-bars and 12 inches wide. The pot-support B rests on the two middle fire-bars, so that the outer fire-bars can be withdrawn without disturbing the pot. The furnace is lined with firebrick, M M, $4\frac{1}{2}$ inches thick, the outer layers N consisting of ordinary brick. The top of the furnace is 29 inches above the floor line. A row of eight furnaces have one stack in common. The distance between the centres of two adjacent furnaces is 2 feet 6 inches.

These furnaces have now been superseded by gas furnaces, using coal gas, with air supply under pressure. The furnaces are circular, and the pots take a charge of about 2,800 ozs. of gold. Furnaces are often placed with their mouths flush with the floor of the melting house, to facilitate charging and manipulating of the crucibles by hand.

A melting furnace in use on many mines on the Rand is shown in plan and elevation in Figs. 187 and 188.[1] The " air supply " pipe accelerates the rate of melting. The fuel is coke. The stack is in this case about 50 feet high. Crucible tilting furnaces fired by coke, oil or gas are now coming into use for melting. They work more quickly than the stationary furnaces, but are more expensive to install. One of their advantages is that the pot remains in the furnace during pouring and recharging, and is less cooled

Fig. 189.—Tilting Gas Furnace.

down than pots which are lifted out before being poured. The reduction in the amount of alternate heating and cooling gives a longer life to the crucible. Solid fuel is less used than gas or oil for these furnaces. A gas tilting furnace is shown in Fig. 189. The crucible is held in place by three projecting bricks, and the gas nozzles are seen on the left. The flue is not shown. Electric furnaces are beginning to be used for melting metal.

The Crucibles.—The bullion is melted in either graphite or clay crucibles, or in graphite pots lined with clay. Detachable clay liners are sometimes

---

[1] Schmitt, *Rand Metallurgical Practice*, vol. ii., p. 185.

used, or a clay pot may be fitted inside a graphite guard pot. The size of the crucibles and the weight of the charges of bullion vary greatly, but in extraction mills, as a general rule, a gold-charge does not exceed 400 ozs., and a silver-charge 1,200 ozs. in weight. In mints and refineries, much larger crucibles are employed, holding different amounts up to 6,000 ozs. of metal.

Melting the Bullion.—All crucibles must be thoroughly annealed before being used ; otherwise, the contained moisture being suddenly converted into steam when the crucible is heated rapidly, the pots crack. The crucible is kept on a shelf near the flue, for as many days or weeks as convenient, before being used. It is then placed on the top of the furnace or in the ashpit for a few hours, when it will probably be safe to hold it over the open furnace by means of the crucible tongs, until it becomes gradually warm. After a few minutes, the crucible being turned round at intervals, it can be lowered rim downwards upon the burning fuel, and as soon as the rim becomes red-hot, the crucible is quite safe, and may be turned over and placed in position for the reception of the gold. With Salamander crucibles, a less degree of care in annealing will suffice, as they are well annealed before being sold. The crucible rests on a firebrick about 3 inches thick, which is laid on the

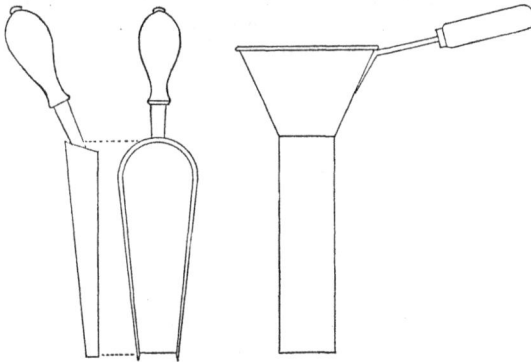

Fig. 190.—Charging-scoop and Shoot.

bars of the grate. If the firebrick were omitted, the bottom of the pot, resting directly on the fire-bars, would be too cold, while a layer of fuel, if placed below the pot, would soon be burnt out, and could not readily be replaced, so that the pot would sink down to the bars. The fuel is built up round the pot until it reaches to its rim, or the top of the muffle, and the fire urged until the whole pot is at a full red heat. Borax is then thrown into the crucible by means of a scoop to slag off metallic oxides and so assist the metal to melt. As soon as the borax is melted, the introduction of the bullion is commenced. The safest way to do this is to use the shoot shown in Fig. 190, which is held in position, its lower edge being inside the crucible, with the left hand, while the metal is transferred to it in a scoop by the right hand. In this way the melter avoids all danger of loss which might be encountered if the metal scrap were wrapped in paper and added by the tongs. Large pieces of metal are added with the crucible tongs. The cover, which must also have been previously well annealed, is kept on the crucible as much as possible. The fuel is pushed down with the poker to avoid scaffolding, and fresh pieces of coke added when required. The crucible is not allowed to become more than two-thirds full at any time, but more metal is added when the first supply has been melted down, and the operation repeated until the pot is sufficiently full of molten material.

REFINING OR TOUGHENING.[1]

Methods of Refining.—The processes in use are as follows :—

    1. Volatilisation.
    2. Oxidation (*a*) by air blowing or roasting.
            (*b*) by " bessemerising."
            (*c*) by nitre.
            (*d*) by metallic oxides.
            (*e*) by cupellation.
    3. Chlorination.
    4. Sulphurisation.
    5. The use of iron or carbon.

The method to be used depends partly on the composition of the bullion, and partly on the means at the disposal of the operator. Cupellation belongs properly to the metallurgy of silver and lead, and need not be dealt with here. (See, however, Tavener's process, p. 391.) The other operations, except roasting, are usually carried out in crucibles, although reverberatory or tilting furnaces are sometimes used for work on a large scale.

    1. **Refining by Volatilisation.**—The volatile metals are partly removed by melting the bullion. Mercury is almost all volatilised when the gold is melted. Mercury boils at 357°, sulphur at 445°, arsenic at 616°, selenium at 690°, cadmium at 770°, and zinc at 910°. Pure gold melts at 1,064°, but its melting point is lowered by the presence of impurities, and the temperature of molten bullion in an ordinary furnace may be taken as between 1,000° and 1,100°. At this temperature part of each of the elements mentioned above is volatilised, and the higher the temperature and the longer the time during which the metal is kept molten, the smaller will be the proportion of the volatile elements retained by the gold, but they cannot be entirely removed by heat alone.

The losses of gold and silver by volatilisation are small. At 1,100° their vapour pressures are insignificant, and are independent of the presence of the impurities named. If, however, a large proportion of volatile elements are present, and the mixture boils, gold and silver are carried off mechanically as spray owing to the bursting of the bubbles, and the losses may be serious. The losses are increased by the passage of currents of gases over the surface of the molten metal, and consequently it is better to cover the metal with charcoal, bone ash, fluxes, or a crucible lid. The use of dust-chambers is desirable.

The losses by volatilisation or as spray are higher in some of the other refining processes, so that dust-chambers are the more necessary when these are used.

    2. **Refining by Oxidation**—(*a*) *By Means of Air.*—The use of a blast of air on the surface of molten gold or silver is probably the oldest of the refining processes. The method appears to be described in the Book of

---

[1] The word " refining " is a convenient one to denote the removal of impurities from gold and silver. In refineries this is called " toughening," the word " refining " being used to denote the separation of gold from silver and copper, which is often called " parting " in metallurgical literature.

Ezekiel,[1] to which the date B.C. 593 is assigned, and the principle is the same as in cupellation, which was practised by the ancients before B.C. 500, and perhaps as early as B.C. 2500.[2] In cupellation, the oxides of the base metals are dissolved and slagged off by litharge, but if a blast of air is used when lead is not present in large quantities, it is necessary to add other fluxes, as many oxides are almost infusible by themselves. In 1580, Ercker recommended the metallurgist to melt brittle gold "with good Venetian borax and drive it before the bellows till it endureth the blowing." [3] The method is still in use at some mills to raise the fineness of low-grade bullion, but does not seem to be employed regularly in refineries. If the oxides of iron, etc., are not slagged off, the dross collects much gold which is difficult to separate by heat alone. If large quantities of base metals are present, the slag soon covers the surface of the metal, and prevents the access of the air. It is, therefore, necessary to skim off slag at frequent intervals, keeping the middle of the charge free from slag to enable oxidation to proceed.

In certain cases, the same object may be attained by granulating the bullion, roasting the granulations spread out on trays at a red heat, with frequent stirring, and melting the product with borax and sand. After two or three repetitions of such treatment, the bullion may be refined sufficiently to be sold.[4]

(b) "*Bessemerising*." [5]—This method consists in passing a stream of air or oxygen through molten bullion in clay pots by means of clay pipes similar to those used in Miller's chlorine process. The base metals are oxidised successively in the order zinc, iron, antimony, arsenic, lead, bismuth, nickel, tellurium, copper. The oxidation of these metals, however, proceeds to some extent simultaneously, some copper being oxidised before the last traces of zinc are eliminated from the bullion. The oxides rise to the surface of the metal, and are slagged off with a mixture of borax and sand. Four parts of sand and three parts of borax are enough for slagging off about six parts by weight of base metals. Lead requires less than one-third its weight of sand, zinc and copper an equal weight of sand, and iron needs $1\frac{1}{2}$ times its weight of sand. The borax may be in great part replaced by about half its weight of sand, but in that case the amount of iron in the slag should be kept greater than the amount of zinc. Tin oxide may be slagged off by pearl ashes. The zinc comes off first as a sheet of flame. Then sparks, due to the formation of magnetic oxide of iron, are seen above the charge; afterwards the action proceeds more quietly. The end of the operation is difficult to determine, except by measuring the air passed through or by dipping out part of the metal, casting it, and bending the ingot. If it is tough, only gold, silver, and copper remain unoxidised. A little silver oxidises simultaneously with copper, and if all the copper is removed into the slag, from 10 to 50 per cent. of its weight of silver is also oxidised, and passes into the slag, from which it is recovered by fusion with carbon and iron. If the operation is stopped as soon as the metal is tough, the losses of silver in the slag are small. The losses of gold in the slag are insignificant, and bullion 990 fine in silver and gold may be obtained from metal only 500 fine in an hour or two. The slag prevents loss by projection or volatilisation, and the cost

[1] Ezekiel, chap. xxii. 18-22 ; see also Jeremiah, chap. vi. 28-30.
[2] See Hoover's *Agricola*, p. 465, "Historical Note on Cupellation."
[3] Pettus on Metals (translation of Ercker's book, London, 1686), p. 218.
[4] J. S. MacArthur, *Trans. Inst. Mng. and Met.*, 1905, **14,** 429.
[5] T. K. Rose, *Trans. Inst. Mng. and Met.*, 1905, **14,** 378.

is trifling. This method was used at the Mint to toughen 41,000 ozs. of brittle standard gold in 1905. Oxygen was passed through charges of 1,200 ozs. for 15 minutes. The loss of gold was about the same as in ordinary melting. The method is also used in refining by chlorine gas, see below, p. 463.

The method was also tried on low-grade cyanide bullion on the Rand by C. W. Lee and W. O. Brunton.[1] In this case air was used. The borosilicate slag was skimmed off before the metal was poured, thus avoiding the difficulty of its removal after pouring. The results were successful, but the scarcity of low-grade bullion on the Rand and the success of the use of manganese dioxide in the oxidation and slagging off of the base metals prevent the method from coming into use.

(c) *Oxidation by Nitre.*—This ancient method was described by Ercker in 1580,[2] who explained that the nitre must be " projected upon the gold just before it melts, as it has little effect on molten gold." The method is still in wide use. The bullion is melted in clay crucibles, and a little nitre (potassium nitrate) or sodium nitrate is thrown on to the surface of the metal. Violent bubbling at once ensues, as heat converts nitrates into nitrites with evolution of oxygen. The nitre is pressed down with a stirrer, and the nitrites and undecomposed nitrates oxidise some of the base metals. The nitrates and oxides corrode the pots, and it is better to have a ring of bone ash next the pot, and to throw the nitre into the " eye " of metal in the centre. The oxides are absorbed by the bone ash, which protects the crucible from attack. After a minute or two, the action of the nitre moderates, and it is then, together with the bone ash, skimmed off with a ladle, and the operation repeated as often as necessary. If the metal is allowed to become pasty, so that it can be mixed with the nitre, the action is much more rapid and effective. If too much nitre is added at one time, the charge boils over, and part is lost.

Iron and zinc can be removed in this way, but the oxidation of lead is more tedious, and bismuth, tellurium and copper are very troublesome. The losses by spirting are heavy, and large quantities of both silver and gold are entangled in the dross and skimmed off. In the treatment of cyanide precipitates the amount of these initial losses has been stated by Alfred James to be 10 per cent.,[3] and by J. S. MacArthur to be as high as 25 per cent.[4] There is little to recommend the method, which seems to remain in use from force of habit. Potassium permanganate has also been used as a substitute for nitre.[5]

(d) *Oxidation by Metallic Oxides.*—Black oxide of copper, CuO, was formerly used in certain cases ; it was mentioned by Ercker. The oxide is stirred in with the molten metal, and the whole then allowed to remain in the furnace for a short time before pouring. All base metals are oxidised, the cupric oxide being reduced first to cuprous oxide, and then to metallic copper. The cuprous oxide is dissolved in the metal, and so carries oxygen to all parts of the molten mass. The process is efficacious, but the gold is, of course, contaminated with the reduced copper. The method has been in use at the Royal Mint for toughening brittle standard gold since 1907. In 1908, 34,046 ozs. of gold were melted in 27 charges with 0·5 per cent.

[1] Lee and Brunton, *J. Chem. Met. and Mng. Soc. of S. Africa*, 1907, **7**, 358.
[2] Ercker, *op. cit.*, p. 216.
[3] James, *Trans. Inst. Mng. and Met.*, 1905, **14**, 423.
[4] MacArthur, *ibid.*, p. 427.
[5] W. R. Dowling, *J. Chem. Met. and Mng. Soc. of S. Africa*, 1905, **5**, 224.

CuO. The fineness of the gold was reduced from 916·6 to 914·9. There was no loss of gold.

The use of manganese dioxide in the Transvaal in refining gold-zinc-slime from the cyanide process was described by Johnson and Caldecott in 1902.[1] Some silver is oxidised and passes into the slag, especially if copper is present. As a cheaper alternative, the author proposed ferric oxide,[2] but this has apparently not been tried, although in certain cases enough ferric oxide is already present in the calcined slime to enable tough bullion of good quality to be produced without the addition of manganese dioxide.[3] In these cases, the oxides are reduced to lower oxides, but not to metals, so that the bullion is not contaminated by the products.

In 1914, W. A. Caldecott[4] treated gold-zinc slime from the cyanide process with a large proportion of manganese dioxide, in order to remove as much silver as possible. The slime was acid-treated and thoroughly calcined, and then fused with 30 per cent. of manganese dioxide and similar quantities of sand and borax. The final product contained on the average gold 938 parts and silver 55 parts per 1,000. Apparently about half the silver in the slime passed into the slag, together with the greater part of the base metals. About 4,000 ozs. of ductile bullion suitable for use in coinage was obtained in this way.

3. **Chlorination.**—Sal-ammoniac, NH_4Cl, is sometimes used to remove lead from gold bullion. When much lead is present, alternate additions of nitre and sal-ammoniac have been recommended. It is probable that the sal-ammoniac acts by decomposing basic compounds of lead which resist the action of nitre. Cupric chloride acts like gaseous chlorine, chloridising base metals and being reduced to cuprous chloride which is volatilised. The fineness of bullion is slightly raised by this agent, but it is not suitable for ordinary refining, although it may be used for toughening brittle standard gold or high-grade bullion containing traces of impurities. Gaseous chlorine is better. Its use is described under the heading of Miller's process of Parting, p. 450. In 1871, 40,000 ozs. of brittle standard gold were toughened at the Mint by the use of chlorine gas.[5] The charges were about 1,100 ozs. each, and the time of passage of the chlorine varied from five to seven minutes. The amount of impurities removed was from 0·1 to 0·3 per 1,000.

An old method of removing traces of impurities from brittle gold was to make repeated small additions of powdered corrosive sublimate (mercuric chloride). After each addition the door of the furnace must be at once closed, as dense poisonous fumes arise which must not be breathed by the workers. Volatile chlorides of zinc, copper, antimony, bismuth, etc., are formed and pass off, carrying with them some gold, of which there is an appreciable loss. A little corrosive sublimate sprinkled on the surface of molten gold will completely toughen every part of it without being mixed with it by stirring, even although the crucible contains several hundred ounces of the metal.

In 1866, 8,800 ozs. of brittle standard gold were toughened in this way at the Mint. It was divided into 14 charges of about 650 ozs. each, and from 12 to 16 ozs. of corrosive sublimate were added to each pot, the total amount

[1] Johnson and Caldecott, *J. Chem. Met. and Mng. Soc. of S. Africa*, July, 1902, **3**, 46.
[2] T. K. Rose, *ibid.*, Oct. 1902, **3**, 59.
[3] H. A. White, *J. Chem. Met. and Mng. Soc. of S. Africa*, Sept. 19, 1914, p. 51.
[4] T. K. Rose, Presidential Address, p. 4, *Bull. Inst. Mng. and Met.*, April, 1915.
[5] *Second Annual Report of the Mint*, 1871, p. 34.

used being 182·3 ozs. troy, or about 2 per cent. The tough bars produced weighed 8,762 ozs., the average fineness being 918·9, or 2·3 per 1,000 higher than the brittle bars. The actual loss of gold was 6·657 ozs., or 0·756 per 1,000.

4. **Sulphurisation** is described by Agricola [1] as a method for separating copper from gold. It is said to be practised in the United States Mints; [2] the following is a brief account :—It is effected in plumbago crucibles, and has for its main object the elimination from retorted metal of iron, when, as sometimes happens, it is present in large quantities. The metal is kept just above its melting point, the temperature being as low as possible in order to avoid unnecessary waste of sulphur by volatilisation. Sulphur is sprinkled round the edges of the molten mass, and stirred in with a graphite stirrer. If sulphur is added near the middle, particles of gold are lost by projection. Sulphide of iron is formed with great energy, and sulphide of silver also, but the latter is not produced rapidly until nearly all the iron has been already converted into sulphide. The gold is unaffected by the sulphur and subsides to the bottom. It is not usually cast by pouring, but allowed to solidify in the pot, a better separation between the gold and the matte being thus effected. The pot is turned out as soon as solidification has taken place, and the matte is broken off with a hammer, the gold being remelted and cast into a bar. The small quantity of gold taken up by the matte is separated by melting with metallic iron.

When retorted metal is infusible from the presence of large quantities of iron free from carbon, it may be refined by melting with galena, according to H. L. Sulman; [3] with pyrite, according to T. C. Cloud; [3] or with sulphur, according to W. McDermott. [3]

5. **The Use of Iron and Carbon.**—Iron is used to remove arsenic, antimony, sulphur, etc. The molten metal is stirred briskly with an iron rod for a few minutes. Antimonides, arsenides or sulphides are formed and separate from the metal.

Carbon is used to assist iron to melt and to remove oxygen from bullion.

At the Philadelphia Mint it is found to be profitable to recover the gold from the iron tools used in stirring, dipping, etc. For this purpose they are melted down in a graphite crucible with a little charcoal to make grey-iron, and kept at a white heat for some time, after which the charge is allowed to cool slowly. Under this treatment the gold and silver separate out (an alloy containing three or four parts of silver to one part of gold being better for the purpose than pure gold), and are found at the bottom of the crucible sharply marked off from the surface of the iron, which is now quite freed from the precious metals. [4]

The melting under charcoal is sometimes necessary to render silver bars fit for coinage when they have been treated by nitre. When silver has been raised to a high degree of fineness, it is affected by a peculiar bubbling due to the evolution of oxygen previously absorbed from the nitre. In this case it is necessary to stir continuously with a graphite rod, keeping the surface covered with charcoal powder, until the bubbling ceases. If the metal, while still effervescing, is poured into a mould, it sprouts at the surface, and a shower of extremely minute particles are projected, often to some

---

[1] Hoover's *Agricola*, p. 462.
[2] *Ninth Report Cal. State Mineralogist*, 1889, p. 64.
[3] Sulman, Cloud, and McDermott, *Trans. Inst. Mng. and Met.*, 1905, **14**, 429, 431, 433.
[4] Egleston, *Metallurgy of Gold, Silver and Mercury*, vol. ii., p. 728.

distance from the mould, requiring to be swept up ; if the crucible is covered by a lid, very heavy effervescence ensues when the lid is lifted. In this case the silver ingots formed are not marketable, being brittle, of low density, and covered by heavy efflorescences.

Casting the Ingots.—The operation of refining by one of the methods described above is necessary before gold can be exactly valued. When base metals other than copper are present, segregation occurs and the solidified metal is not uniform in composition. The result is that the exact value of the bullion cannot be determined, and the buyer must make some allowance to guard against loss. Accordingly bullion is usually, but not invariably, toughened in extraction mills, and this is now generally not necessary in refineries, before parting the silver from the gold. The object at a mill is sometimes merely to melt down the bullion obtained, so as to bring it to a marketable form with as little loss as possible. With this object in view, no nitre is used, but the metal is kept covered by a layer of charcoal to prevent the formation of oxides, and the crucible is poured as soon as the charge has been fused and stirred.

When the removal of base metals has been successfully carried out the refined gold should be of a brilliant green colour, and its surface should remain quiet, without showing any iridescent films or other signs of continued oxidation.

When the metal is supposed to be tough, a small sample may be dipped out and made into a thin ingot, which, after it has been cooled in water, is doubled up by hammering and its degree of toughness thus tested. It is then often remelted with copper to make up the standard alloy of the country, and again cast and hammered or cut in two with a shearing machine. The reason for doing this is that impure gold, although it may be tough when unalloyed with copper, may make brittle standard bars.

It is necessary to stir the charge thoroughly before pouring, as the bar must be as homogeneous as possible to insure a correct assay. Since segregation may occur on cooling, assay pieces are often dipped out immediately after stirring. The subject of taking bullion assay pieces is further considered in Chapter XX. The stirring is usually done with a peculiarly shaped graphite rod, made expressly for the purpose. It is annealed carefully and raised to a full red heat before being introduced into the crucible, and is held firmly by a pair of tongs with special concave curved faces to its jaws, so as to fit the round rod. In the case of very small meltings it is sufficient to lift the crucible out of the furnace with the tongs and to give it a rotary motion just previous to pouring. In doing this, the metal must not be allowed to cool too much or the casting will be defective. It is advisable to close the damper wholly or in part, so as to check the draught, when stirring is being done.

Meanwhile the ingot-mould in which the gold is to be cast has been prepared. It is cleaned thoroughly inside by rubbing with emery paper and oil, or with pumice stone, and wiping with an oily rag ; or it may be black-leaded inside, as this prevents contact between the gold and the iron of the mould. It is then warmed by being placed on the top of the furnace ; its temperature must not be sufficiently high to ignite the oil, but it should be too hot to touch with the hand. When the bullion is ready to pour, the mould is placed on a level surface, such as an iron stool, at a height above the floor of about 12 or 18 inches and a little oil mixed with graphite poured into it. Any cheap non-volatile oil will do, whether animal or vegetable.

The crucible is then lifted from the furnace, usually with basket tongs,

28

and when the temperature of the metal has fallen to but little above its melting point, it is poured rapidly but steadily into the mould. The crucible is then held in the inverted position for a short time, and jarred once or twice to cause the last portion of the metal to flow from it. The oil is ignited, and burns on the top of the cast metal, thus keeping it from tarnishing. In small castings, the slag is allowed to flow out and remain on the top of the metal in the mould ; in large castings, the slag is usually skimmed off before pouring. Beads of metal are caught in a large iron tray with raised edges, in the centre of which the mould is placed. If the mould is clean and has been hot enough, and if enough oil has been used, a clean untarnished bar is produced. It is turned out of the mould by inversion of the latter, while still too hot to be handled, and the slag is separated by one or two light taps with a hammer. The bar is then, in many establishments, momentarily dipped into water to assist in the complete removal of the last fragments of the slag, and it is also a favourite practice to dip the bar, first into dilute sulphuric acid, and then into clean water, the bar retaining warmth enough after removal to expel all moisture. This treatment removes all tarnish, and any adherent particles of slag are then chipped off and assay pieces cut from the bar.

In some refineries large crucibles are used, and 3,000 or 4,000 ozs. of metal are refined at once. In this case, several ingot moulds are filled successively from one pot, usually by means of an iron ladle, the weight of the gold bars manufactured being usually either 200 or 400 ozs. each. Silver bars, on the other hand, are made much larger, usually weighing 1,000 or 1,200 ozs. Mixed bars of bullion (containing both gold and silver) are seldom cast of a greater weight than 600 ozs. in mills.

In conducting all these furnace operations the use of a thick pair of mittens made of sacking or asbestos is to be recommended for the protection of the hands. The feet are best protected by wooden clogs.

Losses of Bullion Incurred in Melting.—The losses sustained in melting vary according to the composition of the metal. At the Royal Mint, the loss in melting without refining standard gold is about 0·2 per 1,000, partly recoverable from the "sweep." At the New York Assay Office they are said to vary from 0·5 to 1·5 per 1,000. The losses may be divided into mechanical loss, and loss by volatilisation. The mechanical loss is reduced by care in the conduct of the operation ; it may be due to a number of causes. The crucible may break and its contents fall into the fire, or be scattered over the floor of the melting house when on the point of being poured. To avoid loss in this way, the ashpit may be constructed of a cast-iron tray, which can be easily scraped out. The floor of the room is made of carefully-laid flagstones, or, better still, of iron plates, in which there are no cracks or crannies capable of hiding metal beads. Projection by spirting out of the crucible may be occasioned, especially if certain impurities, such as tellurium or antimony, are present in the bullion. Recovery of any metal lost in the ashes is effected by panning.

The slags formed in the course of refining frequently contain some small shots of metal, which may be recovered by grinding the slag finely and washing down the product in a pan or on a vanner. At the San Francisco Mint, the residue from the vannings is allowed to accumulate, and at intervals dried and fused with borax in an old graphite crucible at a high temperature. The crucible is left in the furnace over night to cool, and is then broken up, and the shots of metal at the bottom picked out. They are chiefly silver, very little gold being thus recovered. Shots of metal can be recovered with greater certainty from the slags by fusion with lead.

The crucibles, stirrers, lids, etc., also contain a certain quantity of gold and silver. After each melting they are scraped, and the scrapings panned, or, better still, calcined and fused with lead ; but, in spite of this treatment, precious metals accumulate in the pots, which, when worn out, are ground up in an Elspass or other Chilian mill and panned, in order to separate the shots of metal. Sometimes mercury is fed into the grinding mill. The tailing from this treatment will often pay for fusion with lead, and subsequent cupellation. Certain refineries treat large quantities of such residues, with which may be included sweepings of the floor of the melting house.

Osmiridium in Gold Bars.—Gold sometimes contains osmium and iridium. As these metals remain together during the treatment, the mixture is commonly called osmiridium. If this is present in perceptible quantities, a fact which is not usually detected until after the bars have been parted, the gold is remelted in a clean crucible and kept fused at a high temperature for about half an hour, when the osmiridium will settle to the bottom of the crucible. This is due to the fact that osmiridium does not seem to form a true alloy with gold, and, being of high density and very infusible, the particles, unfused or partly fused, settle through the liquid gold. The crucible is then gently lifted out of the furnace, and the greater portion carefully but rapidly poured into a mould ; the remainder, which contains almost all the osmiridium, is allowed to cool in the crucible until it has solidified, and then assayed for osmiridium. An alternative plan is to allow the whole charge to solidify in the crucible, and then to cut off the lowest portion, which is set aside. The osmiridium settles better from an alloy chiefly consisting of silver than from pure gold, and the rich bottoms are consequently melted several times with silver, the lowest part being cut off each time. The gold is thus gradually replaced by silver, which eventually forms by far the greater part of the mass. It is then granulated and parted, and the resulting powder of gold and osmiridium is treated with aqua regia, by which the gold is dissolved, and the osmiridium separated as a black powder. Iridium is, however, not invariably separated from the gold bars in which it is contained, and traces can be observed in some of the refined commercial bars met with in London.

PARTING.

Parting is the separation of silver from gold. During the course of the operation the base metals are separated from both, but, as the presence of a high percentage of these base metals is injurious to the successful conduct of the processes which are chiefly in use, a preliminary refining by one of the methods already described is usually necessary. Only about 10 per cent. of base metals is permissible in the alloys when sulphuric acid is used, and somewhat less in the electrolytic process.

The processes of parting may be tabulated as follows :—

    1. Cementation.
    2. Melting with sulphide of antimony.
    3. Melting with sulphur, and precipitation of the gold from the regulus by silver, iron, or litharge.
    4. Boiling in nitric acid.

5. Boiling in sulphuric acid, sometimes called " refining."
6. A combination of these last two methods.
7. The Gutzkow process (modified sulphuric acid process for the treatment of doré silver).
8. Passing chlorine gas through the molten metal.
9. Electrolysis in silver nitrate solution.
10. Electrolysis in gold chloride solution.
11. Dissolving in aqua regia.

The first of these methods was known to the ancients, and the third was described by Theophilus in the 11th century. The sulphuric acid, chlorine gas and electrolytic processes are the only ones now in use.

1. **Cementation.**—In this ancient and obsolete process, gold was freed from silver, copper, etc., contained in it. The method was described by Agatharcides,[1] B.C. 113, but was almost certainly in use at least as early as B.C. 700;[2] it is possibly still in use in some parts of the East. It consists in heating thin leaves or granulations of argentiferous gold embedded in a cement, consisting of two parts of brick-dust, or some similar material, and one of common salt, in pots of porous earthenware. Several other cements have also been used. The temperature used is a cherry-red heat, which is insufficient to melt the gold alloy. After about thirty-six hours' treatment, the greater part of the silver is converted into the state of chloride, and this, together with the cement, can be removed from association with the granulations by washing with water. The gold can in this way be raised to a fineness of 997 or 998. The silver is recovered from the cement by amalgamation with mercury.[3]

2. **Parting by Means of Sulphide of Antimony.**—This process was also used to purify gold which contained only small quantities of silver. The method was first described about the year 1500.[4] The alloy was repeatedly melted with sulphide of antimony, when the gold became alloyed with the antimony and sank to the bottom of the mass, while the silver was converted into sulphide and floated on the top, mixed with the excess of antimony sulphide added. The gold was subsequently refined by cupellation or by a blast of air directed upon it, the antimony being thus oxidised and volatilised. The method is now obsolete, but was in use at the Dresden Mint up to the year 1846, and gold of the fineness 993 was said to be produced in this way.

3. **Parting by Means of Sulphur.**—This method was first described by Theophilus, and was formerly used for the purpose of concentrating the gold contained in auriferous silver in order to obtain a richer alloy. The granulated alloy was melted with sulphur and some of the silver was thus converted into a matte. The gold was then separated from the matte by hammering, or precipitated from the matte and collected in a smaller quantity of silver by fusion with copper, iron, or litharge. The operation was repeated as often as necessary. Usually only a part of the silver was removed in this way, and the enriched alloy of gold and silver was parted with nitric acid. The silver was recovered from the matte by fusion with iron. The

[1] Gowland, *J. Roy. Anthrop. Inst.*, 1912, **42**, 252.
[2] T. K. Rose, Pres. Address, *Bull. Inst. Mng. and Met.*, April, 1915.
[3] For a full account of this interesting process, as well as of the next succeeding two methods, see Percy's *Metallurgy of Silver and Gold*, pp. 356-402; also Hoover's *Agricola*, p. 458, Historical Note.
[4] Hoover's *Agricola*, p. 458; also p. 451.

method was in use in several refineries in Europe at the beginning of the
last century. The employment of sulphur in refining at the United States
Mints has been already noticed, p. 432.

4. **Parting in Nitric Acid.**—The first mention of the use of nitric acid for
parting silver from gold was made by Geber or Albertus Magnus at a date
prior to the fourteenth century.[1] It was probably not used on a large scale
until much later. In Venice, according to an old tradition,[2] some Germans
were employed in separating gold from Spanish silver in the fifteenth and
sixteenth centuries, the art being kept secret. These refiners were not inaptly
named " gold makers " by those who were unacquainted with their methods.
The process was fully described by Biringuccio in his treatise,[3] published
in 1540, and by Agricola [4] in 1556. It was first used in the Paris Mint about
the year 1514, and in London at least as early as 1594, but for a long period
the operations were conducted in secret in both countries, and it is supposed
that this method of refining was not fully practised in England until about
the middle of the eighteenth century. It was superseded by the sulphuric
acid process in London in 1829. It was used in the United States Mints
throughout a great part of the nineteenth century.

Parting by means of nitric acid is conducted on the large scale in the
same general manner as in the assaying of gold bullion. It consists of the
following operations :—

1. Granulation of the alloys.
2. Dissolution of the silver in nitric acid.
3. Treatment of the gold residues, viz. :—Sweetening by washing with
   water, drying, melting, and casting into bars.
4. Precipitation of the silver as chloride by salt solution.
5. Reduction of the silver chloride by zinc and sulphuric acid.

Granulation of the Alloys.—The gold to be parted must be approximately
free from base metals, particularly from those which are not soluble in nitric
acid, such as tin, arsenic and antimony. If these were present they would
form insoluble oxides, which would remain with the gold, so that further
refining operations would be necessary : they would, moreover, cause a
great increase in the consumption of nitric acid, so, if they are present, the
gold is freed from them as far as possible by melting with nitre, etc. Copper,
lead and other metals which are readily soluble in nitric acid are less ob-
noxious, and small percentages of these are allowed to remain, the presence
of copper in particular being advantageous in promoting rapid dissolution
of the alloy. If present in large quantities, however, even these metals
would create difficulties and expense, increasing the consumption of acid.

The bars are melted together to form an alloy which, it was formerly
believed, must contain one part of gold to three parts of silver [5] (hence the
term " inquartation " applied to this process). Generally, however, the

---

[1] Hoover's *Agricola*, p. 460.
[2] Beckmann, *History of Inventions*, vol. iv., p. 578.
[3] Biringuccio, *De la Pyrotechnia*, Florence, 1540, Book iv., chaps. i.-v., pp. 64-71.
[4] Agricola, *De re Metallica*, Hoover's translation, pp. 443-447.
[5] A smaller proportion of silver, however, was used at least as early as the year 1627 in
Paris. Thus Savot observes in his *Discours sur les Medalles Antiques*, Paris, 1627, chap. vii.,
p. 72 :—" S'il n'y a beaucoup plus d'argent que d'or, l'eau n'agira aucunement: de sorte qu'il
faut qu'il y ayt au moins les deux tiers d'argent, et un autre tiers d'or, et encore que l'eau
soit tres-bonne : car, si elle est foible, elle n'operara point." Savot did not seem to regard
this proportion of 1 to 2 as of recent introduction.

proportion of silver used was less, the minimum being $1\frac{3}{4}$ parts of silver to one of gold. At the Philadelphia Mint, the proportion was $2\frac{1}{3}$ to 1.[1] Doré bars containing small quantities of gold were, of course, preferred to bars of fine silver for the purpose of alloying with the argentiferous gold bars. After the "inquarted" alloy has been thoroughly mixed by being stirred while still in the furnace, the metal is poured into copper tanks filled with cold water, which is kept cool in some refineries by means of a stream of water constantly flowing through the tank. The metal is poured with a circular and wavy motion in a thin stream to prevent the formation of lumps ; leafy granules and small hollow spheres are thus formed. The pouring is done either from a crucible or from a ladle, the vessel being filled from the large crucible by dipping, and held about 3 feet above the surface of the water. In the tank is a perforated copper pan, which is lifted out when the pouring is completed, and the granulations allowed to drain.

Dissolving the Granulations.—The granulated metal was heated with nitric acid in cylindrical vessels of earthenware, porcelain, or platinum, covered by closely-fitting lids provided with delivery tubes to carry away the fumes.

The strength of the acid used varied from that of specific gravity 1·41 to that of specific gravity 1·2. About 3 lbs. of acid (sp. gr. 1·2) were used to dissolve each pound of granulations, but of this quantity the amount used in the last boiling (about 20 per cent. of the whole) was available for further use ; more acid is required if there is much copper present. The granulations were boiled in fresh acid three times in succession.

The reactions that occur are partially expressed by the following equations :—

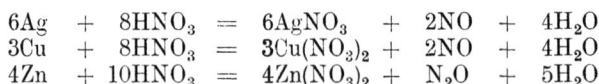

$$6Ag + 8HNO_3 = 6AgNO_3 + 2NO + 4H_2O$$
$$3Cu + 8HNO_3 = 3Cu(NO_3)_2 + 2NO + 4H_2O$$
$$4Zn + 10HNO_3 = 4Zn(NO_3)_2 + N_2O + 5H_2O$$

and similar reactions for other metals. The proportion of nitrous oxide, N_2O, evolved increases towards the end of the operation, as is shown by the diminution in the amount of red fumes which result from the mixture of nitric oxide, NO, with air. It is seen that silver decomposes less than its own weight of nitric acid, while copper and zinc destroy nearly three times their weight of the acid. Nitric acid of specific gravity 1·2 contains about 32 per cent. of anhydrous HNO_3, so that the quantity of acid of this strength theoretically required to dissolve 1 lb. of silver, copper and zinc is about 2·4 lbs., 8·3 lbs., and 7·6 lbs. respectively.

Treatment of the Gold Residue.—The pulverulent gold was " sweetened " by being washed thoroughly in perforated earthenware dishes with boiling distilled water, stirring with a spatula of wood, platinum or porcelain. The gold was thus freed from nitric acid and nitrate of silver, the operation being continued until the washings showed no signs of turbidity on the addition of salt. The sweetened gold was pressed, dried, melted, and cast into bars. The gold thus obtained was usually about 997 or 998 fine, the remainder being chiefly silver.

Treatment of the Silver Solution.—The solution of nitrate of silver was diluted with water, allowed to cool, and then treated with a strong solution of salt regulated so as not to be in large excess, continuous agitation being

---

[1] *Report of the Director of the U.S. Mint* for 1902, p. 121.

kept up by revolving wooden agitators driven by steam power, or by hand paddles. The precipitated chloride of silver was allowed to settle, washed by decantation, and finally in wooden filters lined with linen or some similar material. It was then reduced in lead-lined tanks by means of granulated zinc and water acidulated with sulphuric acid. The reactions involved are as follows :—

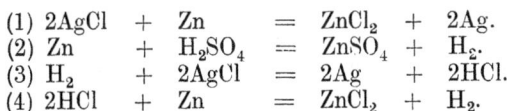

(1) $2AgCl + Zn = ZnCl_2 + 2Ag.$
(2) $Zn + H_2SO_4 = ZnSO_4 + H_2.$
(3) $H_2 + 2AgCl = 2Ag + 2HCl.$
(4) $2HCl + Zn = ZnCl_2 + H_2.$

These reactions explain the fact that, while zinc slowly reduces silver chloride in the presence of water only, the action is quickened by the addition of free acid, by which the zinc is attacked and hydrogen evolved. Nascent hydrogen is a powerful reducing agent, and decomposes silver chloride much more rapidly than zinc does, hydrochloric acid being formed and rendered available for the production of more hydrogen. The result of this is that the action, which is at first slow, becomes more and more rapid as hydrochloric acid accumulates in the solution. The sulphuric acid is only needed to start the reaction. At the San Francisco Mint 1 lb. of acid of 60° B. was added for every 2 lbs. of silver to be reduced. The white chloride of silver gradually turns black-grey as the silver is reduced. Hydrogen is evolved, especially towards the end of the operation.

The dark grey pulverulent silver, after being washed, pressed, dried, and melted into bars, was usually about 998 fine.

5. **Parting in Sulphuric Acid.**—This process superseded the nitric acid method, which is much more expensive, owing to the higher cost of the acid used and of the plant required. The earliest reference to the use of sulphuric acid in parting gold from silver was made by Scheffer in 1753, but the process was not used on the large scale until the year 1802, when it was introduced into France by C. D'Arcet, and worked in a refinery built in Paris for the purpose. It was established in London at the Royal Mint Refinery in 1829 by Mathison, and is still used there, the refinery having been let on lease by the Government since 1852. It has also been used in the United States for many years, but has now been superseded in the mint refineries by the electrolytic process.

The method used varies considerably in different refineries, but essentially consists of the following operations :—

1. Mixing and granulating the alloys.
2. Dissolving the silver from the granulations by means of sulphuric acid.
3. Washing and melting the gold residue.
4. Precipitating the silver from its solution by means of copper.
5. Recovering the copper sulphate by crystallisation or precipitating the copper by electrolysis.

The account given below is a general view of the operations in various refineries, the modifications adopted not being described in most cases.

Mixing and Granulating.—The alloys must be carefully prepared so as to be of suitable composition, as otherwise difficulties are encountered. The most suitable proportion of gold in the alloy is said by Dr. Percy [1] to be from 18 to 25 per cent., including whatever copper there may be present ;

---

[1] Percy, *Metallurgy of Silver and Gold*, p. 471.

but some American writers consider the proportion of one part of gold to two and a half parts of silver to be the most desirable, whilst at a refinery at San Francisco the alloy consisted of two parts of gold to three parts of silver. This proportion was instituted when alloying silver was scarce in California, but the gold thus separated was only 990 fine, containing ten parts of silver, the maximum allowed by law in the gold coins of the United States. If the ordinary proportion of $2\frac{1}{2}$ parts of silver to 1 of gold is used, however, the gold can be obtained about 996 to 998 fine, and the fineness of the gold can be increased to about 999 by fusing it, first with bisulphate of potash and subsequently with nitre. Doré silver containing only a few parts of gold per 1,000, when subjected in the form of bars to the action of the acid, instead of being granulated, yielded gold at San Francisco of 996 fine, after one boiling only.

The amount of base metals present in the alloy is carefully regulated, as their sulphates are little soluble in concentrated sulphuric acid, and consequently are precipitated and interfere with the progress of the operation. Bars of auriferous copper, such as those formed from worn-out amalgamated plates, are added to the parting alloy, as a small amount of copper facilitates the solution of the silver. The proportion of copper must not exceed about 10 per cent., but the usual amount is much less. A small quantity of lead is said to assist in the solution of copper, which is somewhat slowly attacked by concentrated sulphuric acid, and a maximum amount of 5 per cent. of lead does not interfere with the operation. From the economy with which this system of parting can be practised, silver containing only 0·5 part of gold per 1,000 can be separated from it at a profit. At the Vienna Mint, bars are parted containing 0·9 part of gold per 1,000, and at Freiberg bars containing only 0·4 part per 1,000 have been profitably treated.

In England, silver bars are passed through the parting operation, if they contain at least 2 grains of gold per troy pound, or 0·35 part per 1,000, but doré silver is not parted by itself. It is mixed with rich gold alloys.

The parting alloy is usually granulated (by pouring it into water, see above, p. 438), but at a San Francisco Refinery the doré silver is not granulated but melted and cast into bars $\frac{3}{4}$ inch thick, 9 inches wide, and 15 inches long.

Dissolution of the Silver.—This is usually effected in cast-iron kettles, platinum having been abandoned on account of its high cost. The iron used is fine-grained compact white iron, preferably containing 3 or 4 per cent. of phosphorus, which increases the durability, although 2 per cent. only of phosphorus is considered enough by some refiners. The kettle is slowly dissolved by the acid, ferrous sulphate being formed, and, in the course of about two years, the thickness of the vessel is reduced from about 2 inches to from $\frac{1}{4}$ to $\frac{1}{2}$ inch, when it is discarded. The perfect exclusion of air from the interior increases the length of life, and dilute acid must not be allowed to come in contact with the iron, as the latter is freely dissolved by it. The vessels are rectangular or cylindrical, with flat or hemispherical bottoms, the latter being preferred in Europe and the former in America. They are covered with cast-iron lids, about $\frac{1}{2}$ inch in thickness, which are bolted tightly to the vessels, and have bent leaden pipes fitted to them for carrying off the fumes, which consist largely of SO_2. This is sometimes reconverted into sulphuric acid in leaden chambers arranged for the purpose. The cover has also an opening (supplied with a lid made air-tight by a water joint) through which the alloys and acids are added and the operation watched. Heat is supplied by a wood or coal fire (see Fig. 191), or in other ways.

The charge for the pots varies from 200 to 1,000 lbs. of alloy, and the amount of acid required varies from 2 to $2\frac{1}{2}$ times the weight of the alloy,

Fig. 191.—Dissolving Kettles and Hoods used in the Sulphuric Acid Parting Process. New York Assay Office.

depending on the composition of the latter. About one-half of the acid, which is strong commercial acid of 66° B. (sp. gr. 1·85) is added at first, and the temperature cautiously raised to boiling point, when the pot

is closely watched, and, if the ebullition becomes too violent, the temperature
is lowered by regulating the fire and by adding cold acid a little at a time.
The charge is stirred occasionally with an iron tool, particularly towards
the end of the operation, when the undissolved granules of metal must be
freed from the surrounding sediment, consisting of sulphates of the base
metals, and exposed to the action of the acid. The ebullition gradually
subsides and action ceases in about five or six hours, the presence of a greater
proportion of base metals increasing the length of time required. The reactions are as follows :—

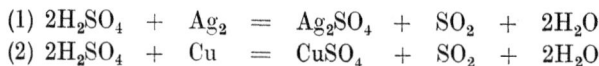

$$(1)\ 2H_2SO_4\ +\ Ag_2\ =\ Ag_2SO_4\ +\ SO_2\ +\ 2H_2O$$
$$(2)\ 2H_2SO_4\ +\ Cu\ =\ CuSO_4\ +\ SO_2\ +\ 2H_2O$$

and similar reactions with tin and lead. The reactions with antimony,
bismuth, zinc and iron are more complicated. It is obvious that 63 parts
of copper decompose as much sulphuric acid as 216 parts of silver. It is
clear, therefore, that an increase in the percentage of copper present necessitates an increase in the amount of sulphuric acid required.

One part of sulphate of silver is soluble in $\frac{1}{4}$ part of boiling concentrated
sulphuric acid, but the solubility rapidly falls off as the temperature and
concentration diminish, so that 180 parts of cold acid of specific gravity
1·08 are required for the same purpose. Sulphate of copper dissolves slightly
in the boiling concentrated acid, but is almost all precipitated in the form of
the white anhydrous salt on cooling. Tin and zinc behave similarly, and
lead makes the solution turbid and milky. The iron would be attacked on
account of the increasing dilution of the acid during the process, owing
to the formation of water, but the latter is usually in great part boiled
off as fast as it forms, or taken up by the anhydrous sulphates.

When the dissolution is complete, the solution is siphoned into a large
lead-lined covered vessel containing hot water (the settling tank) ; a gold
siphon may be used. As an alternative the solution may be ladled into
a settling pot and a few pounds of cold acid of 55° B. added, by which the acid
is cooled and diluted, and some crystals of silver sulphate are formed. These,
falling to the bottom, carry down with them the suspended fine particles
of gold, and so clarify the solution. If much copper is present, however,
this is not necessary, as the slight cooling of the acid is enough to precipitate
some sulphate of copper, which falls to the bottom and adheres to it very
firmly, thus clarifying the liquid and enabling it to be poured off or ladled
out very closely. The clear silver solution is then siphoned off or ladled
out with iron ladles into lead-lined rectangular wooden vats already partly
filled with hot water, in which the precipitation is subsequently effected.

Washing and Melting the Gold Residue.—The residue in the dissolving
pot, if the amount of base metals present is not large, is then boiled twice
more with fresh concentrated sulphuric acid added hot, after which the
gold residue is hard and heavy and rapidly subsides to the bottom, and
the liquors are siphoned into the precipitating vat. The gold is dipped out
with an iron-strainer and transferred to a lead-lined filter-box, where it is
thoroughly washed, first with hot dilute sulphuric acid and subsequently
with boiling water, after which it is pressed, dried and melted. It is sometimes brittle, from the occurrence in it of traces of lead, which is difficult
to separate by sulphuric acid owing to the insolubility and high density of
sulphate of lead.

If the amount of base metals present is very large, the gold residues are ladled into a vessel of hot dilute sulphuric acid and boiled with it by means of steam. In this way, most of the sulphate of silver and the whole of the copper, zinc, iron, etc., remaining with the gold are rapidly dissolved. Care must be taken, however, to add the residues a little at a time, as otherwise the anhydrous sulphate of copper will form lumps, which are only slowly dissolved. The gold is then allowed to settle, and, after the solution has been drawn off, is boiled again with acid if necessary, or if it is already pure enough it is at once washed, dried and melted.

Sometimes no attempt is made to obtain pure gold from auriferous silver in one operation, but the gold is concentrated in a small quantity of silver and then mixed with other alloys rich in gold and parted again. The product of gold thus obtained is purified by heating in a furnace in small iron pots with about half its weight of bisulphate of potash, by which some additional silver is converted into sulphate. The temperature is not raised much above the fusion point of the salt. The fused mass is then boiled in sulphuric acid, and again washed, dried and melted. In the United States these methods were not used, auriferous silver being cast into slabs and parted merely by boiling with sulphuric acid ; fusion with bisulphate of potash was rarely resorted to, but at the United States Mints, the residues were boiled with fresh acid five times in succession.

Precipitation of the Silver.—On pouring the sulphuric acid solution into cold water, most of the silver sulphate is precipitated at once in the form of small crystals, consisting of bisulphate, and the liquid must then be raised to boiling, by means of steam, in order to redissolve them. When the original alloys contain much lead this is not redissolved, and it is, therefore, necessary to let the solution settle and transfer the clear liquid to another vessel. Some particles of gold are usually found in the precipitate thus formed.

The reduction and precipitation of the silver are effected by means of copper, which takes its place in solution. The copper is usually added in the form of scrap while the liquid is being heated up by steam. The precipitation is assisted by constant stirring by means of wooden paddles. In San Francisco, however, the copper was cast into slabs, which were suspended side by side in the solution in a vertical position. The solution should be of about 24° B. ; if it is much more concentrated than this, the precipitation of the silver is imperfect. The end of the reaction is detected by testing with salt solution, and when complete the stirring is stopped, the solution allowed to settle for two hours, and the nearly clear liquid tapped into lead-lined vessels, where further settling of the suspended particles of silver takes place. The precipitate of silver is thoroughly washed with boiling water in wooden filters lined with lead, until the reaction for copper can no longer be obtained with ammonia. Care is taken that no fragments of metallic copper remain with the silver. The metal is then pressed, dried and melted, and is usually from 998 to 999 fine, even without fusion with nitre, in the cases where suspended copper plates have been used for reduction.

Recovery of the Copper.—This has been effected until recently by alternate evaporation and crystallisation of the copper sulphate in lead-lined wooden tanks. The solution, which is still of 24° B., is run from the precipitating tank into the evaporating pan and concentrated to 40° B. by heating with steam ; thence it is transferred to the crystallising tanks, where it is allowed to cool and remain for from ten to twelve days. The mother liquid of 36° B. is then run off and concentrated to 45° B., after which it is again allowed to crystallise, reconcentrated to 55° B., and a third crop of crystals obtained, which contain much iron. The clear acid mother liquor can now be used to dilute the solution of sulphate of silver in the dissolving pot as already described. The excess of acid in surplus liquids is neutralised with oxide of copper, more copper sulphate being thus formed.

The crystals of bluestone are found adhering to the sides and bottom of the tanks. They are detached with copper chisels, redissolved in pure water and recrystallised, the mother liquors being eventually added to the first liquor from the precipitating vats. When the liquors become over-charged with iron, the copper in them is precipitated by means of metallic iron, and they are thrown away or evaporated to get the crystals of sulphate of iron. The bluestone crystals are packed in barrels for the market. One pound of metallic copper with 1·5 lbs. of sulphuric acid of 66° B. will make 4 lbs. of crystallised sulphate of copper.

Owing to the difficulty of selling sulphate of copper, and the large amount of space and time required in obtaining it, it is now considered to be more profitable to precipitate the copper as metal by electrolysis and to concentrate the acid when freed from copper. Both copper and acid can thus be used over again.

Examples of Practice.—In 1902, 984,601 ozs. of standard gold, corresponding to 886,141 ozs. of fine gold, were refined by the process at the San Francisco Mint, at a cost of $57,214, or 6·5 cents per oz. of fine gold, exclusive of superintendence. (This was reduced to 5·1 cents in 1906-7.) The receipts from surplus bullion were $24,949. The amount of sulphuric acid consumed was 1·12 lbs. per standard oz. of gold;[1] this was reduced to 0·78 lb. in 1906-7. The process was discarded at the San Francisco Mint in 1908 and at the New York Assay Office in 1911 in favour of the electrolytic process.

The sulphuric acid parting process was used at the National plant of the American Smelting and Refining Company, at South Chicago, Illinois, in 1906.[2] At this works lead bullion is desilverised by zinc, the zinc scum is distilled in Faber du Faur tilting furnaces, and the residues cupelled in furnaces of the English type. The doré silver left on the test is cast into thick bars of 1,100 ozs. by tilting the test. The bars are parted direct by means of sulphuric acid of 60° B. (sp. gr. 1·71), in tanks of grey cast iron. The acid of 40° B. condensed from the vapours is concentrated in leaden pans. In a large settling tank, slightly warmed, a little gold deposits from the acid solution of sulphates. The solution is then transferred to the large precipitating tank, diluted with water and heated with steam. The silver is precipitated by copper plates (125 plates, each 18 × 8 inches and 1 inch thick), suspended in the solution from iron hooks covered with hard lead. After precipitation the vitriol solution is siphoned off and the silver washed in a vat provided with a false bottom, removed with a wooden shovel and pressed into cakes 10 inches × 10 inches × 6 inches.

The refining of the silver is finished on a cupel hearth, saltpetre being added and the slag removed by means of powdered brick. The silver is then kept fused for twenty minutes under a layer of charcoal, and is cast into iron moulds, previously blackened with a petroleum flame. The silver bars weigh 1,100 ozs. and are 999 fine.

---

[1] *Report of the Director of the U.S. Mint*, 1902, pp. 91, 92.
[2] O. Pufahl, *Eng. and Min. J.*, April 14, 1906, p. 718.

The gold is boiled with several fresh portions of acid, and is then washed,. dried and melted with a little soda in a graphite crucible. It is 995 fine.

The vitriol lye is evaporated down and bluestone obtained by crystallisation.

6. **Combined Process.**—At the Philadelphia Mint a combined process was formerly used, nitric acid and sulphuric acid being employed in succession. The alloys were granulated and digested with concentrated nitric acid for six hours in the same manner as has already been described; the solution was then siphoned off, and the gold washed two or three times. with distilled water, by decantation, subjected to a second boiling with strong nitric acid, and subsequently sweetened in lead-lined filters with boiling water. The gold was then introduced into cast-iron cylindrical kettles and boiled for five hours with strong sulphuric acid, the gold being stirred up with an iron rod every ten or fifteen minutes to prevent agglomeration, and the solution was then ladled out and treated as already described, p. 442. For a charge of 190 lbs. of metal, 175 lbs. of nitric acid were used in the first boiling, and 50 lbs. in the second. Some nitre was added to the sulphuric acid.

The process was introduced in 1866 by A. Mason[1] as an improvement on the nitric acid process, but was more expensive than parting with sulphuric acid alone.

The gold was washed thoroughly and sweetened in wooden filters, boiling distilled water being poured through it until the washings would no longer redden blue litmus paper. The gold was then pressed, dried, melted and cast into bars, which were from 998 to 999 fine. The silver was precipitated. from the washings as chloride by the addition of salt.

The process was much cheaper than the nitric acid process, costing 20 per cent. less for acids, and saving some fuel. The granulations contained 100 parts of gold in 333 of the alloy. After the boiling in nitric acid very little silver was left with the gold.

In the year ended 30th June, 1902, 608,185 standard ozs. of gold, equivalent to 547,367 ozs. fine, were refined by the process at Philadelphia, at a cost of $43,992, or 8 cents per oz. of fine gold, exclusive of superintendence and loss of gold. Operations in the acid refining plant at Philadelphia were finally discontinued in February, 1905, and the electrolytic process. used instead.[2]

7. **The Gutzkow Process for the Treatment of Doré Silver.**—This process of parting by sulphuric acid was invented and patented by F. Gutzkow in 1867, and has been extensively worked in Germany and in San Francisco. It is fully described in Percy's *Metallurgy of Silver and Gold*, p. 479, and only a brief account will be given here. When the patent had expired, Gutzkow introduced and patented several improvements on it, which were used for some time at the Consolidated Kansas City Smelting and Refining Company's Works at Argentine, Kansas, in 1892, but subsequently abandoned, at least in part.

The original Gutzkow process, as employed at the San Francisco Assaying and Refining Works for many years, may be summarised as follows :— The bullion treated was mainly doré silver, consisting of (1) Comstock silver bars or doré bars, usually containing 20 to 100 parts of gold per 1,000 ;

[1] *Production of the Precious Metals in the United States,* 1885, p. 278.
[2] *Report of the Director of the U.S. Mint,* 1904-1905, pp. 60, 86 and 154.

(2) base bars from the Reese River district and from pan-amalgamation of tailing, containing from 100 to 800 parts of silver, and the remainder chiefly copper, with sometimes a little gold. The metal was cast into bars, and parted in that form. The doré bars, when prepared for solution in the acid, weigh about 100 lbs. each, and are 12 inches long, 6 inches broad, and 5 inches thick. The base ingots are melted with fine bars to reduce the average copper contents to 12 per cent., and are cast into bars 1 inch thick, the gold from which is only about 992 fine.

The boiling is done in flat-bottomed thin cast-iron kettles (A, Fig. 192), of which the bottom is only $\frac{3}{4}$ inch thick when new, and $\frac{1}{4}$ inch when worn out. The solution can be rapidly heated, owing to the thinness of the iron kettles, and 200 lbs. of alloy are dissolved in four hours by means of 300 lbs. of sulphuric acid, which comes from the tank, C, and is forced into the kettle through the pipe, f, by the plunger, d. The solution is then siphoned off through the pipe, m, into the tank, E, and diluted with a large quantity of hot mother liquor from a previous crystallisation, which is mainly sulphuric acid of about 58° B.; some water is also added, and the solution partially cooled, so that some crystals of silver sulphate are enabled to separate out and carry down with them the milky precipitate of lead sulphate and any suspended particles of gold; green basic sulphate of iron also settles firmly. The clear solution is then siphoned off into H and cooled to 80° F., and almost all the silver sulphate thus crystallised out. If the acid is concentrated, white soft crystals of bisulphate are formed, which is not desired; if, however, the acid is only at about 58° B., large hard yellow crystals of monosulphate, free from acid but contaminated with copper, are deposited. The mother liquor is pumped back into the tank, E, or to the original acid tank, the device employed for this purpose being to exhaust them of air, so that the acid is sucked up without passing through any valves, which would soon wear out. The crystals of sulphate of silver are transferred to the filtering box, I, with iron shovels, and a hot solution of green vitriol (or ferrous sulphate) of 25° B. run on to them from G. This is at first mainly occupied in dissolving the sulphate of copper, and the first portion of the solution, after passing through the filtering box, is run into a storing vat, where the silver, incidentally dissolved, is precipitated by copper, and the latter subsequently recovered by means of iron. After a time, the copper being dissolved, the silver begins to be reduced, the green solution of iron turning coffee-brown; the reaction is as follows:—

$$2FeSO_4 + Ag_2SO_4 = Fe_2O_3.3SO_3 + 2Ag.$$

The reduction may also be effected by sheets of metallic iron, which is first converted into ferrous sulphate and then into ferric sulphate, the silver being simultaneously reduced to the metallic state. The brown solution of ferric sulphate is boiled with metallic iron in K, in order to regenerate the ferrous salt. The silver is washed, pressed, dried, and melted. The gold from the original dissolving kettle is also washed in a filter, pressed, dried, and melted.

Such was the original Gutzkow process as employed in treating doré bars. Its chief advantage over the ordinary sulphuric acid process was the saving of acid. In 1867, in the ordinary process, none of the acid used was saved, so that it was reduced in amount as much as possible, but did not fall below twice the weight of the silver dissolved. This reduction in the amount of acid used made the finishing of the dissolving a difficult and delicate operation. In the Gutzkow process, however, only the acid decomposed by the silver is lost ; the weight of this is about equal to that of the metal, the rest of the acid being all recovered and used over again in the boiling. Moreover, the long and tedious crystallisation of copper sulphate is avoided,

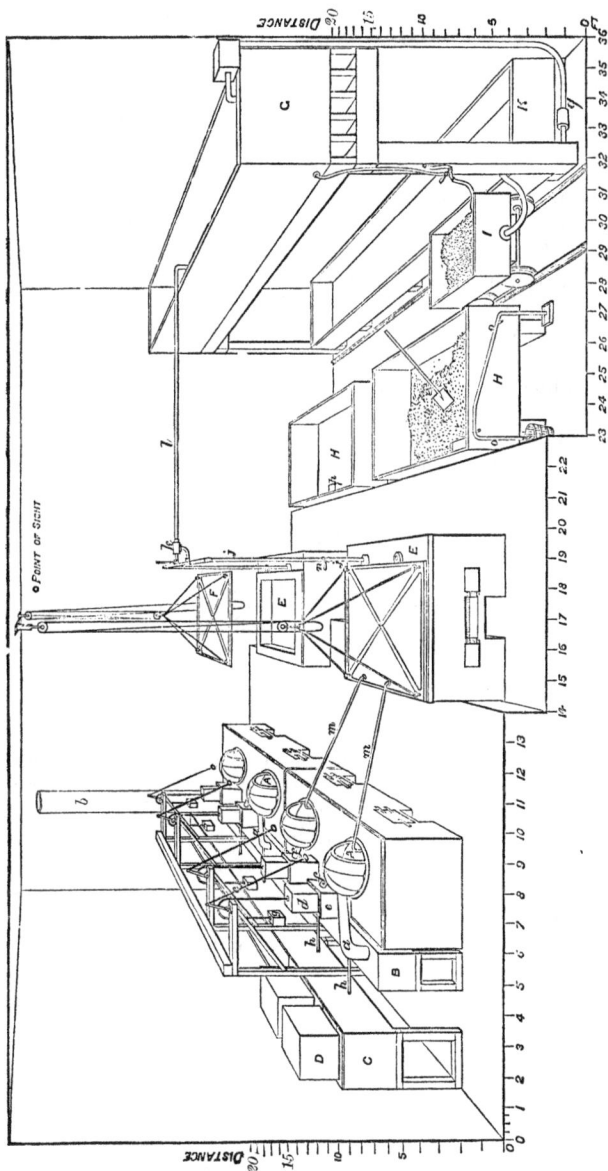

Fig. 192.[1]—Isometric View of Gutzkow Sulphuric Acid Parting Process.

and the space required for the crystallising vats saved. However, several large lead-lined vessels are required for the storage of the various solutions.

[1] Fig. 192 is reproduced by permission of Mr. John Murray from Dr. Percy's *Metallurgy of Silver and Gold*.

As mentioned above, p. 444, the production of sulphate of copper has been discarded in favour of the precipitation of copper by electrolysis, in the ordinary sulphuric acid process. The acid is also carefully recovered, so that the loss is probably less than in the Gutzkow process.

The New Gutzkow Process.[1]—If a large amount of acid is used for the boiling, not only is the silver more completely dissolved and the operation greatly expedited, but the presence of a high percentage of copper does not hinder the parting, as it is kept in solution by the excess of free acid. Thus, for ordinary doré silver, Gutzkow proposed to use four parts of acid to one of bullion; for bars containing 20 per cent. of copper six parts of acid; for still baser bullion, more acid, and so on, never losing more than one part of acid for one of bullion, and recovering the remainder.

The charge for a pot 4 feet in diameter and 3 feet in depth is 400 lbs. of doré silver; the pot is flat-bottomed, with a basin-shaped pocket or well in the centre which is useful for the collection of the gold. The bullion is first attacked by fresh acid of 66° B., run in by gravity from a large tank, and, when most of the silver has been dissolved, mother liquor from a former operation is added, a pitcher-full at a time, until the charge is completely dissolved, which takes from four to six hours. The fire is then moderated and the pot filled with mother liquor to within 1 or 2 inches of the top, when the temperature of the acid will have been so far reduced that only faint fumes are discernible. If no fumes are visible the acid is too cold and some silver sulphate will be precipitated, but otherwise the large excess of acid will keep it in solution. The well-stirred charge is now allowed to settle, which is perfectly accomplished in ten minutes, as the yellowish slowly-subsiding persulphate of iron is transformed to a greenish flocculent compound by the water in the mother liquor, and this settles quickly and carries all suspended matter to the bottom. More iron is dissolved from the kettle than in the ordinary process, owing to the greater dilution of the acid used in boiling.

The solution is now siphoned from the kettle by means of a $\frac{3}{4}$-inch gas pipe into a large cast-iron vessel, only about 1 foot deep, standing in a larger vessel which can be filled with water for cooling the charge. Steam is blown into the still hot acid solution through a lead nozzle, $\frac{1}{8}$ inch in diameter, pointing vertically downwards. This both dilutes and warms the solution, the heating being necessary in order to prevent crystallisation of the silver consequent on the dilution. As soon as the dilution has proceeded sufficiently far to ensure the crystallisation of the hard yellow monosulphate instead of the soft white bisulphate of silver, a point which is found by dipping out small quantities at intervals, and observing their behaviour on cooling, the steam is shut off and the vat cooled with water and left all night. The silver crystals form a coating of about 1 inch thick, which is contaminated with copper sulphate if the mother liquor, by repeated use, has become saturated with it. The mother liquor is now pumped back into the acid storage tank by the creation of a vacuum, and the crystals of sulphate of silver are detached with an iron shovel and thrown into a filtering-box provided with a false bottom. Cold distilled water is sprinkled on the charge, and is allowed to filter through it and flow back into the crystallising vat, until the greater part of the free acid has been removed. The steam is then

---

[1] This is fully described by Gutzkow in the *Eng. and Mng. J.*, Feb. 28, 1891, p. 257, and May 7, 1892, p. 497, from which this account is summarised.

deflected into a " silver filter " where any silver is precipitated that may have been dissolved at the same time as the sulphates of iron and copper. The silver filter is a lead-lined box, partly filled with precipitated copper and provided with a false bottom. The silver separates on the top of the copper as a spongy sheet, a corresponding amount of copper being dissolved. When the crystals of silver sulphate in the first-named filtering box have been completely freed from acid, and from copper and iron sulphates, the stream of water is discontinued. The spongy sheet of silver is then removed from the " silver filter " box and treated with hot water and a few crystals of silver sulphate to dissolve the copper still retained by the sheet. During this whole operation of sweetening the crystals of silver sulphate, only about 3 per cent. of it is dissolved, as it is little soluble in cold water. The copper solution, after passing through the " silver filter," is either run to waste or precipitated by scrap iron in wooden tanks at a nearly boiling temperature.

The crystals are now dried in an iron pan which is placed above a furnace, and, after being mixed with about 5 per cent. of charcoal, they are at once charged into a hot crucible in a melting furnace. The silver sulphate is reduced at a low red heat to metallic silver, carbonic and sulphurous acid gases being evolved. By the time the temperature of melting silver is reached, these gases will have all passed away. The silver is toughened by adding nitre and borax until the so-called " boiling " indicates that the sulphur has all been eliminated, and the metal is then cast into bars.

The gold residue in the dissolving kettle contains undissolved sulphates of lead, iron, antimony, mercury, and often some copper and silver. It is ladled out and boiled with water to dissolve out the sulphates of silver, copper, iron, etc., and, after thorough filtering, it is stirred in a dish with hot water and decanted on to a filter-cloth until the insoluble sulphates of lead, etc., have all been washed off, and the gold is left bright and clean. The gold is stored until enough is collected to make a 200-oz. bar, which is usually brittle. The material collected on the filter-cloth is re-washed once or twice to recover the particles of gold from it, and can then be reduced with charcoal and cupelled. If lead is present in the original alloy, part remains with the gold, and is dealt with in the manner which has been already described, but the greater part is carried off with the silver solution, and is deposited both while the steam is being passed in, and also subsequently during crystallisation of the sulphate of silver, which is coated with it. The sulphate of lead is removed from the crystals by stirring them well in a stream of cold water, by which the insoluble particles of lead and antimony sulphate are carried away ; it can then be collected, reduced and cupelled. Any silver that may be dissolved in the course of this washing is precipitated by copper as before.

The process is seen to differ from the original one in three essential particulars :—1. The solution is diluted with steam instead of with mother liquor, the amount of liquid in use, and consequently the number of lead-lined vats required being thus reduced. 2. The weak silver solution is precipitated at once, instead of being stored in tanks to be used again or to be precipitated at leisure. 3. The silver sulphate is reduced directly with charcoal in a crucible in the furnace. One of the minor advantages of the process is said to be that no stirring is required during the boiling, owing to the large amount of acid used. This saves labour and enables the acid fumes to be more easily condensed, as they are not mixed with air, which in the ordinary way would enter through the aperture left for stirring. The

29

exclusion of air also helps to prolong the life of the iron kettles by checking the attack on them by sulphuric acid.

Titus Ulke, however, announced [1] that the method of reducing sulphate of silver by heating it with charcoal, was tried by the Consolidated Kansas City Smelting and Refining Company, and abandoned, owing to the great volume of objectionable gases evolved, probably accompanied by heavy losses of silver. Gutzkow's original method of reducing sulphate of silver with a solution of ferrous sulphate was at that time used at the company's works at Argentine, and 36,000 ozs. of doré silver (containing on an average 4 per cent. of gold and 95 per cent. of silver) were parted daily at a mean cost (in 1895) of 0·22 cent per oz., not including cost of superintendence and office expenses.[2] The cost in 1892 at this works had been 0·35 cent per oz. These costs corresponded to 5·5 cents and 8·75 cents per oz of refined gold produced. The wages were from $2 to $2½ per day, and sulphuric acid cost one cent per pound.

8. **Miller's Chlorine Process.**—The use of chlorine gas for the purification of molten gold was first proposed by L. Thompson in 1838, and the results of his investigations were published in the *Journal of the Society of Arts* [3] two years later. He stated that " it has long been known to chemists, that not only has gold no affinity for chlorine at red heat, but that it actually parts with it at that temperature, although previously combined. . . . This, however, is not the case with those metals with which gold is usually alloyed. It offers, therefore, at once an easy and certain means of separation."

In 1867, F. B. Miller,[4] the Assayer of the Sydney Mint, applied this property of chlorine to the separation of gold from silver on the large scale, and the process has been in use at Sydney ever since, being particularly suitable for the purpose under the local conditions. Among these conditions may be mentioned the facts that acid is very costly, and that there is a scarcity of silver bullion containing small quantities of gold. The result is that the sulphuric acid parting process would prove very expensive, but the chlorine process can be applied cheaply, as it requires very little acid, and is efficacious in removing small quantities of silver from gold bullion which has not been made up into alloys of definite composition. Before the introduction of the chlorine process no attempt was made to extract the silver from any of the native gold of Australia and New Zealand which was coined at the Sydney Mint. Sovereigns were manufactured containing several per cent. of silver, which replaced part of the copper used as the alloying metal. These sovereigns, some of which are still in existence, can be easily recognised by their pale tint, due to the presence of silver. Such sovereigns have not been manufactured since 1872. Besides separating the silver, the chlorine process removes the small quantities of lead, antimony, etc., which render most of the Australian retorted gold brittle, and so in one operation prepares the gold for coinage. Almost the whole of the gold produced in Australia is now deposited in the mints of Sydney, Melbourne, and Perth. All unrefined bullion received at the mints

[1] Ulke, *Mineral Industry*, 1895, p. 348.

[2] Ulke, *ibid.*, p. 350.

[3] Thompson, *J. Soc. Arts*, 1840-1, [i.], **53,** 16 ; Percy, *Silver and Gold*, p. 402.

[4] Miller, *J. Chem. Soc.*, 1868, **21,** 506 ; and *Trans. Roy. Soc. of New South Wales*, Dec. 1869.

is treated by this process, whatever its composition may be. The average composition of the bullion received in 1913 was as follows :—[1]

	Gold.	Silver.	Base.
Sydney,	874·5	86	39·5
Melbourne,	914·2	43·4	42·4
Perth,	823·8	115·6	60·6

In chlorine refining the furnace used is an ordinary melting furnace, such as has been already described. The tile cover of the furnace has a hole in the centre to allow the chlorine tubes to pass through. Clay crucibles are used, the 4-pint size being generally employed, holding about 600 or 700 ozs. of gold ; they are placed inside graphite pots to prevent loss by cracking. They are glazed inside by melting borax in them to prevent them from absorbing molten chloride of silver. Graphite crucibles are unsuitable, as silver chloride appears to be reduced, presumably by the hydrogen contained in them, as fast as it is formed. The crucibles are covered by loosely fitting lids, through which the clay pipe-stems of about $\frac{3}{16}$-inch bore are passed to the bottom of the crucible for the conveyance of chlorine. The pipe-stem is made red-hot before being introduced into the molten metal, as otherwise it would crack and break off.

The chlorine gas is conveyed in leaden pipes from the generator to the furnaces. All joints and connections consist of well-wired india-rubber tubes, which must be protected from direct radiation from the furnace. Screw compression clamps on these rubber tubes enable the supply of chlorine to be regulated with precision. When the clamps are closed the gas accumulates and forces the acid out of the generator through a safety tube into a vessel placed overhead, and so the further generation of gas is prevented.

The generators being in readiness, the crucibles are slowly heated to redness, and the full charge of 600 or 700 ozs. of bullion introduced and melted, 2 or 3 ozs. of borax being sprinkled on its surface or poured on in a molten state. The chlorine is now allowed to pass slowly through the clay pipe to prevent metal from entering it, and the pipe is plunged to the bottom of the molten metal and kept there by means of a weight attached to it. The full stream of chlorine is now turned on and is heard to be bubbling into the molten metal, by which it is completely absorbed, so that no splashing and projection of the metal occurs. A height of 16 to 18 inches in the safety tube corresponds to and balances a height of 1 inch of gold in the refining crucible. The safety tube acts as an index of the pressure in the generator and of the rate of production of the gas ; any leakage or the exhaustion of the acid is at once indicated by a fall of the liquid in the tube. Fresh acid is added at intervals as it is required.

When the chlorine is introduced, dense fumes at once arise from the surface of the metal owing to the formation of volatile chlorides of the base metals, which are the first to be attacked ; lead gives especially dense fumes, which can be condensed on a cold object held in them. After a time these fumes cease and silver chloride is formed, very little chlorine escaping from the crucible, even if an extremely rapid current is passed into it ; consequently the operation is expedited by every increase in the volume of the current. Towards the end of the operation splashing is more noticeable, and dark

---

[1] *Forty-fourth Annual Report of the Royal Mint,* 1913.

brownish-yellow fumes appear, consisting chiefly of free chlorine. The completion of the refining, however, is indicated by a reddish or brownish-yellow stain containing metallic gold, which is imparted to a piece of white tobacco-pipe when exposed to the action of the fumes for a moment. This stain appears in about one hour and a-half from the start, when 600 ozs. of gold, containing 10 per cent. of silver, are being subjected to treatment. The current of gas is then at once stopped, and the crucible lifted out of the furnace and allowed to cool sufficiently for the gold to solidify. Probably, if the operation were continued after the appearance of the brown stain, appreciable losses of gold by volatilisation would occur.

The chloride of silver, still molten, and floating on the top of the gold, is then poured off into iron moulds, and the crucible inverted on an iron table, when the red-hot cone of gold falls out. This is now fine, and after any adherent chloride of silver has been detached from it by scraping, it simply requires melting into ingots, 98 per cent. of the gold being thus at once rendered available for use. The remainder of the gold is contained in the chloride of silver, partly in the form of entangled shots of metal, but chiefly as a double chloride of silver and gold.

The gold is removed from this by means of sodium carbonate, as follows :—[1] "The argentic chloride is covered by a layer of fused borax, about $\frac{1}{4}$ inch thick, and when all is well fused, the powdered soda is sprinkled on the top of the borax, without stirring, as rapidly as the ensuing action will admit. Occasionally the top layer is dipped with a stirrer slightly underneath the molten argentic chloride, without stirring the latter. When all the necessary soda is added and the action is nearly over, the pot is covered with a lid, and left for about ten to twenty minutes to increased heat, and, when the contents are quite liquid, the pot is lifted out of the fire without previous stirring, and allowed to cool, so as to enable the argentic chloride to be poured off from the gold button at the bottom of the pot.

"The use of 18 ozs. of soda for 230 ozs. of chloride produces a gold button weighing between 30 and 35 ozs., assaying about 920 to 930, and leaves from 0·5 to 1·0 part of gold in 1,000 parts of silver bullion produced.

"To free the argentic chloride from gold, a second treatment with 3 ozs. of soda per pot of 200 ozs. chloride, containing but a minute quantity of gold, will always be found to answer, the only care required being gradual application of the soda and enough heat at the end of the operation."

The time required for the two operations is about half an hour.

"The presence of a large proportion of chloride of copper has been found to prolong the operation considerably on account of oxide of copper being formed on addition of soda, as a much greater heat is required in order to fuse the whole mass. The argentic chloride produced from base gold alloys would contain a large proportion of chloride of copper, etc., and it would be better, therefore, to reduce it direct, and dissolve the reduced metals in acid, to separate gold and silver therefrom."

At the Sydney Mint, according to J. M'Cutcheon, the chloride of silver baled out of the pot is treated with soda ($\frac{1}{2}$ lb. to 350 ozs.), as above, but the portion poured off the gold is partly reduced by granulated zinc, of which 7 ozs. are thrown on the surface of 350 ozs. of molten chloride. This should be equivalent to about 18 ozs. of bicarbonate of soda.

The silver chloride was at first cast into slabs and reduced by zinc plates,

---

[1] *Fourth Annual Report of the Royal Mint*, 1873. Report by A. Leibius, p. 63.

the chloride slabs and zinc being suspended side by side in acidulated water, and connected above by strips of silver.[1] The reduction depends on electrolytic action, hydrogen being generated on the slabs of silver chloride and reducing the silver and copper together. When all action has ceased, the slabs of cupreous silver are lifted out and boiled, first in acidulated water and then in pure water before being melted. As no acid is used the amount of zinc consumed is the theoretical quantity required by the equations—

$$2AgCl + Zn = ZnCl_2 + 2Ag$$
$$CuCl_2 + Zn = ZnCl_2 + Cu.$$

At the Melbourne Mint in the year 1889, the zinc plates employed as

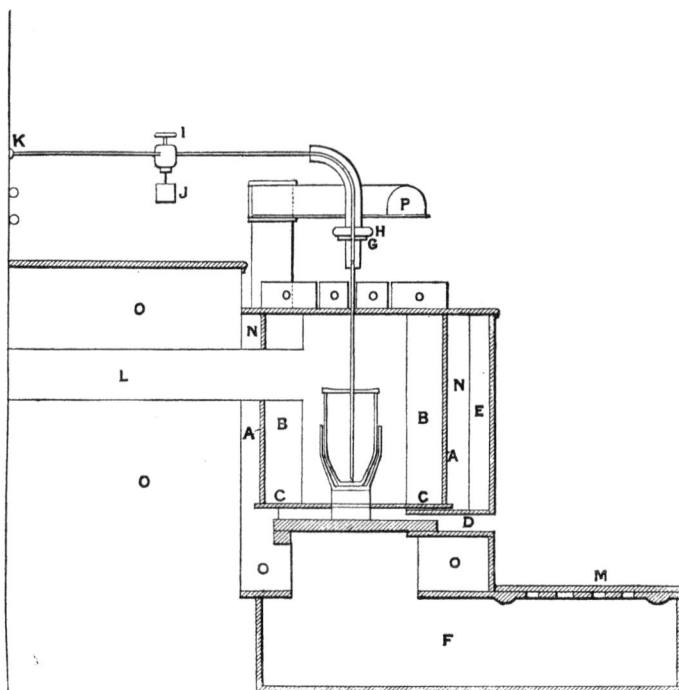

Fig. 193.—Section of Furnace for Chlorine Refining Process.

described above were replaced by sheets of iron,[2] which are now in general use.

The Chlorine Process as now practised at Melbourne.—The following description has been kindly supplied by Francis R. Power, the Assayer at the Royal Mint, Melbourne, by permission of the Deputy Master. It gives the exact methods and apparatus in use in the early part of 1896, and has been amended so as to include all changes introduced subsequently.

Furnaces.—These are thirteen in number. They are built cylindrically (see Fig. 193, in which one of these furnaces is shown in section, with crucible

---

[1] A. Leibius, *Trans. Roy. Soc., New South Wales*, Dec. 1869. The paper is given almost at full length in Percy, *Metallurgy of Silver and Gold*, p. 418.
[2] *Twentieth Annual Report of the Royal Mint*, 1889, p. 126.

and pipe-stem in position), being more compact in this form, more easily cleaned from clinker, and more economical in fuel than the square ones. They are 12 inches in diameter and 21 inches deep. The five fire-bars, $1\frac{1}{2}$ inches square and 18 inches long, are set in a cast-iron box, D, 12 inches by 2 inches, which passes through the brickwork in front of the furnace, the other ends of the fire-bars resting on an iron bar set in the brickwork at the back of the furnace. The bars are 6 inches above the floor. The draught is obtained through a grating in the floor, which covers a portion of the ashpit, over which there slides a cast-iron plate, M, $\frac{1}{4}$ inch thick, for regulating the admission of air, and pivotted in one corner. The flue, L, is 6 inches square, and communicates with a series of five condensing chambers, 8 feet by 8 feet by 5 feet, running the length of the furnaces (42 feet), all communicating and leading to the stack, 80 feet high, common to refining and melting furnaces, which are twenty-one in all. There are three furnace covers, two of them 20 inches by $6\frac{1}{2}$ inches, the third a little smaller, and all are bound with iron. The middle one is perforated by a 1-inch hole, through which the chlorine delivery pipe passes. Glenboig arched firebricks, B, 9 inches by $4\frac{1}{2}$ inches, and tapering from $2\frac{3}{4}$ inches to 2 inches, are used for lining the furnaces, and are set with touching joints in an iron cylinder, A, $21\frac{1}{2}$ inches in diameter, and at least $\frac{1}{4}$ to $\frac{3}{8}$ of an inch thick, which is supported by a cast-iron plate, C, $\frac{5}{8}$ of an inch thick, and 22 inches in diameter, with a 12-inch hole in the centre. This plate is supported by the brickwork which forms the foundation. The ashpit is a cast-iron flanged box, easily cleaned in case of an accident. Round the iron cylinder, concrete, N, is rammed, the front iron plate of the furnace being shifted 2 or 3 inches in, until this is set and then moved out, thus providing an air space, E, which keeps the plates cooler. The furnace top is a plate of cast iron and, in order to facilitate repairs, should be in two pieces for each furnace, halved into one another, the hole being slightly bossed at the edge so that the fire-tiles may run easily on them. One piece has a hole 6 inches in diameter over which the swing ventilating hood, P, is placed by which the pot is covered when removed from the fire. This hood communicates by a passage through the brickwork with the flue. The cylindrical furnace is calculated to last for three years, the square ones lasting only eighteen months and taking three hours to re-line, while the cylindrical ones take one hour. On the other hand, the curved bricks are more expensive than flat ones.

The Crucibles, etc.—The guard pot placed for safety under the white pot and afterwards used for remelting the refined gold, is a plumbago crucible $8\frac{1}{4}$ inches high, 6 inches inside diameter, $\frac{5}{8}$ of an inch thick at the top, and $\frac{3}{4}$ of an inch at the bottom, which is flat inside and stands on a cylindrical firebrick 5 inches in diameter and $2\frac{1}{2}$ inches deep. The clay pots, fitting loosely into the guards, are $10\frac{1}{2}$ inches high, 5 inches in diameter, and $\frac{3}{8}$ of an inch thick at the top, tapering from 1 inch at the bottom. They are covered by a closely fitting lid, dished at the top to catch any globules spirted out by too rapid a current of gas, and perforated by two holes $\frac{3}{4}$ of an inch in diameter. A new pattern of lid has been introduced which has a notch in the edge for the pipe-stem to pass through, the advantage of this being the easy removal of the lid without withdrawing the pipe-stem, as is necessary with the old lids.

The pipe-stem is 24 inches long, tapering from $\frac{3}{8}$-inch thick at the top to $\frac{1}{2}$-inch at the end inserted into the gold, and is wedge-shaped to facilitate the escape of the chlorine when resting on the bottom of the pot. The bore of the pipe-stem is $\frac{1}{8}$-inch in diameter. The thin end of the pipe-stem is attached to the branch delivery pipe by a piece of $\frac{1}{2}$-inch rubber about $2\frac{1}{2}$ inches long, which connects with an ebonite junction, G, 3 inches in length, with a bore of $\frac{1}{10}$-inch, turned with a ring round the middle, which acts as a rest for the 8-oz. weight, H, used as a sinker for the pipe-stem. One end of the ebonite junction is $\frac{1}{2}$-inch in diameter, the other $\frac{3}{4}$-inch; the latter being connected by a stout rubber tube 3 or 4 inches long to a 14-inch lead pipe ($\frac{1}{2}$-inch in diameter) which is attached (by a rubber junction) to a glass stop-cock, I, from the spigot of which a $\frac{3}{4}$-lb. lead weight, J, is suspended to prevent the pressure of gas from blowing it out. The glass stopcocks have replaced the compressor clamps, which were not satisfactory owing to the rubber cutting through, and chlorine leaking past. The rubber joints are sufficiently flexible to allow the pipe-stem to bend down into the pot or

Fig. 194.—Furnaces for Chlorine Process, Melbourne Mint.

to be laid horizontally on a rest when not in use. Each furnace is provided with one glass stopcock to control the flow of gas. The cock is far enough away at the back of the furnace, to be unaffected by the heat when the fire-tiles are removed, and is connected by a $\frac{1}{2}$-inch lead pipe with the main pipe running along the wall. Thinner pipe-stems are found to be as service-able as the above and do not require such careful annealing. The tubes, stopcock, etc., are somewhat diagrammatic in Fig. 193, and can be studied in Fig. 194, which is from a photograph of the furnaces by R. Law. This also shows the crucibles, guard pots, ventilating hoods and chloride cakes.

   The generators [1] are shown in Fig. 195, which is also from a photograph by R. Law. The pressure regulator and the reduction tank are also

---

[1] These generators are now replaced by the Edwards generators described on p. 461.

shown. The generators, eight in number, are three-necked cylindrical stone-ware vessels with domed tops, and having a flange round the middle by which they are supported on the stoneware steam jacket 16 inches high, $16\frac{1}{2}$ inches in diameter, and $\frac{5}{8}$ of an inch thick. The domed vessel is 2 feet high. The three necks have $1\frac{3}{4}$-, $1\frac{3}{8}$-, and $1\frac{3}{8}$-inch holes, the first for charging-in the manganese ore and closed by an india-rubber plug, the second for the pipe leading to the main chlorine pipe, and the third for a branch acid supply tube, $\frac{1}{2}$ an inch in diameter, fitted with a glass stopcock a foot above the neck, between which and the stopcock another $\frac{1}{2}$-inch tube, the overflow, branches. The overflow tube, through which the hot waste of the generators has to pass, is provided with an ebonite stopcock which is turned off during refining. Stout combustion tubing is used. The bottom of the generator is covered with 4 inches of quartz pebbles, to prevent choking of the acid delivery pipe, which reaches to within 1 inch of the bottom of the vessel. 56 lbs. of man-

Fig. 195.—Chlorine Generators, Melbourne Mint.

ganese ore (about 73 per cent. peroxide), broken to $\frac{1}{4}$ to $\frac{1}{2}$ inch cube, is placed on the pebbles, and commercial hydrochloric acid, of specific gravity 1·16, is added as required through the acid delivery pipe by turning the glass stopcock. The acid pipe is of glass, and leads to the eight storage tanks, 20 feet above the floor, which hold 320 lbs. of acid each, and are interconnected by glass tubes luted into the bottoms; the delivery of acid, however, being from one at a time. The gas delivery pipes from the generators all connect with a 1-inch lead pipe, which leads to a distributing vessel with two necks, and partially filled with manganese chloride solution. A pressure gauge of 1 inch glass tube and 15 feet high is luted into the bottom of this vessel, and is fixed to the wall by brackets, 10 to 11 feet of the solution being required to overcome the resistance of about 7 inches of metal in the pots. The pressure in refining is equal to 5 lbs. per square inch. A four-way tube of lead or pottery is passed through the second neck of the vessel, and each arm is connected by thick rubber to glass stopcocks to which $\frac{1}{2}$-inch lead pipes are joined, these pipes leading to three sets of 4, 4, and 5 furnaces respectively, so that the supply of gas can be delivered to a few or all the furnaces, as desired, the subdivision being made for safety in case of a leakage or for convenience if only a few furnaces are in use. All the generators are used whether the quantity of gold to be refined be large or small, the same quantity of acid being run into each. When the flow of chlorine through the gold is stopped the acid in the generators is forced back through the overflow pipe by opening the ebonite tap. It is found necessary to have the main pipe in communi-cation with another two-necked earthenware vessel containing such a quantity of water that when the pressure of gas exceeds the working pressure required, the end of a glass tube, passing to the bottom of the vessel and connected above the neck with an upright 4-inch lead pipe 10 feet high, becomes un-sealed, and the gas escapes through the water in large bubbles, escaping into the air through a glass pipe, inclined at an angle at the top of the lead pipe. When sufficient gas has escaped to reduce the pressure to the working limit the pipe is sealed. Thus the pipe acts automatically in keeping the pressure below such an amount as would endanger the apparatus or cause joints to leak. It is found expedient to cover all the rubber junctions in the generating room with calico and then to paint them. Protected in this way they will last until stopped up by the action of the chlorine which fills them with a lemon-yellow incrustation, at the same time reducing their thickness. All junctions are secured with copper wire where practicable.

Refining Operations.—The guard pot, with the clay pot in it containing 2 or 3 ozs. of fused borax, is placed in the furnace, and is heated gradually until the bottom of the clay pot is dull red. The ingots (of which the larger are slipper-shaped) to be refined, amounting in all to 650 to 720 ozs. in weight, are then placed loosely in the pot, the furnace filled with fuel, and the dampers opened. As soon as the gold is melted, which generally happens in about one and a-half hours, the boraxing of the pots being also effected at the same time, the perforated lid is put on, and the pipe-stem, previously brought carefully to a red heat to prevent cracking or flaking, is pushed to the bottom of the pot. As the pipe is being inserted, the chlorine is gently turned on to avoid stoppage of the passage through the stem by the solidification of metal in it. The supply of chlorine is controlled by the glass stopcock over the furnace, and the amount is adjusted so that the whole of the gas is absorbed and no globules of metal can be thrown up. This can usually be ascertained by feeling the pulsations of the gas through the india-rubber connections as it escapes in bubbles out of the bottom of the pipe-stem. When the gold contains much silver or base metals, the absorption of the chlorine takes place rapidly but gently, very little motion of the contents of the crucible being apparent, but when the gold to be refined is of high assay and also in all cases towards the end of refining, the gas is admitted only in a small stream, and requires careful watching to prevent spirting. When base metals are present in large quantities (over 2 per cent.) dense characteristic fumes of the chlorides of these are given off, and the metal or metals present may be generally identified by the fume or incrustation caused by the condensation of the base chlorides on the pipe or lid.

Gold containing 2 per cent. of silver and 0·5 per cent. of base metal is refined to about 995 fine in one and a-half hours, while that containing 3·5 per cent. of silver and 1·5 per cent. of base metals takes two hours. When larger percentages of silver or base metals are to be dealt with, the time taken is not proportionately longer, because, as mentioned above, a much greater stream of gas may with safety be admitted, though, in all cases, at the beginning the chlorine must be introduced gently on account of there being air in the chlorine mains, and, also, at the end of refining, the supply must be greatly reduced. When nearing completion, the "flame" issuing from the holes in the lid becomes altered in appearance, and much smaller ; it now contains much chlorine mixed with small quantities of the volatile chlorides. The actual completion of the operation is generally known by the appearance of a very characteristic " flame," which is luminous, with a dark brown fringe. In case of doubt, a piece of clean pipe-stem is used as a test. It is placed, cold, for a few seconds in the issuing flame, and if the refining is finished, a clear reddish-brown stain, tending to yellow, is imparted to the test end. This stain consists of ferric oxide and chloride from the oxidation of ferrous chloride, and contains gold and sometimes chloride of silver, and is probably caused by small quantities of chloride of iron retained by the fused gold and non-volatile chlorides, from which it is freed by the unabsorbed chlorine bubbling through. Traces of copper and iron are always found in the refined gold, the bulk of the alloy being silver. As soon as the stain is found to be of the right colour, the current of gas is at once stopped, and the pipe is then withdrawn and the clay pot lifted out of the guard. If the current of gas is continued a loss of gold ensues. The pot is allowed to stand under a hood (to carry off the fumes) until the gold is set, which usually takes place in from five to seven minutes, the fact that solidification has taken place being observed by thrusting a piece of red-hot pipe-stem down through the fused chlorides. The chloride is then poured into a mould provided with a ventilating hood, which, in consequence of the high density of the fumes necessitating a sharp draught to remove them, is connected with the stack. Any borax poured off with the chlorides is allowed to remain, as it is required as a cover for the chlorides in the subsequent fusion for the separation of their gold contents. The pot is then broken, as the cone of gold will not fall out of it soon enough, and the cone of refined gold is remelted in the guard and cast into two flat ingots, 12 inches by 4 inches by $1\frac{1}{2}$ inches, which, when set and still red hot, are placed on a copper lift, dipped in dilute sulphuric acid and then in water, and after removal from the water are still sufficiently hot to dry by their own heat. The broken pots are ground in a small Chilian mill and panned off; and the gold obtained is added to the " end " that is returned at the end of the day. 9,000 ozs., containing up to 10 per cent. of silver and base metals, constitute a day's refining.

An improvement was introduced in 1895, by which a considerable saving of time and material was made. This is the dipping of the fused chlorides and borax from the pot while it is still in the fire, and without previous solidification of the gold, by a small clay crucible, from which they are poured into a covered mould projecting over the furnace, the drops falling back into the pot. This had previously been the practice when the percentage of silver was large, as the silver doubles its bulk on conversion to chloride, and would have overflowed. The last " dip " always contains some gold, and is poured into a separate mould, in which the metal sets at once. The chloride is thence poured into the larger mould, and the gold returned to the pot. The chloride remaining in the pot is then made into a paste with bone ash, after which the refined gold is stirred and cast into ingots, the pot being at once returned to the furnace to be used a second time.

The chlorides, which hold from 5 to 10 per cent. of gold in feathery particles, are remelted during the day in plumbago crucibles holding 300 ozs. of chloride. When fused, 7 per cent. of their weight of bicarbonate of soda is added, cautiously and without stirring, which produces a shower of globules of reduced silver, and these falling through the chlorides carry down nearly all the gold. As one addition of bicarbonate of soda does not entirely free the chlorides from gold, a second addition is made, without removing the crucible from the furnace, ten minutes being allowed after each addition. The pot is then lifted out and placed on one side to allow the metal to set, when the chlorides are poured into a mould 12 inches by 10 inches by 2 inches, practically free from gold and ready for reduction. The silvery button obtained contains from 40 to 60 per cent. of gold, and is refined on the following day.

The silvery ingot and the refined gold contain 99·95 per cent. of the gold issued in the morning for refining, the bulk of the deficiency being in the pots. The amount of gold which goes immediately into work after refining is 97·6 per cent. (an average of thirty days' refining). The amount of gold left in the silver after reduction from the chloride reaches a maximum of 1 part in 10,000, but is usually from $\frac{1}{2}$ to $\frac{1}{3}$ of this quantity.

The cakes of impure chloride, weighing about 250 ozs. each, may be colourless and translucent to brown or chocolate in colour and opaque, the colour depending on the amount of copper salt present. They consist of argentic and cuprous chlorides, with traces of other chlorides and 9 per cent. of chloride of sodium from the decomposition of silver chloride by bicarbonate of soda. When cool, each cake is sewn up in a coarse flannel bag to prevent loss of any silver which may become detached during reduction, and they are then boiled with water in a wooden vat for four or five days. By exposure to air and moisture the cakes become coated with a green deposit, owing to the conversion of the cuprous chloride into cupric oxychloride and hydrated cupric chloride, and the successful removal of a large proportion of the cuprous salt in the vat is due to its solubility in a hot solution of common salt and of hydrated cupric chloride, from which it is redeposited on dilution.

The cakes for reduction are placed alternately with wrought-iron plates $\frac{1}{8}$ inch thick in a cast-iron tank lined with similar plates. The plates are prevented from touching the bags by laths of wood, otherwise the copper would be reduced in the bags, and would be difficult to separate from the silver. The reduction is slow in starting, unless either some liquor is left from a previous operation or some chloride of iron added. The bath is heated by the direct injection of steam (this is absolutely necessary), and the reduction is complete in from two to four days, though sometimes it takes longer. The time may be lessened to twenty-four hours by putting the chloride cakes in metallic connection with the iron plates, or by adding a strong solution of sodium chloride, a method suggested by M. L. Bagge.[1] In the latter case some silver chloride is dissolved, and the silver from it passes into the reduced copper and is lost. The completion of the reduction may be easily ascertained by feeling the cake, when, if any chloride be unreduced, it is left as a hard lump. The reduced silver is taken out of the bags, washed in boiling water for about an hour, and then melted with from 1 to 5 per cent. of nitre. The use of the flannel bags makes the reduced silver of high

---

[1] *Thirty-seventh Annual Report of the Mint*, 1906, p. 127.

standard, as the reduced copper is thus prevented from adhering to the silver cake, from which it was found very hard to detach it without tearing off silver as well. A small percentage of reduced silver and silver chloride is found at the bottom of the tank, its presence being probably due to the solubility of the chloride in solutions of the chlorides of copper, iron, and sodium.

The method now in use at Melbourne for reducing the chloride is due to W. M. Robins. It is as follows :[1]—The chlorides are poured into cakes about ⅜ inch thick, and these are crushed into pieces of about ¼ inch cube, making as little dust as possible. The crushed chlorides, loosely folded in flannel, are placed in open cane baskets, giving layers 3 inches thick. The baskets holding about 350 ozs. each are stacked at one end of a concrete vat of 30 inches cube. At the other end, separated by a perforated wooden diaphragm with ½-inch holes, are some iron plates. The vat is filled with water, and the contents boiled by means of live steam for 24 hours. The copper chloride is dissolved and precipitated by the iron. The residual silver chloride is then placed in loose layers about ¾ inch thick between iron plates in another vat and boiled for 24 hours, when reduction is complete. The silver is then washed, melted in plumbago pots and cast. It averages about 996 fine.

When base gold is refined the amount of copper chloride is large and is probably more easily dissolved in water, so that the resulting silver is finer than if the original bullion contained only a small amount of copper. In the latter case the copper chloride is not easily removed from the chloride cakes.

The silver contained in the gold operated on is distributed in the following manner, the mean results obtained at Melbourne in the five years 1891–95 being given :—

Silver in ingots,	88·77 per cent.
„ left in refined gold,	7·62 „
„ in " sweeps,"	2·00 „
„ unaccounted for,	1·60 „

No silver is now " unaccounted for " since the introduction of Robins' method. The " sweep " from the condensing chambers amounts to about 3 cwts. per annum, and contains an average of 41 ozs. fine gold and 157 ozs. fine silver, which are carried over as globules or volatilised as chloride and condensed.

The mean amount of gold refined per annum at Melbourne during five years (1891-95) was 949,527 ozs., containing gold 937·7, silver 49·6, and base metals (by difference) 12·7. The mean assay of the refined gold for the same period was 995·9, and the mean loss of gold in the refining operations for the same period was 0·175 per thousand. Similar results are still being obtained.[2]

The loss of gold by volatilisation is probably prevented from reaching the large amounts which might be expected from the known volatility of gold chloride by the fact that, during the whole time that the chlorine is being passed, silver and base metals are present, and, by absorbing the gas, protect the gold from its action. Moreover, the cover of slag intercepts any volatilised gold.

---

[1] *Forty-third Annual Report of the Mint*, 1912, p. 103.
[2] *Annual Reports of the Mint*, 1891 to 1913.

The approximate cost of refining per ounce gross weight refined at Melbourne was as follows :—

	In 1894.	In 1895.
Material, . . .	0·1397 of a penny.	0·1215 of a penny.
Wages, . . .	0·1485 ,,	0·1439 ,,
	0·2882 ,,	0·2654 ,,

Half the cost for materials was for hydrochloric acid at £20 15s. per ton.

In the year 1904, 1,074,550 ozs. of rough gold were chlorinated at Melbourne. The average assay before refining was—

Gold, ...	923·0
Silver, ...	42·5
Base, ...	34·5

This was brought up to an average fineness of 994·9, costing approximately per gross ounce treated—

Stores, ...	0·1132 of a penny.
Wear and tear, ...	0·0078 ,,
Treatment of sweep, ...	0·0313 ,,
	0·1523 ,,
Wages, excluding supervision, . . .	0·1222 ,,
Total, ...	0·2745 ,,

Thomas Price, who treated some of the Californian gold bullion by this method on a working scale in his laboratory in San Francisco, stated [1] in 1888 that, with Californian gold, which generally contains more silver than Australian gold, the gold taken up by the chloride of silver amounted to 5 per cent., and even to 10 per cent. of the total weight of the gold. For this reason, and on account of the large amount of silver bullion in the San Francisco market requiring parting, Price considered that the Miller process, while technically successful with Californian gold, was hardly able to compete commercially with the ordinary sulphuric acid process. This opinion was apparently based on an imperfect acquaintance with the chlorine process, but was probably sound at the time when it was expressed.

Edwards' Generator.[2]—In 1901 a new generator, patented by T. Edwards, Pyrites Works, Ballarat, was introduced at the Melbourne Mint. The design is simple, and it has given entire satisfaction. It consists of a wrought-iron flanged cradle (see Fig. 196, which is from a working drawing kindly supplied by F. R. Power), fitted with a lead lining $\frac{3}{8}$ inch thick, cast or hammered from sheet-lead to fit. The lining is supported at the bottom and sides by outside ribs (forming part of the casting), with openings to permit of the free circulation of the hot water used for heating the charge. The water passes between the iron case and the lining through a hole in one of the

---

[1] Price, *Trans. Amer. Inst. Mng. Eng.*, 1888, **17**, 30.
[2] Communicated by F. R. Power.

trunnions on which the generator is suspended, and escapes through a similar hole in the other trunnion. A flat lead-lined iron cover lies on the flange of the lining, and is secured by iron bolts to the flange of the cradle, a gas-tight joint being obtained by a rubber washer 3 inches wide.

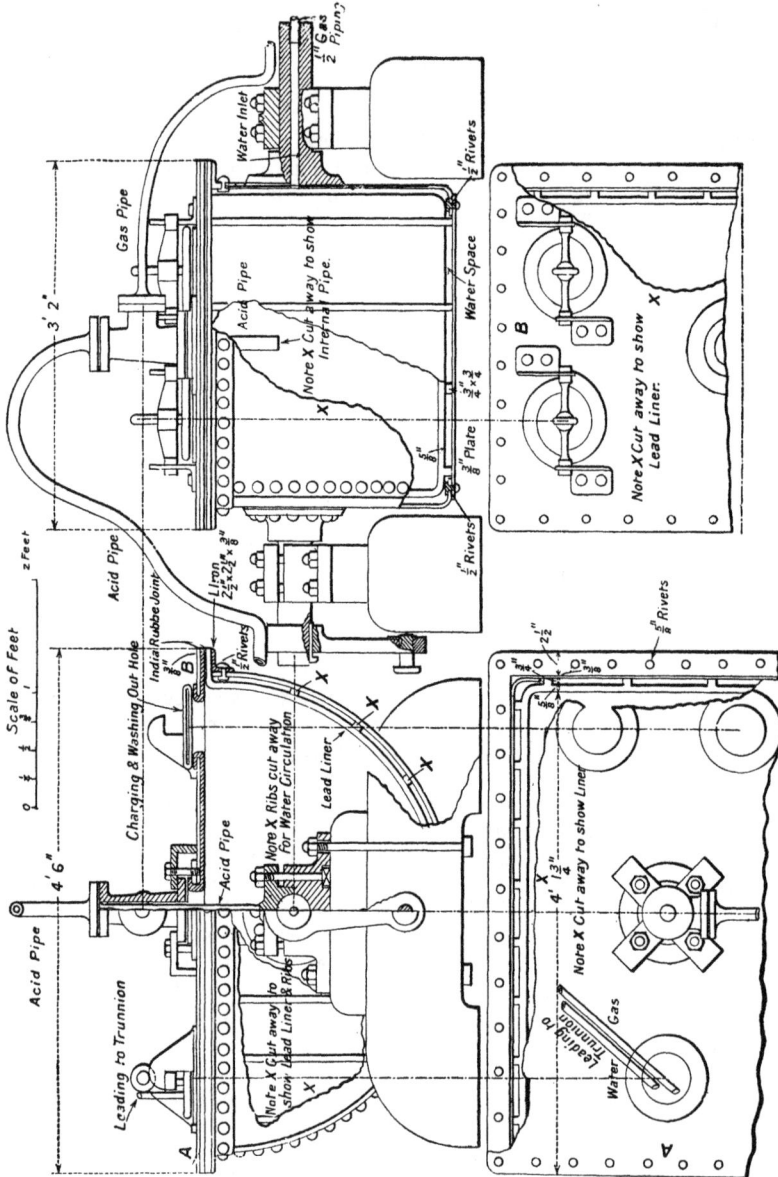

Fig. 196.—Edwards' Chlorine Generator.

There are four openings in the cover, three being 6 inches in diameter (though two of these only are absolutely necessary). These have lead flanges $1\frac{1}{2}$ inches wide and lead-lined iron covers, which are kept in place by stout screws, passing through iron clamps bearing against iron lugs. One opening is used for charging-in and washing out, and the other has attached to its cover two pipes, one for water, which empties the generator of chlorine by forcing it through the second pipe into a second generator, when the charge of the first is exhausted.

The fourth opening, 4 inches in diameter, in the centre of the cover, has a two-way lead casting with flanged openings, fastened over it. Through this the acid is delivered by a siphon pipe, while the chlorine passes through the second opening into the pipe leading to the condensing jar, and thence to the main pipe. The acid delivery, chlorine, and water pipes are bent over to a trunnion, and fastened to it by clips, and have 2 feet of stout rubber tubing, ¾-inch wall, inserted in their length to take up the slight twist produced when the generator rocks. The acid is fed by gravity from a graduated lead jar, 7 feet above, by a lead siphon, having an earthenware tap at the level of the generator to regulate the flow. A crank shaft connecting with the driving arm fixed at one end of the generator gives three complete oscillations per minute, the cover making an angle of 50° with the floor. This ensures the thorough mixing of the charge.

The usual charge is 80 lbs. manganese ore (73 per cent.), 100 lbs. salt, and 10 gallons of water. The acid is allowed to flow down at intervals until about 225 lbs. of chamber acid (sp. gr. 1·83) have been added. To get the maximum yield of chlorine, the charge must be heated nearly to boiling at the end. The generator is easily emptied through a 2-inch pipe burnt into one of the covers. The capacity is about 90 gallons, and with 100 lbs. manganese ore, 125 lbs. salt and 265 lbs. of acid, 1,560 ozs. of base metal and silver have been removed at the Melbourne Mint in one working day.

At the Ottawa Mint, R. Pearson uses the gas given off from *liquid chlorine* contained in cylinders.

Preliminary Refining.[1]—Though the main object of the chlorine process as used at Melbourne is to produce nearly fine gold, which on alloying with copper will be tough, it is frequently employed to enable gold containing much base metal to be partially refined so as to give a correct assay, or to enable a check assay to be made on finer metal, where the appearance of the original clips and ingot make it doubtful if a trustworthy assay is possible. In the case of ingots containing much silver, the operation has to be carefully conducted, otherwise some silver may become converted to chloride. In partial refining it is necessary to vary the pressure of the gas according to the weight of the ingot. The pressure may be reduced automatically from the full pressure as delivered from the generator as follows (see Fig. 197) :—

A is the full pressure main. B is an ebonite plug, with a pin-hole drilled through the centre, inserted in A. C is a glass pressure gauge reaching to the bottom of a Woulff's bottle, D, which contains water, and is of about a litre capacity.

The gas passes from A through B and into D. The tap, E, allows it to pass into the melted metal, through which it bubbles as soon as the pressure in D is sufficiently great to overcome the resistance. Test pieces are taken by dipping from time to time to determine the extent to which the refining has gone.

A preliminary refining with oxygen gas is given to low-grade material

---

[1] Communicated by F. R. Power.

at the Perth Mint. According to A. Ventris[1] the superintendent, zinc, iron, lead, etc., are removed at a lower cost than if chlorine were used.

9. **Electrolysis in Silver Nitrate Solution.**—The method presents some advantages over acid parting, among them being the absence of noxious fumes, except during acid treatment of the gold residues, the small amount of labour and chemicals, the cleanliness and rapidity of working, and the absence of bye-products, so that the loss of metal is reduced to a minimum. In consequence of these advantages, doré silver is generally treated in this way.

The Moebius Process.[2]—This process is in successful operation in several localities in the United States and Germany. It is said[3] to be specially suitable for refining copper bullion containing large proportions of silver and gold with small quantities of lead, platinum, and other metals, but is chiefly used in parting doré silver containing not more than about 10 per cent. of base metals. A similar process is in use in the "silver cells" of

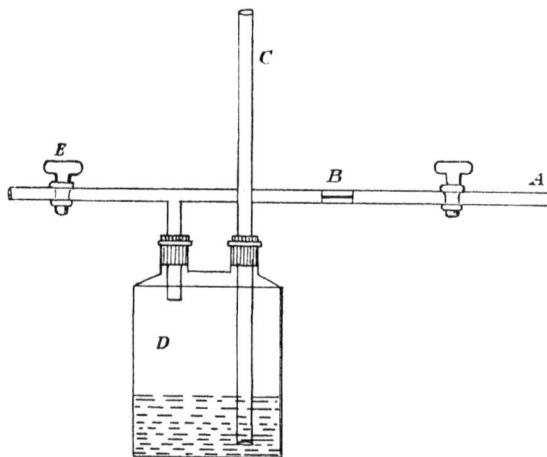

Fig. 197.—Arrangement for Varying Gas Pressure, Melbourne Mint.

the electrolytic processes adopted in the mints of the United States and Canada (see below, p. 466).

The apparatus required consists of a number of earthenware or wooden vats coated with graphite paint. The solution used contains 1 per cent. of nitric acid, which constitutes the electrolyte. This is soon converted into silver and copper nitrate, in which a small percentage of free nitric acid is kept. The anodes consist of plates of bullion of about ½ inch thick, 18 inches long, and 10 inches wide, which are hung in closely woven filter-cloth bags destined to catch the insoluble impurities after the silver, copper, etc., have been dissolved. The anode residue consists of gold, lead (as peroxide), platinum metals, antimony, etc. The cathodes consist of thin rolled plates of pure silver, slightly oiled to prevent adhesion of the deposited metal. These plates are continually scrubbed by a mechanical arrangement

---

[1] Private communication to the author.
[2] Eng. Pat. No. 16,554, Dec. 16, 1884.
[3] Gore, *Electrolytic Separation of Metals* (London, 1890), p. 240.

of wooden brushes, by which their surfaces are kept free from loose crystals of electro-deposited silver. The loose silver falls on to trays placed below, which are removed at intervals, and the silver collected from them. It is necessary to agitate the electrolyte.

The current should have an electromotive force of from one to three volts for each vat, but the potential per cell at the Electrolytic Refining and Smelting Company of Australia at Port Kembla, N.S.W., was eight volts.[1] The current density may be as high as 50 ampères per square foot of anode, but is usually only 20 or 30 ampères per square foot. The copper is not deposited unless the solution becomes too weak in silver or too rich in copper. When too much copper has accumulated in the solution the latter must be removed, the method being as follows :—The bullion anodes are replaced by carbon ones, and a weak current passed until all the silver is deposited. The silver cathodes are then replaced by copper ones, and a strong current passed so as to deposit the copper as rapidly as possible as a loose powder, which falls into a copper box. This box is connected with the cathode to prevent corrosion by the acid which is set free. The liquid thus regenerated is used again as the new electrolyte. Instead of occasional renewal of the whole electrolyte, it is more usual to draw off a portion each day and replace it with solution free from copper.

If no copper were present in the bullion it is clear that there would be no consumption of acid, and it would never be necessary to change the electrolyte. Under the most favourable conditions, therefore, with water power available and the amount of copper in the bullion very small, the cost of parting doré silver would be merely nominal.

At the Pinos Altos Mine, Chihuahua, Mexico,[2] a little plant capable of treating from 3,500 to 4,000 ozs. of doré silver in 24 hours was at work in 1890. The bullion was from 800 to 900 fine in silver, and 25 to 50 in gold, the rest being chiefly copper. The silver anodes were dissolved in about 24 hours, and the melted cathodes were 999 fine. The electric current employed was 170 ampères, the electromotive force being 8 volts. The cost of parting was said to be less than ⅓ cent per gross oz. of bullion.

The Moebius process was also in successful operation at the works of the Pennsylvania Lead Company at Pittsburg from 1886 to 1897 : here it is said that from 30,000 to 40,000 ozs. of doré bullion were refined daily. The doré bullion did not contain more than 2 per cent. of impurities (lead, copper, bismuth, etc.), and was cast into large anode plates, 1·3 cm. thick and weighing 13 to 15 kilogrammes, which are less advantageous than the light thin anodes used at Frankfort. Each anode was dissolved in about 2½ days. When the copper rose to more than 4 or 5 per cent. in the solution, part of the liquid was withdrawn, and the silver precipitated as chloride. The silver produced was from 999 to 999·5 fine, and could be raised by repeated washing to 999·85.[3]

At the Perth Amboy plant of the American Smelting and Refining Company, the process was in use in 1906.[4] The doré bars from electrolytic copper refining were cast into anodes. There were 144 silver cells in 24 divisions. Each cell consumed 260 ampères at 1·75 volts, the total consumption of current being 62 kilowatts and the yield 1,600 ozs. silver per kilowatt in

[1] R. G. Casey, *Mineral Industry*, 1910, p. 223.
[2] G. Maynard, *Eng. and Mng. J.*, May 9, 1891, p. 556.
[3] T. Ulke, *Mineral Industry*, 1895, p. 356.
[4] O. Pufahl, *Eng. and Mng. J.*, Jan. 27, 1906, p. 169.

24 hours, or nearly 100,000 ozs. in all per day. The mean composition of the electrolyte was said to be 10 per cent. free nitric acid, and 17 grammes silver and 35 to 40 grammes copper per litre.

At the Frankfort refinery, where this process is in use,[1] the general arrangements are similar to those noted above. The operation is begun with a very dilute solution of nitric acid and with a current of about 300 ampères per square metre (28 ampères per square foot). As the operation proceeds, copper accumulates in the liquid (the silver anodes being only about 950 fine), and reaches about 4 per cent., the silver being about 0·5 per cent., and the free nitric acid from 0·1 to 1·0 per cent. As the copper accumulates, fresh acid is added, and the current reduced to about 200 ampères per square metre. The E.M.F. required for each cell is 1·4 to 1·5 volts. The anodes are from 6 to 10 mm. thick, and each weighs about 1·5 kilogrammes; they are completely dissolved in about 36 hours, and another plate is at once substituted, the gold slime being cleared out of the anode bag either once or twice a week. The total capacity of the Frankfort refinery (3,200 ozs. per vat in 24 hours) is about 30,000 ozs. per day, the area occupied by the plant being only 40 feet × 45 feet.

At the Guggenheim Smelting Company, New Jersey, a continuously moving endless silver belt was formerly used as cathode, the deposited silver being brushed off by fixed brushes, so that it fell into a separate tank. The capacity was 100,000 ozs. per day, and the cost, exclusive of superintendence, office expenses and royalty, was stated to be about $\frac{1}{8}$ cent per oz.[2]

At Lithgow, in New South Wales,[3] the mud from the electrolytic refining of blister copper, with electrolyte of sulphuric acid, is sifted, washed, heated to a dull red heat and boiled in concentrated sulphuric acid. The residue is then washed, dried and fused with sodium carbonate in a small cupel furnace with the production of bullion which usually contains 12 to 16 per cent. gold and 82 to 86 per cent. silver. This bullion is parted electrolytically in an electrolyte of silver nitrate solution, a silver plate forming the cathode. The gold is caught as a mud in calico bags, and after being heated with nitric acid, is fused, giving a product usually from 990 to 998 fine. The silver is removed from the cathodes by moving wooden arms as fast as it is deposited. The detached crystals are collected and fused, the silver being not less than 996 fine.

The United States Mint Silver Nitrate Process.

In describing the introduction of the gold chloride electrolytic process (described below, p. 474) at the *Philadelphia Mint* in 1902, the Director of the United States Mint and Dr. D. K. Tuttle, the melter and refiner, dwelt on its limitations, and observed [4] that "the ideal refinery plant for a mint would be one in which electrolytic separations are the leading features. The bullion to be parted and refined will be divided into two classes, by selection, and by blending in making up the materials for anodes. The one class will have silver as the predominant metal, but carrying as much gold as will permit its treatment by electrolysis in a silver nitrate bath.

[1] Borchers, *Electric Smelting and Refining*, translated by W. G. McMillan (London 1904), p. 317.
[2] T. Ulke, *Mineral Industry*, 1895, p. 358; *ibid.*, 1900, p. 233.
[3] *Trans. Australasian Inst. Mng. Eng.*, 1911, **15**, 36; *J. Soc. Chem. Ind.*, 1914, **33**, 202.
[4] *Report of the Director of U.S. Mint*, 1901-2, p. 123.

The product from this operation will be fine silver, and the residues will consist of gold, platinum, and other impurities. The other class of bullion is to be largely gold, as at present, selected for electrolytic treatment in a gold-chloride solution. The residues from each process would pass to the other for final treatment. Work is now being done, looking to the realisation of this scheme." In 1905, the Director of the United States Mint stated that[1] " By persistent effort this has now been accomplished, and the energy obtained from a few bushels of coal in the form of an electric current is made to do the work of dray loads of expensive acids. Doré bars are successfully refined in industrial establishments by the Moebius or similar processes, but since in mint practice (and, it may be added, in gold refining generally) silver has to be added to the gold and used as parting material, an economical process must require the minimum percentage of silver in the anodes." The percentage composition adopted was 30 per cent. gold and 70 per cent. silver, copper, lead, etc., or about the same composition as had previously been used for acid parting. The Director proceeds, " The electrolyte used is a 3 per cent. solution of silver nitrate in water, to which is added $1\frac{1}{2}$ per cent. free nitric acid. The tanks are of earthenware, 40 inches by 20 inches and 11 inches deep. In each of these are suspended from conducting-rods 41 anodes and 40 cathodes. The anodes are composed of 300 parts in 1,000 of gold, the remaining 700 parts consisting of silver, copper, and other impurities as parting material. They are cast into bars $7\frac{1}{4}$ inches long by $2\frac{1}{2}$ inches wide and $\frac{3}{8}$ inch thick. The cathodes are strips of fine silver of the same length and width rolled to 0·016 inch thick.

" Eight cells so equipped are connected up in series, and a current with a density of 0·05 ampère per square inch passes through the system. The silver and other soluble metals are extracted from the anode by the combined action of the current and electrolyte, while the gold remains as a chocolate-brown substance sufficiently coherent to retain the original form of the anode. Meanwhile pure silver is deposited in a crystalline but coherent form on the cathode. Heretofore a coherent deposit has not been obtained in the Moebius and other processes in commercial use, being non-adherent crystalline granules, which fall from the cathode to the bottom of the cell.

" The deposit in a coherent form is due to a happy observation of the melter and refiner (Dr. D. K. Tuttle), in which it was discovered that the addition of a very small amount of a colloid, such as gelatine, to the electrolyte changed completely the nature of the deposit, so that the ' vertical system ' of anodes and cathodes became for the first time possible. The cathodes are washed with water, melted without fluxes, and cast into bars."

It may be noted that a similar addition of a colloid substance, usually glue, is made to the bath in the electrolytic refining of copper. It is stated[2] that A. G. Betts was the first to employ this means on a large scale in his lead refining process.

" The anodes retained presistently a small amount of silver, even if subjected to the action of the current after oxygen is freely evolved from their surfaces. . . . If the action of the current be prolonged on the anodes after most of the silver has been dissolved, the nascent oxygen evolved will attack the spongy gold and produce a small but notable quantity of gold tetroxide, soluble in concentrated nitric and sulphuric acids. It is deposited

[1] *Report of the Director of U.S. Mint*, 1904-5, p. 60.
[2] *Electrochemical and Metallurgical Industry*, Oct. 1906, p. 379.

from these solutions on dilution, but, of course, in a finely divided form. It is probably a hydrated oxide, since by simply heating the oxidised anode to 250° " a change occurs, subsequent to which " no gold passes into solution in acids.

" The silver remaining in the anode is removed by a treatment in hot nitric acid, the resulting solution being used to replenish the electrolyte. The gold is then thoroughly washed with water and melted. If platinum is present it will remain with the gold, and we usually pass this through the gold refining cells (Gold Chloride process, see p. 482). The losses in these operations need be very slight if care and cleanliness be used. All accidental sloppages of solutions are mopped with cotton cloths, which are burned and the ashes preserved. The necessary losses should be less than 1 oz. in the thousand."

The process was subsequently introduced at the *Denver Mint*, as described by R. L. Whitehead,[1] who installed the refinery. A summary of the description is given below.

The silver cells are eight Bertuch earthenware tanks, 47 inches long, 26 inches wide and 22 inches deep, arranged in cascades, four on a side. Hard rubber pipes drain from the bottom of one tank and empty into the top of the next below. A hard rubber pump, with a capacity of 18 gallons per minute, forces the electrolyte to a distributing jar placed on a platform above the level of the tanks. The jar has an outlet to each row of the tanks, and is automatically filled by means of a rubber float, which in turn stops and starts the rubber pump, drawing its supply from a tank placed below the level of the last depositing tank. If the automatic mechanism gets out of order, a 2-inch pipe will act as a bye-pass, from the supply to the suction tank. The foundations are of glazed brick, on which are placed slabs of acid-proof Alberene soapstone 3 inches thick. The main floors of the refining room consist of similar slabs 4 feet square and laid with splinted joints. The floor under the depositing tanks and for 18 inches on either side is covered with 18-lb. sheet lead made in the form of a pan, in order to catch any " spill." The electrolyte is 2 per cent. free nitric acid and 3 per cent. silver in solution as nitrate. The system of handling the products is similar to that used in the gold chloride process, see below, p. 483. The gold residue is washed, dried, melted and cast into anodes for the gold cells.

The anodes are suspended by gold hangers from the copper bars which rest on the tops of the tanks, the entire length of the anodes being immersed. The cathodes are rolled sheets of pure silver 0·01 inch thick. The deposited silver is stripped off them, and the cathode straightened so that it is ready for use again. The current density is 20 amperes per square foot, against 7·5 amperes at Philadelphia. There is a 200-ton hydraulic press for pressing silver and copper residue, so as to facilitate handling.

The refinery at *San Francisco* is described by R. L. Whitehead [2] and E. R. Durham.[3] Both the horizontal or Thum system of parting silver and the vertical or Tuttle (Moebius) system were in use in 1909. The tanks for the vertical silver system are of brown earthenware, and are 18 in number, 39 inches long, 19 inches wide and 12 inches deep. These were to be replaced by cells 18 inches deep. The anodes are made up to 300 to 333 parts of gold

[1] Whitehead, *Report of the Director of the U.S. Mint*, 1905-6, p. 59.
[2] Whitehead, *Electrochemical and Metallurgical Industry*, Sept. 1908, p. 355.
[3] Durham, *Trans. Amer. Inst. Mng. Eng.*, Oct. 1911 ; *Eng. and Mng. J.*, Nov. 1911, pp. 901, 950.

per 1,000, not over 100 parts base metal, and the rest silver. The electrolyte contains from 1 to $2\frac{1}{2}$ per cent. free nitric acid, 3 to $4\frac{1}{2}$ per cent. silver as nitrate, and some lead, copper, bismuth and zinc. A solution of glue is added. As the anodes are very base, about 5 per cent. by volume of the electrolyte is drawn off each day and an equal volume of solution very rich in silver is added. This is sufficient to keep the electrolyte pure enough to make a coarse sugary deposit.

The circulation of the electrolyte is by the trough system, the liquid being raised by a centrifugal pump from a lower pump tank to a discharge jar placed above the cells. The liquid flows from a spigot in the bottom of the discharge jar to a trough above the cells, and the flow to each cell or tank is regulated by adjusting the respective openings from the trough. The solution enters at the top of the end of each tank and is drawn from the bottom into another trough leading to the pump tank. Each tank thus has its own individual flow of electrolyte at the speed required by its condition, and at the same time the composition of the electrolyte is kept the same for all tanks. The propeller system of circulation, in which each cell retains its own electrolyte, is considered by R. L. Whitehead to be less advantageous. It was still in use in 1909, according to Durham. The electrolyte having a uniform composition (with trough circulation), practically the same current density is kept in each tank, about 8·3 ampères per square foot of anode surface (Durham).

The distance between anode and cathode centres is $2\frac{1}{8}$ inches. As the dissolving action continues, the anodes are taken out at intervals, and the black sponge, consisting of gold with 10 per cent. of silver and 1 per cent. of base metals, is shaken off into an earthenware jar, washed in a centrifugal machine and melted into anodes for the gold process. After 48 hours the anode cores are removed to the horizontal cells. The cathode deposit of silver is shaken or scraped off the cathodes, which consist of pure silver, washed and dried in a centrifugal machine, and melted. The finished product is about 999·5 fine. The stripping of the deposited silver from the cathodes is facilitated by smearing on them a " dope," consisting of a mixture of silver nitrate, copper nitrate and hydrochloric acid. The detached slime accumulating at the bottom of the cells is removed to the horizontal cells.

There are six horizontal cells. The anode is contained in a basket or tray lined with duck, and consists of a heterogeneous mixture of materials. The cathodes are graphite plates placed at the bottom of the cells. The electrolyte is the same as for the vertical cells, and a current density of 14·3 ampères per square foot of cathode is used. The black gold anode slime passes to the gold cells, and the deposited silver is melted with that from the vertical cells.

The silver in the foul electrolyte is precipitated by copper plates, and returned to the silver anodes. The dissolved copper is recovered in hard lead-lined tanks by deposition on scrap iron. The cement copper contains gold, silver and platinum. The anode residue of gold is washed, and melted into anodes (varying from 920 to 950 fine in gold) for the gold chloride process. There is no boiling in nitric acid. All the products are washed in earthenware centrifugal machines.

The process was introduced at the United States *Assay Office, New York*, in 1912,[1] where, as at the Philadelphia, Denver and San Francisco

[1] *Report of the Director of the U.S. Mint*, 1911-12, p. 49.

Mints, the silver nitrate process was run in conjunction with the gold chloride process. At New York, if the bullion received contains 850 or more gold per 1,000, it is passed directly to the gold cells. If it contains less than 850 per 1,000 of gold, it is passed through both processes.

" In practice the deposits of bullion are so combined that the resulting melt will vary between 250 and 400 parts in 1,000 in gold, 120 parts base and the remaining parts silver. This melt, 3,800 ozs. troy, is poured into anodes 16 inches by 3 inches by $\frac{1}{2}$ inch thick, with a $\frac{1}{4}$-inch hole at one end for suspension in the electrolyte. Allowing the gold contents in the anodes to vary between the above limits reduces the clerical work, and is permitted by the process.

" Each anode is placed inside a muslin bag, tied at the top, and then suspended in the electrolyte by a gold hook from the proper connection. The cells for the silver refining are 47 inches long, 24 inches wide and 24 inches deep, and are made from vitreous acid-proof stoneware. The electrolyte contains about 2 per cent. free nitric acid and 2 per cent. silver nitrate, and this proportion must be practically maintained in order to keep other metals from plating on the cathode. The silver is plated on a thin strip of rolled silver, which is lifted out once in eight hours and the loosely adhering crystals scraped into a dish. The addition of glue to the electrolyte for preventing treeing has been discontinued. The portion of the crystals which fall to the bottom of the tank is regained periodically and added to the fine silver output. This bottom silver runs almost as fine as that which is scraped from the cathodes. The silver and other soluble metals contained in the anodes are gradually dissolved by the action of the electric current while the gold remains as a coherent mass, retaining the shape of the original anode, but of a spongy nature and very easily broken. One of the functions of the bag is to prevent these easily broken anodes from mixing with the silver in the bottom of the cells. In general the bullion contains so much base metal that the anode is very brittle and often breaks prematurely. This necessitates the use of the 'horizontal cell,' in which the anode and cathode lie in a horizontal position. The broken anodes are transferred from the vertical cells to the horizontal ones. The horizontal is slower and requires more energy than the vertical cell, so that it does not replace the vertical type but supplements it.

" A porcelain basin, with a filter bottom, is used in the horizontal cells to hold the anodes, and the cathode is a graphite plate placed about 4 inches below the other pole. The deposited silver is taken out periodically by means of scrapers and added to the fine silver output. The product from the silver cells is from 999 to nearly 1,000 fine, and if any of this product contains more than 1 part of gold in 30,000 parts of silver, it is re-treated.

" The treatment of the foul electrolyte consists in the precipitation of the silver by metallic copper and regaining this copper.

" There are two sets of 16 vertical cells, each in series, across the circuit of 11 volts. The current density is about 6 ampères per square foot, although this varies considerably, according to the condition of the solutions and the nature of the anodes. The cells are placed on a long table built of brick topped with soapstone slabs. The slabs are countersunk, making a $\frac{1}{2}$-inch rim around the whole table, thus preventing drops of the electrolytes from falling on the floor, which is also made of soapstone slabs. The two series of horizontal cells are placed on a somewhat higher table of similar construction. A centrifugal hard rubber pump lifts the foul electrolyte from the level of the cells to that of the (foul electrolyte) precipitating tanks. The 28 horizontal cells are connected two in series across the same circuit as the vertical cells, and they carry a current density of 35 ampères per square foot. The circulation in the vertical cells is obtained by a glass propeller in each cell, a small belted motor furnishing the necessary power, while no circulation is required in the horizontal cells." [1]

There appears to be some difference of opinion as to the relative merits of the horizontal and vertical systems. According to Whitehead,[2] the great drawback to the vertical system is the low current density permissible in obtaining the necessary coherent deposit, amounting to only 7 or 8 ampères per square foot of cathode at the beginning and from 4 to 5 towards the end of the operation, as against 35 to 50 ampères per square foot in the horizontal system. The silver and gold are thus kept in the tanks for about eight days in the former case, as against one or two days in the latter. The anode gold also requires to be boiled in acid after removal from the vertical cells. The vertical cells require only 1 volt as compared with $3\frac{1}{2}$ volts in the horizontal cells.

The following description of the process, as practised in 1913 at the *Royal Mint, Ottawa*, is due to A. L. Entwistle,[3] who is in charge of the refinery :—

Gold bars over 900 fine are passed at once to the gold cells (see p. 484). Gold bars less than 900 fine are alloyed with silver, or, if possible, made up of mixtures of gold deposits of varying fineness, to form an alloy of 40 per cent. gold, 55 per cent. silver and 5 per cent. base metal. From experiment 40 per cent. gold seems to be the maximum amount of gold possible for parting. This mixture is melted in plumbago crucibles capable of holding about 3,000 ozs. The metal is well stirred before pouring, and is cast into closed iron moulds each 15 inches long, 3 inches wide and $\frac{3}{8}$ inch thick. These moulds, 50 in number, are arranged in a carriage on wheels, with four heavy lock screw bolts, two at each end, to tighten them up. At a point 1 inch from the bottom end of the mould there is a tapered pin of $\frac{3}{8}$ inch diameter, which leaves a hole in the bar produced. These bars constitute the anodes for the silver cells and weigh about 100 ozs. each.

The electrolyte, which is a solution of silver nitrate containing 3 to 4 per cent. silver and about 2 per cent. free nitric acid, is contained in an earthenware tank, holding about 400 litres, of the following dimensions, 4 feet long, 2 feet wide, and 2 feet deep. On the sides of the tank, at the top, are placed strips of hard rubber, with grooves cut out, running the length of the tank. In these grooves are placed $\frac{3}{8}$-inch steel rods covered with hard rubber tubes, the ends of which are plugged with the same material. The rods have a strip of silver on top $\frac{5}{8}$ inch wide and $\frac{1}{20}$ inch thick, shaped to fit the rod. The anodes and cathodes are suspended from the strip, which carries the current.

The cathodes are strips of fine silver 21 inches long, 3 inches wide and $\frac{16}{1000}$ inch thick, annealed to within 3 or 4 inches of the top and bent over to fit the rods.

The anodes are suspended by gold hooks slipped through the hole in the end and are enclosed in cotton bags, which are tied at the top round the hook with string. The whole of the anode is immersed in the electrolyte.

---

[1] *Report of the Director of the U.S. Mint*, 1910-12, pp. 48-50.
[2] Whitehead, *Electrochem. and Met. Ind.*, Oct. 1908, p. 409.
[3] Private communication by permission of Dr. J. Bonar, Deputy Master, Ottawa Mint.

Alongside the grooved hard rubber, above mentioned, is a copper rod ¾ inch square, running the length of the tank, to which the silver strips which carry the current are attached by means of brass screws and washers.

The current is divided by splitting the leads from the dynamo and connecting to the copper rod in two places, so that the distribution through the cell is more even.

In the middle of the tank is a glass stirrer or propeller, revolving at about 350 revolutions a minute, and causing an upward current in the liquid. The propeller is driven by a ½ H.P. motor, and the driving pulley

Fig. 198.—Porcelain Filter (washing proceeding), Ottawa Mint.

is grooved for a ¼-inch round belt. This drives a 3-inch fibre pulley, to which is attached the glass stirrer. The bearings for the pulleys are also made of fibre. If the electrolyte is not stirred the deposit is uneven, and short circuiting takes place owing to the deposit " treeing."

The hard rubber-covered steel rods are of two sizes, the short ones 2 feet 3 inches long, and the long ones 4 feet 10 inches long. The tanks are placed close together, and the long rods go across two tanks, and the short ones one tank. Two or more tanks are connected up in series, the current entering through the lines of anodes, which are in parallel, passing to the lines of cathodes, also in parallel, and by means of the silver strips on top of the long rods to the anodes in the next cell, and so on through the whole series.

In each tank there are 9 rows of anodes, 5 in each row, and 9 rows of cathodes, 5 in each row, except in the middle row, where there are only 4 anodes and 4 cathodes, to allow room for the stirrer. The rows of anodes and cathodes are separated by a distance of 2½ inches. The weight of the 44 anodes in the tank is about 4,400 ozs., consisting of gold 1,760 ozs., silver and base metal about 2,640 ozs. There are ten tanks in use, set up in series, standing upon strong wooden tables.

The current used is 350 to 400 ampères at 8 to 9 volts, therefore each cell requires about 0·8 volt. The current density is about 25 to 29 ampères to the square foot. The silver dissolves from the anodes and is deposited upon the cathodes, the base metals being converted into nitrates and going into solution. The deposit of silver is brilliantly crystalline and coherent.

The cathodes are taken out once in 12 hours and scraped with a silver scraper into a large porcelain filter, which is pulled around on a truck. When filled the filter is hoisted up on to a wooden platform by electric power, and the metal is washed free from silver nitrate, copper nitrate, etc., with hot water, and dried in a copper pan. It is then ready for melting and casting into ingots, the fineness being 999·5 or over.

Fig. 198 is from a photograph of a porcelain filter filled with silver ready for washing. Figs. 199 and 200 show in plan and section the porcelain and earthenware filters used by Whitehead at San Francisco.

A little silver is detached from the cathodes in lifting out and falls to the bottom of the tanks. The anodes are enclosed in cotton bags to prevent any small particles of gold from being carried over to the cathodes mechanically; also in case the anode should break, which sometimes happens, the bag prevents the broken pieces from falling to the bottom of the tank.

When separation is complete the gold is left in the original shape of the anode, and is of a dark brown colour, which is easily broken with the fingers and is quite spongy. This is packed

Figs. 199 and 200.—Earthenware and Porcelain Filters (Section and Plan).

into a porcelain filter and washed free from silver nitrate with hot water, allowed to drain, melted without drying, and cast into gold anode plates for further treatment by the gold chloride process. The fineness of the plates is 940 in gold, and the remainder silver, with a little copper, lead, etc., and any platinum that may have been present in the original gold.

If the passage of the electric current is continued after the anode reaches 940 fine in gold, the silver almost ceases to dissolve, and bubbles of gas are evolved at the anode. If the deposit becomes at all soft and mushy the electrolyte requires more silver nitrate, and if it does not adhere to the cathode more nitric acid is required. Racks placed above the tanks are lowered at night, and the cathodes taken out and hung on them to prevent

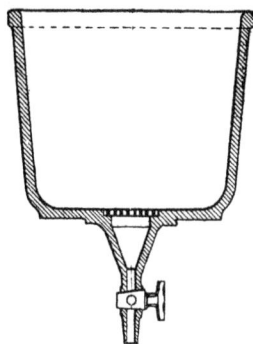

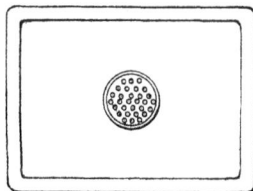

the silver from dissolving and taking up the free acid. The refinery does not work at night.

The anodes are taken out just before the bubbles of gas begin to be evolved, which occurs after about 95 hours ; the men handling the cells bcome very expert in judging when the process is finished.

The cathodes are used until they become too heavy, when they are melted with the crystalline silver and cast into ingots. The cotton bags are used about twice. The electrolyte is used until it becomes foul with impurities. When the specific gravity reaches about 1·320, about half of the solution is drawn off and fresh silver nitrate and water added. The foul solution, which contains from 3 to 6 per cent. silver and a large amount of copper, etc., is run into another tank and nearly all the silver is electro-deposited, iron anodes and silver cathodes being used. There is still a little silver left in solution, which is precipitated with salt solution and the copper is then electro-deposited or recovered with scrap iron, if of sufficient value.

All the washings from the fine silver, etc., are run into a large tank and precipitated with salt solution, and the silver chloride reduced to grey silver with iron plates and a little acid, washed, dried and melted.

The wash-water from the grey silver is run into a wooden tank containing iron plates and allowed to stand for several days before being run down the drain. This tank is cleaned out once a year, and the mud dried and melted. The resultant metal is found to contain about 1 per cent. silver and the rest chiefly copper.

The disadvantage of the process is the length of time for treatment. The average gold received assays between 750 and 800 in fineness. Quite an appreciable amount of gold as received contains 30 per cent. or more of base metal. This material cannot be treated by the silver nitrate process, except a little at a time, and it causes the electrolyte in the silver cells to become foul very quickly. It would be better to treat it by the chlorine process (especially if it is preceded by treatment with oxygen, see p. 429). The chlorine process was to be used instead of the silver nitrate after April, 1915.

10. **Electrolysis in Gold Chloride Solution.**—In this process a solution of gold chloride containing hydrochloric acid is subjected to electrolysis. The anode consists of impure gold. Gold and most other metals are dissolved at the anode, and nearly pure gold is deposited at the cathode, which usually consists of a thin sheet of pure gold.

The process was described in 1863 by Charles Watt, of Sydney, in a communication addressed to the Master of the Mint, London, and now in the Mint Library.[1] Watt used a solution to which he added " from one-fourth to one-fifth its bulk of hydrochloric acid." This he " carefully added " until bubbles of chlorine ceased to be evolved from the anode. He does not mention the heating of the solution nor the addition of chloride of gold. The original solution was to contain 0·5 oz. of gold per pint, or 27 grammes per litre. The voltage used was 1·7, and anodes $\frac{1}{10}$ inch thick were dissolved in 24 hours. This would correspond to a current density of about 330 ampères per square metre. The silver chloride was brushed from the anodes if necessary. The process had the approval of Geo. Gore, but was not adopted at the Sydney Mint, on the ground that no parting process was necessary there. The Miller process was afterwards introduced at Sydney in 1869.

---

[1] T. K. Rose, Pres. Address to the *Inst. Mng. and Met.*, March, 1915.

Watt also described in 1863 the electrolytic parting of gold-silver alloys, " containing not less than three parts of silver to one part of gold," in nitric acid solution. Watt was negotiating " for the introduction of his processes in Europe."

In 1874 E. Wohlwill began experiments in Hamburg with the object of separating gold from platinum.[1] The gold chloride process bearing his name has been in use at the Norddeutsche Affinerie, Hamburg, since 1878, and in 1881 the products were shown at the Electrical Exposition at Paris. The process was introduced at the Deutsche Gold-und-Silber Scheideanstalt, Frankfurt, in 1896, and later at Freiberg-Halsbrücke, in Saxony, and at the Imperial Mint at Petrograd. An identical process worked out independently was adopted at the Philadelphia Mint in 1902, and has since been brought into use at the Denver, San Francisco, New York, and Ottawa Mints, and also as an adjunct of the sulphuric acid process at the London refineries. It was introduced at the Imperial Mints in Austria-Hungary in 1912 for the purpose of refining platiniferous gold.

In the course of Wohlwill's investigations the following results were obtained :— [2]

(1) In a solution of auric chloride or of hydrogen aurichloride, $HAuCl_4$, gold, when used as an anode, behaves like platinum or carbon. No gold is dissolved, but chlorine is given off at the anode.

(2) If, however, hydrochloric acid (even as little as 0·4 gramme per litre) is added, gold is dissolved from the anode under the action of the electric current, and no chlorine is evolved, except as noted below. The chlorides of the alkalies or of ammonium have the same effect as hydrochloric acid.

(3) There is a maximum current density, above which chlorine is given off. The maximum is increased by heating the solution or by the addition of more hydrochloric acid. Below a temperature of 65° the maximum current, which can be used without causing an evolution of chlorine, falls off rapidly, and is said by Wohlwill to be at 50° to 55°, only half of that at 65° to 70°.

The amount of hydrochloric acid required depends on the temperature and on the density of the electric current. At Hamburg from 20 to 50 c.c. of concentrated hydrochloric acid were added per litre, according to the density of the current and other conditions. At New York the solution contains 10 per cent. of free hydrochloric acid.

(4) Gold is dissolved at the anode with the formation of $HAuCl_4$ and $HAuCl_2$. On the assumption that only trivalent gold is formed and exists in the solution, 2·45 grammes of gold should be dissolved and deposited per ampère-hour, but with monovalent gold the amount should be 7·35 grammes. Wohlwill's experiments, and the results obtained in working on a large scale, show that from 2·5 to 3 grammes per ampère-hour are usually deposited, and that the loss at the anode is greater than the amount deposited. It is, therefore, clear that a mixture of trivalent and monovalent gold exists in the solution.

Part of the *anodic loss* is accounted for by the separation of fine particles of gold, which are found for the most part in the silver chloride mud, into which the silver alloyed with the gold of the anode is converted. In ordinary

[1] Wohlwill, *Electrochemical and Metallurgical Industry*, 1904, **2**, 221.
[2] Wohlwill, *Electrochemical and Metallurgical Industry*, 1904, **2**, 221 and 261 ; *Zeitsch. Elektrochem.*, 1898, **4**, 379, 402, 421 ; *J. Chem. Soc.*, 1899, **76**, [ii.], 105.

working with the direct current, the particles of gold in the mud are equal
to about one-tenth of the amount of gold deposited on the cathode. They
are not merely mechanically detached from the anode, but are due to the
formation of aurous chloride at its surface, which subsequently decomposes
into metallic gold and auric chloride. Thus—

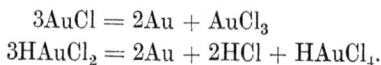

$$3AuCl = 2Au + AuCl_3$$
$$3HAuCl_2 = 2Au + 2HCl + HAuCl_4.$$

In support of this view, Wohlwill has shown that the particles are much
purer than the anode itself, that aurous chloride can be detected in the
solution, and that under certain conditions minute crystals of gold separate
out in all parts of the solution.

A series of experiments, with varying anodic current densities, made by
Wohlwill, show that the formation of monovalent gold ions diminishes as
the current density increases. In experiments with very small current
densities, the non-electrolytic dissolution of gold in the hot acid solution
of $HAuCl_4$ (with formation of $HAuCl_2$) must be allowed for. The equation
is as follows—

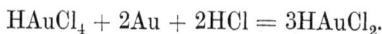

$$HAuCl_4 + 2Au + 2HCl = 3HAuCl_2.$$

This appears to be reversible, a sheet of gold losing weight when the
temperature is raised and gaining it when it is lowered again. After
allowing for this chemical attack, a current density of 1 ampère per square
metre was found to correspond to a cathode deposit of 4·33 grammes and
an anode loss of 6·01 grammes per ampère-hour. It is evident in this case
that 72·9 per cent. of the gold had dissolved in the aurous state, but that
only 38·6 per cent. of the deposited gold was from solution in the aurous
state. This difference corresponds to a large amount of gold thrown down
in the anode mud.

With 1,500 ampères per square metre, 2·48 grammes of gold per ampère-
hour were deposited (as against 2·45 grammes, if pure $HAuCl_4$ had been
present), and in two experiments the losses and gains at the anode and cathode
were nearly equal, thus :—

				Loss at Anode.	Gain at Cathode.
(1)	.	.	.	105·2	104·5
(2)	.	.	.	107·7	105·0

Here, therefore, trivalent gold was present almost exclusively. Such results
are obtained only when the liquid is well stirred.

The existence of $AuCl_4$ ions in the solutions may be regarded as
proved, one piece of evidence in support of this being the reddish-yellow
unstable precipitates of $AgAuCl_4$ which are formed on adding silver nitrate.
The ions of $HAuCl_4$ are H^+ and $AuCl_4^-$. On electrolysis hydrogen is released at the cathode, but is not evolved. It displaces gold in the compounds
$HAuCl_2$ and $HAuCl_4$, so that the gold is precipitated and hydrochloric acid
formed. At the anode the ions $AuCl_2$ and $AuCl_4$ are released and the
nascent chlorine formed by the decomposition of these ions combines
with the gold at the anode. Owing to the cathode deposition and the
migration of the ions containing gold away from the cathode, the
impoverishment in gold of the electrolyte near the cathode is very rapid,
and it is necessary to stir the solution. It is probable that practically no

part of the deposition is due to the electrolysis of $AuCl_3$, dissociated into its ions Au^+ and Cl_3^-; the number of Au ions in existence at any one time, must be inappreciable.

As the presence of monovalent gold causes its deposition in the anode sludge, it is desirable to avoid the formation of such gold, and to favour the dissolution of trivalent gold. This is a strong point in favour of using a high current density, but the need for rapid dissolution and reprecipitation of the gold in order to reduce the charges for interest on the value of the gold in the refinery is an even more cogent reason.

The influence exerted by the *composition of the anode* is of great importance. *Silver* is converted into chloride, but is not dissolved, except in small quantities, and remains adhering to the anode, reducing its effective surface and consequently increasing the anode current density and tending to cause chlorine to be given off. It has been found in practice with an electrolyte containing 5 per cent. of hydrochloric acid, and with the direct current used alone, that when the proportion of silver in the material to be refined exceeds 6 per cent., it is necessary periodically to brush off the deposit of silver chloride adhering to the anodes, in order to avoid the evolution of gaseous chlorine. With the pulsating current (see below, p. 479), this necessity is greatly modified, and gold with 15 to 17 per cent. silver can be refined. Lead and bismuth have to be reckoned as silver in respect of these proportions.

The dissolved silver is in part precipitated with the gold on the cathode, and particles of silver chloride floating in the solution also adhere to the cathode and reduce the fineness of the gold. R. Pearson[1] obtained a gold deposit only 996·8 fine, and containing in addition 2·5 per 1,000 of silver in some experimental work.

Pure *platinum* is not dissolved if it forms the anode, but platinum alloyed with gold is dissolved, and is not reprecipitated with the gold on the anode. According to Wohlwill it accumulates in the solution until it reaches a concentration of 4 or 5 per cent. (or about double that of the gold), above which it should not be allowed to rise. This statement must refer to work in a hot bath. The author has found[2] that in a cold solution containing 2 per cent. of platinum and 5 per cent. of gold, some platinum is deposited with the gold, but if the solution is heated no platinum is deposited. In America small amounts of platinum are recovered from the solution, but the amount, if any, electro-deposited with the gold is unknown. R. Pearson[3] found some cathodic gold to contain a distinct trace of platinum, but does not give the analysis of the electrolyte.

In the year 1911-12 gold equivalent to 6,313,845 standard ozs., 900 fine, or 5,682,461 fine ozs., were treated in the refineries at San Francisco, Denver and New York. An amount of 362·25 ozs. of platinum sponge or 0·06 per 1,000 was recovered, and in addition 27·77 ozs. of palladium and 1·18 ozs. of osmiridium.[4] The amounts recovered are very variable, and, according to D. K. Tuttle,[5] are derived only from dental and jewellers' scrap. Entwistle finds[6] that about 10 ozs. of platinum are recovered per 100,000 ozs. of the gold refined at Ottawa, which is received in about equal quantities from Porcupine and the Yukon.

---

[1] *Forty-fourth Annual Report of the Mint*, 1913, p. 182.
[2] T. K. Rose, Pres. Address, *Inst. Mng. and Met.*, Mar. 18, 1915.
[3] Pearson, *loc cit.*
[4] *Report of the Director of the U.S. Mint*, 1912, p. 17.
[5] *Ibid.*, 1903, p. 63.   [6] Private communication, 1914.

Palladium goes into solution with the platinum, but is not allowed to reach a concentration of more than 0·5 per cent. in order to avoid reprecipitation. *Iridium and osmiridium* do not dissolve but accumulate in the anode sludge. *Lead* dissolves in the solution, but the saturation point is soon reached and $PbCl_2$ crystallises out on the walls of the vat and on the cathodes, making the gold impure. The remedy is to add sulphuric acid to the electrolyte in amount equal (equivalent) to that of the hydrochloric acid. Lead sulphate is precipitated and settles to the bottom. *Copper* and other metals accumulate in the solution, but do not interfere with the purity of the deposited gold until they are highly concentrated; copper is said by Dr. Wohlwill to dissolve entirely as divalent ions. When the electrolyte becomes foul, owing to an accumulation of copper or other impurities, it is drawn off, and the copper and other metals are precipitated and recovered. The electrolyte is renewed most conveniently by withdrawing a part of it each day and making good the deficiency with a solution of pure gold chloride.

Composition of the Electrolyte.—In the course of investigations made by the author,[1] the following results were obtained :—

(1) When the amount of free hydrochloric acid in the bath is raised to about 30 per cent., a current of 5,000 ampères per square metre of anode surface can be used, without causing the evolution of chlorine. Anodes of $\frac{3}{8}$ inch thickness are dissolved in about seven hours.

(2) Under these conditions the amount of gold passing into the anode mud is less than 0·1 per cent. of the amount dissolved.

(3) A further result is that the proportion of silver in the anode is of little importance, as the silver chloride flakes off under the action of the heavy current, and the pulsating current described below (p. 479) is unnecessary. Anodes containing 20 per cent. of silver are readily treated.

(4) The formation of a satisfactory *deposit of gold at the cathode* depends on the proportion of gold in solution as chloride and on the temperature. If the current density is not too great, the deposit of gold is yellow and coherent. If the density exceeds a certain maximum, the deposit is dark coloured and pulverulent. With 3 per cent. of gold in solution, a current of 300 ampères per square metre is too great at 20° C., and a black or dark brown powdery deposit is formed. With 8 per cent. of gold in solution, a current of 400 ampères at 20° gives a coherent sheet of gold. With 20 per cent. of gold in solution, a yellow coherent sheet is formed at 25° with 1,500 ampères per square metre, and an almost equally good deposit at 62° with 5,000 ampères per square metre. This deposit is about 998 fine and of perfect quality, as regards its malleability after being melted, even when the anodes are very impure—*e.g.*, gold 780, silver 195, copper 25.

In practice it is preferred that the gold should be deposited on the cathode in the form of coarse crystalline particles, which are easily detached, but are sufficiently adherent to permit of thorough washing without loss. The crystals are not usually detached from the plate, but are melted with it. Silver plates coated with graphite could, however, be used as cathodes. No trouble is encountered from the formation of dendritic deposits, so that the electrodes may be placed near together. An increase in the concentration of the electrolyte gives a more closely adherent deposit. The presence of platinum in the electrolyte is also said to improve the density of the gold deposit,[2] but palladium has the opposite effect.

---

[1] T. K. Rose, *loc. cit.*   [2] J. B. C. Kershaw, *Electrician*, June 3, 1898, p. 187.

The deposited gold is of a high degree of purity, and it has been suggested that the process offers a cheap and ready method of preparing proof gold, 1,000 fine, for use in bullion assaying. The deposited gold is usually from 999·5 to 999·9 fine, and is seldom of less fineness than 999; it is almost always tough.

The electrolyte gradually becomes weaker in gold chloride, owing to the impurities in the anode, even if no evolution of chlorine at the anode is allowed to take place. It is, therefore, necessary to add gold chloride to the bath. This was prepared in Germany by dissolving gold in aqua regia, but the electrolytic method worked out by R. L. Whitehead[1] is now used in the United States mints. The process consists in passing an electric current through a solution of strong hydrochloric acid with anodes of gold 990 fine, and cathodes of rolled sheet gold suspended in a porous cup or cell. A strong current is used, "200 ampères at 5 volts," and much heat is generated. R. Pearson[2] uses only 17 to 20 ampères per square foot at 2·4 volts. Acid is added occasionally to compensate for loss by evaporation, and for the conversion of HCl and Au into $HAuCl_4$. At San Francisco five porcelain tanks are used for the process, each 24 inches by 15 inches and 15 inches deep, under an acid-proof hood. The capacity is 2,000 ozs. of fine gold per day. It is obvious that stirring is unnecessary, and that as the ions, $AuCl_4$, migrate away from the cathodes, there will be little or no tendency for gold to appear inside the porous cups and to be precipitated on the cathodes.

The amount of gold chloride solution to be added to the electrolyte depends on the composition of the anodes. Thus, Wohlwill calculates[3] that if the anode gold contains 15 per cent. of silver, then 11·5 kilogrammes of the alloy must be dissolved chemically for every 100 kilogrammes deposited electrically, or about 10 per cent. If the composition of the material to be refined is gold 83, silver 15, copper 2, then 14·5 per cent. of the alloy must be dissolved chemically. If the composition is gold 90, copper 10, then 20 per cent. of the alloy must be dissolved chemically.

The introduction of the *pulsating current* in 1908 [4] marks a distinct advance in the electrolytic refining process. The following account is based on the claims of the inventor; as will be seen later, these claims are in great part substantiated in practice. The pulsating or asymmetrical current is best obtained by connecting a direct-current dynamo in series with an alternating-current dynamo. If the maximum voltage of the alternating current is less than that of the direct current, the combined current will always have the same direction, but will change in density periodically from a maximum to a minimum. Such a current is said to produce some good effect in electrolytic gold refining, but it is better to use an alternator with a higher maximum voltage than that of the direct-current dynamo. In this case the combined current will be an alternating one symmetrical with respect to the voltage curve of the direct-current dynamo. The voltage curves are shown diagrammatically in Fig. 201. The straight line e_d is the voltage curve of the direct-current dynamo; the curve e_a is the voltage curve of the alternating-current dynamo. The curve E is the resulting voltage of the combined current. It is symmetrical with respect to the straight line e_d.

[1] Whitehead, *Electrochemical and Metallurgical Industry*, Sept. 1908, p. 357.
[2] Private communication to the author.
[3] Wohlwill, *loc cit.*, 1904, **2**, 261.
[4] *German Patent*, No. 207,555, Sept. 22, 1908; *English Patent*, No. 6276, Mar. 16, 1909; *Met. and Chem. Eng.*, Feb. 1910, p. 82.

The total gold deposited on the cathode can be found from Faraday's law, on the supposition that only the direct current is passing, as the electrolytic action caused by the alternating current is *nil*. A direct-current ammeter (which does not indicate the alternating current) indicates the desired direct current in ampères. The Joulean heat effect of the current is, however, given both by an alternating and a direct current, and, therefore, a hot-wire ammeter will indicate the total effective value (or virtual intensity) $I p$ of the pulsating current, according to the formula $I p = \sqrt{(I d)^2 + (I a)^2}$, where $I d$ is the strength of the direct current and Ia the effective value of the alternating current, both measured in ampères. For instance, if the strength of the direct current is to be equal to the strength of the alternating current—*e.g.*, 200 ampères—the hot-wire ammeter must indicate $\sqrt{80,000}$, or 283 ampères. It is obvious that the strength of the alternating current in use can be calculated from the measurements of the direct-current and hot-wire ammeters. According to Wohlwill,[1] the best ratio of the strength of the alternating current to that of the direct current is not more than 1·1 to 1 if the anodes contain 10 per cent. of silver. As the dynamos are connected in series, the armature of each dynamo must be designed to withstand the maximum heating effect of the total combined current.

The total voltage, from which the power consumption is calculated,

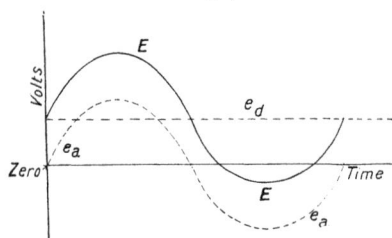

Fig. 201.—Voltage Curve.

is indicated by a hot-wire voltmeter. The degree of voltage depends on the strength of the total current. If, for instance, with anodes containing 10 per cent. of silver, a direct-current density of 1,250 ampères per square metre is used, and an effective density of the alternating current component of 1,375 ampères (in practice less than these amounts are used), then the hot-wire ammeter will indicate a total current density of $\sqrt{1,250^2 + 1,375^2}$, or 1,858 ampères to the square metre. In this case the voltage per cell of the direct-current will be 1·1, the total voltage indicated by the hot-wire instrument 1·4, whence the effective voltage Ea of the alternating current will be calculated as follows :—

$$Ea = \sqrt{1\cdot4^2 - 1\cdot1^2}$$
$$= 0\cdot86.$$

This is the mean effective voltage of the alternating current. Actually the voltage of the alternating current varies from a positive value of 1·2 (*i.e.*, $\sqrt{2} \times 0\cdot86$) to a negative value of 1·2, and the voltage of the pulsating current (E, Fig. 201) from a positive value or maximum of 2·3 to a negative value of 0·1.

If the strength (*i.e.*, effective value) of the alternating current becomes less than 0·707 $\left(i.e., \frac{1}{\sqrt{2}} \right)$ of that of the direct current, its maximum strength will fall below the strength of the direct current, and the combined current will cease to be an alternating current, and will become a mere undulatory

---

[1] Wohlwill, *loc. cit.*

direct current of periodically varying strength, in which case the peculiar advantages of the combined current are lost.

These peculiar advantages are briefly that a stronger direct current can be used without causing an evolution of chlorine at the anode, that the proportion of silver in the anodes may be as much as 15 per cent. or even more, without the necessity arising of scraping the anodes, and that less gold accumulates in the anode mud. When an alternating current is passed through the bath in addition to the direct current, the electric tension of the bath, as indicated on a direct-current voltmeter, falls, and the greater the strength of the alternating current, the greater the reduction in the voltage between the terminals of the refining cell. This may be visualised as a reduction in the resistance of the bath due to depolarisation at the electrodes. Within the usual limits of frequency (that is, for frequencies below 50 periods per second) the value of the frequency has practically no effect on the results.

When the pulsating current is increased beyond a certain limit of density, a slight evolution of oxygen is produced at the anode, occasioning an unimportant loss of current (and so increasing the necessary addition of gold chloride to the bath), and giving the advantage that the falling off of the chloride of silver from the anode is facilitated. When the current density is materially increased beyond this limit, the development of oxygen becomes greater and the loss of current becomes material. The detachment of gas from the anodes in amounts of more than single small bubbles is a sign that loss of current is occurring and that the direct current must be reduced or the alternating current increased.

As an example, if gold bullion containing 10 per cent. of silver is refined by means of the direct current only, an anodic current density of not more than 750 ampères per square metre must be used, and the silver chloride must be scraped off the anode every 45 minutes. If an alternating current is used the strength of which is about 1·1 times that of the direct current, a direct current of 1,250 ampères per square metre of anode can be used, without scraping the anodes (cf. practice at New York, p. 487). Gold containing 20 per cent. of silver can be refined by using an asymmetrical current, the strength of the alternating component of which is to that of the direct current component as 1·7 to 1 with a direct-current density of 1,200 ampères to the square metre (about 700 ampères are used in practice).

The amount of gold passing into the anode mud is diminished by the use of the pulsating current. According to Wohlwill, with the pulsating current, minute particles of gold form in the slime only at the beginning of treatment, when the anode has not yet been covered with silver chloride. At the end of the operation, when the anode finally breaks up, small pieces of gold are mechanically detached. Between these periods the accumulating slime consists almost entirely of chloride of silver. The result is that the pulsating current is an advantage even if the anodes contain very little silver.

By the use of the pulsating current the amount of hydrochloric acid in the bath can be reduced from about 3 per cent. HCl (or, say, 6 or 7 per cent. of acid of sp. gr. 1·19) to a quarter of that amount if the solution is heated to 65°, and the direct-current density is about 1,000 ampères to the square metre. In practice in the United States, the solution is not heated, and about 10 per cent. of commercial hydrochloric acid is added to the bath, with a current density of about 700 ampères per square metre.

Practice in Germany.—The Wohlwill process, as originally used in Germany before the introduction of the pulsating current, may be described

31

as follows :—The anode consists of impure gold, 4 mm. thick, and the electrolyte is a solution of gold chloride, containing 25 to 30 grammes of gold per litre, and from 60 to 70 c.c. of concentrated hydrochloric acid per litre. The liquid is kept at a temperature of from 60° to 70° C. The cathode consists of pure sheet gold, and the current density is from 400 to 500 ampères per square metre. Gold chloride is added from time to time. The anode is stated to be reduced to one-tenth of its thickness in 24 hours, and is then melted down and recast. The electrodes are 3 cm. apart.

With these conditions about 80 per cent. of the gold in the anode is deposited on the cathodes, about 10 per cent. is in the anode mud, and the remaining 10 per cent. is left in the anodes.

The consumption of acid is about 2 per cent. of that which would be required to dissolve gold in the ordinary chemical way. There are practically no fumes, and the cost of electricity in Hamburg is about three pfennige, or one penny per kilogramme of gold.

The platinum is accumulated in solution until it amounts to 50 or 60 grammes per litre, or twice as much as the gold in solution, and is then precipitated with ammonium chloride. In the year previous to May, 1900, the Hamburg refinery treated 2,000 kilogrammes of gold from all parts of the world and accumulated 1½ kilogrammes of platinum, which was worth four times the cost of refining the gold. Iridium is also completely separated from the gold. The costs at Hamburg, exclusive of superintendence and general expenses, are 1s. to 1s. 6d. per kilogramme (32 ozs.) of fine gold, when working fine mud from the Moebius process, and 2s. to 2s. 6d. per kilogramme of argentiferous gold. The cost of a plant for refining gold to the value of £10,000 per day is said to be about £1,000 in Hamburg, including £60 for a dynamo giving 150 ampères at 15 volts, and £250 for a gas engine giving 5 H.P. The space required is about 150 square metres.

Before 1912, all platiniferous gold was sent from the Austrian mints to Frankfort to be refined. The recovery of platinum in 1910-11 was as follows :—[1]

Year.					Gold Refined.	Platinum Recovered.
1910,	.	.	.	.	350·526 kg.	1·703 kg.
1911,	.	.	.	.	1248·173 kg.	9·418 kg.

Practice in America.—The process was introduced at the *Philadelphia Mint* in 1902,[2] owing to the falling off in the amount of doré silver received. It was, however, worked out independently at Philadelphia, and Wohlwill's patents and practice in Germany were not heard of until after all preparations had been made to use the process in America. Seven cells of white porcelain, each 15 inches long, 11 inches wide, and 8 inches deep, are used, filled with a gold solution containing 30 grammes of gold to the litre, to which is added sufficient free hydrochloric acid to suppress the evolution of chlorine at the anodes when the cell is in action. In each cell are 12 anodes in parallel, each 6 inches long, 3 inches wide, and ½ inch thick, and 13 cathodes of fine gold of the same length and width rolled out to 0·01 inch thick. The distance between the cathodes and the anodes is 1½ inches. Circulation of the electrolyte by mechanical means is necessary to secure uniform solution of the

---

[1] Mint Report, Vienna, 1912.
[2] Tuttle, *Report of the Director of U.S Mint*, 1902, p. 121 ; 1905, p. 61.

anodes and deposition of the gold. The anodes are dissolved in about three days. Copper, platinum, lead, etc., remain in the electrolyte. The silver is converted into chloride, and if too much is present it adheres to the anode, forming a protective coating which stops the action. The cells are heated to from 50° to 55° on a bed of sand by means of steam pipes. The seven cells are in series with the dynamo and 100 amperes are used, the difference of potential between the terminals being from $4\frac{1}{2}$ to 5 volts. One attendant with the occasional help of a second is said to be enough to manage the work. About 5,000 ozs. per week are refined with the expenditure of one horse-power.

It is evident from this account that the current density is somewhat less than 33 amperes per square foot of anode, or about 350 amperes per square metre. No alternating current is used. The gold anodes must, from their dimensions, weigh about 6,000 or 7,000 ozs. when new, so that the time of treatment is not less than a week. The cost for acid (20 cents per 1,000 ozs. of deposited gold) and power is trifling, but the labour costs are not inconsiderable, and the length of treatment somewhat great.

The refinery at Philadelphia was closed in 1912, and the work transferred to the Assay Office at New York, where the pulsating current is used.

The gold chloride process, as practised in 1906 at the *Denver Mint*, was as follows :—[1] Gold as low as 800 fine is melted with high-grade gold 990 fine or over to make anodes 940 or 950 fine, the balance being silver and base metals. Most of the anodic gold is from slime from the silver cells (see above, p. 468). If the proportion of silver is over 5 per cent., the silver chloride adheres to the anode and increases the voltage by 25 per cent. If 2 per cent. of copper is also present, the scales of AgCl are detached and in part are thrown against the cathode by the agitation of the electrolyte. They adhere and reduce the deposited gold from 999·9 to 999·5 fine. The electrolyte contains 5 or 6 per cent. of gold as chloride and 5 to 10 per cent. free HCl. The anodes are suspended by means of pure gold strip hangers or hooks from porcelain rods covered with strips of pure gold. There are 14 tanks of Berlin porcelain, each $13\frac{1}{2}$ inches by 18 inches and 12 inches deep. The anodes are 8 inches by $3\frac{1}{2}$ inches. After deposition is complete the cathodes are washed in Berlin porcelain filters, mounted on trucks which are brought close to the tanks, and then raised by an elevator to the filtering platform, after which they are dried all night in a steam drying oven before being melted.

The pulsating current is not used, and the average current density is not given, but 60 amperes per square foot of cathode has been attained. The anodes dissolve in about 36 hours. The temperature of the bath is not given, but the tanks are heated on a sand bath. The electrolyte is circulated by glass propellers in each cell, and difficulties arise on account of differences in the electrolyte in different cells.

The anode mud is boiled in concentrated sulphuric acid, by which the silver chloride (AgCl) is converted into silver sulphate (Ag_2SO_4) and dissolved. The residual gold is melted down for anodes.

The gold in the foul solution is thrown down with ferrous sulphate and ferrous chloride, and filtered off, after which the copper and platinum are precipitated by scrap iron. The clear solution of $FeSO_4$ and $FeCl_2$ is then concentrated by evaporation and used to precipitate the gold as above.

---

[1] Whitehead, *Report of the Director of U.S. Mint*, 1906, p. 57.

The cement copper is melted into anodes and treated electrolytically, yielding an anode mud containing platinum, silver and gold. The plant is capable of refining 100,000 ozs. of gold per week, but is seldom run at its full capacity.

In March, 1913, the pulsating current was installed at the Denver Mint, replacing the old direct-current system.[1] In the new system a direct-current voltage of 1·2 per cell and an alternating-current voltage (maximum) of 1·8 per cell are used. The current density is 65 ampères per square foot, or 700 ampères per square metre. The electrolyte now contains only from 4 to 7 per cent. of hydrochloric acid ; the temperature is 65° C.

Great benefits have resulted from the change, the chief one being that the anodes may now be as low as 800 fine in gold, whereas formerly the lowest permissible fineness was 900. The result is that much gold bullion can be treated direct in the gold cells which formerly passed through the silver cells first. Half the silver cells have accordingly been thrown out of use. For the five months before the change 0·038 lb. of hydrochloric acid and 0·038 lb. of nitric acid were used per oz. of fine gold produced. After the change the amounts were 0·058 lb. of hydrochloric acid and 0·004 lb. of nitric acid per oz. of fine gold. The finely divided and scrap gold in the slime fell from 17·19 per cent. of the fine gold produced to 11·37 per cent. The "cost of operating the refinery" fell from 3·2 cents per oz. of fine gold to 2·0 cents, and the capital cost of the change was paid for in six weeks. The average fineness of the refined gold under each system was 999·5. The average monthly output of the refinery was increased from 146,000 ozs. to 224,000 ozs., and the time of treatment reduced.

The following description of the process as practised at the *Ottawa Mint* in 1913 is given by A. L. Entwistle,[2] who is in charge of the operations. The pulsating current is not used at this refinery :—

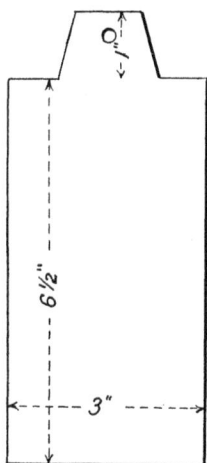

Fig. 202.—Anode Plate.

"Gold deposits of 900 fine and over and the dark brown residue containing 94 per cent. gold from the silver cells are melted in oil furnaces, and cast into open iron moulds of the following inside dimensions :— 6½ inches long, 3 inches wide and ½ inch thick at the top, tapering to ¼ inch at the bottom. At the top of the mould there is a pin which leaves a hole in the plate produced. The anode is ¼ inch thicker at top than bottom, because the metal dissolves quicker at the top than the bottom, and so the whole anode is finished at the same time ; the anodes weigh about 90 to 100 ozs. each. Fig. 202 is a sketch of an anode which is so shaped that only a small portion is left suspended on the hook when it is dissolved. The electrolyte contains about 50 grammes per litre of gold and about 5 per cent. free hydrochloric acid, and is contained in a white porcelain tank 16½ inches long, 11⅝ inches wide and 11 inches deep. The cathodes consist of strips of fine gold 2½ inches wide, 10 inches long and $\frac{15}{1000}$ inch thick, annealed to within 2 inches of the top, and are bent over to fit the rod from which they are suspended.

[1] *Report of the Director of U.S. Mint*, 1913, p. 41.
[2] Private communication, made with the permission of Dr. James Bonar, Deputy Master of the Ottawa Mint.

" The supporting rods are made of $\frac{3}{8}$-inch steel, and are covered with hard rubber tubes, the ends of which are plugged with hard rubber. The current is carried by strips of gold or of fine silver $\frac{1}{16}$ inch thick, which run along the top of the rubber-covered rods. The cells are connected up in series. The anodes are suspended by a gold or silver hook (see below, p. 488) slipped through the hole in the end and immersed, almost up to the hook, in the electrolyte. The cells are kept at 60° by a sand bath placed on an electric heater. The cells are placed in a lead-lined wooden tank, which is enclosed with glass sliding panels and covered with a hood, which has an outlet into the air through the roof, so that the acid fumes do not get into the room.

" The current is divided by splitting the leads from the dynamo and

Fig. 203.—Gold Cells, Ottawa Mint.

connecting to the copper rod, at each end of the bank of cells, in two places, so that the distribution through the cells is as even as possible. In the centre of the cell a hard rubber propeller or stirrer revolves at about 350 revolutions a minute, so that it draws the liquid up. The current is divided between the 11 anodes, which are arranged in four rows, with a space in the middle to allow room for the stirrer. There are also 11 cathodes. Fig. 203 is from a photograph showing four cells set up in series in the above manner. The leads from the dynamo are connected to the two copper rods above the cells, one positive, the other negative, running the whole length of the cells. The cells are shown uncovered, but they are now boxed in. There are 18 cells set up in series, but 6 cells are generally out of use for cleaning purposes.

The cleaning of a set of 6 cells is begun about twice a month. These 6 are cut out of the circuit and the other 6 put in, so that 12 cells are constantly running. The 10 silver cells (see above, p. 471) supply just sufficient gold to keep the 12 gold cells running.

" The current used is 150 ampères at 12 volts for the 12 cells, and the current density is about 100 ampères per square foot. (Reckoning both sides of the anode, it appears to work out at about 50 ampères per square foot.) An anode weighing about 70 ozs. is dissolved in about 42 hours (nearly a week's work), when the residue or top is taken out and replaced by a new anode. When new cathodes are put in the cells the current is lowered to 50 to 100 ampères. If the current is kept up to 150 ampères, the first deposit is soft and mushy, and does not adhere to the cathode, but after two or three hours at the lower current it is quite hard, and then the current is put up to 150 ampères again. If the deposit subsequently becomes mushy and soft, the cell requires the addition of more gold chloride, and if chlorine is evolved at the anodes more hydrochloric acid is required. About a pint of gold chloride solution, containing 20 per cent. gold and 10 per cent. hydrochloric acid, is added to each cell every morning. The evaporation, due to heating, just allows a pint of chloride solution to be added to keep the electrolyte up to within 1 inch of the top of the cell.

" The cathodes are taken out when they are about 60 to 80 ozs. in weight, placed in a porcelain filter, washed free from gold chloride, etc., allowed to drain over night, and melted in plumbago crucibles and cast into iron moulds, the fineness being 999·3 to 999·5.

" The gold in the foul electrolyte and from washing slime, etc., is precipitated with sulphate of iron, allowed to settle, and the liquid siphoned off and run into an earthenware tank containing iron plates, which brings down the platinum and copper. After settling, the liquid is again siphoned off and run into a big wooden tank containing more iron plates, and after settling several days is run from this tank into the drain. The residue in these two tanks is cleaned out at the clean-up, run down with carbonate of soda and borax, and the resultant metal, which consists chiefly of lead, copper, gold and platinum, is assayed and remelted with lead and silver, and cupelled in a cupellation furnace capable of holding 10 lbs. in each charge. The resultant gold, silver and platinum in the button left on the cupel are parted in sulphuric acid. From 100,000 ozs. of fine gold produced, the platinum recovered was only 10 ozs.

" The slimes from the cells are washed free from gold chloride with hot water and then placed in an earthenware tank, with iron plates at about a distance of 3 inches apart and acidified with hydrochloric acid. All the silver chloride is reduced to grey metallic silver in about 72 hours. The iron plates are removed, the grey silver treated with hydrochloric acid and then washed free from iron with water, dried and melted, and cast into 1,000-oz. moulds. These ingots assay about 60 to 70 per cent. gold, and are alloyed with silver, and go through the silver-parting process again.

" The 12 gold cells produce about 4,000 ozs. per week of 44 hours, but about 600 ozs. of this is cast into bars and rolled out to make new cathodes.

" The gold chloride for the gold cells is made by the electro-chemical process, using two cells, exactly similar to those described above, filled with concentrated hydrochloric acid. The anodes (960 fine) and cathodes are of exactly the same dimensions as in the gold cells. The cathodes are suspended in porous jars, 10 inches deep, 12 inches long, and $1\frac{7}{8}$ inches wide, the jars being filled with hydrochloric acid. There are two jars in each cell, and each contains four cathodes. The gold dissolves at the anode, and is prevented from being deposited on the cathode by the porous jar. The cells are connected up in series, the current used for the two cells being 60 ampères at 6 volts, generated by a 1 H.P. motor-generator. After 48 hours the solution contains about 20 per cent. gold and 10 per cent. free acid. These two cells easily supply sufficient gold chloride for the 12 gold cells."

R. Pearson, the Assayer at the Ottawa Mint, in experimenting on the electrolytic production of gold chloride from impure anodes,[1] used bullion assaying gold 922, silver 66, and base metals, chiefly lead, 12 parts per 1,000. In one experiment with an electrolyte containing 28 per cent. of crude hydrochloric acid, he passed through a cell an average current of 83·4 ampères at 2·41 volts, or 17·0 ampères per square foot of anode immersed. The results were as follows :—

Time of Treatment.						Gold in Solution per Litre.
7 hours,	.	.	.	.	.	68·46 grammes.
15 ,,	.	.	.	.	.	139·48 ,,
23 ,,	.	.	.	.	.	203·00 ,,
30 ,,	.	.	.	.	.	263·89 ,,
34 ,,	.	.	.	.	.	305·46 ,,

Chlorine then began to be evolved, and the experiment was stopped. The cost per oz. of gold dissolved was 0·95 cent for acid and power, and 2·19 grammes of gold per ampère-hour were dissolved. Other experiments had similar results.

Pearson also found little difference in results between hot and cold cells with the direct current in the ordinary refining process. He obtained good results in each case with a current density of 14·6 ampères per square foot of anode (157 ampères per square metre), but inferior results with a density of 21 ampères per square foot. In the latter case the cathode gold from the hot cell contained traces of lead and platinum. The anodes contained from 85·6 to 92·5 per cent. of gold. The current densities used were rather low.

The latest and best equipped electrolytic gold refinery is that at the *New York Assay Office*, which was installed in 1912.[2] The silver process run in conjunction at this refinery has already been described on p. 469. The pulsating current is produced from a combination machine, consisting of a driving motor, a direct-current dynamo, a single-phase alternating-current dynamo and a 110-volt exciter. The latter excites the fields of both the alternator and the direct-current machine, but with a hand rheostat in series with each, thus allowing a relative adjustment of voltage. The fields of the exciter itself are excited by an outside source through an adjustable rheostat. This arrangement gives the necessary large range of adjustments to suit every condition that might arise. The claims made for the pulsating current have been in great part substantiated in practice. Anodes containing gold 910, silver 70, base metals 20 are treated without heating the electrolyte with an effective current density of 70 ampères per square

---

Forty-fourth Annual Report of the Mint, 1913, p. 181, and private communication to the author.
[2] B. P. Wirth, *Report of the Director of U.S. Mint*, 1912, p. 50.

foot, or 750 ampères per square metre, and the amount of finely divided gold in the slime is small.

There are four sets of 12 cells each, placed on two soapstone-topped tables. Each cell is 19 inches by 15 inches and 12 inches deep, and is made of royal Berlin porcelain. The anodes are shaped as at Ottawa and are 8 inches long, 3 inches wide and $\frac{1}{2}$ inch thick, weighing about 75 ozs.

The electrical connections are similar to those at Ottawa, but the anodes are hung from the silver-covered rods by gold hooks, which do not dip into the electrolyte. The silver strips replace gold strips which are used at the older refineries. The relative advantages are not stated. At the Ottawa Mint (see above, p. 485), according to A. L. Entwistle,[1] silver strips and hooks were substituted experimentally for gold. They were as efficient as gold at the beginning, but after being in use for a short time they became covered with a thin coating of silver chloride and the resistance rose. With silver strips and hooks a cell required 1·5 volts at 150 ampères, as against 1 volt at 150 ampères with gold strips and hooks. The cost of the extra power is probably less than the interest on the gold, and the formation of silver chloride on the silver hooks can be prevented for a time by gold-plating them, but cleaning the silver strips would be too expensive. The weight of gold in strips and hooks for one cell is about 26 ozs.

At New York the electrolyte consists of a 10 per cent. solution of free hydrochloric acid containing 30 grammes of gold as chloride per litre. The liquid is circulated as at Ottawa by means of hard rubber propellers driven by a motor, and the cascade system previously introduced at San Francisco by R. L. Whitehead[2] is not in use. The trough or cascade system has the advantage that the electrolyte remains uniform in composition, and that gold chloride can be added when required in one operation for all the tanks, and similarly foul solution can be drawn off conveniently. Also the necessary number of assays of the solution is much less when the cascade method is used. For a description of the trough system, see above, p. 469. It is to be introduced at New York.

At San Francisco the anodes are $\frac{3}{4}$ inch apart, and require 1 volt per cell to maintain a current of 450 ampères. This gives a current density of 105 ampères per square foot of cathode surface, and an output of gold per 24 hours per tank of 650 to 675 ozs. The best run for one week with 20 tanks in circuit was 13,500 ozs. per day.[3]

At New York the platinum is precipitated from the foul solution by sal-ammoniac as platinum-ammonium chloride, which is carefully washed, and on heating yields spongy platinum. Palladium is also recovered by a chemical process.

The anode slime, which consists chiefly of silver chloride and finely divided gold, is washed thoroughly, the chloride reduced to metallic silver by zinc, and the whole melted and cast into anodes for the silver cells.

Relative Advantages of the Electrolytic and Sulphuric Acid Methods of Refining Gold.—The electrolytic method avoids acid fumes and yields a pure product always free from elements which cause brittleness. The refined gold is invariably tough, and is well fitted for use in coinage or for the manufacture of wares. Gold from the sulphuric acid process sometimes contains

[1] Private communication, dated Jan. 7, 1914.
[2] Whitehead, *Electrochem. and Met. Ind.*, Sept. 1908, p. 356.
[3] E. B. Durham, *Trans. Amer. Inst. Mng. Eng.*, 1911, **42**, 874.

lead and probably tellurium, and less reliance can be placed on its quality and fitness for use in coinage. The cost of supplies, labour and power is two or three cents per oz. of gold for the electrolytic process in the United States mints, as against six or eight cents for the sulphuric acid process in the same establishments. Moreover, the value of the platinum and palladium recovered in the electrolytic process is considerable (say 0·5 cent per oz. of gold), whilst in the sulphuric acid process these metals are not recovered. On the other hand, in the sulphuric acid process, the gold is quickly brought to account, and the difficulties of periodical stock-taking at short intervals of time are much less than in the electrolytic process. The chief defect of the latter is the greater length of the time of treatment, so that if the loss of interest on the gold is taken into account, a grave disadvantage is disclosed. Taking the total time of treatment from the importation of the gold into the refinery to its exportation at a fortnight, and the gold locked up in the solutions, cathodes, connections, anode mud, etc., as 50 per cent. of the gold undergoing treatment, the interest charge at 5 per cent. would be about 6 cents per oz. of fine gold. It is obvious from such figures that if the gold undergoing treatment is losing interest, the work should be carried on for 24 hours a day with thin anodes and as high a current density as possible.

11. **Aqua Regia Process.**[1]—This process was introduced at the Pretoria Mint after the Miller process had been tried and abandoned owing to the alleged difficulty of treating the gold bullion extracted by the cyanide process. In the aqua regia process the gold is dissolved and precipitated. It is made very difficult if the silver exceeds 100 parts per 1,000, and at Pretoria bullion was not treated if the silver exceeded 50 parts per 1,000. Mill gold with 80 to 110 parts of silver, and cyanide gold if it contained more than 50 parts, were melted with gold obtained by the chlorination process to reduce the silver to less than 50 parts, and granulated. Charges of 500 grammes of the granulations were then placed in each of 40 boiling flasks of 3,250 c.c. capacity, and treated with a mixture of 6 parts HCl to 1 part HNO_3. Some gold was always left undissolved to avoid loss of acid. The flasks were heated on sand baths, and at the end of the day the solution of gold was poured into porcelain vessels holding 100 litres each.

Next morning the clear liquid was siphoned off, leaving the silver chloride at the bottom of the vessel, and transferred to a tank containing a solution of ferrous chloride. The gold thus precipitated was separated and washed, and the ferrous chloride regenerated by the addition of iron. The gold could not be toughened on melting, even by repeated additions (up to 40 times) of a few grammes of $CuCl_2$, and it was necessary to pass chlorine through it. The gold thus obtained was from 996 to 999 fine. About 300 kilogrammes of gold per month were thus refined at a cost, including subsequent coinage, of 1s. 10½d. per standard oz., 916·9 fine. The loss on the transaction was 5½d. per oz., as the Mint bought gold at £3 16s. 9½d. per standard oz., with a deduction of 4d. for refining. The floor of the refinery was slated and drained to carefully constructed gutters, and no liquids were thrown away until they were declared free from gold. The method of testing was to take 20 litres, add some acetate of lead, pass sulphuretted hydrogen, allow to settle, decant, roast the precipitate, and cupel the lead. In spite of all precautions, however, the loss of gold was heavy. This cumbrous and inconvenient process was brought to an end in 1899 by the South African War.

---

[1] Begreer, *The Metallurgy of Gold on the Rand* (Freiberg, 1898), p. 123.

CHAPTER XIX.

THE ASSAY OF GOLD ORES.

Introduction.—The assay of gold ores was probably practised by the Romans,[1] and was elaborated in the Middle Ages. The German assayers had already reached a fair degree of proficiency in the year 1500. Since then the progress in the art has been comparatively moderate.

The assay of gold ores is almost universally conducted in the *dry* way —*i.e.*, by furnace methods. Exceptions will be noted later. The plan of operation is to concentrate the precious metal in a button of lead either (1) By fusion in a crucible ; or, more rarely, (2) By scorification. The button of lead obtained by either method is then subjected to *cupellation*, by which the lead is oxidised and removed, and the resulting bead of precious metal is weighed. In these operations silver and the metals of the platinum group remain with the gold, and are subsequently separated by *inquartation* and *parting*, and, in the case of platinum and its allies, by further special methods.

The exact method to be used in any particular case varies with the composition of the ore. As a general rule, gold ores are better assayed by the fusion process, so that a comparatively large quantity of material may be operated on. Very rich ores may be assayed by scorification, the errors arising from the small amount of material used in the process being less important in their case.

The assay by scorification was preferred to pot-fusion by the German assayers of the 16th century. When fusion in a crucible was recommended, it was only as a preliminary to scorification.[2] Agricola and Ercker describe many of the methods and precautions given in this chapter, such, for instance, as the treatment of cupriferous ores with nitric acid, the use of a salt cover in the crucible, and of " proof-Centners," which correspond to assay-tons. Ercker also mentions the forerunner of the buck-board, and instances assay-offices in which 200 assays of ores were made in a week.

Sampling the Ore and Preparation of the Sample for Assay.—The value of an assay depends largely on the care with which a sample of the ore is selected. Sampling should be, as far as possible, automatic, and independent of the will or judgment of the assayer. The sampling of ore in place in mines is described in books on mining.[3] The taking of running samples is referred to in the section on stamp mill practice (see p. 176).

---

[1] See Hoover's translation of Agricola, 1912, p. 219, note 1.

[2] Agricola, *De Re Metallica*, Basle, 1556, lib. vii. For an account of the Probier-büchlein, published in Germany between 1510 and 1556, see Hoover's *Agricola*, appendix B, p. 612 ; they are the earliest treatises on assaying known to exist. See also Ercker, *Allerfurnemisten Mineralischen Eerzt u. Bergwerks Arten*, Frankfort, 1580, Book i., chap. 10 ; and Book ii., chap. 8. *Laws of Nature in Assaying Metals*, by Sir John Pettus, London, 1686, is a translation of Ercker's work.

[3] *The Sampling and Estimation of Ore in a Mine*, by T. A. Rickard, New York, 1904. See also remarks by Cordner-James, *Trans. Inst. Mng. and Met.*, 1902, **10,** 152 ; and M. H. Burnham, *Trans. Inst. Mng. and Met.*, 1901, **10,** 204 ; also Charleton, *ibid.*, 1901, **9.** 203-231, containing a bibliography on p. 225.

Heaps of ore, placer deposits and impounded tailing are sometimes sampled in the same way as vat or bin charges by driving-in iron pipes[1] or cheese-taster samplers at regular intervals, and mixing the samples withdrawn in the pipes. A form of sampling iron suitable for very dry material is described by Richards.[2] It consists of two iron tubes, each with nearly half its circumference cut away. One tube fits inside the other, and is fixed to a wooden handle by a **T**-piece, so that it can be rotated inside the outer tube. The lower part of the outer tube is not cut away, and is sharpened at the end or drawn out to a point. The tubes are driven into the ore with their openings arranged as shown in the cross-section, a (Fig. 204), and the inner tube is then rotated, passing through the position b (Fig. 204), until it is filled with ore. It is then replaced in its original position, and both tubes withdrawn. The outer tube must also be clamped to a cross-handle, to facilitate its withdrawal.

When the ore to be sampled can be moved, it is reduced in bulk either by hand or by automatic machines, and is generally crushed finer between

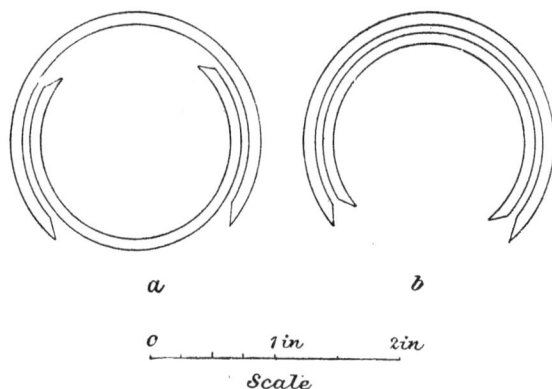

Fig. 204.—Sampling Tube for Dry Ore.

each successive reduction of bulk in accordance with requirements based on theoretical considerations.[3]

If the ore is in course of removal by shovelling, every second, fifth, or tenth shovelful may be set aside as a sample, or the whole heap may be piled up to form a perfect cone, each shovelful being thrown as nearly as possible on the exact apex of the cone, so that the ore runs down on all sides. The heap may be made into a new cone to ensure thorough mixing, or may be at once flattened into a circular cake, and divided into four quarters, along two diameters at right angles to each other. Two opposite quarters are removed and mixed, and the operation repeated if the ore is fine enough. The ore is sometimes made into the form of an annular ring before being coned.[4] "Coning and quartering" is usually preferred to the method of setting aside particular shovelfuls (see, however, Argall, op. cit. infra).

---

[1] W. W. Taylor, Eng. and Mng. J., 1897, **63**, 160; also H. E. Nicholls, Trans. Inst. Mng. and Met., 1905, **14**, 195.
[2] Richards, Ore Dressing, vol. ii. (1903), p. 845.
[3] Smith, The Sampling and Assay of the Precious Metals, p. 90.
[4] Smith, op. cit., p. 98; E. A. Wraight, Assaying, p. 67.

Automatic samplers give better results than hand sampling, and their working costs are smaller. They are divisible into two classes—(1) those which continuously take a part of a moving stream of ore, and (2) those which momentarily take the whole stream of ore at regular intervals of time. The former class of machine is the cheaper, but is considered by Richards[1] and Argall[2] to be useless, "because the values are never evenly distributed across the stream." In many machines a narrow scoop passes steadily across the stream of ore at regular intervals of time, and this is considered to be equivalent to the method used in the machines of class (2). The Bridgman,[3] M'Dermott, Vezin,[4] Collom,[5] Brunton,[6] Snyder,[5] Johnson,[7] and Foster-Coolidge[7] samplers are all well known. Considerations in sampling and a description of the Brunton sampler, which gives highly satisfactory results, are given by D. W. Brunton.[8]

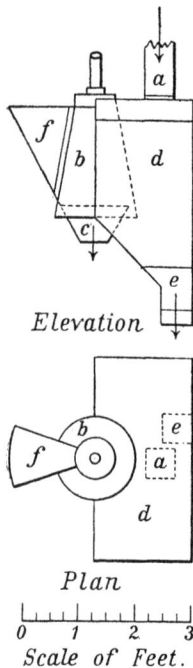

Elevation

Plan

Scale of Feet.

Fig. 205.—Vezin Sampler.

The *Vezin sampler* is one of the few machines that give accurate results, and is stated by Argall[2] to be almost universally employed in Colorado. It consists of two sheet-steel truncated cones, *b*, *c* (Fig. 205), with their bases bolted together. The ore falling from a shoot, *a*, into a hopper, *d*, is carried outside the cone, *c*, and is delivered from the hopper through another shoot, *e*. A scoop, *f*, made of sheet-steel, with a sector-shaped opening above, is rivetted or hooked on to the cone, *b*. The angle of the sector may subtend any desired portion of the circumference of a circle, such as one-tenth (36°), or one-sixteenth (22$\frac{1}{2}$°). Both cones and the scoop are rotated at about 25 or 30 revolutions per minute by bevelled gearing, and when the scoop comes below the shoot, *a*, the whole stream of ore falls into the scoop, and is led into the interior of the cone, *c*, and thence to a separate truck or bucket. In this way a sample is taken about every two seconds. A Vezin sampler of about 3 feet in diameter, and requiring a fall of about 6 feet, is trifling in cost, and treats 30 or 40 tons an hour.

Argall recommends[9] the following course of procedure in sampling telluride ores containing 10 or 15 ozs. per ton :—From 100 short tons crushed to 1 inch cubes, take 20 tons. Crush this to $\frac{1}{4}$ inch, and take 2 tons. Crush this to 8 mesh (= $\frac{1}{16}$ inch), and reduce by "riffling" to 250 lbs. Dry and crush the sample to 30 mesh (= $\frac{1}{80}$ inch), and riffle down to 15 lbs. Crush by a sample-grinder to 90-100 mesh, and riffle down to 1 lb. This is crushed on the buck-board to pass 120 mesh (= $\frac{1}{250}$ inch), and divided for assay.

[1] Richards, *op. cit.*, p. 846.
[2] Argall, *Trans. Inst. Mng. and Met.*, 1902, **10**, 235.
[3] Bridgman, *Trans. Amer. Inst. Mng. Eng.*, 1891, **20**, 416.
[4] Argall, *Trans., Inst. Mng. and Met.*, 1902, **10**, 240, 241, with working drawings.
[5] Richards, *Ore Dressing*, pp. 846-848.
[6] Brunton, *Trans. Amer. Inst. Mng. Eng.*, 1884, **13**, 639 ; 1909, **40**, 580 ; Hofman, *Metallurgy of Lead*, 1893, p. 55.
[7] Hofman, *Mineral Industry*, 1902, **11**, 429.
[8] Brunton, *Trans. Amer. Inst. Mng. Eng.*, 1909, **40**, 567-596.
[9] Argall, *op. cit.*, p. 238.

Poorer and less "spotty" ores may be sampled somewhat more simply, and without such fine preliminary crushing.

In the assay office the sample, however obtained, is further reduced in bulk by an automatic machine, similar in principle to the large samplers, such, for instance, as the Bridgman laboratory sampler,[1] or by the implement known as the *riffle* or *sampling tin* (Fig. 206).[2] This consists of a series of troughs arranged side by side and fastened at equal distances from each other (the width of the spaces being equal to that of the troughs) by strips of metal soldered on to their ends. An even stream of ore being let fall from a shovel on to this sampler, half is retained in the troughs, while half passes through. Careful experiments have proved that each half is representative of the whole. Repetition of the process reduces the sample to any required extent. A sampler with troughs 1 inch wide is suitable for treating materials which include lumps of not more than $\frac{1}{4}$ to $\frac{1}{3}$ inch in diameter. For finely ground materials a convenient width for the troughs is $\frac{3}{8}$ inch.

The *split shovel* (see Fig. 207)[3] is similar in principle to the riffle, and in the *Jones sampler*,[4] instead of troughs and spaces, two sets of troughs are substituted sloping at a high angle in opposite directions. The troughs clear themselves, discharging half the ore at each end of the machine.

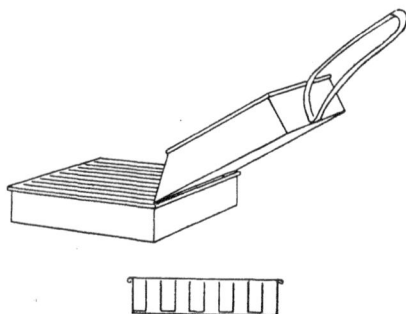

Fig. 206.—Sampling Tin.

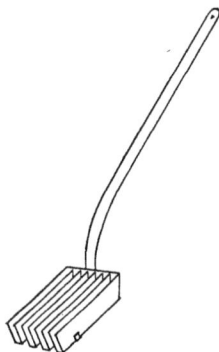

Fig. 207.—Split Shovel for Sampling.

When the sample has been reduced to from 5 to 20 lbs. in weight it is crushed through a 20-mesh sieve, and from 1 to 5 lbs. selected for the assay sample.[5] This sample is used for the estimation of moisture, and is then dried and finely crushed. Opinions differ as to the fineness of the sieve to be used. No one is in favour of a sieve coarser than 80 mesh, and 100 mesh is more common, but some assayers use a 200-mesh sieve. The author's experience is that all ores require crushing to 100 mesh, and that finer crushing may be dispensed with in the case of simple ores; but that complex ores, such as tellurides, should be crushed as finely as possible.[6]

---

[1] Hofman, *Metallurgy of Lead* (1899), p. 59.
[2] E. A. Smith, *The Sampling and Assay of the Precious Metals*, p. 122.
[3] Smith, *op. cit.*, p. 99.
[4] Described by P. Argall, *Trans. Inst. Min. and Met.*, 1902, **10**, 271 ; *Eng. and Mng. J.*, 1903, **76**, 729.
[5] The views of various authorities as to weights of samples and fineness of preliminary crushings in the sampling of large lots of ore are given in Richards, *Ore Dressing* (1st ed., 1903), p. 848.
[6] For further details as to sampling and the preparation of the sample for assay, see Smith, *op. cit.*, pp. 87-126.

The implements employed for crushing samples of ore are various. A small *rock-breaker*, with reciprocating jaws, similar to those used on the large scale, and worked by hand or steam power, is useful for breaking down large lumps, which otherwise may be broken by a hammer. For finer pulverisation a small pair of steel-faced high-speed *rolls* may be used if steam power is available. The *sample-grinders*, resembling coffee mills, and the *disc-grinders*, sold by makers of mining machinery, are more widely used. Such machines are adopted at large smelting or sampling works, where great numbers of samples are crushed daily. In smaller works or offices the *buck-board* (Fig. 208) is most suitable. It is a smooth plate of iron about 2 feet square with a 1-inch rim surrounding it on two or three sides. On this a *bucking hammer* is worked—a heavy piece of iron 15 to 20 lbs. in weight, with a large smooth curved face and a handle 30 inches long. It is moved about on the iron plate (on which the ore is spread) with both hands, one holding the handle, the other pressing the head downwards, the curved face being below, while an oscillatory movement is imparted by the handle. The instrument is very effective if the ore has previously been broken down to the size of coarse sand in a mortar. The *radial bucking-plate*[1] has been devised as a substitute for the buck-board. The *pestle and mortar* are of value in breaking down samples from the size of nuts to that of coarse sand. In grinding down siliceous material, so as to enable it to pass through a fine sieve, the pestle and mortar are far inferior to the buck-board.

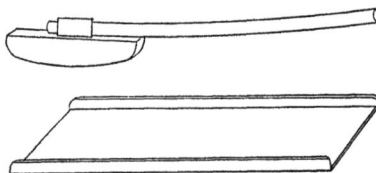

Fig. 208.—Buck-board and Hammer.

The prepared sample may be mixed and divided by the Bridgman divider,[2] or by a sampling tin (see p. 493), and is stored in tin boxes or glass jars, which should be labelled by numbers, none of which are ever repeated. Before weighing out the powdered ore for assay, the whole sample should be turned out into a wide bowl or on to glazed paper, or, better still, rubber cloth (which does not crack and wear out like paper), and thoroughly mixed with a spatula. The sample should never be mixed by shaking, and care should be taken to avoid jarring it after mixing, as metallics in that case tend to settle to the bottom from their superior density, and a fair sample cannot easily be obtained. For the same reason the part of the mxied sample which is taken for assay should not be hastily shovelled on to the balance pan from the top of the pile, but a vertical slice should be taken, some of the lowest layer being carefully scraped up from the rubber cloth.

" *Metallics.*"—In many ores, both gold and auriferous silver occur native in grains or threads. These " metallics " are not readily reducible to a fine state of division, and, though a part always passes through the sieve, some of the larger pieces which have resisted abrasion fail to do so. In some assay offices part of the pulverised ore is thrown back into the mortar with

[1] Smith, *The Sampling and Assay of the Precious Metals*, p. 116.
[2] Hofman, *Metallurgy of Lead*, 1899, p. 72.

the metallics, and grinding is continued until everything passes the sieve. This is a dangerous practice, as it is impossible to ensure the even distribution of metallics through the sample. The smaller pieces which pass through the sieve in the first instance constitute an unavoidable evil which is increased by every piece of metal that follows them. The safer plan is to cupel by themselves the whole of the metallics left on the sieve, and calculate their value per ton of ore independently of the result obtained by the ordinary assay. The total value of the ore is found by adding these two results together.

Assay by means of the Blowpipe.[1]—This method, though less exact than the furnace methods, is of importance, because in prospecting expeditions it is possible by its means not only to detect the gold and silver in any ore, but also to determine its amount quantitatively with fair accuracy. On such expeditions it is difficult to carry the cumbrous apparatus required to make an ordinary assay. The amount of powdered ore taken is usually 100 milligrammes, and this is mixed with borax and about 1 gramme of granulated lead. The whole is wrapped in paper and heated on charcoal in the reducing flame of a blowpipe until the fusion is complete, and then for a short time with the oxidising flame. The lead is then separated from the slag and heated on a bone-ash cupel until it is all converted into litharge. The nearly spherical bead of silver and gold thus obtained is weighed or its diameter carefully measured on an ivory scale, which at once gives the percentage amount in the ore. The gold is usually separated from the silver by parting in nitric acid, and is weighed or melted and measured, but Richards states [2] that the silver can be distilled off by the blowpipe, leaving a bead of pure gold, which can be measured (see also " Assays in Field Work," p. 515).

FUSION OR CRUCIBLE PROCESS OF ASSAY.

This process is divided into four parts, viz. :—(1) Fusion ; (2) Cupellation ; (3) Parting ; (4) Weighing the gold.

(1) **Fusion.**—The object of this operation is to concentrate the precious metals in a button of lead, while the whole of the remainder of the ore forms a fusible slag with suitable fluxes, in which lead sinks. The fusion is made in clay (or rarely iron) crucibles in a wind furnace, similar to, but usually smaller than, that used for melting bullion described on p. 423. Coke is generally used for fuel. Gas and petroleum furnaces are also in wide use. Long-flaming bituminous coal is also used, generally in furnaces of the reverberatory type, especially in the United States and in the Transvaal, where it is displacing coke as fuel.[3]

In Colorado it is usual to make fusions in a muffle furnace, similar to that described below under cupellation, p. 529. The temperature required is about the same as that used in scorification. The advantages claimed are greater cleanliness and neatness and more uniformity in the conditions, the temperature of a muffle being more easily kept constant and uniform than that of an ordinary fusion furnace. Six or eight fusions can be made at one time in a large muffle.

In weighing the materials for a crucible charge the use of a set of *assay-ton* weights saves much labour in calculation. The assay-ton is a weight

---

[1] For a full description of this method, see Plattner's *Blowpipe Analysis*, translated by Prof. Cornwall, 8th ed., New York, 1902, pp. 350-384.
[2] Richards, *J. Franklin Institute*, June, 1896.
[3] A. M'A. Johnston, *Rand Metallurgical Practice*, vol. i., p. 297.

which contains as many milligrammes as a ton contains ounces. Thus an English or long ton of 2,240 lbs. (used in England and Australia) contains 32,666 Troy ounces, so that the corresponding assay-ton must weigh 32,666 mg. or 32·666 grammes. If the weight of the resulting bead of gold (or silver) from an assay-ton of ore is 1·5 milligrammes, then the ore contains 1·5 ozs. of gold per statute ton. If the value per short ton of 2,000 lbs. (used in North America and Africa) is required, the weight of the assay-ton is 29·166 grammes, since there are 29,166 Troy ounces in 2,000 lbs. avoirdupois. If grain weights are preferred, the weight of the assay-ton may conveniently be taken as 326·66 grains (or 291·66 grains for the short ton) ; the weight of the resulting bead of gold in hundredths of a grain then gives the value of the ore in ounces per ton. Sets of weights ranging from $\frac{1}{5}$ A.T. (assay-ton) to 4 A.T. can be bought, or they can be made up from an ordinary box of decimal weights. In the following pages this system is used, one A.T. being equal to 32·666 grammes.

If grammes or grains are used as units, tables are necessary to convert the results obtained by the balance into ounces per ton. The amount of the unit used in weighing should not be lost sight of in the course of conversion. Thus if 50 grammes of ore are taken, and the resulting gold weighs 1·75 milligrammes, the unit of weight being 0·05 milligramme, the balance error corresponds to nearly 8 grains per ton, and the report should be 1 oz. 3 dwts. per ton, not 1 oz. 2 dwts. 20·8 grains as would probably be given in a table.

The weight of ore taken for assay varies according to circumstances. One A.T. is a suitable quantity if the amount of gold is not less than 0·5 oz. per ton, and the balance for the final weighings is sensitive to 0·02 milligramme or less. With poor residues, 2, 4, or even 12 A.T. may be taken, and with very rich ores $\frac{1}{2}$ A.T. may suffice. If more than 3 A.T. of ore are taken, the charge is divided between two or more pots, and the resulting lead buttons scorified down and cupelled together.

The charge is made up of litharge, charcoal, and suitable fluxes, together with metallic iron. *Litharge* or red lead [1] is added in the proportion usually of one and a-half or two parts to two of ore. The amount of lead reduced should be from 25 to 30 grammes per A.T. of ore.

The amount of the *charcoal* added varies with the reducing power (percentage of ash, etc.) of the particular sample which is employed, as well as with the degree of oxidation of the ore. If too little charcoal is used, the amount of reduced lead will be too small, and if too much charcoal is added, the charge remains pasty and emits bubbles of combustible gas (CO). Generally from 1 to $1\frac{1}{2}$ grammes per A.T. of ore are enough, but in some highly basic oxidised or roasted pyritic ores as much as 3 grammes of charcoal powder are required for 1 A.T. of ore, as the oxides must always be completely reduced to the lowest state of oxidation (*e.g.*, FeO) compatible with keeping them in the slag. This point is attained when nearly all the litharge (about 90 per cent.) is reduced to lead. Otherwise some of the gold is oxidised

---

[1] It is sometimes asserted that litharge (PbO) is less suitable than red lead (Pb_3O_4). The latter requires more charcoal to reduce it, and, according to Ricketts and Miller of Columbia University, it is objectionable on the grounds that it oxidises silver. (*Notes on Assaying*, by Ricketts and Miller, p. 39, New York, 1897.) This drawback, however, appears to be shared by litharge. In some comparative experiments made by the author there was no difference in results, so that it would seem that either litharge or red lead can be used with equal advantage. The amounts of silver and gold, if any, in the litharge must be determined by a separate experiment. Red lead contains more oxygen than litharge.

and retained in the slag. If ores contain much sulphur no charcoal is used, and nitre may even be added to burn off the excess of sulphur, but in that case the pot is liable to boil over. With very large quantities of sulphides it is better to increase the quantity of soda, or, if the formation of a matte cannot be prevented in this way, to treat the matte separately, or to roast the ore before fusion. Flour and argol (acid potassium tartrate) may be used as reducing agents instead of charcoal. One gramme of charcoal is capable of reducing 25 to 30 grammes of lead from litharge, and its place may be taken by $2\frac{1}{2}$ grammes of flour or about 4 grammes of argol. These equivalents, however, are approximate, and vary with the sample of the reducing agent. The reducing power of each sample is tested beforehand.

Sodium carbonate is used to flux silica and to take up sulphur, while *borax* is valuable in basic ores to prevent corrosion of the crucible, and render the slag more liquid. The relative amounts required are judged from the appearance of the ore in the first place, and afterwards modified according to the success of the fusion. From 1 to $1\frac{1}{2}$ A.T. of sodium carbonate, and from $\frac{1}{4}$ to $\frac{1}{2}$ A.T. of borax to 1 A.T. of ore are the amounts usually required. Even when the ore is entirely composed of silica, some borax is added. The most convenient form is borax glass. Bicarbonate of soda is less convenient than the normal carbonate. If it is used the water and excess carbon dioxide must be driven off by slow and cautious heating.[1]

Silica is used only for ores full of lime, baryta, compounds of the base metals, etc., or generally whenever the ore does not contain much quartz. It aids fusion in these cases, and protects the crucible from corrosion. From $\frac{1}{4}$ to $\frac{1}{2}$ A.T. of silica to 1 A.T. of ore is generally enough. About two parts of glass may be used instead of one part of silica. *Fluorspar* is added to the charge when the ore contains sulphates of barium or calcium, and in the fusion of cupels. Like borax, it increases the fluidity of almost any charge, but it attacks the crucible, and care must be taken to avoid deficiency of silica when it is used. In general it may be noted that for basic impurities an acid flux is used and for an acid gangue a basic flux. A cover of *common salt* is generally used to protect the charge from the variable oxidising or reducing action of the atmosphere in the furnace. Iron is added to all charges in the shape of large nails or hoop-iron, or even scrap. Sulphur, arsenic, etc., are thus kept out of the lead button. The charge for any given ore is made up by the assayer according to his judgment and experience. Much may be learnt as to the nature of the ore by the examination, with a lens, of unbroken lumps, and after panning, of pulverised ore. The following are some typical charges :—

TABLE XLII.

	Quartzose Ore.	Oxidised Ore (with 25 per cent. iron oxide).	Pyritic Ore (with 25 per cent. iron sulphide).
Ore, ...	1 A.T.	1 A.T.	1 A.T.
Litharge, . . .	1 ,,	1 ,,	1 ,,
Charcoal, . . .	1·2–1·5 grms.	2·0 grms.	0·5–1·0 grm.
Sodium carbonate, .	$1\frac{1}{2}$ A.T.	1 A.T.	1 A.T.
Borax, . . .	5–10 grms.	10 grms.	10 grms.

With salt covers and hoop iron.

[1] J. Bettel, *J. Chem. Met. and Mng. Soc. of S. Africa*, 1899, **2**, 467.

Methods of Operation.—The ore is weighed accurately to 0·01 gramme unless very poor when a less exact approximation is sufficient. The charcoal is always weighed with great care ; the litharge is best measured by a ladle or shot measurer ; the fluxes may also be measured out by a ladle more rapidly than they can be weighed. The various ingredients are thoroughly mixed together on rubber cloth or in the crucible in which the fusion is to take place. Part of the borax is often kept separate and used as a cover, being put on the top of the rest of the charge after transference to the crucible.

The crucible is carefully annealed in the ashpit of the furnace before using. It is lowered into a hollow in the fuel of the furnace made by piling the coke round an old pot and then carefully withdrawing the latter. Both ordinary and basket tongs are useful. The fire should be at a low red heat on charging in—that is, at a temperature of 600° or 700°—and should not be urged at first. The objects to be obtained are as follows :—

(1) The gold is to be brought into such a condition that it will be readily taken up by the molten lead before the subsidence of the latter to the bottom of the pot.

(2) The mixture is to be finally brought to a state of quiet fusion, with low viscosity, so that the reduced lead may subside completely and collect into one button.

(3) There must be no mechanical losses.

Reduction of the lead begins below a red heat, and is probably almost complete before borax glass and sodium carbonate begin to melt (at about 700°) and to attack the gangue. In an ordinary charge which was heated in a muffle at 640° for twenty minutes, and then withdrawn and allowed to cool, it was found that fusion had begun though no effervescence had occurred. No lead was visible until after the charge had been panned, when a large amount of metallic lead in fine particles was recovered. The lead begins to collect into globules visible to the naked eye as soon as the charge begins to work and effervesce, but does not sink until the effervescence has proceeded for several minutes. The slower the melting (*i.e.*, the more the " fritting " stage is prolonged) the more chance is afforded of bringing the lead into intimate contact with all parts of the pasty mass and of collecting the gold from the ore. Also the slower the melting, the less chance there is that part of the charge will be projected from the pot. The final temperature need not, as a rule, be above 1,100°, and may be lower in certain cases.

The influence of the rate of heating on the result is very great. In an experiment designed to test this, four charges of a pyritic ore were made up in accordance with the table given above (p. 497). Two charges were heated very slowly, and remained in the furnace for fifty minutes before fusion was complete. The other two charges were heated as rapidly as possible, and were in the furnace for only twenty minutes. The results were as follows :—

TABLE XLIII.

	Heated Slowly.	Heated Rapidly.
	Ozs. per Ton.	Ozs. per Ton.
(1) Gold,	0·130	0·075
Silver,	4·42	2·33
(2) Gold,	0·125	0·092
Silver,	4·28	2·71

The mean deficiencies in the charges heated rapidly were thus :—Gold 34 per cent., silver 42 per cent., but the results were also less regular than when slow heating was used.

The effervescence is due chiefly to the escape of carbon dioxide from the sodium carbonate as it unites with silica, and is of value in moving about the lead globules.

After a lapse of from 40 to 50 minutes the charge is in a state of tranquil fusion, with the exception perhaps of slight action round the sides or next the iron, if any is used. The crucible is then lifted out of the fire by the tongs, the nails withdrawn and any lead adhering to them shaken off into the pot. The pot is allowed to cool and broken by a hammer to extract the lead button, or the charge may be at once poured into an iron mould. The mould must be cleaned, blackleaded, and warmed before being used.

If the charge does not fuse completely, so that the slag is pasty or has lumps in it, it is advisable to recommence the assay, making such alterations in the charge as experience suggests, having regard to the considerations mentioned above.

When the mould is quite cold its contents are readily separated from it if the precautions mentioned above are taken, and the slag is detached from the lead button with a hammer on an anvil. The slag should be glassy and homogeneous ; if it is streaked it is probable that the fusion has not been perfect. It is green and transparent if the ore is nearly pure quartz ; black and opaque if much iron is present ; but is red if the ore contains much copper.

The lead button should be soft and malleable. If it is hard or brittle it may contain sulphur, arsenic, antimony, copper, etc. Sulphur and arsenic are kept from entering the lead by the addition of iron, and may form with the iron a matte or speiss, which separates as hard blackish-grey or white layers found just above the lead. They are richer than the slag, and may often yield appreciable quantities of gold on further treatment by scorification or roasting and fusion. If the quantity of sodium carbonate is large enough, the sulphur is retained in the slag in the form of sodium and ferrous sulphides. Antimony makes the lead hard, white, brittle and sonorous ; according to Rivot, it is not a source of loss if forming less than 1 per cent. of the lead button (see under Cupellation below). The lead must be completely freed from the slag, very small quantities of the latter interfering seriously with the cupellation, forming a scoria and occasioning loss.

Roasting before Fusion.—Ores containing large quantities of sulphur, arsenic or antimony may be sometimes roasted with advantage as a preliminary to fusion. Roasting is effected in shallow circular clay dishes, in a muffle, or in the crucibles in which the fusion is afterwards performed, The temperature must be kept low at first and the ore frequently stirred with an iron wire or spatula, to prevent fritting, and to expose fresh surfaces to the air. The roasting takes place in two stages : at first, sulphur dioxide, arsenious oxide (As_2O_3), and antimonious oxide (Sb_2O_3) are formed and volatilised, the sulphur burning with a blue flame. The formation of lumps is most to be feared during the first few minutes of the operation, and can scarcely be prevented if much sulphide of antimony is present ; in this case an equal weight or more of pure silver sand is mixed with the crushed ore before charging it into the muffle.

After a time the blue flame disappears, the odour becomes less strong, and sulphates, arsenates and antimonates form. By raising the temperature sulphates are decomposed, but arsenates and antimonates are stable at high temperatures and cause loss of silver in the fusion. To prevent their formation the ore should be roasted in a coke furnace, starting to heat it very gradually and admitting a limited supply of air. In all cases the roasting is nearly complete when the glow caused by stirring is shown only by a few specks of ore; the temperature may then be raised to a strong red heat without danger of fusion. The operation is complete when the ore remains of a uniform colour on stirring. Arsenates and antimonates may be in part removed by re-roasting with powdered charcoal in a covered dish.

James recommends [1] that ores consisting chiefly of sulphide of antimony should be mixed with twice their weight of silica, and roasted at a very low temperature. The sand prevents the charge from " balling." Sulman prefers [2] to mix the charge with carbon instead of sand. In roasting this mixture in a reducing atmosphere, some 96 or 97 per cent. of the antimony is volatilised as sulphide at a fair red heat without loss of gold. C. O. Bannister [3] suggests the addition of charcoal powder towards the end of the ordinary operation of roasting.

The fusion of roasted ores requires more charcoal powder than raw ores, the amount needed being sometimes as much as 3 grammes per A.T. of ore. In general, roasting is to be deprecated, owing to the unavoidable loss by dusting. The results are usually lower than those obtained by other methods, and the operation requires the expenditure of much time and attention. It is useful as a second method in cases of special importance. Before charging concentrate on to a roasting dish, it is advisable to sprinkle sand over the dish to prevent fritted sulphides from adhering to it.

Cleaning the Slag.—The slags of very rich ores may retain enough gold to necessitate further treatment. The slag is roughly crushed and fused with 1 A.T. of litharge, 15 to 20 grains of charcoal and a little sodium carbonate, the same crucible being used again, or the slag may be fused and a mixture of litharge and charcoal thrown on to its surface. If any regulus or speiss forms during the first fusion it must be preserved with special care and re-fused, charcoal being reduced or omitted. The button of lead from the second fusion is cupelled with the first button, or alone; the slag from the second fusion is almost always poor enough to be thrown away.

H. L. Sulman points out [2] that it is not necessary to pour and regrind the slag before " washing " it. Instead of this he uses a method similar to that adopted in " cleaning " scorification slag (see below, p. 516). When the crucible contents have fused and become quite tranquil, he stirs in first a little more oxide of lead and then a pinch of carbon, and thus makes a quick " washing " form part of the ordinary operation.

The Treatment of Basic Ores.—The difficulty of roasting *arsenical and antimonial* ores, which has been discussed above, p. 499, may be avoided [4] by taking ore, 1 A.T.; red lead, 1,000 grains; sodium carbonate, 500 grains; potassium ferrocyanide, 550 grains, with a cover of salt or borax. The button is scorified, together with the matte formed, before cupellation. E. A. Smith [5] found this method unsatisfactory and recommends the

[1] James, *Trans. Inst. Mng. and Met.*, 1901, **9**, 353.
[2] Sulman, *ibid.*, p. 355.
[3] Bannister, *Trans. Inst. Mng. and Met.*, 1907, **16**, 94.
[4] Ricketts, *Notes on Assaying*, New York, 1887, p. 77.
[5] Smith, *Trans. Inst. Mng. and Met.*, 1901, **9**, 334; *The Sampling and Assay of the Precious Metals*, p. 222.

oxidation of sulphide of antimony by nitre, the charge being made up as follows in the case of an ore containing 75 per cent. of stibnite, with gangue mainly consisting of silica :—

Ore, ...	20 to 25 grms.
Red lead, ...	80 ,,
Sodium carbonate, ...	60 ,,
Borax, ...	10 to 20 ,,
Nitre, ...	14 to 20 ,,

With cover of powdered salt.

The button should weigh from 45 to 50 grammes. The fusion is effected in large crucibles at a low temperature. Sodium carbonate forms a fluid slag with oxides of antimony. Another method given by Smith is to digest the ore with concentrated hydrochloric acid, add tartaric acid, filter, dry and scorify, or else fuse with red lead, charcoal and fluxes. An alternative plan is to scorify these ores direct, as well as those containing much copper if they are rich enough.

If they are poor, ores containing *copper* may be treated in three ways,[1] so that each method may serve as a check on the others, viz. :—(a) Fusion with much PbO : the lead button becomes cupriferous, and should be scorified together with the matte ; (b) roasting, followed by fusion and scorification ; (c) treatment with nitric acid, by which all the sulphur and copper is removed. The silver dissolved in the liquid is then precipitated with a solution of common salt, of which a large excess should be avoided. The addition of a solution of lead acetate followed by hydrochloric acid assists in the collection of the silver chloride. The insoluble residue is dried, and can now be readily fused and cupelled. By treatment (c) the lead button is kept free from copper, the presence of which in the lead obtained by methods (a) and (b) renders cupellation difficult and unsatisfactory.

According to J. Loewy,[2] copper ores containing more than 6 per cent. of copper yield lead buttons containing too much copper for satisfactory cupellation. Accordingly he adds silica to ores containing more than 6 per cent. of copper.

Zinc ores may be roasted, and fused with a larger quantity of borax and sodium carbonate than usual with pyritic ores. Instead of roasting, it is easier to desulphurise blende with metallic iron in the fusion, the zinc being volatilised.[3]

According to Hall and Popper,[4] litharge should not be left in the slag of zinciferous ores; the amount of sodium carbonate should be from four to five times that of the ore. A suitable charge for a zinciferous ore was found to be as follows :—

Ore, ...	$\frac{1}{3}$ A.T.
Sodium carbonate, ...	$1\frac{1}{3}$,,
Borax glass, ...	$\frac{1}{2}$,,
Litharge, ...	$\frac{4}{5}$,,

[1] Percy, *Metallurgy of Gold and Silver*, p. 247.
[2] Loewy, *Chem. Zeit.*, 1911, **35,** 278.
[3] Smith, *loc. cit.*
[4] Smith, *op. cit.*, p. 232.

Smith also recommends[1] a crucible fusion with nitre for zinciferous ores, the complete oxidation of 1 part of zinc blende requiring from 1·5 to 2 parts of nitre.

Lead ores should not be roasted, but fused with plenty of iron. E. A. Smith[2] recommends the direct fusion of *bismuth* ores with comparatively large quantities of sodium carbonate and borax, using a low temperature.

Telluride ores are not roasted but fused with excess of litharge at a moderate temperature. Charges for two siliceous American ores containing tellurides are given as follows :—[3]

TABLE XLIV.

	Poor Ore with little Pyrite	Rich Ore with much Pyrite.
Ore, ...	1 A.T.	0·2 A.T.
Sodium carbonate, . . .	1 ,,	1 ,,
Sand, ...	½ ,,	½ ,,
Litharge, ...	6 ,,	4 ,,
Argol, ...	2 grms.	1 grm.
Borax glass, ...	10 ,,	...
Common salt, ...	Cover.	Cover.

The slag is remelted with 1 A.T. of litharge and 2 grammes of argol. The lead button is cupelled direct.

Tindall points out[4] that a most important point in the assay of telluride ores is to crush the ore finely. Cripple Creek ores are crushed through 120-mesh sieves if they are poor. Richer ores are crushed through 150-mesh sieves, and very rich ores through 200-mesh sieves. Lodge states[5] that at Cripple Creek a flux is used made up as follows :—

$$
\left.
\begin{array}{lll}
K_2CO_3, & . & . & . & . & . & 3\cdot07 \text{ kilos.} \\
Na_2CO_3, & . & . & . & . & . & 2\cdot7 \quad ,, \\
\text{Borax glass,} & . & . & . & . & 2\cdot55 \quad ,, \\
\text{Flour,} & . & . & . & . & . & 0\cdot45 \quad ,, \\
\text{Litharge,} & . & . & . & . & . & 13\cdot6 \quad ,, \\
\end{array}
\right\} \text{Mix well.}
$$

Take about 65 or 70 grammes—that is, 2⅓ A.T.—of this mixture to ½ A.T. of ore, and fuse in a muffle furnace. The slag will be glassy and brittle.

Hillebrand and Allen[6] found that the crucible assay is perfectly satisfactory for telluride ores from Cripple Creek. The charge giving the best results on ores containing from 15 to 20 ozs. of gold per ton was as follows :—

Ore, ... 1 A.T.
Sodium bicarbonate, ... 1 ,,
Litharge, ... 6 ,,
Fused borax, ... 10 grms.
Salt, ... Cover.

[1] Smith, *loc. cit.*
[2] Smith, *Trans. Inst. Mng. and Met.*, 1901, **9,** 346.
[3] C. H. Fulton, *J. Amer. Chem. Soc.*, 1898, **20,** 586.
[4] Tindall, *Trans. Inst. Mng. and Met.*, 1901, **9,** 354.
[5] R. W. Lodge, *Notes on Assaying* (New York, 1905), p. 131.
[6] Hillebrand and Allen, *Bull. U.S. Geol. Survey*, series E (1905), No. 253 ; *Chem. News,* 1906, **93,** 100, 109.

The losses of gold in the slag were generally small. For the losses by cupel absorption, see below, p. 508. In general the lead buttons should be large, to facilitate the removal of the tellurium in cupellation.

For assay of tellurides by scorification see p. 517.

(2) **Cupellation.**—This operation is conducted in a muffle furnace, the construction of which is shown in Figs. 210, 211, p. 529. The fire is lighted, a little bone-ash is sprinkled on the floor of the muffle to prevent its corrosion by litharge in case of the upsetting of a cupel, and the cupels introduced as soon as a bright red heat is attained. The cupels are cleaned by gentle rubbing or blowing before being charged in, and are again cleaned with bellows before the lead is added. They are placed in the furnace one by one, or, better, charged in together on a tray, see below, p. 532.

Cupels are little cups made of bone-ash, and are either round or square. In their manufacture the bone-ash is finely powdered so that it will pass a 40-mesh sieve, then slightly moistened with water (to which a little carbonate of potash is sometimes added), put into a mould (Fig. 216), and compressed by the blows of a mallet, or, better still, by means of a screw press, so as to cohere firmly. The mould is preferably made of steel, as gun-metal wears sooner, and an uneven surface is disadvantageous. It is kept clean, bright and smooth, and the sides should be tapered upwards slightly to facilitate the removal of the cupel without injury. Cupels must be dried very carefully and slowly, but completely, as otherwise cracks appear when they are heated, and loss is thereby occasioned. The cupels at the Royal Mint are made several months before being used, and are dried slowly on shelves at some distance from the furnaces. More rapid drying has been tried on many occasions with less satisfactory results. Nevertheless fair results, but not the best, can be obtained when using cupels only a few days' old. If too much water is used in mixing the bone-ash, the cupels lose part of their porosity ; if too little is used they are too soft and crumble readily. About 6 or 7 per cent. of water answers very well. Care must be taken to preserve cupels from the access of nitrous or other acid fumes. These are absorbed by the bone-ash and given out in the furnace so that they are liable to cause spitting. Lengthened heating of an hour or more in the muffle before charging in the lead buttons will prevent loss from this cause. It is always advisable to heat cupels for 15 minutes after they have attained the temperature of the muffle before charging in the lead buttons.

Bone-ash of good quality is supplied by several makers. It may be prepared [1] by carefully calcining clean knuckle-bones. The crude bone-ash is crushed through a 30-mesh sieve, and recalcined. Caustic lime and carbonate of lime are next removed by lixiviation with ammonium chloride, and the bone-ash, after thorough washing and drying, is graded by sieving, and is then ready for use. The old German assayers of the 16th century used a mixture of two parts of wood ashes and one part of bone-ash, chiefly on account of the scarcity of bone-ash.[2] For this mixture, it was necessary to use well-boiled beer, or white of egg or a solution of glue, to make the cupel coherent, but pure bone-ash coheres if water alone is used to moisten it. Sometimes the surface of the cupels consisted of pure bone-ash.

In order to test the effect of the size of the particles of bone-ash on the absorption of silver and gold, the author made the following experiments.

---

[1] Bettel, *J. Chem. Met. and Mng. Soc. of S. Africa*, 1899, **2**, 599.
[2] L. Ercker, *Erszt und Bergwerksarten*, Frankfort, 1580, Bk. i., Chap. v.

Cupels were made from (*a*) "coarse" bone-ash, between 20 and 40 mesh; (*b*) "fine" bone-ash passing a 120-mesh sieve; (*c*) "ordinary" bone-ash, with the following screen analysis :—

On 20-mesh sieve,	.	.	.	.	0	per cent.
,, 30-mesh ,,	.	.	.	.	0·25	,,
,, 40-mesh ,,	.	.	.	.	3·30	,,
,, 60-mesh ,,	.	.	.	.	21·10	,,
,, 80-mesh ,,	.	.	.	.	9·45	,,
,, 100-mesh ,,	.	.	.	.	14·95	,,
,, 120-mesh ,,	.	.	.	.	19·20	,,
,, 140-mesh ,,	.	.	.	.	3·90	,,
Through 140-mesh ,,	.	.	.	.	27·85	,,
					100·00	,,

Charges of 0·5 gramme silver were cupelled with 4 grammes lead, with and also without 0·04 gramme copper. The average losses of silver by absorption were found to be as follows :—

				Loss of Silver per 1,000.	
				In Presence of Copper.	In Absence of Copper.
Coarse bone-ash,	.	.	.	10·45	9·52
Ordinary bone-ash,	.	.	.	11·42	9·10
Fine bone-ash,	.	.	.	9·75	8·90

The coarser the bone-ash, the more readily the prills were detached from the cupel. The loss of gold, similarly tested, was identical in the three cases. These results appear to show that the size of the particles of bone-ash is not of much importance, if they are small enough to pass a 20-mesh sieve.

Magnesia, or rather a mixture of magnesia, lime and some silica, was introduced a few years ago, as a substitute for bone-ash. It is strongly calcined, and does not absorb acid fumes. The cupels require to be made under high pressure, and should be baked at a high temperature before use. They are then as hard and strong as fireclay. Magnesia absorbs and holds less litharge than bone-ash does. For the absorption of gold by these cupels, see below (p. 509).

The cupels having been sufficiently heated in the muffle, the lead buttons obtained as described on p. 499 are charged in by the tongs (A, Fig. 209). The buttons collapse and lose their shape in a few seconds if the temperature is sufficiently high, but the molten mass formed is covered by a black crust for some time later. The crust then breaks up, and the brilliant surface of a liquid bath is seen. The muffle door should be closed immediately after the charging in is completed, and kept closed until all the assays have thus "uncovered." The door is then opened a little way to let in a current of air by which the lead is oxidised, and the litharge, floating to the edge of the bath, is absorbed by the cupel, together with the oxides of other base metals, which are not taken up by bone-ash if they are in a state of purity. Uncombined metals are not readily absorbed by the cupel, and only traces of gold and silver are carried into it by the litharge (see below, p. 508).

As the cupellation advances the lead bath is reduced in size by oxidation and absorption of the oxides; reddish patches float slowly over its surface, appearing earlier in the richer assays; it becomes more convex and brighter; the red spots move more quickly and finally whirl round with great speed and then disappear; moving iridescent bands take their place for a moment and then disappear likewise, and the bead becomes suddenly much duller in appearance, thus indicating that the cupellation is at an end. The cupels may then be removed from the muffle, provided the ore is poor or has little silver in it. It must, however, be remembered that silver absorbs oxygen when molten and gives it off suddenly when solidifying, so that if the bead weighs more than 0·01 gramme (Rivot) little fountains of metal are thrown up, and some part may be projected out of the cupel. This "sprouting," "spitting" or "vegetation" may take place in argentiferous-gold beads if the gold does not exceed one-third of the silver (Levol). At the Royal Mint it is found that a still larger percentage of gold does not prevent spitting unless a trace of copper is present. Where spitting is to be feared, therefore, the cupel containing the silver bead is covered with a red-hot empty cupel, and is withdrawn gradually from the muffle so that the bead may cool slowly. Sprouting is avoided in this way. When the beads are very large (0·2 gramme or more) the muffle is closed and luted up and the fire allowed to die down as in the assay of silver bullion.

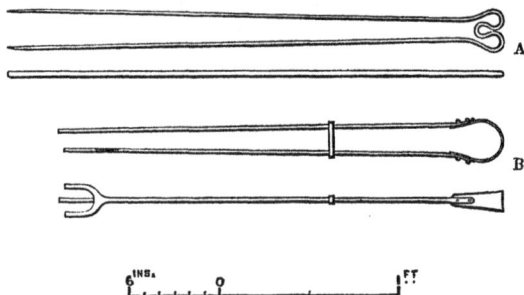

Fig. 209.[1]—Assay Furnace Tools, Royal Mint.

When cooled the beads often "flash"—*i.e.*, brighten suddenly at the moment of solidification. This is due to the fact that the latent heat of fusion being released raises the temperature of the bead enormously, the metal having been in a state of surfusion at a temperature many degrees below its melting point. The flashing of small beads can rarely be observed.

The proper *temperature* for cupellation of lead buttons obtained from the fusion of gold ores is the same as those for silver ores.

Vauquelin, perhaps the earliest writer on the temperature of cupellation, observes[2] that the lead is to be charged in when the cupels are seen to be "rouge légèrement blanc," and that during cupellation the fumes of litharge rise and wind about in the interior of the muffle. This appears to correspond to a temperature of the muffle walls of about 900°. He proceeds to state that the heat is too great if the colour of the cupels is white, the fused metal is seen with difficulty, and the fumes, scarcely visible, rise rapidly to the vault of the muffle without winding about (say 1,000°). If the fumes appear

---

[1] Reproduced from Percy's *Metallurgy of Silver and Gold*, by kind permission of Mr. John Murray.

[2] Vauquelin, *Manuel Complet de l'Essayeur*, Paris, l'an vii. (1800). New edition, Paris, 1836, p. 45.

heavy and dark, moving sluggishly in a direction almost parallel to the bottom of the muffle (say 800°), the temperature is too low. These correct observations have been repeated by Mitchell, Percy and most other writers since Vauquelin. Litharge scales form on the cupel if the temperature is too low.

The lowest temperature at which litharge is absorbed by a cupel, according to Fulton[1] and the author,[2] is about 840°. This is accordingly the minimum temperature for the cupel. The air above the cupel may be cooler. Good soft lead uncovers when the air in the muffle is at a temperature of 675° only, according to the author's measurements, but the temperature of the lead bath itself must be above 840°, or, according to Fulton, may vary from 797° to 862°. As soon as the lead uncovers and has begun to oxidise rapidly, its temperature rises, and the temperature of the muffle may be allowed to fall, until it is only 770° (Fulton), but it should be raised again at the finish of cupellation. Fulton observed that silver cupellation finishing as much as 77° below the melting point of silver yielded pure silver beads, but recommends a minimum temperature of 910°, corresponding to a temperature for the walls and floor of the muffle of somewhat above 850°, a cherry-red heat. When antimony, iron, etc., are present, the temperature must be much higher, and time is saved in all cases by using a higher temperature at first, but the losses are then greater. The muffle should be at a full red heat, the cupel dull red, and the melted lead much more luminous than the cupel ; the fumes should rise slowly to the crown of the muffle. Many assayers prefer to finish the cupellation at a temperature low enough to enable crystals of litharge to form in a ring round the cupel—that is, at about 800°—but it is doubtful whether there is any advantage in this. The temperature must be rapidly reduced at the finish by bringing the cupel towards the front of the muffle, and even then the feathers are not readily formed unless the cupel is almost saturated with litharge. It is probably better that the litharge should be completely absorbed.

If the assays are long in uncovering they may sometimes be started by dropping on them a little charcoal powder wrapped in tissue paper, or, still better, by placing a piece of charcoal near them. If one freezes before completion it is restarted in the same way, or fresh lead is added or the temperature is raised, but the results are not good. Spirting or " spitting " of the lead may occur, due to the use of raw cupels, or of cupels which have been made with too much pressure, or to the presence of sulphur, arsenic, etc., in the lead.

The bead thus obtained should be well rounded and bright, loosely adherent to the cupel and slightly crystalline although malleable. If it contains lead it is more globular and brittle and its surface is very brilliant, while it does not adhere at all to the bone-ash. If copper is present in large quantities, the bead adheres firmly to the cupel, and in extreme cases its surface is blackened. Rhodium and iridium occasion black patches at the bottom of the bead ; platinum makes the surface of the bead crystalline and rugose if it is present in sufficient quantity to prevent flashing.[3] If the bead cracks on being flattened Van Riemsdijk recommends a second cupellation with some more lead and 10 per cent. of cupric chloride. In this way the metals with volatile chlorides are eliminated.

[1] Fulton, *Western Chemist and Metallurgist*, 1908, **4**, 47.
[2] T. K. Rose, *Trans. Inst. Mng. and Met.*, 1909, **18**, 463.
[3] For the effect of small quantities of the platinum metals on the microscopic structure of the bead, see C. O. Bannister and G. Patchin, *Trans. Inst. Mng. and Met.*, 1914, **23**, 163.

Influence of Base Metals on Cupellation [1]—*Iron.*—Its oxide is not readily fusible with lead oxide : the button is long in melting, and a brown scoria is sometimes left on the cupel, which may entangle lead globules and so contain gold. The lead may contain about 4 per cent. of iron, without the formation of any scoria. The cupel is stained dark red, and is moderately corroded.

Zinc burns with a brilliant green flame at first and volatilises ; it forms a voluminous white or pale yellow scoria. The cupel is deeply corroded.

Tin gives a brown scoria if it constitutes more than 3 per cent. of the lead button. In all cases it floats to the top, oxidises, and forms a floating scoria which delays cupellation, but is subsequently carried off by the litharge. The cupel is stained pale brown.

Copper carries gold into the cupel and is usually not wholly removed from the bead. The cupel is stained green if a small quantity of copper is present, and dark brown or black if the quantity is large. If the quantity of lead is not enough to remove the whole of the copper from the prill, the latter spreads out and adheres to the cupel.

Nickel and Cobalt are not so easily carried into the cupel as copper : they form a black scoria.

Antimony does not interfere if less than 2 per cent. is present. If 4 per cent. is present, a slight yellow scoria is formed, and the cupel is stained dark brown.

Arsenic at first burns with a blue flame and causes spitting. After a few minutes the flame turns greenish-white, and becomes very brilliant. The cupel is stained pale brown, and if 4 per cent. of arsenic is present, there is much pale brown scoria.

Manganese causes black stains and deep corrosion of the cupel, and forms a black scoria. The lead boils and spits.

Chromium gives a brick-red stain and a black scoria on the cupel, and *aluminium* a grey scoria ; both these metals delay the course of cupellation. Chromium causes deep corrosion.

Cadmium causes a black sooty ring to form inside the cupel near its margin. When cupellation begins, the action is very violent, a yellowish-red flame is seen, followed by much spitting and the appearance of large red-hot scales floating on the lead. Oxide of cadmium is not absorbed by the cupel.

Tellurium is absorbed quietly, but whitish fumes appear, which give the bath a greenish-white appearance. The cupelled bead is sometimes, but not often, subdivided. The cupel has a brown stain in a ring round the litharge stain, but there is no scoria.

S. W. Smith has shown [2] that tellurium is slowly removed into the cupel during cupellation, and if there is not enough lead to carry it off, so that towards the end the tellurium becomes equal in amount to the gold or gold plus silver, then the surface tension of the globule breaks down completely and the alloy spreads over a wider area, " wets " the cupel and is completely absorbed. The lead should be 80 to 120 times as much as the tellurium to obtain good results in cupellation. The loss of gold by cupel absorption is then small.

[1] See "Notes on Cupellation and Parting," by T. K. Rose, *J. Chem. Met. and Mng. Soc. of S. Africa*, 1905, **5,** 165. The effects noted below are due to the presence in the lead of 4 per cent. of the base metal in most cases.

[2] S. W. Smith, *Trans. Inst. Mng. and Met.*, 1908, **17,** 463.

Selenium gives brown fumes, which change to blue fumes at a little distance above the cupel. The effect on the cupel is similar to that of tellurium, but the brown stain is less pronounced.

Molybdenum causes slight boiling of the lead. The cupel is stained pale brown, and very deeply corroded, but there is no scoria.

Vanadium gives a dark red stain, and causes remarkably deep corrosion of the cupel, but leaves no scoria.

Thallium gives a pale brown, and *bismuth* an orange stain. Neither of these metals leaves a scoria.

Silicon gives a voluminous yellow or white scoria, which delays cupellation, and is left on the cupel.

Losses in Cupellation.—The loss during cupellation by volatilisation was proved to occur by Makins, but is insignificant. The absorption by the cupel is more serious. Rivot states that gold is oxidised to some extent at a red heat in the presence of antimonic oxide, litharge, or cupric oxide, and that it is the oxidised part which is absorbed by the cupel.

Still heavier losses may be caused by large amounts of scoria, in which the whole of the gold and silver may be retained in the form of minute beads, impossible to collect except by scorification with fresh lead. Small losses are sometimes caused by spitting, but these are usually trifling, as the spitting ceases before the lead bath becomes rich in gold and silver.

In some experiments made by the author,[1] the exact losses due to the presence of base metals were determined. In each case 25 grammes of lead were cupelled with 1 milligramme of gold, 4 milligrammes of silver, and 1 gramme of the impurity. Bone-ash cupels made at the Royal Mint were used. The results were as follows :—

Nickel and chromium caused total loss in each case, due to scoria.
Silicon occasioned losses amounting to between 40 and 50 per cent., due to scoria.

		Per cent.		Per cent.	
Tin, .	.	Loss, gold, 2·0 ;	silver,	13·9.	Slight ring of scoria.
Zinc, .	.	,, 9·3	,,	17·6.	Much scoria.
Cadmium, .		,, 3·5	,,	13·1.	Ring of black feathery scoria.
Arsenic,	.	,, 3·9	,,	16·3.	Much scoria.
Antimony, .		,, 5·3	,,	13·2.	No scoria.
Manganese, .		,, 13·6	,,	24·3.	Scoria.
Iron, .	.	,, 4·0	,,	16·6.	No scoria.
Copper,	.	,, 10·0	,,	32·6.	,,
Bismuth,	.	,, 21·8	,,	27·9.	,,
Thallium, .		,, 23·1	,,	34·4.	,,
Molybdenum,		,, 11·0	,,	26·2.	,,
Vanadium,		,, 7·7	,,	21·7.	,,
Tellurium, .		,, 55·8	,,	67·9.	,,
Selenium, .		,, 54·1	,,	64·5.	,,

The losses in cupellation under the same conditions as to temperature, air supply, etc., and using similar cupels, but with pure instead of impure lead, were 1·2 per cent. of gold and 12·8 per cent. of silver. In no case was proof afforded of any loss by volatilisation. Practically the whole of the

---

[1] T. K. Rose, *loc. cit.*

missing gold and silver was recovered by fusing the cupels when tellurium and selenium were present. The percentage losses observed in these experiments appear very large, but it must be remembered that the absolute losses were very small. It will be observed that the formation of a scoria was not necessarily accompanied by a heavy loss of gold.

The percentage losses are much smaller when the beads are larger. Thus E. A. Smith[1] found that the loss on 5 grains of silver (= 324 mgs.) when cupelled with 175 grains of bismuth was 4·6 per cent., against a loss of 1·88 per cent. when lead was substituted for bismuth, and Ricketts[2] observed a loss of 2·3 per cent. on the crucible assay of a lead ore containing 2,260 ozs. of silver per ton.

Lodge records[3] that the loss in cupelling 200 milligrammes of gold with 10 grammes of lead ranged from 0·15 per cent. at 775° to 2·99 per cent. at 1,075°. The addition of 0·05 gramme of copper increased the loss of gold from 0·15 per cent. to 0·25 per cent. In these cases the gold beads were apparently weighed without being parted.

The absorption by the cupel varies with the temperature and with the quality of the bone-ash, as well as with the amount of the impurities present. Crosse[4] observed a loss of 5 per cent. in cupelling small amounts of pure gold with test lead. T. L. Carter[5] found losses up to 10 per cent. of gold with a particular set of cupels. The author compared the absorption of gold and silver at about 900° (the temperature of the air near the cupel) by magnesia cupels with those of two samples of bone-ash cupels with the following mean results :—[6]

TABLE XLV.

Charges.	Losses per Cent. in Cupellation.		
	Magnesia Cupels.	Bone-ash Cupels, "A."	Bone-ash Cupels, "B."
Gold, 0·001 gramme, . . . Silver, 0·004 ,, . . . Lead, 25 grammes, . .	Gold, 1·9 Silver, 8·3	1·2 12·2	7·8 27·1
Gold, 0·001 gramme, . . . Silver, 0·008 ,, . . . Lead, 10 grammes, . .	Gold, 1·4	0·9	5·2
Gold, 0·5 gramme, . . . Silver, 1·25 grammes, . . . Lead, 4 ,, . .	Gold, 0·060	0·055	...
Silver, 0·5 gramme, . . . Lead, 4 grammes, . .	Silver, 1·21	0·99	...

Generally speaking, the absorption by magnesia cupels is less than by bone-ash. Whatever cupels are used, the absorption should be tested frequently.

[1] E. A. Smith, *J. Chem. Soc.*, 1894, **65**, 624.
[2] Ricketts, *Notes on Assaying*, 1897, p. 110.
[3] Lodge, *Notes on Assaying*, 1905, p. 142.
[4] Crosse, *J. Chem. Met. and Mng. Soc. of S. Africa*, 1898, **2**, 325.
[5] Carter, *Eng. and Mng. J.*, May 24, 1902, p. 728.
[6] T. K. Rose, *J. Chem. Met. and Mng. Soc. of S. Africa*, 1905, **5**, 165.

Losses of gold, and especially of silver, are greater at a high temperature than at a low one. This is illustrated by comparing the following mean results of experiments, in which the bone-ash cupels " A " were used, with the results in the table given above :—

TABLE XLVI.

Charges.		Losses in Cupellation at 700°.[1]
Gold, 0·001 gramme, Silver, 0·006 ,, Lead, 25 grammes,	·⎱ ·⎰	Gold, 0·45 per cent. Silver, 8·6 ,,
Gold, 0·001 gramme, Silver, 0·010 ,, Lead, 25 grammes,	·⎱ ·⎰	Gold, 0·39 ,, Silver, 4·5 ,,
Gold, 0·010 gramme, Silver, 0·080 ,, Lead, 25 grammes,	·⎱ ·⎰	Gold, 0·074 ,, Silver, 2·5 ,,

(3) **Inquartation and Parting.**—The bead of silver and gold obtained by cupellation is squeezed between pliers, or flattened with a hammer on a clean anvil, to loosen the bone-ash adhering to its lower surface, and is then cleaned with a brush of wires or stiff bristles. It is then weighed, the silver removed by solution in nitric acid, and the weight of the residual gold taken, when the difference between the two weighings represents the silver. If the bead contains more than one-fourth its weight of gold, more silver is added to it, as otherwise some of the silver will remain undissolved, being protected from the action of the acid by the outer layers of gold. The amount of silver to be added is calculated from the (approximately) known composition of the bead, or guessed from its colour. A pale yellow bead always contains more than 60 per cent. of gold, but a perfectly white bead may not " part " completely. The addition of the silver is sometimes effected in the case of small beads by fusion on charcoal by the blowpipe, but it is better to cupel the bead with the additional silver, wrapped in as small a piece of lead foil as possible. This is called " *inquartation.*" The resulting bead is cleaned, flattened by a hammer, and dropped into boiling nitric acid of about specific gravity 1·10.

Even if the gold is less than one-fourth the weight of the bead, additional silver is sometimes required. When the amount of gold is small, it is convenient to use a large proportion of silver in parting. Suitable proportions of silver for different weights of gold are as follows :—

Weight of Gold.						Ratio of Silver to Gold.
Less than 0·1 mg.,	.	.	.	.	.	20 or 30 to 1
About 0·2 mg.,	.	.	.	.	.	10 to 1
,, 1·0 ,,	.	.	.	.	.	6 to 1
,, 10 mgs.,	.	.	.	.	.	4 to 1
More than 50 mgs.,	.	.	.	.	.	2¼ to 1

With these proportions, the gold does not break up if boiling acid is used.

---

[1] This was the temperature of the air near the cupels.

If much less silver is added, parting is incomplete or is not complete except after prolonged (20 to 30 minutes) boiling in acid.[1]

A. Whitby considers [2] that for 1 mg. of gold about 4 mgs. of silver should be used, and for 0·1 mg. of gold about 1·2 mgs. of silver. He states that the beads break up entirely or partially if larger amounts of silver are used. It is probable, however, that he is referring to parting which has been begun by dropping the beads into acid at a temperature below the boiling point.

The *parting* is effected in a test tube or a porcelain crucible if the bead is small, in a " parting flask " if large. Nitric acid of specific gravity about 1·10 is used for flattened beads containing a large proportion of silver, but if the ratio of silver to gold does not exceed four or five to one, acid of specific gravity 1·20 may be used (made by mixing equal volumes of strong acid and water). There is little advantage in using more dilute acids than that of specific gravity 1·20 for any alloys. The freedom of the nitric acid from chlorine is ensured by adding to it a little spent acid containing some nitrate of silver. In all cases the acids must be previously heated to boiling, as the gold does not in that case break up into such fine particles as if colder acid is used. The acid attacks the bead instantly and violently, turning it black and giving off nitrous fumes. Little beads containing only a small proportion of gold are dissolved in a few seconds, and decanting may be at once proceeded with, but the boiling is usually continued for a minute or two. If the amount of gold is large, the boiling is continued for some minutes. Ten minutes boiling is enough for 10 mgs. gold and 40 mgs. silver. The acid is then poured off, the residue washed twice with boiling water by decantation, and if the bead is very large fresh acid is added of specific gravity 1·20, but with the beads from almost all gold ores no second treatment with acid is required. If a second acid is used, it is boiled for a few minutes, when almost all the silver will have been dissolved. A very small amount of silver, weighing from 0·05 to 0·1 per cent. of the gold, obstinately resists the action of the acid, and remains as a surcharge which may be neglected in almost all ores. The gold usually remains as a single piece if it weighs less than 0·1 mg., even if the proportion is only one part of gold to 40 to 50 parts of silver. If the ore is rich, the gold sometimes, though rarely, breaks up if hot acid is used, and invariably breaks up if the parting is begun with cold acid. The finer particles may float and be lost in decantation. Particles floating on the surface may be sunk by touching with a glass rod or by a drop of water let fall on them. Continued heating of the gold in acid makes it agglomerate to some extent, so that it is easier to wash.

After the final decantation of the acid and washing twice with hot water, the water is drained off and the porcelain crucibles dried at a gentle heat, and then gradually raised to dull redness. The gold which was previously black and soft, being in a fine state of division, now resumes its usual yellow colour, and hardens so that it can be removed to the pan of a balance and weighed. At a bright red heat the glaze in many porcelain crucibles softens and the gold sticks to it. Excessive heating must, therefore, be avoided; an ordinary Bunsen flame answers very well.

If a parting flask or test tube is used for the boiling, the parted gold

---

[1] T. K. Rose, *Loc. cit.* If the addition of silver is known by previous work on the ore to be necessary, it may be added to the charge in the crucible or to the lead during the original cupellation. In this case, if it is essential to determine the silver in the ore, the exact amount of silver added must be allowed for.

[2] Whitby, *J. Chem. Met. and Mng. Soc. of S. Africa*, 1905, **5**, 256.

is transferred to an unglazed Wedgwood crucible. To effect this the flask is filled with water, and the crucible placed over its mouth. On inverting both together the gold falls into the crucible, and the flask is removed in such a way as not to disturb the precious metal. The water which has filled the crucible is then poured off, and the crucible heated as before.

The chief difficulty in parting by the last-named method is encountered in transferring the gold from the glass vessel to the Wedgwood crucible ; minute particles of gold may adhere to the glass and are then left behind. The loss of these is the cause of the low results occasionally observed when this method is used. If the glazed porcelain crucibles are used for both boiling and annealing, no transference from one vessel to another of the gold in its soft state is necessary, so that the source of error mentioned above is avoided. On the other hand, the difficulty of boiling the acid in a small porcelain crucible without sustaining loss by projection may prevent the assayer from continuing the boiling for a sufficient length of time, and some silver may thus be left undissolved ; in consequence of this the results obtained are sometimes too high by as much as 1 or 2 per cent. of the weight of the gold. Such errors would, however, be inappreciable in the assay of poor ores, and are the less serious for the reason that all the other sources of error (unclean slags, absorption by cupel, etc.) tend to make the result too low. The danger of loss by projection is avoided by using watch-glass covers for the crucibles. The latter are heated on clean asbestos sheets, not on sand baths. By assaying an accumulated stock of parted gold derived from ores, it was found at the Royal Mint by J. Phelps that the amount of silver retained by the gold when parted in crucibles as described above (p. 511) is about 0·2 per cent.

To prevent "bumping," one or two small pieces of capillary glass tube or of clay pipe or other porous body are sometimes put into the acid together with the alloy to be parted, but, as a general rule, it may be assumed that parting is finished when bumping begins.

Whatever method of parting is used when large pieces of silver or copper containing very little gold are parted, very finely divided gold may remain suspended in the liquids, and may thus be lost in the course of decantation. Loss by decantation after parting may be reduced by adding a globule of mercury free from gold, and stirring with a glass rod until all the black particles of gold have been absorbed. The spent acid is then poured off and fresh nitric acid of density 1·2 added. On gently warming, the mercury is slowly dissolved and the gold remains as a coherent spongy mass. It is washed with nitric acid and then with water, glowed, re-heated with concentrated nitric acid, washed, re-heated, and weighed. A little mercury remains with the gold.

By the operation of parting, silver, palladium and some platinum are removed in solution, but part of the platinum, and all the rhodium, iridium, etc., remain with the gold. If the presence of these metals is suspected they must be looked for and removed by special methods (vide infra, p. 523).

(4) **Weighing the Gold.**—The balance should turn with $\frac{1}{200}$ milligramme at most, and there is an advantage in using still more sensitive balances. A unit of 0·005 mg. corresponds to 2·4 grains per ton if 1 A.T. of ore has been taken, and 0·2 grain per ton if 12 A.T. have been taken. Great care must be exercised in placing the balance, to avoid vibration from machinery, traffic, etc. In certain cases the foundation must be laid at a depth of many feet and a column built for the support of the balance. The column should

admitted to the muffle above the block, *e*, only, and is withdrawn through the graphite tube, *h*, which connects with the iron tube, *k*. This is furnished with a damper, *l*, and is connected with the chimney. There is no other opening in the muffle, so that the draught through it is quite independent of the draught through the furnace. This is an important feature which should be adopted in all muffles. The muffle rests on a bed of fireclay and pieces of firebrick, which covers the cast-iron girder-plate, *m*. (The retention of this iron plate supporting the muffle was a curious survival of the ancient arrangement described by Ercker in 1580, in which the muffle consisted of an arch only, without a bottom, and rested on an iron sole plate on which the cupels were placed directly.) The top of the muffle is covered with a thick layer of fireclay and graphite to check radiation and to preserve the muffle. The clearance between the muffle and the sides of the furnace is $3\frac{1}{4}$ inches,

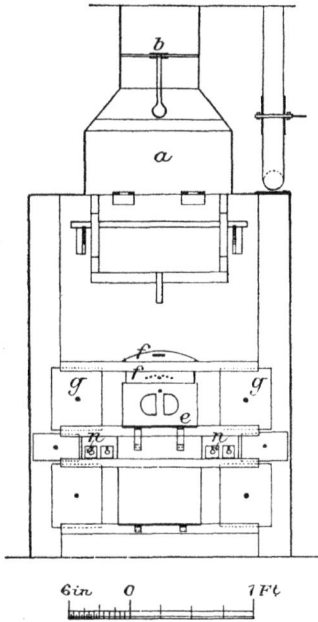

Fig. 210.—Coke Assay Furnace, Royal Mint—Front Elevation.

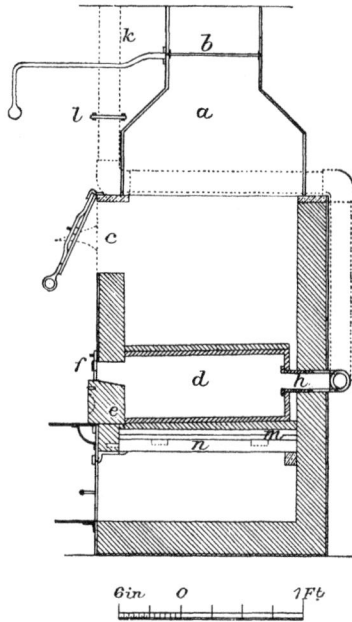

Fig. 211.—Coke Assay Furnace, Royal Mint — Sectional View of Side Elevation.

but at the back it is only 1 inch, and this space is filled with fireclay to prevent overheating.

There are eleven fire-bars, *n*, *n*, but only the three outer ones on each side are covered with fuel. On the other five bars rests the cast-iron girder-plate (Fig. 212), which is flat on the upper surface, but is strengthened with ribs on the under surface in order to prevent buckling. Charcoal, anthracite, or coke can be used in this furnace. The use of coke as fuel in assay furnaces is now exceptional, and charcoal and anthracite are not now used.

Figs. 213 and 214 are sectional diagrams and Fig. 215 a photograph of one of the gas muffle furnaces in use at the Royal Mint. The furnace walls

34

consist of firebrick bound with iron and covered with a layer, about $2\frac{1}{2}$ inches thick, of lagging consisting of a mixture of magnesia and asbestos to check radiation. The muffle, A, is $14\frac{1}{2}$ inches long, $8\frac{1}{4}$ inches wide and 5 inches high, inside measurements with fireclay walls about $\frac{3}{8}$ inch thick. The mouth,

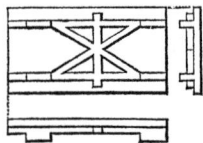

Fig. 212.[1]—Girder Plate to support Muffle, Royal Mint.

B, is closed by a firebrick and by a sliding sheet of mica above it (not shown in the diagram). Air enters through holes in the mica plate and passes out at the back of the muffle through the tube, C, which has a sliding damper and leads into the main flue. The fuel is ordinary coal gas, supplied by a row of Bunsen burners, D, and complete combustion of the gas may be aided by clay fire-balls covering the muffle, as shown in Fig. 213. The flues, E E, carry off the waste gases, and no chimney stack or forced draught is required. The muffle fits closely against the back wall of the furnace, so that the back is not hotter than the rest of the furnace. The back 12 inches of the muffle is used for cupellation, and is of nearly uniform temperature. The muffles last for several months in continuous work.[2]

Petroleum furnaces are used when coal, coke and gas are alike difficult to obtain. Electric resistance furnaces, though generally regarded as unsatisfactory, have been adopted at one of the United States mints, apparently with success.

Furnaces to burn soft coal are sometimes used in places where good coke is very expensive. One type differs little from coke furnaces in

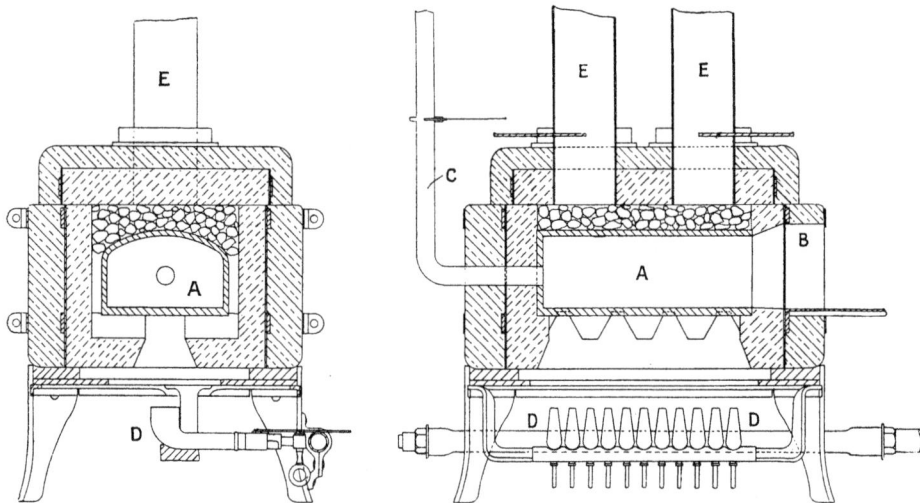

Figs. 213 and 214.—Sections of Assay Muffle Furnace.

construction, but has less space between the muffle and the side walls. The flame of the coal is instrumental in heating the muffle, a comparatively

---

[1] Reproduced from Percy's *Metallurgy of Silver and Gold*, by kind permission of Mr. John Murray.

[2] For descriptions of some other muffle furnaces, see E. A. Smith, *The Sampling and Assay of the Precious Metals*, pp. 20-33.

thin bed of fuel being employed, not reaching to the bottom of the muffle. On the Rand, coal-fired reverberatory furnaces are now much used for heating muffles.[1]

The cupels in use at the Royal Mint consist of bone-ash and are in sets of four, the outer margin being square, as shown in Fig. 216. At some assay offices, larger numbers of cupels are made in one block. The object is to facilitate charging in and withdrawal, but the difficulty of maturing the cupels increases with the size of the block. If cupel trays are used in the furnace, single cupels can be rapidly ranged on the tray by hand before it is placed in the muffle.

Fig. 215.—Muffle Furnace.

The cupel tray used at the Royal Mint is shown in plan and section in Fig. 217. It is $11\frac{3}{4}$ inches long and 6 inches wide, and holds cupels for 72 assays, the cupels fitting loosely inside the rim. The tray is made of "Salamander" graphite, and lasts for some weeks. Iron trays were found to interfere with cupellation and to be rapidly destroyed. Fireclay trays soon break. The tray is sprinkled with bone-ash, and the cupels placed on it before it is charged into the furnace. It is charged in and withdrawn by an

---

[1] A. M'A. Johnston, *Rand Metallurgical Practice*, vol. i., p. 297.

iron " peel " with two flat prongs, 1¼ inches wide, and of the same length as the tray, which slide into the grooves underneath the tray. The furnace tools formerly in use are shown in Fig. 218, where *a* represents the cupel

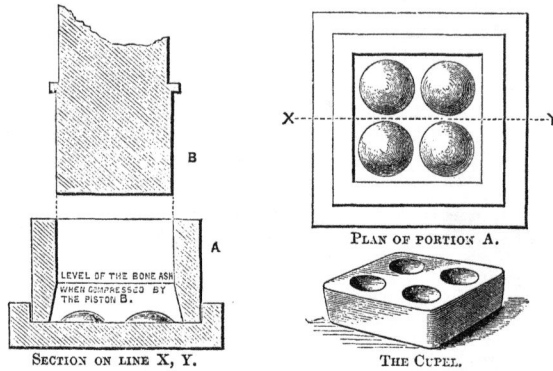

SECTION ON LINE X, Y.   THE CUPEL.

Fig. 216.—Cupel Mould, Royal Mint.

tongs and *b* the tongs used for charging in the lead packets. These tongs are now seldom used in bullion assaying at the Mint.

Method of Operation.—The muffle is brought to a uniform orange-red heat before the cupellation is begun. The cupels are cleaned by a pair of hand bellows just before the assay pieces are charged in. For the latter operation a nickel charging tray, devised by Henry Westwood, of the Birmingham Assay Office, in 1893, is now used. The tray consists of a plate perforated by 72 holes. Underneath is a non-perforated sliding plate. The assay pieces are put in order in the compartments and the tray is placed in position, some help in guiding it being given by stops which touch the front cupels. The sliding plate is then withdrawn and the lead packets fall through the holes into the cupels. " Uncovering " is completed in about two minutes, and during this space of time the muffle is kept closed. The door is then opened, and the draught started through the muffle by opening the damper.

Distinct stages may be noted in the action which now takes place on the cupel. Almost immediately the surface of the molten metal becomes covered with greasy-looking drops of litharge, which are rapidly absorbed by the porous cupel and replaced by others. They pass over the surface at first slowly, but as the operation continues move with greater rapidity.

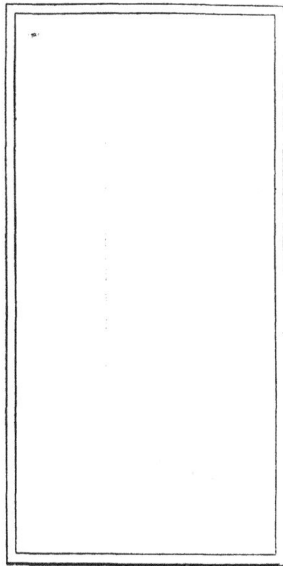

Fig. 217.—Graphite Cupel Tray, Royal Mint.

In from twelve to fifteen minutes the metal suddenly becomes uniformly dull and glowing except for iridescent bands, produced by extremely thin films of fluid litharge, which are seen to pass over it. On the disappearance of these bands a bright liquid globule of a greenish tint is left, but the cupels are not withdrawn from the furnace until the expiration of another fifteen to twenty minutes so that the last traces of lead may be oxidised and absorbed. The completion of cupellation takes place first in the front rows and proceeds regularly backwards.

The cupels are withdrawn from the furnace while the assay pieces are still fluid, and " flashing " ensues in a few seconds. " Flashing " is most marked in the purer buttons, in which but little copper or lead remains. If at least 50 parts of copper per 1,000 of gold were originally present the buttons are never sufficiently freed from it for " sprouting " to take place. If the cupels are withdrawn carelessly the fluid metal moves over the rough surface of the cupel, and minute beads may be separated from the main mass. To avoid this, and losses by sprouting, some assayers do not remove the assays from the furnace until the buttons or " prills " have become solid.

The " flashing " of gold assays was shown by Van Riemsdijk[1] to be due to solidification after superfusion. The temperature of the fluid metal falls until a certain point is reached, when the button solidifies, and the sudden disengagement of the latent heat of fusion reheats it to its true melting point—viz., 950°—and a peculiarly intense light is emitted which rapidly fades as the temperature again falls. A sudden jar at any moment causes the flashing to occur instantly. If the alloy contains a minute quantity of

Scale, 1 inch = 1 foot.

Fig. 218.[2]—Old Assay Furnace Tools, Royal Mint.

iridium, rhodium, ruthenium, osmium or osmium-iridium, the tendency of the cupelled metal to preserve its liquid state below the melting point, and therefore to flash during the final solidification is entirely prevented. The presence of metals of the platinum group (except platinum and palladium) in ingots of commercial gold can be detected by means of this characteristic.

The buttons (which are of the form represented at a, Fig. 220) are removed from the cupels, after cooling, by a pair of sharp-nosed pliers, cleaned by means of a stiff brush or by immersion in warm dilute hydrochloric acid, and are placed in the compartments corresponding to their cupels in the tray (d, Fig. 219). If the bone-ash is not completely removed from their lower surface it is of little moment, since bone-ash is readily dissolved by nitric acid on parting. The surface of the cupels must be carefully examined for minute beads of metal due to spirting of the lead bath, which sometimes happens if there is too strong a draught. If any such beads are found in a cupel the fact is noted and the assay repeated.

If traces of lead remain in the button it is more globular, separates more

---

[1] Van Riemsdijk, *Chemical News*, 1880, **41**, 126.
[2] Reproduced from Percy's *Metallurgy of Silver and Gold*, by kind permission of Mr. John Murray.

easily from the bone-ash of the cupel, and has a brilliant steely surface. The effect of the presence of other metals is discussed on p. 507.

Temperature.—The exact temperature suitable for cupellation can only be ascertained by practice, and the varying light of the day may occasion error in judging the degree of heat. The remarks on temperature in the cupellation of buttons from ores (p. 505) apply here. Care should be taken to ensure that the heat is so high before "charging in" that the chilling which necessarily takes place during this operation shall not cool the muffle below the requisite temperature. It is of more consequence that the muffle should be uniformly hot throughout than that any absolute degree should be attained, as the checks used eliminate uniform errors due to high temperature.

The temperature of an assay muffle was measured by J. Prinsep, Assay Master of the Mint at Benares, by observing differences in the behaviour of a number of silver-gold and gold-platinum alloys when heated.[1] He "made trials in different parts of the same (muffle) furnace. The disparity of heat," he remarks, "is greater than might be supposed." His results were as follows:—

				Maximum Alloy Melted.	Corresponding Temperature.[2]
Front of muffle,	.	.	.	pure silver.	960°.
Middle	,,	(average),	.	silver, 70. gold, 30.	1,025°.
Back	,,	(,,),	.	silver, 50. gold, 50.	1,045°.

In 1892, the temperature of a coke-heated muffle at the Royal Mint was measured by the author by means of the Le Chatelier pyrometer.[3] The muffle was 15 inches long and 6½ inches wide, and it was found that the temperature gradually rose from about 1,050° to 1,080° in passing from front to back, whilst along the sides the temperature was 1° or 2° higher than in the middle line. The mean temperature of the muffle was about 1,070°. With gas furnaces and the present arrangements of the muffle the variation in temperature is much less according to recent measurements.

4. **Preparation of the Assay Buttons for Parting.**—The buttons are flattened by a hammer weighing about 7 lbs., with a convex face, on an anvil kept quite bright and clean and used for this purpose only. A heavy blow is first delivered on the middle of the button, the diameter of which is thereby increased to nearly that of a threepenny piece. Two lighter blows are then given on opposite sides of the disc so as to elongate it, giving it the form shown at *b*, Fig. 220.

After being annealed in the iron tray, *f*, Fig. 219, which is placed in the muffle by the tool, *g*, and left until it is red hot, the flattened buttons are passed in succession through a pair of jeweller's adjusting rolls which are used for this purpose only, and are kept clean and bright. The oil is removed as completely as possible from the rolls before they are used, as otherwise the first fillets come out thinner than the remainder. The rolls are adjusted

[1] Prinsep, *Phil. Trans. Roy. Soc.*, 1828, p. 79.
[2] Roberts-Austen and Rose, *Proc. Roy. Soc.*, 1902, **71**, 162.
[3] Rose, *J. Chem. Soc.*, 1893, **63**, 707.

so that one passage through them reduces the buttons to the required thickness, which is about 0·25 millimetre, or 0·01 inch, or about the thickness of an ordinary visiting card. The " fillets " (c, Fig. 220) thus obtained should all be of uniform size and thickness, with " wire edges," as ragged edges expose them to loss during the boiling. After being rolled they are replaced in the tray, *f*, and annealed at a dull red heat. In some offices, the buttons

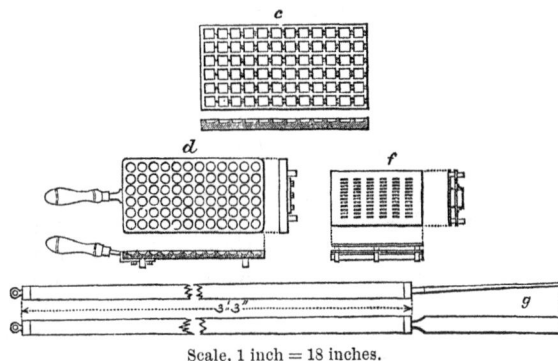

Scale, 1 inch = 18 inches.

Fig. 219.[1]—Button Trays, Royal Mint.

and fillets are annealed by a blowpipe on charcoal instead of in the furnace. They must not be made too hot, as that entails a loss of gold in the parting acid. It is best to raise them to a low red heat. The object of the first annealing is to soften the buttons and facilitate their passage through the rolls, while that of the second is to enable the fillets to be rolled into " cornets " or spirals, *d*, between the finger and thumb, or round a glass rod, and also to put the metal into a suitable physical condition for parting. Unannealed fillets tend to break up in nitric acid. Care is taken to leave that which was formerly the lower side of the button outside, for a reason given below (p. 552). This face is easily recognised, as it is less brilliant than the other.

5. **Parting.** — This was formerly effected by boiling with nitric acid in glass " parting flasks." Platinum boiling trays save time, and are now used whenever possible. The silver is dissolved by the acid, which should be free from chlorine in any form, sulphuric and sulphurous acids, or sulphides from which sulphuric acid may be formed. These substances dissolve gold in the presence of boiling nitric acid. It is sometimes

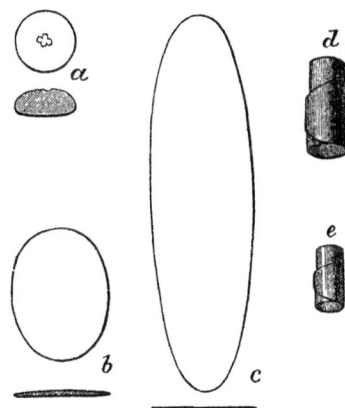

Scale, full size.

Fig. 220.—Stages in working a Gold Bullion Assay Piece.

stated that the acid must be free from nitrous fumes, but this is not necessary, as silver protects gold from the action of nitric and nitrous acids, and as soon as parting begins great quantities of nitrous fumes are generated.

---

[1] Reproduced from Percy's *Metallurgy of Silver and Gold*, by kind permission of Mr. John Murray.

When the flasks are used, 2 ozs. of nitric acid of specific gravity 1·2 are put into each flask and raised to boiling point. A cornet is then introduced and boiling continued for fifteen or twenty minutes (*i.e.*, for about ten minutes after nitrous fumes cease to be given off). Hot distilled water is then added, the solution of nitrate of silver decanted off, and the flask washed by filling with hot water, and decanting. Two ounces of hot nitric acid of specific gravity 1·3 are now poured in and the boiling continued for fifteen or twenty minutes, a piece of fireclay or capillary glass tube being added to prevent bumping; after which decantation and washing is twice performed. Another boiling with acid of specific gravity 1·3 is recommended by Chaudet with the object of dissolving out the last traces of silver and leaving the gold quite pure. This practice has been adopted by many assayers, but is useless and causes loss of gold. If any small particles of gold have become detached from the cornet, time must be allowed for them to settle before each decantation. After the last decantation the flask is filled with hot water, the top covered by a small porous crucible, and the whole is carefully inverted; the pure gold, which is of a dark brown colour and exceedingly fragile, falls through the liquid and rests in the crucible, the water which enters with it being afterwards poured off. The crucible is dried and then annealed at a red heat over gas or in the muffle, when the gold shrinks greatly, though still preserving its shape, hardens and regains its ordinary pale yellow colour. It can now be weighed.

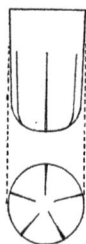

Fig. 221. Platinum Cup for Parting.

When the platinum boiler is used the cornets are put on platinum pins, as at the Sydney Mint, or more usually into platinum cups, one of which is shown in Fig. 221. These cups are supported in a platinum tray (which holds 144 cups at the Royal Mint) and the whole lowered by a platinum hook into a platinum vessel containing about 60 ozs. of hot nitric acid. Some attention must be paid to the temperature of the acid. At the Royal Mint, acid of specific gravity 1·20 (in which, however, a small quantity of silver is already dissolved) is used, and the temperature at the moment of introduction of the tray is from 95° to 100° C. At higher temperatures the action is so vigorous that the vessel may boil over. When the ratio of 2·5 parts of silver to 1 part of gold is used, some care is necessary in putting the assay pieces into the acid. A temperature of 90° should not be departed from widely; if the acid is colder than this the cornets tend to break up. Cornets containing 2 parts of silver to 1 of gold are less delicate, and can be put into acid of sp. gr. 1·26, even if it is cold, or at any temperature up to boiling point, without showing any signs of breaking up. They can also be put into hot acid of sp. gr. 1·32 without injury, but are not safe in cold acid of the same strength. They break up in acid of sp. gr. 1·42, whether it is hot or cold. Cornets containing larger proportions of silver may not break up if the first acid is weaker, say of sp. gr. 1·2.

Boiling is kept up gently for about thirty minutes, and the tray is then withdrawn, drained, washed by dipping vertically in and out of a vessel of hot distilled water, drained again, similarly washed in a second vessel of water, and placed in a second platinum boiler filled with boiling nitric acid of specific gravity 1·20 to 1·22, free from silver. In this the cornets remain for a period of about thirty minutes, when they are drained and washed as before, and are then ready for annealing.

The platinum tray is dried on a hot plate, and annealed over gas or in a muffle kept specially for the purpose, and free from lead fume, particles of bone-ash, etc. The temperature should be as high as possible, consistent with the safety of the cornets. They fuse at 1,064° C. (having the same melting point as pure gold), at which temperature the muffle appears orange-red. If annealed at a low temperature, the cornets are rough in texture, dull and fragile, being crushed easily between the finger and thumb. In this condition they adhere to the platinum, and, in detaching them, fragments are often left sticking fast inside the cups. If annealed properly the cornets are smooth, lustrous and hard, showing signs of incipient fusion under a magnifying glass, and only yielding to considerable force exercised by the finger and thumb. Under these conditions they can always be detached from the platinum entire, and do not readily absorb moisture and gases from the air. After boiling, the cornets are very soft and fragile and dark brown in colour; on being annealed they shrink and harden, and regain the ordinary yellow colour of gold.

Relative Advantages of Parting in Flasks and in Platinum Boilers.—The use of platinum trays and boilers effects a great saving of time in decanting and washing, as one operation takes the place of as many as 144. If the standard of an alloy is unknown, so that it is not certain that the cornet will remain entire in the acid, it must, of course, be boiled separately, as, if one cornet in the tray breaks up, fragments may adhere to a number of others. The manipulation of the platinum tray is easier than that of parting flasks, and, in addition, the treatment of the cornets is more uniform, so that the correction afforded by the use of checks is more trustworthy. On the other hand, if platinum or palladium is present in an individual cornet, it imparts a straw-yellow or orange colour to the acid; but where a number of cornets are boiled together, it is obviously impossible to say from which the colour is derived, so that less information on the subject is obtained.

6. **Weighing the Cornets.**—The balance must readily indicate differences of 0·05 per 1,000, or $\frac{1}{40}$ milligramme, with a weight of $\frac{1}{2}$ gramme in each pan. The " checks " or " proofs " (*vide infra*) are weighed first, and their mean excess or deficiency in weight applied as a correction to all the cornets worked with them. This may be done by means of a light rider. The " weighing in " correction (p. 527) is also allowed for, care being taken to observe that this correction represents a definite weight of the original alloy of gold, not of fine gold. It follows that in the assay of a gold alloy 500 fine, half the weighing-in correction must be used in correcting the result. With gold 900 fine, nine-tenths of the weighing-in correction is to be applied, and proportionate amounts with gold of other finenesses. The report is at once indicated by the marks on the weights without further calculation.

The weighing is usually by substitution, the proof cornets being followed in the same pan by the ordinary cornets, so that if the proof cornets are of the same mass as the others, the weight in the other pan is a mere counterpoise, and its error, if any, is immaterial. When proof cornets are of different mass from the ordinary cornets, allowances are made, the surcharge being taken as proportional to the weight of the cornet.[1]

The weights are the $\frac{1}{2}$ gramme (or sometimes some other weight, such as

---

[1] For various rapid methods of weighing, intended to save calculation and to avoid mistakes when cornets are of very different weights, see G. Foord, *Proc. Roy. Soc. of Victoria*, Nov. 1875; A. O. Watkins, *Chem. News*, 1912, **106**, 248, 259.

5 grains, 7 grains, or 10 grains), which is stamped "1,000," and decimal subsidiary weights stamped 900, etc., down to 0·5. The stamped numbers denote the number of $\frac{1}{2}$ milligrammes ("millièmes") contained in the weight. Ordinary weights in the gramme system may, of course, be used, each milligramme corresponding to 2 per 1,000 in the assay system. The report finally made gives the number of parts (in millièmes and tenths) of pure gold in 1,000 parts of the alloy.

The weights usually supplied even by the best makers differ from the standard weights by varying amounts, which sometimes exceed 0·1 per 1,000. These errors cannot in general be disregarded. In many sets of weights a cumulative error of 0·5 per 1,000 or more may be introduced. It is necessary to allow for the errors, or still better, to adjust the weights by carefully polishing them with rouge or with fine emery, or by rubbing them on a ground glass plate if they are too heavy, or by plating them with gold if they are too light. It is not advisable to gild weights unless they consist of platinum. Riders should be used instead of small weights. For further remarks on weights see above (p. 513).

Auxiliary Balance for Weighing Cornets.—Robert Law, of the Melbourne Mint, has invented[1] an auxiliary balance which gives the approximate weight of the cornets without the use of any weights. The balance has a pan to hold the cornet attached to one end of the beam, and the other end is prolonged as a pointer, the position of which (when it comes to rest), with reference to a curved scale, denotes the percentage of gold present. The cornet is then transferred to the pan of an ordinary balance, and the necessary weights, already determined by the auxiliary balance within narrow limits, are placed on the other pan. For weighing large numbers of cornets of widely differing weights, this balance is said to be convenient, saving both time and wear of the balance and weights. It has been in use since 1896 at the Australian Mints. The scale gives the weight of cornets from unrefined gold between 700 and 1,000 fine. The balance is not required in the assay of refined gold bullion or in other cases where the probable composition of the gold is already known approximately.

Surcharge.—The gold cornet does not actually contain the whole of the gold present in the original alloy and nothing else. Gold is lost by (a) volatilisation; (b) absorption by the cupel; (c) solution in the acid. On the other hand, the cornet always retains (1) some silver; (2) occluded gases. The algebraical sum of these losses and gains is called the "surcharge," since the cornet usually weighs more than the gold originally present in the assay piece; if the reverse is the case, the work is less accurate. The various losses and gains are discussed in detail below.

Losses of Gold.—G. H. Makins[2] found gold and silver in the proportion of about 1 to 9 in the dust taken from a flue used only in gold and silver cupellation, but did not attempt to ascertain the percentage loss of gold by volatilisation. He also showed that large amounts of gold were dissolved by nitric acid in the course of assaying, and attributed the dissolution of gold to the presence of nitrous acid; but he supposed that it would not be dissolved in the weaker acid, where nitrous acid was formed in larger quantities, owing to the protective action exercised by undissolved silver, which formed the positive element in the gold-silver couple. The fact that gold is dissolved

---

[1] R. Law, *J. Chem. Soc.*, 1896, **89**, 526.
[2] Makins, *J. Chem. Soc.*, 1860, **13**, 77.

by nitric acid in parting operations has been well known to bullion assayers for many years. It is, however, very slight, except when strong acid of specific gravity 1·42 or more is used. F. P. Dewey has published the results of some exact experiments on the subject.[1] In 1872 A. H. Allen showed [2] that gold dissolved in nitric acid is not precipitated on dilution with water.

As the result of experiments made by the author at the Royal Mint in 1893,[3] it was found that in the assay of standard gold (916·6 fine) with the ordinary surcharge of 0·4 to 0·8, the loss of gold is about 0·4 per 1,000, of which about 82 per cent. is absorbed by the cupel, 8 per cent. is dissolved in the acid, and the remaining 10 per cent., which is unaccounted for, was put down as having probably been volatilised. These ratios, however, vary considerably, as a hot fire increases the loss by absorption, while by prolonged boiling in acid, and especially by annealing the fillets at a high temperature, the amount of gold dissolved in the acids is increased.

These results were obtained when using the ratio for cupellation of gold 1 part, silver 2·75 parts, copper one-twelfth part, with the old-fashioned coke furnaces, muffle open to the fuel, and strong acid of specific gravity 1·32.

With the ratio, gold 1 part, silver $2\frac{1}{6}$ parts, copper one-twelfth part, with isolated muffle, gas-fired, the losses of gold have been more recently determined by the author, the details being as follows :—

Volatilisation.—The fumes from the muffle were passed through an iron tube of $1\frac{1}{2}$ inches diameter, with a right-angled bend near the furnace. Much litharge was deposited at and before the bend, and a small quantity, diminishing rapidly in amount with the distance from the furnace, in the straight portion of the tube. In one experiment 35 batches of assays were worked successively in May, 1910, a batch consisting of 72 assay pieces, each consisting of 4 grammes of lead, about 0·5 gramme of gold and about 1 gramme of silver. The amount of litharge volatilised and condensed was 76 grammes, and this contained 0·2143 gramme silver and 0·0012 gramme gold. The proportions of the metals present in the furnace which were volatilised and condensed were thus 0·7 per cent. of lead, 0·008 per cent. of silver, and 0·0001 per cent. of gold. The loss of gold by volatilisation of one part in a million is, of course, inappreciable in the assay. In other experiments similar results were obtained, the gold volatilised ranging downwards from 0·0001 per cent. to 0·00006 per cent. It is, of course, possible that a part of the volatilised gold was not condensed, but it is difficult to believe that as much as one-half could have escaped.

The amount of gold *absorbed* by the cupel varies from 0·35 to 0·40 per 1,000, in the cupellation of gold 1 part, silver $2\frac{1}{6}$ parts, lead 8 parts. The loss may run up to 0·5 per 1,000 of gold if the furnace is hotter than usual. The gold *dissolved* in the acid varies with the strength of the acid and with the ratio of gold to silver. With a ratio of 1 to 2·17 the amount of gold dissolved is less than with the ratio 1 to 2·5. With acid of specific gravity 1·20 and the ratio of 1 to 2·17, the amount of gold dissolved in parting is certainly less than 0·005 per 1,000. With acid of specific gravity 1·30, this is increased to 0·03 per 1,000 if the boiling is maintained for five hours after the ordinary parting is completed.

[1] Dewey, *J. Amer. Chem. Soc.*, 1910, **31**, 318.
[2] Allen, *Chem. News*, 1872, **25**, 85.
[3] T. K. Rose, *J. Chem. Soc.*, 1893, **63**, 710.

The loss of gold in assaying in present practice at the Mint may thus be stated as :—

Volatilisation,	.	.	.	.	.	0·001 per 1,000.
Cupel absorption,	.	.	.	.	.	·400 ,,
Dissolution in parting acid,	.	.	.	.	·005 ,,

It is clear therefore that the loss by absorption in the cupel is alone worthy of consideration.

An increase in the percentage of copper in the assay piece is accompanied by an increase in the loss, as is shown in the following table, which gives the relative surcharges obtained in the parting assay of gold-copper alloys of different standards ; it is compiled from a number of results obtained at the Royal Mint :—

TABLE LI.

Standard of Alloy.	Surcharge on the Assay Piece per 1,000.
916·6	+ 0·6
750	+ 0·5
625	+ 0·4
500	·0
375	− 0·6

The amounts of lead used were in accordance with those given in column B of Table L, p. 528. Table LI, as already explained, only gives relative surcharges ; the absolute amounts vary with the treatment.

The influence of the temperature of the furnace on the surcharge was pointed out by Roberts-Austen.[1] The following table is compiled from the results of his experiments :—

TABLE LII.

Temperature of Furnace.	Composition of Alloy.	Surcharge.
(a) Slightly lower than usual, . . .	Gold, 916·7	+ 0·05
(b) Ordinary temperature, . . .	Copper, 83·3	− 0·10
(c) Slightly higher than usual, . . .		− 0·37

The loss of gold was found to be 0·645 per 1,000 in series (a), and 0·723 per 1,000 in series (b).

Rössler[2] has shown that the loss of gold in cupellation increases with the amount of lead used and decreases as the amount of silver is increased.

The author obtained the following results at the Royal Mint in 1914, by cupelling under precisely similar conditions as to temperature and draught the stated amounts of lead with 0·5 gramme gold, 1·054 grammes silver and 0·046 gramme copper :—

---

[1] See Percy, *Metallurgy of Silver and Gold*, p. 275.
[2] Rössler, *Dingl. Poly. J.*, **206**, 185.

TABLE LIII.

Weight of Lead.	Loss of Gold in Cupellation.
1 gramme.	0·18 per 1,000.
2 grammes.	0·26 ,,
4 ,,	0·50 ,,
8 ,,	0·80 ,,
20 ,,	1·90 ,,

Silver Retained in the Cornet.—It has been found at the Royal Mint that after boiling in the first acid the amount of silver retained is about 2·5 to 3·0 parts per 1,000. The amount left undissolved by the second acid varies with the length of time of its action. Under normal conditions with a surcharge of 0·6 to 0·8, the silver left in the cornets is from 1·0 to 1·2.

By continuing to boil in the second acid kept at about specific gravity 1·25 by occasional additions of water, the following results were obtained by the author in 1911 :—

TABLE LIV.

	Surcharge per 1,000.	Fineness of Cornets.
After 30 minutes in first acid, . . .	+ 3·03	996·79
,, 5 ,, second acid, . .	1·18	998·59
,, 10 ,, ,, . .	0·96	998·81
,, 20 ,, ,, . .	0·78	999·03
,, 30 ,, ,, . .	0·59	999 07
,, 40 ,, ,, . .	0·41	999·27
,, 50 ,, ,, . .	0·30	999·38
,, 1 hour in ,, . .	+ 0·23	999·45
,, 2 hours in ,, . .	− 0·04	999·66
,, 5 ,, ,, . .	− 0·27	999·77

After each hour, fresh acid free from silver was substituted for the previous acid. The results show that very little gold is dissolved in acid of specific gravity 1·25, when parting cornets of the ratio of 1 to $2\frac{1}{6}$, the total amount of gold dissolved in five hours in such acid being about 0·32 per 1,000.

If the acid continues to boil until its constant strength is reached (sp. gr. 1·42), gold is dissolved to a considerable extent (0·5 per 1,000). If much gold is dissolved the results are not uniform. When checks are used perfectly satisfactory results are obtained with surcharges of from + 0·2 to + 1·0 per 1,000 or even more.

Effect of Varying the Proportion of Silver to Gold.—The views of various authors on the best proportion to be used have been given above (p. 527). The following are some of the results of a series of experiments made by the author in the years 1904-5. In each case the results are the means of a number of closely concordant assays worked together under similar conditions. Except where stated otherwise, the cornets were parted by boiling in nitric acid for two half-hours, the first acid being of sp. gr. 1·26, and the second acid of sp. gr. 1·32. In all cases 0·5 gramme of gold, 0·045 gramme of copper and 4 grammes of lead were used :—

Ratio of silver to gold, .	.	2·71	2·50	2·00
		Per 1,000.	Per 1,000.	Per 1,000.
Surcharge, . . .	.	+ 0·292	+ 0·208	+ 0·062
Loss of gold, . .	.	0·474	0·547	0·593
Silver retained by cornet,	.	0·766	0·755	0·655

When the ratio of 2 parts of silver to 1 part of gold is used, it is seen that the amount of silver retained by the cornet is less, the loss of gold greater and the surcharge less than when more silver is used.

If the surcharges are higher, the relation of the results remains approximately the same as shown below :—

Ratio of silver to gold, .	.	.	2·75	2·00
			Per 1,000.	Per 1,000.
Surcharge—Series (a), .	.	.	+ 0·608	+ 0·466
,, ,, (b), .	.	.	+ 0·873	+ 0·705

If less than 2 parts of silver to 1 of gold are used, the surcharge begins to rise again, as follows :—

Ratio of silver to gold,	2·17	2·00	1·90	1·75
Surcharge, per 1,000,	+ 1·05	+ 1·03	+ 1·15	+ 1·60

If the ratio of silver to gold is lower than 1·75 to 1, the cornets are not properly parted by boiling in two acids in the ordinary way. By boiling for half an hour each in three acids of sp. gr. 1·26, 1·32, and 1·41 respectively, the following results were obtained :—

Ratio of silver to gold, .	2·00	1·90	1·75	1·50	1·25	1·00
Surcharge, per 1,000, .	+ 0·80	+ 0·80	+ 0·87	+ 1·52	+ 3·60	+ 533·0

Among other results, it was found that the uniformity of the assays was greater when the ratio of about 2 to 1 was used than in the case of other ratios. This led to the adoption of the ratio of $2\frac{1}{6}$ to 1 for bullion assays at the Royal Mint.

Occluded Gases.—Graham[1] proved that certain cornets retained twice their volume of gases (mainly carbon monoxide) in occlusion after annealing. This would amount to two parts by weight in 10,000. According to Varrentrapp, the gas retained varies with the temperature at which annealing takes place. Recent experiments at the Royal Mint show that under ordinary working conditions the amount of gases absorbed by cornets is practically *nil*.

Checks or Proofs.—Since the losses and gains detailed above are dependent on so many conditions, it is always necessary to subject check-pieces of known composition to the same operations as the alloys under examination. The use of checks in the Royal Mint was prescribed by law as early as the fourteenth century.[2] Standard trial plates (916·6 fine) were made and used for

---

[1] Graham, *Phil. Trans. Roy. Soc.*, 1866, p. 533.
[2] *Fourth Annual Report of the Mint*, 1873, p. 38.

this purpose. Pure gold is now in general use for check assays except in special cases, such as at mints and hall marking offices.[1] If 1,000 parts of pure gold were taken originally as a check, and the weight of the resulting cornet is 1,000·3, then 0·3 must be deducted from all the other results. No appreciable error is caused, provided that the standard of the alloys under examination is not much below 1,000. For example, if an alloy is 900 fine, the amount to be deducted may be assumed to be nine-tenths of 0·3 per 1,000 or 0·27, the error in deducting 0·3 being in this case one-tenth of the surcharge. Experiments show that the assumption made above is justified, and that the loss of gold is proportional to the amount of gold present if the composition of the assay pieces is similar. The errors introduced by the use of different weights are usually much greater, unless they are carefully adjusted (see above, p. 538).

When the bullion to be assayed contains a large proportion of a base metal or metals, check assay pieces are made up of similar composition. Duplicate check assays for each variety of bullion in the batch are required. In cases where the composition of the bullion is not already approximately known and cannot be judged by considerations arising from its colour, hardness, origin or other data, the effect of the base metals on the surcharge is ignored. The assays are in that case less exact, and the unknown base metals will usually cause an increase in cupel absorption, so that the results will be too low.

The variation of surcharge caused by differences of temperature and draught in various parts of the muffle is determined by placing check assay pieces in suitable positions in the batch. The exact number of proofs and their positions is a matter for experiment with each furnace in use, and varies with the size of the batch of assays. At least two proofs, and preferably three, are used for a few assays, and from four to six proofs are usually enough for a large batch, such as 72, as used at the Royal Mint. With careful work the surcharge throughout the batch does not vary on account of differences of temperature and draught in the furnace.

It has been assumed above that absolutely pure gold is used for proofs, but pure gold is not always available for the purpose. The method of preparation of proof gold in use at the Royal Mint is given below. Since the assay of an alloy only gives the relative fineness of proof gold and the alloy, it follows that, if the proof gold is not quite pure, the amount found in the alloy will be in excess of the truth. If a sample of proof gold is less pure than the finest yet obtained, an allowance is made. Thus, if it is 999·9 fine, a deduction of 0·1 is made from all results of assays checked by it. This deduction is readily proved to be a very close approximation to the correct one.

Lastly, the " weighing-in " correction is applied. If the original weight taken was, say, 1,000·4 (recorded as + 4), it is sufficient to deduct 0·4 from the final weighing. In this correction 0·4 of alloy is reckoned as fine gold, but the error is inappreciable when alloys differing but little from pure gold are under examination. In the case of an alloy 900 fine, nine-tenths of the weighing-in correction must be added to or deducted from the weight of the cornet. If the alloy is 500 fine, one-half of the weighing-in correction must be applied, and in general if the weighing-in correction is x and the weight of the cornet is y, the correction to be applied in weighing out is $\dfrac{xy}{1,000}$.

---

[1] Roberts-Austen and Rose, *Proc. Roy. Soc.*, 1900, **67**, 105 ; 1902, **71**, 161.

Preparation of Pure Gold.—The purest gold obtainable is required for use as standards or check pieces in the assay of gold bullion. The following method of preparing it is now in use at the Royal Mint. About 20 ozs. or 620 grammes of gold assay cornets from the purest gold which can be obtained are dissolved in a large flask in about two litres of aqua regia consisting of a mixture of 400 c.c. of nitric acid and 1,600 c.c. of hydrochloric acid. The acid is added little by little, and heat is not applied at first, or the action will be too violent. When action is at an end and all the gold dissolved, the solution is decanted into a large porcelain basin, the undissolved silver chloride being separated, and the excess of acid driven off on a water bath, the sides of the basin being kept cool to prevent gold chloride from creeping up the sides. At the finish the blackish-red liquid is raised to 100° for a few minutes and then allowed to solidify. It should solidify at about 70°. The solid consists of $HAuCl_4$. It is dissolved in distilled water and diluted to about 20 litres ($4\frac{1}{2}$ gallons) in a cylindrical glass jar. More silver chloride is precipitated and is allowed to settle clear, an operation which takes from 3 to 5 days. The clear liquid is then siphoned off from the silver chloride into an equal volume of distilled water, when it remains clear, no further separation of silver chloride taking place. The mixture is siphoned into a saturated solution of SO_2 in distilled water. This is conveniently made by passing the gas coming from a siphon of liquid SO_2 through two wash bottles and then into distilled water. The amount of SO_2 solution required is about 60 litres, into which the 40 litres of gold solution is siphoned in a thin stream with constant stirring.

Under these conditions the yellow-coloured gold solution is instantly decolourised, owing to the reduction of $HAuCl_4$ to $HAuCl_2$, thus :—

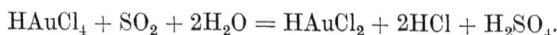

$$HAuCl_4 + SO_2 + 2H_2O = HAuCl_2 + 2HCl + H_2SO_4.$$

After a few minutes gold begins to be precipitated in a fine state of division, the action being represented by the following equations :—

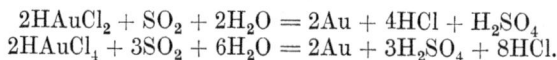

$$2HAuCl_2 + SO_2 + 2H_2O = 2Au + 4HCl + H_2SO_4$$
$$2HAuCl_4 + 3SO_2 + 6H_2O = 2Au + 3H_2SO_4 + 8HCl.$$

The precipitation is effected in five large cylindrical glass vessels.

The gold is allowed to settle for a few hours, after which the clear solution is siphoned off and the precipitate is washed by decantation in a porcelain basin, and is then transferred to a large flask of a capacity of 3 or 4 litres and repeatedly shaken with cold distilled water. The mouth of the flask is closed by a watch-glass held in position by the finger during the shaking. The water is frequently changed at first, but is left unchanged for several hours after the first few days. After washing for a week with cold water, the gold may be treated with strong ammonia with occasional shaking for 24 hours and the washing with cold water resumed. (It is found, however, that if this is done, a trace of ammonium sulphate is expelled from the gold on melting.) Finally, the flask is boiled for a few days, with occasional changes of water, until the presence of chlorides in the water can no longer be detected in a long test tube by means of silver nitrate, after the gold has been boiled in the unchanged water for five or six hours. The entire operation of washing occupies two or three weeks, distilled water being used throughout.

then not be allowed to touch any part of the superstructure. Balances are made with more care than weights, which are almost always inaccurate when received from the manufacturer. Small weights often differ from their stated value by 5 per cent., or even more. For this reason riders of 5 mgs. and 0·5 mg. should be used, the smallest weight used in the pan being 5 mgs. The errors on riders are reduced according to their position on the beam. At a distance of one-tenth of the half-beam length from the centre, the error of the rider is reduced to one-tenth of its full amount. The rider must fit the beam so that it always lies in a plane perpendicular to it. The value of each division on the ivory scale is made, as nearly as possible, equal to 0·1 mg., 0·05 mg., 0·01 mg., or 0·005 mg., according to the sensitiveness of the balance, and the divisions are divided into 10 parts by the eye with the aid of a lens. The rider is always placed exactly on a subdividing mark on the beam, which should be nicked.

If riders are used, as suggested above, the difficulties due to errors in weights are reduced ; but, if great accuracy is required, the riders must be examined, and, if necessary, adjusted by the assayer. It is necessary that the riders should be in concordance with the assay-ton weights. It may generally be assumed that the assay-ton weights and the 1 grm. or 0·5 grm. bullion assay weight are in agreement when received from the manufacturer. The errors on weights of 1 A.T. (29·166 grms.) or of 1 grm. rarely exceed 0·01 per cent. The assumption that these weights are correct involves a possible error of only 0·02 per cent., and this corresponds to only 0·1 grain of gold per ton on a 20-dwt. ore, and 0·01 grain of gold per ton on a 2-dwt. ore, amounts which are inappreciable. It is enough, therefore, to see that the rider bears the correct ratio to the gramme weight. This must be ascertained by building up a set of weights from 1 mg. to 1,000 mgs.

The weighings will be sufficiently accurate if they are made on a balance in which the divisions on the ivory scale are each equal to 0·005 mg. These divisions are divided into ten parts by the eye. Care must be taken to avoid unequal heating of the balance case. This can be watched if two thermometers reading to 0·01° C. are kept in the case—one near each end of the beam. In reading the indications of the balance, parallax is avoided by applying the eye to a pin-hole in a card fixed at about 12 inches from the ivory scale. A swinging ivory scale fixed to the beam and passing across a thread in a telescope gives still more exact readings. The position of rest of the balance is given by the formula, $\dfrac{l_1 + 2\,l_2 + l_3}{4}$, or, with greater exactness, by $\dfrac{l_1 + 3\,l_2 + 3\,l_3 + l_4}{8}$, where l_1, l_2, l_3, l_4 are successive positions of arrest (end of swing) of the balance, l_1 and l_3, on one side of the middle of the scale, and l_2 and l_4 on the other. If the middle of the scale is taken as zero, l_2 and l_4 will be negative.

The riders must be of platinum or aluminium. Brass riders increase in weight, and gilded brass riders oxidise and increase in weight still more rapidly. Some gilded brass riders at the Royal Mint increased in weight from 5·0 mgs. to 5·025 mgs. in six months. Platinum riders, if too light, may be gilded in a cyanide bath, but light aluminium riders must be rejected. Any rider may be reduced in weight by light rubbing on a sheet of ground glass, or by other mechanical means. The error on a rider can in this way be made less than 0·005 mg. without difficulty, but, if greater accuracy is desired, it saves time to determine the error of the weight and to apply it as a correction to the result of each assay.

33

The gold can be weighed by the method of substitution by which the bias of the balance is eliminated. The principle of this method is to counterpoise the gold, then to remove it from the pan, and compensate for the loss of weight by a rider. If direct weighing is preferred, the position of rest of the balance when the pans are empty is determined, the gold is put in the pan, a rider put on the beam in such a position as to equalise approximately the effect of the weight of the gold, and then the position of rest again determined. The method of swings is always used to determine the position of rest, disregarding the first swings. An example is appended.

A. *Position of Rest of Empty Balance.*—The swings, after the first two, are in succession $+ 3\cdot4$ divisions of the index, $- 2\cdot7$ divisions, $+ 3\cdot2$ divisions. (Plus divisions are always reckoned on the side where the weights are placed, minus divisions on the side where the gold is added.) The position of rest or true zero is $\frac{1}{2}\left(\frac{3\cdot4 + 3\cdot2}{2} - 2\cdot7\right) = + 0\cdot3$ division. This determination is repeated after the parted gold assay pieces have all been weighed.

B. *Value of the Division.*—A weight of, say, $0\cdot1$ milligramme (by means of a rider) is placed on the weight side, and the previous work repeated. The swings are $- 2\cdot1$, $+ 0\cdot4$, $- 2\cdot0$, and the new position of rest is $\frac{1}{2}\left(+ 0\cdot4 - \frac{2\cdot1 + 2\cdot0}{2}\right) = - 0\cdot8$, and $1\cdot1$ division $= 0\cdot1$ milligramme. This value does not alter much from day to day.

C. *Weight of Parted Gold.*—The gold is found to require about 1 milligramme to counterbalance it, and the swings are now $- 4\cdot6$, $+ 0\cdot2$, $- 4\cdot4$. The new position of rest is $\frac{1}{2}\left(0\cdot2 - \frac{4\cdot6 + 4\cdot4}{2}\right) = - 2\cdot15$. The gold, therefore, weighs $1 - \left(2\cdot45 \times \frac{0\cdot1}{1\cdot1}\right)$ milligramme—that is, $0\cdot777$ milligramme.

By weighing in this way results correct to $0\cdot002$ mg. are easily obtained by using the best assay balances in which the value of the scale division is $0\cdot005$ mg., and still greater accuracy is attainable by weighing the gold in each pan in succession and by taking means of a number of observations.

Examination of Assay Materials.—The fluxes are usually free from the precious metals, but the litharge, red-lead and granulated lead contain a small amount of silver and less gold (see p. 75). These are estimated by fusion with charcoal (or in the case of granulated lead by scorification) and cupellation. To prevent errors due to " salting," which may occur accidentally in an office where gold bullion or residues are handled, it is advisable to run a full blank charge at least once a week.

Examination of the Cupel.—When rich ores are assayed, appreciable quantities of gold are carried into the cupel, especially if certain base metals (see above, p. 508) are present in the lead button. To assay the cupel, all clean bone-ash is detached and thrown away, and the remainder crushed so as to pass through an 80-mesh sieve, and the charge made up as follows :—

Cupel, ...	. 20	grammes.
Fluorspar, ...	. 15	,,
Sand, ...	. 15	,,
Sodium carbonate, . . .	. 20	,,
Borax, ...	. 10	,,
Litharge, ...	. 25	,,
Charcoal, ...	. 1	,,

Salt cover and hoop iron.

Or, according to Lodge,[1]

Cupel, ...	30 grammes.
Litharge, ...	50 ,,
Borax glass, ...	20 ,,
Sand, ...	5 ,,
Argol, ...	$2\frac{1}{2}$,,

According to the author's experience, when fluorspar is omitted the slag is pasty, and loss of lead tends to occur in pouring. The fusion is made in a clay crucible, and the resulting button of lead cupelled. The slag of the original fusion may be cleaned by adding it to the charge, when less fluxes will be required.

E. A. Smith gives the following charge for fusing magnesia cupels :—[2]

Cupel (magnesite), . . .	40 to 60 grammes.
Sodium carbonate, ...	20 ,,
Borax, ...	20 ,,
Litharge, ...	40 ,,
Silica, ...	15 ,,
Argol, ...	2·5 ,,

Assays in Field-work.—Rough determinations of the value of gold ores by the methods described above may be made in prospecting or examining expeditions with simple apparatus, using a blacksmith's forge fire as a source of heat for fusion and cupellation. S. K. Bradford describes[3] a portable assay outfit which can be packed in a valise 26 inches long. He measures the cupelled beads of gold-silver and of parted gold by Wagoner's method (see below, p. 518). The fusion and cupellation are carried out as usual. See also "Assay by means of the Blowpipe," p. 495.

Assay by Scorification.

As stated above, this process is especially applicable to (a) complex ores, (b) very rich ores, (c) ores which are mainly valuable for their silver contents. The losses are small and the operations are easy to conduct, and need not be varied much for different classes of ore. For these reasons the process is preferred to the crucible process by some assayers ; if the ore is poor several assays are made and the lead buttons scorified together. The chief disadvantage of the process lies in the small quantity of ore that can be treated, so that the presence of one or two metallic particles of gold may cause the result to be erroneous.

Scorification is conducted in a muffle at a much higher temperature than that required for cupellation. It must be high enough to melt litharge when contaminated by silica and oxides of copper, iron, manganese, etc. A temperature of 1,050° to 1,100° C. is usually enough. The charge is placed in a *scorifier*, a shallow circular fireclay dish 2 to 3 inches in diameter, which is charged in by the tongs shown at B, Fig. 209. The charge consists of about 50 grains of ore (or $\frac{1}{10}$ A.T.), 500 to 1,000 grains of granulated lead, and a

[1] Lodge, *Notes on Assaying*, 1905, p. 57.
[2] E. A. Smith, *The Sampling and Assay of the Precious Metals*, p. 180.
[3] Bradford, *Trans. Amer. Inst. Min. Eng.*, 1910, **41**, 561.

few grains of borax glass.[1] The ore is mixed with half the granulated lead, the mixture put in the scorifier and smoothed down, the rest of the lead spread over evenly and the borax put on the top. An addition of some litharge to the cover is often made. The amount of lead to be used varies with the nature of the ore. Ricketts and Miller give the following table [2] as a guide :—

TABLE XLVII.

Character of Gangue.	For One Part of Ore take	
	Parts of Granulated Lead.	Parts of Borax Glass.
Quartzose,	8 to 10	0
Basic,	8 to 10	0·25 to 1·00
Galena,	6	0·15
Arsenical,	16	0·10 to 0·50
Antimonial,	16	0·10 to 1·00
Fahlerz,	12 to 16	0·10 to 0·15
Pyrite,	10 to 15	0·10 to 0·20
Blende,	10 to 15	0·10 to 0·20
Telluride,	16 to 18	0·10
	and a cover of litharge.	

The borax lessens the corrosion of the scorifier and renders the slag more liquid, but its quantity is kept as low as possible to prevent the slag from completely covering the bath of metal too soon. After charging in, the door of the muffle is closed until fusion takes place. As soon as the lead is melted the door is opened, and a current of air allowed to pass over the bath of metal. Some of the ore is now seen to be floating on the surface of the lead, and is rapidly oxidised, partly by the air and partly by the litharge which immediately begins to form. The sulphur, arsenic, antimony, etc., are thus soon eliminated, while copper, iron, and other bases oxidise and slag off with the borax, and the silica and other acids form fusible compounds with the litharge. Effervescence and spirting may occur, especially if the scorifier has not been well dried by warming before it is used. The slag soon forms a ring completely encircling the bath of metal. As oxidation of the lead proceeds, the litharge flows to the sides and increases the quantity of slag until at length the ring closes completely over the metal, leaving a flat uniform surface. This usually happens after from thirty to forty minutes. The slag should be " cleaned " before withdrawal. This is done by placing 3 grains of charcoal powder wrapped in tissue paper on the surface of the slag, with a pair of cupel tongs, and closing the muffle door. A number of globules of lead are formed by the reduction of the litharge, and these, falling through the slag, extract and carry down with them any gold and silver which it may still contain, and concentrate them in the molten lead below. The fusion being quiet again, the charge is poured into an iron mould with the scorifying tongs, and the lead button cleaned from slag with a hammer. If the

[1] By the use of a flatter scorifier than is usually made, or even by using a roasting dish, ½ A.T. of ore may be taken with a charge of 75 grammes of lead. E. H. Simonds, *Cal. Mines and Minerals*, 1899, p. 226.

[2] Ricketts and Miller, *Notes on Assaying*, 1897, p. 96.

button is too large to cupel at once it is re-scorified in the same dish, fresh lead being added if it is not soft and malleable.

The slag from the scorifier should always be separately re-treated in the case of very rich ores, but seldom contains much gold.

The losses incurred in this process are chiefly due to improper temperatures. If the muffle is too cold at first, gold is retained by the slag. This initial low temperature is often indicated by the occurrence of white patches of sulphate of lead on the surface of the slag after pouring.[1] If the slag is pasty, borax is added, but the slag is then rich. It is better to begin again with less ore or more lead. When extremely rich sulphides are assayed, results are often obtained which do not agree well. Stetefeldt[2] recommends in all such cases that the sulphides be attacked by nitric acid and the residues dried and scorified.

Lodge states[3] that the losses of gold in scorifying are greater than in cupelling. He accordingly recommends the use of the former method only when absolutely necessary, as in the assay of zinc residues from the cyanide process, and of copper ingots or bars and material rich in copper. No experiments seem to be on record showing the relative losses of gold when the crucible and scorification methods are used on ordinary quartzose or pyritic ores.

Telluride ores often yield erroneous results, usually ascribed to volatilisation of the gold, but according to the author's experience, the vanished gold can be found in the slag or cupel, and has not been volatilised (see also p. 10). The tellurium should not be allowed to enter the lead in large amounts or should be removed again by scorification with litharge (S. W. Smith) before the lead is passed to the cupel (see above, p. 507).

Lodge[4] recommends the mixture in the scorifier to be covered with litharge when rich tellurides are being assayed. S. W. Smith gives the reason for this[5]—viz., that at temperatures of 700° to 900° litharge is reduced by tellurium which passes into the slag, thus—

$$2PbO + Te = Pb_2O + TeO.$$

Lodge considers the pot assay and scorification assay to be equally good for tellurides. Tindall[6] and Fulton are in favour of the crucible method.

Detection of Gold in Minerals.—The detection of gold in loose alluvial ground has been already described (p. 101). A similar method may be employed in the case of auriferous quartz after grinding it. If the concentrate obtained in either case contains sulphides, these are collected, roasted or treated with nitric acid and re-ground. The light particles of oxide of iron can now be separated from any gold that may be present by washing. " Colour " may often be obtained thus when none could be seen after the first concentration. The washing is made easier by removing from the concentrate the magnetic oxides and iron from the grinding tools by means of a magnet. All the finest particles of gold are lost in the process of washing, and consequently many auriferous ores cannot be made to " show colour."

---

[1] Percy, *Metallurgy, Silver, and Gold*, p. 244.
[2] Stetefeldt, *Lixiviation of Silver Ores*, New York, p. 106.
[3] Lodge, *Notes on Assaying*, p. 113.
[4] Lodge, *Mineral Industry*, 1899, **8**, 397.
[5] S. W. Smith, *Trans. Inst. Mng. and Met.*, 1908, **17**, 473.
[6] Tindall, *Trans. Inst. Mng. and Met.*, 1901, **9**, 354.

518   THE METALLURGY OF GOLD.

The following method devised by Wm. Skey, analyst to the Geological Survey of New Zealand, is said by him to give good results. The sample of ore is carefully roasted, then digested with an equal volume of an alcoholic solution of iodine for a length of time varying from twenty minutes to twelve hours, the longer time being allowed if the ore is poor. A piece of Swedish filter paper is then saturated with the clear supernatant liquid and afterwards burnt to an ash ; if gold is present in the ore the ash is coloured purple, and the colouring matter can be quickly removed by bromine. This method is said to show the presence of as little as 2 dwts. gold per ton in certain ores, but is not uniformly successful. A solution of iodine in potassium iodide is said to be better than the alcoholic solution. Bromine or chlorine may be substituted for iodine. A mixture of 5 to 10 parts of bromine with 100 of water may be used to 100 parts of ore. The ore must be fine enough to pass an 80-mesh sieve and should be reground after roasting. After leaving the mixture to stand for some hours with occasional stirring, the liquid is filtered and the excess of bromine evaporated from the clear solution, which may then be tested with stannous chloride. Dr. Don[1] found these methods defective, and was forced to use the ordinary crucible assay when examining material containing small amounts of gold, taking samples of 4·48 lbs. of ore. L. Wagoner[2] determines the weight of small beads by fusing them and measuring their diameter under the microscope. The error in reading is about 0·001 mm. or 6·36 per cent. on a gold bead weighing 0·001 mg.

Special Methods of Assay.

1. **Amalgamation Assay.**[3]—This is useful in determining the amount of "free" gold present in the ore capable of being extracted by mercury. The sample of ore is crushed finely enough to pass through a 60- or 80-mesh sieve, made into a paste with a little water, an equal weight of mercury added, and the whole ground in an iron mortar with a pestle for from two to four hours, small additional amounts of water or mercury being added from time to time according to the appearance of the triturated mass. The consistency of the mass must be such that the globules of mercury do not sink in it but are broken up into very small particles. A little sodium amalgam dissolved in the mercury prevents it from flouring. The grinding is continued until the particles of ore are all impalpably fine. A machine for the purpose, called the "arrastra" mortar, has a large pear-shaped muller loosely fitting the inside of the mortar, and capable of being revolved in it by means of a handle. Complete amalgamation is performed in this machine much more rapidly than in an ordinary mortar. When the operator judges that amalgamation is complete, enough water is added to reduce the mass to a thin pulp and stirring is continued for a few minutes to collect the mercury at the bottom. The contents of the mortar are then "washed down" in a pan, the mercury collected and distilled, and the residue, consisting of gold, silver and base metals, scorified with test-lead, cupelled and parted. If it is not necessary to estimate the silver extraction, the mercury may be dissolved with nitric acid, and the gold cupelled and weighed.

The sample may, of course, be panned before amalgamation. In that

---

[1] Don, *Trans. Amer. Inst. Mng. Eng.*, 1897, **27**, 565.
[2] Wagoner, *Trans. Amer. Inst. Mng. Eng.*, 1901, **31**, 798.
[3] See also *Trans. Amer. Inst. Mng. Eng.*, 1905, **35**, 399.

case Charleton recommends [1] treatment in a weighed golden dish, the surface of which has been amalgamated.

2. **Assay by Chlorination.**—Plattner's method for the assay of roasted pyrites consists in placing the mineral moistened with water in a glass cylinder, 200-250 millimetres deep and 20-30 millimetres in diameter, and introducing a current of chlorine gas at the bottom. When the odour of chlorine is noticed above the ore, a cover is put on, the stream of chlorine stopped and the whole left for twenty-four hours, after which the reaction is complete if chlorine is still in excess. Boiling water is now run through the ore until all soluble salts have been washed out, and the gold contained in the solution is precipitated by ferrous sulphate, collected, cupelled and parted. This method fails if more than about 20 per cent. of silver is present, as the chloride of silver formed encrusts the gold and protects it from the action of the chlorine.[2] Balling recommends the addition of common salt to dissolve the chloride of silver, but found that the telluride ores of Nagyag yielded only 85 per cent. of their silver and 92 per cent. of their gold when successively treated with chlorine and sodium chloride.[3]

Another method is to place the completely roasted sample in a stoppered bottle with enough water to make the whole of the consistency of thin mud. The ore and water should together occupy about two-thirds of the bottle. Bleaching powder and a thin glass bulb filled with dilute sulphuric acid are then added and the bottle securely closed. As cork is attacked by chlorine, glass or vulcanite stoppers are better, and the screw-stoppered soda water bottles are most convenient; if corks are used they must be wired down. The bottle is then shaken so as to break the sulphuric acid bulb and mix its contents with the bleaching powder, when chlorine is evolved. The bottle is now left for several hours in a warm place, being shaken occasionally by hand to mix its contents. At the end of a period of eight to twelve hours the bottle is opened, and if excess of chlorine is still present the liquid is separated from the ore and the latter washed thoroughly by filtration or decantation. The liquid and washings, whether clear or muddy, are warmed to expel free chlorine, and an excess of ferrous sulphate is then added to them. The precipitate is collected, scorified with lead, and cupelled. In all cases it is better to keep the first liquid separate from the washings, which should be concentrated by evaporation, since, if this is not done, the precipitate of gold may be too fine to settle and will pass through filter-paper. Bromine may be used instead of the materials generating chlorine. The quantities of chemicals required will be such as are sufficient to generate a volume of chlorine equal to twice the capacity of the bottle used, or to make a solution of 2 per cent. of bromine in water. Only finely divided gold comparatively free from silver is extracted by this method.

3. **Assay of Metallic Copper and of Copper Matte containing Gold and Silver.**—Two methods are in use—a furnace method and a mixed wet and dry method.[4]

(*a*) *Furnace Method.*—Ten scorifications are made each of 0·1 A.T. of the sample, 50 grammes test lead (using half as cover) and 1 gramme borax

[1] Charleton, *Trans. Inst. Mng. and Met.*, 1901, 9, 69.
[2] See results of Hofman and Magnuson on p. 313.
[3] Balling, *L'Art de l'Essayeur* (Paris, 1881), p. 433.
[4] A. H. Ledoux, *Trans. Amer. Inst. Mng. Eng.*, Oct. 1894, and 1904, 33; Peter's *Modern Copper Smelting* (7th ed., 1895), p. 67. For full bibliography see E. A. Smith, *The Sampling and Assay of the Precious Metals*, p. 320.

glass as cover. The lead buttons are cupelled separately and the beads weighed together, and parted. The cupels are re-fused in five lots of two each with 90 grammes of litharge, 50 grammes each of soda and borax, and 3 grammes argol. The results are higher than those obtained by the mixed method.

(b) *Mixed Wet and Dry Method.*—Weigh out 1 A.T. of the auriferous material, place it in a beaker of a litre capacity, and add gradually enough nitric acid of specific gravity 1·14 to dissolve it completely ; heat until red fumes cease to come off, dilute to 600 c.c. with water, and add sodium chloride solution and then 10 c.c. of a concentrated solution of lead acetate and 1 c.c. of concentrated sulphuric acid, and allow the lead sulphate to settle. The precipitate is filtered off and washed. It contains the gold which has been collected and carried down by the sulphate of lead. The filter-paper and precipitate are dried, the paper burned, and the ash and lead sulphate scorified with test-lead. The button is cupelled and the gold with any trace of silver it may contain is weighed and then parted. The difficulty of the method is partly due to the fact that gold is dissolved by solutions of nitrate of copper containing nitric acid.

The silver is determined on a separate sample, which is dissolved and treated as before, except that a slight excess of sodium chloride, or, better still, of sodium bromide (Whitehead[1]) is added after the sulphuric acid and before the lead acetate. The solution is well stirred, and the precipitate allowed to settle and treated as before. The use of the lead acetate is to cause the precipitates of gold and silver to settle quickly, and to enable them to be filtered effectively. Sodium bromide is used instead of the chloride, on account of the greater insolubility of the silver salt.

Fulton gives a crucible method for the assay of copper matte, thus :—[2]

Matte, ...	0·25 A.T.
Silica, ...	10·5 grammes.
Litharge, ...	67 ,,
Sodium carbonate, ...	24 ,,
Potassium nitrate, ...	5 ,,

For disused battery plates, Johnston[3] gives a fusion charge, as follows :—

Drillings, ...	0·5 A.T.
Sulphur, ...	0·25 ,,
Borax, ...	0·4 ,,
Glass, ...	0·4 ,,
Sodium carbonate, ...	1·75 ,,
Litharge, ...	1 ,,
Charcoal, ...	1·5 grammes.

4. **Assay of Purple of Cassius.**—One part of Purple of Cassius is fused with three parts of carbonate of soda, cooled and dissolved in water. The gold remains undissolved, and is collected on a filter and cupelled after incineration.[4]

---

[1] C. Whitehead, *Chem. News*, 1892, **66,** 19.
[2] Fulton, *Fire Assaying*, p. 109 ; Johnston, *Rand Metallurgical Practice*, vol. i., p. 317.
[3] Johnston, *Rand Metallurgical Practice*, vol. i., p. 319.
[4] *Ann. de Pharmacie*, vol. xxxix.

5. **Assay of Graphite Crucibles.**—Graphite crucibles, stirrers, etc., contain considerable quantities of gold and silver after use in melting bullion. They may be crushed with water in the Elspass mill or other Chilian mill (*q.v.*), and the products washed down in order to separate shots of metal. The residues, even after roasting, yield only a moderate proportion of their values when treated by amalgamation, and can still be profitably smelted. The sampling of the crushed material before amalgamation is difficult, and must be carried out with great care. The dried sample is roasted and scorified, or fused in pots, but J. Loewy obtains correct results by scorifying without previous roasting.[1] He takes 5 grammes of the material, 60 grammes of lead, and a little borax or glass powder.

Scorification has been discarded at the Royal Mint and a method of pot fusion adopted. The method is to estimate the moisture in a large sample, then to roast thoroughly at a high temperature so as to remove all the graphite, and to make up a charge for fusion as follows :—

Roasted material, ...	20 grammes.[2]
Litharge, ...	60 ,,
Sodium carbonate, ...	30 ,,
Borax, ...	30 ,,
Charcoal, ...	1·5 ,,

The lead buttons weigh about 33 grammes. The slags are cleaned by fusion with litharge and charcoal. The method is due to S. W. Smith.

6. **Estimation of Gold in Dilute Solution**—*Chloride Solutions.*—A. Carnot has shown[3] that the rose colour produced in a chloride solution in presence of arsenate of iron is very sensitive, and can be used for colorimetric estimation of small quantities of gold. To attain this end a neutral solution of chloride of gold of known strength is prepared, and some drops of solution of arsenic acid added slowly to it. Then after a time two or three drops of dilute ferric chloride and some hydrochloric acid are added. If the liquid is not acidulated, a flocculent purple precipitate forms ; if it is too acid the reaction fails, and only a faint blue colour is seen. The liquid is made up to 100 c.c. with distilled water, a pinch of zinc dust added, and the mixture shaken in a flask. A colour is produced, varying from rose to purple, according to the amount of gold present. The solution is clear, can be filtered unchanged, and kept without alteration for some time. If more than one milligramme of gold is present (1 in 100,000), the colour becomes too intense for small differences to be noticeable ; if less than one-tenth of this amount is present (1 in 1,000,000), the colour becomes too faint. Between these proportions the amount of gold present in a liquid can be determined by comparison with a series of prepared coloured solutions.

For more dilute solutions of chloride of gold, the test described on p. 65, depending on the use of stannous chloride, may be used. A number of precipitates are prepared from solutions of gold containing known amounts, and compared with that given by the solution to be estimated. Suitable

---

[1] Loewy, *J. Chem. Met. and Mng. Soc. of S. Africa*, 1898, **2**, 205.
[2] This is an approximate weight. The actual weight taken is that of the roasted material derived from 25 grammes of the original moist sample.
[3] Carnot, *Compt. rend.*, 1883, **97**, 105 ; *J. Chem. Soc.*, 1884, **46**, 115.

volumes of liquid from which the test precipitates are obtained are as follows :—

TABLE XLVIII.

Strength of Solution. Parts of Water Present to each Part of Gold.	Volume of Solution used to obtain a Precipitate.
1 to 5 millions.	100 c.c.
5 to 20 ,,	500 ,,
20 to 50 ,,	1,000 ,,
50 to 100 ,,	3,000 ,,

If more gold is present than one part per million, the colour of the precipitate is too intense for accurate measurement. In comparing the colours of liquids, Veley's *tintometer* is generally useful.

Vanino and Seemann[1] recommend the reduction of gold chloride by hydrogen peroxide in the presence of caustic soda or potash. The reaction is complete in a few minutes in the cold. In dilute solutions, the hydrogen peroxide is destroyed by heating the solution, and hydrochloric acid is added to agglomerate the precipitate. A solution containing 3 parts of gold in a million gives a slight red colour with a bluish shimmer.

Estimation of Gold in Cyanide Solution.—The amount of gold in the solution is often determined by evaporating a known bulk to dryness in lead basins, or in porcelain basins with the addition of litharge mixed with the liquid before evaporation is commenced. The determination is finished by reducing the lead by fusion in a crucible with charcoal, and cupelling the lead button. This method is said by A. M'A. Johnston[2] to be the most accurate one. He gives 20 A.T. of solution as a convenient amount, and adds to it 10 grammes of a mixture of litharge, charcoal and silica. Care is taken to avoid spitting, and the heat is reduced towards the end of the operation, or the dried residue may cake on the dish, when it is necessary to moisten it with dilute nitric acid. The dry residue is fused with 30 grammes of a mixture of 100 parts litharge, 25 parts sand and 3 parts charcoal.

Another method, given by Johnston,[3] is to take 584 c.c. (20 A.T.) of solution, boil, add 30 c.c. of saturated lead acetate solution and 3 grammes of zinc shavings, and again boil for half an hour ; add 25 c.c. of hydrochloric acid and keep hot until the zinc is dissolved. The spongy lead is washed, pressed into a ball, dried and cupelled. Gold can also be determined as follows :—Add an excess of a solution of silver nitrate, filter, reduce the precipitate with zinc and hydrochloric acid ; filter, dry, cupel and part the button.

At St. John del Rey Mine,[4] solutions of silver in sodium cyanide and of sodium plumbite are added, the silver and gold are precipitated by zinc, and the filtered precipitate is calcined gently in a pot and fused with red lead, flour and fluxes. The lead button is cupelled and the bead parted. The results are said to be accurate.

---

[1] L. Vanino and L. Seemann, *Ber.*, 1899, **32,** 1968-72 ; *J. Chem. Soc.*, 1899, **76**, [ii.], 578.
[2] A. M'A. Johnston, *Rand Metallurgical Practice*, vol. i., p. 311.
[3] Johnston, *op. cit.*, p. 312.
[4] D. M. Levy and H. Jones, *Bull. Inst. Min. and Met.*, May, 1915.

7. **Assay by Electrolysis.**—If the gold is in solution, potassium cyanide is added in the proportion of 2·5 grammes per 100 c.c. of solution, and a current is passed through the solution between platinum electrodes. The current density should be about 0·3 ampère per square decimetre, the fall of potential 2·7 to 4 volts, and the temperature 50° to 60° C. The gold is deposited in about 1½ hours,[1] and is weighed together with the cathode in the usual way.

According to F. M. Perkin and W. C. Prebble,[2] it is better to use a solution of ammonium thiocyanate instead of potassium cyanide. With a current of 0·4 to 0·5 ampère per square decimetre, deposition is complete in 1½ to 2 hours at ordinary temperatures. The best way of removing the deposited gold from platinum electrodes is by means of a solution of potassium cyanide containing an oxidising agent such as hydrogen peroxide, sodium peroxide, or an alkali persulphate.

8. **Assay of Black Sand. Determination of Gold in the Presence of Platinum, Iridium, etc.**—These metals remain with the gold in great part when the material is subjected to the ordinary assay by fusion, cupellation and parting. For the assay of such material as black sand (see p. 183), methods have been proposed by C. Toombs,[3] Jas. Gray,[4] and A. F. Crosse.[5]

Toombs roasts black sand until free from sulphur, and fluxes as follows :—

Black sand, ...	1 A.T. (weight before roasting).
Sodium carbonate, . . .	2 ,,
Borax, ...	1½ ,,
Litharge, ...	1¼ ,,
Flour, ...	4 grammes.

The cupelled bead is parted with nitric acid of specific gravity 1·12, and the gold is dissolved in 10 per cent. aqua regia, leaving iridium undissolved.

Gray fuses and cupels as usual, but with the addition of excess of silver. The bead is parted in sulphuric acid and the gold and platinum dissolved in aqua regia, leaving iridium, etc., in the residue. The filtrate is made alkaline with caustic soda and the gold precipitated by H_2O_2, thus :—

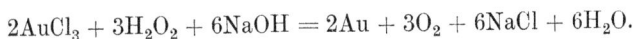

$$2AuCl_3 + 3H_2O_2 + 6NaOH = 2Au + 3O_2 + 6NaCl + 6H_2O.$$

The platinum may be precipitated with H_2S.

Crosse adds silver and melts the cupelled bead with three times its weight of cadmium (see below, p. 546), and parts in nitric acid, 1 : 3. The residue is fused with acid sulphate of potassium and treated with water to remove rhodium and palladium. The undissolved portion is treated with aqua regia, which leaves iridium, etc., unattacked. Gold is precipitated in the filtrate by SO_2, and platinum by magnesium and hydrochloric acid.

[1] A. Classen, *Quantitative Analysis by Electrolysis*, translated by B. B. Boltwood, 1903, p. 207.
[2] Perkin and Prebble, *Electro-chemist and Metallurgist*, Feb. 1904, pp. 490-494.
[3] Toombs, *J. Chem. Met. and Mng. Soc. of S. Africa*, 1913-14, **14**, 4 ; see also C. B. Horwood, *Mng. and Sci. Press*, Nov. 8, 1913, p. 724.
[4] Gray, *J. Chem. Met. and Mng. Soc. of S. Africa*, 1913-14, **14**, 2.
[5] Crosse, *ibid.*, p. 373.

A. M'A. Johnston[1] gives charges for concentrate or black sand as follows :—

(a) *Unroasted Concentrate, etc.*

Concentrate,	0·5 A.T.
Sodium carbonate,	0·5 ,,
Borax,	0·3 ,,
Litharge,	2·5 ,,
Silica,	0·5 ,,

Wash slag after fusion with litharge and charcoal.

(b) *Roasted Concentrate, etc.*

Concentrate,	1 A.T. (weight before roasting).
Sodium carbonate,	1 ,,
Litharge,	2 ,,
Borax,	0·5 ,,
Silica,	0·5 ,,
(Partly covering roasting dish.)	
Flour,	10 to 12 grammes, but varying with completeness of roasting.

The metallics, separated on a 200-mesh sieve, are in either case treated separately.

---

[1] A. M'A. Johnston, *Rand Metallurgical Practice*, vol. i., p. 313.

CHAPTER XX.

THE ASSAY OF GOLD BULLION.

Introduction.—The assay of gold bullion, as described in this chapter, has for its sole object the estimation of the percentage of gold present in the alloy, all other constituents being disregarded. In the first instance, the simple case of the assay of gold alloys containing appreciable quantities of only copper and silver will be dealt with. Refined gold ingots and the alloys used for coinage, and for almost all jewellery, come under this head. The effect of large quantities of other impurities and the precautions thereby rendered necessary will be discussed later.

The method universally employed is that of cupellation and subsequent parting. The gold bullion is cupelled with silver and lead, by which the greater part of the base metals present is removed as oxides dissolved in litharge, and an alloy of gold and silver left on the cupel. This is "parted" with nitric acid, which dissolves the silver and leaves the gold unattacked.

In the following pages the practice at the Royal Mint is described, but the same description would apply, with very slight alterations, to the methods used at other mints and assay offices.

The "parting assay" was first mentioned in a decree of King Philippe of Valois, published in the year 1343.[1] The methods of procedure in the 16th century have been described by Agricola[2] and by Ercker,[3] and those in the 17th century have been briefly described by Savot[4] and by J. Reynolds,[5] and more fully in the *Compleat Chymist*. In 1666 Pepys saw the parting assay being practised at the Mint in the Tower of London, and from his description it is clear that the method then employed bears a surprisingly strong resemblance to that of the present day. In 1829 a Royal Commission was appointed in France to examine into all questions relating to the methods of assaying gold and silver. The Committee arrived at the conclusion that the method adopted for assaying gold often overstated the amount of precious metal by 1 part per 1,000.[6] The Mint Conference held in Vienna in 1857,[7] resulted in the almost universal adoption of a more uniform method of manipulation.

The degree of accuracy now attained in most assay offices reduces the probable error in the report of an assay to about 0·1 per 1,000, but, to prevent the error from rising above this amount, all weighings must be correct to 0·05 per 1,000, which is not always the case in ordinary bullion assays.

---

[1] *First Annual Report of the Royal Mint*, 1870, p. 103.
[2] Hoover's *Agricola*, p. 247.
[3] Pettus' *Ercker* ("The Laws of Art and Nature in Assaying, etc., Metals"), bk. ii., Chap. xv.
[4] Savot, *Discours sur les Medalles Antiques* (Paris, 1627), p. 72.
[5] Reynolds, *A New Touchstone for Gold and Silver Wares* (London, 1679), p. 362.
[6] *Report of the Select Committee on the Royal Mint*, 1837, Appendix B, p. 123.
[7] *Kunst- und Gewerbeblatt Baiern*, 1857, p. 151.

The system may be conveniently regarded as comprising six distinct operations, viz. :—

1. Selection of the sample.
2. Preparation of the assay piece for cupellation.
3. Cupellation.
4. Preparation of the assay piece for parting.
5. Parting and annealing the cornets.
6. The final weighing and reporting.

1. **Selection of the Sample.**—Alloys of gold with either silver or copper or with both are practically uniform in composition if they have been melted and well mixed, and the ingot has not been pickled after casting. In these cases a single outside cut is representative of the composition of the whole of the ingot. The cut must, of course, be clean, and is usually taken in the middle of one of the ends of the ingots at the bottom. A gouging tool is used, worked either by hand or by steam or electric power. Chisel cuts from the corners are also sometimes taken.

When other metals are present, the solidified ingot is not uniform in composition, and a dip-sample is taken. The metal is melted in a plumbago crucible with a covering of borax, stirred vigorously, and one or more dip-samples taken by an iron ladle, or better, by a plumbago spoon. The sample may also be taken by a plumbago stirrer, the foot of which has a hollow cavity,[1] or by a little plumbago crucible fastened with iron wire to an iron rod. The sample is dipped out with a borax cover, and poured into an iron mould or allowed to cool in the dipper. Granulation by pouring into water may result in partial oxidation of some of the base metals.

Moderately pure tough ingots are sometimes sampled by means of outside cuts, even if there is no certainty that some base metal besides copper is not present. Cuts from a top corner and an opposite bottom corner are then taken, or drillings are taken from the top and bottom of the ingot. The results are not always trustworthy (see p. 433).

The sampling of gold wares is difficult, because the outside is usually finer than the interior, in consequence of the pickling of the wares after manufacture. " It is the usual practice to remove the ' colour ' from gold wares by a preliminary scraping or by ' buffing ' before scraping to obtain the sample proper." [2] The sampling of " base bullion " belongs to the metallurgy of lead or copper.[3]

2. **Preparation of the Assay Piece for Cupellation.**—If necessary, the assay piece is "flatted" on a clean anvil by means of a hammer with a rounded face, weighing about 11 lbs., and a portion, weighing about 0·5 gramme, is obtained by cutting with shears and filing. The metal to be cut is held firmly between the fore-finger and thumb of the left hand, and care is taken to keep the plane of the piece of gold perpendicular to the cutting faces of the shears, otherwise damage is done to the latter. Only clean portions of metal must be used.

When the assay is reported to $\frac{1}{10000}$ part, it is evident that the balance used must clearly indicate a difference in weight of 0·1 per 1,000 or 0·05

---

[1] See *J. Chem. Met. and Min. Soc. of S. Africa*, 1897, **2**, 794. The use of iron tools is apparently not approved by the Committee appointed by this Society to consider the matter. Iron would absorb sulphur, etc., from the sample

[2] E. A. Smith, *The Sampling and Assay of the Precious Metals*, p. 324.

[3] For description of methods, see E. A. Smith, *op. cit.*, chap. xx.

milligramme. It is convenient to have the balance adjusted so that one subdivision on the ivory scale traversed by the pointer corresponds to this quantity. The difference in weight of the piece of alloy from 0·5 gramme, or " 1,000," is noted down in parts per 10,000 as the " weighing-in correction." The weighed piece is wrapped in pure lead foil together with the silver necessary for parting, and some copper unless it is present in the assay piece. The lead packets are put in order in the numbered compartments of the wooden tray shown at c, Fig. 219, p. 535, their position being noted on the assay paper. The corners of the packets are squeezed down so as to fit the cupels by pliers with concave rounded faces specially designed for the purpose, and the assays are then ready to be charged into the furnace.

It was formerly considered necessary for the metals to be present in the proportion of 1 of gold to 3 of silver, but, as early as the year 1627, Savot relates that the proportion of 1 to 2 was used, and strong acid employed in the boiling, " quand on veut faire quelque essay curieux et exact." [1] Both Chaudet and Kandelhardt recommend the proportion of 1 to $2\frac{1}{2}$, on the ground that less silver is then retained by the cornet than in any other case. If much more than three parts of silver are present the gold breaks up in the acid. Pettenkofer [2] found that the proportion of 1 to 1·75 could be employed if the assays were boiled in concentrated nitric acid for some time. At the Royal Mint the proportion used was formerly 1 to 2·75, but is now 1 to $2\frac{1}{6}$ (see also p. 542 in section on " Surcharge "). The test silver should be assayed for gold, but the presence of a small quantity, say one part or less in 100,000, does not matter if pieces of the same silver are used for " proof " and ordinary assays.

The amount of lead used varies with the proportion of base metals present. For gold 900 fine and upwards, eight times its weight of lead is used at the Mint and answers very well, but the copper is not completely removed in the course of cupellation. The following table shows the proportions recommended by D'Arcet, [3] Cumenge & Fuchs, [4] and Kandelhardt [5] respectively :—

TABLE XLIX.

Gold in 1,000 Parts: the Alloying Metal being Copper.	Amount of Lead employed for One Part of Alloy.		
	D'Arcet.	Cumenge & Fuchs.	Kandelhardt.
1,000	1	1	8
900	10	14	16
800	16	20	20
700	22	24	24
600	24	28	24
500	26	32-34	28
400	34	32-34	28
300	34	32-34	32
200	34	32-34	32
100	34	30	32
50	34	28	32
0	...	11	32

[1] Savot, *Discours sur les Medalles Antiques* (Paris, 1627), p. 73.
[2] Pettenkofer, *Bergwerksfreund*, 1849, **12**, 6.
[3] Pelouze et Frémy, *Traité de chimie générale*.
[4] Cumenge et Fuchs. *Encyclopædie Chimique*, vol. iii., L'or, p. 154.
[5] Kandelhardt, *Gold-Probirerfahren*, Berlin, p. 3.

Kandelhardt's table is modified to make it uniform with the others. It must be remembered that, although the above table gives the quantities of lead which will remove the greater part of the copper present in the alloy during cupellation, the last 2 or 3 parts per 1,000 are obstinately retained by the gold, and cannot be entirely eliminated even by a second cupellation with fresh lead. Instead of attempting to remove all the copper present in an alloy of low standard by one operation, using large quantities of lead as above, it saves time and gives more uniform results if a smaller amount of lead is used in two successive cupellations. For one part of gold bullion 400 fine, 16 parts of lead are enough if added in this way. The object in view is not to remove all the copper by cupellation, but to obtain a well-formed, clean and bright button suitable for parting. Some copper must be retained by the assay piece in order to prevent " sprouting " at the moment of solidification of the cupelled button. For this reason, the amount of copper in the assay piece charged into the furnace should be not less than 50 parts per 1,000 of gold.

In the cupellation of triple alloys of gold, silver and copper, somewhat less lead than that given above is used in France and in this country. E. A. Smith recommends the quantities given in column A [1] in the following table. In column B are given the quantities used by the author at the Royal Mint.

TABLE L.

Standard.		Amount of Lead for One Part of Alloy.	
Gold in Parts per 1,000.	Carats.	A.	B.
916·6	22	8	8
750	18	15	8
625	15	18	12
500	12	20	16
375	9	24	16

In this case, the 22 carat contains a little silver, and the other standards usually from 100 to 125 parts of silver per 1,000. The lead in column B is in the last three cases added in two parts, the first addition consisting of 8 parts, and the second addition (after the first cupellation is complete) consists of the remainder.

3. **Cupellation.**—A coke muffle furnace formerly in use at the Royal Mint is shown in front and side elevation in Figs. 210 and 211. It consists of an outer casing of wrought-iron plates about ⅛ inch thick, united by angle iron. This casing is connected with a chimney 60 feet high by means of a wrought-iron hood and flue, a, which is provided with a damper, b. The lining consists of Stourbridge firebricks. Fuel is introduced through the opening c.

The mouth of the fireclay muffle, d, is closed by the graphite block, e, and the sliding iron plate, f, when cupellation or scorification is proceeding. For annealing and other purposes the sliding doors, g, g, are used. Air is

---

[1] Smith, op. cit., p. 326.

The gold is then dried in a covered porcelain basin and melted in a clay pot which has previously been washed out with molten borax. The pot is covered and heated in a gas furnace in an oxidising atmosphere. No flux is used, but it is advantageous, at any rate in some cases, to bubble oxygen through the molten gold for a minute or two before pouring. The gold is poured in an iron mould, which is neither smoked nor oiled, but rubbed with powdered graphite, and then brushed clean with a stiff brush. The ingot is cleaned by brushing and heating in hydrochloric acid, dried, and rolled out. The rolls must be clean and bright, and free from grease. The surface of the rolled gold plate may be again cleaned by scrubbing with fine sand and ammonia and also with hydrochloric acid, but if it is left rough, the plate readily accumulates impurities on its surface.

The addition of hydrobromic acid to the clear solution of gold chloride has been tried without throwing more silver out of solution or affecting the quality of the gold. An additional precaution is to remove the gases taken up by the gold during the process of melting by heating it to redness *in vacuo*, but the consequent improvement in fineness, if any, cannot be detected by assay. The method has also been tried of melting the precipitated gold in a silica tube *in vacuo* and allowing it to solidify before admitting the air. On the solidification of the gold the tube breaks. There seems to be no improvement in the fineness of the gold due to this treatment.

Fine gold is made a little purer by scraping it just before it is assayed, and also by heating it to redness, as suggested by S. W. Smith. This is probably due to the removal from the surface of impurities such as grease, moisture, dust, condensed vapour, etc.[1]

Oxalic acid is often used instead of sulphurous acid as the precipitant, and is stated by Krüss to be the best precipitant if platinum is present. Platinum is not precipitated from its solution as tetrachloride by sulphurous acid, and tellurium, though reduced by sulphurous acid at the same time as gold, would not be present in the solution under the conditions named. Sulphurous acid acts in cold solutions, but if oxalic acid is used, it is better to warm the solution and to leave it to stand for three or four days.

J. W. Pack[2] uses aluminium to precipitate the gold from its solution as chloride. F. Mylius[3] extracts the gold chloride with ether, but the method gives poor results, judging from an examination of the gold prepared in this way at the Hamburg Mint. At the Melbourne Mint, fine gold is prepared by passing chlorine gas through molten gold previously purified by precipitation and washing.[4] The results are highly satisfactory.

Limits of Accuracy in Gold Bullion Assay.—Attention may here be drawn to the errors introduced by the lack of delicacy of the very finest assay balances in ordinary use. It has elsewhere been shown by the author[5] that by weighing in the way indicated above, errors not greater than 0·15 per 1,000 may be introduced. It is, therefore, clear that this amount represents the limit of accuracy when such balances are used. By weighing correctly to 0·01 per 1,000, however, and performing all other operations with scrupulous care, then in the determination of gold in high-standard alloys of gold and copper, or of gold, silver and copper, whether pure or contaminated

[1] T. K. Rose, *J. Inst. of Metals*, 1913, **10**, 160.
[2] Pack, *Mng. and Sci. Press*, Mar. 7, 1908, p. 324.
[3] Mylius, *Zeitsch. anorg. Chem.*, 1911, **70**, 203.
[4] F. R. Power and R. Law, *Forty-fourth Annual Report of the Mint*, 1913, p. 138.
[5] T. K. Rose, *J. Chem. Soc.*, 1893, **63**, 700-714; see also J. Phelps, *ibid.*, 1910, **97**, 1272.

with traces of lead, bismuth, zinc, antimony, nickel, and some other elements, the error does not exceed ± 0·02 per 1,000, if the mean of three results is taken.

Parting with Sulphuric Acid.—The use of sulphuric acid of 66° B. instead of nitric acid for parting is recommended by some assayers on the ground that the losses of gold by dissolution in nitric acid are variable, while sulphuric acid does not dissolve gold. The inconveniences suffered by the use of sulphuric acid are that (1) lead and platinum are left undissolved by it ; (2) violent bumping of the liquid occurs during ebullition ; and (3) sulphate of silver is not very soluble in water, and the washing is consequently done with dilute sulphuric acid. However, it is stated that less silver is left undissolved in the cornets than in the parting by nitric acid if the proportion of gold to silver is between 1 to 2 and 1 to 3.

Preliminary Assay.—If the composition of an alloy is quite unknown, a preliminary assay is necessary in order to determine the right quantities of silver, copper, and lead to be added. This determination may be made by the touchstone, by considerations of the colour and hardness of the alloy, or by cupelling 0·1 gramme of it with 0·25 gramme of silver and 2 grammes of lead, and parting the button in a flask. Simple cupellation with lead gives satisfactory results if silver is absent or insignificant in quantity ; according to Frémy this method, in which parting is dispensed with, is accurate to 3 millièmes if carefully performed with proofs.

Assay of Gold by means of Cadmium.—Balling has shown [1] that cadmium may be substituted for silver in the operation of parting. The ½ gramme of gold alloy is placed in a porcelain crucible in which a little fragment of potassium cyanide has been previously fused in order to protect the metal from the air. Cadmium is then added in the proportion of 2½ to 1 of gold. If silver is present in addition, the combined weight of cadmium and silver must be 2½ times that of the gold. The whole is fused and then cooled and plunged into hot water to clean the button, which is then parted in nitric acid (specific gravity 1·2), boiled in water for some minutes, dried and weighed. The silver, if any, can be estimated by precipitation as chloride from the acid solution, or, better, by titration with sulphocyanide. By this method the losses of gold and silver incidental to cupellation are entirely avoided. A similar method, employing zinc in place of cadmium, had previously been recommended by von Jüptner. [2]

Alloys of Gold, Silver and Copper.—These may be assayed by the method just given, the copper being estimated as difference ; or the gold may be estimated as usual, and other assay pieces cupelled with enough lead to remove all the copper. The buttons thus obtained contain silver and gold only, and the proportion of silver is found by difference. The method of double cupellation, by which the button of silver and gold is weighed and then subjected to inquartation and parting, is less accurate than that just given, and both are inferior to cadmium parting and sulphocyanide titration.

The cupellation designed to remove the copper is made with less lead than the quantities given on p. 527. If little gold is present, half the amount of lead there given is used, with increasing proportions as the amount of gold present increases. The temperature of cupellation must also be lower than

---

[1] Balling, *Oestr. Zeitsch. für Berg. und Httnwesn.*, 1879, p. 597.
[2] Jüptner, *Zeitsch. anal. Chem.*, 1879, p. 104.

for gold, approximating more to that used for silver. Proofs of similar composition must be used and the operations require much practice before the necessary skill is acquired. The difficulty is that part of the silver is lost by cupel absorption and part of the copper is retained in the prill. It is almost impossible to arrange that these two amounts shall be equal. Accordingly the method should not be used. The cadmium parting already described, followed by determination of the silver in solution by titration with sulphocyanide or by precipitation as chloride, should always be preferred. If less than 33 per cent. of gold is present originally, the alloy may be parted at once without cupellation and the silver estimated as above.

Effects of the Presence of other Metals on Gold Bullion Assay.—The effects on cupellation are the same as those given under ore assay (p. 507). In general, if a scoria is formed owing to the presence of large quantities of antimony, arsenic, cobalt, nickel, iron, tin, zinc, or aluminium, there is a loss of gold. The alloy should in that case be scorified with lead as a preliminary to cupellation. If mercury is present gold is carried off in the form of spray, and lost as the mercury boils off. The presence of tellurium is sometimes indicated by the formation of numbers of minute beads of precious metal dispersed over the surface of the cupel. Tellurium compounds are best analysed in the wet way (see p. 553).

If members of the platinum group are present they remain undissolved by the parting acid, and hinder the solution of the silver, and the assay is consequently rendered unreliable. The treatment of the alloys is discussed below (p. 549). The effects of the presence of small quantities of various metals on the surcharge in the ordinary parting assay is shown below. The table is the result of experiments made by the author.[1] The presence of 5 per cent. bismuth does not affect the surcharge. All assay pieces contained 1,000 parts of gold, 2,500 parts of silver, and 91 parts of copper, other metals being added in the proportions indicated.

Metal added.	None.	Antimony, 50 Parts.	Zinc, 70 Parts.	Tellurium, 33 Parts.	Iron, 50 Parts.	Nickel. 50 Parts.
Surcharge,	+ 0·42	+ 0·28	+ 0·35	+ 0·07	+ 0·27	+ 0·22

These differences indicate the necessity of employing special checks containing these metals if such be present in the alloys.

ASSAY OF VARIOUS GOLD ALLOYS.

It is often impracticable to apply the ordinary parting assay to the examination of low-standard alloys of gold with other metals. These are then tested by various other methods, of which a summary is given below, the alloys being grouped in four series for convenience :—

    A. Alloys requiring scorification.
    B. Amalgams.
    C. Alloys containing members of the platinum group.
    D. Tellurium compounds.

A. Scorification of Alloys.—Alloys of gold containing *arsenic* or *antimony* are reduced to a fine powder and scorified with thirty parts of lead and a half part of borax. If the slag becomes pasty towards the end of the

---

[1] T. K. Rose, *J. Chem. Soc.*, 1893, **63**, 706.

operation more borax is added, a little at a time. If the lead button obtained is hard, a second scorification is necessary, with the addition of more lead. There is some loss of gold in the slag.

Iron or Manganese Alloys.—The operation is tedious and difficult with these alloys, as they are difficult to fuse, having higher melting points than pure gold,[1] and the oxides of iron do not form easily fusible compounds with the litharge. An extremely high temperature is required; ten parts of lead, one of borax, and one of silica usually suffice.

Cobalt and Nickel.—Twenty parts of lead are used, but no borax at first, so that the oxidation of the nickel may not be hindered. A very high temperature and the subsequent addition of two parts of borax are necessary. Several successive scorifications are required as nickel and cobalt are difficult to oxidise.

Zinc.—Oxide of zinc does not form a fusible mixture with litharge, and the slag is only rendered pasty by borax, unless it is added in large quantities. Gold is lost in the slag, but the loss is minimised by slagging off the zinc as rapidly as possible. Use 15 to 20 parts of lead and two to three parts of silica, with a little borax. Instead of silica, caustic soda may be used (W. A. C. Newman).

Tin.—Twenty parts of lead are required; oxide of tin is rapidly formed but the slag is not easily fusible. Large amounts of borax are necessary, or still better, borax mixed with caustic potash or soda which forms a fusible stannate with SnO_2.

Aluminium.—Alloys containing this metal cannot be assayed by scorification and cupellation. As soon as fusion takes place in the muffle, aluminium floats to the top of the bath, being of low density, and is rapidly oxidised, producing alumina which forms an exceedingly infusible scoria not easily removed by litharge. The production of the latter, moreover, is checked by the scum. Caustic soda would form a fusible aluminate.

Edward Matthey observes[2] that the removal of aluminium by digestion in hydrochloric acid, and the collection of the residual gold, does not yield satisfactory results. The process he recommends is as follows:—Accurately weighed portions of 50 grains each of the alloys are fused with litharge, under a flux of potassium carbonate and borax with a small proportion of powdered charcoal, and the resulting slag re-fused with a further small quantity of litharge and powdered charcoal. The lead buttons containing all the gold (the aluminium having combined with the fluxes employed) are cupelled, and the resulting gold cupelled with silver and parted with nitric acid in the usual manner. The assays must be worked with checks or standards of fine gold and pure aluminium.

In the majority of the preceding cases it is better to analyse the alloys by wet methods (see p. 553).

B. Amalgams.—The alloy is placed in a weighed porcelain crucible and gradually heated so as to drive off the mercury. After the greater part of the mercury has been driven off the temperature is raised to a full red heat which is maintained for half an hour. About 0·1 per cent. of mercury still remains in the gold after this treatment and can only be completely removed by cupellation and parting. Checks must be used, as the loss of gold in the operation may amount to 1 per 1,000.

---

[1] Frémy, *Ency. Chim.*, vol. iii., L'or, p. 147.
[2] Matthey, *Phil. Trans. Roy. Soc.*, 1892, **183, A,** 647.

A better method is to dissolve the mercury in nitric acid mixed with an equal volume of water. A porcelain crucible or basin should be used and gentle heat applied. If the action is not too violent the gold does not break up, but is left as a spongy coherent mass. It is washed, first with nitric acid, and finally with water, and ignited. After ignition, further treatment with concentrated nitric acid for a few moments is necessary. The gold still contains some mercury, which is removed by cupellation and parting. Mercury may be made use of to collect very finely divided precipitated or parted gold into a coherent mass, the mercury being removed with nitric acid.

C. Platinum Group.—Cupellation must be performed at a higher temperature than usual, and the iridescent bands are seen to remain longer, although they are less numerous.

Platinum-Gold Alloys.—The button obtained by cupellation is dull and crystalline. If the alloy contains as much as 7 or 8 per cent. of platinum the cupellation proceeds slowly, brightening is only obtained at a very high temperature and the button appears flattened, and has a rough crystalline surface and a grey colour. If more than 10 per cent. is present, brightening does not occur at all, and the other features just mentioned are more strikingly exhibited. On parting, the platinum is partly dissolved with the silver, but the assay-piece must be boiled in acid for a long time, and the parting is incomplete. When the ordinary parting assay is used the results are not satisfactory if more than 1 or 2 parts of platinum are present per 1,000 of alloy. It is necessary, if more platinum is present, to add fine gold in accordance with the rules given in the section below on the assay of alloys of gold, silver, platinum and copper.

Edward Matthey recommends[1] the following method for alloys of platinum 900, and gold 100 parts :—50 grains of each of the alloys are taken and treated with an excess of nitrohydrochloric acid which gradually dissolves the whole. The resulting solutions of platinum-gold chloride are then evaporated nearly to dryness to drive off the free acid and diluted with distilled water to about 20 c.c., a degree of strength ascertained by experiment to be the best for the precipitation of the gold. The metallic gold is thrown down by means of crystals of oxalic acid, and is carefully washed, dried and weighed. It is necessary to use checks.

Alloys of Gold, Silver, Platinum and Copper.—In this case it is necessary to determine in the first instance the approximate composition of the alloy and afterwards to make an exact assay.

In the *approximate assay*,[2] the amount taken is usually 50 milligrammes. This is cupelled with 1 gramme of lead at a very high temperature, and if the button is flat it is again cupelled with more lead. When a rounded button has been produced its weight is accepted as that of the gold, silver and platinum together. The button is cupelled with twice its weight of silver and parted in concentrated sulphuric acid, which dissolves the silver and leaves the gold and platinum as a residue. The gold is parted from the platinum in nitric acid, but it is necessary for the amount of gold to be at least ten times that of the platinum, in order that the result may be satisfactory. This proportion of gold is accordingly added with $2\frac{1}{2}$ times its weight of silver, and the cupelled button is parted in nitric acid. The cornet represents the original

[1] Matthey, *op. cit.*, p. 649.
[2] T. K. Rose, *The Precious Metals*, p. 272.

gold contents (after allowing for the amount added) and the platinum is taken by difference.

Thus, to take an example :—

Weight of alloy taken for assay, . .	100 half-milligrammes.
Weight after cupellation (Au + Pt + Ag), .	95 ,,
To this is added 100 mg. of silver, and the whole is cupelled. The weight after second cupellation and parting in sulphuric acid (Au + Pt) is . . .	75 ,,

To this 750 half-milligrammes of gold and $2\frac{1}{2} \times 825$ or 2,062·5 half-milligrammes of silver are added, and the whole is cupelled.

Weight after third cupellation and parting in nitric acid = 800 half-milligrammes, of which 750 had been added in the form of pure gold. The approximate composition of this alloy is thus determined to be as follows :—

Gold, ...	50
Platinum, ...	25
Silver, ...	20
Copper, ...	5
	100

Final or Exact Assay.—Errors may occur owing to the following causes :—

(1) In cupellation, copper is retained and is liable to be weighed as silver. A second cupellation with fresh lead would remove the copper but occasion a loss of silver.

(2) In the sulphuric acid parting some platinum may be dissolved in the strong acid.[1] This is prevented by using slightly diluted acid for the first boiling—*e.g.*, 100 parts by volume of concentrated acid, diluted with 20 parts of water. The second boiling may be in concentrated acid, or similarly diluted acid. On the other hand, part of the silver remains undissolved in sulphuric acid.[2] This can be prevented (a) by increasing the proportion of silver to 8 or 10 times that of the gold and platinum taken together, a course which results in the cornet breaking up ; or (b) by adding gold to make its ratio to platinum at least 10 to 1, and adding an amount of silver equal to three times the amount of platinum and gold combined.[3]

(3) In the nitric acid parting, part of the platinum remains in the cornet undissolved. The amount, however, is small (1 or 2 parts per 1,000) if the ratio of gold to platinum is 10 to 1, and that of silver to gold and platinum combined about 3 to 1.[4] The errors are avoided or reduced to small proportions by the use of proofs or check assay pieces, the composition of which is determined by the approximate assay, and by the adoption of the following scheme of operations, which was worked out by M. Forest at the Paris Mint.[5]

---

[1] A. Steinmann, *J. Soc. Chem. Ind.*, 1911, **30**, 1216.
[2] J. F. Thompson and E. H. Miller, *J. Amer. Chem. Soc.*, 1906, **28**, 1115.
[3] H. Carmichael, *J. Soc. Chem. Ind.*, 1903, **22**, 1324.
[4] Carmichael, *loc. cit.* ; also Thompson and Miller, *loc. cit.*
[5] *6me Rapport par l'Administration des Monnaies et Médailles*, 1901, p. xxix. See also E. A. Smith, *The Sampling and Assay of the Precious Metals*, pp. 410-428.

Take 0·5 gramme (" 1,000 " in gold bullion assay weights) of the alloy for the sulphuric acid parting if the platinum does not exceed 100 millièmes, and less weight if a greater proportion of platinum is present. Gold is added, if necessary, to give a ratio of ten parts of gold to one of platinum.

The amount of lead required for cupellation is about double the amount of lead which would be required in a gold assay (see p. 527), so that in a case in which the copper amounts to 150 per 1,000, about 15 grammes of lead would be used for 0·5 gramme of the alloy. In cases where the platinum exceeds 200 per 1,000 of the alloy, d'Arcet and Chaudet recommend a second cupellation with 1 or 2 grammes of lead.

If there is a surcharge in the proofs of more than one or two parts per 1,000 in the first cupellation, showing retention of copper, the prills are recupelled with fresh lead.

Taking as an example a case in which the approximate assay of the alloy is—

Gold, ...	500
Platinum, ...	250
Silver, ...	100
Copper, ...	150
	1,000

the weight taken for assay will be 0·25 gramme.

The cupelled button or prill in this case requires the addition of 0·5 gramme of gold, and is cupelled with 3 parts of silver to 1 part of gold and platinum taken together—that is to say—

$$
\begin{aligned}
\text{Gold + platinum} &= 0·75 \text{ gramme.}\\
\text{Silver already present} &= 0·025 \quad ,,\\
\text{Silver required to be added} &= 2·225 \text{ grammes.}
\end{aligned}
$$

The amount of lead required for this second cupellation is 5 grammes.

The button is worked as described above in the assay of gold bullion and the cornet boiled in slightly diluted sulphuric acid for ten minutes. This is a delicate operation, as the boiling point of sulphuric acid (325° C.) is so high that glass vessels are liable to be cracked by cold draughts. Moreover, the acid boils explosively with heavy bumping, and boiling sulphuric acid is terribly corrosive and inflicts dangerous wounds. It is advisable to use silica ware or a platinum boiler, if possible, but if glass is used it must be protected from the air by being wrapped in asbestos, and heated at the sides more than at the bottom. Sometimes pieces of carbon are added or a capillary tube,[1] to assist the acid to boil quietly. After boiling, the vessel is allowed to cool, as otherwise it would crack during decantation. The liquid is then decanted into a dry vessel, or into a considerable bulk of cold water, to avoid danger from the heat generated by mixing sulphuric acid and water. The decantation must be as complete as possible, to avoid heating on the addition of the washing water, which should be warm in order to dissolve the sulphate of silver. After washing twice, the cornet is again boiled for ten minutes in sulphuric acid and washed, dried and annealed. Its weight less 0·5 gramme, the amount of gold added, gives the amount of gold + platinum, in this case 750 per 1,000.

---

[1] Smith, *op. cit.*, p. 419.

For the determination of the gold, 0·25 gramme or 500 half-milligrammes of the original alloy is taken, containing approximately—

> 250 gold
> 125 platinum
> 50 silver
> 75 copper.

This is cupelled with 0·5 gramme of pure gold, making 10 parts of gold to 1 part of platinum, and 2·25 grammes of silver and 5 grammes of lead. The resulting button is worked and parted in nitric acid as described in the assay of gold bullion. If, as generally happens, there is a surcharge in the check assays of more than 1 or 2 parts per 1,000, showing the retention of platinum, the inquartation and parting is repeated. The final weight is then 1,250, and after deducting 1,000 and doubling the remainder, the result 500 is obtained.

If necessary as small a quantity as 0·1 gramme may be taken for assay. This is a convenient amount for alloys containing 900 to 950 platinum and the remainder gold.

The cupellation requires a higher temperature in proportion as the percentage of platinum is greater, and alloys containing over 50 per cent. of platinum cannot be freed from lead without the help of the oxygen-gas blowpipe. The dissolved platinum colours the nitric acid solutions brown. The cornets are dark coloured or grey if they contain platinum.

Palladium-Gold Alloys.—The palladium is dissolved in parting if the weight of silver is at least three times that of the gold, yielding an orange coloured solution. Matthey recommends double parting. Separation may also be effected by fusion with six to eight parts of potassium bisulphate, and dissolving out the dark brown palladious sulphate by boiling water.[1] A second fusion is usually necessary to render the gold residue quite pure.

As in the case of platinum, palladium may be separated from gold by dissolving the alloy in aqua regia, evaporating to dryness, taking up with water and precipitating the gold by oxalic acid from a very dilute solution : the palladium remains in solution.

Rhodium- and Iridium-Gold Alloys.—Iridium, if present, always sinks to the bottom of the cupelled button as it is very dense (specific gravity = 21·38), and is not usually fused at the temperature of the muffle, but occurs in the state of fine black crystalline particles. Hence, when the button is rolled into a cornet with the lower face outwards (p. 535) iridium occurs as black sooty spots or streaks which are seen by a lens to fill up depressions in the surface of the gold.

Both rhodium and iridium are almost insoluble in aqua regia. If gold alloys containing both of them are parted in the ordinary way with nitric acid, only a small quantity of rhodium goes into solution with the silver. The residue consisting of the gold, the iridium and most of the rhodium may then be attacked by dilute aqua regia when the gold is dissolved together with only traces of the other metals. These may be separated by evaporating the solution to dryness and heating to dull redness, when the reduced metals being no longer alloyed may be completely separated by dissolving the gold in aqua regia.

---

[1] H. Rose, quoted by E. Cumenge and E. Fuchs, *L'or dans le laboratoire* (Frémy, *Ency. Chim.,* p. 177.

According to d'Hennin,[1] iridium may be separated from gold by fusing it with fluxes, the charge being made up as follows :—

Iridic gold, ...	12·5 grains.
Sodium arsenate, ...	3 „
Black flux (a deflagrated mixture of nitre and tartar), ...	18 „
Ordinary flux (consisting of borax, cream of tartar, charcoal and litharge), . .	20 „

The iridium forms a speiss with the iron and the arsenic, and the lead button formed at the bottom of the fused mass contains all the gold.

The estimation of osmium, iridium, and ruthenium in gold bullion has also been studied by Riche, Leidié, and Quennessen,[2] assayers at the French Mint.

D. Tellurium Alloys.—These must be treated by wet methods. The alloy may be dissolved in nitrohydrochloric acid, and the solution containing both gold and tellurium evaporated with a large excess of hydrochloric acid until no more chlorine is given off, when both gold and tellurium are readily precipitated by a current of sulphur dioxide gas. On attacking the precipitate with dilute nitric acid the tellurium is dissolved in the state of tellurous acid, and the gold residue may be dried and weighed, and its purity ascertained by inquartation and parting. The greater part of the tellurium may be removed from gold-tellurium alloys by boiling in nitric acid, and the residue can be cupelled and parted with very little loss of gold.

Wet Methods of Assay of Gold Alloys, Compounds, etc.—Assays or complete analyses of gold bullion, natural minerals, etc., can be made by the ordinary chemical methods given in books on quantitative analysis. From 1 to 5 grammes of bullion are usually enough, but a much larger amount is necessary if the alloy is nearly pure gold. M. Forest[3] takes 300 grammes of gold bullion when examining it for small quantities of impurities. In general the residue left after prolonged action of nitric or sulphuric acid is not sufficiently pure to weigh as gold, and complete solution in aqua regia is usually necessary. From the solution the gold may be precipitated by (a) ferrous sulphate, (b) sulphurous acid, (c) oxalic acid, (d) sulphuretted hydrogen, (e) ammonium sulphide, followed by the addition of hydrochloric acid. The following remarks may be of value in aiding the chemist in his choice of a precipitant in any particular case.

Nitric acid must always be expelled from the solution by warming with successive additions of hydrochloric acid. The acid solution must not be heated too strongly or loss of gold chloride by volatilisation occurs. Some other chlorides escape more freely. Ferrous sulphate and sulphurous acid act well in strongly acid (HCl) solutions ; oxalic acid, sulphuretted hydrogen, and ammonium sulphide best in presence of small quantities of HCl. The solution should be dilute (say 1 part of gold in 300 of water), so that other metals may not be carried down by the gold. Sulphate of iron gives a very finely divided precipitate which is difficult to wash by decantation without loss ; precipitation is slow in cold solutions. Oxalic acid causes plates and scales to form which are readily washed and are very pure ; it acts best in

[1] d'Hennin, *Dingl. Poly. J.*, **137**, 443.
[2] *Huitième Rapport par l'Administration des Monnaies et Médailles*, 1903, p. xxix.
[3] Forest, *Neuvième Rapport par l'Administration des Monnaies et Médailles*, 1904, p. xxxi.

boiling liquids, but a temperature of 80° for forty-eight hours suffices ; in the cold or in the presence of much hydrochloric acid or alkaline chlorides the action is very slow and partial ; a large excess of the precipitant must be present. Oxalic acid is used for solutions containing metals of the platinum group, which are not precipitated by it. Alkaline oxalates act better than the free acid.

Sulphurous acid is an excellent precipitant for most solutions. It acts rapidly and completely in the cold, and does not readily precipitate other metals, except tellurium. Sulphuretted hydrogen is used in the absence of all other metals whose sulphides are insoluble in hydrochloric acid.

In all cases careful consideration must be given to the nature of the base metals present, and the precipitant which will not render any of them insoluble must be selected.

Other Methods of Bullion Assay.—Among methods which have been proposed at various times, and which may still be of service occasionally in particular cases, may be mentioned :—(1) The trial by the touchstone (a method still extensively used by jewellers) ; the assay by means of considerations as to (2) the colour, and (3) the density of alloys ; (4) spectroscopic assay. A brief description of each of these methods is appended.

1. **Trial by the Touchstone.**—This is the oldest method of assay. It is described by Theophrastus, about 300 B.C.,[1] and the methods in use in Germany in the 16th century are fully detailed in Agricola.[2] The assay consists in rubbing the gold bullion to be tested on a hard dark-coloured smooth stone, and comparing the appearance and colour of the streak with those made by carefully prepared touch needles of known composition. The effect of the action of nitric acid and dilute aqua regia on these streaks is also noted. Touchstones usually consist of Lydianstone or of silicified wood, and black or dark green stones are best. Only alloys of gold and copper or of gold and silver can be thus tested. The trial is more sensitive for alloys below 750 fine than for higher standards. The amount of gold in alloys between 700 and 800 fine can be determined correct to 5 parts per 1,000.

2. **Colour and Hardness of Alloys.**—These properties form a guide to the composition of copper-gold alloys, an increase of copper corresponding to a heightening of the colour and an increase of the hardness as tested with shears or a knife. On heating the alloy to redness in air, the degree of blackening of the surface is a further indication of the percentage composition, if compared with plates of known fineness.

3. **Density of Gold-copper Alloys.**—The determination of the fineness of these alloys by taking their densities was investigated by Roberts-Austen at the Royal Mint in 1876.[3] He showed that the densities found by experiments were nearly equal to those obtained by calculation on the assumption that the union of the two metals was accompanied neither by contraction nor expansion. The alloys examined ranged from 860 to 1,000 fine, and were made into discs which were all compressed to the same extent. The conclusion arrived at was that the fineness of large masses of gold can be deduced from their densities correct to $\frac{1}{10000}$ part. In the case of individual coins the results are only approximate.

---

[1] See Hoover's *Agricola*, p. 252, note 37.
[2] *Op. cit.*, pp. 252-260.
[3] Roberts-Austen, *Seventh Annual Report of the Mint*, 1876, p. 41.

4. **Assay by means of the Spectroscope.**—This method of determining the composition of gold-copper alloys was investigated by Lockyer and Roberts-Austen.[1] The arc spectrum of pure gold was shown to be altered by successive additions of copper, and near the English standard (916·6 fine) a difference of 2 or 3 parts in 10,000 in composition could be readily detected, but the amount of metal volatilised is so small that it cannot be made to represent with certainty the average composition of the mass, which is never perfectly homogeneous. When operating on a slip of alloy formed of

Silver,	.	.	.	.	.	708
Copper,	.	.	.	.	.	254
Gold,	.	.	.	.	.	38
						1,000

the spectra of copper and silver alone were visible (Cupel). In an alloy of gold and copper containing from 200 to 250 parts in the 1,000 of the precious metal, the gold spectrum is barely visible. On the other hand, in an alloy of gold and copper containing traces only (·01 per cent.) of the latter, the copper spectrum was distinctly shown.

The detection of traces of gold in alloys or ores by means of the spectroscope, though sometimes attempted, is not remarkable for its delicacy or certainty. One method of procedure [2] is to dissolve the auriferous material in aqua regia, evaporate off the nitric acid, and pass induction sparks through the surface film of liquid, when the spectrum shows some narrow bands and some nebulous bands. The latter only are seen if a drop of the solution is placed in a Bunsen flame. The method may sometimes be useful when complex minerals are being examined.

---

[1] Lockyer and Roberts-Austen, *Phil. Trans. Roy. Soc.*, 1874, **164**, [ii.], 495, and *Mint Report*, 1874, p. 38.
[2] Frémy, *Ency. Chim.*, vol. iii., L'or. p. 134.

CHAPTER XXI.

STATISTICS OF GOLD PRODUCTION.

Annual Production of the Gold Mines of the World.—The production of gold in ancient times cannot be closely estimated, but, judged from a modern standpoint, it was probably very small. In the middle ages, however, between the fall of Rome and the discovery of America, the production was far smaller than before, and Jacob observes that in this period "the precious metals were sought not by exploring the bowels of the earth, but by the more summary process of conquest, tribute, and plunder." Even after the exploitation of the New World began, the output of gold was for many years much too small to satisfy the cupidity of the conquerors. The development of the mining industry was prevented by the ruin and destruction of the natives, and by the almost incessant irregular warfare waged against the Spaniards in America in the 16th century, first by the Dutch and later by the English. Fifty years elapsed after Columbus discovered America, before the annual production of gold reached £1,000,000, and even at the end of the 17th century Soetbeer [1] estimates that it was only £1,500,000. The discovery and working of the rich Brazilian placers during the next half century raised the annual product to over £3,500,000 in the period 1740-1760 (Soetbeer), but as these deposits became exhausted, the output again fell off, and in the period 1810-1820 had again sunk to about £1,500,000 per annum. The gradual development of the Siberian placers was the main cause of the subsequent steady increase in production up to an average of £7,500,000 per annum in the period 1841-1850 (Soetbeer), and this was followed by a sudden rise consequent on the discoveries in California and Australia. The maximum output from the rich placers of these countries was reached in 1852, when the world's production of gold is estimated by Sir Hector Hay to have been £36,500,000. [2] After falling to £21,000,000 in 1862 (Hay), the output remained nearly stationary until about the year 1888, when, from various causes mentioned below, the production again began to increase, and in 1899 reached £63,000,000. After a sudden fall to £53,000,000 in 1900 owing to the Boer War, the increase in production was resumed, and in 1911 the output was £98,000,000, the greatest amount on record.

The increase since 1888 is due (1) to the discovery and development of new districts, and (2) to the progress in the art of metallurgy. The goldfields in South and West Africa, in West Australia, on the Yukon, at Cripple Creek in Colorado, in Mexico, Canada, India, and several other countries, have all

---

[1] Soetbeer, *Materialen, etc.*, 1879, p. 1.

[2] *First Report of the Commission on the Precious Metals* (C. 5,099), 1887, p. 308.

been developed since 1888. The production in Africa alone rose from £900,000 in 1888 to £43,500,000 in 1912. With modern practice, the development, working out and abandonment of gold deposits is far more rapid than in former times, and some of these fields already show signs of approaching exhaustion.

The increase of production due to improved metallurgical methods is difficult to estimate. Most of it must certainly be put down to the account of the cyanide process.

As regards the future production, it is obviously impossible to predict what will be the effect of the opening up of goldfields as yet undiscovered, or of the invention of new processes for treating ore of lower grade than that which can now be dealt with profitably. Excluding the Transvaal, the production of the rest of the world has been falling since 1909, and that of the Transvaal has fallen since 1912. In the existing conditions a continued fall in the world's output appears to be indicated.

TABLE LV.—GOLD PRODUCTION OF THE WORLD,
1848-1913.

Year.	Value in £ Millions.	Year.	Value in £ Millions.
1848	13·5	1881	20·6
1849	17·5	1882	20·2
1850	18·5	1883	19·6
1851	24·0	1884	19·1
1852	36·5	1885	19·5
1853	31·0	1886	20·4
1854	25·4	1887	21·7
1855	27·0	1888	22·6
1856	29·5	1889	24·9
1857	26·6	1890	23·2
1858	24·8	1891	26·8
1859	24·9	1892	30·1
1860	23·8	1893	32·4
1861	22·8	1894	37·2
1862	21·4	1895	40·9
1863	21·3	1896	41·5
1864	22·6	1897	48·5
1865	24·0	1898	59·0
1866	24·2	1899	63·0
1867	23·2	1900	53·7
1868	24·0	1901	53·4
1869	24·2	1902	61·1
1870	23·8	1903	67·2
1871	23·3	1904	70·5
1872	22·0	1905	79·2
1873	22·3	1906	81·7
1874	21·5	1907	83·9
1875	22·1	1908	89·5
1876	22·3	1909	93·7
1877	23·4	1910	94·7
1878	22·1	1911	97·9
1879	20·8	1912	97·5
1880	21·2	1913	93·3

The table on p. 557 gives the production of gold in the world in millions of pounds sterling for each year from 1848 to 1913.[1] The figures for the years 1848 to 1851 are quoted from A. Del Mar's *History of the Precious Metals*, p. 179, and the *Westminster Review*, Jan., 1876. Those for the years 1852 to 1885 are on the authority of Sir Hector Hay (*Report of the Commission on the Precious Metals*, 1887 (C. 5099), p. 308). Those for the years 1886 to 1899 are from the Reports of the Director of the United States Mint. Those for the years 1900 to 1911 are from the Reports of the Chief Inspector of Mines, Home Office (Mines and Quarries, Part IV., annually). Those for the years 1912 and 1913 are from the *Mineral Industry*. These authorities are the best available. Where the original estimates are in dollars or kilogrammes, they have been converted into the value in pounds sterling.

The figures in the following table giving the relative production for the years 1900 to 1911 in the British Empire and in Foreign Countries are from *Mines and Quarries* (Home Office annual publication), Part IV. The figures for the years 1912 and 1913 are from the *Mineral Industry* :—

TABLE LVI.—RELATIVE PRODUCTION OF GOLD.

Year.	British Empire.		Rest of World.		Total Kilos.
	Kilos.	Percentage.	Kilos.	Percentage.	
1900,	188,491	47·94	204,705	52·06	393,196
1901,	184,854	47·28	206,171	52·72	391,025
1902,	232,507	51·94	215,137	48·06	447,644
1903,	284,837	57·92	206,835	42·08	491,672
1904,	306,133	59·31	209,994	40·69	516,127
1905,	342,005	58·95	238,082	41·05	580,087
1906,	364,189	60·84	234,447	39·16	598,636
1907,	372,149	60·52	242,583	39·48	614,732
1908,	392,080	59·84	263,250	40·16	655,338
1909,	395,281	57·62	290,748	42·38	686,029
1910,	393,152	56·69	300,224	43·31	693,376
1911,	410,477	57·25	306,388	42·75	716,865
1912,	436,985	61·22	276,882	38·78	713,867
1913,	432,549	63·34	250,362	36·66	682,911

The curves in Fig. 222 are based on these figures.

The production of gold contained in or obtained from ore raised in the individual countries of the world during the year 1911 are given in the table on p. 559.[2]

---

[1] The value of 1 oz. troy of fine gold is approximately £4 4s. 11·454d. or £4·2477273. The logarithm of this number is 0·6281566. The value of 1 kilogramme or 32·1507267 ozs. of fine gold is £136·56756, logarithm 2·1353475. In United States currency these values are— 1 oz. $20.671452, 1 kilogramme $664.6025.

[2] *Mines and Quarries*, 1911, Part iv., Home Office, Cd. 7,217.

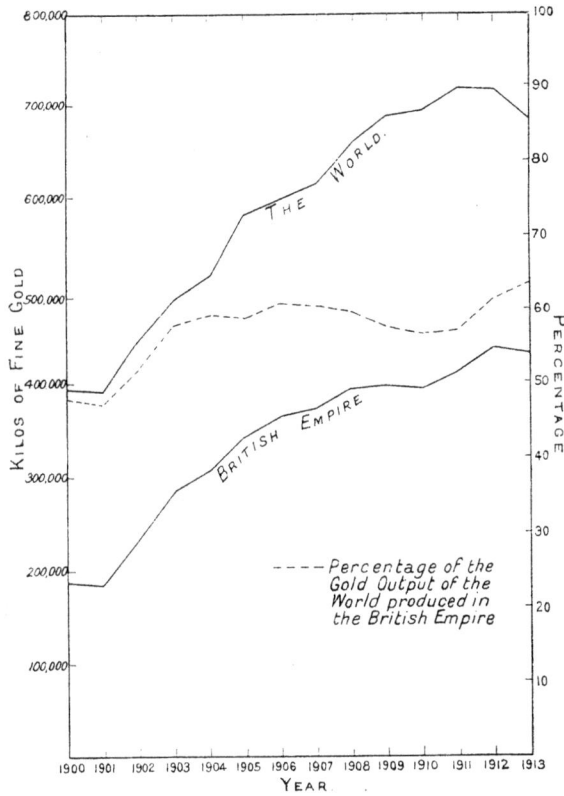

Fig. 222.--Gold Production of the World.

TABLE LVII.

British Empire.

Output in Kilos.

Great Britain and Ireland, ...	11
Canada, ...	14,717
Newfoundland, ...	71
British Guiana, ...	1,342
South Africa, ...	256,642
Rhodesia, ...	19,573
Swaziland, ...	460
Bechuanaland Protectorate, . . .	170
Gold Coast, ...	7,899
Australia, ...	77,346
New Zealand, ...	14,148
Papua, ...	455
British Borneo, ...	879
India, ...	16,388
British Protected Malay States, . .	89
Federated Malay States, . . .	287
Total for British Empire, . . .	410,477

Rest of World.

Output in Kilos.

Europe—

France,	2,726
Germany,	117
Austria-Hungary,	3,467
Italy,	29
Portugal,	76
Servia,	421
Turkey,	1
Sweden,	11
Russia (for 1910),	64,958

North America—

United States,	145,784
Mexico,	42,775

South America—

Argentine,	435
Bolivia,	33
Brazil,	3,428
Chile,	1,112
Colombia,	4,766
Costa Rica,	1,370
Cuba,	30
Ecuador,	416
French Guiana,	2,994
Dutch Guiana,	798
Honduras,	1,361
Nicaragua,	1,400
Panama,	121
Peru,	741
Uruguay,	106
Venezuela,	1,278

Africa—

Abyssinia,	364
Belgian Congo,	913
Egypt,	158
French West Africa,	55
German East Africa,	375
Portuguese East Africa,	93
Madagascar,	2,850

Asia—

China,	4,987
Indo-China,	113
Japan,	4,681
Formosa,	1,752
Korea,	4,348
Dutch East Indies,	4,655
Philippine Islands,	290

Total for rest of world,	306,388

Total **Output of the World** in 1911, 716,865 kilogrammes, or 23,047,728 ozs.

Of the total, South Africa contributed 34·8 per cent., United States 20·4 per cent., Australia 10·8 per cent., Russia 9 per cent., and Mexico 6 per cent., these five countries contributing 81 per cent. of the output between them.

The product of the United States for 1913, excluding the Philippine Islands and Porto Rico, is divisible as follows :—[1]

California, ...	$20,241,300
Colorado, ...	18,109,700
Alaska, ...	15,201,300
Nevada, ...	11,977,400
South Dakota, ...	7,214,200
Arizona, ...	4,101,400
Utah, ...	3,570,300
Montana, ...	3,320,900
Oregon, ...	1,477,900
Idaho, ...	1,244,300
New Mexico, ...	892,000
Washington, ...	657,500
North Carolina, ...	115,200
Other States, ...	52,900
Total, ...	$88,176,300

The product of Australia in 1913 is divisible as follows :—[2]

	Ozs. Fine.
Western Australia, ...	1,314,043
Victoria, ...	434,932
Queensland, ...	265,735
New South Wales, ...	149,657
Tasmania, ...	33,400
South Australia, ...	7,689
Total, ...	2,205,456

The product of the Witwatersrand alone in 1913 was 8,430,998 ozs. fine of the value of £35,812,605,[3] or 38·4 per cent. of the output of the world.

The total product of the chief gold-producing countries for the years 1850-1913 was as follows :—[4]

Australia and New Zealand (from 1851), .	£640,607,000
United States (from 1849), . . .	722,294,000
Canada (from 1862), ...	63,994,000
India (from 1880), ...	42,381,000
West Africa (from 1880), . . .	12,050,000
Transvaal (from 1884), ...	401,000,000
Rhodesia (from 1898), ...	25,282,000
Other countries, ...	543,613,000
Total, ...	£2,451,221,000

[1] *Report of the Director of the U.S. Mint*, 1914, p. 197.
[2] *Annual Report of the Mint*, London, 1913, p. 150.
[3] *Report of the Transvaal Chamber of Mines.*
[4] *The Statist*, 28th February, 1914.

Cost of Production of Gold.—In view of the fact that the value of money is measured in almost all gold-producing countries by means of the weight of the metal itself, it would be of special interest to estimate the average cost per ounce of its production. Complete data, however, are lacking. In the leading goldfield of the world, the Witwatersrand, the cost of production per fine ounce of gold at those mills which were working more or less continuously seems to have been about £3 10s. in 1892, and Hatch and Chalmers estimated it for the mines situated on the outcrop of the Main Reef at about £2 14s. in 1895. Later details are as follows for the Witwatersrand :—[1]

TABLE LVIII.

Year.	Tons of Ore Milled.	Cost per Ton Milled.	Gold Produced, Ozs.	Working Cost per Oz. of Gold.
1908	18,196,589	18s. 0d.	6,782,538	£2 9 6
1913	25,628,432	17s. 11d.	8,430,998	2 14 4

In spite of the improvements in mining and metallurgical practice, it is clear that the cost of producing an ounce of gold in the Transvaal has been increasing of late years, owing to the fall in the grade of the ore and to the rise in prices, which involves an increase in the cost of labour and supplies.

If, however, the results on the Witwatersrand can be taken as an index of the industry of gold production generally, it is clear that prices must rise much more—that is to say, the value of gold must depreciate considerably further—before any perceptible check is given to the output from that cause. The slight falling off of output during the last two or three years appears to be due entirely to the partial exhaustion of the existing goldfields.

Consumption of Gold.—Many attempts have been made at various times to estimate the amount of gold used annually for purposes other than that of additional coinage. In 1831, Jacob [2] estimated the annual amount of gold " converted into utensils and ornaments " at about £4,500,000, while the production was under £2,000,000, without counting, however, imports from India and China, which were supposed to be considerable. In 1881, Dr. Soetbeer estimated the annual industrial consumption of gold in the world, after making deductions for old material employed, as amounting to 84,000 kilogrammes or £11,500,000, and in 1885 he put it at 90,000 kilogrammes or £12,300,000, against a production of £21,000,000. In 1891, the same authority gave the industrial consumption of gold added to the amount hoarded and that exported to the East as 120,000 kilogrammes or £16,400,000, against a production of £24,000,000. In 1894, Ottomar Haupt [3] estimated the recent industrial consumption at only 270,000,000 francs or £10,700,000 per annum, but the exports to the East are not included in this amount. The Director of the United States Mint [4] estimated the industrial consumption for 1893 at $50,500,000 or about £10,300,000, for the year 1903 at

[1] *Reports of Transvaal Chamber of Mines.*
[2] Jacob, *History of the Precious Metals*, vol. ii., p. 322.
[3] Haupt, *Arbitrages et Parités*, 8th ed., Paris, 1894.
[4] *Production of the Precious Metals*, 1893, p. 53. *Report of the Director of the U.S. Mint*, 1904, p. 48 ; 1911, p. 24.

$76,350,000 or about £15,700,000, and for the year 1910 at $111,848,500 or about £23,000,000. Doubtless the industrial consumption of gold is increasing in amount with the increase of population and wealth, but nevertheless it is fair to assume that no such rapid growth has taken place of late years in consumption as has been already pointed out to have taken place in production.

The export of gold to India and certain other countries, however, puts a different complexion on the matter. The gold absorbed by India and Egypt in particular is in great part hoarded by individuals, who make no use of it for purposes of currency, and it may be classed as consumption equally with that used in the arts in the Western world. The Director of the United States Mint gives the following amounts[1] as having been diverted from monetary use during the eleven years, 1900 to 1910 :—

Industrial consumption,	.	.	.	$958,000,000
India,	.	.	.	433,000,000
Egypt,	.	.	.	146,000,000
Japan,	.	.	.	69,000,000
South America, .	.	.	.	343,000,000
Mexico,	.	.	.	28,500,000
Total, .	.	.	.	$1,977,500,000

The total of about £400,000,000 represents nearly one-half of the production of the period. The remainder was mainly added to the stock of gold in banks and treasuries,[2] very little being required for additions to money in actual circulation, in spite of the increase of population and wealth, owing to the gradual change in the habits of the people in the management of their monetary transactions.

The world's stock of gold in 1914, excluding India and the East, is put by R. A. Lehfeldt[3] at 12,410 tons or £1,695,000,000,[4] and the annual increase in this stock at about 3 per cent. The increase is generally regarded as having a beneficial effect on the trade of the world, and as tending to cause a rise in prices, although there is no general agreement as to the extent of its effects in the latter direction.

---

[1] *Report of the Director of the Mint for* 1911, Washington, 1912.
[2] *Loc. cit.*, for details.
[3] Lehfeldt, *J. Chem. Met. and Mng. Soc. of S. Africa*, Sept. 1419, p. 52.
[4] Compare estimated stocks of £34,000,000 in 1492 and £300,000,000 in 1850.

BIBLIOGRAPHY.

It has been found impracticable to enumerate the articles and paragraphs relating to the metallurgy of gold, which have from time to time appeared in the various periodicals, as very little search results in the accumulation of thousands of such references. In general, therefore, only the names of some of the publications which contain important or interesting matter on the subject are given; but a few exact references on special points have been added, and many others occur in the footnotes to the text.

PERIODICAL LITERATURE.

American Journal of Science. New York. From 1818.
Anales de la Mineria Mexicana, ó sea ; Revista de Minas. Mexico. From 1861.
Anales de Minas. Madrid. From 1841.
Annalen der Berg- und Hüttenkunde. Salzburg, 1802-5.
Annales des Mines. Paris. From 1816.
Annuaire du Journal des Mines de Russie. Petrograd. First published in 1840.
Annuaire des Mines et de la Métallurgie Françaises. Paris. First published in 1876.
Annual Reports of the Californian State Mineralogist. Sacramento. From 1881 to 1890. First Biennial in 1892.
Annual Reports of the Deputy-Master of the Royal Mint. London. From 1870.
Annual Reports of the Director of the United States Mint. Washington.
Annual Reports on Gold Mining. Victoria, British Columbia. From 1875.
Australian Mining Standard. Sydney and Melbourne. From 1872.
Berg- und Hüttenmannisches Jahrbuch. Vienna. From 1866.
Berg- und Hüttenmannische Zeitung. Freiberg. From 1842.
Biennial Reports of the Nevada State Mineralogist. From 1871.
Boletin oficial de minas. Madrid, 1844-5.
Bulletin de l'Association amicale des anciens élèves de l'École des Mines. Paris. From 1869.
Bulletin de la Société de l'Industrie Minérale. Quarterly. St. Etienne. From 1855.
Canadian Mining Journal. Toronto. From 1880.
Dingler's Polytechnisches Journal. From 1815.
Economic Geology. London. From 1906.
Electrochemical and Metallurgical Industry. New York. From 1903 to 1909.
Engineering and Mining Journal. New York. From 1866.
Jahrbuch für Berg- und Hüttenwesen. Freiberg. From 1827.
Journal of the Chemical Society. London. From 1841.
Journal of the Chemical, Metallurgical and Mining Society of South Africa. Johannesburg. From 1894.
Journal of the American Chemical Society. Easton, U.S.A. From 1879.
Journal of the Society of Chemical Industry. London. From 1882.
Journal of the Geological Society. Quarterly. London. From 1815.
Journal of the Institute of Metals. London. From 1909.
Journal des Mines de Freiberg. Koehler. Freiberg, 1788-1793.
Journal des Mines de Russie. 1832 to 1835.
La Mineria. Mexico, 1843.
Metallurgical and Chemical Engineering. New York. From 1910.
Mineral Industry. New York. Annually from 1892.
Mines and Minerals, formerly *Colliery Engineer.* Monthly. Scranton, Pa., U.S.A. From 1897.
Mining and Scientific Press. San Francisco. From 1865.
Mining and Smelting Magazine. London, 1862-65.

Mining Journal. London. From 1836.
Mining and Engineering World. Chicago. From 1890.
Mining Magazine, formerly *Pacific Coast Miner.* Monthly. New York. From 1904.
Mining Magazine. London. Monthly. From 1909.
Mining Review. Denver, Colorado, 1873-76.
Mining World. London. From 1871.
National Physical Laboratory. Collected Researches. London.
Nouveau Journal des Mines de Freiberg. Koehler & Hoffmann. Freiberg, 1795-1804.
New Zealand Government Mining Journal. Wellington. From 1897.
Oesterreich Zeitschrift für Berg- und Hüttenwesen. Otto Freihern. Vienna. From 1853.
Ontario Bureau of Mines, Annual Reports. From 1892.
Pacific Miner.
Precious Metals of the United States, Annual Reports on. Washington. From 1880.
Proceedings of the California Academy of Sciences. San Francisco.
Proceedings of the Canadian Mining Institute. Montreal.
Proceedings of the Colorado School of Mines. Golden City, Colorado.
Proceedings of the International Congress of Applied Chemistry.
Proceedings of the Colorado Scientific Association.
Reports of the Mining Commissioner of New Zealand. Wellington, New Zealand. From 1871.
Reports of the South African Association for the Advancement of Science. Johannesburg. From 1903.
Revista Minera y Metalurgica. Madrid. From 1850.
Revue de Métallurgie. Paris. From 1904.
Revue Universelle des Mines. Paris. From 1857.
School of Mines Quarterly. Columbia, U.S. From 1879.
Scientific American. New York. From 1846.
Silliman's American Journal of Science and the Arts. New Haven and New York. From 1816.
South African Mines, Commerce, and Industries. Johannesburg.
Transactions of the American Institute of Mining Engineers. Philadelphia. From 1871.
Transactions of the Australasian Institute of Mining Engineers. Sydney. From 1893.
Transactions of the Canadian Mining Institute.
Transactions of the Institution of Mining Engineers. Newcastle-on-Tyne. From 1852.
Transactions of the South African Association of Engineers. Johannesburg. From 1902.
Transactions of the Institution of Mining and Metallurgy. London. From 1892.
United States Geological Survey Bulletins. New York.
United States Bureau of Mines, Annual Reports and Bulletins. From 1910.
Zeitschrift für anorganische Chemie. Berlin. From 1826.
Zeitschrift für Metallographie. Berlin. From 1911.

GENERAL METALLURGY OF GOLD.

Geber, the Works of. Translated by R. Russell. London, 1686.
Biringuccio. De la Pirotechnia. Venice, 1540. French translation, Rouen, 1627.
Agricola (Georgius). De re metallica. Bâle, 1556. Reprinted, Bâle, 1657. Translated 1912, see below.
Michaelis (Johannis). De Oro. Leipzig, 1630.
Barba (Alphonzo). Arte de los metales. Madrid, 1639. French translations, Paris, 1751, and La Haye, 1782.
Schlüter. Principles of Metallurgy and Assaying. Brunswick, 1738.
Vargas (Perez de). Traité singulier de métallurgie. Translated from the Spanish. Paris, 1743.
Lewis (Wm.) Commercium Philosophico-Technicum. London, 1763.
Valerius. Grundriss der Metallurgie. Ulm, 1768.
Cramer. Principes de métallurgie et docimasie. Blankenburg, 1774.
Jars. Voyages métallurgiques. Paris, 1774-81.
Karsten. System der Metallurgie. Breslau, 1818.
Kiessling. Die Metallurgie. Dresden, 1841.
Ansted (D. T.) The Gold Seekers' Manual. London, 1849.
Landrin (H.) Traité de l'or. Paris, 1850 and 1863.
Rammelsberg. Lehrbuch der chemischen Metallurgie. 1850.
Phillips (J. A.) Gold Mining and Assaying. London, 1852.
Phillips (J. A.) *Encyclopædia Metropolitana.* Article on " Metallurgy." London, 1854.

Rivot (L.) Principes généraux du traitement des minerais métalliques. Paris, 1859, nouv. ed., 1871-73.

Kerl (B.) Die Rammelsberger Hüttenprozesse. Clausthal, 1861.

Kustel (G.) Processes of Gold and Silver Extraction. San Francisco, 1863.

Kerl (B.) Handbuch der metallurgischen Hüttenkunde. Clausthal, 1865.

Phillips (J. A.) The Mining and Metallurgy of Gold and Silver. London, 1867.

Overman (F.) A Treatise on Metallurgy. New York, 1868.

Crookes (Wm.) and Röhrig (E.) Treatise on Practical Metallurgy. Translated from the German. London, 1868-70.

Blake (W. P.) Report on the Precious Metals. Washington, 1869.

Raymond (R. W.) Mineral Resources West of the Rocky Mountains. 7 vols. Washington, 1869-74.

Raymond (R. W.) Mines, Mills, and Furnaces of the Pacific States. New York, 1871.

Schiern (F.) Sur l'origine de la tradition de fourmis qui ramassent l'or. Copenhagen, 1873.

Makins (G. H.) Manual of Metallurgy, particularly Precious Metals. 2nd edition. London, 1873.

Greenwood (W. H.) Manual of Metallurgy, vol. ii. London, 1875.

Kerl (B.) Grundriss der Metallhüttenkunde. Leipzig, 1880.

Simonin (L.) L'Or et l'Argent. Paris, 1880.

Percy (John). Metallurgy of Silver and Gold, vol. i. London, 1880.

Ryan (J.) Gold Mining in India. London, 1881.

Lock (A. G.) Gold : Its Occurrence and Extraction. With a Bibliography. London, 1882.

Egleston (T.) The Progress of the Metallurgy of Gold and Silver in the United States. New York, 1882.

Balch (W. R.) Mines, Miners, and Mining Interests of the United States in 1882. Philadelphia, 1882.

Restrepo. Estudio sobre las minas de oro y Plata de Colombia. Bogota, 1884.

Egleston (T.) Metallurgy of Silver, Gold, and Mercury in the United States. 2 vols. London, 1887-90.

Balling (C.) Grundriss der electrometallurgie. Stuttgart, 1888.

Frémy. L'Or dans le laboratoire. Encyclopædie Chimique, vol. iii. Cahier 16e. Paris, 1888.

Lock (G. W.) Practical Gold Mining. London, 1889.

Juptner (H.) Traité pratique de chimie métallurgique. Paris, 1891.

Frémy. L'Or dans les centres de travail et de l'industrie. Ency. Chim., vol. v. Cumenge and Fuchs. Paris, 1891.

Phillips (J. A.) and Bauerman (H.) The Elements of Metallurgy. London, 1891.

Raymond (R. W.) Gold and Silver : Report of the 11th Census of the U.S. New York, 1892.

Hatch (F. H.) and Chalmers (J. A.) Gold Mines of the Rand. London, 1895.

De la Coux (H.) L'or. Gites aurifères—Extraction de l'or. Paris, 1895.

Launay (L. de). Les Mines d'Or du Transvaal. Paris, 1896.

Becker (H.) L'Or, Minerais aurifères et auro-argentifères. Paris, 1896.

Schnabel (C.) Translated by H. Louis. Handbook of Metallurgy. 2 vols. London, 1898. 2nd edition, 1905.

Coignet (F.) Traitement des Quartz Aurifères. Bulletin de la Société L'Industrie Minérale. Tome xii.-xiii. 1898-9.

Begreer (B. W.) Metallurgy of Gold on the Rand. Freiberg, 1898.

Eissler (M.) The Metallurgy of Gold. 5th edition. London, 1900.

Wade (E. M. and M. L.) A Compendium of Gold Metallurgy (pocket book). Los Angelos, 1901.

Hatch (F. H.) The Kolar Goldfield. Memoirs of the Geological Survey of India. Calcutta, 1901.

Charleton (A. G.) Gold Mining and Milling in Western Australia. London, 1903.

Clark (D.) Australian Mining and Metallurgy. Sydney, 1904.

Levat (E. D.) L'Industrie aurifère. Paris, 1905.

Anonymous. West Australian Metallurgical Practice. Kalgoorlie, 1906.

Hauser (H.) L'Or. 2nd edition. Paris, 1907.

Crane (W. R.) A Treatise on Gold and Silver. New York, 1908.

Rose (T. K.) The Precious Metals, comprising Gold, Silver and Platinum. London, 1909.

Fulton (C. H.) Principles of Metallurgy. New York, 1910.

Roberts-Austen (Sir W. C.) Introduction to the Study of Metallurgy. 6th edition. London, 1910.

Austin (L. S.) Metallurgy of the Common Metals. 3rd edition. New York, 1911.

Lang (Herbt.) Metallurgy. 3 vols. New York, 1912.

Le Chatelier (H.) Introduction a l'Etude de la Métallurgie. Paris, 1912.

Prost (Eugene). Cours de Métallurgie des Métaux autre que la Fer. Paris, 1912.

Rose (T. K.) Article "Gold" in Thorpe's Dictionary of Applied Chemistry, vol. ii. London, 1912.

Encyclopœdia Britannica. Article "Gold." 11th edition. London, 1912.

Agricola (Georgius). De re Metallica. Translation from Latin edition of 1556, by H. C. and L. H. Hoover. London, 1912.

Borchers (Relf). Fortschritte der Edelmetallurgerei während der letzten Jahrzehnte, Saale, 1913.

Hofman (H. O.) General Metallurgy. New York, 1913.

Rand Metallurgists. Text-Book of Rand Metallurgical Practice. 2 vols. 2nd edition. 1913.

Power (F. D.) Pocket-Book for Miners and Metallurgists. London, 1913.

Smith (S. W.) Roberts-Austen: a Record of His Work. London, 1914.

Gowland (Wm.) Metallurgy of the Non-Ferrous Metals. London, 1914.

CHAPTERS I., II., AND III.—PROPERTIES OF GOLD, ITS ALLOYS AND COMPOUNDS.

Budelius (R.) De monetis. Coloniæ Agrip, 1591.

Savot. Discours sur les Medalles Antiques. Paris, 1627.

Potier (M.) Philosophica Chemica. Francfort, 1648.

Borrichius. Hermetes Ægyptiorum et Chemicorum Sapientia. Copenhagen, 1674.

Gobet. Les anciens mineralogistes du royaume de France. Paris, 1679.

Gellert (C. E.) Metallurgic Chemistry. Translation. London, 1796.

Hatchett (C.) Wear of Coins. *Phil. Trans. Roy. Soc.*, 1803, p. 43.

Hatchett (C.) Experiences et observations sur l'or. Paris, 1804.

D'Arcet. L'art de dorer le bronze. Paris, 1818.

Schmieder. Geschichte der Alchemie. Halle, 1832.

Boue (P.) Traité d'orfèvrerie, bijouterie, et jouaillerie. Paris, 1832.

Gmelin (L.) Translated by H. Watt. Handbook of Chemistry. Chapter on Gold. Vol. vi., pp. 200-251. London, 1851.

Ansell (G. F.) A Treatise on Coining. London, 1862.

Rossignol (J. P.) Les métaux dans l'antiquité. Paris, 1863.

Ronchaud (L. de). Dictionnaire des antiquités grecques et romaines. Article, "aurum."

Mommsen. Histoire de la monnaie romaine. Paris, 1868.

Lepsius. Les métaux chez les Egyptiens. Paris, 1877.

Wright (C. R. A.) Metals. London, 1878.

Cripps. Old English Plate. London, 1878, 1891.

Lenormant (F.) La Monnaie dans l'antiquité. 3 vols. Paris, 1878.

Gee (G. E.) The Goldsmith's Handbook. London, 1879.

Pollen (J. H.) Ancient and Modern Gold and Silversmith's Work. London, 1879.

Noback (Fr.) Münz-, Maass- und Gewichtsbuch. Leipzig, 1879.

Douan. Inoxydation, dorure, et platinage des métaux. Paris, 1880.

Brandis. Das Münz-mass und Gewichtwesen in Vorder-Asien bis auf Alexander den Grossen. Berlin.

Wagner (A.) Gold, Silber, und Edelsteine. Handbuch für Gold-, Silber-, Bronze-Arbeiter und Juweliere. Vienna, Leipzig and Pesth, 1881 and 1895.

Wheatley and **Delamotte.** Art Work in Gold and Silver. London, 1881.

Bloxam (C. L.) Metals: their Properties and Treatment. London, 1882.

Achiardi (T.) Metalli. Milan, 1883.

Chaffers (W.) Hall Marks on Gold and Silver Plate. London, 1883.

Kenyon (R. L.) The Gold Coins of England. London, 1884.

Roberts-Austen (W. C.) *Cantor Lectures, Soc. of Arts.* London, 1884, 1888, 1892, 1897, and 1901.

Berthelot (M.) Origines de l'alchemie. Paris, 1885.

Streeter (E. W.) Gold: the Standards of all Countries. London, 1885.

Kopp (H.) Die Alchemie in älterer und neuerer Zeit. Heidelberg, 1886.

Schaefer (H. W.) Die Alchemie, etc. Flensborg, 1887.

Riche (A.) Monnaie, Médailles et Bijoux. Paris, 1889.

Brannt (W. T.) Metallic Alloys. London, 1889.

Guettier (A.) Practical Guide for the Manufacture of Metallic Alloys, 1865. Trans lated from the French by A. A. Fesquet. New York, 1890.

Smith (E. A.) Dental Alloys. London, 1897.

Hiorns (A. H.) Mixed Metals. London, 1901. Metallography. 1902.

Griffiths (A. B.) Dental Metallurgy. London, 1903.

Osmond (F.) and Stead (J. E.) Microscopic Analysis of Metals. London, 1904. 2nd edition, 1914.

Desch (C. H.) Metallography. London, 1913.

Guertler (W.) Metallographie. Berlin, 1913.

Gulliver (G. H.) Metallic Alloys. 2nd edition. London, 1913.

Law (E. F.) Alloys and their Industrial Applications. 2nd edition. London, 1914.

Rosenhain (W.) Introduction to the Study of Physical Metallurgy. London, 1914.

CHAPTER IV.—MODE OF OCCURRENCE AND DISTRIBUTION OF GOLD.

Holzchul. Remarques sur l'or des mines de Saxe. Penig, 1805.

Atkinson (S.) Discoverie and Historie of the Gold Mines in Scotland. Edinburgh, 1825.

Miers (J.) Travels in Chile and La Plata. London, 1828.

Dupont (S. C.) De la Production des Métaux précieux au Mexique : considérée dans ses rapports avec la métallurgie, etc. Paris, 1843.

Tchihatchef (Pierre de). Voyage Scientifique dans l'Altai Oriental. Paris, 1845.

Papers relating to the Discovery of Gold in Australia. 2 vols. London, 1852-57.

Calvert (J.) The Gold Rocks of Great Britain and Ireland. London, 1853.

Clarke (W. B.) Researches in the Southern Gold Fields of New South Wales. Sydney, 1860.

Davison (S.) Geognosy of Gold Deposits in Australia. London, 1861.

Rosales (H.) Essay on the Origin and Distribution of Gold in Quartz Veins. Melbourne, 1861.

Brown (J. R.) Mineral Resources of the United States. Washington, 1867.

Lovell (J.) Gold Fields of Nova Scotia. Montreal, 1868.

Smyth (R. Brough). Gold Fields of Victoria. Melbourne, 1869.

Mackay (J.) Report on the Thames Gold Fields. Wellington, N.Z., 1869.

Cotta (B. von). Treatise on Ore Deposits. Translated. New York, 1870.

Bateman (A. W.) South African Gold Fields. *The Times*, Sept. 28, 1874.

Domeyko (J.) Ensayo sobre los Depositos Metaliferos de Chile. Santiago, 1876.

Bain (A. G.) Gold Regions of S.E. Africa. London and Cape Colony, 1877.

Church (J. A.) The Comstock Lode : its Formation and History. New York, 1879.

Jenney (W. P.) Mineral Resources of the Black Hills of Dakota. Washington, 1880.

Ball (Prof. B.) Diamonds, Coal, and Gold in India : their Occurrence and Distribution. London, 1881.

Blake (W. P.) Geology and Mineralogy of California. Sacramento, 1881.

Jervis (G.) Dell'Oro in Natura. Rome, 1881.

Emmons (S. F.) and Becker (G. F.) Precious Metals: being vol. xiii. of U.S. Census Reports of 1880. Washington, 1885.

Handbook of New Zealand Mines. Wellington, 1887.

Ferguson (A. M.) All about Gold in Ceylon and S. India. Colombo, 1888.

Liversidge (J.) Minerals of New South Wales. London, 1888.

Posewitz (T.) Borneo : its Geology and Mineral Resources. London, 1892.

Lock (C. G. W.) Economic Mining. London, 1895.

Phillips (J. A.) Treatise on Ore Deposits. 2nd edition. Edited by H. Louis. London, 1896.

Schmeisser (K.) and Vogelsang (K.) The Gold Fields of Australasia. Translated by H. Louis. London, 1898.

Anderson (J. W.) The Prospector's Handbook : a Guide for the Prospector and Traveller in search of metal-bearing and other valuable minerals. 9th edition. London, 1902.

Truscott (S. J.) The Witwatersrand Gold Fields. 2nd edition. London, 1902.

Posepny (F.) The Genesis of Ore Deposits. New York, 1902.

MacLaren (J. M.) The Occurrence of Gold in Great Britain. *Transactions of the Institution of Mining Engineers*, 1902, 25, 435-508.

Fuchs (E.) and Launay (L. de). Traité des gîtes minéraux et métallifères. 2 vols. Paris, 1903.

Launay (L. de). Les richesses minérales de S. Afrique et de Madagascar. Paris, 1903.
Lindgren (W.) and **Ransome (F. L.)** Geology and Gold Deposits of Cripple Creek District, Colorado. United States Geological Survey. Professional Paper No. 54. 1906.
Lock (C. G. W.) Mining in Malaya for Gold and Tin. London, 1907.
Burrows (A. G.) Porcupine Gold Area, Ontario. Part II. *Twentieth Annual Report of the Ontario Bureau of Mines.* Toronto, 1911.
Thomas and **Macalister.** Geology of Ore Deposits. London, 1911.
Cahen (E.) and **Wootton (W. O.)** Mineralogy of the Rarer Metals. London, 1912.
Hayes (C. W.) and **Lindgren (W.)** Contributions to Economic Geology. Washington, 1912.
Maclaren (J. M.) Gold : its Geological Occurrence and Geographical Distribution. London, 1912.
Hatch (F. H.) Geological Survey of the Witwatersrand and other Districts in the Southern Transvaal. London, 1913.
Simpson (E. S.) and **Gibson (C. G.)** Geology and Ore Deposits of Kalgoorlie; Part I. *Geol. Survey of Western Australia,* Bull. 42.
Tyrrell (J. B.) Gold of the Klondyke. London, 1913.
Emmons (S. F.) and others. Emmons' Volume on Ore Deposits. New York, 1915.

CHAPTERS V. AND VI.—PLACER MINING.

Moneeram. Native Account of Washing for Gold in Assam. *J. Asiat. Soc. Bengal,* 1838, **7,** 621.
Abbott (Capt. J.) Account of the process employed for obtaining gold from the sand of the River Beyass : with a short account of the gold mines of Siberia. *J. Roy. Asiat. Soc. Bengal,* 1847, **16,** 266-272.
Delesse. Gisement et exploitation de l'or en Australie. Paris, 1853.
Report of the Royal Commission appointed to inquire into the best methods of removing sludge from the gold fields. Melbourne, 1859.
Radde (Gustav). Reisen im Süden von Ost-Sibirien in den Jahren, 1855-59. Petrograd, 1863.
Debombourg (G.) Gallia aurifera. sur Études les alluvions aurifères de la France. Lyons, 1868.
Christy (S. B.) Ocean Placers of San Francisco. *Proc. Cal. Acad. Sci.,* August, 1878.
Egleston (T.) Hydraulic Mining in California. London, 1878.
Whitney (J. D.) Auriferous Gravels of the Sierra Nevada of California. Cambridge, U.S., 1880.
Hammond (J. H.) Auriferous Gravels of California and the Methods of Drift Mining. *Prod. of Gold and Silver in U.S. for* 1881. Washington, 1882.
Bowie, Jr. (A. J.) Practical Treatise on Hydraulic Mining in California. New York, 1885.
Bowie. Jr. (A. J.) Mining Debris in Californian Rivers. San Francisco, 1887.
Hammond (J. H.) Auriferous Gravels of California. *Ninth Annual Report California State Mining Bureau,* pp. 105-138, 1889.
Gould (E. S.) Practical Hydraulic Formulæ for the Distribution of Water through long Pipes. New York, 1891.
Kirkpatrick (T. S. G.) Hydraulic Gold Miner's Manual. New York, 1891.
Wagenen (T. F. van). Manual of Hydraulic Gold Mining. New York, 1891.
Levat (E. D.) Mémoire sur l'Exploitation de l'Or en Siberie Orientale. Paris, 1896.
Johnson (J. C. F.) Getting Gold. A Practical Treatise for Prospectors. London, 1897.
M'Kay (A.) Older Auriferous Drifts of Central Otago, New Zealand Mines Dep. Wellington, 1897.
Batz (Baron René de). Les Gisements Aurifères de Siberie. Paris, 1898.
Wilson (E. B.) Hydraulic and Placer Mining. London and New York, 1898.
Grothe (A.) Gold Dredging in the United States. *Mineral Industry,* 1899, 8, 326-336.
Barbour (T. J.) Gold Dredging in California. *Proceedings of the California State Miners' Association.* San Francisco, 1900.
Schrader (F. C.) and **Brooks (A. H.)** Preliminary Report of the Cape Nome Gold Region. *Professional Papers U.S. Geological Survey.* Washington, 1900.
Longridge (C. C.) Hydraulic Mining, in four parts. London, 1903 and 1912.
Longridge (C. C.) Gold Dredging. London, 1905.
Weatherbe (D'Arcy.) Dredging for Gold in California. San Francisco, 1907.
Prelini (C.) Dredgers and Dredging. New York, 1912.
Winston (W. B.) and **Janin (C.)** Gold Dredging in California. 1912.

Aubury (L. E.) Gold Dredging in California. *Californian State Min. Bureau*, Bull. 57. 1913.

Earl (T. C.) Gold Dredging. London, 1913.

Purington (C. W.) Hydraulic Elevator Work on Anvil Creek, Nome, Alaska. London, 1913.

Longridge (C. C.) Gold and Tin Dredging and Mechanical Excavators. 3rd edition. London, 1914.

CHAPTERS VII., VIII., AND IX.—ORE CRUSHING AND AMALGAMATION.

De Born. Amalgamation des minerais d'or et d'argent. Vienna, 1786. English translation by R. E. Raspe, 1791.

Sonneschmied. L'Amalgame espagnol. Leipzig, 1811.

Sonneschmied. Traité sur l'amalgamation. Ronneburg, 1811.

Rivot. Nouveau procédé de traitement des minerais d'or et d'argent (a manuscript in the archives of the École des Mines de Paris). Paris, 1818.

Ortmann. Kurze Geschichte der Amalgamation in Sachsen. Freiberg, n.d.

Lawson (G.) Improvements in Amalgamation. *Trans. Nova Scotia Inst.* 1866.

Hague (J. D.) Gold Mining in Colorado. *Report on the Fortieth Parallel*, vol. iii. Washington, 1870.

Keith (N. S.) Amalgamated Copper Plates. *Eng. and Mng. J.*, 1871, **11**, 270.

Blake (W. P.) Mining Machinery. New Haven, 1871.

Fonseca. Mémoire sur l'amalgamation Chilienne. Paris, 1872.

Bergmann (E. von). Die Anfänge des Geldes in Ægypten. Vienna, 1872.

Sonneschmidt (F.) Tratado de la Amalgamacion de Mexico. Mexico, 1876.

Egleston (T.) Californian Stamp Mills. London, 1880.

Egleston (T.) Treatment of Gold Quartz in California. London, 1881.

Randall (P. M.) Quartz Operators' Handbook. New York, 1888.

M'Dermott and Duffield. Gold Amalgamation and Concentration. London and New York, 1890.

Charleton (A. G.) The Choice of Coarse and Fine Crushing Machinery and Processes of Ore Treatment. *Trans. Fed. Inst. Mng. Eng.*, 1892-94. Seven Papers.

Louis (H.) Handbook of Gold-Milling. London, 1894.

Rickard (T. A.) Variations in Gold-Milling. New York, 1895.

Rickard (T. A.) The Stamp Milling of Gold Ores. New York and London, 3rd edition, 1901.

Lock (C. G. W.) Gold Milling : Principles and Practice. London, 1901.

Adams (W. J.) Hints on Amalgamation and the General Care of Gold Mills. Chicago, 1901.

Richards (R. H.) Ore Dressing. 2 vols. New York, 1903.

Horner (J. G.) Modern Milling Machines : Their Design, Construction, and Working. New York, 1906.

Tinney (W. H.) Gold Mining Machinery. New York, 1906.

Caldecott (W. A.) Development of Heavy Gravitation Stamps. *J. Chem. Met. and Mng. Soc. of S. Africa*, 1910, **10**.

M'Farren (H. W.) Practical Stamp Milling and Amalgamation. London, 1910.

Rand Metallurgists. Rand Metallurgical Practice. 2 vols. London, 1912. 2nd edition. London, 1913.

Mar (A. del). Stamp Milling. San Francisco, 1912.

Preston (E. B.) Californian Gold Mill Practices. *Californian State Min. Bureau*, Bull. 46. 1913.

CHAPTER XI.—CONCENTRATION IN GOLD MILLS.

Gaetzschmann (M. F.) Die Aufbereitung. (Mechanical Concentration of Ores.) Leipzig, vol. i., 1864 ; vol. ii., 1872.

Rittinger (P. von). Lehrbuch der Aufbereitungskunde. Berlin, 1867, 1870, and 1873.

Smyth (Sir W. W.) Dressing, or the Mechanical Preparation of Gold Ores. Lectures on Gold. *Mining Journal*, 1873.

Schmidt (A. W.) Der Schlamfänger auch Kornfänger genannt. Dillenburg, 1877.

Habermann (J.) Comparison of the Salzburg Table, the Rittinger Table, and the Hand Buddle. *Oesterr. Zeitsch. für Berg- und Hüttenwesen*, 1879, No. 8.

Cazin (F. M. F.) Dynamical Metallurgy or Mechanical Ore Concentration. *Mining Record*, 1881-2.

Callon (J.) Lectures on Mining. Translated by Le Neve Foster and W. Galloway, vol. iii. London, 1886.

The Settling of Solid Particles in Liquids. Bulletin No. 36, *United States Geological Survey.* Washington, 1886.

Ore Dressing in California. *Sixth Report of the Cal. State Min.*, 1886. This is a full account of the machines actually at work.

Lock (C. G. W.) Mining and Ore Dressing Machinery. London, 1891.

Kunhardt (W. B.) The Practice of Ore Dressing in Europe. New York, 1891.

Commans (R. E.) Concentration and Sizing of Crushed Ore. *Proc. Inst. Civil Eng.*, 1894.

Rosales (H.) Report on the Loss of Gold in the reduction of Auriferous Veinstone in Victoria. Melbourne, 1895.

De La Goupillière (H.) Cours a'exploitation des Mines (chapter on Ore Dressing). Paris, 1896-7.

Lock (C. G. W.) Gold Milling. London, 1901.

MacLaren (J. M.) Queensland Mining and Milling Practice. *Queensland Geological Survey.* Brisbane, 1901.

Davies (E. H.) Machinery for Metalliferous Mines. London, 1902.

Richards (R. H.) Ore Dressing. New York, 1903.

Foster (C. Le Neve). Ore and Stone Mining. 7th edition. London, 1910.

Richards (R. H.) Text Book of Ore Dressing. New York, 1911.

Hoover (T. J.) Concentration of Ores by Flotation. London, 1913.

CHAPTER XIII.—ROASTING OF GOLD ORES.

Plattner (C. F.) Die metallurgische Röstprozesse. Freiberg, 1856.

Kustel (G.) Roasting of Gold and Silver Ores. San Francisco, 1880.

Christy (S. B.) Losses of Gold in Roasting. *Trans. Amer. Inst. Mng. Eng.*, 1888.

Adams (W. H.) Pyrites : Practical Methods for Extraction of Gold, etc. New York, 1892.

CHAPTER XIV.—CHLORINATION.

Whelpley and Storer. Method of separating Metals from Sulphurets. Boston, 1866.

Küstel (G.) Concentration and Chlorination. San Francisco, 1868.

Egleston (T.) Leaching Gold and Silver Ores in the West. New York, 1883.

Egleston (T.) Leaching Gold Ores containing Silver. London, 1886.

O'Driscoll (F.) Notes on the Treatment of Gold Ores. London, 1889.

France (Ch. D.) Extraction par voie humide du cuivre, de l'argent, et de l'or. Paris, 1897.

Wilson (E. B.) The Chlorination Process. London and New York, 1897.

Thomson (F. A.) and **Goodall (S. L.)** The Portland Mill. (Description of the Chlorination Plant.) *Mines and Minerals*, Oct. and Nov. 1904.

CHAPTERS XV., XVI., and XVII.—CYANIDE PROCESS.

Scheidel (A.) The Cyanide Process. Sacramento, 1894. 2nd edition, 1901.

Reunert (T.) Diamonds and Gold in South Africa. London, 1894.

Gaze (W.) Handbook of Practical Cyanide Operations. London, 1898.

Eissler (M.) The Cyanide Process. 3rd edition. London, 1902.

Fulton (C. H.) The Cyanide Process. Bulletin No. 5, *South Dakota School of Mines.* Rapid City, Dakota, 1902.

James (A.) Cyanide Practice. 3rd edition. London, 1903.

Uslar (M.) and **Erlwein (G.)** Cyanid Prozesse zur Goldgewinnung. Halle, Germany, 1903.

Bosqui (F. L.) Practical Notes on the Cyanide Process. 3rd edition. New York, 1904.

Miller (A. S.) The Cyanide Process. 2nd edition. London, 1906.

Rickard (T. A.) Recent Cyanide Practice. San Francisco, 1907.

Wilson (E. B.) The Cyanide Process. 4th edition. New York and London, 1908.

Bain (H. Foster). More Recent Cyanide Practice. San Francisco, 1910.

Clennell (J. E.) Cyanide Handbook. New York, 1910.

M'Cann (F.) Beneficio de metales de plata y oro por cyanuracion. 1910.

572 THE METALLURGY OF GOLD.

Clennell (J. E.) Chemistry of Cyanide Solutions resulting from the Treatment of Ores. 2nd edition. New York, 1910.

Julian (H. F.) and Smart (E.) Cyaniding Gold and Silver Ores. 2nd edition. London, 1911.

MacFarren (H. W.) Cyanide Practice. New York and London, 1912.

M'Cann (F.) Cyanide Practice in Mexico. San Francisco, 1913.

Von Bernewitz (M. W.) Cyanide Practice, 1910-1913. San Francisco, 1913.

Park (Jas.) The Cyanide Process of Gold Extraction. Auckland and Melbourne, 1896. London, 5th edition, 1913.

Megraw (H. A.) Details of Cyanide Practice. London, 1914.

Megraw (H. A.) Practical Data for the Cyanide Plant. San Francisco, 1910. London, 1914.

CHAPTER XVIII.—REFINING AND PARTING OF GOLD BULLION.

Goddard (Jonathan). Experiments on Refining Gold with Antimony. *Phil. Trans. Roy. Soc.*, 1676.

Egleston (T.) Parting Gold and Silver in California. New York, 1877.

Gore (G.) Art of Electro-Metallurgy. New York, 1884.

Zoppeti. L'electrolisi in metallurgica. Milan, 1885.

Egleston (T.) Parting Gold and Silver by means of Iron at Lautenthal. New York, 1885.

Egleston (T.) The Separation of Silver and Gold from Copper at Oker. Washington, 1885.

Egleston (T.) Treatment of Gold and Silver at the United States Mint. London, 1886.

Gumbinner (Sven). Parting Gold and Silver in the United States. *Ninth Report of the Cal. State Min.*, 1889, pp. 62-104. This is a complete account of the methods in use in the United States.

Watt (A.) Electro-Metallurgy Practically Considered. London, 1889.

Ulke (Titus). Parting and Refining Gold and Silver. *Mineral Industry*, 1895, pp. 343-366.

Iles (M. W.) Notes on the Moebius Process for Parting Gold and Silver Ores. *Mineral Industry*, vol. viii., 1899 ; also vols. ii., iv., and v.

M'Millan (W. G.) Treatise on Electro-Metallurgy. London, 1899.

Watt (A.) and Philip (A.) Electro-Plating and Electro-Refining of Metals. London, 1902.

Langbein (G.) Translated by W. T. Brannt. Complete Treatise on the Electro-Deposition of Metals. Philadelphia, 1902 and 1913.

Borchers (W.) Translated by W. G. M'Millan. Electric Smelting and Refining. 2nd edition. London, 1904.

Clark (D.) Gold Refining. London, 1909.

Allmand (A. J.) Principles of Applied Electro-Chemistry. London, 1912.

CHAPTERS XIX. AND XX.—ASSAYING.

Ercker (L.) Allerfurnemisten Mineralischen Eerzt und Bergwerks Arten. Frankfort, 1580. Another edition, Frankfort, 1629.

Carranza (A.) El Ainstamieto i Proporcion de las Monedas de Oro, Plata i Cobre i la reduccion distos Metales a su Debida estimaccion. Madrid, 1629.

Bedrock (Wm.) A New Touchstone for Gold and Silver Wares. London, 1651. 2nd edition, 1679.

Le Febure (N. R.) Compleat body of chymistry. 1670.

Pettus (Sir John). "Fleta Minor." Laws of Nature in Assaying Metals. Translated from the German of L. Ercker. London, 1686.

Cramer (J. A.) Elementa artis docimasticæ. Lugduni Batavorum (Leyden), 1741.

Symonds (W.) Essay on the Weighing of Gold, etc. London, 1756.

Cramer, M.D. (J. A.) Elements of the Art of Assaying Metals. Translated from Latin by C. Mortimer, M.D. 2nd edition. London, 1764. An account of this book by Prof. Austin of Michigan University is given in the *Eng. and Mng. J.*, July 28, 1904, p. 144.

Pouchet. Le nouveau titre des matières d'or et d'argent. Rouen, 1789.

Aldridge (W. J.) The Goldsmith's Repository, containing Treatise on Assaying. London, 1789.

Vauquelin. Manuel de l'Essayeur. Paris, An. vii. (1800).

Becquerel. Gold and Electricity. *Ann. Chim. Phys.*, 1823, **24** ; also Article by **Oersted,** d°, 1828, **39**, 274.

Chaudet. L'Art de l'essayeur. Paris, 1835.

Bodemann (Th.) Anleitung zur Berg- und Hüttenmannischen Probirkunst. Clausthal, 1845.

Berthier. Traité des essais par la voie séche. Paris, 1847.

Pettenkofer. *Bergwerksfreund*, vol. xii. (1849). Article on Gold Bullion Assaying.

Ryland (A.) Assay of Gold and Silver Wares. London, 1852.

Watherston (J. H.) The Gold Valuer. London, 1852.

Plattner (C. F.) Probirkunst. Freiberg, 1853. 6th edition. Leipzig, 1897.

Terrell (A.) Atlas de Chemie analytique minerale. Paris, 1861.

Bodemann and **Kerl.** Treatise on Assaying. Translated by W. A. Goodyear. New York, 1868.

Domeyko (J.) Tratado de Ensayes, tanto por la via seca como por la via humeda. Chile, 1873. 5th edition. Mexico and Paris, 1889.

Phillips (J. S.) Explorers and Assayers' Companion. San Francisco, 1879.

Attwood (G.) Practical Blowpipe Assaying. London, 1880.

Chapman (E. J.) Assay Notes. Practical instructions for the determination by furnace assay of Gold and Silver in rocks and ores. Toronto, 1881.

Balling (C.) Probirkunde. Przibram, 1879.

Balling (C.) L'Art de l'essayeur. Paris, 1881.

Kerl (B.) Metallurgische Probirkunst. Translated by W. T. Brannt (*Assayers' Manual*). London, 1883. Translated by Garrison. 2nd edition. Philadelphia, 1889.

Jagnaux (Raoul). Traité pratique de analyses chimiques et d'essais industriels. Paris, 1884.

Rössler (H.) Article on Gold Bullion Assaying. *Dingler's Polyt. Journ.*, 1884, **206**.

Black (J. G.) Chemistry of the Gold Fields. Dunedin, N.Z., 1885.

Rivot (L. E.) Docimasie Traité d'Analyse des substances minérales. Tome v. Paris, 1886. (Containing useful notes on assaying complex ores).

Mitchell (W.) Manual of Practical Assaying. Edited by Wm. Crookes. 6th edition. London, 1888.

Hiorns (A. H.) Practical Metallurgy and Assaying. London, 1888.

Frémy. L'or dans le Laboratoire. Cumenge and Fuchs. *Ency. Chim.*, vol. iii., c. 16e. Paris, 1888.

Ross (W. A.) Blowpipe Analysis. London, 1889.

Brown and **Griffiths.** Manual of Assaying of Gold, Silver, etc. London, 1890.

Lieber (O. M.) Assayer's Guide. New York.

Plattner. Blowpipe Analysis. Enlarged by **Richter (Th.)** Translated by H. B. Cornwall. New York, 1890.

Riche (A.) L'Art de l'essayeur. Paris, 1892.

Campredon (L.) Guide pratique du chimiste, metallurgiste et de l'essayeur. Paris, 1898. 2nd edition, 1909.

Aaron (C. H.) Manual of Assaying. 3rd edition. San Francisco, 1900.

Brown (W. L.) Manual of Assaying. Gold, Silver, etc. 9th edition. Chicago, 1900.

Simonds (E. H.) Practical Course in the Fire Assaying for Gold, Silver, etc. San Francisco, 1900.

Ricketts and **Miller.** Notes on Assaying. 3rd edition. 1900.

Merritt (W. H.) Field Testing for Gold and Silver Ores. London, 1900.

Merritt (W. H.) Gold and Silver Ores, what is their value ? New York, 1901.

Miller (A. S.) Manual of Assaying. 1901.

Rhead (E. L.) and **Sexton (A. H.)** Assaying and Metallurgical Analysis. London, 1902. 2nd edition, 1911.

Lindgren (W.) Tests for Gold and Silver Ores in Shales from Western Kansas. Bull. No. 202, *U.S. Geol. Survey.* Washington, 1902.

Macleod (W. A.) and **Walker (C.)** Metallurgical Analysis and Assaying. London, 1903.

Lodge (R. W.) Notes on Assaying and Metallurgical Analysis. New York, 1904.

Philips (H. J.) Gold Assaying. London, 1904.

Argall (P. H.) Western Mill and Smelter Methods of Analysis. New York, 1905.

Low (A. H.) Technical Methods of Ore Analysis. New York and London, 1905.

Buskett (E. W.) Fire Assaying. New York, 1907.

Barr (J. A.) Testing for Metallurgical Processes. 1910.

Seamon (W. H.) A Manual for Assayers and Chemists. New York, 1910.

Smith (J. R.) Modern Assaying. Philadelphia, 1910.

Furman (H. van F.) A Manual of Practical Assaying. New York. 8th edition, 1911.

Sutton (F.) Systematic Handbook of Volumetric Analysis. 10th edition, 1911.

Fulton (C. H.) Fire Assaying. 1st edition, 1907. 2nd edition. New York, 1912.
Beringer (J. J. and C.) Manual of Assaying. 13th edition. London, 1913.
Lord (M. W.) and Demorest (D. J.) Metallurgical Analysis. New York, 1913.
Collins (G. E.) Systematic Testing in the Evolution of Mill Practice. London, 1914.
Smith (E. A.) The Sampling and Assay of the Precious Metals. London, 1914.
Wraight (E. A.) Assaying in Theory and Practice. London, 1914.
Park (J.) Practical Assaying. London, 1914.

CHAPTER XXI.—STATISTICS OF GOLD PRODUCTION AND CONSUMPTION.

Jacob (Wm.) A History of the Precious Metals. 2 vols. London, 1831.
Humboldt (A. von). Fluctuations in the Supplies of Gold. London, 1839.
Blake (W. P.) The Production of the Precious Metals. London and New York, 1869.
Soetbeer (A.) Edelmetall Production. Gotha, 1879.
Del Mar (A.) A History of the Precious Metals (containing a Bibliography). London, 1880. New York, 1902.
O'Brien. Treatise on Gold and Silver. London, 1884.
Soetbeer (A.) Materialen zur Erläuterung und Beurteilung der wirtschaftlichen Edelmetallverhältnisse und der Währungsfrage. Berlin, 1886.
Roswag (C.) L'Argent et L'Or, Production, consommation et circulation des métaux précieux. 2 vols. Paris, 1889-90.
Soetbeer (A.) Litteraturnachweis über geld und Münzwesen. Berlin, 1892.
Welton (W. S.) Practical Gold Mining : its commercial aspects. London, 1902.
Launay (L. de). Translated by O. C. Williams. The World's Gold. London, 1908.
Annual Reports of the Director of the Mint (United States).
Annual Reports of the Royal Mint (London).
Reports of Chamber of Mines of Transvaal.
Mining Statistics of Western Australia (Government Gazette).
Whitaker's Almanac. London. Annually.
Canadian Department of Mines—Annual Statements of Production of Gold and Silver. From 1907.
Statistical Abstracts for Foreign Countries. Annually from 1876. Board of Trade, London.
Statistical Abstracts for British Empire. Annually from 1905. Board of Trade, London.
Mines and Quarries. General Report with Statistics. Part IV., Colonial and Foreign Statistics. Annually from 1900. Home Office, London.
Rapport au Ministre des Finances par l'Administration des Monnaies et Médailles. (Annual Reports of the Mint, Paris.)
Mineral Industry. New York. Annually since 1892.

INDEX.

ERRATA.

Page 51. Footnote 2.—For Von Pelabon *read* H. Pelabon.
  ,, 53.   ,, 2.--For Von Maey *read* E. Maey.
  ,, 271. Last line but one.—For Clark and Stansfield's *read* Clarkson and Stanfield's.
  ,, 356. Footnote 1.—For Power *read* O. P. Powell.

www.ingramcontent.com/pod-product-compliance
Lightning Source LLC
Chambersburg PA
CBHW062009190326
41458CB00009B/3025